# Advances in Remote Sensing Technology and the Three Poles

# Advances in Remote Sensing Technology and the Three Poles


*Edited by*

*Manish Pandey*
*Chandigarh University, Department of Civil Engineering, Mohali, India*

*Prem C. Pandey*
*School of Natural Sciences, Shiv Nadar Institution of Eminence, Center for Environmental Sciences & Engineering, Greater Noida, Uttar Pradesh, India*

*Yogesh Ray*
*National Centre for Polar and Ocean Research, Headland Sada, Vasco-da-Gama, Goa*

*Aman Arora*
*Bihar Mausam Seva Kendra, Planning and Development Department, Bihar, India*

*Shridhar D. Jawak*
*University Centre in Svalbard, Longyearbyen, Norway*

*Uma K. Shukla*
*Banaras Hindu University, Institute of Science, Varnasi, Uttar Pradesh, India*



WILEY

This edition first published 2023


*Registered Office(s)*
John Wiley & Sons, Inc., 111 River Street, Hoboken, NJ 07030, USA
John Wiley & Sons Ltd, The Atrium, Southern Gate, Chichester, West Sussex, PO19 8SQ, UK

*Editorial Offices*
9600 Garsington Road, Oxford, OX4 2DQ, UK
The Atrium, Southern Gate, Chichester, West Sussex, PO19 8SQ, UK

For details of our global editorial offices, customer services, and more information about Wiley products visit us at www.wiley.com.

Wiley also publishes its books in a variety of electronic formats and by print-on-demand. Some content that appears in standard print versions of this book may not be available in other formats.



*Library of Congress Cataloging-in-Publication Data*
Names: Pandey, Manish (Assistant professor), editor.
Title: Advances in remote sensing technology and the three poles / edited by Manish Pandey [and five others]
Description: Hoboken, NJ : John Wiley & Sons, 2023.
Identifiers: LCCN 2022028873 | ISBN 9781119787723 (hardback) | ISBN 9781119787730 (pdf) | ISBN 9781119787747 (epub) | ISBN 9781119787754 (ebook)
Subjects: LCSH: Remote sensing--Polar regions. | Remote sensing--Hindu Kush-Himalayan Region.
Classification: LCC G70.5.P73 .A38 2023 | DDC 621.36/780911--dc23/eng20221013
LC record available at https://lccn.loc.gov/2022028873

Cover image: © LEONELLO CALVETTI/Getty Images
Cover design by Wiley

Set in 9.5/12.5pt STIXTwoText by Integra Software Services Pvt. Ltd, Pondicherry, India
Printed and bound by CPI Group (UK) Ltd, Croydon, CR0 4YY

C9781119787723_051222

Indra Bir Singh (1943–2021)
Department of Geology,
University of Lucknow,
Lucknow, Uttar Pradesh, India

Prof I.B. Singh was an eminent scholar of international repute, a dedicated geoscientist, and an ideal teacher. He was born on 8 July 1943 in Lucknow, Uttar Pradesh, India. Prof Singh breathed his last in the morning of 11 February 2021 after a brief illness. He completed his secondary education from the Lucknow Christian College in 1956. For higher education, he joined the Lucknow University, from where he obtained a BSc (1961) and MSc (1962) in Geology. This was the time when his classmates noticed in him an exceptional ability to look at the subject of Geology in a more common way relating to natural processes at work. After completing his post graduation, he joined the Oil and Natural Gas Corporation of India for a brief period and being unsatisfied with the job he left. He went to Germany to obtain a Dr.Rer.nat. degree from Technical University, Stuttgart, Germany in 1966 under the supervision of H. Aldinger. He then worked as a research associate at Senckenberg Institute, Wilhelmshaven, Germany in 1966. He spent two years (1967–1968) as a Post-doctoral fellow at the Oslo University, Norway. Later, he returned to Senckenberg Institute, Wilhelmshaven, as Alexander von Humboldt Fellow and worked from 1969 to 1972 on modern shallow marine sediments.

In 1972, Prof Singh returned to India and started working in the Department of Geology, University of Lucknow, from where he retired as Head of the Department in 2008. He brought out qualitative changes in teaching and research of the Department. He headed the department in the most democratic way and raised it to the level of Centre of Advance Studies in Geology. Returning to India, he started working on the sedimentary sequences of the Himalaya and central India ranging in age from the Precambrian to Holocene. Applying his experience of working on modern sediments of marine and fluvial origin, Prof Singh was able to interpret the depositional environments of the Himalayan rock sequences precisely in terms of physical processes and age. He reinterpreted the Krol belt of Himalaya as Upper Proterozoic, which had been considered to be of Mesozoic age for over a century. This study changed the stratigraphy and evolutionary history of the Himalaya. He made many significant contributions for understanding rock sequences of the Kashmir, Kachchh, Gondwana, and Himalaya, east coast delta. These studies provided an in-depth understanding of the depositional processes based on field-based Facies Analysis.

In the early 1990s, Prof Singh established a very strong group with his students and adopted a multiproxy approach including Facies Analysis supported by OSL dating, geochemistry, and isotopic signatures to study the Ganga Plain of Himalayan Foreland Basin. He worked on the landform evolution, architectural element analysis of channel bars. and floodplain deposits. His group identified the contribution of interfluves (doab) processes in the fluvial domain which was a new discovery. This study provided insights into the nature of river systems, chronology to the Late Quaternary landform evolution, tectonic events, and climate changes in the Ganga Plain. He emphasized geoarchaeological aspects of the Ganga Plain and has been able to establish palaeovegetation, human settlement patterns, and agricultural activity, particularly the domestication of rice. Working on different aspects, Prof Singh guided 15 doctoral theses and published about 200 research papers in journals of national and international repute. His students have now established themselves as leaders in their own right and are a tribute to the training he imparted. With Prof H.E. Reineck, he co-authored the book, "Depositional Sedimentary Environments," published in 1973. This is a classic book on depositional environments and has been translated into Russian and Chinese. With A.S.R. Swamy, he also wrote the book entitled "Delta Sedimentation: East Coast of India."

Prof Singh was elected as Fellow of Indian National Science Academy, New Delhi in 1995. He was a recipient of the National Mineral Award, Government of India (1996) and National Award for excellence in Earth System Science in 2013. He also received the L. Rama Rao Birth Centenary Award of the Geological Society of India. He was honored with Fellow of Alexander von Humboldt Foundation, Germany in 1988–1989. Prof Singh was visiting Professor at Louisiana State University, USA (1984–1986) and at the University of Erlangen–Nuremberg, Germany (1998–1999). He has served as a board member of governing bodies on several committees dealing with research and teaching and is easily placed among those few who have impacted Indian geosciences, research, and teaching in a fundamental way and with indigenous resources and ideas.

# Contents

## About the Editors

**Dr Manish Pandey** currently works at the University Center for Research & Development (UCRD), Chandigarh University located in Mohali, Punjab, India. He earned his graduation (Geography honors) and post-graduation (Geography) from the University of Allahabad located in Allahabad, Uttar Pradesh, India. He has been awarded a research grant as Junior Research Fellow (JRF) and Senior Research Fellow (SRF) for carrying out his doctoral research by the Council of Scientific and Industrial Research (CSIR), Ministry of Human Resource Development, Government of India. After earning his PhD degree in the field of Geomorphology, he has been engaged in post-doctoral research (at different research positions) for more than five years. His research interests are in Geography, Fluvial and Glacial Geomorphology, Glaciology, and Remote Sensing & Geoinformatics (GIS). Recently, he discovered his new area of interest in the application of artificial intelligence, machine learning, and deep learning algorithms in the domains of natural hazards, and how their application can be extended for exposure of land to future natural hazards. His simple interest is in understanding the process–form relationship in diverse environmental settings. He is an experienced research associate with a demonstrated capability of working in the research industry, skilled in Cartography, Geomorphology, well versed in GIS packages like ArcGIS, QGIS, ERDAS Imagine, and Data Analysis, and is a strong research professional with a Doctor of Philosophy (PhD) in Fluvial Geomorphology from Banaras Hindu University. His exposure to glaciological field work and training by India's elite government institutions like the Geological Survey of India, and geospatial training provided to him by institutions like ISRO, has infused some very important skills in the respected fields of research needed to carry out this project to finality. Dr Manish has been in the field training groups carrying out research in the study of the Himalayan Foreland Basin deposits, ancient Neogene Siwalik sequences and their modern analogs like the Gangetic Foreland Basin sediments facies, to understand the role of synsedimentary processes in the evolution of one of the world's most important foreland basin systems on the planet. He has published high-quality peer reviewed research articles in national/international scientific journals and books including *Ecological Indicators*, *Science of the Total Environments*, *Advances in Space Research*, *Frontiers in Earth Science*, etc.

**Dr Prem C. Pandey** received PhD from the University of Leicester, United Kingdom, under Commonwealth Scholarship and Fellowship Plan. He did his Post-Doctoral from the Department of Geography and Human Environment, Faculty of Exact Sciences, Tel Aviv University Israel. Currently, he is working as Assistant Professor at the School of Natural Sciences, Center for Environmental Sciences & Engineering, Shiv Nadar Institution of Eminence (erstwhile, Shiv Nadar University), Uttar Pradesh, India. Previously, he has been associated with Banaras Hindu University India as a SERB-NPDF fellow. He received his BSc and MSc degrees (Environmental Sciences) from Banaras Hindu University and his M.Tech degree (Remote Sensing) from Birla Institute of Technology, India. He has worked as a Professional Research fellow on remote sensing applications in the National Urban Information System funded by the NRSC Government of India. He has been a recipient of several awards including Commonwealth Fellow United Kingdom, INSPIRE fellow GoI, MHRD-UGC fellow GoI, Malviya Gold Medal from Banaras Hindu University, SERB-NPDF from the Government of India, and Young Investigator Award. Dr Pandey is working on three projects related to Monitoring of wetlands/chilika lakes, mainly focusing on ramsar sites along with other natural resources-based research work funded by the NGP and SERB Government of India. Dr Pandey is also working with science collaborators in real-time disaster monitoring in the Himalayan regions. He has published more than 45 peer reviewed journal papers , 6 edited books, several book chapters, and presented his work at national and international conferences. He is a serving member (associate editor) of the editorial board for Geocarto International Journal, Taylor & Francis, and acted as guest editor for *Remote Sensing*, MDPI. Additionally, he is also a member of ISG (Indian Society of Geomatics), ISRS (Indian Society of Remote Sensing), IUCN-CEM (2017–2025), Society of Wetland Scientists (2021–2022), SPIE, and AAG. Dr Pandey focuses his research on remote sensing for natural resources including forestry, agriculture, urban studies, environmental pollutant modeling. and climate change.

**Dr Yogesh Ray** is presently working as Scientist-E at National Centre for Polar and Ocean Research, Ministry of Earth Sciences (Government of India) Goa, India. He earned his PhD from Wadia Institute of Himalayan Geology, MoU with HNB Garhwal University. He has published several papers in peer-reviewed journals and chapters in edited volumes. Research interests lie in Clastic Sedimentology, Geomorphology, Himalayan Geology, and the evolution of the Antarctic landscape in the Pliestocene-Holocene. Actively involved with the Indian Antarctic programme. Dr Ray participated in the 33rd, 35th, 37th, and 40th Indian Scientific expeditions to Antarctica (ISEA) and was entrusted with the responsibility of "Voyage Leader" during the 37th and 40th ISEA. He was bestowed with the Young Researcher Award 2010, Ministry of Mines, Government of India, Foundation day best research paper award 2010–2011 from the Wadia Institute of Himalayan Geology, Dehradun, India, the Shri PV Dehadrai Memorial Gold Medal, and Prof MS Srinivasan Gold Medal from Banaras Hindu University, Varanasi, India.

**Dr Aman Arora** has completed his doctorate (PhD) in Geography, specializating in Remote Sensing (RS) and Geographic Information System (GIS) and has more than twelve years experience in different public and private organizations. He also holds a master's degree as well as a postgraduate diploma in RS & GIS. Dr Aman Arora has core expertise in change detection analysis, urban planning, network analysis, flood frequency analysis, hydrometeorological trend analysis, and spatial modeling. His current research interests are in the fields of risk map analysis for different natural hazards by utilizing satellite images and advanced statistical algorithms including machine learning models in GIS environment. He had received awards and travel grants from different organizations/institutes of international repute such as the National Science Foundation, USA; United Nations Office for Outer Space Affairs, Vienna, Austria; Council of Scientific Industrial Research, India; and Sun Yat-sen University, China; for research work presentations, participation in conferences, and training programs.

In his current role as a Scientific Officer/Scientist (RS & GIS) at Bihar Mausam Sewa Kendra, Planning & Development Department, Government of Bihar, Dr Aman Arora is leading his team in providing support to others by performing accurate and timely delivery of weather maps to the stakeholders and officials for Bihar. Also, he and his team are actively involved in monitoring, assessment, and forecast of hydrological extreme events (floods/droughts) and meteorological extreme events (heat waves/cold waves).

In addition to his professional and academic achievement, Dr Aman Arora is an active International Rated Chess Player recognized by The Fédération Internationale des Échecs (FIDE), Switzerland.

**Dr Shridhar D. Jawak** is currently working as a Senior Adviser in Remote Sensing at the Svalbard Integrated Arctic Earth Observing System (SIOS), Longyearbyen, Norway. He is on the advisory/evaluation board of three European projects focusing on Earth observation activities. He has chaired more than 14 sessions in international conferences, published more than 40 peer-reviewed articles, and presented more than 100 conference presentations in the past 12 years. He has participated in three summer expeditions to Antarctica and one field campaign to Svalbard during his doctoral research. He has acted as a co-principal investigator (Co-PI) of four projects peer-reviewed by national referees. He has guided/co-guided around 90 master's students for their master's thesis/summer training projects in the field of Earth observation and Remote Sensing in Polar Regions. He is the recipient of five prestigious awards/fellowships: 1) Emerging leadership grant: 2022 by the Arctic Frontiers; 2) International Mentorship Award: 2021 by the Association of Polar Early Career Scientists (APECS); 3) Indian National Geospatial Award: 2018 by the Indian Society of Remote Sensing (ISRS); 4) International Arctic Science Committee (IASC) Fellow: 2017; and (5) Recipient of Young Geospatial Scientist: 2017 by the Geospatial World Forum. His research interests include remote sensing of the cryosphere, specifically focusing on the usage of multi-satellite (SAR/Optical/LiDAR) and airborne data for spatiotemporal changes in the cryosphere of the Arctic, Antarctic, and Himalayas.

**Professor Uma Kant Shukla** is currently an Alexander von Humboldt Fellow at the Center for Advanced Study in Geology, Institute of Science, Banaras Hindu University, Varanasi. He is a sedimentologist with 30 years of doctoral and postdoctoral research experience. His research interest hinges on facies analysis of ancient and modern deposits of fluvial, marine, and lake origin. For more than one decade, he has been using modern tools such as architectural element analysis, palaeocurrent, trace fossil, etc., in the study of the Himalayan Foreland Basin deposits, including ancient Neogene Siwalik sequences and modern Gangetic Foreland Basin sediments to generate facies models for various depositional domains, and to evaluate the role of synsedimentary tectonics and palaeoclimate influencing the mode of sedimentation through Neogene-Quaternary times. Study of modern fluvial processes has been helpful to understand the river dynamics in the past. The incised valley system of Gangetic Plain Rivers has also been compared to the Stuttgart Formation (Carnian, Late Triassic) of Germany, which is believed to have been formed under similar climatic settings. Palaeolake deposits of Ladakh and paleoflood sediments have been investigated and the interpretation of palaeoclimate and tectonic evolution of the Trans-Himalayan terrain. Glacial history of the Suru-sub-Basin of Kashmir Himalaya has been studied and an inventory of paleoclimatic response of glaciers has been proposed.

Recently, it has been realized that Geoarchaeological studies of ancient settlements may have societal implications and can help masses to understand existing myths and traditions in a more scientific way. Therefore, a collaborative attempt has been made involving experts from the Archaeology and Geography Departments of BHU to unravel the settlement history of Varanasi city and its possible evolution with the dynamics of River Ganga through the ages. This work has resulted in an authored book titled *Varanasi and Ganga*, published by Aryan Books International, New Delhi. Study of the Cretaceous Lameta and Bagh Formation of Central India, Precambrian sequences of Kumaun Lesser Himalaya, and the Vindhyan basin has helped to understand the processes of sedimentation, facies models, and palaeogeographic reconstruction of these basins. The Permian-Triassic Boundary in Spiti Himalaya has been studied and a major catastrophe has been deduced.

Professor Shukla has, to his credit, authored more than 90 national and international peer reviewed papers, book chapters, and has presented his research findings in different conferences and seminars, both in India and abroad. He has been a member of expert committees in MoES, Government of India, and SERB, a statutory body under the Department of Science and Technology, Government of India.

## Notes on Contributors

***Avinash Kumar Pandey***
Department of Chemistry, GLA University, Chaumuhan, Mathura, Uttar Pradesh, India
Email ID: avinash.pandey@gla.ac.in

***Benidhar Deshmukh***
Discipline of Geology, School of Sciences, Indira Gandhi National Open University, New Delhi, India.
Email ID: bdeshmukh@ignou.ac.in

***Maneesh Kuruvath***
Kuruvath House, Muttathukulangara, Vellikulangara P.O Thrissur, Kerala, India
Email ID: maneeshkuruvath@gmail.com

***Manish Pandey***
University Center for Research & Development (UCRD), Chandigarh University, Mohali, Punjab, India
Email ID: manish07sep@gmail.com

***Rahul Devrani***
Wadia Institute of Himalayan Geology, Dehradun-248001, India; University School of Environment Management, Guru Gobind Singh Indraprastha University Delhi, India
Email ID: rahuldevrani18@gmail.com

***Rahul Mohan***
ESSO National Centre for Polar and Ocean Research, Ministry of Earth Sciences, Government of India, Headland Sada, Vasco-Da-Gama, Goa, India
Email ID: rahulmohan@ncpor.res.in; rahulmohan@ncpor.res.in

***Rohit Kumar***
Discipline of Geology, School of Sciences, Indira Gandhi National Open University, New Delhi, India.
Email ID: rohitkumargeologist@gmail.com

***Romulus Costache***
Department of Civil Engineering, Transilvania University of Brasov, Brasov, Romania
Email ID: romuluscostache2000@yahoo.com; romulus.costache@icub.unibuc.ro

***Satarupa Mitra***
University Center for Research & Development (UCRD), Chandigarh University, Mohali, Punjab, India
Email ID: ruupa86@gmail.com

***Yogesh Ray***
ESSO-National Centre for Polar and Ocean Research, Ministry of Earth Sciences, Government of India, Headland Sada, Vasco-Da-Gama, Goa, India
Email ID: yogeshray@gmail.com; yogesh@ncpor.res.in

***Aditya Mishra***
Hemvati Nandan Bahuguda Garhwal University, Srinagar, Uttarakhand, India
Email ID: mishraaditya557@gmail.com

***Aljasil Chirakkal***
Indian Institute of Science Education and Research Mohali, Manauli, Punjab, India
Email ID: jasilchirakkal@gmail.com

***Alvarinho J. Luis***
ESSO-National Centre for Polar and Ocean Research, Ministry of Earth Sciences, Government of India, Headland Sada, Vasco-Da-Gama, Goa, India
Email ID: alvjuis@gmail.com

***Aman Arora***
Bihar Mausam Seva Kendra, Planning and Development Department, Government of Bihar, Patna, Bihar, India
Email ID: aman.jmi01@gmail.com

***Ambili Anoop***
Indian Institute of Science Education and Research
Mohali, Manauli, Punjab, India
Email ID: anoop.ambili85@gmail.com

***Anand K. Singh***
ESSO-National Centre for Polar and Ocean Research,
Ministry of Earth Sciences, Govt. of India, Headland Sada,
Vasco-Da-Gama, Goa, India
Email ID: singhaaks@gmail.com; aks@ncpor.res.in

***Anant Pande***
Department of Endangered Species Management, Wildlife
Institute of India, Dehra Dūn, India
Email ID: anantpande1984@gmail.com; anant@wii.gov.in

***Anil Kumar***
Wadia Institute of Himalayan Geology, Dehradun,
India
Email ID: anilwihg@gmail.com

***Antonio Bonaduce***
Nansen Environmental and Remote Sensing Centre, Norway
Email ID: Antonio.Bonaduce@nersc.no

***Anupam Sharma***
Birbal Sahni Institute of Palaeosciences, Lucknow, India
Email ID: anupam110367@gmail.com

***Archana Singh***
ESSO-National Centre for Polar and Ocean Research,
Ministry of Earth Sciences, Govt. of India, Headland Sada,
Vasco-Da-Gama, Goa, India
Email ID: archanasingh@ncpor.res.in; archana.singh199101@gmail.com

***Ashutosh Venkatesh Prasad***
ESSO-National Centre for Polar and Ocean Research,
Ministry of Earth Sciences, Government of India,
Headland Sada, Vasco-Da-Gama, Goa, India
Email ID: venkatesh.ashu@gmail.com

***Astha Dangwal***
Discipline of Geology, School of Sciences, Indira Gandhi
National Open University, New Delhi, India.
Email ID: asthamahima@gmail.com

***Avinash Kumar***
National Centre for Polar and Ocean Research, Ministry
of Earth Sciences (Govt. of India) Headland Sada, Vasco-da-Gama, Goa, India; India & Department of Geography,
University of Calgary, Calgary, AB, Canada
Email ID: avinash@ncaor.gov.in; kumaravinash13@gmail.com

***Chandra Prakash Singh***
Department of Endangered Species Management, Wildlife
Institute of India, Dehra Dūn, India
Email ID: cpsingh@sac.isro.gov.in

***Devendra Singh***
Botanical Survey of India, Acharya Jagadish Chndra Bose
Indian, Botanic Garden, Howrah, India
Email ID: singhdrds@rediffmail.com

***Devojit Bezbaruah***
Department of Applied Geology, Dibrugarh University,
Dibrugarh, Assam, India
Email ID: devojit.bezbaruah@gmail.com

***Divya David T.***
ESSO-National Centre for Polar and Ocean Research,
Ministry of Earth Sciences, Government of India,
Headland Sada, Vasco-Da-Gama, Goa
Email ID: divya@ncpor.res.in

***Giribabu Dandabathula***
Regional Remote Sensing Centre (West), NRSC, Indian
Space Research Organization, Jodhpur, India
Email ID: dgb.isro@gmail.com

***Harish Chandra Nainwal***
Hemvati Nandan Bahuguda Garhwal University,
Srinagar, Uttarakhand, India
Email ID: nainwalhc@yahoo.co.in

***Jagriti Mishra***
School of Environmental Sciences, Jawaharlal Nehru
University, New Delhi, India; Civil Engineering Research
Institute for Cold Region, Sapporo,
Japan
Email ID: jagritimp@gmail.com

***Juhi Yadav***
ESSO-National Centre for Polar and Ocean Research,
Ministry of Earth Sciences, Government of India,
Headland Sada, Vasco-Da-Gama, Goa, India
Email ID: juhi@ncpor.res.in

***K. P. Krishnan***
ESSO-National Centre for Polar and Ocean Research,
Ministry of Earth Sciences, Government of India,
Headland Sada, Vasco-Da-Gama, Goa, India
Email ID: kpkrishnan@gmail.com

***Kasturi Mukherjee***
Associate Professor, Department of Geography, Adamas
University, Kolkata. India
Email ID: kasturi.mukherjee@adamasuniversity.ac.in

**_Keshava Balakrishna_**
Department of Civil Engineering, Manipal Institute of Technology, Manipal Academy of Higher Education, Manipal, Karnataka, India
Email ID: k.balakrishna@manipal.edu

**_Kiledar Singh Tomar_**
ESSO-National Centre for Polar and Ocean Research, Ministry of Earth Sciences, Government of India, Headland Sada, Vasco-Da-Gama, Goa, India
Email ID: kdsingh108@hotmail.com; kiledarsingh108@gmail.com

**_Kiran Singh_**
Department of Geography, Central University of Punjab, Bathinda, Punjab, India
Email ID: kiran.singhgeo@gmail.com; kiran.singh@cup.edu.in

**_KN Prudhvi Raju_**
Banaras Hindu University, Varanasi, Uttar Pradesh, India
Email ID: knpraju1954@gmail.com

**_Krishna G. Misra_**
Birbal Sahni Institute of Palaeosciences, 53-University Road, Lucknow, India
Email ID: kg_misra@bsip.res.in

**_M. Javed Beg_**
ESSO-National Centre for Polar and Ocean Research, Ministry of Earth Sciences, Government of India, Headland Sada, Vasco-Da-Gama, Goa, India
Email ID: javed.beg@gmail.com; mjbeg@ncpor.res.in

**_Mallikarjun Mishra_**
Banaras Hindu University, Varanasi, Uttar Pradesh, India
Email ID: mallikarjungeobhu2016@gmail.com

**_Mehta Bulbul_**
Indian Institute of Science Education and Research Mohali, Manauli, Punjab, India
Email ID: bulbulmehta11@gmail.com

**_Mondip Sarma_**
Department of Applied Geology, Dibrugarh University, Dibrugarh, Assam, India
Email ID: mondip_sarma@yahoo.com

**_Neelam Verma_**
Amity School of Earth & Environment Sciences, Amity Education Valley Gurugram, Manesar, Panchgaon, Haryana, India
Email ID: neeelverma@gmail.com

**_Nguyen Thuy Linh_**
Institute of Applied Technology, Thu Dau Mot University, Binh Duong Province, Vietnam
Email ID: nguyenthuylinh@tdmu.edu.vn

**_Nivedita Mehrotra_**
Birbal Sahni Institute of Palaeosciences, Lucknow, India
Email ID: nivedita_mehrotra23@hotmail.com

**_ON Bhargava_**
Honorary Professor, Geology Department, Panjab University, Chandigarh, INSA Honorary Scientist
Email ID: onbhargava@gmail.com

**_Parv Kasana_**
Department of Geology, University of Delhi, Chattra Marg, Delhi, India
Email ID: kasanaparv009@gmail.com

**_Prabhat Ranjan_**
Central Pollution Control Board, Ministry of Environment, Forest and Climate Change, Parivesh Bhawan, East Arjun Nagar, Shahdara, Delhi, India
Email ID: prabhatranjan.jnu@gmail.com

**_Pradeep Srivastava_**
Indian Institute of Technology, Roorkee; Wadia Institute of Himalayan Geology, Dehradun, India
Email ID: pradeep71s@rediffmail.com

**_Prateek Gantayat_**
Lancaster Environment Centre, Lancaster University, UK
Email ID: gantayat.prateek@ymail.com

**_Pratik Dash_**
Assistant Professor, Department of Geography, Khejuri College, West Bengal, India
Email ID: pratik.bidyut@gmail.com

**_Praveen K. Mishra_**
Wadia Institute of Himalayan Geology, Dehradun, Uttarakhand, India
Email ID: pk.geobhu@gmail.com

**_Prem Chandra Pandey_**
Center for Environmental Sciences & Engineering (CESE), School of Natural Sciences (SoNS), Shiv Nadar Institution of Eminence, Gautam Buddha Nagar, Uttar Pradesh, Delhi, India
Email ID: prem26biT@gmail.com

**Prodip Mandal**
Department of Geography, Cooch Behar Panchanan Barma University, Cooch Behar, West Bengal, India

**R. Shankar**
The Institute of Mathematical Sciences, Chennai, Tamil Nādu, India
Email ID: shankar@imsc.res.in; shankar.chennai@gmail.com

**Rahul Devrani**
Wadia Institute of Himalayan Geology, Dehradun, India; University School of Environment Management, Guru Gobind Singh Indraprastha University Delhi, India
Email ID: rahuldevrani18@gmail.com

**Rahul Raj**
Centre for Korean Studies, School of Language, Literature & Culture Studies Jawaharlal Nehru University, New Delhi, India
Email ID: rahulrajkorean@gmail.com

**Ravi S. Maurya**
Birbal Sahni Institute of Palaeosciences, 53 University Road, Lucknow, India
Email ID: ravishankarmaurya94@gmail.com

**Richard Davy**
Nansen Environmental and Remote Sensing Centre, Norway
Email ID: Richard.Davy@nersc.no

**Rohit Srivastava**
ESSO-National Centre for Polar and Ocean Research, Ministry of Earth Sciences, Government of India, Headland Sada, Vasco-Da-Gama, Goa, India
Email ID: rohits@ncpor.res.in; rohitphy@gmail.com

**Roshin P. Raj**
Nansen Environmental and Remote Sensing Centre, Norway
Email ID: roshin.raj@nersc.no

**Saeid Janizadeh**
Department of Watershed Management Engineering and Sciences, Faculty in Natural Resources and Marine Science, Tarbiat Modares University, Tehran, Iran
Email ID: janizadehsaeid@modares.ac.ir

**Sagar F. Wankhede**
Department of Geoinformatics, Mangalore University, Mangalore, Karnataka, India
Email ID: swankhede436@gmail.com

**Sajeed Zaman Borah**
Techno Canada Inc., Barmer, Rajasthan, India
Email ID: sajeedborah28@gmail.com

**Sandhya Misra**
Birbal Sahni Institute of Palaeosciences, 53 University Road, Lucknow, India
Email ID: sandhyabsip@gmail.com; sandhya.sharma@bsip.res.in

**Sangita Kumari**
Department of Civil Engineering, Indian Institute of Technology Bombay, Powai, Maharashtra, India
Email ID: singh.sangita15@gmail.com

**Santosh K. Shah**
Birbal Sahni Institute of Palaeosciences, Lucknow, India
Email ID: santoshkumar_shah@bsip.res.in

**Shailendra Saini**
ESSO-National Centre for Polar and Ocean Research, Ministry of Earth Sciences, Government of India, Headland Sada, Vasco-Da-Gama, Goa, India
Email ID: shailendra.saini@gmail.com; shailendra@ncpor.res.in

**Shraban Sarkar**
Department of Geography, Cooch Behar Panchanan Barma University, Cooch Behar, West Bengal, India
Email ID: shraban.sarkar@yahoo.com

**Shridhar D. Jawak**
Svalbard Integrated Arctic Earth Observing System (SIOS), SIOS Knowledge Centre, Longyearbyen, Svalbard, Norway
Email ID: shridhar.jawak@gmail.com

**Sukumar Parida**
Physical Research Laboratory, Navrangpura, Ahmedabad, India
Email ID: paridasukumar@gmail.com

**Shyam Ranjan**
School of Environmental Sciences, Jawaharlal Nehru University, New Delhi, India
Email ID: shyamrockranjan@gmail.com

**Siddhi Garg**
Wadia Institute of Himalayan Geology, Dehradun, Uttarakhand, India
Email ID: g.siddhi86@gmail.com

***Som Dutt***
Wadia Institute of Himalayan Geology, Dehradun, Uttarakhand, India
Email ID: somduttmudgil1@gmail.com

***Sourav Chatterjee***
National Centre for Polar and Ocean Research, Ministry of Earth Sciences (Government of India) Headland Sada, Vasco-da-Gama, Goa, India
Email ID: sourav@ncpor.res.in

***Surajit Ghosh***
International Water Management Institute, Colombo, Sri Lanka
Email ID: surajitghosh.ind@gmail.com

***Swathi M.***
ESSO-National Centre for Polar and Ocean Research, Ministry of Earth Sciences, Government of India, Headland Sada, Vasco-Da-Gama, Goa, India
Email ID: mswathi@ncpor.res.in

***Takuya Inoue***
Civil Engineering Research Institute for Cold Region, Sapporo, Japan
Email ID: inouetakuya@hiroshima-u.ac.jp

***Tapos Kumar Goswami***
Department of Applied Geology, Dibrugarh University, Dibrugarh, Assam, India
Email ID: taposgoswami@gmail.com

***Uday Sharama***
ESSO-National Centre for Polar and Ocean Research, Ministry of Earth Sciences, Government of India, Headland Sada, Vasco-Da-Gama, Goa, India
Email ID: uday@ncaor.gov.in; udaysharmaofficial@gmail.com

***Uma Kant Shukla***
Center for Advanced Study in Geology, Institute of Science, Banaras Hindu University, Varanasi, India
Email ID: shukla_umakant@yahoo.com

***Upendra Baral***
Key Laboratory of Continental Collision and Plateau Uplift, Institute of Tibetan Plateau Research, Chinese Academy of Sciences, Beijing China; Kathmandu Center for Research and Education, Chinese Academy of Sciences, Tribhuvan University, Kirtipur
Email ID: upendrabaral@gmail.com

***Uttam Pandey***
1-Birbal Sahni Institute of Palaeosciences, Lucknow-226007, India; 2-Graduate School of Environmental Studies, Nagoya University, Nagoya, Japan
Email ID: uttampandeygeo@gmail.com

***Varun Narayan Mishra***
Centre for Climate Change and Water Research (C3WR), Suresh Gyan Vihar University, Jaipur, Rajasthan, India
Email ID: varun9686@gmail.com

***Yadav Ankit***
Indian Institute of Science Education and Research Mohali, Manauli, Punjab, India
Email ID: ankitvrishm22@gmail.com

# Foreword

The polar regions of the world comprise the Antarctic, the Arctic, and the regions of Hindukush-Karakorum-Himalaya-Tibet jointly referred to as the three poles of our planet. These distinct geographical locations, separated from each other are the natural laboratories of the earth to study and quantify the effects of climate change. Additionally, the regions of the three poles are remote and have extreme conditions, and possess challenges to study in detail. With the advent of satellites and remote sensing technology, these regions can be studied like never before.

This book provides the latest information about the techniques, advances, and wide-ranging applications of remote sensing on all three poles. The degradation of polar ice sheets and the Himalayan glaciers to natural and anthropogenic forcing on local, regional, and global scales is elucidated in dedicated chapters. Additionally, chapters related to flora and fauna, landscape changes, and multidisciplinary research activities undertaken at Indian research stations in the Antarctic and Arctic are also briefly touched up.

All the editors have done a commendable task to pool up the present understanding related to the recent advances in the field of remote sensing along with the case studies from the experts and the leading scholars. I am sure the book will serve a useful purpose for academicians, researchers, and students.

M. Ravichandran

# Preface

Poles are the most sensitive to climate change and their impacts on different components of the earth system in the polar regions are becoming prominently more visible. Owing to the intense focus of the research community on the study in all the three polar regions, and the allocation of proportionately high funds for research by international and national organizations of various countries in these sensitive regions of our planet, we now have a better understanding of the three pole environments.

While working on a theme paper that involved the Himalayas, the Arctic, and the Antarctic, members of the present group of editors have noticed that, though there is an abundance of research work going on in different spheres of the three poles using geospatial data and technology, it is dispersed and not compiled. And also, there is a dearth of books that systematically account for research works relating to themes on any of the spheres of our planet, such as remote sensing technology. Aiming at filling that gap, we, in this publication, seek to provide an insight into advancements in geospatial techniques, relating to multidisciplinary study of the three poles, namely Antarctica, the Arctic, and the Himalayas. This book provides both traditional as well as advanced geospatial techniques used in lithospheric, atmospheric, hydrospheric, biospheric, and anthropospheric contexts of the three poles along with their strengths, limitations, and gap areas.

The importance of this book is also because there is geological evidence that proves the existence of several supercontinents like Columbia, Rodinia, Pangea, and Gondwana, which provide clues about the connectedness of all the continental landmasses at some points in time of the history of planet Earth. This book brings together research works on different aspects of all the spheres of all the three poles into one place. Even though the breaking up of the most recent supercontinent Pangea and the drifting away of continents have created the present-day Himalayas, the Arctic, and Antarctica, they are connected through different subsystems like atmospheric and oceanographic components. The two distinctive poles, the North Pole and the South Pole of our planet, are situated in the Arctic and Antarctica respectively. Whereas the Arctic is an ocean surrounded by continents, in contrast, Antarctica is a continent surrounded by oceans and due to its immense altitude and being the youngest orogeny on Earth, the Himalaya is fondly referred to as the third pole. Not only are these landmasses peculiar in their own physical, climatic, and ecosystem components, but are interconnected by the teleconnections of the atmosphere, hydrosphere, lithosphere, biosphere, and cryosphere through different feedback systems.

Recent developments in satellite remote sensing, geoinformatics, and landscape evolution modeling techniques have made it possible to trace the harsh effects of climate change on the three poles, two of them being the icy continent surrounded by oceans and the partially frozen ocean surrounded by continents. The increased resolution of the satellite data has aided in the quantification of ever-changing landforms and surface processes. This book attempts to understand the subtle link between climate change and its effects on the cryospheric and related processes. There are a total of five subsections in the book that aim to include chapters dealing with the Quaternary geology and geomorphology of Antarctica, Arctic, and the Himalaya, GPS, geodesy, geodynamics, glacier monitoring, glacier dynamics, sea–ice interaction with the continent, hydrology aquatic and terrestrial floral and faunal dynamics, etc., depending upon the availability of contributions and the book size-related constraints.

The five sections, Section I to Section V, comprise original and review research articles on various aspects of our planet's systems under the boundaries of the three poles regions. Section I, entitled "Earth Observation (EO) and Remote Sensing (RS) Applications in Polar Studies" includes six chapters that encompass a critical review of past, present, and future satellite missions, their data characterizes, and availability; data accuracy assessments, and various software packages, tools, add-ons for morphometry and landscape evolution modeling; and finally reviews of various spectral indices used for identification and assessment of the health of respective elements of identification (EOI) for all the spheres of our planet, such as lithosphere, hydrosphere, biosphere, atmosphere, and anthroposphere. Section II of this book has five chapters dedicated to "Antarctica: The Southernmost Continent having the South Pole, Environment and Remote Sensing" dealing with glacial dynamics, terrestrial quaternary deglaciation, Antarctic biodiversity relating to geospatial technology, prospects of Bryophytes in the Larsemann Hills, and sea-ice variability relating to physical forcing. Section III, focusing on "Himalayas: The Third Pole Environment and Remote Sensing" has 13 chapters covering all the spheres, i.e., lithosphere, hydrosphere, biosphere, atmosphere, and anthroposphere. In Section IV, five chapters on "The Arctic: The Northernmost Ocean Having the North Pole Environment and Remote Sensing" focus on gaps in polar research, glacier facies evaluation with high-resolution satellite products, supraglacial lakes impact on Greenland Ice Sheets dynamics, and aerosol variation over space and time in the polar regions. The last section, Section V, has two chapters covering research collaboration efforts among national and international polar research organizations, and an overview of the multi-disciplinarity of the National Antarctic Programs of India.

Since the book covers a very wide spectrum of the research scope, we could not include chapters on all the aspects of all the spheres of the three poles. For example, there is an absence of chapters on lithospheric, atmospheric, and anthropospheric contexts in Section II dealing with Antarctic Environments. Similarly, Section IV lacks chapters on topics falling within the lithosphere, biosphere, and anthroposphere. Though Himalayan environments are dealt with in Section III that also covers chapters on all the five spheres, there is great scope for including quality chapters on different specific topics, e.g., relating modeling and remote sensing of different elements of those five systems of our planet. Hopefully, these topics will be covered in separate books under this special series called "Advancements in Remote Sensing Technology and The Three Poles."

*Dr Manish Pandey*
*Dr Prem Chandra Pandey*
*Dr Yogesh Ray*
*Dr Aman Arora*
*Dr Shridhar D. Jawak*
*Prof Uma Kant Shukla*

## List of Acronyms

**Chapter 1**
TPR: Third Pole Region
NPR: Northern Polar Region
SPR: Southern Polar Region
AMOC: Atlantic Meridional Overturning Circulation
ENSO: El Niño Southern Oscillation
EMS: Electromagnetic Spectrum
RADAR: Radio Detection and Ranging
HFT: Himalayan Frontal Thrust
GEE: Google Earth Engine
MHD: Mahalanobis Distance classifier

**Chapter 2**
EO: Earth Observation
NASA: National Aeronautics and Space Administration
HEO: High Earth Orbit
GSO: Geosynchronous Orbit
LEO: Low Earth Orbit
GRACE: Gravity Recovery and Climate Experiment
ICESat: Ice, Cloud, and land Elevation Satellite

**Chapter 3**
ALOS: Advance Land Observing Satellite
ASF: Alaska Satellite Facility
ASTER GDEM: Advanced Spaceborne Thermal Emission and Reflection Radiometer Global Digital Elevation Model
DEM: Digital Elevation Model
DGM: Digital Ground Model
DGPS: Differential Global Positioning System
DHM: Digital Height Model
DSM: Digital Surface Model
DTM: Digital Terrain Model
EGM: Earth Gravitational Model
EI: Error Index
ERDAS: Earth Resources Data Analysis System
GCP: Ground Control Point
GPS: Global Positioning System
ICESat:Ice, Cloud and land Elevation Satellite
IDEM: Intermediate Digital Elevation Model
InSAR: Interferometric Synthetic Aperture Radar
JAROS: Japan Resources Observation System Organization
JAXA: Japan Aerospace Exploration Agency
LiDAR: Light Detection and Ranging
LPS: Leica Photogrammetric Suite
MAE: Mean Absolute Error
ME: Mean Error
MERIT: Multi Error Removed Improved-Terrain
METI: Ministry of Economy, Trade and Industry
PALSAR: Phased Array L-band Synthetic Aperture Radar
PPP: Public Private Partnership
RMSE: Root Mean Square Error
RPC: Rational Polynomial Coefficient
RTC: Radiometric Terrain Correction
RTK-GNSS: Real Time Kinematic-Global Navigation Satellite System
SAR: Synthetic Aperture Radar
SOI: Survey of India
SPI: Stream Power Index
SRTM: Shuttle Radar Topography Mission
TWI: Topographic Wetness Index
UTM: Universal Transverse Mercator
WGS: World Geodetic System

**Chapter 4**
GIS: Geographical Information System
DEM: Digital Elevation Model

GUI: Graphical User Interface
LiDAR: Light Detection and Ranging
ALM: ACME-Accelerated Climate Modeling for Energy
TAK: Topographic Analysis Toolkit
ITC: Inter Tropical Convergence
ILWIS: Integrated Land and Water Information System
CERL: Construction Engineering and Research Laboratory
TAS: Terrain Analysis System
SAGA: System for Automated Geoscientific Analyses
TIN: Triangular Irregular Network
TAPES: Terrain Analysis Programs for Environmental Sciences

**Chapter 5**
LEM: Landscape Evolution Modeling
GEM: Global Environmental Monitoring
TIN: Triangular Irregular Network

**Chapter 6**
LiDAR: Light Detection and Ranging
UAVs: Unmanned Aerial Vehicles/Uncrewed Aerial Vehicles
VIS: Visible
NIR: Near Infra-Red/Near Infrared
MIR: Middle Infrared
RS: Remote Sensing
EO: Earth Observation
SPOT: Satellite Pour l'Observation de la Terre
RADAR: Radio Detection and Ranging
MODIS: Moderate Resolution Imaging Spectroradiometer
LAI: Leaf Area Index
EoD: Elements of Detection

**Chapter 7**
AIS: Antarctic Ice Sheet
ASTER: Advanced Spaceborne Thermal Emission and Reflection Radiometer
COMNAP: Council of Managers of National Antarctic Programs
DEM: Digital Elevation Model
DInSAR: Differential SAR Interferometry
EAIS: East Antarctic Ice Sheet
GCP: Ground Control Point
GRD: Ground Range Detected
HH: Horizontal transmit and horizontal receive
InSAR: Interferometric SAR
IW: Interferometric Wide
MIMC: Multi Image Multi Chip
PRG: Polar Record Glacier
PS-InSAR: Permanent Scatterer Interferometry
REMA: Reference Elevation Model of Antarctica
SAR: Synthetic Aperture Radar
SLC: Single Look Complex

**Chapter 8**
LGM: Last Glacial Maximum
EAIS: Eastern Antarctic Ice Sheet
Ka: Kilo years
Ma: Million Years
BP: Before Present
CRN: Cosmogenic Radionuclides
OSL: Optically Stimulated Luminescence
DML: Dronning Maud Land
MIS: Marine Isotope Stage

**Chapter 9**
APIS: Antarctic Pack Ice Seal program
CMFRI: Central Marine Fisheries Research Institute
CSIR: Council of Scientific and Industrial Research
ERDAS: Earth Resources Data Analysis System
ESRI: Environmental Systems Research Institute
ETM: Enhanced Thematic Mapper
GIS: Geographic Information System
MLC: Maximum Likelihood Classification
NIO: National Institute of Oceanography
OBIS: Ocean Biodiversity Information System
RPAS: Remotely Piloted Aircraft Systems
SPOT: Satellite pour l'Observation de la Terre
UAS: Unmanned Aerial System
UAV: Unmanned Aerial Vehicle
VHR: Very High Resolution
VTOL: Vertical Take-Off and Landing
WII: Wildlife Institute of India

**Chapter 10**
BSIP: Birbal Sahni Institute of Palaeobotany
CAL: Central National Herbarium
GSI: Geological Survey of India
IITM: Indian Institute of Tropical Meteorology
KM: Kilometer
KV: Kilowatt
LH: Larsemann Hills
NCPOR: National Centre for Polar and Ocean Research
NHO: National Hydrographic Office
OS: Schirmacher Oasis
SAC: Space Application Centre
SEM: Scanning Electron Microscope
SOI: Survey of India
WII: Wildlife Institute of India

**Chapter 11**
ACW: Antarctic Circumpolar Wave
AMJ: April, May, June
AT: Air Temperature

ECMWF: European Centre for Medium-Range Weather Forecasts
ENSO: El Niño–Southern Oscillation
JAS: July, August, September
JFM: January, February, March
NSIDC: National Snow and Ice Data Center
OND: October, November, December
SAM: Southern Annular Mode
SIC: Sea Ice Concentration
SIE: Sea Ice Extent
SMMR: Scanning Multichannel Microwave Radiometer
SSM/I: Special Sensor Microwave Imager
SSMIS: Special Sensor Microwave Imager/Sounder
SST: Sea Surface Temperature

**Chapter 12**
DZ: Detrital Zircons
ONGC: Oil and Natural Gas Corporation
OSL: Optically stimulated luminescence

**Chapter 13**
ARU: Aru Valley
GTA: Green Top of Aru
MS: Mean sensitivity
SD: Standard deviation
AC-1: First-order autocorrelation
EPS: Expressed Population Signal

**Chapter 14**
ICESat-2: Ice, Cloud, and land Elevation Satellite-2
GDEM: Global Digital Elevation Model
SAR: Synthetic Aperture Radar
NASA: National Aeronautics and Space Administration
ATLAS: Advanced Topographic Laser Altimeter System
GLAS: Geoscience Laser Altimeter System
NSIDC: National Snow and Ice Data Centre
GT: Ground Track
DAAC: Distributed Active Archive Centre
RGT: Reference Ground Tracks
ATBT: Algorithm Theoretical Basis Document
CWC: Central Water Commission (India)
REDD: Reducing Emissions from Deforestation and forest Degradation
SRTM: Shuttle Radar Topography Mission
ACE: Altimeter Corrected Elevation
GCPs: Ground Control Points
TIN: Triangulated Irregular Network

**Chapter 15**
EHE: Extreme Hydrological Event
GEE: Google Earth Engine
GIS: Geographical Information System
CHRS: Center for Hydrometeorology and Remote Sensing
DEM: Digital Elevation Model
NBSS-LUP: National Bureau of Soil Survey and Land Use planning
NW: North West
NE: North East
RUSLE: Revised Universal Soil Loss Equation
MMF: Morgan-Morgan-Finney
SWAT: Soil and Water Assessment Tool
WEPP: Water Erosion Prediction Project
USLE: Universal Soil Loss Equation
LOF: Lake Outburst Flow
LULC: Land Use Land Cover
NE–SW: North East-South West

**Chapter 16**
HCO: Holocene Climate Optimum
IPCC: Intergovernmental Panel on Climate Change
ISM: Indian summer monsoon
GHGs: Greenhouse gases
NDVI: Normalized difference vegetation index
NEH: Northeast Himalayas
ENSO: El Niño-Southern Oscillation
IOD: Indian Ocean Dipole
NAO: North Atlantic Oscillation
MAP: Mean annual precipitation
MWP: Medieval Warm Period
LIA: Little Ice Age
ACC: Abrupt climate change

**Chapter 17**
AHP: Analytical Hierarchy Process
AUC: Area Under Curve
CI: Convergence Index
CR: Consistency Ratio
DEM: Digital Elevation Model
FHI: Flood Hazard Index
LULC: Land Use Land Cover
MCDA: Multi Criteria Decision Analysis
ROC: Receiver Operating Characteristics
TPI: Topographic Position Index
TWI: Topographic Wetness Index

**Chapter 18**
SRTM DEM: Shuttle Radar Topography Mission and Digital Elevation Model.
MBT: Main Boundary Thrust.
HFT: Himalayan Frontal Thrust.
A.S.T.M: American Standard Test Sieve Series.
CT: coarse truncation.
FT: fine truncation
C.M.: "C"–Coarser one percentile value in micron and "M" median value in micron on log-probability scale

**Chapter 19**
DEM: Digital Elevation Models
UARC: Upper Alaknanda River catchment
GIS: Geographical Information System

**Chapter 20**
CH: Central Himalaya
EH: Eastern Himalaya
GlaThiDa: Glacier Thickness Database
GPR: Ground Penetrating Radar
GSI: Geological Survey of India
HP: Himachal Pradesh
IHR: Indian Himalayan region
J&K: Jammu and Kashmir
UK: Uttarakhand
WH: Western Himalaya

**Chapter 21**
GLOF: Glacial Lake Outburst Floods
HHC: High Himalayan crystalline
IMD: Indian Meteorological Department
ISM: Indian Summer Monsoon
ISZ: Indus Suture Zone
ITCZ: Inter Tropical Convergence Zone
KF: Karakoram Fault
KFZ: Karakoram Fault Zone
KMC: Karakoram Metamorphic Complex
KPC: Karakoram Plutonic Complex
LLOF: Landslide lake outburst floods
MIS: Marine Isotopic Stage
SSZ: Shyok Suture Zone
SWD: Slack water deposits
TRMM: Tropical Rainfall Measuring Mission
TSS: Tethys sedimentary succession

**Chapter 22**
USLE: Universal Soil Loss Equation
RUSLE: Revised Universal Soil Loss Equation
MUSLE: Modified Universal Soil Loss Equation
LULC: Land Use-Land Cover
ANSWERS: The Areal Non-point Source Watershed Environment Response Simulation
CREAMS: Chemical Runoff and Erosion from Agricultural Management System
SLEMSA: Soil Loss Estimation Model for Southern Africa
EPIC: Erosion Productivity Impact Calculator
GLEAMS: Groundwater Loading Effects of Agricultural Management System
MADALUS: Mediterranean Desertification and Land Use
KYERMO: Kentucky Erosion Model
WEPP: Water Erosion Prediction Project
LISEM: Limburg Soil Erosion Model
EUROSEM: European Soil Erosion Model
GAMES: Guelph Model for Evaluating the Effects of Agricultural Management Systems on Soil Erosion and Sedimentation
SWAT: Soil and Water Assessment Tool
EROSION 2D/3D = 2D Rainfall Erosion Model
DEM: Digital Elevation Model
LC: Lower Chambal basin
NBSS-LUP: National Bureau of Soil Survey and Land Use Planning
CHRS: Center for Hydrometeorology and Remote Sensing
ASF: Alaska Satellite Facility
RG: Rishiganga basin

**Chapter 23**
ISM: Indian summer monsoon
WD: Western Disturbances
LGM: Last Glacial Maximum
YD: Younger Dryas
HCO: Holocene Climatic Optimum
MWP: Medieval Warm Period
LIA: Little Ice Age
RWP: Roman Warm Period
DACP: Dark Ages Cold Period
LULC: Land-Use/Land-Cover Change
MSS: Multispectral Scanner
TM: Thematic Mapper
ETM +: Enhanced Thematic Mapper Plus

**Chapter 24**
LULC: Land Use/Land Cover
LULCC: LULC changes
MSS: Multispectral Scanner System
OLI: Operational Land Imager
FCC: False Colour Composite
MLC: Maximum-Likelihood Classification
OA: Overall Accuracy

**Chapter 25**
GMT: Global mean temperature
SOECO: Societal, ecological, and economical
GLOFs: Glacial lake outburst floods
HK: Hindu-Kush

**Chapter 26**
SAR: Synthetic Aperture Radar
SI: Snow and Ice
GPR: Ground Penetrating Radar
MSS: Multispectral
WV-2: WorldView-2
WV-3: WorldView-3
DN: Digital Number
ISODATA: Iterative Self-Organizing Data Analysis

VHR: Very High-Resolution
USA: United States of America
NIR: Near Infrared
SWIR: Shortwave Infrared
MHD: Mahalanobis Distance
MD: Minimum Distance
ENVI: Environment for Visualizing Images
FLAASH: Fast Line-of-Sight Atmospheric Analysis of Spectral Hypercubes
MODTRAN4: Moderate Resolution Atmospheric Transmission 4
O1: Operator 1
O2: Operator 2
O3: Operator 3
O4: Operator 4
TERCAT: Terrain Categorization

**Chapter 27**

$\sigma_{xx}$: longitudinal stress (Pa)
$\sigma_{yy}$: lateral stress (Pa)
$\tau_{xy}$: Shear stress (Pa)
$\sigma_1$: prinicipal stress (Pa)
$\sigma_2$: principal stress (Pa)
$\sigma_v$: Square of von Mises stress (Pa)
Rxx: von Mises stress (Pa)
$\tau_c$: Critical stress (Pa)
d: Crevasse depth (m)
$\rho_i$: Ice density ($kg/m^3$)
b: Water depth in crevasse (m)
$K_I$: Stress intensity ($Pam^{0.5}$)
$K_{IC}$: Critical stress intensity ($Pam^{0.5}$)
$H_L$: Lake height (m)
$H_{Li}$: Reference lake height (m)
$A_{Li}$: Reference lake area ($m^2$)
$\zeta$: Difference between lake and channel height (m)
w: Channel width (m)
$f_R$: Channel roughness parameter
S: Channel slope
g: Acceleration due to gravity ($m/s^2$)
$H_c$: Channel height (m)
L: Latent heatof fusion (kJ/kg)
SWR: Short wave radiation ($W/m^2$)
alb: Albedo
$h_{lake}$: lake depth (m)
$\sigma$: Stefan Boltzmann constant ($W / m^2/k^4$)
$LWR_{in}$: Incoming longwave radiation ($W/m^2$)
H: Sensible heat flux ($W/m^2$)
LE: Latent heat flux ($W/m^2$)

**Chapter 28**

AO: Arctic Ocean
AOO: Arctic Ocean Oscillation
AW: Atlantic Water
BSO: Barents Sea Opening
CMIP: Coupled Model Intercomparison Project
ECV: Essential Climate Variable
FS: Fram Strait
GMSL: Global mean sea-level
GRACE: Gravity Recovery and Climate Experiment
NwASC: Norwegian Atlantic Slope Current
SLP: Sea-level pressure
SSH: Sea surface height

**Chapter 29**

AOD: Aerosol Optical Depth
AERONET: Aerosol Robotic Network
BC: Black Carbon
DB: Deep Blue
DJF: December-January-February
DT: Dark Target
ECMWF: European Centre for Medium-Range Weather Forecasts
EOS: Earth Observing System
ERA: ECMWF Reanalysis
IFS: Integrated Forecast System
JJA: June, July, August
MAM: March, April, May
MODIS: MODerate resolution Imaging Spectroradiometer
OC: Organic Carbon
POLAERNET: POLar AERosol NETwork
RH: Relative Humidity
SH: Specific Humidity
SON: September, October, November

**Chapter 30**

ISEA: Indian Scientific Expedition to Antarctica
ISESO: Indian Scientific Expedition to the Southern Ocean
HSM: Historic Site and Monuments
NCPOR: National Centre for Polar and Ocean Research
CEP: Committee for Environmental Protection
COMNAP: Council of Managers of National Antarctic Programs
SCAR: Scientific Committee on Antarctic Research
CCAMLR: Commission for the Conservation of Antarctic Marine Living Resources
DML: Dronning Maud Land
DROMLAN: Dronning Maud Land Air Network
MADICE: Mass Balance, Dynamics, and Climate of the Central Dronning Maud Land Coast, East Antarctica
SONIC: Schirmacher Oasis Nippon (Japan) India Coring
SIWHA: Sea Ice and Westerly Winds during the Holocene in coastal Antarctica
GeoEAIS: Geological Exploration of the Amery Ice Shelf
IAOFA: Integrated Atmospheric Observation Facility for Antarctica

**Chapter 31**

ACAP: Arctic Contaminants Action Program
AFoPS: Asian Forum of Polar Sciences
AMAP: Arctic Monitoring and Assessment Programme
APECS: Association of Polar Early Career Scientists
ASSW: Arctic Science Summit Week
CAFF: Conservation of Arctic Flora and Fauna
EPB: European Polar Board
EPPR: Emergency Prevention, Preparedness and Response
FARO: Forum of Arctic Research Operators
IAG: Information Advisory Group
IASC: International Arctic Science Committee
IASSA: International Arctic Social Science Association
IPCC: Intergovernmental Panel on Climate Change
IPRN: Indian Polar Research Network
MoES: Ministry of Earth Sciences
NCPOR: National Centre for Polar and Ocean Research
NySMAC: Ny-Ålesund Science Managers Committee
PAME: Protection of the Arctic Marine Environment
PEI: Polar Educators International
RICC: Research Infrastructure Coordination Committee
RiS: Research in Svalbard
RSWG: Remote Sensing Working Group
SAON: Sustaining Arctic Observing Networks
SDMS WG: SIOS data management system working group
SDWG: Sustainable Development Working Group
SIOS: Svalbard Integrated Arctic Earth Observing System
SOAG: Science Optimization Advisory Group
SSF: Svalbard Science Forum
UArctic: University of Arctic

# Section I

## Earth Observation (EO) and Remote Sensing (RS) Applications in Polar Studies

# 1

# The Three Poles

## Advances in Remote Sensing in Relation to Spheres of the Planet Earth

*Manish Pandey[1,2,*], Prem C. Pandey[3], Yogesh Ray[4], Aman Arora[5], Shridhar Digambar, Jawak[6], and Uma Kant Shukla[7]*

[1] *University Center for Research & Development (UCRD), Chandigarh University, Mohali 140413, Punjab, India*
[2] *Department of Civil Engineering, University Institute of Engineering, Chandigarh University, Mohali 140413, Punjab, India*
[3] *School of Natural Sciences, Shiv Nadar Institution of Eminence, Greater Noida, Uttar Pradesh 201314, India*
[4] *National Centre for Polar and Ocean Research, Ministry of Earth Sciences, Government of India, Vasco-da-Gama, India*
[5] *Bihar Mausam Seva Kendra, Planning and Development Department, Government of Bihar, Patna 800015, Bihar*
[6] *SIOS – Svalbard Integrated Arctic Earth Observing System, University Centre in Svalbard, Longyearbyen, Norway*
[7] *Center for Advanced Study in Geology, Institute of Science, Banaras Hindu University, Varanasi 221005, India*
[*] *Corresponding author*

## 1.1 Introduction

Planet Earth functions as an integrated, complex, adaptive system (Steffen et al., 2020) composed of dynamic and interdependent sub-systems making life possible and sustaining it (Mahaffy et al., 2018). The scale of operation of the systems/sub-systems of the Earth vary from interplanetary through intra-planetary to sub-planetary levels. Miller and Miller (1982) have included everything necessary about this planet as a system when they state, "The Planet Earth, from its center to the outer limits of its atmosphere, including everything in and on it, is a mixed living and nonliving system within the solar system, the Milky Way galaxy, and, ultimately, the universe." Among the sub-systems of the Earth System are the geosphere, hydrosphere, cryosphere (a sub-system of the hydrosphere as per some classifications), atmosphere, biosphere, and anthroposphere (Williams and Ferrigno, 2012). These systems and sub-systems are connected through various feedbacks via exchange of matter and energy (O'Neill and Steenman-Clark, 2002; Reid et al., 2010). For instance, through the biogeochemical cycle, which propounds cyclic transformation and transport of matter and energy on Earth (Rodhe, 1992), a connection among the geosphere, biosphere, atmosphere, and hydrosphere has been established (Schlesinger et al., 2011). Similarly, the part of the hydrosphere, the oceans, transports and recycles sediment, nutrients, and temperature through the global conveyor belt of ocean circulation (Yamazaki and Trigg, 2016). Another example of global transfer of energy and matter through the component of atmosphere is explained through the application of the general circulation model (Corby et al., 1972), which explains the movement of atmospheric layers [layers of mechanical mix of various gases, e.g., $N_2$, $O_2$, Ar, $CO_2$, and different types of aerosol particles, e.g., black carbon, biomass burning aerosols, mineral dust, sea salt (Masson-Delmotte et al., 2021)], facilitating the transfer of matter and energy among the spheres of our planet. A brief concept of the Earth as a system and its different components, together with their connectedness through various feedbacks and loops, is described in Section 1.1.1. The structure of this chapter follows the form of: (a) the Planet Earth as a system and its sub-systems are described; (b) the role of the three poles and the Three Pole Regions (TPR) in interacting sub-systems of Planet Earth is briefly described; (c) how remote sensing technology has facilitated the advancement in research of our planet in the Three Poles Regions for mapping/monitoring/quantitative analysis of sub-systems; (d) advances in remote sensing technology in relation to process-form studies in the TPR; (e) the aim of the book covered in all its five sections; (f) overview of all the chapters in the book; and (g) summary and recommendations.

*Advances in Remote Sensing Technology and the Three Poles*, First Edition. Edited by Manish Pandey, Prem C. Pandey, Yogesh Ray, Aman Arora, Shridhar D. Jawak, and Uma K. Shukla.

### 1.1.1 Earth as a System and Components of the Earth System

The Earth, which evolved over a period of 4.6 Ga (giga-annum), functions as one planetary system composed of interconnected individual parts as sub-systems or components (Condie, 2011). In system theory lexicon, the spheres of the Earth are called components or sub-systems. These sub-systems, namely geosphere (lithosphere comprising of crust, and solid part of the upper mantle, upper and lower mantle, and core), hydrosphere (including limnosphere, oceanosphere, and cryosphere), atmosphere, biosphere, and anthroposphere are in the process of continuous change and work interconnectedly to form one entity known as the Earth System.

### 1.1.2 Role of the "Three Poles" and the Three Poles Regions in the Earth System

In the executive summary of the IPCC summary for policy-makers, the role of TPR is stated in a very concise and accurate manner, and states, "The polar regions are losing ice, and their oceans are changing rapidly. The consequences of this polar transition extend to the whole planet and are affecting people in multiple ways." Through the global ocean conveyor belt carrying warm waters from the tropical regions to the colder polar regions, the North Polar Region (NPR) and the South Polar Region (SPR) help to redistribute temperatures over the globe and play a crucial role in maintaining the horizontal temperature and pressure gradients. These zones, through interaction of the matter–energy couplet at the hydrosphere–atmosphere interface, help in maintaining the quasi-equilibrium suitable for sustaining life forms on the planet. The third pole, through its role in origin of the monsoon (Dell'Osso and Chen, 1986) and hosting the third most extensive glaciers and frozen water sources (Zhang et al., 2015), plays a crucial role in the Earth System functioning.

#### 1.1.2.1 Defining the Three Poles, Three Poles Regions, and Their Geographical Extent

The point of intersection of the Earth's rotational axis, at an axial tilt angle of about 23.4365472133° [as on 1 January 2021 (Bertrand and Legendre, 2021)], varies between 22.5° and 24.5° during a year cycle, with its surface defining two poles: The Geographical south and the Geographical north poles. These poles are two points located on the Antarctica continent, and the Arctic Ocean (Figures 1.1A, B, D, and E). The area of the globe enclosed by the 66.5° N parallel defines the North Polar Region (NPR), whereas the area south of the 66.50° S parallel demarcates the South Polar Region (SPR) (Strahler, 2013).

Yao et al. (2012a) state that since the area encompassing Tibet and the surrounding mountain ranges, owing to it high altitude above sea level being able to sustain frozen water, has the largest inventory of glaciers and snowfields after the North and South polar regions, scientists refer to it as the "Third Pole." The average elevation of the "largest and the highest mountain region," called the Third Pole, is over 4000 m above sea level (a.s.l.). This third pole hosts all the peaks in the world above 7000 m a.s.l., some of them including "fourteen world-acknowledged mountains over 8000 m a.s.l., such as the Mt Everest (8848.5 m a.s.l.), Nanga Parbat (8125 m a.s.l.), K2 (8611 m a.s.l.), Annapurna (8078 m a.s.l.), Gosainthān (8027 m a.s.l.), Kanchenjunga (8586 m a.s.l.). (Yao et al., 2012a). Figures 1.1C and F show the position of the Third Poles on the globe and on the map, respectively.

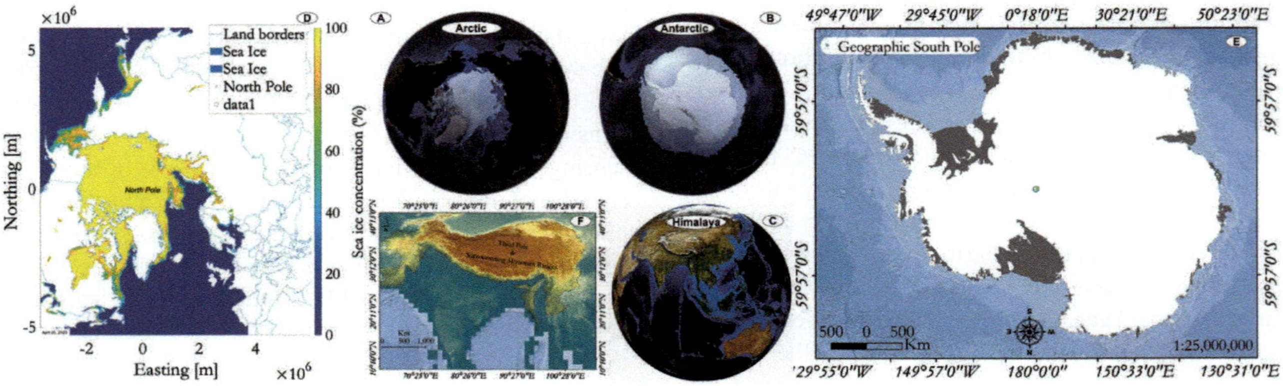


Figure 1.1 Location map of the three poles and surrounding regions. Photo credit for A, B, and C, NASA. Figures 1.1A, B, and C show the location of the north pole, south pole, and third pole and surrounding regions, respectively, on the 3D globe. Figures 1.1D, E, and F are 2D maps of the TPR.

#### 1.1.2.2 Interaction Among Components of the Earth System and Role of the Three Poles

The life supporting conditions existing on Earth, as per the "Goldilocks effect" statement (Lawton, 2001), which examines, "why, in general, terms, Earth's natural blanket of atmospheric $CO_2$ and distance from the Sun make the planet 'just right' for life, neither too hot (like Venus) nor too cold (like Mars)," warrants the working of our planet's components/subcomponents in a complex interlinked fashion but also in synchrony. By definition of a system, all the components and sub-components of the Earth System are interdependent and interact in a very complex way to make the Earth function in a way that works to sustain life; the interaction pathways are illustrated in Figure 1.2. This figure shows that all five components: geosphere, hydrosphere, atmosphere, biosphere, and anthroposphere are interconnected with each other in one way or the other. The anthroposphere, one of the components of the Earth System, is connected to all the other four components at different spatiotemporal scales. All the connections among various components/subcomponents affect every other component. For instance, paleoclimatic records reveal that long-term continuous variability in Earth System parameters has brought intermittent change, like shutting down of the Atlantic Meridional Overturning Circulation (AMOC) (Rahmstorf, 1995). Similarly, mathematical models, theoretically, suggest a phenomena wherein dynamics cause a flip for topographic cell numbers resulting from variations in the angular velocity of the planet (Williams, 1988a, 1988b). Coupled working of the ENSO, El Niño, and La-Nina sub-systems on an interannual timescale presents another example of this interconnectedness of the systems/sub-systems of Planet Earth. Unfortunately, climate system modeling has not advanced to the stage where it can accurately predict whether "global warming will fundamentally alter the ENSO dynamics in the future" (Schellnhuber, 2009). The interference of the anthropospheric component in the cryospheric component in the Arctic, Antarctic, and Greenland, evaluated by Notz (2009), presents the path of interconnectivity of these two sub-systems. Convergence and divergence (Stern, 2018), subduction (Davies, 1992), negative buoyancy force of the slab (Conrad and Lithgow-Bertelloni, 2002), under the umbrella of Plate Tectonics, are some of the mechanisms through which the different sub-systems categorized under the geosphere interconnect with each other (Alvarez, 1982; Storey, 1995; Anderson, 2006; Maruyama et al., 2007; Sobolev et al., 2007; Condie, 2013). The signatures of connectivity of components, and "innumerable intertwined processes" (Schellnhuber, 2009) through which they are interconnected, are among some of the most important and intriguing research topics in the field of Earth System Science (Falkowski et al., 2000; Collins et al., 2005; Jöckel et al., 2005; Gruber and Galloway, 2008; Warner et al., 2008; Flato, 2011; Huang et al., 2022; Sokol et al., 2022; Zhai et al., 2022).

During the early phases of the origin of the Earth, collision of a Mars-sized celestial body with our planet lead to evaporation of a proto-ocean, changing the parameters of motion of the planet itself, formation of the moon and establishing the interaction pathway between the celestial bodies' impact and other components of the Earth (Figure 1.2). The role the Tibet plateau, part of the Third Pole Region (TPR), proposed in theories of the origin and working of the monsoon system, highlights the significant role of the third pole and its connectivity with the atmospheric component and the lithospheric component (Spicer et al., 2003).

Connectivity of oceanosphere and atmosphere in the NPR in the changing climate scenario has extensively been reviewed by Taylor et al. (2018). In this work, they have recommend that the Arctic Ocean surface turbulent fluxes are the nature of episodic exchange and fluctuations for arctic climate change prediction improvements.

Pöschl and Shiraiwa (2015) present an extensive review of multiphase chemical interactions at the atmosphere–biosphere interface at the Anthropocene temporal scale that shed light on health relationships influenced by the multiphase chemical reactions taking place at the interface horizon.

The ozone hole over the North Pole and its healing has been in the news during the last two to three decades (Solomon, 1988, 2019; Peter, 1994; Solomon et al., 2016). The full account of why there is a decreasing trend in the stratospheric ozone thickness is not completely constrained to date (Peter, 1994). Here the relationship between the atmospheric component and Antarctica presented a new theme of research in 1985 when the ozone hole was discovered. The emergence of healing signs in the ozone hole after implementation of the Montreal Protocol, because of which many governments the world over "phased out the production of many of the human-made compounds responsible for stratospheric ozone destruction" has widely been reported (Varotsos, 2002; Weatherhead and Andersen, 2006; Hofmann and Montzka, 2009; Solomon et al., 2016; Kuttippurath and Nair, 2017). It signifies the link between stratospheric components of the atmosphere and anthroposphere. It also reveals the connectedness of the anthropic activities anywhere on the globe and polar regions.

In global-scale studies, different programs, commissioned by a host of international institutions, led to the evolution of concepts of the anthroposphere, and helped in defining/setting the planetary boundaries, and have been overviewed under the umbrella of "International Geosphere–Biosphere Programme (IGBP)" during 1986

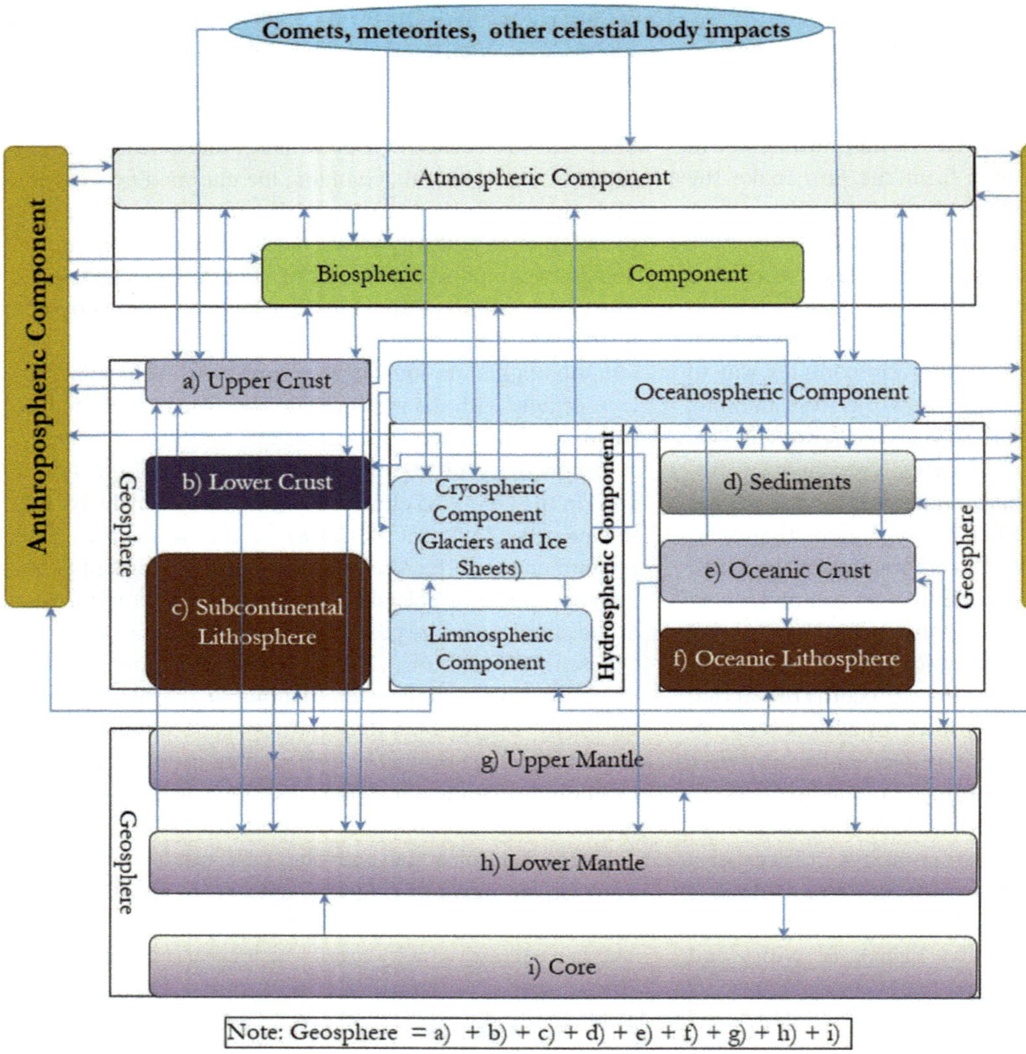


**Figure 1.2** Interaction pathways among sub-systems of Planet Earth, based on a description given in Condie, 2011 / with permission of Elsevier other related references are shown by arrows.

(starting of IGBP), 2015 (Seitzinger et al., 2015). According to Schellnhuber (2009), the Earth's system operates at the Holocene timescale (starting at ≈ 10 ka) and the surface of our planet is characterized by a distinct distributional pattern of "ice sheets, wind regimes, ocean currents, biomes, and deserts," which cumulatively form environmental facies.

### 1.1.3 Advancement of RS Technologies in Relation to Their Application in the Three Poles Regions

#### 1.1.3.1 Remote Sensing Technology Advancements

Over the last seven decades, since the first successful launch of the first artificial satellite, SPUTNIK-1 in 1957, remote sensing technology has witnessed tremendous advances. The advancement has been seen throughout different domains of the field. Development in the field of remote sensing sensor technology can be grouped into the following: i) electromagnetic spectrum (EMS) related; ii) sensor mounting platform related; iii) advancement in digital image processing algorithms and methods; and iv) development in computing power and affordability of high-end computers.

During the earlier phases of remote sensing technology, sensors covered broadbands within specified regions, such as the visible range (380–740 nm) of the EMS. Gradually, the number of spectral bands increased in the same wavelength region, in the sensors on board satellites during the 1980s. A look at the spectral bands of the sensors in different satellites under the series of Earth-Observation

Satellite Missions like the Landsat Program reveals that the three cameras under the spectral bands, i.e., blue-green (475–575 nm), orange-red (580–680 nm), and red to near-infrared (690–830 nm) in the Return Beam Vidicon (RBV) sensor and four cameras with spectral bands: i) green (0.5–0.6 μm); ii) red (0.6–0.7 μm); iii) near-infrared (0.7–0.8 μm); and iv) near-infrared (0.8–1.1 μm) fitted in the Multispectral Scanner (MSS) sensor was upgraded into hyperspectral mode in Landsat 8 and 9 (Roy et al., 2014; Masek et al., 2020). Operational Land Imager (OLI) records in 9 bands and Thermal Infrared Sensor (TIRS) captures energy fluxes in 2 bands in Landsat 8. Operational Land Imager 2 (OLI-2) and Thermal Infrared Sensor 2 (TIRS-2) sensors on board Landsat 9 carry the same number of bands as Landsat 8, but there are certain improvements in the TIRS band (Hair et al., 2018; McCorkel et al., 2018). In the evolutionary history of the satellite program, it has been observed that the four resolutions, i.e., spatial, temporal, spectral, and radiometric, have also been improved upon in the later phases of these programs (Markham et al., 2004; Jafarbiglu and Pourreza, 2022). All weather data capturing capability in the microwave region of the EMS provides RADAR remote sensing as a special privilege. There have been important advancements in RADAR remote sensing in the last few decades.

To achieve high spectral resolutions, while still opting for high spectral resolutions necessary for finer detectability of specific objects of detection, space engineers and scientists have implemented shifting orbits of the platforms from high earth orbit (HEO) through medium or mid-earth orbit (MEO) to low and very low earth orbits (LEO and VLEO) (Montenbruck et al., 2002; Crisp et al., 2020). From HEO, MEO, to LEO satellite platforms producing RS products, the remote sensing community has now shifted its attention to data from drone-based (also called unmanned aerial vehicles or uncrewed aerial systems)/LiDAR-based sensors accruing tremendous benefits in terms of ultra-high spatial resolution, maintaining high spectral resolution and desired repeatability of surveys. But there are also some limitations that constrain earth observation/modeling research works (Bhardwaj et al., 2016). Figure 1.3D shows the Earth observation platforms and their position in space orbits.

Since the data obtained from high spatial and spectral resolutions contain complex information about the object of detection, and come in large storage sizes, preprocessing and readying it for final analyzes and assessment needs both enhanced algorithms/methods and high-end computing capabilities to be available to the scientific communities/end users of these image products (Peral et al., 2018; Orynbaikyzy et al., 2019; Alvarez Lopez et al., 2022; Cruz et al., 2022; Hanif et al., 2022; Munawar et al., 2022; Wassie et al., 2022; Yu et al., 2022).

Radar altimetry has proved to be one of the most important advancements in the field of remote sensing, being actively applied for monitoring and mapping of water resources, topographic variations due to a variety of aggradational/degradational processes, ocean surfaces, etc. (Abdalla et al., 2021). Figure 1.4 summarizes the evolution of different radar satellite constellation missions. Readers are advised to consult Benveniste (2011) for developing an understanding of the working principles of radar altimetry. Utmost significance and future prospects in the field of radar altimetry can be understood by the fact that despite operational RADAR altimetry missions already in place in their respective orbits, e.g., Cryosat 2, Sentinel 3A and 3B, Sason 3, and Saral, work on future RADAR altimetry missions like Sentinel 3C and 3D, Sentinel 6A M. Freilich, Jason-CS, SWOT (Surface Water and Ocean Topography), Sentinel 6B, Sentinel 3 NG Topo, and Sentienl-6 NG is in full throttle by the concerned agencies.

#### 1.1.3.2 Role of Remote Sensing (RS) in Mapping/Monitoring/Quantitative Analysis of Sub-Systems of Our Planet in the Three Poles Regions

Application of remote sensing technology and their products in the NPR and SPR can be grouped into the following categories: i) geospheric component and remote sensing; ii) hydrospheric component and remote sensing; iii) atmospheric component and remote sensing; iv) biospheric component and remote sensing; and v) anthropospheric component and remote sensing.

The geospheric component related measurements and changes therein being mapped, monitored, and predicted using remote sensing products obtained from different platforms like satellites, specifically designed airplanes, helicopters, drones, and LiDARs, are some of the ongoing themes of research in TPR.

Mapping, gradual monitoring, and prediction of hydrospheric components like ice sheet dynamics, ice volume changes, glacier ice sheet melt patterns and drainage evolution in the supraglacial areas, sea-level change monitoring and their effects/consequences on coastal communities and other areas, water-related hazards such as floods, tsunamis, and storm surges, sea route monitoring for maintaining connectivity to the research stations in the northern and southern polar regions, ocean current tracking, sea surface temperature, and biological flourishing in the upper ocean layers like phytoplankton blooms, atmospheric phenomena and applications of remote sensing, SAR interferometry, feature tracking, scatterometry, altimetry, and gravimetry are themes that dominate the main areas that scientists are incessantly working on to quantify and predict (Drewry, 1975; Frezzotti, 1993, 2002; Liu and Bromwich, 1993; Vaughan et al., 1993; Bishop et al.,

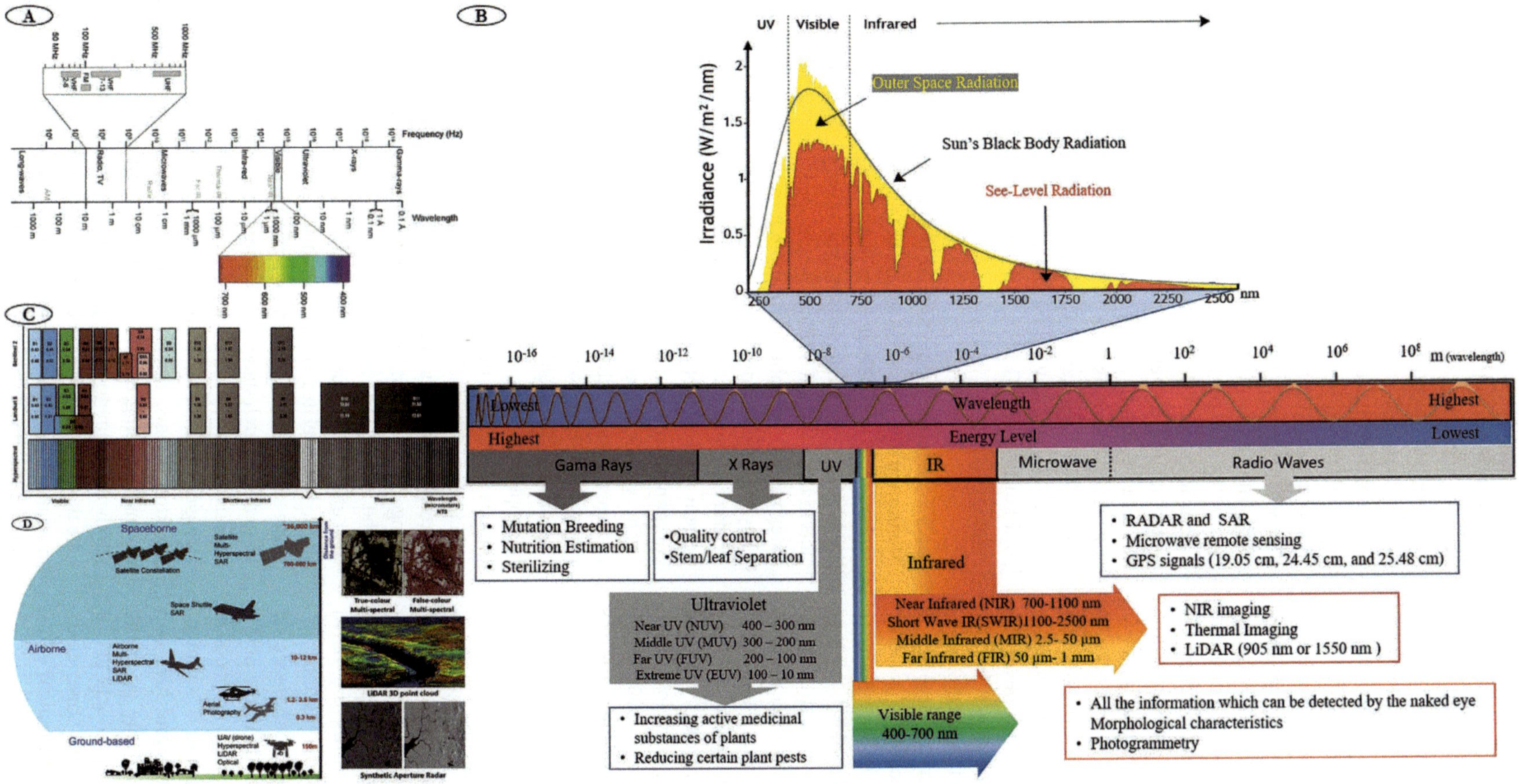


**Figure 1.3** (A) Range of electromagnetic spectrum (EMS) radiated from the Sun reaching the Earth's outer atmosphere. (B) Detailed portrayal of the EMS with wavelength bands in various spectral bands and some of their applications. (C) A comparative illustration of hyperspectral bands of Sentinel-2 and Landsat-8 presented as examples only. Low frequency (left) to high frequency (right) arrayed EMS and different bandwidths therein for which sensors used in remote sensing instruments on board various types of RS platforms have been developed, are presented in (A) and (D). Optical sensors on board most of the RS platforms cover the visible bandwidth along with some parts of the infrared region. For LiDAR sensors development, mostly the visible to near-infrared (NIR) bandwidth range have been the target regions to date. The RADAR sensors use bandwidth slots from microwave regions. In (D), various RS platforms are sketched. Source for (A) Penubag: https://commons.wikimedia.org/wiki/File:Electromagnetic-Spectrum.png; Creative Commons Attribution-Share Alike 2.5 Generic; Figure 1.3B is after Jafarbiglu and Pourreza, 2022 / Public Domian CC BY / Elsevier; Figure 1.3C Lechner et al., 2020 / Public Domian CC BY / Elsevier and Figure 1.3D are Lechner et al. (2020), ELSEVIER, CC BY 4.0.

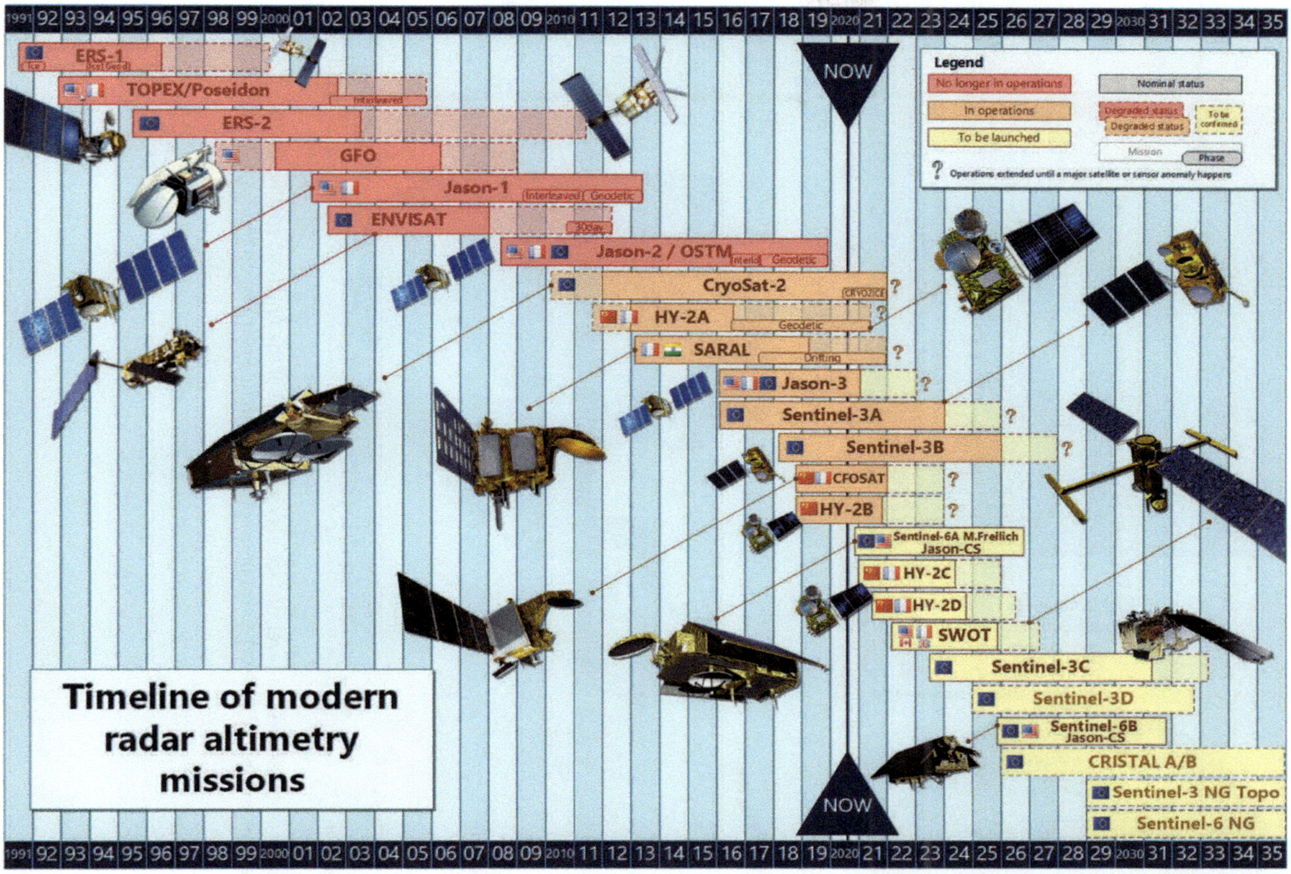


**Figure 1.4** Evolutionary history of satellite altimetry. *Source*: Abdalla et al., 2021 / with permission of Elsevier.

1996; Harris et al., 1999; Jezek and Onstott, 1999; Frezzotti et al., 2002, 2004; Montenbruck et al., 2002; Chao and Gick, 2004; Liu and Jezek, 2004; Markham et al., 2004; Thomas et al., 2004; Arthern et al., 2006; Williams et al., 2006; Picard et al., 2007; Scambos et al., 2007; Bindschadler et al., 2008; Chen et al., 2008; Sandau, 2010; Benveniste, 2011; Palm et al., 2011; Berrocoso et al., 2012; Bromwich et al., 2012; Siingh et al., 2012; Rignot et al., 2013; Wesche et al., 2013; Gernez et al., 2014; Pijoan et al., 2014; Roy et al., 2014; Vieira et al., 2014; Arcalís-Planas et al., 2015; Gorodetskaya et al., 2015; Zmarz et al., 2015; Bhardwaj et al., 2016; Calviño-Cancela and Martín-Herrero, 2016; Scheinert et al., 2016; Seo et al., 2016; Greene et al., 2017; Hugentobler and Montenbruck, 2017; Hui et al., 2017; Hair et al., 2018; McCorkel et al., 2018; Peral et al., 2018; Pour et al., 2018; Pudełko et al., 2018; Zanutta et al., 2018; Zhou et al., 2018; Beiranvand Pour et al., 2019; Lu et al., 2019; Orynbaikyzy et al., 2019; Cardenas et al., 2020; Crisp et al., 2020; Dąbski et al., 2020; Dirscherl et al., 2020, 2021; Ferreira et al., 2020; Fudala and Bialik, 2020; González et al., 2020; Lechner et al., 2020; Masek et al., 2020; Pereira et al., 2020; Salvatore et al., 2020; Tabibi et al., 2020; Abdalla et al., 2021; Alvarez-Vanhard et al., 2021; Groh and Horwath, 2021; Haran et al., 2021; Sun et al., 2021; Allahvirdi-Zadeh et al., 2022; Alvarez Lopez et al., 2022; Cruz et al., 2022; Hanif et al., 2022; Jafarbiglu and Pourreza, 2022; McAllister et al., 2022; Munawar et al., 2022; Pina and Vieira, 2022; Prasad et al., 2022a, 2022b; Wassie et al., 2022; Xu et al., 2022; Yu et al., 2022). Some of the research output products of continental-scale remote sensing are presented in Figure 1.5.

Some topics on which remote sensing (and related aspects) have been applied on a continental scale are:

- Snow extent and water equivalent
- Essential satellite elements of an observing system
- Daily, all-weather, global imaging
- High-resolution all-weather polar ice dynamics imaging
- Surface temperature and albedo change
- Ice topography/elevation/thickness change
- Gravity and mass distribution and exchange
- Position monitoring/navigation
- Telecommunications for data relay
- Solid precipitation
- Ice sheets (topographic change, mass change, and sea level)
- Gravity and the geoid

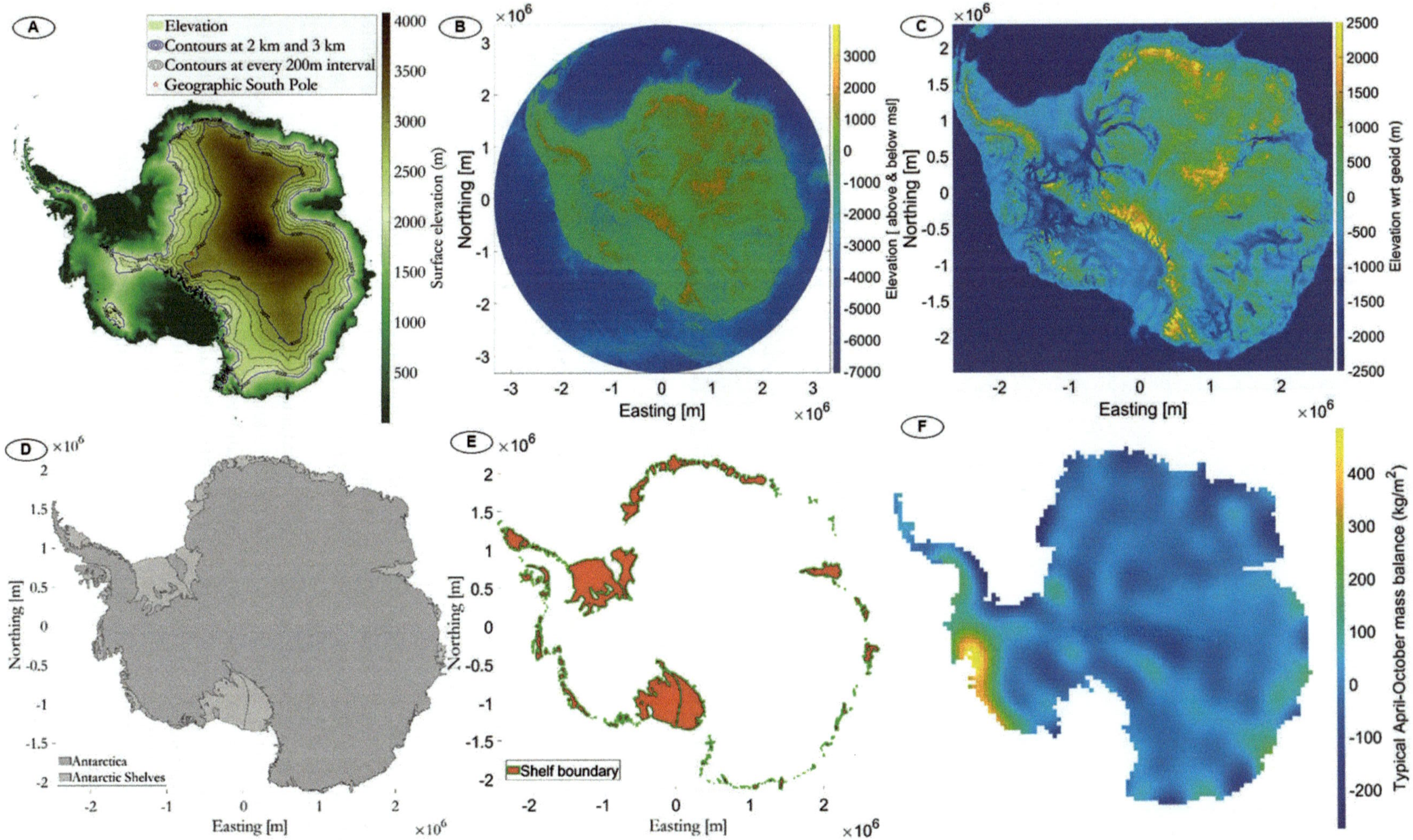


**Figure 1.5** Some examples of studies carrying out measurement, mapping, and modeling of: (A) above sea-level surface elevation model of the Antarctic continent; (B) bathymetric and surface elevation data combined to provide the idea of surface (both above and below surface) distribution of the Antarctic Polar Region (AnPR); (C) visualization of the elevation distribution of the AnPR; (D) and (E) distribution of the shelf zone.

**Table 1.1** Some references on research in all five sphere domains of the three poles.

| Components | Antarctica | Arctic | Himalaya |
|---|---|---|---|
| Geosphere | (Pour et al., 2018) | (Bedini, 2011; Ekholm, 1996; Franke et al., 2020) | (Quincey et al., 2005) |
| Hydrosphere | (Palm et al., 2011; Dirscherl et al., 2020) | (Barry et al., 1989; Wadhams, 1990; Johannessen et al., 1999; Greuell and Knap, 2000; Thomas, 2001; Cooper and Smith, 2019; Wake, 2019) | (Azam et al., 2021; Bhambri and Bolch, 2009; Song et al., 2014) |
| Atmosphere | (Gorodetskaya et al., 2015; Singh et al., 2012) | (Morison et al., 2018; Engelmann et al., 2021; Tjernström et al., 2004) | (Guleria et al., 2012; Kumar et al., 2018; Barros et al., 2004) |
| Biosphere | (Arcalís-Planas et al., 2015; Malenovský et al., 2017) | (Stow et al., 2004; Rogers et al., 2019) | (Sharma et al., 2016; Gairola et al., 2013; Mohapatra et al., 2019) |
| Anthroposphere | (Hui et al., 2017; Cincinelli et al., 2017) | (Kumpula et al., 2011, 2012) | (Mishra et al., 2020; Khanal et al., 2022; Pandey and Sharma, 2021; Pandey et al., 2021) |

The subcomponent-wise research papers cited in Table 1.1 will give readers some idea of the types of research into TPR wherein remote sensing is applied.

## 1.2 Aim of the Book and Its Five Sections

This book aims to present high-quality research utilizing the potential of advancement in remote sensing technology over the last 5–6 decades conducted in the Three Poles Regions (TPRs), gathered together in an edited book format. Being engaged in the research and having known the well-established concept of connectedness of the three poles, the editors have noticed a lack of polar research material gathered together and this book attempts to fill this gap.

The contents proposed for this book have been thoughtfully partitioned into five sections. The first section, Section I, deals with papers on general topics applicable to all the poles, such as remote sensing data and platforms, softwares, tools, add-ons used for processing the RS datasets and extracting valuable information from them. Starting from south to north, the next three sections, Sections II, III, and IV, have been apportioned to the three poles: the Antarctic, the Himalaya (third pole), and the Arctic. The last section, Section V, has been proposed for chapters on common topics related to national/international collaborations in polar research, polar data sharing policies and agreements, and nature of job availability in polar research domain. This book's scope is very wide covering the research spectrum including all the spheres of our planet around the TPR. It became impossible to cover all the papers on all the proposed themes and topics and remain a small volume. An overview of all the papers included herein is briefly presented below.

## 1.3 Overview of the Contributing Chapters Covering Research About Different Aspects of the Sub-Systems of Our Planet in the Three Poles Regions

Section I, entitled "Earth Observation (EO) and Remote Sensing (RS) Applications in Polar Studies," contains six chapters which attempt to give a general idea about the advancement in remote sensing technology, salient features of the past, present, and future satellite data products, and an overview of the softwares, tools, and add-ons used for processing the datasets requisitioned from different remote sensing platforms. In Chapter 1, by Pandey et al. (2022), an overview on softwares and other tools cover those applicable in two aspects: 1) morphometry, and 2) landscape evolution, glacier dynamics, and ice sheet dynamics.

Chapter 2, by Mishra et al. (2022b), gives a brief introduction to various orbits in which we can launch satellites and how they are used for various purposes. They also discuss various active earth observation missions, both present and future, and their available sources of data and policies. The application of such data in polar study has also been explored. Future missions include: JPSS-2 (joint polar satellite system in 2022); JPSS-3 and JPSS-4 in 2026 and 2031 respectively; SWOT-surface water ocean topography, February 2022.

In Chapter 3, Mandal and Sarkar (2022) assess the accuracy of freely available Digital Elevation Models (DEM) in the Darjeeling-Sikkim Himalayas. They have included SRTM (Shuttle Radar Topography Mission), ASTER (Advanced Spaceborne Thermal Emission and Reflection Radiometer), GDEM (Global Digital Elevation Model), and DEMs. This chapter presents an overall understanding of how DEM accuracy behaves in the third pole environmental setting.

Chapter 4, by Mitra et al. (2022b), presents an overview of morphometric software packages, tools, and add-ons that are currently used by the geoscience community. This chapter is of great value to both undergraduate researchers and scientists working in the interdisciplinary field who have a desire to enter the field of morphometry.

In a similar pursuit, Chapter 5, by Mitra et al. (2022a), attempts to compile most of the popularly used models and tools in the field of landscape evolution modeling (LEM). With the availability of global digital elevation models at as high as 30 m spatial resolution, free of any cost to the scientific community, and affordability of computers powerful enough to handle large computations, younger-generation scientists are pouring their skills into LEM techniques.

Chapter 6, by Mishra et al. (2022c), compiles all the popular spectral indices used to quantify different objects, also known as elements of detection (EoD), in various spheres of our planet. For example, spectral indices like normalized difference mud index (NDMI), bare soil index, ferrous mineral ratio, clay mineral ratio, used for quantifying the respective objects of the lithosphere; normalized difference snow index, normalized index debris index, normalized difference water index, for hydrospheric objects; NDVI (Normalized Difference Vegetation Index), GNDVI, EVI (Enhanced Vegetation Index), for biospheric element quantification; cloud index, and its normalized version NDCI (Normalized Difference Cloud Index) for atmospheric elements; and burn area index, built-up index, for anthropospheric components have been reviewed upon for their application in the fields of interest.

Section II of the book entitled, "Antarctica: The Southernmost Continent having the South Pole, Environment and Remote Sensing," includes five chapters.

Chapter 7, by Tomar et al. (2022), has used radar satellite data from Sentinel-1 satellite to monitor glacier dynamics of the Polar Record Glacier (PRG) located in eastern Antarctica over the period from 2016 to 2020. Ice front position movement and glacier velocity has been calculated using 18 scenes of the HH-Polarization, 5.6 cm wavelength SAR dataset. The study discovered that the PRG moved >2 m/day during the study period. Also, the glacier front position shifted ~1750 m during the study period. Bedcover topographic patterns have significant control over the glacier velocities in the Antarctic glaciers.

Chapter 8, by Sharma et al., (2022), reviews terrestrial deglaciation studies based on cosmogenic radionuclides dating, radiocarbon dating, and luminescence dating. The authors have observed that several coastal regions and inland nunataks of Dronning Maud Land and Princess Elizabeth Land were already ice-free before the Last Glacial Maximum (LGM). And the initiation of deglaciation around the Lambert Glacier-Amery Ice Shelf started around ~20–18 ka years BP and deglaciation in other areas of Eastern Antarctica started after ~12 ka, as evident from regions of Enderby Land, MacRobertson Land, and Wilkes Land.

Pande et al. (2022), in Chapter, 9, have detailed the utility of Unmanned Aerial Vehicles and satellite imagery for wildlife monitoring in Antarctica, with examples. They have also briefly discussed the seabird monitoring being conducted under the Indian Antarctic Program and give recommendations on upscaling the techniques being currently used.

Singh (2022a), in Chapter 10, has documented the detailed taxonomic identification keys used for monitoring of 12 species and infraspecific taxa of Bryophytes (one liverwort and 11 mosses) of the Larsemann Hills belonging to 5 families and 7 genera. Out of those 12 species, one species, *Guembelia longirostris* (Hook.) Ochyra & Zamowiec is recorded for the first time from East Antarctica, earlier reported from West Antarctica; and the *Bryum argenetum* Hedw. var. argenetum has been recorded for the first time in the Larsemann Hills, which extends the range of distribution of other species on various islands of the Larsemann Hills.

Swathi et al. (2022), in Chapter 11, using 42 years of long measurements (1979–2020) of passive-microwave satellite data, have conducted sea ice extent (SIE) analyzes on monthly, interannual, decadal, and seasonal timescales for the Southern Ocean.

Section III, entitled "Himalayas: The Third Pole Environment and Remote Sensing," contains 13 chapters on different aspects of the third pole of our planet, the Himalayas.

Chapter 12, by Bhargava (2022), explores the unsolved problems in different domains of Earth Sciences in the third pole region. Brief descriptions are provided of the problems and research gaps in the fields of: a) Stratigraphy and Palaeontology; b) Sedimentology; c) Tectonics and structure; d) Magmatism and geochronology; e) Metamorphism; f) Mineral deposits; g) Palaeomagnetic studies; h) Glaciological studies; and i) Geomorphological studies.

Pandey et al. (2022), in Chapter 13, have assessed glacier fluctuations through tree-ring analysis, field survey, and remote sensing. This is the first study that approaches the Kolahoi glacier fluctuation through the aid of satellite remote sensing and tree-ring chronology using the Himalayan Blue Pine (*Pinus wallichiana*). It opens the way to extending glacier fluctuation tracking periods with tree-ring analysis.

In Chapter 14, Dandabathula (2022) has rigorously tested recently launched (September 2018) ICESat-2 Photon Data for its applicability in the third pole regions.

The study domains tested by the author includes: i) annual water levels variations in mountain lakes; ii) inland water level detection in mountainous regions; iii) longitudinal profiling of rivers; iv) terrain profiling; v) lake ice phenology study; vi) tree height estimation; and vii) generation of DEMs. This chapter vindicates the findings of other researchers who conducted their work in different parts of the world and reported wide applicability ICES at 2 Photon Data for various purposes.

In Chapter 15, Kumar et al. (2022c) estimated variation in the soil loss rate induced by a particular Extreme Hydrological Event (EHE) occurring on 3 August 2012. Using Google Earth Engine (GEE), Geographical Information System (GIS), and ALOS-PALSAR Digital Elevation Model (DEM), they derived inputs to run the RUSLE model. The authors report average soil loss rate in the Assiganga River Basin increased from 2.64 tons/ha/month in May 2012 (pre-event) to 3.35 tons/ha/month in October 2012 (post-event), even with less rainfall.

Mehta et al. (2022), in Chapter 16, have comprehensively reviewed climate–vegetation interaction in an anthropogenically intervening state of environmental conditions based on instrumental and remote sensing derived data. They have attempted to track down the impact of various anthropogenic drivers on vegetation in the Indian Himalayas. As for assessing the ecosystem sensitivity to climate variability, they have followed empirical approaches including pollen records, carbon isotope (δ13C), and phytolith assemblages.

Dash et al. (2022), in Chapter 17, conducted their work in the Bhagirathi River Basin, modeling flash flood susceptibility using nine flash flood related explanatory variables based on a multicriteria decision making (MCDM) approach in a GIS environment. The applied model resulting in a flood hazard index performed well with 78% accuracy in terms of area under curve (AUC). The study found that "distance-from-river" (priority ~29.5%) is the most influential predictor variable (also called the explanatory variable) following geomorphology (20.7%), elevation (15.7%), and slope (12.2%).

In Chapter 18, Sarma et al. (2022) describe their work on the role of the Himalayan Frontal Thrust (HFT), and have interpreted the evolutionary history of part of Himalaya in Arunachal Pradesh, located in the North-Eastern part of India. The authors have used digital elevation models and intensive field visits for documentation and mapping of the sedimentary features and processes thereof.

In Chapter 19, Devrani et al. (2022) have assessed the sensitivity of the most popularly used different digital elevation datasets, e.g., ALOS-PALSAR DEM, ASTER GDEM, COOPER DEM, CARTOSAT, NASA DEM, SRTM DEM, and topographic sheet (TOPOSHEET at 1:50,000 scale) derived-DEM for river profile extraction. The study concludes that among 30 m spatial resolution digital elevation models, freely available for non-commercial use, SRTM DEM is the most reliable for extraction of stream longitudinal profiles.

In Chapter 20, Mishra et al. (2022a) present a brief review of glacier ice thickness estimation methods using various geophysical methods glaciers in different parts of the Indian Himalaya. Geophysical methods used for ice thickness estimation in the Indian Himalaya included in this review include: i) gravity surveys; ii) magnetic surveys; iii) electrical resistivity surveys; iv) seismic surveys; and v) ground penetrating radar surveys. The authors have presented their GPR survey results conducted over the Satopanth Glacier in 2016 and reported ice thickness of 38–50 m near a glacier snout. Summary of results from GPR surveys conducted in the Indian Himalaya by different researchers makes this chapter very useful for those working in the cryospheric environment of the third pole.

In Chapter 21, Kumar et al. (2022a) present an overview of the physical landscape of the Ladakh Himalaya, in relation to the forces sculpturing the landforms, such as: i) earth surface processes related to atmospheric forces (coupling of monsoon and westerlies); and ii) continuum northward movement of the Indian plate against the batholith as a rigid body. The main aim of the study is to understand the paleoclimate variability, hydrological changes in the major rivers, glaciation in the Ladakh, and human migration for trading via the Silk Route.

In Chapter 22, Kumar et al. (2022b) present a review of remote sensing and GIS-based soil loss estimation models such as USLE (Universal Soil Loss Equation) and the USLE family models, e.g., RUSLE-1 (Revised Universal Soil Loss Equation), RUSLE-2, and MUSLE (Modified Universal Soil Loss Equation). Furthermore, they simultaneously compared the soil loss results obtained from the RUSLE model applied over two topoclimatologically distinct drainage sub-basins, upper (Rishiganga) and marginal (lower Chambal) Ganga River basins, which are parts of the Ganga River Basin. The result shows the mean rate of soil loss of 9.82 ton/ha/yr in the Rishiganga and 1.58 ton/ha/yr in the lower Chambal basin.

In Chapter 23, Mishra et al. (2022) present a review on the role of wetlands as potential zones preserving the human-plant-climate interaction signals archived in the form of spore/pollen in pond or lacustrine environments of the western and eastern Himalaya. These natural proxies help to decode past monsoon and other climatic shifts at geological timescales.

In Chapter 24, Mishra (2022d) has attempted to quantify the changes in landuse land cover in the mountainous drainage basin of the Alaknanda River during the period covering about four and a half decades from 1976 to 2020. The author has used Landsat 2-Multispectral Scanner System (MSS), and Landsat 8-Operational Land Imager (OLI) datasets and supervised the (MLC) method to achieve landuse land cover (LULC) maps. The study estimated that the agricultural land in the basin has decreased by 1.92%, whereas settlement areas have increased by 4.47% during the study period. This study vindicates results of other studies reporting increasing anthropogenic interference and effects of climate change becoming clearly visible.

Section IV, entitled, “The Arctic: The Northernmost Ocean having the North Pole, Environment and Remote Sensing,” hosts five chapters, each focusing on different aspects of the north polar region.

Chapter 25, by Ranjan et al. (2022), presents a synoptic overview of the three poles from the cryosphere change perspective together with their association with hydrological ecological, societal, and economic conditions.

Chapter 26, by Jawak et al. (2022), explores the application of ultra-high-resolution satellite optical imagery, WorldView-3 MSI (Multispectral: 1.24 m GSD at Nadir), and shortwave infrared (SWIR) (3.7 m GSD at Nadir) for quantification of glacier surface facies and important supraglacial features, that are highly important to increase the accuracy of distributed mass balance modeling results. The two classification algorithms, the Minimum Distance classifier (MD) and the Mahalanobis Distance classifier (MHD) protocols, were executed in ENVI 5.3.

In Chapter 27, Gantayat (2022) has discussed three methods that simulate the evolution of SGLs in the GrIS. These models are called SLING (Leeson et al., 2012), SRLF (Banwell et al., 2012a) and SRLFCI (Koziol et al., 2017). All these methods were applied over different regions located in the southwest GrIS.

Chatterjee et al. (2022), in Chapter 28, summarize their recent findings using the remotely sensed sea-level estimates in the Arctic Ocen (AO) and further assess the representation of the same in the new generation climate models participating in the Coupled Model Intercomparison Project-6 (CMIP6). Satellite altimetric observations from platforms like GEOSAT (NASA) in 1985 to SENTINEL 6 (ESA/NASA/NOAA/Eumetsat) in 2020, have been overviewed and their results are discussed in this chapter.

In Chapter 29, Srivastava (2022) has analyzed and present spatial and seasonal variations in mid-visible aerosol optical depth derived from the MODerate resolution Imaging Spectroradiometer (MODIS) sensor on board the Terra satellite over two of the polar regions, i.e., the Arctic and the Antarctic. They also emphasize and highlight the importance of year-round *in-situ* aerosol optical property measurements over polar regions. India is developing the POLar AERosol NETwork (POLAERNET), an *in-situ* network of aerosol measurements, to address these challenges. The study has also analyzed global fire activities that accentuates the global warming.

Section V, entitled, “The Research Institutions on the ‘Three Poles’, Data Pools, Data Sharing Policies, Career in Polar Science Research and Challenges,” explores research collaborations in the polar research regions and the multidisciplinary nature of polar research.

In Chapter 30, Singh et al. (2022a) present an overview of multidisciplinary research activities under the aegis of the “Indian Antarctic Programme” starting in 1981/82 and completing over four decades of uninterrupted research expeditions. The research work in the Antarctic covers myriad fields incorporating seismology, crustal and ice sheet dynamics, paleoclimate reconstruction, quaternary geological deglaciation history, meteorology, space weather, and environmental monitoring and assessment, to name a few. Therefore, the research in the Antarctic Programmes of India is truly multidisciplinary in nature. In addition, expedition vessel-based research named, “Indian Scientific Expeditions to the Southern Ocean” (ISESO), started in 2004, aimed at exploring the Southern Ocean and its surroundings, is ongoing.

Chapter 31, by Singh et al. (2022b), presents a detailed overview of the coordination platforms engaged in promoting and facilitating international cooperation in addressing scientific questions and activities and ensuring optimum utilization of the infrastructure in the Arctic region. This chapter also throws light on, and discusses, the details of Arctic research coordination among the international and national communities.

## 1.4 Summary and Recommendations

In the wake of rapid developments in the last two to three decades in the fields of remote sensing technology, big data science, and computing, the need is to study the three poles of our planet as they play very important roles in (in)directly governing the processes in global cycles, such as biogeochemical cycles, hydrological cycles, and global ocean circulation conveyor belts. The sections in this book includes chapters on topics within the broad domain of lithospheric, hydrospheric, cryospheric, atmospheric, biospheric, and anthropospheric components of the Earth’s systems.

## References


Abdalla, S., Abdeh Kolahchi, A., Ablain, M. et al. (2021). Altimetry for the future: building on 25 years of progress. *Adv. Sp. Res.* 68: 319–363. doi:10.1016/j.asr.2021.01.022.

Allahvirdi-Zadeh, A., Wang, K., and El-Mowafy, A. (2022). Precise orbit determination of LEO satellites based on undifferenced GNSS observations. *J. Surv. Eng.* 1 8(1): 1–22. doi:10.1061/(ASCE)SU.1943-5428.0000382.

Alvarez Lopez, Y., Garcia-Fernandez, M., Alvarez-Narciandi, G. et al. (2022). Unmanned aerial vehicle-based ground-penetrating radar systems: a review. *IEEE Geosci. Remote Sens. Mag.* 2–22. doi: 10.1109/MGRS.2022.3160664.

Alvarez, W. (1982). Geological evidence for the geographical pattern of mantle return flow and the driving mechanism of plate tectonics. *J. Geophys. Res. Solid Earth* 87: 6697–6710. doi: 10.1029/JB087iB08p06697.

Alvarez-Vanhard, E., Corpetti, T., and Houet, T. (2021). UAV and satellite synergies for optical remote sensing applications: a literature review. *Sci. Remote Sens.* 3: 100019. doi: 10.1016/j.srs.2021.100019.

Amir, A.Z. and El-Mowafy, A. (2021). Precise orbit determination of LEO satellites based on undifferenced GNSS observations. *J.Survey. Eng.* 18(1): 1–22.

Anderson, D.L. (2006). Speculations on the nature and cause of mantle heterogeneity. *Tectonophysics* 416: 7–22. doi: 10.1016/j.tecto.2005.07.011.

Arcalís-Planas, A., Sveegaard, S., Karlsson, O. et al. (2015). Limited use of sea ice by the Ross seal (*Ommatophoca rossii*) in Amundsen Sea, Antarctica, using telemetry and remote sensing data. *Polar Biol.* 38: 445–461. doi: 10.1007/s00300-014-1602-y.

Arthern, R.J., Winebrenner, D.P., and Vaughan, D.G. (2006). Antarctic snow accumulation mapped using polarization of 4.3-cm wavelength microwave emission. *J. Geophys. Res.* 111: D06107. doi: 10.1029/2004JD005667.

Azam, M.F., Kargel, J.S., Shea, J.M. et al. (2021). Glaciohydrology of the Himalaya-Karakoram. *Science* 80: 373. doi: 10.1126/science.abf3668.

Barros, A.P., Kim, G., Williams, E. et al. (2004). Probing orographic controls in the Himalayas during the monsoon using satellite imagery. *Nat. Hazards Earth Syst. Sci.* 4: 29–51. doi: 10.5194/nhess-4-29-2004.

Barry, R.G., Miles, M.W., Cianflone, R.C. et al. (1989). Characteristics of Arctic sea ice from remote-sensing data and their relationship to atmospheric processes. *Ann. Glaciol.* 12: 9–15. doi: 10.3189/S0260305500006893.

Bedini, E. (2011). Mineral mapping in the Kap Simpson complex, central East Greenland, using HyMap and ASTER remote sensing data. *Adv. Sp. Res.* 47: 60–73. doi: 10.1016/j.asr.2010.08.021.

Beiranvand Pour, A., Park, Y., Crispini, L. et al. (2019). Mapping listvenite occurrences in the damage zones of Northern Victoria Land, Antarctica using ASTER satellite remote sensing data. *Remote Sens.* 11: 1408. doi: 10.3390/rs11121408.

Benveniste, J. (2011). Radar altimetry: past, present and future. In: *Coastal Altimetry*, 1–17. Springer-Verlag, Berlin, Heidelberg, Germany. doi: 10.1007/978-3-642-12796-0_1.

Berrocoso, M., Torrecillas, C., Jigena, B. et al. (2012). Determination of geomorphological and volumetric variations in the 1970 land volcanic craters area (Deception Island, Antarctica) from 1968 using historical and current maps, remote sensing and GNSS. *Antarct. Sci.* 24: 367–376. doi: 10.1017/S0954102012000193.

Bertrand, P. and Legendre, L. (2021). The living earth: our home in the solar system and the universe. In: *Earth, Our Living Planet: The Earth System and Its Co-Evolution with Organisms* (ed. P. Bertrand and L. Legendre), 1–47. Springer, Gewerbrestrasse, Cham, Switzerland. doi: 10.1007/978-3-030-67773-2_1.

Bhambri, R. and Bolch, T. (2009). Glacier mapping: a review with special reference to the Indian Himalayas. *Prog. Phys. Geogr. Earth Environ.* 33: 672–704. doi: 10.1177/0309133309348112.

Bhardwaj, A., Sam, L., Bhardwaj, A. et al. (2016). LiDAR remote sensing of the cryosphere: present applications and future prospects. *Remote Sens. Environ.* 177: 125–143. doi: 10.1016/j.rse.2016.02.031.

Bhargava, O. N. (2022). Some unresolved problems in the Himalaya: A synoptic view. In: *Advancements in Remote Sensing Technology and the Three Poles.* (ed. M. Pandey, P. Pandey, A. Arora, Y. Ray, S. Jawak, U. Kant Shukla), 1st edn, 193–202. John Wiley & Sons, Inc., Chichester, UK.

Bindschadler, R., Vornberger, P., Fleming, A. et al. (2008). The landsat image mosaic of Antarctica. *Remote Sens. Environ.* 112: 4214–4226. doi: 10.1016/j.rse.2008.07.006.

Bishop, J. L., Koeberl, C., Kralik, C. et al. (1996). Reflectance spectroscopy and geochemical analyses of Lake Hoare sediments, Antarctica: implications for remote sensing of the Earth and Mars. *Geochim. Cosmochim. Acta.* 60: 765–785. doi: 10.1016/0016-7037(95)00432-7.

Bromwich, D.H., Nicolas, J.P., Hines, K.M. et al. (2012). Tropospheric clouds in Antarctica. *Rev. Geophys.* 50: RG1004. doi: 10.1029/2011RG000363.

Calviño-Cancela, M. and Martín-Herrero, J. (2016). Spectral discrimination of vegetation classes in ice-free areas of Antarctica. *Remote Sens.* 8: 856. doi: 10.3390/rs8100856.

Cardenas, C., Casassa, G., Aguilar, X. et al. (2020). From space to earth: physical and biological impacts of glacier dynamics in the marine system by means of remote sensing at Almirantazgo Bay, Antarctica. In: *2020 IEEE Latin American GRSS & ISPRS Remote Sensing Conference (LAGIRS)*, 308–312. IEEE, Danvers, MA, U.S.A. doi: 10.1109/LAGIRS48042.2020.9165686.

Chao, C. and Gick, R. (2004). Long-term evolution of navigation satellite orbits: GPS/GLONASS/GALILEO. *Adv. Sp. Res.* 34: 1221–1226. doi: 10.1016/j.asr.2003.01.021.

Chatterjee, S., Raj, R. P., Bonaduce, A., and Davy, R. (2022). Arctic Sea level change in remote sensing and new generation climate models. In: *Advancements in Remote Sensing Technology and the Three Poles*, (ed. M. Pandey, P. Pandey, A. Arora, Y. Ray, S. Jawak, U. Kant Shukla), 1st edn, 390–400. John Wiley & Sons, Inc., Chichester, UK.

Chen, J.L., Wilson, C.R., Tapley, B.D. et al. (2008). Antarctic regional ice loss rates from GRACE. *Earth Planet. Sci. Lett.* 266: 140–148. doi: 10.1016/j.epsl.2007.10.057.

Cincinelli, A., Scopetani, C., Chelazzi, D. et al. (2017). Microplastic in the surface waters of the Ross Sea (Antarctica): occurrence, distribution and characterization by FTIR. *Chemosphere*. 175: 391–400. doi: 10.1016/j.chemosphere.2017.02.024.

Collins, N., Theurich, G., DeLuca, C. et al. (2005). Design and implementation of components in the Earth System Modeling Framework. *Int. J. High Perform. Comput. Appl.* 19: 341–350. doi: 10.1177/1094342005056120.

Condie, K.C. (2011). Earth systems. In: *Earth as an Evolving Planetary System*, 1–10. Elsevier, Oxford, UK. doi: 10.1016/B978-0-12-385227-4.00010-9.

Condie, K.C. (2013). *Plate Tectonics & Crustal Evolution*, 1–485. Elsevier, Pergamon Press, Inc., New York, U.S.A.

Conrad, C.P. and Lithgow-Bertelloni, C. (2002). How mantle slabs drive plate tectonics. *Science* 80(298): 207–209. doi: 10.1126/science.1074161.

Cooper, M. and Smith, L. (2019). Satellite remote sensing of the Greenland Ice Sheet Ablation zone: a review. *Remote Sens.* 11: 2405. doi: 10.3390/rs11202405.

Corby, G.A., Gilchrist, A., and Newson, R.L. (1972). A general circulation model of the atmosphere suitable for long period integrations. *Q.J.R. Meteorol. Soc.* 98: 809–832. doi: 10.1002/qj.49709841808.

Crisp, N.H., Roberts, P.C.E., Livadiotti, S. et al. (2020). The benefits of Very Low Earth Orbit for earth observation missions. *Prog. Aerosp. Sci.* 117: 100619. doi: 10.1016/j.paerosci.2020.100619.

Cruz, H., Véstias, M., Monteiro, J. et al. (2022). A review of synthetic-aperture radar image formation algorithms and implementations: a computational perspective. *Remote Sens.* 14: 1258. doi: 10.3390/rs14051258.

Dąbski, M., Zmarz, A., Rodzewicz, M. et al. (2020). Mapping glacier forelands based on UAV BVLOS operation in Antarctica. *Remote Sens*. 12: 630. doi: 10.3390/rs12040630.

Dandabathula, G. (2022). Applications of ICESat-2 photon data in the third pole environment. In: *Advancements in Remote Sensing Technology and the Three Poles*, (ed. M. Pandey, P. Pandey, A. Arora, Y. Ray, S. Jawak, U. Kant Shukla), 1st edn, 213–229. John Wiley & Sons, Inc., Chichester, UK.

Dash,P., and Kasturi Mukherjee S.G. (2022). Flash flood susceptibility mapping of a himalayan river basin using multi-criteria decision-analysis and GIS. In: *Advancements in Remote Sensing Technology and the Three Poles*, (ed. M. Pandey, P. Pandey, A. Arora, Y. Ray, S. Jawak, U. Kant Shukla), 1st edn, 257–267. John Wiley & Sons, Inc., Chichester, UK.

Davies, G.F. (1992). On the emergence of plate tectonics. *Geology* 20: 963. doi: 10.1130/0091-7613(1992)020<0963:OTEOPT>2.3.CO;2.

Dell'Osso, L. and Chen, S.-J. (1986). Numerical experiments on the genesis of vortices over the Qinghai-Tibet plateau. *Tellus A* 38A: 236–250. doi: 10.1111/j.1600-0870.1986.tb00468.x.

Devrani, R., Kumar, R., Kuruvath, M., Kasana, P., and Shailendra Pundir, M. P. (2022). Himalayan river profile sensitivity assessment/analysis by validating of DEMs and comparison of hydrological tools. In: *Advancements in Remote Sensing Technology and the Three Poles*, (ed. M. Pandey, P. Pandey, A. Arora, Y. Ray, S. Jawak, U. Kant Shukla), 1st edn, 283–298. John Wiley & Sons, Inc., Chichester, UK.

Dirscherl, M., Dietz, A.J., Dech, S. et al. (2020). Remote sensing of ice motion in Antarctica: a review. *Remote Sens. Environ.* 237: 111595. doi: 10.1016/j.rse.2019.111595.

Dirscherl, M., Dietz, A.J., Kneisel, C. et al. (2021). A novel method for automated supraglacial lake mapping in Antarctica using Sentinel-1 SAR imagery and deep learning. *Remote Sens.* 13: 197. doi: 10.3390/rs13020197.

Drewry, D.J. (1975). Terrain units in eastern Antarctica. *Nature* 256: 194–195. doi: 10.1038/256194a0.

Ekholm, S. (1996). A full coverage, high-resolution, topographic model of Greenland computed from a variety of digital elevation data. *J. Geophys. Res. Solid Earth* 101: 21961–21972. doi: 10.1029/96JB01912.

Engelmann, R., Ansmann, A., Ohneiser, K. et al. (2021). Wildfire smoke, Arctic haze, and aerosol effects on mixed-phase and cirrus clouds over the North Pole region during MOSAiC: an introduction. *Atmos. Chem. Phys.* 21: 13397–13423. doi: 10.5194/acp-21-13397-2021.

Falkowski, P., Scholes, R.J., Boyle, E. et al. (2000). The global carbon cycle: a test of our knowledge of earth as a system. *Science* 80(290): 291–296. doi: 10.1126/science.290.5490.291.

Ferreira, V.G., Yong, B., Seitz, K. et al. (2020). Introducing an improved GRACE global point-mass solution: a case study in Antarctica. *Remote Sens* 12: 3197. doi: 10.3390/rs12193197.

Flato, G.M. (2011). Earth system models: an overview. *WIREs Clim. Chang.* 2: 783–800. doi: 10.1002/wcc.148.

Franke, S., Jansen, D., Binder, T. et al. (2020). Bed topography and subglacial landforms in the onset region of the Northeast Greenland Ice Stream. *Ann. Glaciol.* 61: 143–153. doi: 10.1017/aog.2020.12.

Frezzotti, M. (1993). Glaciological study in Terra Nova Bay, Antarctica, inferred from remote sensing analysis. *Ann. Glaciol.* 17: 63–71. doi: 10.3189/S0260305500012623.

Frezzotti, M. (2002). Snow megadunes in Antarctica: sedimentary structure and genesis. *J. Geophys. Res.* 107: 4344. doi: 10.1029/2001JD000673.

Frezzotti, M., Gandolfi, S., La Marca, F. et al. (2002). Snow dunes and glazed surfaces in Antarctica: new field and remote-sensing data. *Ann. Glaciol.* 34: 81–88. doi: 10.3189/172756402781817851.

Frezzotti, M., Pourchet, M., Flora, O. et al. (2004). New estimations of precipitation and surface sublimation in East Antarctica from snow accumulation measurements. *Clim. Dyn.* 23: 803–813. doi: 10.1007/s00382-004-0462-5.

Fudala, K. and Bialik, R.J. (2020). Breeding colony dynamics of southern elephant seals at Patelnia Point, King George Island, Antarctica. *Remote Sens.* 12: 2964. doi: 10.3390/rs12182964.

Gairola, S., Procheş, Ş., and Rocchini, D. (2013). High-resolution satellite remote sensing: a new frontier for biodiversity exploration in Indian Himalayan forests. *Int. J. Remote Sens.* 34: 2006–2022. doi: 10.1080/01431161.2012.730161.

Gantayat, P. (2022). Supraglacial lake filling models: Examples from Greenland. In: *Advancements in Remote Sensing Technology and the Three Poles*, (ed. M. Pandey, P. Pandey, A. Arora, Y. Ray, S. Jawak, U. Kant Shukla), 1st edn, 381–389. John Wiley & Sons, Inc., Chichester, UK.

Gernez, P., Reynolds, R.A., and Stramski, D. (2014). Within-day variability of particulate organic carbon and remote-sensing reflectance during a bloom of *Phaeocystis antarctica* in the Ross Sea, Antarctica. *Int. J. Remote Sens.* 35: 454–477. doi: 10.1080/01431161.2013.871598.

González, R., Toledano, C., Román, R. et al. (2020). Characterization of stratospheric smoke particles over the Antarctica by remote sensing instruments. *Remote Sens.* 12: 3769. doi: 10.3390/rs12223769.

Gorodetskaya, I.V., Kneifel, S., Maahn, M. et al. (2015). Cloud and precipitation properties from ground-based remote-sensing instruments in East Antarctica. *Cryosph.* 9: 285–304. doi: 10.5194/tc-9-285-2015.

Greene, C.A., Gwyther, D.E., and Blankenship, D.D. (2017). Antarctic mapping tools for Matlab. *Comput. Geosci.* 104: 151–157. doi: 10.1016/j.cageo.2016.08.003.

Greuell, W. and Knap, W.H. (2000). Remote sensing of the albedo and detection of the slush line on the Greenland ice sheet. *J. Geophys. Res. Atmos.* 105: 15567–15576. doi: 10.1029/1999JD901162.

Groh, A. and Horwath, M. (2021). Antarctic ice mass change products from GRACE/GRACE-FO using tailored sensitivity kernels. *Remote Sens.* 13: 1736. doi: 10.3390/rs13091736.

Gruber, N. and Galloway, J.N. (2008). An Earth-system perspective of the global nitrogen cycle. *Nature* 451: 293–296. doi: 10.1038/nature06592.

Guleria, R.P., Kuniyal, J.C., Rawat, P.S. et al. (2012). Validation of MODIS retrieval aerosol optical depth and an investigation of aerosol transport over Mohal in north-western Indian Himalaya. *Int. J. Remote Sens.* 33: 5379–5401. doi: 10.1080/01431161.2012.657374.

Hair, J.H., Reuter, D C., Tonn, S.L. et al. (2018). Landsat 9 thermal infrared sensor 2 architecture and design. In: *IGARSS 2018-2018 IEEE International Geoscience and Remote Sensing Symposium*, 8841–8844. IEEE. doi: 10.1109/IGARSS.2018.8518269.

Hanif, A., Muaz, M., Hasan, A. et al. (2022). Micro-doppler based target recognition with radars: a review. *IEEE Sens. J.* 22: 2948–2961. doi: 10.1109/JSEN.2022.3141213.

Haran, T., Bohlander, J., Scambos, T. et al. (2021). *MODIS mosaic of Antarctica 2003–2004 (MOA2004) image map, version 2.* Digital media, National Snow and Ice Data Center, Boulder, CO: NASA Natl. Snow Ice Data Cent. Distrib. Act. Arch. Cent. https://doi. org/10.7265/N5ZK5DM5

Harris, A.J.L., Wright, R., and Flynn, L.P. (1999). Remote monitoring of Mount Erebus volcano, Antarctica, using polar orbiters: progress and prospects. *Int. J. Remote Sens.* 20: 3051–3071. doi: 10.1080/014311699211615.

Hofmann, D.J. and Montzka, S.A. (2009). Recovery of the ozone layer: the ozone depleting gas index. *Eos, Trans. Am. Geophys. Union* 90: 1. doi: 10.1029/2009EO010001.

Huang, Y., Wang, Y., and Ziehn, T. (2022). Nonlinear interactions of land carbon cycle feedbacks in Earth System Models. *Glob. Chang. Biol.* 28: 296–306. doi: 10.1111/gcb.15953.

Hugentobler, U. and Montenbruck, O. (2017). Satellite orbits and attitude. In: *Springer Handbook of Global Navigation Satellite Systems*, 59–90. Cham: Springer International Publishing. doi: 10.1007/978-3-319-42928-1_3.

Hui, F., Zhao, T., Li, X. et al. (2017). Satellite-based sea ice navigation for Prydz Bay, East Antarctica. *Remote Sens.* 9: 518. doi: 10.3390/rs9060518.

Jafarbiglu, H. and Pourreza, A. (2022). A comprehensive review of remote sensing platforms, sensors, and applications in outcrops. *Comput. Electron. Agric.* 197: 106844. doi: 10.1016/j.compag.2022.106844.

Jawak, S. D., and Sagar Filipe Wankhede, K.B. (2022) High-resolution remote sensing for mapping glacier facies in the Arctic. In: *Advancements in Remote Sensing Technology and the Three Poles*, (ed. M. Pandey, P. Pandey, A. Arora, Y. Ray, S. Jawak, U. Kant Shukla), 1st edn, 371–380. John Wiley & Sons, Inc., Chichester, UK.

Jezek, K.C. and Onstott, R.G. (1999). The role of remote sensing in the environmental monitoring of Antarctica 1. *Polar Geogr.* 23: 55–70. doi: 10.1080/10889379909377664.

Jöckel, P., Sander, R., Kerkweg, A. et al. (2005). Technical note: the Modular Earth Submodel System (MESSy): a new approach towards Earth System Modeling. *Atmos. Chem. Phys.* 5: 433–444. doi: 10.5194/acp-5-433-2005.

Johannessen, O.M., Shalina, E.V., and Miles, M.W. (1999). Satellite evidence for an Arctic sea ice cover in transformation. *Science* 80(286): 1937–1939. doi: 10.1126/science.286.5446.1937.

Khanal, S., Pokhrel, R.P., Pokharel, B. et al. (2022). An episode of transboundary air pollution in the central Himalayas during agricultural residue burning season in North India. *Atmos. Pollut. Res.* 13: 101270. doi: 10.1016/j.apr.2021.101270.

Kumar, A., Singh, N., Anshumali, et al. (2018). Evaluation and utilization of MODIS and CALIPSO aerosol retrievals over a complex terrain in Himalaya. *Remote Sens. Environ.* 206: 139–155. doi: 10.1016/j.rse.2017.12.019.

Kumar, A., and Rahul Devrani, P. S. (2022). Landscapes and paleoclimate of the Ladakh Himalaya. In: *Advancements in Remote Sensing Technology and the Three Poles*, (ed. M. Pandey, P. Pandey, A. Arora, Y. Ray, S. Jawak, U. Kant Shukla), 1st edn, 308–320. John Wiley & Sons, Inc., Chichester, UK.

Kumar, R., and Rahul Devrani, B.D. (2022). A review of remote sensing and GIS-based soil loss models with a comparative study from the upper and marginal Ganga River basin. In: *Advancements in remote sensing technology and the three poles*, (ed. M. Pandey, P. Pandey, A. Arora, Y. Ray, S. Jawak, U. Kant Shukla), 1st edn, 321–339. John Wiley & Sons, Inc., Chichester, UK.

Kumar, R., Devrani, R., Dangwal, A., Deshmukh, B., and Dutt, S. (2022). Extreme hydrological event induced temporal variation in soil erosion of the Assiganga River basin, NW. Himalaya. In: *Advancements in Remote Sensing Technology and the Three Poles*, (ed. M. Pandey, P. Pandey, A. Arora, Y. Ray, S. Jawak, U. Kant Shukla), 1st edn, 230–246. John Wiley & Sons, Inc., Chichester, UK.

Kumpula, T., Forbes, B.C., Stammler, F. et al. (2012). Dynamics of a coupled system: multi-resolution remote sensing in assessing social-ecological responses during 25 years of gas field development in Arctic Russia. *Remote Sens.* 4: 1046–1068. doi: 10.3390/rs4041046.

Kumpula, T., Pajunen, A., Kaarlejärvi, E. et al. (2011). Land use and land cover change in Arctic Russia: ecological and social implications of industrial development. *Glob. Environ. Chang.* 21: 550–562. doi: 10.1016/j.gloenvcha.2010.12.010.

Kuttippurath, J. and Nair, P.J. (2017). The signs of Antarctic ozone hole recovery. *Sci. Rep.* 7: 585. doi: 10.1038/s41598-017-00722-7.

Lawton, J. (2001). Earth system science. *Science* 80(292): 1965–1965. doi: 10.1126/science.292.5524.1965.

Lechner, A.M., Foody, G.M., and Boyd, D.S. (2020). Applications in remote sensing to forest ecology and management. *One Earth* 2: 405–412. doi: 10.1016/j.oneear.2020.05.001.

Liu, H. and Jezek, K.C. (2004). A complete high-resolution coastline of Antarctica extracted from orthorectified radarsat SAR imagery. *Photogram. Eng. Remote Sens.* 70: 605–616. doi: 10.14358/PERS.70.5.605.

Liu, Z. and Bromwich, D.H. (1993). Acoustic remote sensing of planetary boundary layer dynamics near Ross Island, Antarctica. *J. Appl. Meteorol.* 32: 1867–1882. doi: 10.1175/1520-0450(1993)032<1867:ARSOPB>2.0.CO;2.

Lu, Y., Shao, Q., Yue, H. et al. (2019). A review of the space environment effects on spacecraft in different orbits. *IEEE Access* 7: 93473–93488. doi: 10.1109/ACCESS.2019.2927811.

Mahaffy, P.G., Krief, A., Hopf, H. et al. (2018). Reorienting chemistry education through systems thinking. *Nat. Rev. Chem.* 2: 0126. doi: 10.1038/s41570-018-0126.

Malenovský, Z., Lucieer, A., King, D.H. et al. (2017). Unmanned aircraft system advances health mapping of fragile polar vegetation. *Methods Ecol. Evol.* 8: 1842–1857. doi: 10.1111/2041-210X.12833.

Mandal, P., and Sarkar, S. (2022) Assessing the accuracy of Digital Elevation Models in Darjeeling-Sikkim Himalayas. In: *Advancements in Remote Sensing Technology and the Three Poles*, (ed. M. Pandey, P. Pandey, A. Arora, Y. Ray, S. Jawak, U. Kant Shukla), 1st edn, 36–48. John Wiley & Sons, Inc., Chichester, UK.

Markham, B.L., Storey, J.C., Williams, D.L. et al. (2004). Landsat sensor performance: history and current status. *IEEE Trans. Geosci. Remote Sens.* 42: 2691–2694.

Maruyama, S., Santosh, M., and Zhao, D. (2007). Superplume, supercontinent, and post-perovskite: mantle dynamics and anti-plate tectonics on the Core–Mantle Boundary. *Gondwana Res.* 11: 7–37. doi: 10.1016/j.gr.2006.06.003.

Masek, J.G., Wulder, M.A., Markham, B. et al. (2020). Landsat 9: empowering open science and applications through continuity. *Remote Sens. Environ.* 248: 111968. doi: 10.1016/j.rse.2020.111968.

Masson-Delmotte, V., Zhai, P. et al. (eds.) (2021). IPCC, 2021: Climate Change 2021:the Physical Science Basis. *Contribution of Working Group I to the Sixth Assessment Report of the Intergovernmental Panel on Climate Change.* Cambridge, UK: Cambridge University Press. Available at: https://www.ipcc.ch/report/ar6/wg1/downloads/report/IPCC_AR6_WGI_Full_Report.pdf.

McAllister, E., Payo, A., Novellino, A. et al. (2022). Multispectral satellite imagery and machine learning for the extraction of shoreline indicators. *Coast. Eng.* 174: 104102. doi: 10.1016/j.coastaleng.2022.104102.

McCorkel, J., Montanaro, M., Efremova, B. et al. (2018). Landsat 9 thermal infrared sensor 2 characterization plan overview. In: *IGARSS 2018-2018 IEEE International Geoscience and Remote Sensing Symposium*, 8845–8848. IEEE. doi: 10.1109/IGARSS.2018.8518798.

Mehta, B., Yadav, A., Aljasil, C., Ambili, A.; and Mishra, P. K. (2022). Understanding the present and past climate-human-vegetation dynamics in the Indian Himalayas: A comprehensive review. In: Manish Pandey, Prem Pandey, Aman Arora, Yogesh Ray, Shridhar Jawak, Uma Kant Shukla (Eds.), *Advancements in Remote Sensing Technology and the Three Poles*, 1st edn. Chichester, UK: John Wiley & Sons, Inc., pp. 247–256.

Miller, J.G. and Miller, J.L. (1982). The earth as a system. *Behav. Sci.* 27: 303–322. doi:10.1002/bs.3830270402.

Aditya Mishra, H. C., and Nainwal, R. S. (2022). Glacier ice thickness estimation in Indian himalaya using geophysical methods: A brief review. In: Manish Pandey, Prem Pandey, Aman Arora, Yogesh Ray, Shridhar Jawak, Uma Kant Shukla (Eds.), *Advancements in Remote Sensing Technology and the Three Poles*, 1st edn. Chichester, UK: John Wiley & Sons, Inc., pp. 299–307.

Mishra, J., and Takuya Inoue, A.P. (2022). Continuous satellite missions, data availability, and nature of future satellite missions with implications to Polar Regions. In: *Advancements in Remote Sensing Technology and the Three Poles*, (ed. M. Pandey, P. Pandey, A. Arora, Y. Ray, S. Jawak, U. Kant Shukla), 1st edn, 24–35. John Wiley & Sons, Inc., Chichester, UK.

Mishra, M., Singh, K. K., Pandey, P. C., Devrani, R., Pandey, A. K., Prudhvi Raju, K. N., Ranjan, P., Arora, A., Costache, R., Janizadeh, S., and Nguyen Thuy Linh, M.P. (2022). Spectral indices across remote sensing and sensor relating to the three poles: An overview of applications, challenges, and future prospects. In: *Advancements in Remote Sensing Technology and the Three Poles*, (ed. M. Pandey, P. Pandey, A. Arora, Y. Ray, S. Jawak, U. Kant Shukla), 1st edn, 83–116. John Wiley & Sons, Inc., Chichester, UK.

Mishra, P.K., Rai, A., and Rai, S.C. (2020d). Land use and land cover change detection using geospatial techniques in the Sikkim Himalaya, India. *Egypt. J. Remote Sens. Sp. Sci.* 23: 133–143. doi: 10.1016/j.ejrs.2019.02.001.

Mishra, V. N. (2022). Investigation of land use/land cover changes in Alaknanda River Basin, Himalaya during 1976 – 2020. In: *Advancements in Remote Sensing Technology and the Three Poles*, (ed. M. Pandey, P. Pandey, A. Arora, Y. Ray, S. Jawak, U. Kant Shukla), 1st edn, 351–356. John Wiley & Sons, Inc., Chichester, UK.

Misra, S., Sharma, A., Ravi, S., and Maurya, K. G. M. (2022). Wetlands as potential zones to understand spatiotemporal plants-human-climate interaction: A review on palynological perspective from western and eastern Himalaya. In: *Advancements in Remote Sensing Technology and the Three Poles*, (ed. M. Pandey, P. Pandey, A. Arora, Y. Ray, S. Jawak, U. Kant Shukla), 1st edn, 340–350. John Wiley & Sons, Inc., Chichester, UK.

Mitra, S., Devrani, R., Pandey, M., Arora, A., Costache Romulus, S.J. (2022). Landscape modelling, glacier & ice sheet dynamics, and the three poles: A review of models, softwares and tools. In: *Advancements in Remote Sensing Technology and the Three Poles*, (ed. M. Pandey, P. Pandey, A. Arora, Y. Ray, S. Jawak, U. Kant Shukla), 1st edn, 58–82. John Wiley & Sons, Inc., Chichester, UK.

Mitra, S., Pundir, S., Devrani, R., Arora, A., Pandey, M., and Romulus Costache, S. J. (2022). An overview of morphometry software packages, tools, Add-ons. In: *Advancements in Remote Sensing Technology and the Three Poles*, (ed. M. Pandey, P. Pandey, A. Arora, Y. Ray, S. Jawak, U. Kant Shukla), 1st edn, 49–57. John Wiley & Sons, Inc., Chichester, UK.

Mohapatra, J., Singh, C.P., Tripathi, O.P. et al. (2019). Remote sensing of alpine treeline ecotone dynamics and phenology in Arunachal Pradesh Himalaya. *Int. J. Remote Sens.* 40: 7986–8009. doi: 10.1080/01431161.2019.1608383.

Montenbruck, O., Gill, E., and Lutze, F. (2002). Satellite orbits: models, methods, and applications. *Appl. Mech. Rev.* 55: B27–B28. doi: 10.1115/1.1451162.

Morison, J., Wilkinson, J., Alkire, M. et al. (2018). The north pole region as an indicator of the changing Arctic Ocean. *Arctic* 71: 1–5. Available at: https://www.jstor.org/stable/26646182.

Munawar, H.S., Hammad, A.W.A., and Waller, S.T. (2022). Remote sensing methods for flood prediction: a review. *Sensors* 22: 960. doi: 10.3390/s22030960.

Notz, D. (2009). The future of ice sheets and sea ice: between reversible retreat and unstoppable loss. *Proc. Natl. Acad. Sci.* 106: 20590–20595. doi: 10.1073/pnas.0902356106.

O'Neill, A. and Steenman-Clark, L. (2002). The computational challenges of Earth-system science. *Philos. Trans. R. Soc. London. Ser. A Math. Phys. Eng. Sci.* 360: 1267–1275. doi: 10.1098/rsta.2002.1000.

Orynbaikyzy, A., Gessner, U., and Conrad, C. (2019). Crop type classification using a combination of optical and radar remote sensing data: a review. *Int. J. Remote Sens.* 40: 6553–6595. doi: 10.1080/01431161.2019.1569791.

Palm, S.P., Yang, Y., Spinhirne, J.D. et al. (2011). Satellite remote sensing of blowing snow properties over Antarctica. *J. Geophys. Res.* 116: D16123–D16123. doi: 10.1029/2011JD015828.

Pande, A., Anand, A., and Shailendra Saini, K. S. (2022). Geospatial tools for monitoring vertebrate populations in Antarctica with a note on the ecological component of the

Indian Antarctic program. In: *Advancements in Remote Sensing Technology and the Three Poles*, (ed. M. Pandey, P. Pandey, A. Arora, Y. Ray, S. Jawak, U. Kant Shukla), 1st edn, 144–154. John Wiley & Sons, Inc., Chichester, UK.

Pandey, U., Shah, S. K., and Mehrotra, N. (2022). Fluctuations of Kolahoi glacier, Kashmir valley, its assessment with tree-rings of Pinus wallichiana and comparable satellite imageries and field survey records. In: *Advancements in Remote Sensing Technology and the Three Poles*, (ed. M. Pandey, P. Pandey, A. Arora, Y. Ray, S. Jawak, U. Kant Shukla), 1st edn, 203–212. John Wiley & Sons, Inc., Chichester, UK.

Pandey, P. C., & Sharma, L. K. (Eds.). (2021). Advances in Remote Sensing for Natural Resource Monitoring. John Wiley & Sons.

Pandey, P. C., Koutsias, N., Petropoulos, G. P., Srivastava, P. K., and Ben Dor, E. (2021). Land use/land cover in view of earth observation: Data sources, input dimensions, and classifiers—a review of the state of the art. *Geocarto Int.* 36(9), 957–988.

Peral, E., Im, E., Wye, L. et al. (2018). Radar technologies for earth remote sensing from CubeSat platforms. *Proc. IEEE.* 106: 404–418. doi: 10.1109/JPROC.2018.2793179.

Pereira, F., Marques, J.S., Heleno, S. et al. (2020). Detection and delineation of sorted stone circles in Antarctica. *Remote Sens.* 12: 160. doi: 10.3390/rs12010160.

Peter, T. (1994). The stratospheric ozone layer: an overview. *Environ. Pollut.* 83: 69–79. doi: 10.1016/0269-7491(94)90024-8.

Picard, G., Fily, M., and Gallee, H. (2007). Surface melting derived from microwave radiometers: a climatic indicator in Antarctica. *Ann. Glaciol.* 46: 29–34. doi: 10.3189/172756407782871684.

Pijoan, J., Altadill, D., Torta, J. et al. (2014). Remote geophysical observatory in Antarctica with HF data transmission: a review. *Remote Sens.* 6: 7233–7259. doi: 10.3390/rs6087233.

Pina, P. and Vieira, G. (2022). UAVs for science in Antarctica. *Remote Sens.* 14: 1610. doi: 10.3390/rs14071610.

Pöschl, U. and Shiraiwa, M. (2015). Multiphase chemistry at the atmosphere–biosphere interface influencing climate and public health in the anthropocene. *Chem. Rev.* 115: 4440–4475. doi: 10.1021/cr500487s.

Pour, A.B., Park, Y., Park, T.-Y.S. et al. (2018). Regional geology mapping using satellite-based remote sensing approach in Northern Victoria Land, Antarctica. *Polar Sci.* 16: 23–46. doi: 10.1016/j.polar.2018.02.004.

Prasad, P., Dai, B.K., and Ramakrishna, B.N. (2022a). Gyro sensor calibration of ISRO's remote sensing satellites. *Aerosp. Syst.* 5: 11–19. doi: 10.1007/s42401-021-00118-6.

Prasad, P., Loveson, V.J., Chandra, P. et al. (2022b). Evaluation and comparison of the earth observing sensors in land cover/land use studies using machine learning algorithms. *Ecol. Inform.* 68: 101522. doi: 10.1016/j.ecoinf.2021.101522.

Pudełko, R., Angiel, P., Potocki, M. et al. (2018). Fluctuation of glacial retreat rates in the eastern part of Warszawa Icefield, King George Island, Antarctica, 1979–2018. *Remote Sens.* 10: 892. doi: 10.3390/rs10060892.

Quincey, D.J., Lucas, R.M., Richardson, S.D. et al. (2005). Optical remote sensing techniques in high-mountain environments: application to glacial hazards. *Prog. Phys. Geogr. Earth Environ.* 29: 475–505. doi: 10.1191/0309133305pp456ra.

Rahmstorf, S. (1995). Bifurcations of the Atlantic thermohaline circulation in response to changes in the hydrological cycle. *Nature* 378: 145–149. doi: 10.1038/378145a0.

Ranjan, S., Pandey, M., and Raj, R. (2022) Hydrological changes in the Arctic, the Antarctic, and the Himalayas: A synoptic view from the cryosphere change perspective. In: *Advancements in Remote Sensing Technology and the Three Poles*, (ed. M. Pandey, P. Pandey, A. Arora, Y. Ray, S. Jawak, U. Kant Shukla), 1st edn, 359–370. John Wiley & Sons, Inc., Chichester, UK.

Reid, W.V., Chen, D., Goldfarb, L. et al. (2010). Earth system science for global sustainability: grand challenges. *Science* 80(330): 916–917. doi: 10.1126/science.1196263.

Rignot, E., Jacobs, S., Mouginot, J. et al. (2013). Ice-shelf melting around Antarctica. *Science* 80(341): 266–270. doi: 10.1126/science.1235798.

Rodhe, H. (1992). 4 modeling biogeochemical cycles. In: *Global Biogeochemical Cycles International Geophysics Book Series* (ed. G.V. Wolfe, S.S. Butcher, R.J. Charlson, et al.), 55–72. doi: 10.1016/S0074-6142(08)62687-X.

Rogers, A., Serbin, S.P., Ely, K.S. et al. (2019). Terrestrial biosphere models may overestimate Arctic $CO_2$ assimilation if they do not account for decreased quantum yield and convexity at low temperature. *New Phytol.* 223: 167–179. doi: 10.1111/nph.15750.

Roy, D.P., Wulder, M.A., Loveland, T.R. et al. (2014). Landsat-8: science and product vision for terrestrial global change research. *Remote Sens. Environ.* 145: 154–172. doi: 10.1016/j.rse.2014.02.001.

Salvatore, M.R., Borges, S.R., Barrett, J.E. et al. (2020). Remote characterization of photosynthetic communities in the Fryxell basin of Taylor Valley, Antarctica. *Antarct. Sci.* 32: 255–270. doi: 10.1017/S0954102020000176.

Sandau, R. (2010). Status and trends of small satellite missions for Earth observation. *Acta Astronaut.* 66: 1–12. doi: 10.1016/j.actaastro.2009.06.008.

Sarma, M., Zaman, S. Borah, D. B., and Tapos Kumar Goswami, U. B. (2022). The role of Himalayan frontal thrust in the upliftment of Kimin Formation and the migration of sedimentary basin in Arunachal Himalaya, around Bandardewa, Papumpare District, Arunachal Pradesh. In: *Advancements in Remote Sensing Technology and The Three Poles*, (ed. M. Pandey, P. Pandey, A. Arora,

Y. Ray, S. Jawak, U. Kant Shukla), 1st edn, 268–282. John Wiley & Sons, Inc., Chichester, UK.

Scambos, T.A., Haran, T.M., Fahnestock, M.A. et al. (2007). MODIS-based Mosaic of Antarctica (MOA) datasets: continent-wide surface morphology and snow grain size. *Remote Sens. Environ.* 111: 242–257. doi: 10.1016/j.rse.2006.12.020.

Scheinert, M., Ferraccioli, F., Schwabe, J. et al. (2016). New Antarctic gravity anomaly grid for enhanced geodetic and geophysical studies in Antarctica. *Geophys. Res. Lett.* 43: 600–610. doi: 10.1002/2015GL067439.

Schellnhuber, H.J. (2009). Tipping elements in the Earth System. *Proc. Natl. Acad. Sci.* 106: 20561–20563. doi: 10.1073/pnas.0911106106.

Schlesinger, W.H., Cole, J.J., Finzi, A.C. et al. (2011). Introduction to coupled biogeochemical cycles. *Front. Ecol. Environ.* 9: 5–8. doi: 10.1890/090235.

Seitzinger, S.P., Gaffney, O., Brasseur, G. et al. (2015). International Geosphere–Biosphere Programme and Earth system science: three decades of co-evolution. *Anthropocene* 12: 3–16. doi: 10.1016/j.ancene.2016.01.001.

Seo, M., Kim, H.-C., Huh, M. et al. (2016). Long-term variability of surface albedo and its correlation with climatic variables over Antarctica. *Remote Sens.* 8: 981. doi: 10.3390/rs8120981.

Sharma, M., Areendran, G., Raj, K. et al. (2016). Multitemporal analysis of forest fragmentation in Hindu Kush Himalaya: a case study from Khangchendzonga Biosphere Reserve, Sikkim, India. *Environ. Monit. Assess.* 188: 596. doi: 10.1007/s10661-016-5577-8.

Sharma, U., and Yogesh Ray, M.P. (2022). Terrestrial deglaciation signatures in East Antarctica. In: *Advancements in Remote Sensing Technology and the Three Poles*, (ed. M. Pandey, P. Pandey, A. Arora, Y. Ray, S. Jawak, U. Kant Shukla), 1st edn, 128–143. John Wiley & Sons, Inc., Chichester, UK.

Siingh, D., Chate, D.M., and Ali, K. (2012). Time-elapsed evolution of aerosol size distributions by snow particles after the passage of blizzards over the Maitri (Antarctica). *Int. J. Remote Sens.* 33: 962–978. doi: 10.1080/01431161.2010.542206.

Singh, A. K., Ray, Y., Saini, S., and Rahul Mohan, M. J. B. (2022) Multi-disciplinary research in the Indian antarctic programme and it's international relevance. In: *Advancements in Remote Sensing Technology and The Three Poles*, (ed. M. Pandey, P. Pandey, A. Arora, Y. Ray, S. Jawak, U. Kant Shukla), 1st edn, 415–419. John Wiley & Sons, Inc., Chichester, UK.

Singh, A., and Divya David, T. K. P. K. (2022). Indian and international research coordination in the arctic. In: *Advancements in Remote Sensing Technology and the Three Poles*, (ed. M. Pandey, P. Pandey, A. Arora, Y. Ray, S. Jawak, U. Kant Shukla), 1st edn, 420–426. John Wiley & Sons, Inc., Chichester, UK.

Singh, D. (2022). Bryophytes of Larsemann Hills, East Antarctica and future prospects. In: *Advancements in Remote Sensing Technology and the Three Poles*, (ed. M. Pandey, P. Pandey, A. Arora, Y. Ray, S. Jawak, U. Kant Shukla), 1st edn, 155–178. John Wiley & Sons, Inc., Chichester, UK.

Sobolev, A.V., Hofmann, A., Kuzmin, D.V. et al. (2007). The amount of recycled crust in sources of mantle-derived melts. *Science* 80(316): 412–417. doi: 10.1126/science.1138113.

Sokol, N.W., Slessarev, E., Marschmann, G.L. et al. (2022). Life and death in the soil microbiome: how ecological processes influence biogeochemistry. *Nat. Rev. Microbiol.* 20: 1–16. doi: 10.1038/s41579-022-00695-z.

Solomon, S. (1988). The mystery of the Antarctic Ozone "Hole." *Rev. Geophys.* 26(1): 131–148. doi: 10.1029/RG026i001p00131.

Solomon, S. (2019). The discovery of the Antarctic ozone hole. *Nature* 575: 46–47. doi: 10.1038/d41586-019-02837-5.

Solomon, S., Ivy, D.J., Kinnison, D. et al. (2016). Emergence of healing in the Antarctic ozone layer. *Science* 80(353): 269–274. doi: 10.1126/science.aae0061.

Song, C., Huang, B., Ke, L. et al. (2014). Remote sensing of alpine lake water environment changes on the Tibetan Plateau and surroundings: a review. *ISPRS J. Photogramm. Remote Sens.* 92: 26–37. doi: 10.1016/j.isprsjprs.2014.03.001.

Spicer, R.A., Harris, N.B.W., Widdowson, M. et al. (2003). Constant elevation of southern Tibet over the past 15 million years. *Nature* 421: 622–624. doi: 10.1038/nature01356.

Srivastava, R. (2022). Spatio-temporal variations of aerosols over the Polar regions based on satellite remote sensing. In: *Advancements in Remote Sensing Technology and the Three Poles*, (ed. M. Pandey, P. Pandey, A. Arora, Y. Ray, S. Jawak, U. Kant Shukla), 1st edn, 401–411. John Wiley & Sons, Inc., Chichester, UK.

Steffen, W., Richardson, K., Rockström, J. et al. (2020). The emergence and evolution of Earth System Science. *Nat. Rev. Earth Environ.* 1: 54–63. doi: 10.1038/s43017-019-0005-6.

Stern, R.J. (2018). The evolution of plate tectonics. *Philos. Trans. R. Soc. A Math. Phys. Eng. Sci.* 376: 20170406. doi: 10.1098/rsta.2017.0406.

Storey, B.C. (1995). The role of mantle plumes in continental breakup: case histories from Gondwanaland. *Nature* 377: 301–308. doi: 10.1038/377301a0.

Stow, D.A., Hope, A., McGuire, D. et al. (2004). Remote sensing of vegetation and land-cover change in Arctic Tundra Ecosystems. *Remote Sens. Environ.* 89: 281–308. doi: 10.1016/j.rse.2003.10.018.

Strahler, A. (2013). *Introducing Physical Geography*, 6th ed. Boston, MA: J. Wiley & Sons.

Sun, X., Wu, W., Li, X. (2021). Vegetation abundance and health mapping over southwestern antarctica based on worldview-2 data and a modified spectral mixture analysis. *Remote Sens.* 13: 166. doi: 10.3390/rs13020166.

Swathi, M., Yadav, J., and Avinash Kumar, R. M. (2022). Antarctic sea ice variability and trends over the last four decades. In: *Advancements in Remote Sensing Technology and the Three Poles*, (ed. M. Pandey, P. Pandey, A. Arora, Y. Ray, S. Jawak, U. Kant Shukla), 1st edn, 179–190. John Wiley & Sons, Inc., Chichester, UK.

Tabibi, S., Geremia-Nievinski, F., Francis, O. et al. (2020). Tidal analysis of GNSS reflectometry applied for coastal sea level sensing in Antarctica and Greenland. *Remote Sens. Environ.* 248: 111959. doi: 10.1016/j.rse.2020.111959.

Taylor, P., Hegyi, B., Boeke, R. et al. (2018). On the increasing importance of air–sea exchanges in a thawing arctic: a review. *Atmosphere* Basel 9: 41. doi: 10.3390/atmos9020041.

Thomas, R., Rignot, E., Casassa, G. et al. (2004). Accelerated sea-level rise from West Antarctica. *Science* 80(306): 255–258. doi: 10.1126/science.1099650.

Thomas, R.H. (2001). Remote sensing reveals shrinking Greenland ice sheet. *Eos, Trans. Am. Geophys. Union* 82: 369–369. doi: 10.1029/01EO00226.

Tjernström, M., Leck, C., Persson, P.O.G. et al. (2004). The summertime Arctic atmosphere: meteorological measurements during the Arctic Ocean Experiment 2001. *Bull. Am. Meteorol. Soc.* 85: 1305–1322. doi: 10.1175/BAMS-85-9-1305.

Tomar, K. S., Kumari, S., and Ashutosh Venkatesh Prasad, A. J. L. (2022). Glacier dynamics in East Antarctica: A remote sensing perspective. In: *Advancements in Remote Sensing Technology and the Three Poles*, (ed. M. Pandey, P. Pandey, A. Arora, Y. Ray, S. Jawak, U. Kant Shukla), 1st edn, 119–127. John Wiley & Sons, Inc., Chichester, UK.

Varotsos, C. (2002). The southern hemisphere ozone hole split in 2002. *Environ. Sci. Pollut. Res.* 9: 375–376. doi: 10.1007/BF02987584.

Vaughan, D.G., Mantripp, D.R., Sievers, J. et al. (1993). A synthesis of remote sensing data on Wilkins Ice Shelf, Antarctica. *Ann. Glaciol.* 17: 211–218. doi: 10.3189/S0260305500012866.

Vieira, G., Mora, C., Pina, P. et al. (2014). A proxy for snow cover and winter ground surface cooling: mapping *Usnea* sp. communities using high resolution remote sensing imagery (Maritime Antarctica). *Geomorphology* 225: 69–75. doi: 10.1016/j.geomorph.2014.03.049.

Wadhams, P. (1990). Evidence for thinning of the Arctic ice cover north of Greenland. *Nature* 345: 795–797. doi: 10.1038/345795a0.

Wake, B. (2019). A drift in the Arctic. *Nat. Clim. Chang.* 9: 733–733. doi: 10.1038/s41558-019-0597-3.

Warner, J.C., Perlin, N., and Skyllingstad, E.D. (2008). Using the Model Coupling Toolkit to couple earth system models. *Environ. Model. Softw.* 23: 1240–1249. doi: 10.1016/j.envsoft.2008.03.002.

Wassie, Y., Mirmazloumi, S.M., Crosetto, M. et al. (2022). Spatio-temporal quality indicators for differential interferometric synthetic aperture radar data. *Remote Sens.* 14: 798. doi: 10.3390/rs14030798.

Weatherhead, E.C. and Andersen, S.B. (2006). The search for signs of recovery of the ozone layer. *Nature* 441: 39–45. doi: 10.1038/nature04746.

Wesche, C., Jansen, D., and Dierking, W. (2013). Calving fronts of Antarctica: mapping and classification. *Remote Sens.* 5: 6305–6322. doi: 10.3390/rs5126305.

Williams, D.L., Goward, S., and Arvidson, T. (2006). Landsat. *Photogram. Eng. Remote Sens.* 72: 1171–1178. doi: 10.14358/PERS.72.10.1171.

Williams, G.P. (1988a). The dynamical range of global circulations? Part II. *Clim. Dyn.* 3: 45–84. doi: 10.1007/BF01080901.

Williams, G.P. (1988b). The dynamical range of global circulations? Part I. *Clim. Dyn.* 2: 205–260. doi: 10.1007/BF01371320.

Williams, R.S., Jr. and Ferrigno, J.G. (2012). State of the Earth's cryosphere at the beginning of the 21st century: glaciers, global snow cover, floating ice, and permafrost and periglacial environments. *Director* 508: 344–6840. Available at: https://pubs.usgs.gov/pp/p1386a.

Xu, D., Tang, X., Yang, S. et al. (2022). Revisiting ice flux and mass balance of the Lambert Glacier–Amery Ice Shelf system using multi-remote-sensing datasets, East Antarctica. *Remote Sens.* 14: 391. doi: 10.3390/rs14020391.

Yamazaki, D. and Trigg, M.A. (2016). The dynamics of Earth's surface water. *Nature* 540: 348–349. doi: 10.1038/nature21100.

Yao, T., Thompson, L.G., Mosbrugger, V. et al. (2012a). Third Pole Environment (TPE). *Environ. Dev.* 3: 52–64. doi: 10.1016/j.envdev.2012.04.002.

Yao, T., Thompson, L., Yang, W. et al. (2012b). Different glacier status with atmospheric circulations in Tibetan Plateau and surroundings. *Nat. Clim. Chang.* 2: 663–667. doi: 10.1038/nclimate1580.

Yu, T., Li, M., Li, W. et al. (2022). Polarimetric calibration technique for a fully polarimetric entomological radar based on antenna rotation. *Remote Sens.* 14: 1551. doi: 10.3390/rs14071551.

Zanutta, A., Negusini, M., Vittuari, L. et al. (2018). New geodetic and gravimetric maps to infer geodynamics of Antarctica with insights on Victoria Land. *Remote Sens.* 10: 1608. doi: 10.3390/rs10101608.

Zhai, Y., Han, Y., Lu, H. et al. (2022). Interactions between anthropogenic pollutants (biodegradable organic nitrogen and ammonia) and the primary hydrogeochemical component Mn in groundwater: evidence from three polluted sites. *Sci. Total Environ.* 808: 152162. doi: 10.1016/j.scitotenv.2021.152162.

Zhang, G., Yao, T., Xie, H. et al. (2015). An inventory of glacial lakes in the Third Pole region and their changes in response to global warming. *Glob. Planet. Change* 131: 148–157. doi: 10.1016/j.gloplacha.2015.05.013.

Zhou, Y., Yan, G., Zhao, J. et al. (2018). Estimation of daily average downward shortwave radiation over Antarctica. *Remote Sens.* 10: 422. doi: 10.3390/rs10030422.

Zmarz, A., Korczak-Abshire, M., Storvold, R. et al. (2015). Indicator species population monitoring in Antarctica with UAV. *Int. Arch. Photogramm. Remote Sens. Spat. Inf. Sci.* 40: 189–193.

# 2

# Continuous Satellite Missions, Data Availability, and Nature of Future Satellite Missions with Implications to Polar Regions

*Jagriti Mishra*[1,4,*], *Takuya Inoue*[2], *and Avinash Kumar Pandey*[3]

[1] *School of Environmental Sciences, Jawaharlal Nehru University, New Delhi 110067, India*
[2] *Graduate School of Advanced Science and Engineering, Hiroshima University, Higashi Hiroshima, Japan*
[3] *Institute of Applied Science and Humanity, GLA University, Mathura, Uttar Pradesh, India*
[4] *Civil Engineering Research Institute for Cold Region, Sapporo, Japan*
[*] *Corresponding author*

## 2.1 Introduction

According to NASA (National Aeronautics and Space Administration), a *satellite* is a body that moves in a curved path around another body in space. This curved path is called an *orbit* and can be either circular or elliptical. Satellites can be natural or man-made. The Earth is a natural satellite revolving around the Sun and the moon. The first man-made satellite was Sputnik which was launched in October 1957 by Russia. Man-made satellites have different shapes and sizes depending on the purpose they are designed to fulfil, like communication, Earth Observation (EO), navigation, etc. Depending on their purpose, the satellites are installed with respective sensors. Recent advancements in satellite sensor technologies and storage and processing technologies for big-data have aided Remote Sensing (RS) researchers to make use of satellite data for various purposes like forest cover estimation, climate change studies, forest fire damage estimation, prediction of typhoons, etc. EO and RS technologies have especially aided research in the polar regions where human intervention is minimal and often expensive. EO missions have contributed to the availability of multisensor, multitemporal, multiresolution, and multifrequency data that make it possible to study the changes in ice sheets in polar regions, mapping the unknown polar regions (Wulder et al., 2019), mapping of ice flow (Mouginot et al., 2017), study of limate change, etc. In this chapter, we introduce the reader to different satellite missions (note: mainly Earth Observation satellites that relate to the scope of this chapter and book), data available from satellites, different platforms available for accessing the satellite data, and future satellite missions with implications to the polar regions.

### 2.1.1 Types of Orbit

The orbit is the circular or elliptical path in which a body (moon, satellite, etc.) travels around another body. This circular or elliptical path is caused by gravity. A path is circular when the distance from the Earth is continually the same, whereas an elliptical path changes its distance from the Earth (Riebeek and Simmon, 2009). The distance between a satellite and the Earth is the height of orbit which determines the speed at which the satellite will circle the Earth. Satellites closest to the Earth will have a higher speed due to gravity. Satellites taking an elliptical orbit change their distance from the Earth at different times; hence changing their speed. When a satellite is closest to the Earth, i.e., at its perigee, it is the fastest, whereas when it is at its pogee (farthest from the Earth), it is the slowest. The elliptical path of planets around the Sun and variation in speed of objects due to gravity was first introduced by Kepler. The height of orbit a satellite will be placed at depends on the purpose the satellite is designed for.

Other factors determining the shape of the orbit of a satellite are *eccentricity* and *inclination*. Eccentricity is the measure of the shape of an orbit. When eccentricity ($e$) is zero, the satellite follows a perfectly circular orbit around the Earth. As eccentricity ($e$) increases, the shape of the orbit becomes more elliptical. The value of eccentricity ($e$) for any satellite can be 0 or more, but is always less than 1.

*Advances in Remote Sensing Technology and the Three Poles*, First Edition. Edited by Manish Pandey, Prem C. Pandey, Yogesh Ray, Aman Arora, Shridhar D. Jawak, and Uma K. Shukla.

When $e = 1$, the orbit becomes parabolic; when $e > 1$, the orbit becomes hyperbolic, and when $e \geq 1$, the satellite enters escape orbit and will eventually exit our solar system. The *inclination* is the measure of the angle an orbit possesses relating to the Earth's equator. An inclination of $90^0$ will put the satellite almost above the poles, and when the inclination is zero, the satellite orbits the Earth directly above the equator (Figure 2.1).

#### 2.1.1.1 High Earth Orbit (HEO)

The satellites in High Earth Orbit (HEO) revolve at about 35,780 km away from the surface of the Earth. Satellites in this orbit take the same time as the Earth to finish one revolution, i.e., 23 hours, 56 minutes, and 4 seconds ($ESA_a$, 2021). The satellites in this orbit appear to be stationary as they move at the same speed as Earth; hence the satellites are always above the same place on the Earth. Therefore, this orbit is often known as the Geo-Synchronous Orbit (GSO). When a GSO satellite has zero eccentricity and zero inclination, it orbits directly above the Earth's equator and said to be in the Group on Earth Observations (GEO). Tele-communication and weather monitoring satellites are often put in this orbit. However, this orbit does not efficiently cover areas at higher latitudes.

#### 2.1.1.2 Medium Earth Orbit (MEO)

Medium Earth Orbit (MEO) satellites are closer to the earth and hence faster. MEO encompasses a wide range of orbits between LEO (Low Earth Orbit: discussed in Section 2.2.2.5) and HEO. Navigation satellites like GPS (Global Positioning System) and the European Galileo System are placed into this orbit. The two famously known MEOs are the semi-synchronous and the Molniya orbits.

#### 2.1.1.3 Semi-Synchronous Orbit

The satellites in the Semi-Synchronous Orbit orbit have almost zero eccentricity, i.e., they follow an almost circular orbit. The satellites take 12 hours to complete an orbit.

#### 2.1.1.4 Molniya Orbit

The Molniya Orbit was invented in Russia. The Molniya orbit is extremely eccentric (0.72) and has a high inclination ($63.4^0$) (Kidder and Vonder Haar, 1990). This highly eccentric and inclined orbit enables satellites to maximize the observation time over higher latitudes (Figure 2.2). They move very fast when they are close to the Earth and then slow down as they move away from the Earth. The satellites in this orbit also take 12 hours to complete an entire revolution; however, they spend two-thirds of their time over one hemisphere. This orbit is used by Russian Communication satellites.

#### 2.1.1.5 Low Earth Orbit (LEO)

Low Earth Orbit (LEO) lies comparatively much closer to the Earth's surface, at 200–1000 km above the surface. This orbit is generally used for satellite imaging, as satellites in this orbit can capture high-resolution images given their closer proximity to the Earth's surface ($ESA_a$, 2021). The satellites in this orbit take around 90 minutes to orbit the Earth. The International Space Station (ISS) uses this orbit.

#### 2.1.1.6 Polar Orbit and Sun-Synchronous Orbit

Satellites with Polar and Sun-Synchronous orbits travel from north to south rather than from east to west. They approximately pass over the poles with an occasional deviation of about 20 to 30 degrees. The Sun-synchronous orbit is a type of Polar orbit in which the satellite is in sync with the Sun, i.e., the satellite always visits the same spot at the same local time ($ESA_a$, 2021). Landsat is a polar sun-synchronous satellite.

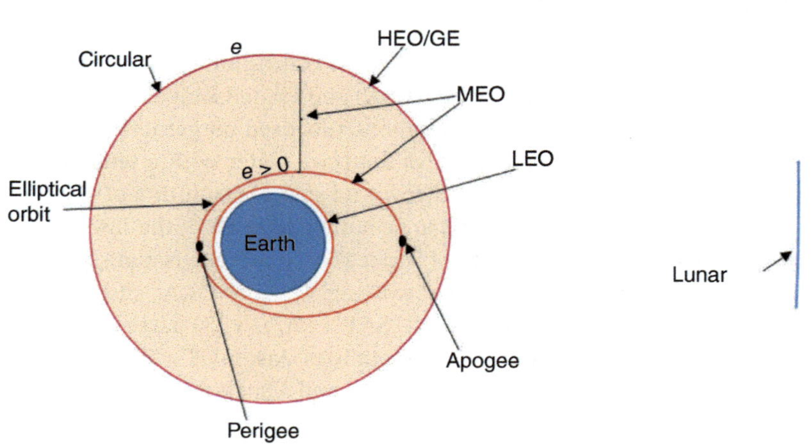


**Figure 2.1** Depiction of various orbits, Apogee (farthest point from Earth) and Perigee (nearest point to Earth). Image Adapted from NASA.

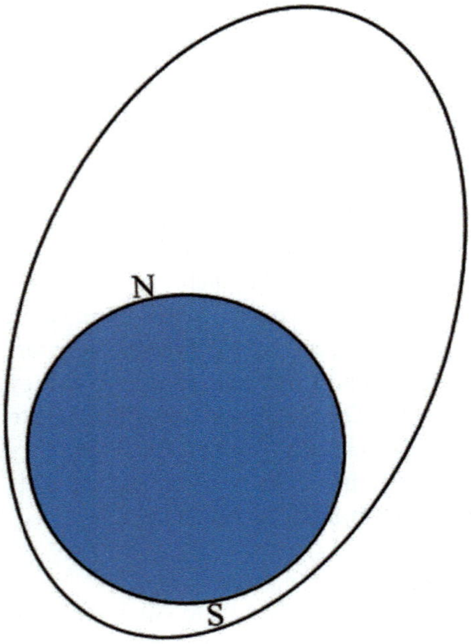


**Figure 2.2** Representation of Molniya Orbit.

#### 2.1.1.7 Lagrange's Point

Langrage's points are also known as L-points. These are positions in space where the combined gravitational force of Earth and Sun create positions where an orbiting satellite can remain stable and connected to the Earth ($ESA_a$, 2021). Usually, a satellite orbiting the sun will gradually move very far from the Earth, losing all contact with the Earth. There are five such known L-points.

## 2.2 Satellite Missions and Data Availability

Earth Observation Satellites (EOS) mainly collect images by recording bands of the electromagnetic spectrum at various spatial and temporal resolutions. Usually, these satellites are placed in Sun-Synchronous or Polar orbits.

The Civilian EOS was started in the 1970s by the USA, when it began its civilian EO mission with Landsat in 1972. They were then followed by Europe and Japan. Landsat 1 was the only EO mission data available to the public in 1972. Now there are over 59 publicly open EO agencies (Dowman and Reuter, 2017). The Landsat mission has the longest continuous data record of planet Earth (Emery and Camps, 2017; $NASA_a$, 2021), having shown human-scale processes such as urban growth for over 40 years.

The Sentinel mission by Copernicus is targeted to continuously gather EO data by missions like Envisat (2002–2012) ($ESA_b$, 2021). Sentinel-1 was launched in 2014 equipped to perform radar imaging of land and ocean regardless of the weather conditions. The Sentinel satellite constellation aims to cover sea-ice zones and European coastal areas on a daily basis and a bi-weekly coverage of the entire world's landmasses ($ESA_b$, 2021).

GRACE (Gravity Recovery and Climate Experiment) launched two satellites in tandem in 2002 (NASA, 2022b). These satellites captured a detailed measurement of the Earth's gravitational field. GRACE EO data have been employed in studies for societal benefits like variations in water availability (Ahmed et al., 2014). The GRACE mission ended in 2017, but was followed by GRACE- FO (GRACE Follow On) in 2018 (Landerer et al., 2020).

In 2003, NASA launched ICESat (Ice, Cloud, and Land Elevation Satellite) aiming to measure ice-sheet mass balance, height of clouds and aerosols, and land and vegetation characteristics. The ICESat mission ended in 2009, having expanded our knowledge of stratospheric clouds common over polar areas. The ICESat mission is a touchstone in EO, especially over the polar regions. ICESat Bridge and ICESat-2 (2018~) are now continuing this mission.

Table 2.1 provides details of various satellite missions and platforms available to access those data. Also, Table 2.2 provides details of tools and platforms for accessing remotely sensed data.

## 2.3 Future Satellite Missions

In 2022, JPSS-2 (Joint Polar Satellite System) is scheduled to be launched, which will collect terrestrial, atmospheric, and oceanic data from across the globe. JPSS-3 and JPSS-4 are scheduled to be launched in 2026 and 2031, respectively (https://www.jpss.noaa.gov). With a temporal resolution of 14 times a day, these satellites cover the entire globe twice a day.

Also, in February 2022, NASA launched SWOT (Surface Water and Ocean Topography). It was conjointly developed by NASA and France's Centre National D'Etudes Spatiales (CNES) with contributions from the Canadian Space Agency (CSA) and United Kingdom Space Agency. The SWOT mission is focused on gathering information about the Earth's surface water with a temporal resolution of 15–25 days and spatial resolution of 120 km.

Scheduled to be launched in 2025, the Jason-CS (Jason Continuity of Service) or Sentinel 6 is a mission that will continue the existing USA (NOAA and NASA) and Europe (EUMETSAT, ESA, CNES) partnership program of Sentinel hub satellites. Jason-CS will have a temporal resolution of 10 days and a spatial resolution of 315 km. Table 2.3 provides details of other relevant space missions.

**Table 2.1** Satellite missions and platforms for accessing the data. The table has utilized data available from (Chawla et al. 2020). We only cover the satellite clusters active to date in this manuscript.

| Satellite Mission | Country/ Organization | Type Of Sensor | Temporal Resolution (Days) | Spatial Resolution (m) | Period | Common Applications | Data Availability | | | |
|---|---|---|---|---|---|---|---|---|---|---|
| | | | | | | | Platform | Fee | Registration | Download |
| Landsat | National Aeronautics and Space Administration (NASA) and United States Geological Survey (USGS) – USA | Multispectral Scanner system | 16 | 60 | 1972–Present | Remote sensing, crop health detection, water body detection, water quality analysis, managing forest fires, Human Population census, exploring changes in land cover and land use. | Google Earth and Google Earth Engine | Free | Required | No: web-based |
| | | | | | | | Earth Explorer and other USGS/EROS data inventories like: GloVis, M2M API | Free | Required | No: web-based |
| | | | | | | | LandsatLook Viewer | Free | Required | No: web-based |
| | | | | | | | Web-enabled Landsat Data (WELD) | Free | Required | No: web-based |
| | | | | | | | Landsat Commercial Cloud Data Access from Amazon Web Services (AWS) | Free | AWS account required (Charges apply) | No: web-based |
| | | | | | | | WorldView | Free | Required | No: web-based |
| Sentinel | Copernicus Mission by Europe | Multi-Spectral Imager<br>SAR Radar Altimeter (SRAL) | 5 | 60 | 2014–Present | Maritime monitoring, land monitoring, border surveillance for security. | Copernicus Open Access Hub<br>CREODIAS, MUDI web services,ONDA, sabl∞ | Free | Required | No: web-based |
| | | | | | | Management of natural disasters, man-made emergency situations, and humanitarian crises. | Earth Online – Earth Observation information discovery platform | Free | Required | No: web-based |
| | | | | | | | Google Earth Engine | Free | Required | No: web-based |
| | | | | | | | Data Hub Software (DHuS) | Free | No | Required |

*(Continued)*

**Table 2.1** (Continued)

| Satellite Mission | Country/ Organization | Type Of Sensor | Temporal Resolution (Days) | Spatial Resolution (m) | Period | Common Applications | Data Availability | | | |
|---|---|---|---|---|---|---|---|---|---|---|
| | | | | | | | Platform | Fee | Registration | Download |
| GRACE (Gravity Recovery and Climate Experiment) GRACE-Follow-On (FO) | NASA–USA and German Aerospace Center (Deutsches Zentrum für Luft- und Raumfahrt) (DLR) – Germany | K- Band Ranging twin-satellite system | 30 | $4 \times 10^5$ | 2002–2017 2018–Present | Monitoring the loss of ice mass from ice sheets, understanding factors responsible for sea-level rise and ocean circulation, monitoring global groundwater resources, monitoring areas where dry soils are contributing to drought, monitoring changes in the solid Earth. | NASA GRACE Data Analysis Tool (DAT) | Free | No | Required |
| Geostationary Operational Environmental Satelite (GOES) | NASA-USA and National Oceanic and Atmospheric Administration (NOAA) – USA | GOES Imager | – | $4 \times 10^3$ | 1975–Present | Improved hurricane track and intensity forecasts, increased thunderstorm and tornado warning lead time, earlier warning of lightning ground strike hazards, detection of heavy rainfall and flash flood risks, monitoring of smoke and dust, monitoring air quality, aviation route planning. | GOES Image Viewer | Free | No | No: web-based |
| RapidEye | European Space Agency (Germany), Planet | Multi-Spectral Imager | 1 | 6.5 | 2008–Present | Identifying vegetation encroachment, assessing forest status, monitoring illegal logging and deforestation, agricultural monitoring, | Earth Online – Earth Observation information discovery platform | Free | Required | No: web-based |

| | | | | | | | | | | |
|---|---|---|---|---|---|---|---|---|---|---|
| Ziyuan 3 | People's Republic of China | Multispectral Camera | 4–5 | 5.8 | 2012–Present | environmental monitoring, resource management, disaster relief, urban planning, national security. | – | – | – | – |
| Ice, Cloud and Land Elevation Satellite (ICESat)<br>ICESat-2<br>IceBridge | NASA- USA | Geoscience Laser Altimeter System (GLAS) | 180<br>91<br>– | 65 | 2003–2010<br>2018–Present<br>– | Surface water monitoring, modeling of sensitive bird habitats, identify and track icebergs in shipping lanes, mitigate volcanic events. | National Snow & Ice Data Centre (NSIDC) – Distributed Active Archive Center (DAAC) | Free | No | No: web-based |
| | | | | | | | HDF View | Free | No | No: web-based |
| Joint Polar Satellite System (JPSS):<br>National Oceanic and Atmospheric Administration's (NOAA) Polar Orbiting Environmental Satellites (POES)<br>NOAA-20 (Formerly JPSS-1) | National Oceanic and Atmospheric Administration (NOAA) and the European Organisation for the Exploitation of Meteorological Satellites (EUMETSAT) | Advanced Very High-Resolution Radiometer (AVHRR)<br>Visible Infrared Imaging Radiometer Suite (VIIRS) | 14 times a day<br>16 | $1.1 \times 10^3$<br>375 | 2009, 2012–Present<br>2011–Present | Weather forecasting to help meteorologists to issue an accurate and complete picture of Earth's weather, predicting severe weather conditions like hurricanes, blizzards, etc. that helps save lives. | NOAA CLASS: Advanced Very High-Resolution Radiometer: Comprehensive Large Array-Data Stewardship System | Free | No | No: web-based |
| | | | | | | | Earth observation data Centre, NERC | Free | No | No: web-based |
| Terra | NASA – USA | Advanced Spaceborne Thermal Emission and Reflection Radiometer (ASTER) | 16 | 15 | 1999–Present | Air quality monitoring, volcanic ash monitoring, monitoring oil spills and their effect on surrounding environment, tracking wild fires. | 2.1 GLoVis | 2.2 | 2.3 Required | 2.4 No: web-based |
| | | | | | | | ASTER/AIST | Free | Required | No: web-based |
| | | | | | | | Earth Explorer | Free | Required | No: web-based |
| | | | | | | | AppEEARS | Free | Required | No: web-based |
| | | | | | | | LPDAAC Data Pool | Free | Required | No: web-based |

*(Continued)*

Table 2.1 (Continued)

| Satellite Mission | Country/ Organization | Type Of Sensor | Temporal Resolution (Days) | Spatial Resolution (m) | Period | Common Applications | Data Availability | | | |
|---|---|---|---|---|---|---|---|---|---|---|
| | | | | | | | Platform | Fee | Registration | Download |
| Terra, Aqua | NASA – USA | Moderate Resolution Imaging Spectroradiometer (MODIS) | 16 | 250 | 2000–Present | Aerosol retrieval, air quality monitoring, volcanic ash monitoring, monitoring oil spills and its effect on surrounding environment, tracking wild fires, river analysis and modeling. | NASA's Earth Data Search | Free | Required | No: web-based |
| | | | | | | | Google Earth Engine | Free | Required | No: web-based |
| Cryosat-2 | European Space Agency (ESA) | Synthetic Aperture Radar(SAR)/ Interoferometric Radar Altimeter-2 (SIRAL-2) | 369 (30-day pseudo subcycle) | $15 \times 10^3$ | 2010–Present | River analysis and modeling, climate monitoring, environmental analysis, maritime security monitoring, marine meteorology. | CryoSat User Tool | | | |
| | | | | | | | CryoSat Software Routines | | | |
| | | | | | | | CryoSat Matlab routines | Paid | | Downloaded on system |
| Satellite for Argos and ALtiKa (SARAL) | ISRO (Indian Space Research Organization) and CNES (Space Agency of France) | Altika | 35 | $8 \times 10^3$ | 2013–Present | Climate monitoring, environmental analysis. | Meteorological & Oceanographic Satellite Data Archival Centre-Space Applications Centre, ISRO | Free/ Paid | Required | No: Download URLs provided |
| MetOp-A/B/C | European Organisation for the Exploitation of Meteorological Satellites (EUMETSAT) As part of the Initial Joint Polar System (IJPS) shared with the US NOAA | Advanced SCATterometer (ASCAT) | 1–2 | $25 \times 10^3$ | 2007–Present | Weather forecasting, climate modeling, monitoring atmospheric composition like aerosols, etc. to help determine air quality. | Earth Observation Portal – EUMETSAT (www.eumetsat.int) | Free | Required | No: Some datasets available via proposal review |

| | | | | | | | | | | |
|---|---|---|---|---|---|---|---|---|---|---|
| RADARSAT-2 | Canadian Space Agency | SAR C-band | 24 | 3–100 | 2007–Present | Ice-edge detection, ice-type discrimination, marine surveillance, disaster management like response during calamities by providing near real-time data, forest-type mapping. | European Space Agency: https://earth.esa.int | Free | Required | No: web-based |
| | | | | | | | Earth observation Portal: https://directory.eoportal.org/web/eoportal/home | Free | Required | No: web-based |
| ALOS-2 | Japan Aerospace Exploration Agency: Earth Observation Research Center | Phased Array L-band Synthetic Aperture Radar (PALSAR-2) | 14 | – | 2014–Present | Ice-edge detection, ice-type discrimination, marine surveillance, disaster management like response during calamities by providing near real-time data, forest-type mapping, providing continuous data for cartography, regional observation, disaster monitoring and management, environmental monitoring. | ALOS-2 tool: https://www.eorc.jaxa.jp/ALOS-2/en/doc/pal2_tool.htm | Free | Required | No: web-based |
| CORONA, ARGON, and LANYARD (declassified military intelligence satellites) | Central Intelligence Agency Directorate of Science and Technology | Telescopic camera system 2.5 | – 2.6 | Ground resolution of 12.9 m 2.7 | 1960–1972 2.8 | Reconnaissance and production of maps for US intelligence agencies (to spy during the Cold War), digital terrain modeling, mapping forest cover change. | National Archive Records and Administration (NARA) | Free | Required | No: web-based |
| | | | | | | | National Air Survey Center Corporation (NASCC) | Free | Required | No: web-based |

**Table 2.2** Other useful sources for satellite data access (data utilized from Lubin, et al., 2009).

| | |
|---|---|
| British Antarctic Survey | https://www.bas.ac.uk/data/uk-pdc |
| Australian Antarctic Data Centre | https://data.aad.gov.au |
| Alaska Satellite Facility | https://asf.alaska.edu |
| Arctic and Antarctic Research Institute | http://www.aari.ru |
| Antarctic Meteorology Research Center | https://amrc.ssec.wisc.edu |
| Geographic Information System | ArcGIS (Paid), QGIS, Planet stories (Paid), EOS Land Viewer |

**Table 2.3** Future satellite missions (ISRO, 2021; $ESA_c$, 2022; $NASA_c$, 2022; OSCAR, 2022).

| Mission | Agency |
|---|---|
| Sentinel 6 B | NASA and ESA |
| Climate Absolute Radiance and Refractivity Observatory Pathfinder (CLARREO Path.) | NASA |
| Earth Surface Mineral Dust Source Investigation (EVI-4) (EMIT on ISS) | NASA |
| Geostationary Carbon Cycle Observatory (EVM-2) (GeoCarb) | NASA |
| Geosynchronous Littoral Imaging and Monitoring Radiometer (EVI-5) (GLIMR) | NASA |
| Investigation of Convective Updrafts (EVM-3) (INCUS) | NASA |
| Tropospheric Emissions: Monitoring of Pollution (EVI-1) (TEMPO) | NASA |
| Total and Spectral Solar Irradiance Sensor-2 (TSIS-2) | NASA |
| Time-Resolved Observations of Precipitation structure and storm Intensity with a Constellation of Smallsats (EVI-3) (TROPICS) | NASA |
| SpaceX series | SpaceX |
| Biomass,<br>EarthCARE,<br>Euclid,<br>Plato,<br>Copernicus | ESA |
| High-Resolution Satellite–1 | ISRO |
| Methane Remote Sensing Mission (MERLIN) | CNES and DLR |

## 2.4 Applicability of Satellite Products in Three Poles Regions

Climate change is one of the leading causes of disasters in the polar regions and Himalayan mountains. The rise in temperature is leading to a decrease in polar ice sheets (Hogg et al., 2017; Slater et al., 2021). In the Himalayan region, the glacier sheet is melting leading to an increase in freeze-thaw cycles. This is producing a higher amount of suspended sediment and flow leading to floods like those in Uttarakhand-India on 7 February 2021 (https://www.preventionweb.net/news/view/76491). The Earth lost about 28 trillion tonnes of ice during 1994–2017 (Slater et al., 2021). Ice sheets perform functions like reflecting the Sun's rays so that the ocean temperature is maintained. Shrinking sea ice leads to higher absorption of sun rays by the oceans and atmosphere, which thereby is speeding up sea ice melt as well as causing a rise in sea levels (CPOM: Slater et al., 2021). Satellite imagery data along with numerical models can help us determine the extent of ice sheet depletion, understand and combat the causes, and thereby be better prepared or to entirely eliminate disasters caused by sea-level rise or glacier melting. Microwave remote sensing is one of the most efficient ways to study polar ice. This cannot only tell the sea ice apart from open waters, but can also distinguish first-year ice sheet (sea ice from autumn to winter that melts in the summer) and multi-year ice sheet (sea ice typically closer to the poles that withstands successive melt seasons) (Lubin and Massom, 2007).

Also, supraglacial lakes in the Arctic and Antarctic regions form when meltwater collects on the surface of a glacier, ice shelf, or ice sheet (Moussavi et al., 2020). Supraglacial lakes increase the absorption of the Sun's rays by reducing the ice-surface albedo. They are also precursors to the collapse of floating ice-shelves (Chudley et al., 2019; Stokes et al., 2019; Moussavi et al., 2020). Remote sensing can help us identify these lakes and understand the effect of climate change on the poles. Satellites such as Landsat and Sentinel provide free access to data and so are a popular choice among scientists for such studies. Researchers have also used images from CORONA, ARGON, and LANYARD to study changes in glacier masses to help us determine formation of supraglacial lakes (Bolch et al., 2008; Lamsal et al., 2011).

Satellite imagery has also assisted and expanded our understanding of unique polar climatic features like Polynyas. Global Land Ice Measurements from Space (GLIMS) (http://www.glims.org) is a project dedicated to studying changing glaciers, especially mountain glaciers that are one of the immediate and indispensable climate change indicators (Kääb, 2005). GLIMS majorly uses ASTER (Advanced Spaceborne Thermal Emission and Reflection Radiometer) data, which consists of three subsystems VNIR (Visible Near Infrared, i.e., a telescope that only acquires a stereo pair image), SWIR (ShortWave Infrared or fixed aspheric refracting telescope), and TIR (Thermal Infrared).

Remotely sensed data provide us with a birds-eye view of planet Earth and also aids us in the mapping of physically inaccessible locations like glaciers and rockglaciers (Kääb, 2002; Toutin, 2008). Also, satellite data have helped us estimate snow depth allowing us to predict the snowmelt (Lievens et al., 2019), which in turn can aid in the prediction of events such as floods in the high mountainous regions and other wet places.

Another major natural disaster that earth observing systems can help with is surveillance of earthquakes. Along with shaking, earthquakes also pose a risk of triggering tsunamis, landslides, flooding, etc. (Tilloy et al., 2019). Earthquakes cause displacement of the Earth's surface causing deformation in the crust. Satellite data can help us to determine polar and Himalayan earthquake events. GNSS (Global Navigation Satellite System) and InSAR (Interferometric synthetic aperture radar) data is being used extensively to predict earthquakes (Elliott, 2020). Remotely sensed data can also help us study landslides, especially in the Himalayan region where landslides are a leading cause of disasters. Scientists have used ASTER DEM data to study landslides in the Himalayan region (Toutin, 2008).

## 2.5 Challenges and Limitations

Since the inception of the Earth Observation missions in the 1970s, the EO data and sensors have advanced tremendously in terms of spatial and temporal resolution and sensors. Most of the EO missions are operated by governments., but there are also several commercial organizations collecting data (DigitalGlobe, Planet, WorldDEM$^{TM}$, Google Earth, Microsoft Bing, etc.). Most of these are high-resolution datesets with a global sample distance of less than 5 meters. One of the challenges existing with the availability of such extensive data at such varying resolutions is the validation of that data. Data validation comprises geometric accuracy and accuracy of the content (Dowman and Reuter, 2017). This is usually achieved by sampling several selected areas and collecting data in the field, for comparing the ground data with satellite data. Several locations across the world are selected in order to validate the data collected by the satellites, for example, the Globeland30 land cover dataset (Chen et al., 2015).

An additional concern is the standardization of the data. Different EO missions use different sensors for data collection; hence the data comes in various resolutions and formats. To receive the maximum benefit out of this ever-increasing enormous dataset, it is necessary to standardize the data. Initiatives are being taken to realize the essentiality of the availability of standardized data (INSPIRE: EC, 2007).

Another pressing issue that scientists face is the limitation of access to data, i.e., availability of free data and accessibility by any user, alias, open data (Harris and Baumann, 2015, 2021). While there are several datasets available free of charge (e.g., Landsat, Sentinel), a lot of data like DEMs by Airbus Defense and Space, data by Planet, etc. are available only for payment (Dowman and Reuter, 2017). While EO missions by countries like the USA, Japan, and the European nations are easily accessible, countries like China, India, Russia, etc. actively send satellites for Earth Observation missions; however, they do not allow access to this data by the research community outside of their own government organizations (or allow only limited access). A report published by CODATA (Committee on Data for Science and Technology) for the GEO (Group on Earth Observations) secretariat in France, emphasizes the need for open EO data (GEO, 2015). They uncover and signify the benefits of open EO data to society, such as interdisciplinary research and innovation opportunities, improving societal welfare (GEO, 2015); for example, participation of the California Planet to provide free data to the UN (United Nations) for achieving sustainability goals.

## 2.6 Summary

This chapter introduces the reader to various present and future satellite missions. We also discuss various applications of remotely sensed data in the polar and Himalayan studies; for example, detection of lakes, polynyas, mapping of glaciers and rock glaciers, etc. Global Earth Observation datasets are widely available and can be accessed via several commercial and government organizations, with many of them providing datasets free of charge. Efforts are being made toward increasing the availability of ready-to-use, free EO datasets. The extent and availability of remotely sensed data make it a promising approach to better understand our poles and the high-altitude mountains and mountain glaciers in the Himalayan Region.

## Acknowledgments

The authors want to thank the editors and two anonymous reviewers who helped improve the earlier versions of this chapter. Jagriti Mishra wishes to thank Ramanujan fellowship no. RJF/2021/000168 for financial assistance.

## ORCiD IDs

Jagriti Mishra: https://orcid.org/0000-0002-8561-8293

Takuya Inoue: https://orcid.org/0000-0003-1423-8186

Avinash Kumar Pandey: https://orcid.org/0000-0002-3324-5309

## References

Ahmed, M., Sultan, M., Wahr, J. et al. (2014). The use of GRACE data to monitor natural and anthropogenic induced variations in water availability across Africa. *Earth-Science Reviews* 136: 289–300. doi: 10.1016/j.earscirev.2014.05.009.

Bolch, T., Buchroithner, M., Pieczonka, T. et al. (2008). Planimetric and volumetric glacier changes in the Khumbu Himal, Nepal, since 1962 using Corona, Landsat TM and ASTER data. *Journal of Glaciology* 54(187): 592–600. doi: 10.3189/002214308786570782.

Chawla, I., Karthikeyan, L., and Mishra, A.K. (2020). A review of remote sensing applications for water security: quantity, quality, and extremes. *Journal of Hydrology* 585: 124826. doi: 10.1016/j.jhydrol.2020.124826.

Chen, Jun, Chen, Jin, Anping Liao, et al. (2015). Global land cover mapping at 30 m resolution: a POK-based operational approach. *ISPRS Journal of Photogrammetry and Remote Sensing* 103: 7–27. ISSN 0924-2716, https://doi.org/10.1016/j.isprsjprs.2014.09.002.

Chudley, T.R., Christoffersen, P., Doyle, S.H. et al. (2019). Supraglacial lake drainage at a fast-flowing Greenlandic outlet glacier. *Proceedings of the National Academy of Sciences* 116(51): 25468–25477. doi: 10.1073/pnas.1913685116.

Dowman, I. and Reuter, H.I. (2017). Global geospatial data from Earth observation: status and issues. *International Journal of Digital Earth* 10(4): 328–341. doi: 10.1080/17538947.2016.1227379.

EC (2007). Directive 2007/2/EC of the European Parliament and of the Council of 14 March 2007 Establishing an Infrastructure for Spatial Information in the European Community (INSPIRE). ELI. http://data.europa.eu/eli/dir/2007/2/oj (accessed June 2021).

Elliott, J.R. (2020). Earth observation for the assessment of earthquake hazard, risk and disaster management. *Surveys in Geophysics* 41(6): 1323–1354. doi:10.1007/s10712-020-09606-4.

Emery, W. and Camps, A. (2017). Chapter 1: The history of satellite remote sensing. In: *Introduction to Satellite Remote Sensing* (ed. W. Emery and A. Camps), 1–42. Elsevier. https://doi.org/10.1016/B978-0-12-809254-5.00001-4.

$ESA_a$ https://www.esa.int/Enabling_Support/Space_Transportation/Types_of_orbits#:~:text=A%20low%20Earth%20orbit%20(LEO,very%20far%20above%20Earth's%20surface (accessed June 2021).

$ESA_b$ https://www.esa.int/Enabling_Support/Operations/Envisat (accessed June 2021).

$ESA_c$ https://www.esa.int/Enabling_Support/Operations/Current_and_future_missions (accessed April 2022).

GEO (2015). The value of open data sharing. https://earthobservations.org/documents/geo_xii/GEO-XII_09_The%20Value%20of%20Open%20Data%20Sharing.pdf (accessed June 2021).

Harris, R. and Baumann, I. (2015). Open data policies and satellite Earth observation. *Space Policy* 32: 44–53. doi: 10.1016/j.spacepol.2015.01.001.

Harris, R. and Baumann, I. (2021). Satellite Earth observation and national data regulation. *Space Policy* 56: 101422. doi: 10.1016/j.spacepol.2021.101422.

Hogg, A.E., Shepherd, A., Cornford, S.L. et al. (2017). Increased ice flow in Western Palmer Land linked to ocean melting: ice flow in Western Palmer Land. *Geophysical Research Letters* 44(9): 4159–4167. doi: 10.1002/2016GL072110.

ISRO, accessed June 2021. https://www.isro.gov.in/spacecraft/earth-observation-satellites,

Kääb, A. (2002). Monitoring high-mountain terrain deformation from repeated air- and spaceborne optical data: examples using digital aerial imagery and ASTER data.

*ISPRS Journal of Photogrammetry and Remote Sensing* 57(1–2): 39–52. doi: 10.1016/S0924-2716(02)00114-4.

Kääb, A. (2005). Combination of SRTM3 and repeat ASTER data for deriving alpine glacier flow velocities in the Bhutan Himalaya. *Remote Sensing of Environment* 94(4): 463–474. doi: 10.1016/j.rse.2004.11.003.

Kidder, S.Q. and Vonder Haar, T.H. (1990). On the use of satellites in molniya orbits for meteorological observation of middle and high latitudes. *Journal of Atmospheric and Oceanic Technology* 7(3): 517–522. Available from: https://doi.org/10.1175/1520-0426(1990)007<0517:OTUOSI>2.0.CO;2

Lamsal, D., Sawagaki, T. and Watanabe, T. (2011). Digital terrain modelling using Corona and ALOS PRISM data to investigate the distal part of Imja Glacier, Khumbu Himal, Nepal. *Journal of Mountain Science* 8(3): 390–402. doi:10.1007/s11629-011-2064-0.

Landerer, F.W., Flechtner, F.M., Save, H. et al. (2020). Extending the global mass change data record: GRACE Follow-On instrument and science data performance. *Geophysical Research Letters* 47: e2020GL088306. https://doi.org/10.1029/2020GL088306

Lievens, H., Demuzere. M., Marshall, H-P. et al. (2019). Snow depth variability in the Northern Hemisphere mountains observed from space. *Nature Communications* 10(1): 4629. doi: 10.1038/s41467-019-12566-y.

Lubin, D., Ayres, G., and Hart, S. (2009). Remote sensing of polar regions: lessons and resources for the International Polar Year. *Bulletin of the American Meteorological Society* 90(6): 825–835. doi: 10.1175/2008BAMS2596.1.

Lubin, D. and Massom, R. (2007). Remote sensing of Earth's polar regions: opportunities for computational science. *Computing in Science and Engineering* 9(1): 58–71. doi: 10.1109/MCSE.2007.16.

Mouginot, J., Rignot. E., Scheuchl. B. et al. (2017). Comprehensive annual ice sheet velocity mapping using Landsat-8, Sentinel-1, and RADARSAT-2 Data. *Remote Sensing* 9(4): 364. doi: 10.3390/rs9040364.

Moussavi, M.. Pope, A., Halberstadt, A.R.W. et al. (2020). Antarctic supraglacial lake detection using Landsat 8 and Sentinel-2 Imagery: towards continental generation of lake volumes. *Remote Sensing* 12(1): 134. doi:,10.3390/rs12010134.

$NASA_a$ https://landsat.gsfc.nasa.gov/about (accessed June 2021).

$NASA_b$ https://grace.jpl.nasa.gov (accessed June 2021).

$NASA_c$ https://eospso.nasa.gov/future-missions (accessed April 2022).

OSCAR https://space.oscar.wmo.int/satelliteprogrammes (accessed April 2022).

Riebeek, H. and Simmon, R. (2009). *Catalog of Earth Satellite Orbits: Orbits Catalog – NASA*. https://earthobservatory.nasa.gov/features/OrbitsCatalog Accessed June 2021.

Slater, T., Lawrence, I.R., Otosaka, I.N. et al. (2021). Review article: Earth's ice imbalance. *The Cryosphere* 15(1): 233–246. doi:10.5194/tc-15-233-2021.

Stokes, C.R., Sanderson, J.E., Miles, B.W. et al. (2019). Widespread distribution of supraglacial lakes around the margin of the East Antarctic Ice Sheet. *Scientific Reports* 9(1): 13823. doi: 10.1038/s41598-019-50343-5.

Tilloy, A., Wulder, Malamud, B.D., Winter, H. et al. (2019). A review of quantification methodologies for multi-hazard interrelationships. *Earth-Science Reviews* 196: 102881. doi: 10.1016/j.earscirev.2019.102881.

Toutin, T. (2008). ASTER DEMs for geomatic and geoscientific applications: a review. *International Journal of Remote Sensing* 29(7): 1855–1875. doi: 10.1080/01431160701408477.

Wulder, M.A., Loveland, T.R., Roy, D.P., Kive, et al. (2019). Current status of Landsat program, science, and applications. *Remote Sensing of Environment* 22: 127–147. doi: 10.1016/j.rse.2019.02.015.

# 3

# Assessing the Accuracy of Digital Elevation Models for Darjeeling-Sikkim Himalayas

*Prodip Mandal[1] and Shraban Sarkar[1,*]*

[1] *Department of Geography, Cooch Behar Panchanan Barma University, Cooch Behar, West Bengal 736101, India*
[*] *Corresponding author*

## 3.1 Introduction

Digital Elevation Model (DEM) presents the elevation of the Earth's surface in digital form by an array of grids or lists of three-dimensional (3D) coordinates. DEM is also known as the Digital Height Model (DHM), Digital Ground Model (DGM), Digital Surface Model (DSM), and Digital Terrain Model (DTM). These terms originate from different countries (Li et al., 2004). More specifically, DSM represents the Earth's surface elevation including all objects (i.e., vegetation and man-made features), while DTM contains elevation of the bare Earth's surface (Martha et al., 2010). DEM is often used as a generic term for DSM and DTM (Hirt, 2014), only representing height information without defining surface characteristics (Peckham and Jordan, 2007).

In recent years, DEM has become an essential data in various research fields, including geomorphology (Boulton and Stokes, 2018), oceanography (Libina and Nikiforov, 2020), climatology (Daly et al., 2002, 2008), and biodiversity (Sesnie et al., 2008). The various topographic and hydrologic parameters such as slope, aspect, curvature, drainage networks, Topographic Wetness Index (TWI), Stream Power Index (SPI), etc. can be extracted from DEM. These parameters are a prerequisite of different studies such as flood inundation and management (Sanders, 2007; Sampson et al., 2015; Araújo et al., 2019); landslide susceptibility assessment (Sarkar et al., 2013a; 2016); glacial mass balance (Berthier et al., 2006; Bolch and Kamp, 2006); landform mapping and analysis (Weibel and Heller, 1990); vegetation mapping (Kellndorfer et al., 2004; Volařík, 2010; Simard et al., 2011; Korets et al., 2016; O'Loughlin et al., 2016); terrain morphology (Spark and Williams, 1996); hydrology (Moore et al., 1991); meteorology (Zhang et al., 2014); volcanology (Vassilopouloua et al., 2002; Grosse et al., 2012); and soil depth estimation (Sarkar et al. 2013b). The quality of the aforesaid research outputs depends upon the resolution and accuracy of DEM (Zhang and Montgomery, 1994; Januchowski et al., 2010; Gómez-Gutiérrez et al., 2011).

The different sources obtain DEM at various resolutions. The DEM can be prepared after interpolating contour lines (Taud et al., 1999), spot heights from topographic maps (Wilson and Gallant, 2000), and ground control points collected by a GPS receiver. However, the stereo-photographic method (San and Suzen, 2005; Hohle, 2009), airborne laser scanning (Favey et al., 2003), radar interferometry, and radar altimetry (Kervyn, 2001) are widely used to produce the DEM. All these methods have some limitations and DEMs prepared by these techniques are not free from errors. Errors in DEM can be categorized into three types: i) gross errors which emerge during data collection processes (Rodgriguez et al., 2006); ii) systematic errors caused by orientation difference of the stereo images; and iii) random errors whose cause is unknown. The levels of error vary with terrain conditions (Holmes et al., 2000). The quality of DEM is also determined by the data receiving sensors and algorithms used in the DEM generation process (Hebeler and Purves, 2009).

There are different types of free global and quasi-global level DEMs with varying resolutions available in different geospatial data portals, although high-resolution DEM comes under proprietary databases. Open source Synthetic Aperture Radar (SAR)-based 12 m TerraSAR-X add-on for

*Advances in Remote Sensing Technology and the Three Poles,* First Edition. Edited by Manish Pandey, Prem C. Pandey, Yogesh Ray, Aman Arora, Shridhar D. Jawak, and Uma K. Shukla.

Digital Elevation Measurement (TanDEM-X), 12.5 m Phased Array L-band Synthetic Aperture Radar (PALSAR) from Advanced Land Observing Satellite (ALOS) and 30 m Shuttle Radar Topography Mission (SRTM) DEM, and optical stereo-image-based 30 m Cartosat-1 and 30 m Advanced Spaceborne Thermal Emission and Reflection Radiometer Global DEM (ASTER GDEM) are widely used.

The accuracy of different types of DEMs has been tested throughout the world. It is evident that the physiography and land use types of a region determine the DEM quality (Hawker et al., 2019; Uuemaa et al., 2020). Comparison of LiDAR DEM with ASTER, AW3D30, MERIT, TanDEM-X, SRTM, and NASADEM in Estonia, China, New Zealand, and Norway shows slope has a strong influence on DEM accuracy (Uuemaa et al., 2020). The effect of topography on DEM accuracy has also been noted in northern (Kramm and Hoffmeister, 2019) and central Chile (Podgórski et al., 2019). After considering a LiDAR-derived elevation dataset as reference, TanDEM-X is found to be more accurate than MERIT and SRTM in areas without forest cover, in 32 different places on the globe (Hawker et al., 2019). With respect to RTK-GNSS-based ground survey points, the 0.4 arcsec resolution TanDEM-X Intermediate Digital Elevation Model (IDEM04) has an 1.03 m Root Mean Square Error (RMSE) in open terrain in Kruger National Park of South Africa (Baade and Schmullius, 2016). TanDEX-X also provides better accuracy in urban open spaces of south Florida (Zhang et al., 2019a). With respect to the GPS collection elevation points, vertical accuracy of TanDEM-X is satisfactory (RMSE of $\pm 1.1$ m) in less vegetated areas in North America (Wassel et al., 2018). A similar kind of comparison shows that the accuracy of TanDEM-X is not good enough (standard deviation of error $<3.5$ m) in the Central Andean Plateau (Purinton and Bookhagen, 2017). However, a high agreement between TanDEM-X and reference points collected by Differential Global Positioning System (DGPS), with a mean absolute vertical error (MAE) of 0.53 m, and an RMSE of 0.73 m, has been reported in Patagonia (Pasquetti et al., 2019). Even in southeastern Tibet, a 12 m resolution TanDEM-X DEM was shown to be more efficient and promising to detail geomorphological analysis in comparison to 1 arcsec ASTER and SRTM DEM (Pipaud et al., 2015).

In the Indian subcontinent, both the horizontal and vertical accuracy have been tested in the plain as well as on Himalayan terrain (Martha et al., 2010; Deo and Rao, 2014; Agarwal et al., 2020). The accuracy of various multi-sensor open-source DEMs has been compared in parts of the Karakoram region (Kumar et al., 2020) and snow/glacier covered western Himalayas (Snehmani et al., 2013; Pandey et al., 2017). The vertical and planimetric accuracies of Cartosat-1 DSM was estimated by comparing it with checkpoints obtained with a DGPS in the Higher Himalayas (Martha et al., 2010). In Siwalik Hills, open source DEMs (ASTER and SRTM) and their derivatives have been evaluated using Cartosat-1 DEM and Survey of India (SOI) topographical map-based heights (Mukherjee et al., 2013). Reference elevation data that have been collected through DGPS have been compared in previous studies with TanDEM-X in flat and hilly terrain and found significant deviation with topographical changes (Deo and Rao, 2014). The horizontal accuracy of ASTER and SRTM DEM were also evaluated with respect to the Cartosat-1 DEM in the Ganga plain (Rawat et al., 2013).

The literature review is revealing the accuracy of DEM with respect to high precision DGPS collected checkpoints have rarely been tested in forest-tea garden covered Darjeeling-Sikkim Himalayas. This section of the Himalayas falls under active seismic zone (BIS, 2002) and has a long history of earthquake, catastrophic rainfall, and associated landslide events (Martha et al., 2015; Sarkar et al., 2016). Lives, economy, and environment of the region are disrupted every year due to geohazards. Therefore, the area is considered as one of the most hazard prone regions of the Himalayas. A detail terrain analysis from a high-quality DEM may be useful for geohazard mitigation and management. The accuracy level of multisensor DEMs in Darjeeling-Sikkim Himalayas is still unknown. In this perspective, the present study has assessed the vertical and horizontal accuracy of Cartosat-1 DEM, TanDEM-X, and ALOS PALSAR DEM in a part of Darjeeling-Sikkim Himalayas.

## 3.2 Study Area

The test site (a scene of Cartosat-1 Stereo images covering 27.5 km × 27.5 km, central coordinates 27°00′00.32″N and 88°28′30.52″E) is located in Darjeeling-Sikkim Himalayan region. The physiography of the area is highly diversified and characterized by flat Terai-Dooars in the south, narrow Sub-Himalayas in the middle, and rough wide Lesser Himalayas in the north (Figure 3.1). Elevation ranges from 78 m in the south to 2189 m in the north. The Tista is the trunk river and flows in a southerly direction through the middle of the study area. The region is highly dissected by many small rivers/streams that joined to the River Tista from both sides. The entire region is wrapped with very dense to sparse forest, tea gardens, and built-up areas. Kalimpong is the major town with district headquarter located in the north-eastern part of the test site. A group of small settlements (i.e., Kalijhora, Berik Dara, Rambhi, etc.) have developed along the National Highway-10, which passes through the bank of River Tista.

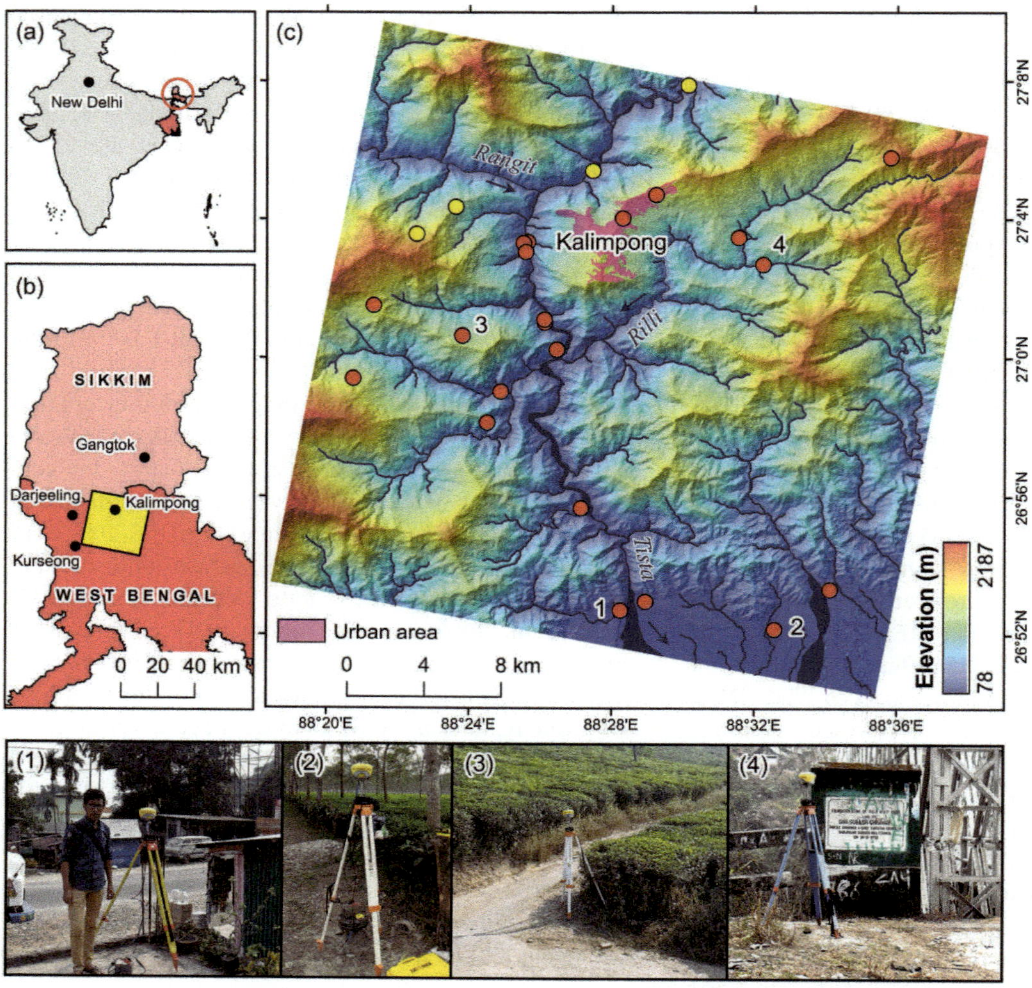


**Figure 3.1** Location map: (a) India; (b) northern part of West Bengal and Sikkim state, the yellow demarcation on the map showing the coverage area of the study DEMs; and (c) the area under study (part of Darjeeling-Sikkim Himalayas). Location of GCPs have been marked in red (collected by DGPS) and yellow (collected by handheld GPS receiver) dots. ALOS PALSAR DEM has been used to show the elevation of the study area. Inset numbers (1, 2, 3 & 4) in (c) indicating locations of field photos (1–4).

## 3.3 Materials and Methods

To achieve our objectives, the present study has used a set of multi-sensor DEMs and reference elevation points with UTM (Zone 45 North) projection. Details of data has been presented in Table 3.1.

### 3.3.1 Generation of Cartosat-1 DEM and Orthoimage

A DEM has been prepared by a photogrammetric technique from a Cartosat-1 stereoscopic satellite image (Path: 0585, Row: 0272) through automatic image-matching using ERDAS IMAGINE Photogrammetry (® ERDAS, Inc.). The Rational

**Table 3.1** Details of data used in the study.

| Data Type | Source | Year | Spatial Resolution | Purpose |
|---|---|---|---|---|
| Satellite imagery | Cartosat – 1 (Stereo) | 2011 | 2.5 m | Digitization of reference stream network, preparation of DEM with 10 m cell size, automatic extraction of stream network from DEM |
| DEM | TanDEM-X | 2010–2015 | 12 m | Automatic extraction of stream network |
| | ALOS PALSAR | 2016 | 12.5 m | Automatic extraction of stream network |
| Field survey | DGPS | 2019–2020 | | Three dimensional coordinates for GCPs |

Polynomial Coefficients (RPCs) supplied by the data provider have used during block preparation. Automated correlation between the Aft and Fore stereo pair have been performed based on 26 GCPs (22 collected by DGPS and 4 by a handheld GPS receiver) and auto tie points. Stereo correlation is a computational and statistical procedure utilized to derive a DEM automatically from a stereo pair of registered images (Lang and Welch, 1999). In the present study, the stereo pair was correlated with an RMSE of 0.853512, which is acceptable (Figure 3.2). Finally, a 10 m cell size DEM has been extracted (Figure 3.3).

The extracted Cartosat-1 DEM has been used to orthorectify the 2.5 m resolution Cartosat-1 Aft satellite image (Figure 3.3).

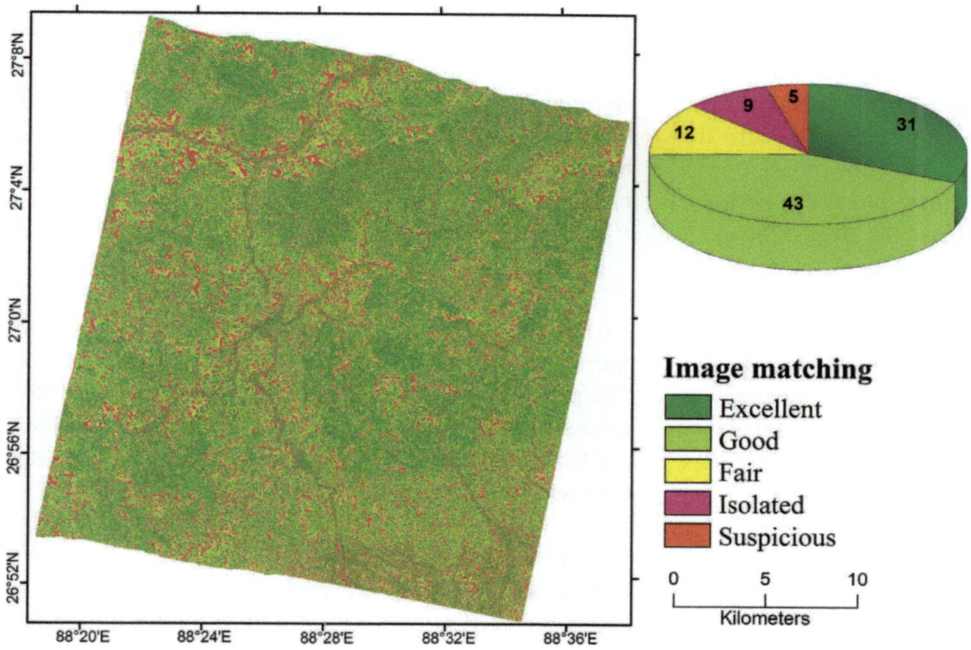


**Figure 3.2** Spatial and graphical representation of image matching performance during DEM extraction from Cartosat-1 stereo images, showing its quality.

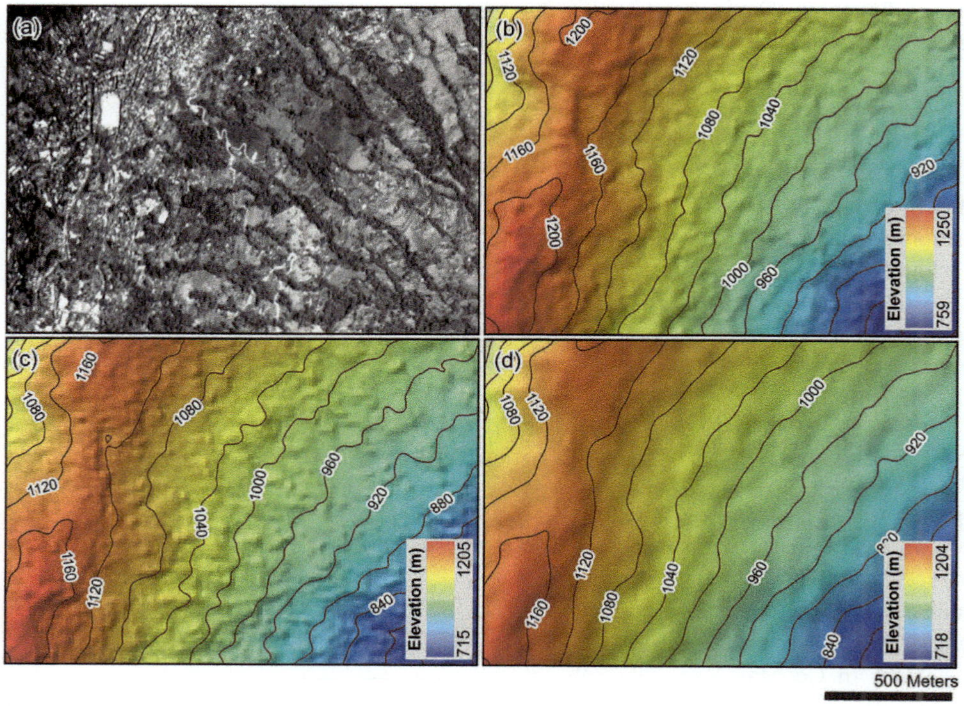


**Figure 3.3** A preview of optical satellite image, optical DEM, and InSAR DEM around Kalimpong town: (a) Cartosat-1 satellite image (2.5 m): (b) Cartosat-1 DEM (10 m): (c) TanDEM-X (12 m); and (d) ALOS PALSAR (12.5 m).

### 3.3.2 TanDEM-X

TanDEM-X is a German Earth observation mission, using spaceborne SAR (Synthetic Aperture Radar) interferometry. The SAR system is capable of scanning the Earth in all weather conditions and around the clock using the X-band signal. TanDEM-X receives altitude and position data from its twin satellite. Both the satellites acquired data from December 2010 to January 2015 to create the global DEM (Zink et al., 2014; Rizzoli et al., 2017). The TanDEM-X DEM has a resolution of 12 m, that is freely available from the German Aerospace Center (Deustches Zentrum für Luft- und Raumfahrt: DLR). Though the mission was launched as a Public Private Partnership (PPP) between DLR and Airbus Defense and Space, the DLR is the main authority for providing TanDEM-X data to the scientific community after critically examining the scientific proposal given by the researchers (Figure 3.3).

According to the mission specification, the TanDEM-X product's absolute horizontal and vertical accuracy is 10 m, while the relative vertical accuracy is 2 m for slopes with a slope of ≤20% and 4 m for slopes greater than 20% (Wessel, 2016).

### 3.3.3 ALOS PALSAR

PALSAR is a L-band Interferometric SAR (InSAR) radar installed on the Japanese satellite ALOS. PALSAR was jointly developed by Japan Aerospace Exploration Agency (JAXA), Japan Resources Observation System Organisation (JAROS), and Japanese Ministry of Economy, Trade & Industry (METI). The ALOS satellite was launched in 2006 and operated until 12 May 2011. The L-band SAR system is capable of scanning the Earth in all weather conditions and around the clock.

Alaska Satellite Facility (ASF) provides radiometrically terrain corrected (RTC) SAR data products at both 12.5 m (RT1) and 30 m (RT2) resolutions for the major part of the globe (ASF Product Guide, 2015) (Figure 3.3).

### 3.3.4 DGPS Survey for Obtaining Ground Control Points (GCPs)

The DGPS survey in static mode has been conducted to collect high precision GCPs. The 3D coordinates of GCPs were measured by the triangulation method in the selected base station using two rovers. The instruments installed at three different sites received satellite signals for at least 4 hours. The instrumental setup was repeated and placed at 22 locations to collect GCPs. After processing the signals received at 22 stations in GeoMax Geo Office 3.6 software (© GeoMax AG), the coordinates of GCPs were obtained.

### 3.3.5 Datum Transformation

To compare vertical accuracy between different DEMs, it is required to transform them into a similar vertical datum. Cartosat-1 DEM, TanDEM-X, and DGPS-based GCPs are referenced to WGS84 ellipsoid (Gruber et al., 2012), while ALOS PALSAR is referenced to the EGM96 geoid (orthometric height) (Lemoine et al., 1997; Farr et al., 2007; Tachikawa et al., 2011). Approximate orthometric height (H) can be obtained by subtracting geoid height (N) from ellipsoid height (h) and may expressed as H ≈ h − N (Johnson, 2004). This approach has been used to transformed Cartosat-1 DEM, TanDEM-X, and GCPs into orthometric height.

### 3.3.6 Accuracy Assessment Methods

Earlier studies revealed that both vertical and spatial accuracy of DEMs varies with land use (Hawker et al., 2019). The present study has assessed the vertical accuracy with reference to the GCPs (Ground Control Points) for different land use classes, while spatial accuracy of DEMs has been assessed with reference to the stream networks.

#### 3.3.6.1 Vertical Accuracy

Highly accurate reference elevation data is required to assess the vertical accuracy of a DEM. Vertical accuracy has been calculated by subtracting the reference elevation data (GCPs) from the DEMs used in this study. This study has used 22 high-precision GCPs (obtained by DGPS survey) as the reference data.

Equation 3.1 has been used for error estimation:

$$\Delta h_i = h_{DEM(i)} - h_{GCP(i)} \tag{3.1}$$

In Equation 3.1, $\Delta h_i$ is the vertical height error of pixel $i$, $h_{DEM(i)}$ is the elevation of pixel $i$ in DEM, and $h_{GCP(i)}$ is the elevation of GCP falling within pixel $i$.

The differences in elevation have been measured statistically, i.e., Mean Error (ME), Mean Absolute Error (MAE), Root Mean Square Error (RMSE), and skewness (Grohmann, 2018; Hawker et al., 2019).

The ME, MAE, and RMSE values have been calculated as in Equations 3.2 to 3.4:

$$ME = \frac{1}{n}\sum_{i=1}^{n}\left[h_{DEM(i)} - h_{GCP(i)}\right] \tag{3.2}$$

$$MAE = \frac{1}{n}\sum_{i=1}^{n}\left[\left|h_{DEM(i)} - h_{GCP(i)}\right|\right] \tag{3.3}$$

$$RMSE = \sqrt{\frac{1}{n}\sum_{i=1}^{n}\left[h_{DEM(i)} - h_{GCP(i)}\right]^2} \tag{3.4}$$

The *ME*, *MAE* values near zero, and a small *RMSE* value indicate a good match (or higher-level accuracy) between reference elevation points (*GCPs*) and *DEM* elevation.

#### 3.3.6.2 Spatial Accuracy

Stream networks manually digitized from Cartosat-1 ortho-image have been used as references and compared with stream networks automatically extracted from Cartosat-1 DEM, TanDEM-X, and ALOS PALSAR to assess the spatial accuracy. This comparison has been carried out between 30 m influence zone (buffer) of the manually and automatically extracted stream networks (Martha et al., 2010).

The degree of stream mismatch has been calculated by an Error Index (EI) in Equation 3.5 as follows:

$$EI = \frac{A_{DEM} - \left(A_{Ref} \cap A_{DEM}\right)}{A_{Ref}} \tag{3.5}$$

In Equation 3.5, $A_{Ref}$ is the area of the reference stream buffer, $A_{DEM}$ is the area of the stream buffer automatically extracted from DEMs, and $A_{Ref} \cap A_{DEM}$ is the common area between them. *EI* values range between 0 and 1, where lower values indicating a higher degree of match of the stream networks.

## 3.4 Results and Discussion

### 3.4.1 Vertical Accuracy Assessment: Comparison of DEMs With Reference to GCPs

Comparison of Cartosat-1 DEM, TanDEM-X, and ALOS PALSAR with reference to GCPs are presented in Figure 3.4. From Table 3.2, it is clear that there is a strong agreement between topography and errors in Cartosat-1 DEM. The higher level of error in Cartosat-1 DEM has been noted in high altitude areas of the Sikkim-Darjeeling Himalayan terrain. However, no such association has previously been observed in the case of TanDEM-X and ALOS PALSAR. It is evident that the performance of automatic image matching technique between optical stereo pairs fails in steep Himalayan terrain (Martha et al., 2010). That may be the reason for the higher topographic influence on Cartosat-1 satellite image. It is also noted that accuracy of InSAR DEMs highly depends on topography in the Andes mountains in northern (Kramm and Hoffmeister, 2019) and central Chile (Podgórski et al., 2019).

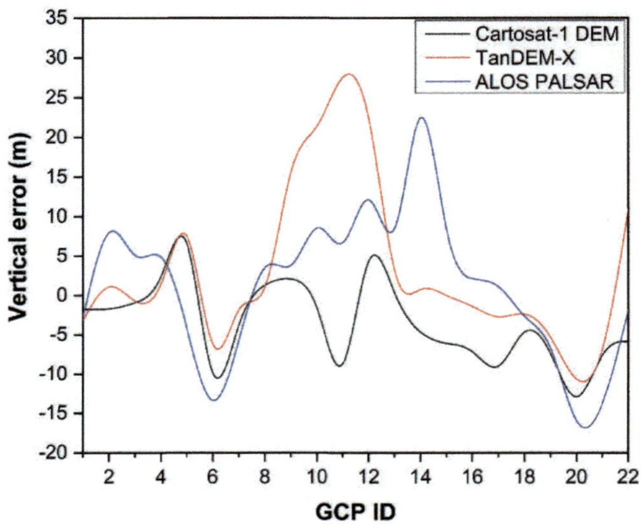


Figure 3.4 Vertical errors of the DEMs with reference to GCPs.

The error statistics (Table 3.3) of these three DEMs reveals that the Cartosat-1 DEM is more accurate than TanDEM-X and ALOS PALSAR. The RMSE for the Cartosat-1 DEM of 8.23, which is very close to the earlier research, has been conducted using on similar data in the northwestern Himalayas (Snehmani et al., 2013). However, the obtained ME, MAE, and RMSE for TanDEM-X in the Darjeeling-Sikkim Himalayas are much greater than the global accuracy level (Wessel et al., 2018).

The DGPS survey was challenging in the Himalayas due to the lack of accessibility and low satellite signals. That may be one of the reasons behind the mismatch in elevation between reference GCPs and DEMs used in this study.

### 3.4.2 Vertical Accuracy of DEMs for Different Land Use Classes

Since the image matching perfection of Cartosat-1 stereo images and scanning performance of SAR system depend upon land use practices, the vertical accuracy of DEMs has been assessed for four land use classes, i.e., dense vegetation, open forest, tea garden, and built-up area. The descriptive error statistics of DEMs with reference to GCPs collected from different land use classes show a clear variation in DEMs' accuracy (Table 3.4 and Figure 3.5).

#### 3.4.2.1 Dense Forest

About 41% of total GCP reference points, collected from the dense forests, have been used to find out the deviation between reference and DEMs elevation. TanDEM-X is not providing a satisfactory level of accuracy in dense forest cover areas (Table 3.4). The RMSE value of the TanDEM-X is 15.32 m, which is very much higher than its global level of accuracy, and ranges from 1.0 to 8.7 m irrespective of terrain and land use/land cover diversity (Vassilaki and Stamos, 2020). Globally, in dense forest areas, TanDEM-X performs with RMSE ±1.8 m in compression to kinematic GPS points (Wessel et al. 2018), which is lower than the present study. The X-band SAR wave is unable to fully

**Table 3.2** Vertical error in Cartosat-1 DEM, TanDEM-X, and ALOS PALSAR with reference to the elevation of GCPs.

| | | UTM Coordinates | | Elevation (m) | | | | | | | | | |
|---|---|---|---|---|---|---|---|---|---|---|---|---|---|
| GCP ID | Land Use Class | Easting | Northing | GCP | Geoid Undulation | Transformed GCP | Cartosat-1 DEM | Cartosat-1 DEM Error | TanDEM-X | TanDEM-X Error | ALOS PALSAR | ALOS PALSAR Error |
| Point 1 | Built-up | 646071.03 | 2973842.09 | 103.58 | −48.07 | 151.64 | 149.94 | −1.71 | 148.51 | −3.14 | 149.07 | −2.58 |
| Point 2 | Built-up | 647231.82 | 2974315.10 | 105.74 | −47.90 | 153.64 | 151.79 | −1.85 | 156.42 | 2.79 | 165.61 | 11.97 |
| Point 3 | Tea garden | 653240.90 | 2972912.67 | 120.41 | −47.99 | 168.39 | 167.40 | −0.99 | 166.78 | −1.61 | 170.99 | 2.59 |
| Point 4 | Built-up | 647454.30 | 2998152.06 | 1181.33 | −45.49 | 1226.83 | 1227.25 | 0.42 | 1225.94 | −0.89 | 1234.35 | 7.52 |
| Point 5 | Dense forest | 658308.01 | 2998152.06 | 1736.00 | −44.58 | 1780.57 | 1794.70 | 14.12 | 1795.23 | 14.66 | 1776.32 | −4.25 |
| Point 6 | Built-up | 645923.75 | 2994738.94 | 1133.65 | −45.78 | 1179.43 | 1161.54 | −17.89 | 1165.90 | −13.53 | 1161.71 | −17.72 |
| Point 7 | Built-up | 651313.26 | 2993769.46 | 661.19 | −45.76 | 706.94 | 705.33 | −1.61 | 708.54 | 1.60 | 701.59 | −5.36 |
| Point 8 | Open forest | 652466.01 | 2992349.03 | 576.61 | −45.76 | 622.59 | 624.31 | 1.72 | 619.13 | −3.46 | 628.96 | 6.37 |
| Point 9 | Dense forest | 641521.45 | 2993412.99 | 187.80 | −46.08 | 233.88 | 236.28 | 2.40 | 253.45 | 19.56 | 235.08 | 1.20 |
| Point 10 | Dense forest | 641329.94 | 2993352.48 | 189.26 | −46.09 | 235.35 | 236.28 | 0.93 | 255.25 | 19.90 | 247.14 | 11.78 |
| Point 11 | Dense forest | 641439.24 | 2992881.80 | 179.22 | −46.18 | 225.40 | 209.46 | −15.93 | 255.07 | 29.68 | 228.18 | 2.78 |
| Point-12 | Open forest | 644166.80 | 2979279.32 | 156.74 | −37.50 | 194.24 | 204.91 | 10.67 | 220.76 | 26.52 | 211.54 | 17.30 |
| Point 13 | Open forest | 655789.31 | 2975068.43 | 159.58 | −47.18 | 206.75 | 205.84 | −0.91 | 204.08 | −2.67 | 207.18 | 0.43 |
| Point 14 | Dense forest | 642935.57 | 2987689.19 | 254.49 | −47.14 | 301.64 | 296.42 | −5.22 | 303.44 | 1.81 | 334.36 | 32.72 |
| Point 15 | Tea garden | 638559.92 | 2988370.49 | 969.40 | −48.04 | 1017.43 | 1011.15 | −6.28 | 1017.16 | −0.27 | 1020.82 | 3.39 |
| Point 16 | Tea garden | 634427.30 | 2989936.87 | 1302.16 | −47.91 | 1350.07 | 1343.72 | −6.35 | 1348.80 | −1.27 | 1351.82 | 1.75 |
| Point 17 | Dense forest | 640365.77 | 2985439.79 | 204.34 | −47.51 | 251.85 | 240.61 | −11.24 | 248.39 | −3.46 | 253.80 | 1.95 |
| Point 18 | Dense forest | 639759.62 | 2983780.39 | 682.48 | −44.76 | 727.24 | 724.64 | −2.59 | 725.60 | −1.64 | 723.78 | −3.46 |
| Point 19 | Open forest | 633522.80 | 2986056.97 | 1197.53 | −47.51 | 1245.04 | 1238.60 | −6.43 | 1240.38 | −4.66 | 1240.35 | −4.69 |
| Point 20 | Built-up | 642366.27 | 2989137.28 | 195.74 | −44.86 | 240.60 | 224.31 | −16.29 | 228.12 | −12.48 | 221.46 | −19.14 |
| Point 21 | Dense forest | 642361.88 | 2989113.52 | 196.27 | −44.87 | 241.13 | 235.26 | −5.87 | 231.35 | −9.78 | 225.86 | −15.28 |
| Point 22 | Dense forest | 642345.02 | 2989267.25 | 188.72 | −44.83 | 233.55 | 232.56 | −0.99 | 244.72 | 11.18 | 231.83 | −1.71 |

**Table 3.3** Error statistics for different types of DEMs with reference to 22 GCPs (units are in meters).

| DEM | ME | MAE | RMSE | Skewness |
|---|---|---|---|---|
| Cartosat-1 | −3.27 | 6.02 | 8.23 | 0.05 |
| TanDEM-X | 3.13 | 8.48 | 12.10 | 0.87 |
| ALOS PALSAR | 1.25 | 8.00 | 11.26 | 0.57 |

**Table 3.4** Error statistics for different types of DEMs with reference to GCPs collected from different types of land use classes (units are in meters).

| Land Use Class | No. of GCP collected | DEM | ME | MAE | RMSE | Skew-ness |
|---|---|---|---|---|---|---|
| Dense forest | 9 | Cartosat-1 | −2.71 | 6.59 | 8.54 | 0.41 |
| | | TanDEM-X | 9.10 | 12.41 | 15.32 | 0.06 |
| | | ALOS PALSAR | 2.86 | 8.35 | 12.86 | 1.16 |
| Open forest | 4 | Cartosat-1 | 1.26 | 4.93 | 6.30 | 0.39 |
| | | TanDEM-X | −1.07 | 4.32 | 4.56 | 1.07 |
| | | ALOS PALSAR | 4.85 | 7.20 | 9.51 | 0.45 |
| Tea gardens | 3 | Cartosat-1 | −4.54 | 4.54 | 5.19 | 0.71 |
| | | TanDEM-X | −1.05 | 1.05 | 1.20 | 0.52 |
| | | ALOS PALSAR | 2.58 | 2.58 | 2.66 | −0.03 |
| Built-up | 6 | Cartosat-1 | −6.49 | 6.63 | 9.95 | −0.69 |
| | | TanDEM-X | −4.28 | 5.73 | 7.74 | −0.46 |
| | | ALOS PALSAR | −4.21 | 10.71 | 12.35 | −0.008 |

penetrate dense vegetation cover; rather it backscatters from the vegetation canopy (Hawker et al., 2019; Zhang et al., 2019b). This may be one of the reasons for poor TanDEM-X data quality in the respective areas. While a good agreement of elevation with reference GCPs were found in Cartosat-1 DEM (RMSE 8.54 m), the level of accuracy for Cartosat-1 DEM is closer to the earlier research (RMSE ±9.6 m) carried out in the Manali region of northwestern Himalayas (Snehmani et al., 2013).

#### 3.4.2.2 Open Forest

In open forest cover areas, the elevation of TanDEM-X is closer to the reference GCPs. Among the three DEMs, the lowest RMSE was found in TanDEM-X (4.56 m), followed by Cartosat-1 DEM (6.30 m), and ALOS PALSAR (9.51 m). The TanDEM-X is much more appropriate in open areas (Hawker et al., 2019; Zhang et al., 2019a). In the open vegetation cover areas, the SAR antenna can record the intensity of the radiation that is backscattered from the Earth's surface (Grohmann, 2018). The reflected incident energy from the ground makes a perfect elevation model.

#### 3.4.2.3 Tea Garden

Based on three reference GCPs, the present study found that TanDEM-X and ALOS PALSAR DEM data quality in tea garden areas are comparatively better than the Cartosat-1 DEM. In tea garden areas there is no obstacle to penetrating the vegetation canopy by X & L-band. Therefore, energy backscattered from the ground surface represents the Earth's surface elevation perfectly. It is reported that globally, in low vegetation cover areas, TanDEM-X offers a good quality of data, RMSE is ±1.1 m (Wessel et al., 2018).

#### 3.4.2.4 Built-up Area

After comparing the DEMs, and six reference GCPs collected from Sevoke, Kalimpong, and 29th mile Baluwakhani settlement areas, the study established that TanDEM-X has the lowest RMSE 7.74 m, followed by Cartosat-1 DEM (9.95 m), and ALOS PALSAR (12.35 m). Moreover, all the DEMs poorly represent the surface elevation in areas covered with built-up/settlement. The negative skewness in descriptive error statics represents the asymmetric distribution of errors (Table 3.4).

After verifying the TanDEM-X data from developed areas throughout the globe, ±1.4 m RMSE was reported (Wessel et al. 2018), which is lower than the present study. The higher degree of mismatch in elevation may be the reason for strong backscattering from the buildings' top and tree canopies. While optical stereo images, i.e., Cartosat-1, failed to make DTM, rather it provided DSM. Less number of trees in the settlement area may slightly improve the DEM data quality.

### 3.4.3 Spatial Accuracy Assessment: Comparison of DEMs With Reference to Stream Networks

The influence zone of automatically extracted stream networks from DEMs has been compared with the reference streams. This comparison has been done on streams flowing from west to east, north to south, northeast to southwest, and northwest to southeast, as proposed by Martha et al. (2010) (Figure 3.6). The EI obtained after intersecting the influence area is presented in Table 3.5. The result shows the case of north-south oriented valley streams extracted from Cartosat-1 DEM that matches well with reference stream networks, whereas it is poor for west-east oriented valleys. A good agreement between TanDEM-X and ALOS PALSAR extracted and reference streams were noted for west-east, north-south, and northeast-southwest oriented stream

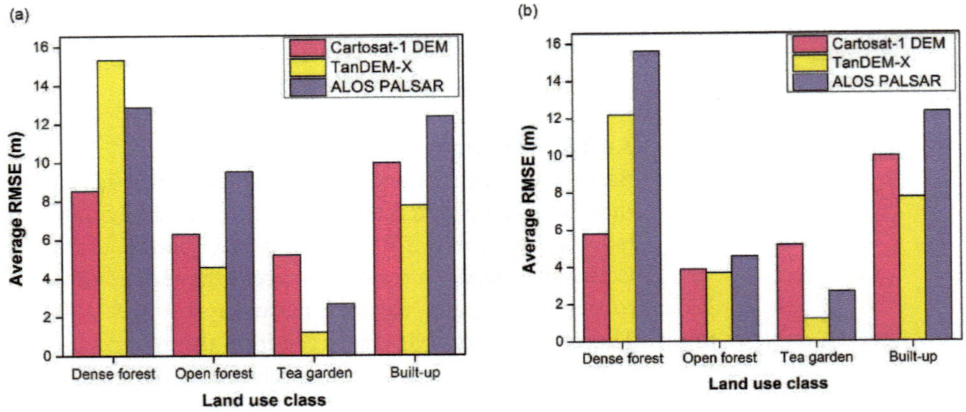


**Figure 3.5** Average RMSE for different land use classes: (a) after considering all GCPs; (b) after excluding four GCPs (three from dense forest and one from open forest cover area) containing the highest level of vertical errors. RMSE for dense forest cover area may reduce up to 5.79 in Cartosat-1 DEM, while all the DEMs provide good accuracy in open forest cover areas with RMSE 3.69–4.57.

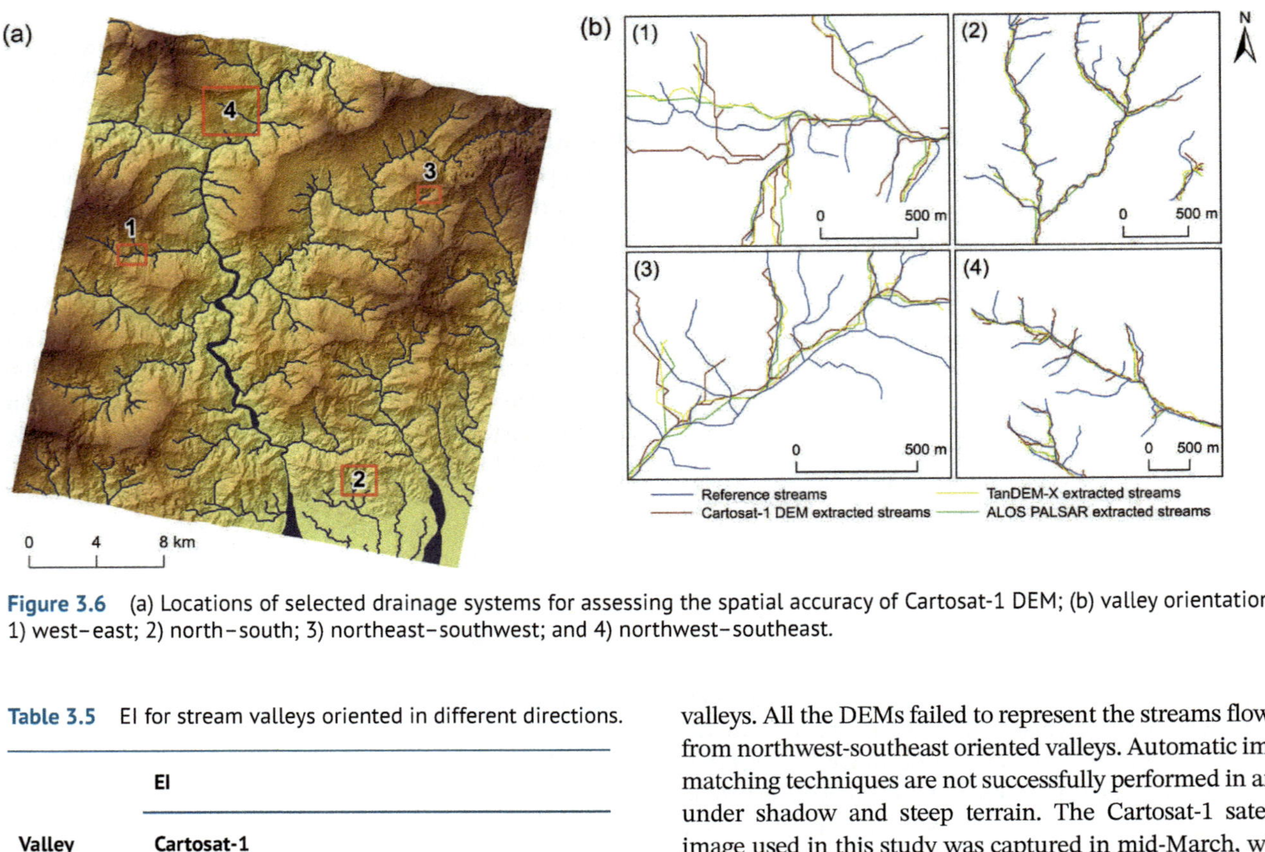


**Figure 3.6** (a) Locations of selected drainage systems for assessing the spatial accuracy of Cartosat-1 DEM; (b) valley orientations: 1) west–east; 2) north–south; 3) northeast–southwest; and 4) northwest–southeast.

**Table 3.5** EI for stream valleys oriented in different directions.

| | EI | | |
|---|---|---|---|
| **Valley Orientation** | **Cartosat-1 DEM** | **TanDEM-X** | **ALOS PALSAR** |
| West-east | 0.40 | 0.25 | 0.26 |
| North-south | 0.17 | 0.19 | 0.22 |
| Northeast-southwest | 0.36 | 0.29 | 0.28 |
| Northwest-southeast | 0.41 | 0.40 | 0.46 |

valleys. All the DEMs failed to represent the streams flowing from northwest-southeast oriented valleys. Automatic image matching techniques are not successfully performed in areas under shadow and steep terrain. The Cartosat-1 satellite image used in this study was captured in mid-March, when the angle of the Sun is oblique over the Himalayas. The oblique angle casts a long shadow on the northern slope of the mountain. That may be one of the reasons for high EI in west-east oriented valleys. In the north-south oriented valleys, sunlight may also pass the mountain ranges. This gives a good match between the stereo pair and produces a high-quality DEM. Similar findings were reported from the

Garhwal Himalayas by Martha et al. (2010). The InSAR-based TanDEM-X and ALOS PALSAR have no such limitations (Figure 3.7). Therefore, they are performing well in multi-oriented stream valleys.

## 3.5 Conclusions

After comparing the Cartosat-1 DEM, TanDEM-X, and ALOS PALSAR with respect to the collected GCPs and stream networks of the Darjeeling-Sikkim Himalayas for assessing and understanding the level of DEM accuracy, the study has come up with the following observations and conclusions:

1) A comparative study has been carried out between 10 m resolution Cartosat-1 DEM, 12 m TanDEM-X, and 12.5 m resolution ALOS PALSAR to assess the vertical and spatial accuracy with reference to the 22 high-precision GCPs and manually digitized stream networks from Cartosat-1 orthoimage, respectively.
2) The results show that accuracy of DEM depends on topography. A higher mismatch has been noted between elevation of GCPs and DEMs in high altitude areas and vice versa. Cartosat-1 DEM has an MAE of 6.02 m and an RMSE of 8.23 m, which is comparatively lower than ALOS PALSAR and TanDEM-X.
3) It is also observed that the variation in elevation between GCPs and DEMs is higher in areas under dense forest cover. In such a situation, Cartosat-1 DEM performs well compared with others. In the areas covered by tea gardens, elevation of TanDEM-X matches the GCPs appropriately.
4) The spatial accuracy of DEMs has been assessed by EI derived after comparing stream influence areas flowing in different directions. In north-south oriented valleys automatically extracted stream networks from DEMs matche well with the reference streams. Whereas, in the case of northwest-southeast oriented valleys, a higher degree of mismatch has been noted.

The study suggests that Cartosat-1 DEM is performing well in Sikkim-Darjeeling Himalayan terrain. Further studies are required with more well distributed GCPs to understand the accuracy of DEM from a better perspective.

ORCID (Prodip Mandal): https://orcid.org/0000-0003-0118-3331
ORCID (Shraban Sarkar): https://orcid.org/0000-0001-5846-0972

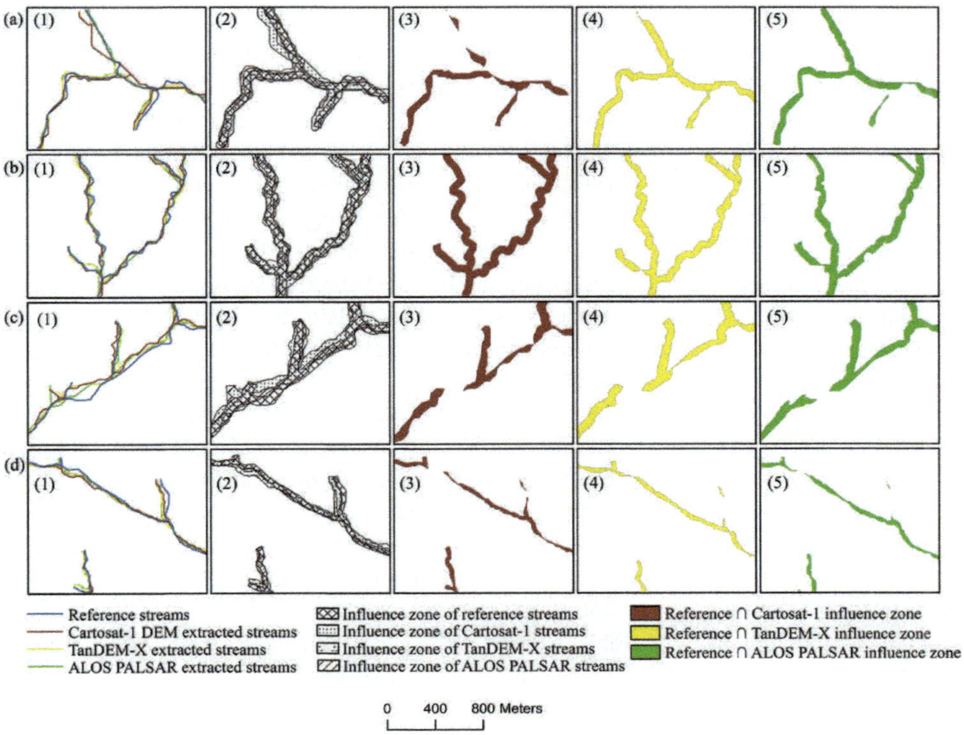


**Figure 3.7** Comparison between reference and automatically extracted stream (Column 1) influence areas (Column 2), intersection of influence areas (Columns 3–5) for: (a) west–east; (b) north–south; (c) northeast–southwest; and (d) northwest–southeast oriented valleys.

## Acknowledgments

The first author is grateful to University Grants Commission, New Delhi, India for providing the fellowship [UGC-Ref. No. 3279/(SC) (NET-JAN 2017)] for carrying out the research work. The authors are thankful to Dr Alokananda Ghosh (Department of Geography, Tehatta Sadananda Mahavidyalaya, Tehatta, India) for her constructive comments and suggestions to improve the manuscript. Authors report no conflict of interest.

## References

Agarwal, R., Sur, K., and Rajawat, A.S. (2020). Accuracy assessment of the CARTOSAT DEM using robust statistical measures. *Modeling Earth Systems and Environment* 6(1): 471–478. doi: 10.1007/s40808-019-00694-9.

Araújo, P.V.N., Amaro, V.E., Silva, R.M. et al. (2019). Delimitation of flood areas based on a calibrated a DEM and geoprocessing: case study on the Uruguay river, Itaqui, southern Brazil. *Natural Hazards & Earth System Sciences* 19(1): 237–250. doi:10.5194/nhess-19-237-2019.

ASF (2015). ASF radiometrically terrain corrected ALOS PALSAR products. Product guide: 2–12. https://asf.alaska.edu/wp-content/uploads/2019/03/rtc_product_guide_v1.2.pdf.

Baade, J. and Schmullius, C. (2016). TanDEM-X IDEM precision and accuracy assessment based on a large assembly of differential GNSS measurements in Kruger National Park, South Africa. *ISPRS Journal of Photogrammetry & Remote Sensing* 119: 496–508. doi: 10.1016/j.isprsjprs.2016.05.005.

Berthier, E., Arnaud, Y., Vincent, C. et al. (2006). Biases of SRTM in high-mountain areas: implications for the monitoring of glacier volume changes. *Geophysical Research Letters* 33(8): L08502. doi: 10.1029/2006GL025862.

BIS (2002) Seismic zonation map of India. IS: 1893 (Part: I).

Bolch, T. and Kamp, U. (eds.) (2006). Glacier mapping in high mountains using DEMs, Landsat and ASTER data 37–48. Available at: http://web.unbc.ca/~bolch/publications/BolcKamp06_GeoRaum.pdf.

Boulton, S.J. and Stokes, M. (2018). Which DEM is best for analyzing fluvial landscape development in mountainous terrains? *Geomorphology* 310: 168–187. doi: 10.1016/J.GEOMORPH.2018.03.002.

Daly, C., Gibson, W.P., Taylor, G.H. et al. (2002). A knowledge-based approach to the statistical mapping of climate. *Climate Research* 22(2): 99–113. doi: 10.3354/cr022099.

Daly, C., Halbleib, M., Smith, J.I. et al. (2008). Physiographically-sensitive mapping oftemperature and precipitation across the conterminous United States. *International Journal of Climatology* 28: 2031–2064. doi: 10.1002/joc.1688.

Deo, R. and Rao, Y. (2014). Vertical accuracy assessment of TanDEM-X DEM over Indian test areas. In: *EUSAR 2014; 10th European Conference on Synthetic Aperture Radar*, 1–4.

Farr, T., Rosen, P.A., Caro, E. et al. (2007). The Shuttle Radar Topography Mission. *Reviews of Geophysics* 45(2): 1–33 (Paper number 2005RG000183). doi: 10.1029/2005RG000183.

Favey, E., Geiger, A., Gudmundsson, G.H. et al. (2003). Evaluating the potential of an airborne laser-scanning system for measuring volume changes of glaciers. *Geografiska Annaler: Series A Physical Geography* 81(4): 555–561.

Gómez-Gutiérrez, Á., Schnabel, S., Contador, F.L. et al. (2011). Testing the quality of open-access DEMs and their derived attributes in Spain: SRTM, GDEM and PNOA DEM. *Proceedings of Geomorphometry 2011 Conference, Redlands, CA, USA*. 53–56.

Grohmann, C.H. (2018). Evaluation of TanDEM-X DEMs on selected Brazilian sites: comparison with SRTM, ASTER GDEM and ALOS AW3D30. *Remote Sensing of Environment* 212: 121–133. doi: 10.1016/j.rse.2018.04.043.

Grosse, P., van Wyk de Vries, B., Euillades, P.A. et al. (2012). Systematic morphometric characterization of volcanic edifices using digital elevation models. *Geomorphology* 136(1): 114–131. doi: 10.1016/j.geomorph.2011.06.001.

Gruber, A., Wessel, B., Huber, M. et al. (2012). Operational TanDEM-X DEM calibration and first validation results. *ISPRS Journal of Photogrammetry & Remote Sensing* 73: 39–49. doi: 10.1016/j.isprsjprs.2012.06.002.

Hawker, L., Neal, J., and Bates, P. (2019). Accuracy assessment of the TanDEM-X 90 Digital Elevation Model for selected floodplain sites. *Remote Sensing of Environment* 232: 111319. doi: 10.1016/j.rse.2019.111319.

Hebeler, F. and Purves, R.S. (2009). The influence of elevation uncertainty on derivation of topographic indices. *Geomorphology* 111(1–2): 4–16. doi: 10.1016/j.geomorph.2007.06.026.

Hirt, C. (2014). Digital terrain models. In: *Encyclopedia of Geodesy* (ed. E. Grafarend), 2–7. Cham: Springer. doi: 10.1007/978-3-319-02370-0_31-1.

Höhle, J. (2009). DEM generation using a digital large format frame camera. *Photogrammetric Engineering & Remote Sensing* 75(1): 87–93. doi: 10.14358/PERS.75.1.87.

Holmes, K.W., Chadwick, O.A., and Kyriakidis, P.C. (2000). Error in a USGS 30-meter digital elevation model and its impact on terrain modeling. *Journal of Hydrology* 233(1–4): 154–173. doi: 10.1016/S0022-1694(00)00229-8.

Januchowski, S.R., Pressey, R.L., VanDerWal, J. et al. (2010). Characterizing errors in digital elevation models and estimating the financial costs of accuracy. *International Journal of Geographical Information Science* 24(9): 1327–1347. doi: 10.1080/13658811003591680.

Johnson, A. (2004). *Plane and Geodetic Surveying*. London: CRC Press.

Kellndorfer, J., Walker, W., Pierce, L. et al. (2004). Vegetation height estimation from shuttle radar topography mission and national elevation datasets. *Remote Sensing of Environment* 93(3): 339–358. doi: 10.1016/j.rse.2004.07.017.

Kervyn, F. (2001). Modelling topography with SAR interferometry: illustrations of a favourable and less favourable environment. *Computer Geosciences* 27: 1039–1050.

Korets, M.A., Ryzhkova, V.A., Danilova, I.V. et al. (2016). Vegetation cover mapping based on remote sensing and digital elevation model data. *International Archives of the Photogrammetry Remote Sensing and Spatial Information Sciences: ISPRS Archives* 41: 699–704. doi: 10.5194/isprsarchives-XLI-B8-699-2016.

Kramm, T. and Hoffmeister, D. (2019). A relief dependent evaluation of digital elevation models on different scales for Northern Chile. *ISPRS International Journal of Geo-Information* 8(10): 430. doi: 10.3390/ijgi8100430.

Kumar, K.S., Kumar, A., Khanduri, V.P. et al. (2020). Indigenous knowledge for disaster solutions in the hilly state of Mizoram, Northeast India. In: *Techniques for Disaster Risk Management and Mitigation* (ed. P.K. Srivastava, S.K., Singh, U.C., Mohanty, et al.), 23–32. USA: John Wiley & Sons, Inc.

Lang, H.R. and Welch, R. (1999). *Algorithm Theoretical Basis Document for ASTER Digital Elevation Models, Version 3.0 Jet Propulsion Laboratory*. Available at: http://eospso.gsfc.nasa.gov/eos_homepage/for_scientists/atbd/docs/ASTER/atbd-ast-14.pdf.

Lemoine, F.G., Smith, D.E., Kunz, L. et al. (1997). The development of the NASA GSFC and NIMA joint geopotential model. July: 461–469. doi: 10.1007/978-3-662-03482-8_62.

Li, Z., Zhu, C. and Gold, C. (2004). *Digital Terrain Modeling: Principles and Methodology*. New York: CRC Press.

Libina, V.N. and Nikiforov, L.S. (2020). Digital elevation model of the bottom in the operational oceanography system. *Oceanology* 60(6): 854–860.

Martha, T.R., Babu Govindharaj, K., and Vinod, K. (2015). Damage and geological assessment of the 18 September 2011 Mw 6.9 earthquake in Sikkim, India using very high resolution satellite data. *Geoscience Frontiers* 6(6): 793–805. doi: 10.1016/j.gsf.2013.12.011.

Martha, T.R., Kerle, N., Van Westen, C.J. et al. (2010). Effect of sun elevation angle on DSMs derived from Cartosat-1 data. *Photogrammetric Engineering & Remote Sensing* 76(4): 429–438. doi: 10.14358/PERS.76.4.429.

Moore, I.D., Grayson, R.B., and Ladson, A.R. (1991). Digital terrain modelling: a review of hydrological, geomorphological, and biological applications. *Hydrological Processes* 5(1): 3–30. doi: 10.1002/hyp.3360050103.

Mukherjee, S., Joshi, P.K., Mukherjee, S. et al. (2013). Evaluation of vertical accuracy of open source Digital Elevation Model (DEM). *International Journal of Applied Earth Observation & Geoinformation* 21(1): 205–217. doi: 10.1016/j.jag.2012.09.004.

O'Loughlin, F.E., Paiva, R.C.D., Durand, M. et al. (2016). A multi-sensor approach towards a global vegetation corrected SRTM DEM product. *Remote Sensing of Environment* 182: 49–59. doi: 10.1016/j.rse.2016.04.018.

Pandey, P., Manickam, S., Bhattacharya, A. et al. (2017). Qualitative and quantitative assessment of TanDEM-X DEM over western Himalayan glaciated terrain. *Geocarto International* 32(4): 442–454. doi: 10.1080/10106049.2016.1155655.

Pasquetti, F., Bini, M., and Ciampalini, A. (2019). Accuracy of the TanDEM-X digital elevation model for coastal geomorphological studies in Patagonia (South Argentina). *Remote Sensing* 11(15): 1767. doi: 10.3390/rs11151767.

Peckham, R.J. and Jordan, G. (2007). Digital terrain modelling. In: *Development and Applications in a Policy Support Environment Series: Lecture Notes in Geoinformation and Cartography*, 1–313. Berlin-Heidelberg: Springer-Verlag.

Pipaud, I., Loibl, D., and Lehmkuhl, F. (2015). Evaluation of TanDEM-X elevation data for geomorphological mapping and interpretation in high mountain environments: a case study from SE Tibet, China. *Geomorphology* 246: 232–254. doi: 10.1016/j.geomorph.2015.06.025.

Podgórski, J., Kinnard, C., Petlicki, M. et al. (2019). Performance assessment of TanDEM-X DEM for mountain glacier elevation change detection. *Remote Sensing* 11(2): 187.

Purinton, B. and Bookhagen, B. (2017). Validation of digital elevation models (DEMs) and comparison of geomorphic metrics on the southern Central Andean Plateau. *Earth Surface Dynamics* 5(2): 211–237. doi: 10.5194/esurf-5-211-2017.

Rawat, K.S., Mishra, A.K., Sehgal, V.K. et al. (2013). Comparative evaluation of horizontal accuracy of elevations of selected ground control points from ASTER and SRTM DEM with respect to CARTOSAT-1 DEM: a case study of Shahjahanpur district, Uttar Pradesh, India. *Geocarto International* 28(5): 439–452. doi: 10.1080/10106049.2012.724453.

Rizzoli, P., Martone, M., Gonzalez, C. et al. (2017). Generation and performance assessment of the global TanDEM-X digital elevation model. *ISPRS Journal of Photogrammetry & Remote Sensing* 132: 119–139. doi: 10.1016/j.isprsjprs.2017.08.008.

Rodríguez, E., Morris, C.S., and Belz, J.E. (2006). A global assessment of the SRTM performance. *Photogrammetric Engineering & Remote Sensing* 72(3): 249–260. doi: 10.14358/PERS.72.3.249.

Sampson, C.C., Smith, A.M., Bates, P.D. et al. (2015). A high-resolution global flood hazard model. *Water Resources Research* 51(9): 7358–7381. doi: 10.1002/2015WR016954.Received.

San, B.T. and Süzen, M.L. (2005). Digital elevation model (DEM) generation and accuracy assessment from ASTER stereo data. *International Journal of Remote Sensing* 26(22): 5013–5027. doi: 10.1080/01431160500177620.

Sanders, B.F. (2007). Evaluation of on-line DEMs for flood inundation modeling. *Advances in Water Resources* 30(8): 1831–1843. doi: 10.1016/j.advwatres.2007.02.005.

Sarkar, S., Roy, A.K., and Martha, T.R. (2013a). Landslide susceptibility assessment using information value method in parts of the Darjeeling Himalayas. *Journal of the Geological Society of India* 82(4): 351–362. doi: 10.1007/s12594-013-0162-z.

Sarkar, S., Roy, A.K., and Martha, T.R. (2013b). Soil depth estimation through soil-landscape modelling using regression kriging in a Himalayan terrain. *International Journal of Geographical Information Science* 27(12): 2436–2454. doi: 10.1080/13658816.2013.814780.

Sarkar, S., Roy, A.K., and Raha, P. (2016). Deterministic approach for susceptibility assessment of shallow debris slide in the Darjeeling Himalayas, India. *Catena* 142: 36–46. doi: 10.1016/j.catena.2016.02.009.

Sesnie, S.E., Hagell, S.E., Otterstrom, S.M. et al. (2008). SRTM-DEM and LANDSAT ETM+ data for mapping tropical dry forest cover and biodiversity assessment in Nicaragua. *Revista Geográfica Acadêmica* 2: 53–65. Available at: http://www.fs.fed.us/global/iitf/pubs/ja_iitf_2008_sesnie001.pdf.

Simard, M., Pinto, N., Fisher, J.B. et al. (2011). Mapping forest canopy height globally with spaceborne lidar. *Journal of Geophysical Research: Biogeosciences* 116(4): 1–12. doi: 10.1029/2011JG001708.

Snehmani, Singh, M., Gupta, R.D., and Ganju, A. (2013). Extraction of high resolution DEM from Cartosat-1 stereo imagery using rational math model and its accuracy assessment for a part of snow covered NW Himalaya. *Journal of Remote Sensing & GIS* 4(2): 23–34. Available at: http://stmjournals.com/index.php?journal=JoRSG&page=article&op=view&path[]=3020.

Spark, R.N. and Williams, P.F. (1996). Digital terrain models and the visualization of structural geology. *Computer Methods in the Geosciences* 15(C): 421–446. doi: 10.1016/S1874-561X(96)80032-0.

Tachikawa, T., Hat, M., Kaku, M. et al. (2011). Characteristics of ASTER GDEM version2. In: *Geoscience & Remote Sensing Symposium (IGARSS), 2011 IEEE International*, 3657–3660.

Taud, H., Parrot, J.F., and Alvarez, R. (1999). DEM generation by contour line dilation. *Computers & Geosciences* 25(7): 775–783. doi: 10.1016/S0098-3004(99)00019-9.

Uuemaa, E., Ahi, S., Montibeller, B. et al. (2020). Vertical accuracy of freely available global digital elevation models (ASTER, AW3D30, MERIT, TanDEM-X, SRTM, and NASADEM). *Remote Sensing* 12(21): 3482. doi: 10.3390/rs12213482.

Vassilaki, D.I. and Stamos, A.A. (2020). TanDEM-X DEM: comparative performance review employing LIDAR data and DSMs. *ISPRS Journal of Photogrammetry & Remote Sensing* 160: 33–50. doi: 10.1016/j.isprsjprs.2019.11.015.

Vassilopoulou, S., Hurni, L., Dietrich, V. et al. (2002). Orthophoto generation using IKONOS imagery and high-resolution DEM: a case study on volcanic hazard monitoring of Nisyros Island (Greece). *ISPRS Journal of Photogrammetry & Remote Sensing* 57(1–2): 24–38. doi: 10.1016/S0924-2716(02)00126-0.

Volařík, D. (2010). Application of digital elevation model for mapping vegetation tiers. *Journal of Forest Science* 56(3): 112–120. doi: 10.17221/74/2009-jfs.

Weibel, R. and Heller, M. (1990). *A framework for digital terrain modelling*. In: *Fourth International Symposium on Spatial Data Handling*, Zurich, Switzerland, 219–229.

Wessel, B. (2016). TanDEM-X ground segment: DEM products specification document. EOC, DLR, Oberpfaffenhofen, Germany, Public Document TD-GS-PS-0021. Available at: https://tandemx-science.dlr.de.

Wessel, B., Huber, M., Wohlfart, C. et al. (2018). Accuracy assessment of the global TanDEM-X digital elevation model with GPS data. *ISPRS Journal of Photogrammetry & Remote Sensing* 139: 171–182. doi: 10.1016/j.isprsjprs.2018.02.017.

Wilson, J.P. and Gallant, J.C. (2000). Secondary topographic attributes. In: *Terrain Analysis: Principles and Applications* (ed. J.P. Wilson and J.C. Gallant), 87–131. New York: John Wiley & Sons.

Zhang, C., Lin, H., Chen, M. et al. (2014). Scale matching of multiscale digital elevation model (DEM) data and the weather research and forecasting (WRF) model: a case study of meteorological simulation in Hong Kong. *Arabian Journal of Geosciences* 7(6): 2215–2223. doi: 10.1007/s12517-014-1273-6.

Zhang, K., Gann, D., Ross, M. et al. (2019a). Comparison of TanDEM-X DEM with LiDAR data for accuracy assessment in a coastal urban area. *Remote Sensing* 11(7): 876. doi: 10.3390/RS11070876.

Zhang, K., Gann, D., Ross, M. et al. (2019b). Accuracy assessment of ASTER, SRTM, ALOS, and TDX DEMs for Hispaniola and implications for mapping vulnerability to coastal flooding. *Remote Sensing of Environment* 225: 290–306. doi: 10.1016/j.rse.2019.02.028.

Zhang, W. and Montgomery, D.R. (1994). Digital elevation model grid size, landscape representation, and hydrologic simulation. *Water Resources Research* 30(4): 1019–1028.

Zink, M., Bachmann, M., Bräutigam, B. et al. (2014). TanDEM-X: the new global DEM takes shape. *IEEE Geoscience & Remote Sensing Magazine* 2(2): 8–23. doi: 10.1109/MGRS.2014.2318895.

# 4

# An Overview of Morphometry Software Packages, Tools, and Add-ons

*Satarupa Mitra[1], Shailendra Pundir[2], Rahul Devrani[3,4], Aman Arora[5], Manish Pandey[1,6,*], Romulus Costache[7,8], and Saeid Janizadeh[9]*

[1] *University Center for Research & Development (UCRD), Chandigarh University, Mohali, Punjab, India*
[2] *Department of Geology, Centre of Advanced Study, Kumaun University, Nainital, India*
[3] *Wadia Institute of Himalayan Geology, 33 GMS. Road, Dehradun 248001, India*
[4] *Delhi School of Climate Change & Sustainability, Institution of Eminence, University of Delhi, Delhi 110007, India*
[5] *Bihar Mausam Seva Kendra, Planning and Development Department, Government of Bihar, Patna 800015, Bihar, India*
[6] *Department of Civil Engineering, University Institute of Engineering, Chandigarh University, Mohali, Punjab, India*
[7] *Department of Civil Engineering, Transilvania University of Brasov, 5, Turnului Str, Brasov 500152, Romania*
[8] *Danube Delta National Institute for Research and Development, 165 Babadag Street, Tulcea 820112, Romania*
[9] *Department of Watershed Management Engineering and Sciences, Faculty of Natural Resources and Marine Science, Tarbiat Modares University, Tehran 14115-111, Iran*
[*] *Corresponding author*

## 4.1 Introduction

The role of morphometry has been of great interest to the community of geomorphologists working to understand the role of climate and tectonics in the evolution of topography, high relief, deep gorges, valleys, and exhumation of deep crustal rocks (Aravind et al., 2022; Baillie and Norbu, 2004; Clark et al., 2004; Devrani et al., 2015; Jamieson et al., 2004; Pundir et al., 2021; Thiede et al., 2004). Geomorphometry incorporates quantitative analysis, characterization, and modeling of the topography of landforms. The morphometric information about the surface features became separated from the broader field of geomorphology and was accepted as an independent field mainly during the second half of the 20th century. It followed precursory works in this field, such as by Horton (1932) on quantitative evaluation of the processes that shape stream networks and drainage basins (Szypuła, 2017). Gradually, geomorphometry became a sophisticated area of study, owing to the complexities and varieties of terrain surfaces on Earth, and even more so for the analysis of remotely sensed data (Franklin, 2020; Sofia, 2020). The concept of "surface" encompasses discernible surfaces within several different fields of study, such as topography, land cover, bathymetry, water tables, tree canopies, soil horizons, and lithological and sedimentary units of the sub-surfaces. This relationahip of geomorphometry to various fields makes it a multidisciplinary area of research. Through the measurements of surficial forms and features, geomorphometry also measures the evolution of such features, the processes that influence the evolution, including the process-form relationships and interactions which control the evolution (Bishop et al., 2018; Cavazzi et al., 2013; Lecours et al., 2016).

The features on the Earth's surface keep changing on a temporal scale due to the impact of various processes that come under the broad categories of atmospheric, surface, and tectonic processes (Keller and Pinter, 1996). It affects the local biosphere, members of which need to adapt to the changing geomorphology to survive. Therefore, parallel evolution of the methods of analysis and modeling of the morphological parameters is also essential to keep up with constant change for the benefit and safeguarding of the inhabitants of the biosphere. Therefore, as an initial step toward studying landscape evolution, morphology, and the morphometry of the landscape needs to be studied first. Integration of geomorphometric variables was developed to quantify topographic surface features conditions and used at appropriate scales (Sofia, 2020; Szypuła, 2017). It is essential to analyze remotely sensed data to produce viable results (Franklin, 2020). Also, the significance of geomorphometric analysis is enhanced as the increasing population plays a pivotal role as an influencing factor on landforms, making it necessary to understand the degree of anthropogenic influence on the landform of any particular area (Cao et al., 2020).

*Advances in Remote Sensing Technology and the Three Poles,* First Edition. Edited by Manish Pandey, Prem C. Pandey, Yogesh Ray, Aman Arora, Shridhar D. Jawak, and Uma K. Shukla.

A broad categorization of the primary applications of geomorphometry as provided by Bishop et al. (2018) and Wilson (2018) is the following: i) Digital Terrain Modeling; ii) depiction of topographic properties based on multiple scales utilizing mathematical or statistical methods; iii) analysis and mapping of geomorphometric features; and iv) modeling of the evolution of the topography. The factors that play a part in these applications are the geomorphometric indices, and knowledge of these characteristics is the basis of geomorphometry. Availability of remote-sensing data, invention of faster computers and development of better computing methods has increased the possibilities of geomorphometric analyzes for a variety of research and practical implementations and will continue to improve and enhance in the future.

This chapter provides a comprehensive overview of various geomorphometric tools, developed and incorporated in the past few decades, to enlighten the readers on the diversity of their implementation and usefulness. This collective analysis aims to aid beginners and professionals in this field and encourage future developments. This chapter focuses on the computational aspects of the tools, including a tool's primary purpose and the geomorphometric variables incorporated within it, while the explanation of the concepts behind the individual geomorphometric variables is beyond the scope of this chapter. Geomorphometric tools have been organized here based on two broad categories and the respective sub-categories as shown in Figure 4.1.

Categorization based on user interfaces is a functional category; for example, add-on tools are the ones that can be run within another GIS software, either as an add-on or a plug-in to that software, which either comes with the software or can be downloaded and installed separately. The input and output are through the user interface of the leading software. The morphometric analysis only forms a part of the more significant GIS analysis of remote-sensing data for such tools. Then comes the stand-alone tools, which are developed for the sole purpose of morphometric analysis and do not incorporate any other functions. The input and output of such tools are through the user interface. Also, the final purpose of the first two categories is the availability of the visualization of the input and output through graphical user interfaces. The third sub-category is the tools that do not incorporate any user interface, and the input and output are not represented in any visual format. The data processing is done through codes within a coding base only for quantification. Such coding bases can be traditional command prompts where codes can be executed using functions from a coding library or package, or a coding base designed particularly for a specific function or a software package comprising several functions, which means that a software package would have its very own coding base. With the general availability or accessibility of the morphometric tools, all tools can be categorized as either Open-Source or Proprietary. Open-Source ones are available free of cost and, in most cases, can be easily downloaded and installed by any user. Proprietary tools are the ones that need to be either purchased or are available from the developer upon request. Such tools may not be made Open-Source, mainly because of advanced functions and features that require expensive maintenance and might be required only by professionals and are therefore purchased by organizations and institutions where its members can use them as required.

The contents of this chapter are for the benefit of professionals and newcomers in the field of morphometric analysis of remote-sensing data with the hope that the broader content of the overview provided will aid in the development and implementation of new and improved methods as well as tools in terms of all aspects, such as functions available, better user interfaces, better visualizations, and easier availability and accessibility.

## 4.2 Overview of Morphometry Tools and Toolboxes

As mentioned earlier, the morphometric tools and toolboxes are classified based on their computational functionality and availability. However, each tool's morphological processes incorporated within them are more or less

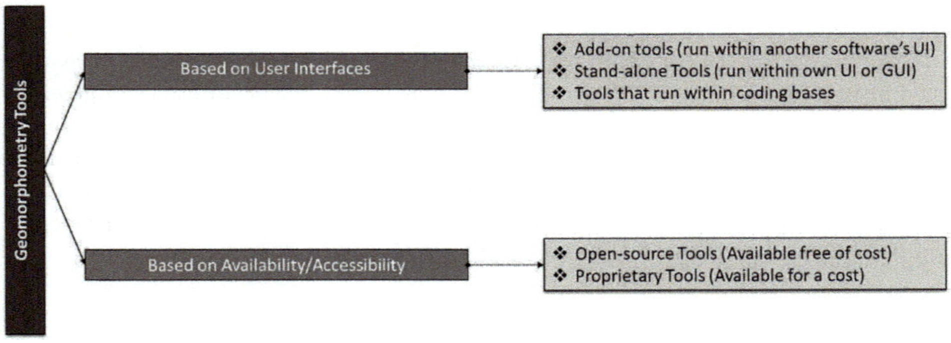


**Figure 4.1** Categories of morphometric tools.

similar in terms of the underlying mathematical and statistical concepts that govern the morphological processes and morphometric quantifications. The general concept in computer languages is that the morphometric quantifications are incorporated within the tools as methods of a class representing the broader geomorphological processes within which the sub-process occurs. After running each function, the program gives outputs as values for each method, representing the values for the morphometric measurement for which the method is assigned. As the scope of this chapter is only to provide a review of all tools and toolboxes, details of each method and functions within those methods have not been included. For further details on any tool, the readers should refer to the reference literature associated with each tool. The following paragraphs provide an overview of the tools based on their functionality.

The Gradient Metrics toolbox is an ArcGIS ADD-ON that contains various gradient and geomorphometric metrics used for surface analysis (Evans et al., 2010, 2016). This tool has been available on GitHub. (https://github.com/jeffreyevans/GradientMetrics) from January 2018. The Morphometric parameters used in this tool are mean slope, slope impedance, slope gradient, surface area ratio, surface relief ratio, and many more. The processes included in the Gradient Metrics toolbox are slope/aspect transformations, heat load index, compound topographic index, and more. This tool is compatible with Python3 and ArcGIS Pro and can be used through these user interfaces; and is Proprietary for use.

The TOPOTOOLBOX is a MATLAB-based program mainly used to analyze digital elevation models (DEMs) used in morphometry analysis (Schwanghart and Kuhn, 2010). It is an open code library/base in MATLAB and is used as a tool that works by using a set of MATLAB functions to perform topological analysis and modeling in earth surface sciences using GIS and non-GIS environments. It utilizes the concept of Object-Oriented Programming by encapsulating information on spatial referencing, flow topology, and geometry as individual objects, each having specific attributes to be implemented. It also requires the use of the Image Processing Toolbox in MATLAB. The Morphometric parameters used in this tool are the GRIDobj methods which calculate various topographical features, such as: "aspect," "curvature," "elevate minima," "hill shade," "hyps curve," "identify flats," and "roughness." Similarly, the FLOWobj class includes methods that calculate various flow features, such as: "drainage basins," "flow distance," "flow convergence," and "flow path extract." The processes included in the TOPOTOOLBOX were developed to support a combination of spatial and non-spatial numerical analysis, involving gridded Digital Elevation Model (DEM) analysis and material fluxes and spatial variability of water, sediment, chemicals, and nutrients. Typical objects used in DEM analysis are surface flow paths and the resulting stream network. It is a user interface tool, and the primary goal of TopoToolbox is to provide numerical tools for DEM analysis, which are in the form of functions that can be used as a library in custom-made scripts or functions for advanced analysis or modeling purposes. The tool's other features introduce a novel method for storing flow directions as topologically ordered vectors of indices and enables calculation of flow-related attributes such as flow accumulation. It is possible to visually explore and interact with the DEMs through the Graphical User Interfaces (GUIs). The limitations of the tool include medium-sized DEMs, and memory constraint is imposed by MATLAB's primary use of main memory. The latest version of the tool is TOPOTOOLBOX2, an enhanced version of TOPOTOOLBOX (Schwanghart and Scherler, 2014). It reduced memory overhead of up to 33% of that required by the previous version and achieved approximately 20 times faster speed than other conventional algorithms. The TopoApp provides rapid access to several essential and advanced functions of TopoToolBox2. The other model part of the tool is TTLEM (TopoToolbox Landscape Evolution Model) (Campforts et al., 2017), which is a spatially explicit, raster-based LEM to study fluvially eroding landscapes in TopoToolbox2, and is an Open-Source tool for use.

The TecDEM software is implemented in MATLAB, which applies tectonic geomorphology indices to digital elevation models (DEMs) (Shahzad and Gloaguen, 2011a, 2011b). The significant features of the tool are meant for applying complex tectonic geomorphologic aspects to DEM data on a global scale. It is used for extracting and investigating stream profiles. Users are free to write their MATLAB functions according to their requirements. The tool uses standard parameters implemented here: watershed delineation, stream profile generation, steepness, flow direction determination, and knickpoint selection. The processes incorporated in the tool are: pre-processing of the drainage network, determining the flow directions and contributing areas, delineation of the drainage network, extraction of the watershed, analysis of the stream profile, determination of concavity and steepness index, determination of hack index, preparation of isobase map, incision map, surface roughness map, drainage density map, basin asymmetry, and basin hypsometry. A GUI is provided within MATLAB, which consists of six menus (File, Prepare, Display, Analyse, Results, and Help) and a textbox that keeps a log of the processes that can be saved for reference, and is Proprietary for use.

The Whitebox Tools Geospatial Analysis Tools (GAT) 3.0.8 is an advanced Open-Source geospatial data analysis platform (Schwanghart and Scherler, 2014). It provides better flow directions than TopoToolBox, which enforces flow convergence in flat terrain. The tool is written in Java, Groovy, and Python where users can add custom tools using a JVM language. It contains more than 385 tools to perform spatial analysis on raster datasets. Some examples of the tools are GIS tools, Image processing tools, Hydrology tools, Terrain Analysis tools, and LiDAR tools, and they are Proprietary for use.

The Terrain Analysis Using Digital Elevation Models (TauDEM) is a set of tools for analysing terrain using digital elevation models (Schwanghart and Scherler, 2014). The significant feature of the model is that it can hold the entire DEM in the main memory, and it partitions the DEM into sub-domains that are processed in parallel to manage the main memory capacity better. The other features are currently packaged as an extendable component: a toolbar plugin for ArcGIS and MapWindow (a GIS data viewer). It is Open-Source for use.

ArcGeomorphometry is a software tool for geomorphometric analysis of DEMs within GIS (Rigol-Sanchez et al., 2015). The significant features of the tool are that it comprises Python (Numpy) functions for processing land surface features and variables. The morphometric parameters used involves slope, curvature, aspect, openness, and relief. The user interface of the tool consists of a graphical menu that presents the processing functions within ArcGIS software. The ArcGeomorphometry GUI can be integrated along with ArcToolbox, and all the functions can be incorporated into ArcGIS. It is Proprietary for use.

TerraceM tool is mainly used to analyze marine and lacustrine terraces using high-resolution topography (Jara-Muñoz et al., 2015). The tool's significant feature is a MATLAB GUI meant to analyze and increase the efficiency and estimate the precision of shoreline angle elevations of wave-cut terraces in marine and lacustrine environments using high-resolution topography. The Morphometric parameters used are that it incorporates processes related to terrace morphology such as river incision, scarp diffusion, and eolian diffusion. The Processes incorporated in the tool can process a series of profiles from several sites in an efficient and structured workflow. The results obtained from TerraceM can be exported in Google Earth and ESRI Shapefile formats. Mapping Toolbox of TerraceM exports results in ESRI shapefile and Google Earth format. The latest version of the tool is TerraceM-2 (Jara-Muñoz et al., 2019), which provides new methodologies and includes the processing speed and mapping capabilities of TerraceM. TerraceM-2 demonstrates the advantage of an integrated mapping and modeling approach. The tool's limitations are that it can be run on standard desktops installed with MATLAB version 2011b or a higher version only. It is Proprietary for use.

ACME Land Module, ALM (ACME-Accelerated Climate Modelling for Energy) (Wang et al., 2017) is an earth systems component of ACME and is used to simulate aspects of the land surface such as ecosystem dynamics, biogeochemistry, biophysics, heterogeneity, and hydrological cycle. Sub-models within ALM are organized based on carbon-nitrogen cycles, water balance, and energy balance. It is Proprietary for use.

Topographic Analysis Toolkit (TAK) is a Matlab function used for topographic analysis (Forte and Whipple, 2019). The significant features TAK includes are that it is used to build a series of high-level topographic analysis tools for performing a variety of topographic analysis, including the generation of maps of normalized channel steepness and selection and statistical analysis of groups watersheds. TAK can be used without any programming skills. The underlying TopoToolbox functions in TAK require MATLAB and other toolboxes to run. To expand the utility of TAK, the authors have provided compiled versions of the functions that use the free MATLAB runtime environment for users who do not have institutional access to MATLAB or all the required toolboxes. The processes incorporated are analysis of the stream network, basin-averaged analysis, and swath profiles. It is Open-Source for use.

The Transient-Profiler (Jaiswara et al., 2019) is a set of comprehensive and interactive MATLAB tools to extract geomorphic indices from DEM for obtaining landscape evolution period and basin characterization. The tool can be used to analyze river profiles, surface uplift, drainage networks, steepness, and knickpoints. The tool takes the transient basin as an indicator of tectonic influence in bedrock landscapes as tectonic and erosional processes shape the Earth's surface. Transient basins are created as the river basin adjusts to changing tectonic, topological, climatic, and boundary conditions, and bedrock channels act as efficient gradation agents. It can be run as a MATLAB toolbox and is Proprietary for use.

## 4.3 Stand-Alone Tools

ILWIS (Integrated Land and Water Information System) is an Open-Source Geographical Information Systems (GIS) and Remote-Sensing software that was developed by the Faculty of Geo-Information and Earth Observation (ITC). Its development started in 1984 at the International Institute of Geoinformation Science and Earth Observations (ITC) in the Netherlands, and in 2007 it became an Open-Source GIS (Hendrikse, 2000; Hengl et al., 2009). It can be used for hydrology and flow characteristics analyses, e.g., determination of the flow, optimization of the flow, flow indices, sink removal/sink filling, topological optimization, extraction of the drainage network, extraction of the catchment, and statistical parameter extraction. The tool is a standalone GIS package with its own user interface and is Open-Source for use.

GRASS GIS is a free Geographic Information System (GIS) software used for geospatial data management and analysis, image processing, graphics, maps, and other GIS-based analyses. This software was initially developed by the Construction Engineering and Research Laboratory (CERL) of the US Army Corps of Engineers and henceforth modified by CERL until 1996. An international GRASS Development Team has undertaken the task until now (Neteler et al., 2012). It is available as an Open-Source software package with a portable source code base, and the modularity of the software minimizes overheads. Other useful features are: SQL Query Builder, model builder, vector digitizer, and georeferencing tools. It has the ability to perform multi-scale parameterization. Almost all standard morphometric parameters are available for implementation in GRASS GIS. Analysis of all regular morphometric processes can also be implemented in GRASS GIS. The software provides its user interface and it is an Open-Source software.

The LandSerf is a freely available Geographical Information System (GIS) for visualizing and analysis of surfaces. The software was initially developed in 1996 as a platform for scale-based analysis of DEMs (Wood, 2009). It can be used to characterize a surface on multiple scales, which implies that slope, curvature, and type of feature can be measured for a wide range of spatial scales, which enables the user to evaluate the scale dependencies. LandSerf can also represent processes in the form of a script, using its scripting language, LandScript, which can be written in any text editor and can be run from a command-line interface. It can be used for morphometric parameters, e.g., land surface morphometric parameters such as slope, relief, curvature, and others. It is a Java-based, web-based application that requires a working Java Runtime Environment (JRE). It is an Open-Source software.

RiverTools is a user-friendly GIS application to analyse and visualize digital terrain, watersheds, and river networks. The software was initially available in 1998, and many improved versions have been released (Ferrando, 2017; Peckham, 2009). It is a toolkit designed explicitly for extracting hydrological information from DEMs, using specialized grids such as mass-flux grids, d-infinity grids, D8-based grids, finite-difference, and other derived grids. It also implements state-of-the-art algorithms for the computation of geomorphometric quantities. It can be used for the flow parameters like flow rate and topographical parameters like slope and aspect. The RiverTools uses various visualization tools such as density plot tool, shaded relief tool, and extraction tools such as the extraction of a D8 flow grid, basin outlet extraction, and river network extraction. The RiverTools has an interactive point-and-click GUI and is Open-Source.

Terrain Analysis System (TAS) is an essential tool for modeling environmental systems (Lindsay, 2005). TAS is incorporated with the functionalities of standard GIS tools along with a substantial number of sub-programs representing various geomorphological and hydrological processes, which makes it capable of advanced modeling of catchment processes and can therefore be used in the research of geomorphology, watershed modeling, hydrology, climatology, physical geography, and environmental science. TAS is also freely available for download. An array of parameters is implemented in this tool, described in detail in Lindsay (2005). Some of the most significant are slope, aspect, curvature, relief, flow direction, watershed, view shade, visibility index, topographic wetness index, stream power index, landform classification, stream index number, and flow path and stream profile. Flow routing algorithms incorporated in TAS are D8, FD8, Rho8, FRho8, FD8-Quinn, ADRA, and $D_{\infty}$. TAS Graphical User Interface (GUI) allows the display of multiple images simultaneously and also allows the overlaying of various raster and vector images to achieve the desired visualization. TAS operates only on the Windows platform. It is an Open-Source software.

SAGA (System for Automated Geoscientific Analyses) is a geographic information system (GIS) based system first released in 2004 (Conrad et al., 2015). SAGA is preceded by three applications for digital terrain analysis, SARA, SADO, and DiGeM. SAGA provides a comprehensive set of tools ranging from pre-processing DEMs such as filtering and filing procedures and the generation of first- and second-order derivatives, such as slope and curvature. It incorporates more sophisticated and process-oriented terrain parameters, such as altitude over the channel network. It is coded in C++ and organized in modules and comprises an Application Programming Interface and a Graphical User Interface and interfaces for scripting in R and Python. It is divided into different modules allowing raster and vector processes to be run. In windows, the SAGA GIS extension is automatically installed with QGIS. The software is Open-Source.

uDig (Abera et al., 2014) is a "Spatial Toolbox" that provides toolsets for hydrological, geomorphometric, and topographical analysis. JGrasstools is the library of spatial tools utilized by uDig. It can be used to analyze aspect, slope, gradient, curvature, drain directions, flow directions, hill shade, insolation, basin shape, etc., in hydro-geomorphology, DEM manipulation, hillslope attributes quantification, etc. It runs on its own GUI and is Open-Source.

MicroDEM (Guth and Guth, 2006) is capable of computing regional geomorphic parameters and terrain organizations and interprets DEMs in both geometric and rectangular grids and is also capable of reading other types of DEMs and can create statistical graphs for the DEMs. It can create parameter maps representing regional statistics. It can be used for land surface parameters such as slope, aspect, curvature, etc. It runs within its own GUI. It is a microcomputer mapping program that is capable of running only in Windows. The software is Open-Source for use.

## 4.4 Tools that Run within Coding Bases

LandLab is used to build numerical landscape models (Barnhart et al., 2020; Hobley et al., 2017). It is a tool to provide a numerical modeling environment to quantify earth surface dynamics, based on Python language. Landlab is capable of modeling earth surface processes related to geomorphology, glaciology, hydrology, stratigraphy, and other related fields. Robust and reusable components of Landlab can model flows such as glacial ice, volcanic material, landslide debris, water, sediment, etc. Landlab v2.0 contains modularized elements of numerical models where each module represents a single physical process, thereby reducing the necessity to create or reimplement models and contains a suite of utilities that support the numerical methods, input/output, and visualizations. Because of the modular nature of Landlab, it can be used for model comparisons and analysis of sensitivity and epistemic uncertainty. The tripartite structure of Landlab, consisting of a core grid module, a library of process components, and a set of supporting utilities, supports the subdivision of the code into Python modules which can be used within a Python environment independently. Description of processes incorporated in Landlab v1.0 can be obtained from Hobley et al., (2017) while those for Landlab v2.0 are described in Barnhart et al. (2020). Grids in Landlab are represented as Python objects, including methods and functions to process geometric and topological parameters. Elements of the grids consist of nodes, links, cells, corners, faces, and patches. It is Open-Source and distributed as source code and pre-packaged binaries for Windows, Mac, and Linux.

Badlands is a parallel TIN-based landscape evolution model built to simulate topography development in various spaces and times (Salles, 2015; Salles et al., 2018a, 2018b; Salles and Hardiman, 2016). It is an Open-Source python-based code. The coding base has a python front-end which provides a flexible interface and increases the portability of the software. The morphometric parameters used are Transport capacity law, stream-power law, linear and non-linear diffusion, etc. Processes incorporated in the tool are sediment transport and deposition, fluvial incision, hillslope processes, climate change effects, etc. It is an Open-Source software package.

LSDTopoTools is a software package for analysing topography (Mudd et al., 2015). It is an Open-Source software package tool for reproducible topographic analysis on large datasets. LSDTopoTools can be used in geomorphological, hydrological, and ecological applications. It also comes with the facility of providing landscape modeling. The morphometric parameters used are channel profile, hilltop flow, and rock exposure. The processes incorporated in the tool are channel network extraction, channel steepness extraction, hilltop flow routing algorithm to connect hilltops to channel networks, and mapping of rock exposure. It is an Open-Source software package.

Fastscape is a flexible and modular landscape evolution model that is highly connected to the Python scientific ecosystem (Bovy et al., 2020; Braun and Willett, 2013; Cordonnier et al., 2019; Yuan et al., 2019a, 2019b). It is a set of efficient algorithms to simulate various erosion, transport, and deposition processes and represents a set/stack of software components written in Python. The morphometric parameters used are erosion, transport, and deposition. The essential processes used are flow-routing, channel erosion, depression resolving, and hillslope diffusion. FastScape uses a modular interface to make the models and algorithms accessible to a more significant number of users. It can be used as a component of other software as well. Such components can be organized in a stack, giving each one specific scope of function. It is Open-Source for use.

ANUDEM is a program that calculates regular grid digital elevation models (DEMs) with sensible shape and drainage structures from arbitrarily large topographic datasets (Hutchinson et al., 2011). The software can arbitrarily process different input files having different arbitrary sizes. Some of the morphometric parameters used in the software are sink point, point elevation, contour line, streamline, etc. It is Proprietary for use.

TAPES (Terrain Analysis Programs for Environmental Sciences) and TAPES-G (Grid version) are used for terrain analysis. This program computes slope, aspect, upslope contributing area, profile and plan curvature, and other topographic attributes for each cell of a square-grid digital elevation model (DEM) (Gallant and Wilson, 1996). The significant features of the program calculate topographic attributes for each cell of a DEM. The morphometric parameters used in the program are slope, aspect, profile curvature, plan curvature, upslope contributing area, flow path length, etc. and is Proprietary for use.

## 4.5 Conclusion

The quantification of the Earth's surface features lies at the heart of geomorphometry (Wilson, 2018). The incorporation of geomorphometry into GIS for the analysis of remote-sensing data has led to the development of various geomorphometric tools, whose brief overview has been provided in this chapter. From such an overview, it can be gathered that as the diversity of the discipline of geomorphometry is appreciably comparable to the varied functions incorporated in various tools, some are developed for narrow and specific purposes. In contrast, others can be used for broader aspects of spatial analysis. The use of these tools depends on the choice made by the user, depending on the requirement of the user, and availability of required resources at the user-end. In order to keep pace with the advancement of remote-sensing methods, which is increasing the variety of the available data, the future needs a constant improvement of computational methods; not only in terms of data but also in changing climatic, environmental, and geomorphological conditions, geomorphometric analysis, and the usefulness of morphometric analysis. The morphometric tools will never fade and comprehensive knowledge of these tools and their functionalities would be helpful for all earnest users of these tools.

## References

Aravind, A., Rahul, D., Jayangondaperumal, R. et al. (2022). Non-emergent thrust front along the northwest Jammu Sub-Himalaya: evidence from channel profile analysis across the Surin Mastgarh Anticline. *Himal. Geol.* 43: 303–318.

Baillie, I.C. and Norbu, C. (2004). Climate and other factors in the development of river and interfluve profiles in Bhutan, Eastern Himalayas. *J. Asian Earth Sci.* 22: 539–553. https://doi.org/10.1016/S1367-9120(03)00092-0

Barnhart, K.R., Hutton, E.W.H., Tucker, G.E. et al. (2020). Short communication: Landlab v2.0: a software package for earth surface dynamics. *Earth Surf. Dyn.* 8: 379–397. https://doi.org/10.5194/esurf-8-379-2020

Bishop, M.P., Young, B.W., and Huo, D. (2018). Geomorphometry: quantitative land-surface analysis and modeling. In: *Reference Module in Earth Systems and Environmental Sciences. Treatise of Geomorphology* (ed. M.P. Bishop), 162–186. Elsevier Inc. https://doi.org/10.1016/b978-0-12-409548-9.11469-1

Bovy, B., Braun, J., Cordonnier, G. et al. (2020). The FastScape software stack: reusable tools for landscape evolution modelling. EGU General Assembly 2020, Online, 4–8 May 2020, EGU2020-9474. https://doi.org/10.5194/egusphere-egu2020-9474, 2020

Braun, J. and Willett, S.D. (2013). A very efficient O(n), implicit and parallel method to solve the stream power equation governing fluvial incision and landscape evolution. *Geomorphology* 180–181:170–179. https://doi.org/10.1016/j.geomorph.2012.10.008

Campforts, B., Schwanghart, W., and Govers, G. (2017). Accurate simulation of transient landscape evolution by eliminating numerical diffusion: the TTLEM 1.0 model. *Earth Surf. Dyn.* 5: 47–66. https://doi.org/10.5194/esurf-5-47-2017

Cao, W., Sofia, G., and Tarolli, P. (2020). Geomorphometric characterisation of natural and anthropogenic land covers. *Prog. Earth Planet. Sci.* 7: 2–17. https://doi.org/10.1186/s40645-019-0314-x

Cavazzi, S., Corstanje, R., Mayr, T. et al. (2013). Are fine resolution digital elevation models always the best choice in digital soil mapping? *Geoderma* 195–196: 111–121. https://doi.org/10.1016/j.geoderma.2012.11.020

Clark, M.K., Schoenbohm, L.M., Royden, L.H. et al. (2004). Surface uplift, tectonics, and erosion of eastern Tibet from large-scale drainage patterns. *Tectonics* 23: 1–20. https://doi.org/10.1029/2002TC001402

Conrad, O., Bechtel, B., Bock, M. et al. (2015). System for Automated Geoscientific Analyses (SAGA) v. 2.1.4. *Geosci. Model Dev.* 8: 1991–2007. https://doi.org/10.5194/gmd-8-1991-2015

Cordonnier, G., Bovy, B., and Braun, J. (2019). A versatile, linear complexity algorithm for flow routing in topographies with depressions. *Earth Surf. Dyn.* 7: 549–562. https://doi.org/10.5194/esurf-7-549-2019

Devrani, R., Singh, V., Mudd, S.M. et al. (2015). Prediction of flash flood hazard impact from Himalayan river profiles. *Geophys. Res. Lett.* 42: 5888–5894. https://doi.org/10.1002/2015GL063784

Evans, J., Oakleaf, J., Cushman, S. et al. (2010). Geomorphometric and gradient metrics toolbox. ESRI ArcGIS 7.

Evans, J.S., Oakleaf, J., Cushman, S. (2016). Geomorphometry & Gradient Metrics Toolbox [WWW Document]. URL https://www.arcgis.com/home/item.html?id=63ffcecf3b2a45bf99a84cdaedefaccf

Ferrando, A. (2017). RIVERtools: an IDE for RuntIme VERification of MASs, and beyond. *CEUR Workshop Proceedings* 2056: 13–26.

Forte, A.M. and Whipple, K.X. (2019). Short communication: the Topographic Analysis Kit (TAK) for TopoToolbox. *Earth Surf. Dyn.* 7: 87–95. https://doi.org/10.5194/esurf-7-87-2019

Franklin, S.E. (2020). Interpretation and use of geomorphometry in remote sensing: a guide and review of integrated applications. *Int. J. Remote Sens.* 41: 7700–7733. https://doi.org/10.1080/01431161.2020.1792577

Gallant, J.C. and Wilson, J.P. (1996). TAPES-G: a grid-based terrain analysis program for the environmental sciences. *Comput. Geosci.* 22: 713–722. https://doi.org/10.1016/0098-3004(96)00002-7

Guth, P.L. and Guth, S.C. (2006). MICRODEM terrain grid computing: global SRTM geomorphometry. *Computing* 1–6. https://www.usna.edu/Users/oceano/pguth/website/research/srtm_grid/morphometry.htm.

Hendrikse, J. (2000). Geostatistics in ILWIS. *Int. Arch. Photogram.* 33: 365–375.

Hengl, T., Maathuis, B.H.P., and Wang, L. (2009). Geomorphometry in ILWIS. *Dev. Soil Sci.* 33: 309–331. https://doi.org/10.1016/S0166-2481(08)00013-5

Hobley, D.E.J., Adams, J.M., Siddhartha Nudurupati, S. et al. (2017). Creative computing with Landlab: an open-source toolkit for building, coupling, and exploring two-dimensional numerical models of Earth-surface dynamics. *Earth Surf. Dyn.* 5: 21–46. https://doi.org/10.5194/esurf-5-21-2017

Horton, R.E. (1932). Drainage-basin characteristics. *Eos, Trans. Am. Geophys. Union* 13: 350–361. https://doi.org/10.1029/TR013i001p00350

Hutchinson, M., Xu, T., and Stein, J. (2011). Recent progress in the ANUDEM elevation gridding procedure. *Geomorphometry* 2011: 19–22.

Jaiswara, N.K., Kotluri, S.K., Pandey, A.K. et al. (2019). Transient basin as indicator of tectonic expressions in bedrock landscape: approach based on MATLAB geomorphic tool (Transient-profiler). *Geomorphology* 346: 106853. https://doi.org/10.1016/j.geomorph.2019.106853

Jamieson, R.A., Beaumont, C., Medvedev, S. et al. (2004). Crustal channel flows. 2: Numerical models with implications for metamorphism in the Himalayan-Tibetan orogen. *J. Geophys. Res. Solid Earth* 109: 1–24. https://doi.org/10.1029/2003JB002811

Jara-Muñoz, J., Melnick, D., Pedoja, K. et al. (2019). TerraceM-2: a MATLAB® interface for mapping and modeling marine and lacustrine terraces. *Front. Earth Sci.* 7: 1–18. https://doi.org/10.3389/feart.2019.00255

Jara-Muñoz, J., Melnick, D., and Strecker, M.R. (2015). TerraceM: a MATLAB® tool to analyse marine and lacustrine terraces using high-resolution topography. *Geosphere* 12: 176–195. https://doi.org/10.1130/GES01208.1

Keller, E.A. and Pinter, N. (1996). *Active Tectonics: Earthquakes, Uplift, and Landscape*. Upper Saddle River, NJ: Prentice Hall.

Lecours, V., Dolan, M.F.J., Micallef, A. et al. (2016). A review of marine geomorphometry, the quantitative study of the seafloor. *Hydrol. Earth Syst. Sci.* 20: 3207–3244. https://doi.org/10.5194/hess-20-3207-2016

Lindsay, J.B. (2005). The Terrain Analysis System: a tool for hydro-geomorphic applications. *Hydrol. Process.* 19: 1123–1130. https://doi.org/10.1002/hyp.5818

Massaad, J.S., Barbar, A.M., and Ismail, A. (2017). LandSurf: a smart tool for evaluating properties and lands. In: *2017 Sensors Networks Smart and Emerging Technologies*, 1–4. (SENSET) (IEEE). doi: 10.1109/SENSET.2017.8125038.

Mudd, S.M., Grieve, S.W.D., Milodows, D.T. et al. (2015). LSDTopoTools Open source geomorphology. In: *BSG Annual Meeting 2015*, London (GIT HUB), 1. Available at: https://lsdtopotools.github.io/

Neteler, M., Bowman, M.H., Landa, M. et al. (2012). GRASS GIS: a multi-purpose open source GIS. *Environ. Model. Softw.* 31: 124–130. https://doi.org/10.1016/j.envsoft.2011.11.014

Peckham, S.D. (2009). Geomorphometry in rivertools. *Dev. Soil Sci.* 33: 411–430. https://doi.org/10.1016/S0166-2481(08)00018-4

Pundir, S., Adlakha, V., Devrani, R. et al. (2021). Tectonic control over shallow crustal exhumation across the India–Asia Convergent Margin. *Tectonics* 40: 1–34. https://doi.org/10.1029/2021TC006722

Rigol-Sanchez, J.P., Stuart, N., and Pulido-Bosch, A. (2015). ArcGeomorphometry: a toolbox for geomorphometric characterisation of DEMs in the ArcGIS environment. *Comput. Geosci.* 85: 155–163. https://doi.org/10.1016/j.cageo.2015.09.020

Salles, T. (2015). Badlands: a parallel basin and landscape dynamics model. *SoftwareX* 5: 195–202. https://doi.org/10.1016/j.softx.2016.08.005

Salles, T., Ding, X., and Brocard, G. (2018a). pyBadlands: a framework to simulate sediment transport, landscape dynamics and basin stratigraphic evolution through space and time. *PLoS One* 13: 1–24. https://doi.org/10.1371/journal.pone.0195557

Salles, T. and Hardiman, L. (2016). Badlands: an open-source, flexible and parallel framework to study landscape dynamics. *Comput. Geosci.* 91: 77–89. https://doi.org/10.1016/j.cageo.2016.03.011

Salles, T., Pall, J., Webster, J.M. et al. (2018b). Exploring coral reef responses to millennial-scale climatic forcings: insights from the 1-D numerical tool pyReef-Core v1.0. *Geosci. Model Dev.* 11: 2093–2110. https://doi.org/10.5194/gmd-11-2093-2018

Schwanghart, W. and Kuhn, N.J. (2010). TopoToolbox: a set of Matlab functions for topographic analysis. *Environ. Model. Softw.* 25: 770–781. https://doi.org/10.1016/j.envsoft.2009.12.002

Schwanghart, W. and Scherler, D. (2014). Short communication: TopoToolbox 2: MATLAB-based software for topographic analysis and modeling in Earth surface sciences. *Earth Surf. Dyn.* 2: 1–7. https://doi.org/10.5194/esurf-2-1-2014

Shahzad, F. and Gloaguen, R. (2011a). TecDEM: a MATLAB based toolbox for tectonic geomorphology. Part 1: Drainage network preprocessing and stream profile analysis. *Comput. Geosci.* 37: 250–260. https://doi.org/10.1016/j.cageo.2010.06.008

Shahzad, F. and Gloaguen, R. (2011b). TecDEM: a MATLAB based toolbox for tectonic geomorphology. Part 2: Surface dynamics and basin analysis. *Comput. Geosci.* 37: 261–271. https://doi.org/10.1016/j.cageo.2010.06.009

Sofia, G. (2020). Combining geomorphometry, feature extraction techniques and earth-surface processes research: the way forward. *Geomorphology* 355: 107055. https://doi.org/10.1016/j.geomorph.2020.107055

Szypuła, B. (2017). Digital elevation models in geomorphology. In: *Hydro-Geomorphology – Model and Trends*. London: IntechOpen. https://doi.org/10.5772/intechopen.68447[Online].

Thiede, R.C., Bookhagen, B., Arrowsmith, J.R. et al. (2004). Climatic control on rapid exhumation along the Southern Himalayan Front. *Earth Planet. Sci. Lett.* 222: 791–806. https://doi.org/10.1016/j.epsl.2004.03.015

Wang, D., Pei, Y., Hernandez, O. et al. (2017). Compiler technologies for understanding legacy scientific code: a case study on an ACME land module. *Procedia Comput. Sci.* 108: 2418–2422. https://doi.org/10.1016/j.procs.2017.05.264

Wilson, J.P. (2018). Geomorphometry: today and tomorrow. *Peer J Preprints* 6: e27197v1. https://peerj.com/preprints/27197.pdf

Wood, J. (2009). Geomorphometry in LandSerf. *Dev. Soil Sci.* 33: 333–349. https://doi.org/10.1016/S0166-2481(08)

Yuan, X.P., Braun, J., Guerit, L. et al. (2019a). A new efficient method to solve the stream power law model taking into account sediment deposition. *J. Geophys. Res. Earth Surf.* 124: 1346–1365. https://doi.org/10.1029/2018JF004867

Yuan, X.P., Braun, J., Guerit, L. et al. (2019b). Linking continental erosion to marine sediment transport and deposition: a new implicit and O(N) method for inverse analysis. *Earth Planet. Sci. Lett.* 524: 115728. https://doi.org/10.1016/j.epsl.2019.115728

# 5

# Landscape Modeling, Glacier and Ice Sheet Dynamics, and the Three Poles

## A Review of Models, Softwares, and Tools

*Satarupa Mitra[1], Rahul Devrani[2], Manish Pandey[1,3,*], Aman Arora[4], Romulus Costache[5,6], and Saeid Janizadeh[7]*

[1] *University Center for Research & Development (UCRD), Chandigarh University, Mohali 140413, Punjab, India*
[2] *Delhi School of Climate Change & Sustainability, Institution of Eminence, University of Delhi, Delhi 110007, India*
[3] *Department of Civil Engineering, Chandigarh University, Mohali 140413, India*
[4] *Bihar Mausam Seva Kendra, Planning and Development Department, Government of Bihar, Patna 800015, Bihar, India*
[5] *Department of Civil Engineering, Transilvania University of Brasov, 5, Turnului Str, Brasov 500152, Romania*
[6] *Danube Delta National Institute for Research and Development, 165 Babadag Street, Tulcea 820112, Romania*
[7] *Department of Watershed Management Engineering and Sciences, Faculty of Natural Resources and Marine Science, Tarbiat Modares University, Tehran 14115-111, Iran*
* *Corresponding author*

## 5.1 Introduction

A landform can be defined as a physical entity (Schumm and Licthy, 1965) or a geomorphological unit (Evans, 2012) and its definition can be nomenclature from morphological and mathematical perspectives. In the morphological approach, a landform can be defined as a feature on the Earth's surface (Renwick, 1992), produced in a defined system with the action of different geomorphic processes in a time period (Brundsen and Thornes, 1979). In mathematics, a landform can be defined as a derivative based on many functions like structure, process, and time (Davis, 1899), whereas its shape can also be defined by an equation (Torech, 1965). The landform change could be described in terms of the surface geometry variation, where we describe the visual impression of the landform and the surface changes within a time period of a decade to a century (Etzelmuller, 2000). The dynamics of landscape change is often described in terms of parameters like tectonics, climate, hydrology, and erosion where time is restricted for a considerable interval, i.e., geological timescale. Landscape evolution models usually illuminate the long-term processes (Willgoose, 2005). Abstract mathematical relationships between landforms and their governing processes help unravel the understanding of those changes in landforms, both terrestrial and submarine, at different spatiotemporal scales. The evolution of landscape is a subject of intrigue in the geomorphology community as the changing landscape forms can make or break the tranquillity of human settlements, being the cause of an array of geological and meteorological disturbances. Geomorphology studies the characteristics of various landforms and their relationships with the processes that impact their evolution. On coarser spatial and temporal scales, these relationships can be established with relative ease through observations and inferences, whereas on the more acceptable forms of such scales, the relationships are harder to discern (Van Der Beek, 2013). Thus, the solution to such a quandary comes in models, typically identified by three variants of nomenclature: Landscape Evolution Models, Surface Process Models, and Geomorphological Models. A "model" is a reduced representation of reality (Malamud and Baas, 2013), providing the ease of studying a process in a scale or environment that is devoid of specific influences and thus enables testing the effects of various factors on landscape evolution, for the depiction of past landscape formations as well as for prediction of future evolution. Implementation of Landscape Evolution Models can be either laboratory- or field-based, conceptual, statistical, or computer models. Application of the models serves the purpose of representing temporal terrain changes due to factors such as glacial erosion and deposition, sediment transport, fluvial erosion and deposition, debris flow, landslides, surface uplift and subsidence, and regolith production (Malamud and Baas, 2013; Tucker and Hancock, 2010).

In order to acquire an adequate perception of the current scenario of landscape evolution models, it is imperative to have at least a brief understanding of the timeline of the study of landscape evolution, starting from the initial

*Advances in Remote Sensing Technology and the Three Poles,* First Edition. Edited by Manish Pandey, Prem C. Pandey, Yogesh Ray, Aman Arora, Shridhar D. Jawak, and Uma K. Shukla.

dependency on theories to the modern modeling methods. According to Pazzaglia (2003), Louis Agassiz, considered the grandfather of landscape models, while studying the influence of glaciation on landscapes in 1840, contemplated the idea of landform configuration as a function of time, thus laying the foundation for landscape modeling. However, Agassiz's ideas were not implemented by the next generation of geomorphologists. According to Chen et al. (2014), Gilbert (1909) established qualitative and quantitative principles governing all landscapes to develop LEMs. Before the appropriate modeling techniques came into existence, theories on landscape evolution, such as those given by Davis (1899), Penck (1953), and Hack (1960), aided in the study of geomorphology. It was not until the late 1970s that numerical models were developed for simulating geomorphological processes following the advancement of computing techniques. The earliest models simulated landscape evolution using grids of cells, running water through that grid, and changing the elevation of individual cells accordingly depending on accretion or erosion. Advancing through the increase in computing power, the models evolved for the incorporation of more complex physical processes such as fluvial erosion, slope processes, tectonic processes, and the influence of climate and lithology on drainage networks, while further developments led to the use of irregular mesh for a more accurate representation of the terrain and associated processes (Braun and Sambridge, 1997; Coulthard, 2001; Howard, 1994; Tucker et al., 2001; Tucker and Slingerland, 1994).

This chapter aims to provide an investigative review of Landscape Evolution Models and their counterparts in the field of glaciology, Glacier Evolution Models, which are variants of Landscape Evolution Models applied to the interpretation of the evolution of glaciers and glacial landscapes, implementing only those parameters which are considered to be influencing factors for the evolution of landscapes in glacial regions. Following in the footsteps of several reviews on the development and application of Landscape Evolution Models, such as by Chen et al. (2014), Codilean et al. (2006), Martin and Church (2004), Coulthard (2001), Tucker and Hancock (2010), Van Der Beek (2013), and Willgoose (2005), this chapter is dedicated to the assessment of the applicability of the existing models toward the study of the landscape evolution in the three poles, that is, the Arctic, the Antarctic, and the Himalayas, in order to understand the internal working principals of the models for determining their applicability to the glaciological studies of these regions. The chapter proceeds with a descriptive review of various landscape evolution models used until now, finally reaching the goal of summarizing glacier evolution models applied explicitly to the study of glacial regions.

## 5.2 Taxonomy

Taxonomy of landscape evolution models can be categorized into three broad areas, as depicted in Figure 5.1), based on geomorphic processes, the process of modeling, and application areas. This section provides an overview of the models according to their classification criteria. In contrast, the following section provides a brief description of the general characteristics of the models, the physical laws involved in the functioning of those models, and the working principles of the essential processes involved in landscape evolution models.

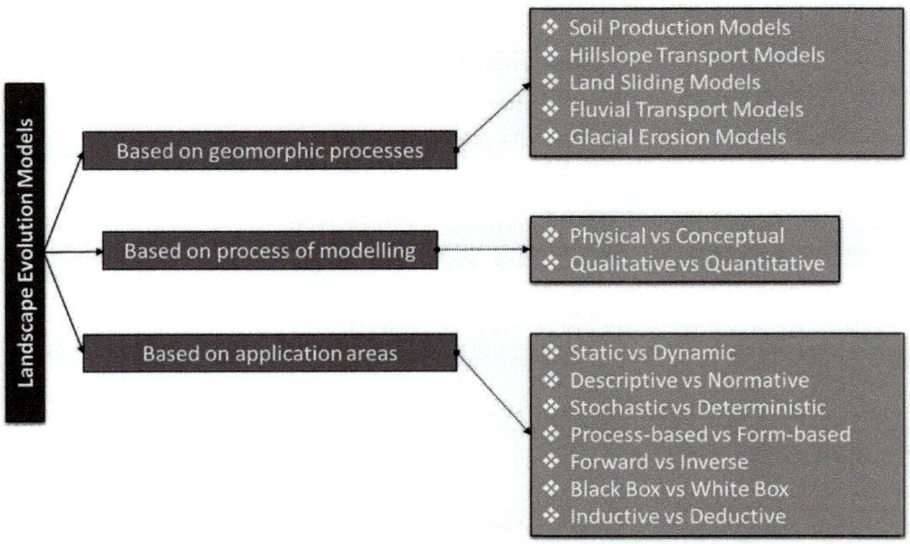


**Figure 5.1** Tree representation of a broad classification of Landscape Evolution Models.

### 5.2.1 Geomorphic Process-Based Models

The first scheme is the scheme of classification of Landscape Evolution Models based on geomorphic processes (Van Der Beek, 2013) given in diagrammatic form in Figure 5.2.

### 5.2.2 Classification Based on Process of Modeling

Next in line of classification of LEMs is that based on the process of modeling (Baas, 2017; Pazzaglia, 2003), depicted in Figure 5.3.

**Qualitative models** can be further sub-categorized into three categories based on geomorphic processes and three categories based on modeling. All such categories are briefly described below.

#### 5.2.2.1 Based on Geomorphic Processes

Qualitative models classified based on geomorphic processes and their brief descriptions are presented in Figure 5.4.

#### 5.2.2.2 Based on Modeling Process

This particular category of quantitative modeling provides the most time-efficient, cost-effective methods of modeling landscape evolution. The constant improvement of such modeling techniques is the beacon of hope for continuous accuracy improvements in the future. Modeling processes-based model characteristics are presented below in schematic form.

**a. Statistical models of observed and simulated data**

such models involve mathematical expressions for process representations, and the results may or may not be displayed as visually representing landscapes and inferences may be drawn from the models based on the results of the quantitative calculations

**b. Computer Simulation Models**

such models display the simulations of landscape evolution processes on the basis of mathematical calculations being employed at the back end of the simulation program

**1) Soil Production Models**

Landscape Evolution models essentially require the inclusion of a soil production term in its quantitative expression as the transportation of soil which shapes landscape characteristics.only follows soil production

**2) Hillslope Transport Models**

Such models describe the movement of sediments along hillslopes with respect to time and is represnted by a diffusion equation governed by a transport law

**3) Land Sliding Models**

Land sliding is one of the requisite processes which influences erosion of hillslopes and us mathematically expressed in landscape evolution models using the implementation of theory if slope stability and landslide probability

**4) Fluvial Transport Models**

Fluvial incision and fluvial transport are the most essential processes which affect landscape evolution and is implemented in LEMs using a stream power law or expressed as transport capacity being function of stream power

**5) Soil Production Models**

Depicts glacial erosion processses that impact the evolution of glacial landscapes and quantitatively incorporates glacial erosion laws and mass conservation equations for ice thickness estimation

Figure 5.2 Classification of Landscape Evolution Models according to geomorphic processes they work for.

**1) Physical Models**

Such models provide scaled replica of the processes involved in landscape evolution, in physical forms, and can be constructed on the basis of expert knowledge without the involvement of any mathematical calculations

**2) Conceptual Models**

Such models represent a process verbally or visually, possibly through pictorial representations and they also do not use any form of mathematical formulation for explaining the processes that produce landscape evolution

**3) Qualitative Models**

Such models do not employ mathematical formulations and represent landscape evolution in terms of factors such as size, shape, and relief of landforms, employing only expert knowledge in the respective aspects

**4) Quantitative Models**

Such models employ numerical quantifications and represents geomorphic processes responsible for landscape evolution in terms of mathematical expressions. As such modelling is based on physical laws along with mathematical laws, therefore correct identification of boundary conditions necessary for the expression of the processes, ensures accuracy of interpretation

Figure 5.3 Classification scheme of LEMs based on process of modeling.

**a. Surrogate Models**

considers a single metric of a landscape, such as mean relief or mean elevation, for evolution modelling

**b. Multi-Process Models**

all the processes that contribute towards landscape evolution of a region, are considered by such models

**c. Coupled Geodynamic Surface Process Models**

such models link multi-process models with geodynamic models

**Figure 5.4** Classification of Qualitative LEMs based on geomorphic processes.

Moving on to the LEM classifications based on application areas (Baas, 2017; Pazzaglia, 2003), these have been described in a dichotomic fashion in Figure 5.5.

## 5.3 Working Principles for Geomorphological Models

Every landscape evolution model, no matter in which category it belongs, follows similar philosophy behind the mathematical quantification of geomorphic processes. Specific physical laws and partial differential equations govern the formulation of geomorphological processes that dominate the LEMs. The two significant goals of landscape evolution modeling are determining which primary variables and which physical laws should be implemented for the landscape evolution under consideration. In this context, it can be noted that landscape altitude and the influence of water are two processes that should be considered as bare necessities for all LEMs (Chen et al., 2014). Therefore, this section provides a brief explanation of the working principles of vital geomorphological processes through a representation of their associated mathematical equations, expressed in a state-variable form, from which the most influential state variable is selected for a particular LEM (Chen et al., 2014; Van Der Beek, 2013):

In general, the continuity equation $\frac{\partial h}{\partial t} = U - \nabla Q_s$ rules the basic level of LEMs, where $(\frac{\partial h}{\partial t})$ is the temporal change in elevation $(\frac{\partial h}{\partial t})$ in dimension [L],$(U)$ is the rock uplift which can be spatiotemporally varying (in $[LT^{-1}]$), and $(Q_s)$ in $[L^3T^{-1}]$ represents the divergence of sediment flux.

### 5.3.1 Soil Production

Production of soil exponentially decreases with soil depth $(H_s)$ and is thus modeled using Equation 5.1.

$$\frac{\partial H_s}{\partial t} = \varepsilon_0 e^{-\alpha H_s} \tag{5.1}$$

where $(\varepsilon_0)$ is the rate of production of soil or rate of erosion at the surface and is expressed in the unit $[LT^{-1}]$, and $(\alpha)$ is a "weathering decay constant" denoted by the unit $[L^{-1}]$.

| | | |
|---|---|---|
| **Static Models** | vs | **Dynamic Models** |
| Represents processes in a state of equilibrium | | Represents processes as a function of changing time |
| **Descriptive Models** | vs | **Normative Models** |
| Describes reality with a simplistic description | | Predicts outcomes based on certain given conditions |
| **Stochastic Models** | vs | **Deterministic Models** |
| Represents the features of a geomorphic system in terms of probabilities and statistics | | Provides precise values quantitatively |
| **Process-based Models** | vs | **Form-based Models** |
| Simulates physical processes that are assumed to be occurring in reality | | Render forms and shapes that are found in geomorphic landscapes |
| **Forward Models** | vs | **Inverse Models** |
| Follow the objective of predicting a final or future state of a geomorphic system | | Determines past conditions such as initial or boundary conditions, based on a current state |
| **Black Box Models** | vs | **White Box Models** |
| Uses "invisible" internal procedures so that they are untraceable. Advanced statistical methods are often involved in such models | | Represents all processes and relationships in transparent and precisely specified ways |
| **Inductive Models** | vs | **Deductive Models** |
| Generalizes and categorizes a system from empirical results | | Simulates consequent forms and processes based on the assumption of a theory or framework |

**Figure 5.5** LEM classifications based on application areas.

### 5.3.2 Hillslope Transport

Short-range hillslope transport processes are depicted by a diffusion equation, which is a combination of a continuity equation and a transport law, which states sediment flux to be dependent on slope gradient in a linear manner. The diffusion equation is Equation 5.2:

$$\frac{\partial h}{\partial t} = -\kappa \nabla^2 h \tag{5.2}$$

where ($\kappa$) is the landscape diffusivity in [$L^2T^{-1}$]. When a hillslope is in a steady state, $\frac{\partial h}{\partial t}$ remains constant from both spatial and temporal aspects, and thus Equation 5.2 is used to predict parabolic hillslopes where the maximum slope gradients and relief linearly calibrate with the erosion rates. However, when it comes to higher slope angles approximately greater than 20 degrees, the hillslopes become more linear, making this equation unsuitable for modeling such slopes. Therefore, to maintain a universality, an empirical and non-linear diffusion law came into existence, which better fits the variability in the morphology of the hillslopes (Equation 5.3):

$$\frac{\partial h}{\partial t} = -\frac{\kappa \nabla^2 h}{1 - \left(\frac{\nabla h}{S_c}\right)^2} \tag{5.3}$$

where ($S_c$) is a threshold hillslope gradient.

### 5.3.3 Land Sliding

Modeling of land-slide occurrence takes into account the variability of the hillslope processes for studies on scales smaller than those considered for Hillslope Transport Models, and Equation 5.4 may be used.

$$h_c = \frac{4c}{\rho g}\left(\frac{sin\beta cos\theta}{1 - cos(\beta - \varnothing)}\right) \tag{5.4}$$

where $h_c$ is the height of the stable state in [L], $c$ is the implicit cohesion in [$ML^{-1}T^{-2}$], $\rho$ is density in [$ML^{-3}$], g is acceleration due to gravity in [$LT^{-2}$], $\beta$ is the topographic slope angle, and $\varnothing$ is the angle of critical friction. For the variance factor, a land-sliding probability $p_l$ is also considered, which depends on mechanical stability and the time difference from the last landslide $\Delta t_{slide}$ (Equation 5.5).

$$p_l = \frac{h}{h_c} + k_0 \frac{\Delta t_{slide}}{dt_{char}} \tag{5.5}$$

where $k_0$ and $dt_{char}$ are constant terms.

### 5.3.4 Fluvial Incision and Transport

Such events are modeled by considering that fluvial incision rates are proportional to either the total stream power, the unit stream power, or shear stress at the base. In this context, a stream-power law is implemented, which represents the formulations of flow velocity, hydraulic geometry, and discharge as functions of the drainage area (A) and stream gradient (S) (Equation 5.6).

$$\frac{\partial h}{\partial t} = KA^m S^n \tag{5.6}$$

where $K$ is a dimensional constant that depicts the incision resistance of the substratum, having the dimension [$L^{(1-2m)}T^{-1}$], and $n$ and $m$ are dimensionless exponents. Several other alternative equations have been explained concerning fluvial incision and transport models (Van Der Beek, 2013).

### 5.3.5 Glacial Erosion

Glacial Erosion considers glacial erosion law, which takes into account erosion due to both quarrying and glacial abrasion, as linearly proportional to the sliding velocity at the base ($u_s$) (Equation 5.7).

$$\frac{\partial h}{\partial t} = K_g u_s \tag{5.7}$$

($u_s = 0$) in case of ice with a cold base, which has been frozen up to the bedrock, and ($u_s != 0$) for temperate and ice with a wet base. $u_s$ depends on basal temperature (based on the assumption of a conductive geo-thermal gradient in the ice), water pressure at the base (proportional to ice overburden), surface slope, and ice thickness ($H_g$), which is calculated using the mass conservation Equation 5.8,

$$\frac{\partial H_g}{\partial t} = \nabla F + M = \nabla\left(H_g W\left[\underline{u_d} + u_s\right]\right) + M \tag{5.8}$$

where $M$ is the mass balance term in [$LT^{-1}$], $W$ is the valley width, $\overline{u_d}$ is the sum of the vertically-averaged velocity of deformation, $u_s$ is the sliding-velocity, and $F$ is the vertically integrated mass-flux [$L^3T^{-1}$] = $H_g * W * \underline{u_d} * u_s$. Equation 5.8 indicates that mass balance and ice dynamics models are required to obtain a glacial erosion model. An ice dynamics model is obtained through calculations using "shallow ice approximations," which assumes that in-plane deformation of ice is negligible.

Besides geomorphological processes, meteorological processes such as wind, rainfall, and snowfall also affect landscape evolution. However, such factors are not directly incorporated into the landscape evolution modeling but instead are implemented into the formulation of climate models. Then they are combined with landscape evolution models to understand the combined effect of such factors. Therefore, the meteorological processes and their effects are not in the scope of this chapter.

Apart from mathematical and physical laws involved in the quantification of geomorphic processes which govern the functioning of landscape evolution models, specific characteristics of the model have to be considered, such as the ones briefly described below (Baas, 2017):

**1) A Physical Basis**

This involves the decision regarding what form or process equations should be selected

**2) Simplicity**

Determining what are the only necessary and sufficient number of processes and parameters for representing the geomorphic systems. Only that many should be used to maintain simplicity

**3) Generality and Richness**

This implies that model formulation should be sufficiently general and versatile to allow application to a variety of areas and situations

**4) Potential for scaling up and down**

The model, being a representation at a smaller scale, should have the ability to be aggregated to represent geomorphology at a larger scale

As opposed to the general description and underlying working principles of landscape evolution models provided in the preceding section, the following sections of this chapter describe the general characteristics encompassing the working principles of Landscape Evolution Models and Glacier Evolution Models (GEMs), including GEMs applied specifically to alpine regions, the Arctic, and the Antarctic, keeping in view their relevance to applications in the three poles, which is one of the objectives of this book.

## 5.4 Landscape Evolution Models

Landscape Evolution Models (LEMs) aid in determining erosion rates and landscape stability over large timescales such as thousands of years (Hancock et al., 2010). In the initial stages, quantitative analysis of digital topographic data required specific domain knowledge and programming skills. However, the development of broad, flexible, Open-Source code bases has caused the spread of the scope of topographic analysis, thus leading to improved modeling (Forte and Whipple, 2019). This section provides an overview of the most widely used LEMs. Some of the earliest LEMs that gained popularity were DEM-based models such as SIBERIA, GOLEM, CASCADE, CHILD, and CAESAR, which provided the ease of user interfaces for the incorporation of multiple geomorphic processes. Later came the popularity of toolboxes and toolkits such as TOPOTOOLBOX and TTLEM. They provide a platform for incorporating many processes to achieve more flexibility in testing a variety of past and future landscape evolution scenarios. Subsequently, some other models, which have demonstrated considerable success, are described briefly. Lastly, this section describes a machine learning-based LEM implementation, which is significant because machine learning is a beacon for future improved prediction accuracy. Mathematical formulations for all the processes applied in each model are based on the mathematical equations described in the previous section. This section and all the subsequent sections' descriptions of the models are listed in a chronological system for better understanding.

Most researchers of LEM have utilized quantitative methods for a single process or various combinations of processes. Therefore, it is prudent to have a platform for landscape evolution modeling that provides such platforms comprising an indefinite number of parameters that can be selectively combined to produce LEMs.

### 5.4.1 DEM-Based Models

In DEM-based models, the landscape can be dynamically adjusted to erosion or deposition, producing a better representation of slope and angle during the simulations (Hancock et al., 2010). The following are brief descriptions of the most popular DEM-based models:

### 5.4.2 SIBERIA

- **References**: Beaumont et al. (1992); Willgoose et al. (1991)
- **Processes Incorporated**: Catchment forms, hydrological processes, tectonic processes, and erosion processes. The linear diffusion equation is incorporated for hillslope transport. The transport-limited stream power equation is applied for the fluvial process. Also, a combination of

these processes is used to represent multiple flow routing, slope weathering, and armoring.

- **Other Features**: Consideration of fixed engineered structure in the study area.
- **User Interface:** SIBERIA uses grids of square-shaped cells for each iteration and provides the averages of erosion rates through annual timescales.

### 5.4.3 GOLEM

- **Reference**: Tucker and Slingerland (1994)
- **Processes Incorporated**: Hillslope soil production is exponentially decreasing with soil depth. The linear diffusion equation is considered for hillslope transport. Both detachment-limited and transport-limited stream power equations are considered for fluvial processes.
- **Other Features**: Coupling other weathering models and sediment production models allows the comparison of arid and humid environments. Incorporating a tectonic model allows the simulation of uplift and understanding of links between tectonics and erosion.
- **User Interface**: GOLEM uses square-shaped grid cells. Water is directed down the line of steepest descent. Works on landscape evolution on a temporal scale of 0.1 million to 10 million years.

### 5.4.4 CASCADE

- **Reference**: Braun and Sambridge (1997)
- **Processes Incorporated**: Basic processes associated with diffusive slopes undergoing simple fluvial erosion and deposition, based on a specific carrying-capacity of the slope.
- **Other Features**: Only model among DEM-based models to incorporate horizontal tectonic movements and the only model that gives real-time graphical output. They are written in Fortran and C programming languages.
- **User Interface**: Mainly used for extensive area simulations with millions of years of temporal scale. It uses an adaptive and irregular mesh grid that is self-adapting and deformable and applies the Triangular Irregular Network (TIN) method to represent erosion and deposition between the natural neighbor nodes of the mesh. Due to adaptability, nodes can be added and moved dynamically during simulations, and the model also allows simulating complicated geometries and easily integrating horizontal tectonic movement, which are difficult to model using a regular mesh.
- **Limitations:** Not Open-Source and can only be used on Linux and Unix systems.
- **Further Modifications**
  - Braun et al. (1999) developed a surface process model with a modified form of the CASCADE algorithm for implementation of glacial erosion, large-scale fluvial processes, and local hillslope processes with the assumption that during glacial erosion, glacial debris is transported by the ice at the ice velocity ($\vec{u}$). The novelty of this model is the combination of fluvial and ice erosions, both spatially and temporally.
  - Tomkin and Braun (2002) developed a numerical surface process model ICE CASCADE based on a spatial discretized version of CASCADE using fluvial processes of large-scale, glacial erosion, and local hillslope processes to find out the effect of alpine glaciation on the characteristics of the relief of a tectonically active orogen. Their results showed the following: that glaciers tend to concentrate the process of erosion near the peaks, fluvial-erosion tends to get reduced in the downstream direction of the glaciers, glaciers with frozen bases tend to increase the relief, and glaciation in the absence of widespread basal freezing tends to reduce relief.
  - Herman and Braun (2006) applied a combination of a 1-D model to obtain the position and height of the drainage divide and the 2D CASCADE model to get the effects of recent glaciations

### 5.4.5 ZScape

- **References**: Densmore et al. (1998); Ellis et al. (1999)
- **Processes Incorporated**: This model defines two distinct sets of processes. One is a tectonic rule set for modeling bedrock displacement, and the other is a geomorphic rule set for modeling the resulting bedrock topography and mass rearrangement on the surface.
- **Other Features**: This considers the 3D aspects of tectonic and surface processes.
- **User Interface**: This is a numerical model which uses a grid-based finite difference method and uses timescales of $10^4$–$10^6$ years.

### 5.4.6 CHILD

- **Reference**: Tucker et al. (2001)
- **Processes Incorporated**: Fluvial processes are elaborately represented by calculating width and depth of a channel within a node, from which a user can derive meandering models, detachment-limited and transport-limited rules of sediment transport, and routines of overbank deposition. It also incorporates a meandering flow model, the Topographic Steering River Meandering (TSRM) model developed by Lancaster (1998), which works on the basis of the continuity of a simplified flow-field going through bends.

- **Other Features**: Written in C++ programming language.
- **User Interface**: CHILD incorporates a simulated environment for exploring consequences of the hypotheses, parameters, and boundary conditions produced by the model. It allows the user to specify different grain sizes and also allows recording the age of the deposits enabling the model in constructing simplified alluvial stratigraphy. Uses Triangular Irregular Network (TIN).

### 5.4.7 CAESAR

- **Reference**: Coulthard (2001) (Developed by Coulthard between 1998 and 2000)
- **Processes Incorporated**: It can model fluvial erosion processes using nine grain sizes. The model can simulate individual events with a greater emphasis on fluvial processes.
- **Other Features**: The model presents channels topography with a divergent flow and does not include tectonic effects.
- **User Interface**: Uses millions of small grid cells (ranging from 1 to 50 m) to represent the landscape. The timescale ranges from 10 to 10,000 years.

### 5.4.8 APERO

- **Reference**: Carretier and Lucazeaut (2005)
- **Features**: The model can investigate how alluvial sedimentations affect catchment responses toward climatic or tectonic changes. It includes different slope processes and river processes competing with each other leading to "transport-limited," "detachment-limited," or "mixed" conditions of transport in the mountains at dynamic equilibrium. The results indicated that adding the alluvial aspect increased long-term mountain denudation.
- **Processes Incorporated**: Different slope processes and river processes that compete with each other in order to investigate the effects of the presence of alluvial sedimentation at range fronts on the catchment responses toward climatic changes or tectonic changes.

### 5.4.9 SIGNUM (Simple Integrated Geomorphological Numerical Model)

- **Reference**: (Refice et al., 2012)
- **Features**: It is a landscape evolution model based on Triangular Irregular Network within the MATLAB environment for simulating the development of topography at various space scales and timescales. It can stimulate hillslope processes, a fluvial incision into bedrocks, spatially varying surface uplift, thrust and faulting, and climate changes.

### 5.4.10 TTLEM (TopoToolbox Landscape Evolution Model) 1.0

- **Reference**: (Campforts et al., 2017)
- **Significant Features**: It is a spatially-explicit, raster-based LEM used to study landscapes which are fluvially eroding, within TopoToolbox2.
- **Processes Incorporated**: Higher-order flux-limiting TVD-FVM (Total Variation Diminishing-Finite Volume Method) is used for solving partial differential equations for river incision and tectonic displacement. A 2D TVD-FVM is able to accurately simulate the evolution of landscapes which are affected by tectonic displacement of a lateral nature.

Below is a brief comparative summary of some of the key features pertaining to a few of the models mentioned above (Coulthard, 2001):

1. CASCADE and GOLEM work better for finer spatial and temporal scales

2. SIBERIA, CAESAR, and, CHILD work better for shorter time scales and smaller spatial scales

3. SIBERIA and GOLEM run a simpler mesh and therefore, their code is simpler. CASCADE and CHILD having irregular mesh, incorporate more complicated codes

4. SIBERIA and GOLEM represent slope processes better and are suitable in modeling the effects of weathering

## 5.5 Other Models

### 5.5.1 DELIM

- **Reference**: (Howard, 1994)
- **Features**: It is a drainage basin simulation model which applies creep and threshold slumping to detachment limited and transport limited fluvial processes

### 5.5.2 EROS

- **Reference**: Crave and Davy (2001)
- **Features**: It is a cellular-automata-based stochastic model used for simulating natural variability of the climatic events using deterministic processes of erosion-sedimentation. The model demonstrated its capability to reproduce complicated geomorphological processes by calculating various geomorphological instabilities that influence the relief dynamics.
- **Processes Incorporated**: The numerical method used in the model calculates the water discharge of the run-off as a function of the return period of basic walking elements and also generates water flux distributions at every point of the system depending on local drainage area and also walker rainfall.

### 5.5.3 Landscape Evolution Model Using Global Search

- **Reference**: (Paik, 2012)
- **Features**: Incorporates a Global D8 or GD8, where D8 refers to deterministic eight-neighbor flow direction retrieval, generating angular uncertainty in determining flow paths. GD8 reduces the uncertainty associated with flow modeling at the local scale. According to the authors, the choice of flow path extraction method is a significant issue in numerical simulation of fluvial landscape evolution. LEGS is the first LEM to implement GD8.

### 5.5.4 eSCAPE

- **Reference**: (Salles, 2019)
- **Features**: A Python-based LEM that can be implemented at both regional and global scales and is Open-Source and cross-platform, is distributed under the GPLv3 license, and is available on GitHub. Processes simulated by this model depend on a simple mathematical representation of the landscape processes to compute the Earth's surface evolution. The model implements a series of iterative algorithms defined in a matrix form and uses the PETSc parallel library for solving the matrix equations. According to the author, eSCAPE is the first LEM to have the ability to solve problems on large and unstructured grids consisting of millions of nodes, thus being able to simulate processes at a global scale.

### 5.5.5 r.sim.terrain 1.0

- **Reference**: (Harmon et al., 2019)
- **Features**: This is a free, Open-Source GIS-based, topographic LEM with dynamic hydrology that can be used for simulating short-term changes in topography for both steady-state as well as dynamic flow regimes across various spatial scales. It incorporates empirical models for the erosion of soil and also a physics-based model for shallow superficial water flow and soil erosion for computing short-term changes in topography.

## 5.6 Combined/Application-Specific Models

In addition to the commonly used models, some researchers have developed Landscape Evolution Models to keep in view particular application aspects or observe the results of taking integrated approaches with more than one model. Some such studies have been described below chronologically.

Garcia-Castellanos et al. (1997) devised a numerical model, developed in C language using the finite-difference method, for the formation and evolution of foreland basins. Temme et al. (2006) devised an algorithm where LEMs are used to deal with sinks, representing depressions in landscapes in DEMs, as the authors noted that for models simulating soil redistribution, realistically dealing with depressions is a problem in every time step. They also tested an application running the LAPSUS model with the new algorithm to compare with conventional runs. Coulthard et al. (2013) developed an integrated application of the CAESAR model and the LISFLOOD-FP, a 2D hydrodynamic model. The authors' objectives for such integration are the expected exclusion of shorter-term hydrodynamic effects, like a flood wave, into LEMs, as models for such non-steady flow require incorporating shorter time steps, such as in seconds and also require more expensive numerical solutions. LEMs are usually developed with steady flow models with longer time steps in several years. The authors inform that LISFLOOD-FP uses a simplified 2D flow model and the CAESAR-LISFLOOD coupled model proved to be faster and more computationally efficient.

## 5.7 Machine Learning Models

Mokarram et al. (2022) constructed an approach called genetic landscape evolution (GLE), for modeling drainage on the basis of the states of optimal-waterway-network for predicting the drainage morphometry parameters and hydrological processes. GLE aims to achieve results such that the system expends minimum total energy.

## 5.8 LEMs Developed for Glaciated Landscapes

The study of glaciers is significant due to their influence on the global climate system, such as the rise of sea level and availability of seasonal water, while also acting as sources of geohazards (Maussion et al., 2019). How have glaciers affected geomorphology and meteorology in the past? Moreover, how they will affect them in the future is a vital matter of speculation. Apart from the standard processes implemented in the construction of LEMs, Glacier Evolution Models (GEM) require the consideration of several other factors that affect and are affected by glacier evolution. According to Oerlemans (1997), hypsometry and mean slope are the most important factors determining a glacier's response to climate change. Unlike LEMs, GEMs are also categorized into physical-based snow models, numerical glacier models, LEMs modified to study glaciated and glacierized terrains, and machine learning models.

In this regard, it is relevant to mention some categorization aspects mentioned by some researchers: Bokhorst et al. (2016) gave three main categories of physically-based snow models:

1) Zero-layer (comprising the soil layer) or single-layer models
2) Intermediately complex models (including physical processes within snow pack)
3) Detailed snow-pack models

Categorization of numerical glacier models according to dimensions:

1) 1D models such as flowline models, where the grid is only one cell wide and include factors such as shape and lateral drag for simulating glacier velocities, such as applied by Nick et al. (2010).
2) 2D models, such as the one used by (Plummer and Phillips, 2003), are preferred for understanding the influence of topography and include both flowline models and plan-view models.
3) 3D models have spatially distributed models for calculating the ice flow pertaining to various depths within the ice column.

As mentioned in the previous sections, some successful GEMs have been listed chronologically to understand their constraints and constrictions.

Braithwaite and Zhang (1999) devised a model for understanding changes in glacier mass balance. Their model used precipitation and temperature for calculating snow and ice melting, snow accumulation, and probable refreezing of meltwater. They considered regular intervals of altitude for the modeling process. The model was applied to 37 glaciers worldwide, and it was inferred that sub-polar glaciers have lower sensitivities toward temperature, while maritime glaciers and tropical glaciers have higher sensitivities toward temperature. Pazzaglia (2003) stated that by comparing some studies on the landscape evolution of glacial regions, it can be inferred that freezing at the bed of a glacier is an important geomorphic process. According to the author, for running models, relief in the landscape of the model increases in the case when glaciers are assumed to be frozen to their bed and otherwise, if the relief is lower in the region, erosion concentrates on the interfluves. Herman and Braun (2008) devised a model where a coarse grid for calculating ice thickness and velocity and a finer grid for calculating evolution of surface topography, were interpolated. This also included an empirical constriction term for simulating reduction in glacier velocities due to friction with the sidewalls of the valley. Jamieson et al. (2008) used a 3D ice sheet model GLIMMER to model landscape evolution under ice sheets. This model is fully-coupled and thermomechanical. GLIMMER contains an erosion component for examining evolution of landscapes underneath ice sheets implementing long timescales. The model can predict ice thickness, basal velocities, and the distribution of areas at pressure melting points. GLIMMER also applies shallow-ice approximation, assuming that normal stress components on topographic and ice surface slopes are ignorable. Putkonen et al. (2008) applied a model for moraine degradation using a proportionality constant and local slope for determining the rate of transport. The authors worked on the concept of glaciers' spatial extent of the characteristic signatures left behind. The characteristics considered in their model were measured for Antarctic, tropical, and alpine moraines, and the agreement between their model predictions and the data collected from the field were quite strong. Hirabayashi et al. (2010) developed a global glacier model named HYOGA, which can compute glacier mass balance considerably. Using HYOGA, the authors developed a 59 years long, that is, from 1948 to 2006 time series of glacier mass balance and glacier area of global scale. Their inference was that most glaciers had lost mass during their chosen study period. MacDougall et al. (2011) assessed parameter sensitivity and transferability of a glacier melt model in a dry subarctic environment. The authors investigated the effect of model tuning with different statistics, and finally, using mass balance data, the parameter transferability of the model was examined by considering two melt seasons and two neighboring glaciers. Egholm et al. (2012) devised a computational framework of a coupled nature, that simulated simultaneous ice flow, water flow, and sediment flow in a glacial landscape evolution model. They inferred that their model had the strength of production of morphological details of the landscape, facilitating direct comparisons with existing landforms and knowledge of the physical processes operating under glaciers. Van Der Beek (2013) tells us that glacial erosion processes were rarely included in landscape-evolution models, even though several 1D models have been developed in the past for representing the evolution of glacial valleys, such as by Harbor (1992), MacGregor et al. (2000), and Oerlemans et al. (1998). The CASCADE model, developed by Braun and Sambridge (1997), eventually modified by Tomkin and Braun (2002), and then Herman and Braun (2008), uses a simple glacial erosion law. Van Der Beek (2013) also infers that glacial-erosion models require both mass-balance models as well as ice-dynamics models. Rawlins et al. (2013) presented an improved pan-Arctic water balance model to simulate soil freeze dynamics and thaw dynamics. With the model, they have evaluated simulated soil temperature and active layer thickness. The new model simulations indicated that there was no systematic soil temperature bias. The authors also inferred that simulated temperatures matched well with summer observations, but the results were mixed in winter, containing both positive and negative biases. Gabbi et al. (2014) compared empirical-based and physically-based glacier surface-melt models for obtaining long-term simulations of response of glaciers. They tested the performance of five empirical glacier-melt models over a multi-decade period. The melt models were coupled with an accumulation model and a model for surface evolution and the obtained seasonal mass-balance measurements were used for calibration. The authors inferred that the melt relationships obtained by them were robust and suitable for long-term modeling and at the same time, more empirical approaches are oversensitive to temperature fluctuations. Huss and Hock (2015) presented a new Global Glacier Evolution Model (GloGEM) for glacial change and sea-level rise detection, considering the challenge of global-scale modeling of roughly 200,000 glaciers on Earth. GloGEM can calculate the mass changes of all glaciers on Earth except the ice-sheet glaciers and can also calculate mass loss caused by frontal-ablations in the case of marine-terminating glacier fronts. Ugelvig et al. (2016) modeled glacial landscape evolution by constructing an erosion law for landscape evolution simulation considering the factor of subglacial quarrying. The authors have applied the physics of quarrying to long-term landscape evolution connecting meter-scale cavities to kilometer-scale glacial landscapes. Applying erosion law of glacial quarrying, their results produced significant large-scale glacial landforms. Their glacial landscape evolution model consisted of the following components: Ice Flow Model, Sliding Law, and Hydrological Model. Their inference was that quarrying was more effective in modeling valleys than higher elevations. Hewitt and Schoof (2017) presented two models based on energy conservation to represent the temperature content and water content of polythermal ice sheets and glaciers. The authors inferred that the two models are simple enough to be included within existing ice-sheet models without the addition of significant modifications.

## 5.9 Some Significant Glacier Evolution Models

**Open Global Glacier Model (OGGM) v1.1**, introduced by Maussion et al. (2019), is Open-Source and coded in Python and provides a modular and numerical framework capable of computing past as well as future glacier changes. The modular framework of OGGM enables the addition of modules of varying capacities to the code database. OGGM can simulate several dozen glaciers on a personal computer compared to other global models, which can take a few hours for a temporal scale of 100 years using supercomputing environments. OGGM runs the following modules: Pre-processing, Flow Lines and Catchments, Climate Data and Mass Balance, Ice Thickness, and Ice Dynamics. In the future, new physical processes and calibration tools may be added.

**GlacierMIP**, introduced by Hock et al. (2019), systematically compares global-scale glacier model results to facilitate model improvements and reduce uncertainties in global glacier projections. The authors have investigated output from six glacier models between 2015 and 2100 by comparing the predicted glacier mass changes and area changes and corresponding sea-level contributions provided by the models.

**ReSCAL-Snow** is a model which was designed for the quantitative study of snow bedforms, wherein the simulation was adapted from a sand-dune model, ReSCAL (Real-Space Cellular Automaton Laboratory) (Rozier and Narteau, 2014), while also inheriting its granular transport capability. Kochanski et al. (2019) added features to Rescal-Snow to simulate other snow processes and fluid processes. They also explained the limitations of Rescal-Snow, stating that the configuration of the cellular automata can set the natural length and timescales of the model. The model cannot resolve centimeter-scale snow features, such as the corners of sharp-edged features.

**CICE**, initially released in 2013, is a sea-ice component to a atmosphere-land global climate model which is compatible with the ocean circulation model called Parallel Ocean Program (POP). Components of CICE include a thermodynamic model, an ice dynamics model, and a transport model. CICE was built to simulate polar sea ice's growth, melting, and movement (Hunke et al., 2015).

**IceFlow** was initially written by Willliam Colgan in Matlab and later translated into Python by Andrew Wickert in 2014 to be used as an automated interface in GRASS GIS software for retrieving input data (Andy Wickert, 2014).

**SedBerg** is a 2D model developed to model the formation, drift, and melt of icebergs by simulating sedimentation in high-latitude glaciated fjords and uses Monte Carlo-based techniques and thus describes the stochastic behavior of iceberg formation and dynamics (Mugford and Dowdeswell, 2010).

**SICOPOLIS** (SImulation COde for POLythermal Ice Sheets) is a 3D dynamic/thermodynamic model for simulating the evolution of large ice sheets and ice caps and was first developed in 1993 with subsequent modifications applied over the years (Greve, 2007). SICOPOLIS has been applied to various problems such as on the past, present, and future glaciation of the Northern Hemisphere, Greenland, Antarctica, and polar ice caps of planet Mars (Ahlkrona, 2011; Greve, 2020; Greve and Blatter, 2016; Greve et al., 2021; Logan et al., 2020; Rückamp et al., 2019).

**Elmer/Ice** is an ice sheet model that uses the Open-Source multi-physics code Elmer and was originally developed for solving local ice-flow problems. Resulting from almost ten years of development, Elmer/Ice gained the capability to solve larger-scale problems and is being used for making projections of the evolution of the entire Greenland Ice Sheet continuing for two centuries into the future (Gagliardini et al., 2013). Favier et al. (2019) applied a coupled framework combining NEMO (Nucleus for European Modeling of the Ocean) and Elmer/Ice.

**SNOWPACK** is numerical snow and land-surface model introduced by Lehning et al. (1999). In this, snow is modeled as a porous medium having three phases (i.e., ice, air, and water), and the volumetric contents of each of the phases characterize the model. The model includes a detailed description of mass exchange and energy exchange between the atmosphere, snow, soil, and vegetation cover. Its applications have covered the following areas: avalanche forecasting, climate change assessment, superimposed ice simulations, permafrost sensitivity studies, and simulation of snow storage.

**Glimmer-CISM** is an Open-Source, 3D, thermomechanical ice sheet model written in Fortran-95, which can run in standalone mode and also be interfaced to global climate models (Rutt et al., 2010).

**CISM** (Community Ice Sheet Model) is a parallel 3D thermomechanical model and is written with an Open-Source code. It solves the equations for momentum balance, temperature evolution, and thickness of ice sheets. This model has been subtantiated for standard test problems, such as experiments with the Ice Sheet Model Intercomparison Project for Higher Order Models (ISMIP-HOM). CISM was also included in the initMIP-Greenland initialization experiment (Lipscomb et al., 2019).

**Icepack**, introduced by (Shapero et al., 2021), is a Python package used for flow modeling of glaciers and ice sheets. The main goal of icepack is to produce tools that can be helpful for both experts and novices. It uses 2D shallow ice and shallow stream models and a 3D hybrid model.

**VarGlaS** (Variational Glacier Simulator), introduced by Brinkerhoff and Johnson (2013), is a higher-order finite-element ice-sheet model. It is built on a finite element framework called FEniCS. Unlike other ice-sheet models, it solves enthalpy equations rather than temperature. Brinkerhoff and Johnson (2013) suggested that VarGLaS predicted a thinning effect within the interior and thickening effect at the margins of the ice sheet.

**Gc2d** is a finite-difference numerical model used to simulate the formation and evolution of valley glaciers in temperate zones and can simulate glacial evolution over millennial timescales. It works on a 2D topographic surface with a pre-specified meteorological setting. Gc2d works by solving equations for ice flux and mass conservation and explicitly integrates glacier and climate simulation components, which gives it the ability to simulate feedback between changing ice surface and climate forcings. The computational simplicity of the model allows the user to quickly determine the effect of new processes (Kessler et al., 2006).

**ISSM** (Ice Sheet System Model), initially developed in 2005, is a thermo-mechanical finite element model for ice-sheet flow simulation at the continental scale, including higher-order stresses and better data assimilation techniques (Larour et al., 2012).

**CryoGrid3** was designed for applications on permafrost and comprises a flexible platform for exploring novel parameterizations applied for a range of permafrost processes (Westermann et al., 2016).

**CVPM** (Control Volume Permafrost Model) is a non-linear heat transfer model used to predict permafrost temperature with depth. CVPM's domain is highly flexible and can be set for various dimensions and temporal and spatial scales. The numerical implementation of CVPM is based on the control volume method.

**ThawLake1D** is used to model the conduction of heat through permafrost systems of lakes, by evaluating the temperature with depth, by calculating the conduction of heat through water based on the incoming solar-radiation. It calculates the thermal conductivity and the specific heat capacity (Matell et al., 2013).

**GRANTISM** (**GReenland & ANTarctic Ice Sheet Model**), developed by Pattyn (2006), is primarily an educational model, using physics similar to scientific ice-sheet models but with a much faster computation time. The model runs fast because of the use of an over-implicit numerical method. The model is classroom suitable as it uses a user-friendly interface and implements the Microsoft ExcelSheet program within itself.

**IcIES** (Ice sheet model for Integrated Earth System Studies) is a large-scale, 3D dynamic/thermodynamic ice-sheet model with shallow ice approximation dynamics (Greve et al., 2011).

**SAFRAN-CROCUS-MEPRA (SCM chain)**, being composed of three separate models, SAFRAN, CROCUS, and, MEPRA, was developed at Meteo-France, Centre d'Etudes de la Neige. SAFRAN is a meteorological analysis model used to compute meteorological variables. CROCUS is a physically-based model used for simulating the evolution of a snow profile, taking as input the weather conditions that SAFRAN calculates. The novelty of CROCUS is in its ability to simulate snow metamorphism in near-surface and buried layers. MEPRA is an expert system that adds new information such as mechanical characteristics to the profile simulated by CROCUS to analyse the mechanical stability of the profile and finally provides a diagnostic risk assessment (Coléou et al., 2009).

**FISOC** (Framework for Ice Sheet – Ocean Coupling) was developed by Gladstone et al. (2021) for performing experiments with coupled ice sheet–ocean modeling. The Regional Ocean Modeling System (ROMS) was coupled with Elmer/Ice for the Marine Ice Sheet-Ocean Model Intercomparison Project (MISOMIP).

**Kudryavtsev Model** uses an approximate analytical solution of the Stefan freeze-thaw problem for calculating the maximum annual thawing depth for a given vertical profile and the mean annual permafrost temperature at the bottom of the thaw layer at a steady state. The model contains snow insulation, vegetation effect, permafrost temperature, and thaw depth modules (Wang et al., 2020).

**Potsdam Parallel Ice Sheet Model (PISM)** was introduced by Winkelmann et al. (2011) to simulate large-scale ice-sheet shelf systems. Calculation of ice-sheet velocities were done by superimposing two shallow stress-balance approximations and ice-shelf margins and were modeled using Neumann Boundary Conditions. Martin et al. (2011) conducted a dynamic-equilibrium simulation of Antarctica for present-day conditions using PISM, and the results showed that different flow regimes in sheets and shelves as well as the zone of transition between them, were well-captured. It supported the superposition approach of Shallow Ice Approximation (SIA) and Shallow Shelf Approximation (SSA) for representing the fast motion of grounded ice. Albrecht et al. (2020a) evaluated PISM's sensitivity to different parameter choices and boundary conditions and provided estimates of various uncertainties. Albrecht et al. (2020b) presented the analysis of the ensembled use of the four parameters of PISM and their effects as a continuation of the previous study. They ran an ensemble of 256 models using four model parameters by varying them using the complete factorial parameter sampling method.

**RIMBAY** is a 3D, thermo-mechanical, ice-dynamics model which uses multi-approximation for comprehensive applications (Thoma et al., 2014).

**MALI** (MPAS-Albany Land Ice) is an ice-sheet model of variable resolution which uses amorphous Voronoi grids represented on a plane or sphere. It is fabricated using the Model for Prediction Across Scales (MPAS) framework and the Albany multi-Physics code base for solving partial differential equations of coupled systems. MALI is the glacier component of the Energy Exascale Earth System Model (E3SM) (Hoffman et al., 2018).

**BISICLES** is an ice-sheet model containing kilometer-scale refinement in resolution and also refinement in higher-order ice-flow physics. The model uses inferences from previous models on deglaciation for initializing the ice-sheet temperature and for recent reconstructions of the ice sheet, for developing the topography of the region and the ice sheet (Matero et al., 2020).

**WAVI** is a model representing vertical shearing within the ice and membrane stresses caused by horizontal stretching and shearing (Arthern et al., 2015).

**GlabTop** (Glacier Bed Topography) is used for modeling the distribution of the thickness and the topography of the bed for large samples of glacier using a GIS. The model is constructed on the basis of an empirical relationship between average basal shear stress and a range of elevations of individual glaciers (Linsbauer et al., 2012).

**PICO** (Potsdam Ice-shelf Cavity mOdel) is based on an ocean-box model that is used for resolving ice-shelf cavities, using a boundary layer melt formation. Also, PICO is implemented as a module of the PISM model (Reese et al., 2018).

**PSU/UofC** is a 2D, finite-element thermo-mechanical flowline model for ice-sheet evolution and is used to complement 3D models by allowing thorough investigation of the parameter space (Parizek et al., 2005).

**f.ETISH** is a fast, thermo-mechanical and coupled ice-sheet/ice-shelf model which is capable of considering critical and complex processes, such as the instability mechanism of the marine-ice sheet (Bulthuis et al., 2017). Here the authors have used the f.ETISH model for understanding the part played by the uncertainties in sub ice-shelf processes for assessing the contributions to sea-level rise for the coming centuries.

**HYOGA** is a glacier model of global scale, that calculates the temporal variations of mass balance for around 24,000 glaciers worldwide, covering 90% of the total glacierized area of the Earth, except the ones in Antarctica and Greenland (Hirabayashi et al., 2013). HYOGA also enables the simulation of future dissipation and future changes in mass and area of glaciers.

Bolibar et al. (2020) introduced a novel technique of simulating and reconstructing annual glacier-wide SMB series using ANN, a DL method. Unlike most other glacier models, the authors created a parameterized model based on data science, which applies physical processes. Their method used deep learning to simulate yearly glacier-wide

SMBs, using topo-climatic predictors. Comparing and cross-validating their non-linear SMB model with other known linear statistical methods, the authors found that deep learning outperformed linear methods using a dataset of 32 alpine glaciers. They thus inferred that deep learning in glacier studies can achieve higher accuracy in capturing non-linearities of the climate and glacier systems and in the construction or simulation of SMB time series of a glacier.

## 5.10 Models Developed for Alpine Regions

Glaciers are responsible for shaping alpine landscapes by creating glacial signatures such as broadening valley bottoms, enhancing local valley relief, generating multiple steps, over-deepening valley floors, and causing tributary valleys to hang. These signatures result from millions of years of erosion caused by advancing and retreating alpine glaciers (MacGregor et al., 2000). Alpine glaciers influence the hydrological cycles of the mountains and their foothill regions (Callegari et al., 2017) and are highly sensitive to climate fluctuations (Rabatel et al., 2005). The Himalayan mountains and the Tibetan Plateau, referred to as the "world's third pole" (Yao et al., 2020), are the focus of the study of alpine glaciers in this chapter. The following part of the section is a chronological listing and description of successful GEMs applied to alpine regions. The understanding of such applications can be oriented toward the Himalayas, a constituent of the three poles.

Harbor (1992) combined a finite element model with an erosion model for the flow of ice through a glacier cross-section in order to investigate the development process of a U-shaped valley, which is a significant glacial landform. Oerlemans (1997) studied the sensitivity of an alpine glacier using a 3D numerical model for ice flow for calculating mass flux of ice along a central flow line. MacGregor et al. (2000) used a 1D finite-difference staggered-grid model of a temperate valley glacier, and for their model, the use of mass balance, ice motion, and ice discharge is similar to other models of ice flow such as those (Oerlemans, 1984; Waddington and Clarke, 1988). Their numerical model was used to study the development of the longitudinal profiles of glaciated valleys. Huggel et al. (2003) studied a modeling approach that considers the glacial environment's current evolution and provides a first-order assessment of the hazards caused by outbursts of glacier lakes by simulating the debris flow from such outbursts using two GIS models based on topography and applying them to two such events in alpine regions. The multiple-flow-direction approach yielded more details about the propagation, whereas the single-flow-direction approach was better suited for process automation. Hock (2005) reviewed various glacier melt processes of alpine glaciers and snow-covered alpine regions and explained two types of glacier melt models: Energy Balance Melt Models and Temperature Index Melt Models. Energy Balance Melt Models consider the following factors: Net Radiation, Global Radiation, Albedo, Longwave Radiation, Turbulent Heat Fluxes, Ice Heat Flux, and Other Heat Fluxes. Pellicciotti et al. (2005) presented a glacier-melt model using a temperature index which incorporated albedo and incoming shortwave radiation, which offered better results than models requiring only temperature data inputs. The model is aimed at lower data requirements and simplicity of computational. Compared to three other temperature index models, their model offered vital improvements. Herman and Braun (2006) studied fluvial response to glaciations choosing an alpine region as a study area, using a 2D LEM. For their model, the authors have considered fundamental equations used in three kinds of models: Geomorphic Models (for Fluvial and Hillslope Erosion), Tectonic Models, and Tectonomorphic Models. Herman and Braun (2008) modeled ice-cap evolution to predict the parts of the landscape affected by glacial erosion, while specifically identifying the complexity of the erosion patterns. Their Erosion Model consists of the following components: Glacial Erosion, Fluvial Erosion, and Hillslope Erosion. Along with this, they employed a Tectonic Model. The authors have also suggested future improvement of the model by including periglacial processes. Oerlemans et al. (2011) used a minimal glacier model for studying the global dynamics of an arctic glacier. The model applies strongly parameterized ice dynamics and a simple law for iceberg calving. Their model predicted the future fate of the Arctic glacier with considerable accuracy. Egholm et al. (2011) modeled glacier-flow in steep terrains, using iSOSIA. which is a depth-integrated second-order shallow-ice approximation. Linsbauer et al. (2013) compared three different methods, having different levels of complexity and different temperature increase scenarios, to model scenarios of the future evolution of alpine glaciers. They noted that the three approaches have similar results for the overall long-term evolution. Li et al. (2015) integrated a glacier-retreat model with a hydrological model, and found that their coupled model was able to provide good estimations of mass balance and runoff and was also able to provide estimate of the runoff components. Egholm et al. (2015) developed landscape evolution models to integrate two important periglacial processes. They also included the factor of subglacial erosion. The authors devised three types of computational experiments using different spatial scales and involving erosion processes of both periglacial and glacial environments. Ramsankaran et al. (2018) used an

Glacier Bed Topography Version 2 (GlabTop2) model, which was parameterized for estimating the thickness distribution and total volume of ice in an alpine glacier. GlabTop2 is an SIA-based spatially distributed approach. They tested their approach to estimating the optimal shape factor of the studied glacier. They inferred that the model combined with the proposed parameterizations had enormous potential for application to data-scarce alpine glaciers for quantifying reliable ice thickness. Réveillet et al. (2018) used a snowpack model called SURFEX/ISBA-Crocus, which is forced with the data from the SAFRAN meteorological reanalysis, as the performance of the model decreased significantly without *in-situ* meteorological measurements. Gao et al. (2018) combined a spatially-distributed hydrological model called FLEX and a glacier-retreat model called Δh-parameterization. Their results showed that the integrated model accurately estimated runoff and reproduced historical variations. Zekollari et al. (2019) modeled the future evolution of glaciers in an alpine region using the mass-balance model, GloGEMflow Model, which is an extended version of the Global Glacier Evolution Model (GloGEM) and is forced with regional climate model (RCM) simulations. Rounce et al. (2020b) developed a calibration method using geodetic mass-balance observations and a Bayesian inverse model for quantifying model parameter uncertainties. Rounce et al. (2020a) calibrated the Open-Source Python Glacier Evolution Model (PyGEM) using Bayesian inference and a new dataset of geodetic mass balance observations, which enabled the recording of spatial variations of mass balance. The Bayesian inference helped in quantifying the uncertainty associated with the parameters of the model.

## 5.11 Models Developed for the Arctic Regio

Snow is a swiftly changing attribute of the Arctic region, and the changes in snow cover and snowpack conditions over time makes the tasks of measurement and prediction more demanding for the purpose of assessing scenarios on the response of snow cover to the evolving Arctic climate for the preparation of strategies for adaptation (Bokhorst et al., 2016). Various models have been developed, specifically for the Arctic region:

Raymond and Harrison (1987) tested the fit of ice-motion models to the evolution of geometry and velocity observations of variegated glaciers in Alaska during their normal flow phase. The authors inferred that the sliding rate directly varies with shear stress and inversely varies with effective everyday stress, but they could not identify any quantitative relationship explaining the spatial pattern and time evolution. Wright et al. (2005) presented a surface-energy/mass-balance model to understand the impact produed by superimposed ice on the mass balance of the arctic glaciers and consider future climate change scenarios. Their model includes calculations for the refreezing of meltwater and the formation of superimposed ice. Their results show that these two factors make a significant contribution to the mass balance. Wadham et al. (2006) combined ground-penetrating radar (GPR) and flow modeling to validate mass-balance measurements and/or infer an Arctic glacier's mass balance/dynamic history. Wright et al. (2007) comparatively assessed different models for refreezing of meltwater superimposed on high arctic glaciers. The authors tested the refreezing algorithms, Surface Energy Balance Model, Englacial Model, Bassford Refreezing Model, and other refreezing approaches for up to the year 2050, considering future climatic scenarios and reached the following conclusions regarding mass-balance modeling. The choice of the meltwater refreezing model is essential, and future predictions of mass balance are independent of the specific meltwater refreezing model used. Bingham et al. (2008) implemented a 3D finite difference flow model to understand seasonal variations in basal sliding distribution at a polythermal Arctic glacier. Their work highlighted the significance of adding basal motion into models for predicting the response of the land ice toward changing climate. Dornes et al. (2008) used a combined modeling approach for evaluating the ablation of snow cover and snow-melt runoff in a sub-Arctic basin. The authors' inference was that the two-model aggregation approach yielded good results but could not correctly represent the evolution of the varying rates of snowmelt. Rye et al. (2010) evaluated and inferred that the European Centre for Medium-Range Forecasts ERA-40 reanalysis performed well with the surface mass-balance model. Pimentel et al. (2017) modeled intra-annual dynamics of a marine-terminating glacier, using a higher-order and hydrologically-coupled flow model and found that sea ice has insignificant influence on the flow speed of glaciers, compared to the drainage of meltwater, which has more influence. Hansen et al. (2020) prepared a 3D model of englacial and subglacial drainage systems, using speleological surveys and exploration of a meltwater conduit with the help of a high-resolution Ground Penetrating Radar (GPR) survey and obtained 3D documentation of the englacial drainage system.

## 5.12 Models Developed for the Antarctic Region

Numerical modeling of the Antarctic Ice Sheet has undergone massive changes, initiated by models focusing on long-term diffusion response toward changes in surface mass balance. New models are incorporating processes that occur at the boundary of the ice sheet with the ocean, thus exhibiting fast and short-term volume changes and, currently, ocean models are being coupled with the study of critical areas of the Antarctic Ice Sheet (Pattyn et al., 2017). Several studies have been done on modeling in the Antarctic region:

Oerlemans (1982) developed a time-dependent model that used vertical mean ice velocity to compute bedrock adjustment and used an empirical and diagnostic relationship in order to derive the dissemination of ice thickness in ice shelves. They inferred that the snowfall rate as well as the ice/snow melt rate is dependent on the sea-level temperature, the surface slope, the distance to open water, and the elevation. Payne et al. (1989) presented a model that represented the growth and then the decay of the Ice Sheet of the Antarctic Peninsula for the last glacial-interglacial cycle. Their representation was used to observe the main controls on the behavior of the ice sheet. Their tests showed that the ice-sheet behavior demonstrated most sensitivity toward sea-level change, accumulation, and basal marine melting, and showed lesser sensitivity toward isostasy, calving rates, how the accumulation was varying spatially, and the parameterization of the ice flow. Lingle and Clark (1985) developed a time-dependent numerical model of the interactions among the Earth and the marine ice sheet to simulate the retreat of a West Antarctic ice stream. Their model was used to compute the solid earth deformations which were caused by changes in loading of ice and water during the ice stream retreat. Huybrechts (1990) presented a 3D thermo-mechanical, time-dependent model of the entire ice sheet of the Antarctic to understand how ice-sheet geometries are affected by the shifts in glacial-interglacial cycles within environmental boundary conditions over a particular timescale. Their model considered a coupled ice shelf, basal sliding, grounding line dynamics, and isostatic bed adjustment. Then the model calculates ice flow for floating and grounded ice on a fine mesh. Simulation of the existing ice sheet indicated that their model gave realistic results. From their experiments, the authors deduced that the Antarctic ice sheet at a steady state might cause approximately 16 m of lowering of the global sea level at the time of maximum glaciation. Macayeal et al. (1996) developed a standard numerical experiment as a test package for development and inter-comparison of different ice-shelf models. They performed an intercomparison of five ice-sheet models and inferred that the differences among the results were being obtained from the equations used, the methods of discretizations being used, the ice-front boundary conditions, and the discretization method. Meneghel et al. (1999) proposed a model of the glacial reduction since the LGM, based on the general morphology and the existence of ice-cored banded moraines. Ritz et al. (2001) presented a thermo-mechanical model for simulation of the ice-sheet evolution of the Antarctic over long periods, predicting grounding line migration by coupling three variants of ice flow observed in Antarctica, such as flow of the ice shelf, the slower inland flow of the ice caused by ice deformation, and the faster flow of ice in ice-stream regions. Their simulations covered four glacial-interglacial cycles and thus indicated that the migration of the grounding line which is induced by sea-level changes acts as the primary factor which governs the ice-volume evolution of the Antarctic. Lythe and Vaughan (2001) introduced a new modeling suite, BEDMAP, which contains a database of measurements of ice thickness over the past 50 years of the Antarctic ice sheet. A suite of digital models of the topography were compiled, using the ice-sheet thickness grids over the grounded ice sheet and ice shelves, the thickness of the water column beneath the floating ice shelves, the bed elevations beneath the grounded ice sheets, and the sub-ice-shelf cavities. BEDMAP grid calculates the total volume of the Antarctic ice sheet, and the total sea-level equivalent. The gridded dataset might be available from the authors. Schmeltz et al. (2002) used a finite element model for coupled ice-stream and ice-shelf flow for the study of the sensitivity of an Antarctic glacier toward the changes in the basal conditions and the ice shelf. The dynamic, geometric, and physical characteristics of the glacier were used as input parameters for the model. By matching model velocity to InSAR observations, they used the model to investigate the effects of small perturbations on an ice flow. Pattyn et al. (2004) applied a 3D, time-dependent, thermo-mechanical ice-sheet model that used higher-order stress gradients for simulating ice flow across a subglacial lake in Antarctica. Benn et al. (2007) developed three models to test a new calving theory. Their model1 showed that strain rates and velocities are dependent on sufficient pressure, and consequently are dependent on the thickness of ice relative to depth of water and also showed that imposing thinning led to acceleration and consequently, terminus retreat and thus, ice shelves were not able to form. Their model2 inferred the independence of ice velocity from changes in ice thickness unless accompanied by changes in gradient of the ice surface, the strong dependence of the velocities on the width of the channel, and the tendency of the calving margins to stabilize at the widenings of the flow units. Their model3 exhibited a combination of the two other models, and its results suggested the sensitivity of the calving glaciers toward imposed changes in thickness, if the basal drag shows flow resistance but remains stable in case most of the resistance comes from lateral drag. Finally, the authors inferred that the model can be used for simulating most of the known range of calving behavior of glaciers and can also provide new ways of modeling the response of the ice sheet toward climate change. Le Brocq et al. (2009)

presented a sub-glacial waterflow model for West Antarctica, considering the effect of a subglacial and thin-film-based water-flow model which included a coupled ice/water flow model of continental scale for the West Antarctic ice sheet (WAIS). They applied their model in both a time-dependent form and a steady-state form and for deriving the water depth at equilibrium for the configuration of the present day and demonstrated the effectiveness of coupling the model with a model of ice flow using a sliding parameter with a variable range and a function of the sub-glacial water depth. Finally, the authors infer that their proposed model demonstrated the ability of reproducing the elevation of the surface and the thermal regime and velocity of the WAIS, which are essential factors in modeling its evolution and has to be addressed for reliable models for prediction. Otero et al. (2010) developed a 3D extension of (Benn et al., 2007) to overcome limitations such as dynamic models being simplistic and in 2D. Here, the authors used a full Stokes model of glacier dynamics, added further refinement of the calculation for depth of crevasses, and eventually introduced a "yield strain rate." According to the authors, an important feature of their results is producing a "calving bay" at the glacier front, providing an additional source of model validation. Pattyn (2010) studied a hybrid ice sheet/ice stream model to determine Antarctic subglacial conditions. Their model used a combination of numerical modeling of the thermodynamics of the ice sheet with prior information obtained through a simple technique of assimilation. Their results showed that 55% of the submerged section of the Antarctic Ice Sheet experiences a pressure melting point and thus reached the inference that sub-glacial flow of water is mostly concentrated under the larger outlet glaciers of Antarctica, and the large sub-glacial lakes are located mostly at the source of the sub-glacial water systems. Kowalewski et al. (2012) used a 2D numerical model applying COMSOL multi-physics, which is used to calculate the rate of vapor diffusion and spatial variability within porous media, in order to understand the role played by sublimation polygons for modulating the sublimation of buried glacier ice in Antarctica. Whitehouse et al. (2012a) worked on a deglacial model on Antarctica utilizing an immense geological and glaciological data database. They performed a reconstruction of the ice sheets of Antarctica for a chosen timescale incorporating a numerical model to generate a physically consistent ice surface across the entire continent. According to their predictions, the Antarctic Ice Sheet contributed approximately 9 m to 10.5 m to the global sea level. Whitehouse et al. (2012a) presented a glacial isostatic adjustment model (GIA) for Antarctica driven by a deglaciation history developed using a new icesheet model. The model's sensitivity was tested with uncertainties in the de-glaciation history, errors that can result in significant misfits. The authors infer that this study provided a new GIA correction for GRACE satellite data for Antarctica, which permitted more accurate constraints on ice-mass change at present. Farinotti et al. (2014) derived glacier-wide estimates of ice-thickness distribution and bedrock topography for an Antarctic glacier using a calibrated model with radio-echo soundings. Additional constraints for the model were obtained from the surface topography and ice-flow measurements. They cross-validated their results with NASA's IceBridge project measurements and found excellent agreement between the two sources. Pollard and Deconto (2012) developed a hybrid, 3D ice-sheet shelf model for long-term continental-scale applications. Their model incorporated a hybrid combination of scaled shallow-ice and shallow-shelf approximations for the flow of ice. It also applied to floating ice shelves, grounding-line migrations, and parameterized ice fluxes at the grounding line. Briggs et al. (2013) configured a glacial systems model (GSM) for ensemble analysis of the past evolution of the Antarctic Ice Sheet, while introducing a large set of model parameters in order to overcome the uncertainties that are associated with modeling of such complex systems. The authors explain that the dynamic core of the GSM comprises of a 3D thermo-mechanically modeled ice sheet, which solves both shallow-ice and shallow-shelf approximation problems. Along with the core component, the model comprises a basal-drag component which is parameterized, an isostatic and viscoelastic adjustment solver, a varied set of climate forcings, a ice-shelf calving and tidewater functionality, along with a component for sub-ice-shelf melt (SSM). Briggs and Tarasov (2013) presented a database comprising of observational data for forcing Antarctic Ice Sheet deglaciation chronologies (AntICEdat). This database is available as a spreadsheet, and the evaluation method uses observational error models, data weighing, and different chronologies can be evaluated and compared. Levermann et al. (2014) estimated the uncertainty of future ice discharge from Antarctica with a combined study of ice-sheet model response, oceanic response, and uncertainty in the climate forcing. They derived the oceanic forcing considering the effect of atmospheric warming from 19 different climate models from the Coupled Model Intercomparison Project (CMIP-5) and also including two ocean models of the EU-project Ice2Sea. The dynamic response of the ice sheet was obtained from experiments from five different Antarctic Ice Sheet Models of the intercomparison project called Sea-Level Response to Ice-Sheet Evolution (SeaRISE). Argus et al. (2014) worked on a model named ICE-6G_C(VM5a) for modeling the deglaciation history of Antarctica, which is a revision of its predecessor model ICE-5G(VM5). Maris et al. (2015) modeled the ice sheet of the Antarctic from the LGM until 2100 using a dynamical ice model, ANICE, while forcing the model with various records of temperature, sea level, and surface mass balance, in order to investigate the significance of such factors for the ice-sheet evolution. Emetc et al. (2018) produced a statistical fracture model for Antarctic glaciers and ice shelves. The authors have presented an alternative statistical method for modeling the most likely fracture zones, applying this method to all the main ice-shelf regions in Antarctica. Their fracture location predictions demonstrated a success rate of 84% on average for grounded ice and 61% in the case of floating ice. They also inferred that the Antarctic ice shelves could be classified based on fractures controlling the fracture locations. Yu et al. (2018) used different ice-flow models in conjunction with ice-shelf-melt scenarios and laws of basal friction in order to model an Antarctic glacier's grounding-line retreat over the next 100 years. They investigated various approximations of the ice-stress balance, ice-shelf melt parameterizations, and basal-friction parameterizations. They simulated the glacier evolution using ice-sheet models of various level of complexity and found that the sensitivity of the grounding line retreat toward ice-shelf melt increased when a full Stokes model was used, a Budd friction was used, or an ice-shelf melt was applied on partially-floating elements. All their simulations suggested a rapid and sustained retreat of the glacier and the uncertainty of their projections was inconsequential in the first 30 years with an overall contribution of 5 mm of sea-level rise, and after that, the contribution to sea-level changes with the model configurations. Quiquet et al. (2018) developed a revision of the GRISLI (Grenoble ice sheet and land ice) ice-sheet model, originally developed by Ritz et al. (2001), focusing only on the Antarctic ice sheet as it features the importance of the dynamics of the grounding line. They performed a full calibration of the model based on 300 ensembled simulations. Their model can perform multi-millennial long-term integrations, using atmospheric and oceanic proxies and can replicate the advances of the grounding line which is expected during glacial periods and the following retreats of the grounding line during terminations. With a rapidly converging initialization method, Le Clec'h et al. (2019) used GRISLI in order to simulate the present-day Greenland ice sheet.

## 5.13 Conclusion and Future Prospects

Throughout the history of modeling in geomorphology, there have been numerous contemplations regarding the methods applied and the corresponding calibre achieved (Baas, 2017). Consequently, the review presented here is not complete in all aspects. However, efforts have been made to cover as much ground as possible on this subject. For the objective of obtaining betterment in modeling, essentially all Landscape Evolution Models aim to obtain an optimal manifestation of a collection of geomorphic processes to be considered, and the viability of such an assemblage depends on its ability to resolve all problems of landscape evolution associated to a particular region and at the same time being independent of spatial-temporal scales. Nevertheless, on account of various constraints, the possibility of reaching such perfection is slim. Such constraints amount to aspects such as a lack of understanding of the exact influence of geomorphological and climatological processes affecting landscape evolution, appropriate numerical expression of the processes, and issues relating to the upscaling and downscaling of such processes (Van Der Beek, 2013). In this regard, Temme et al. (2011) appraised that ideal LEMs do not exist as they are always case-specific and therefore depend on the area of application. A review of general LEMs tells us that most well-established models followed almost the same methods but with different process integrations, which can be modified according to requirements. Nevertheless, extensive efforts are being continued by researchers worldwide on methods to be used for the explication of the uncertainties associated with the modeling processes. Additionally, coupling the interactions between geomorphological and climatological processes is trending in recent times due to the advantage of obtaining the combined effects of the processes on the landscape, thus depicting the natural processes more accurately, leading to a tremendous leap in the study and analysis of landscape evolution models.

In this review, the landscape evolution of the three poles has been considered, as the glaciated terrains of the three poles are vital influencers for climatological conditions worldwide. It has been considered as a purpose for this review that the glacier evolution models described and explained here would guide the future development and improvement of such models.

It is vital to mention as a concluding remark that the application of machine learning to the field of modeling landscapes would be a beacon of hope for the future study of landscape evolution, with the ulterior motive of achieving better accuracy and time efficiency in understanding the history of a landscape or predicting the future of a landscape.

## Acknowledgment

The lead author and MP are thankful to the UCRD, Chandigah University for allowing their lab and library services during the preparation of this manuscript.

## Declaration of Competing Interest

The Authors declare no conflict of interest of any form among them.

## References


Ahlkrona, J. (2011). Implementing higher order dynamics into the ice sheet model SICOPOLIS, 51 pp.

Albrecht, T., Winkelmann, R., and Levermann, A. (2020a). Glacial-cycle simulations of the Antarctic Ice Sheet with the Parallel Ice Sheet Model (PISM). Part 1: Boundary conditions and climatic forcing. *Cryosphere* 14: 599–632.

Albrecht, T., Winkelmann, R., and Levermann, A. (2020b). Glacial-cycle simulations of the Antarctic Ice Sheet with the Parallel Ice Sheet Model (PISM). Part 2: Parameter ensemble analysis. *Cryosphere* 14: 633–656. https://doi.org/10.5194/tc-14-633-2020

Argus, D.F., Peltier, W.R., Drummond, R. et al. (2014). The Antarctica component of postglacial rebound model ICE-6G_C (VM5a) based on GPS positioning, exposure age dating of ice thicknesses, and relative sea level histories. *Geophys. J. Int.* 198: 537–563. https://doi.org/10.1093/gji/ggu140

Arthern, R.J., Hindmarsh, R.C.A., and Williams, C.R. (2015). Flow speed within the Antarctic ice sheet and its controls inferred from satellite observations. *J. Geophys. Res. F Earth Surf.* 120: 1171–1188. https://doi.org/10.1002/2014JF003239

Baas, A.C.W. (2017). Models in geomorphology. In: *International Encyclopedia of Geography: People, the Earth, Environment and Technology* (ed. D. Richardson, N. Castree, M.F. Goodchild, et al.), 59–96. https://doi.org/10.1002/9781118786352.wbieg0882

Beaumont, C., Fullsack, P., and Hamilton, J. (1992). *Erosional control of active compressional orogens*. In: *Thrust Tectonics* (ed. K.R. Clay), 1–18. Dordrecht, Netherlands: Springer. https://doi.org/10.1007/978-94-011-3066-0_1

Benn, D.I., Hulton, N.R.J., and Mottram, R.H. (2007). "Calving laws," "sliding laws" and the stability of tidewater glaciers. *Ann. Glaciol.* 46: 123–130. https://doi.org/10.3189/172756407782871161

Bingham, R.G., Hubbard, A.L., Nienow, P.W. et al. (2008). An investigation into the mechanisms controlling seasonal speedup events at a High Arctic glacier. *J. Geophys. Res. Earth Surf.* 113: 1–13. https://doi.org/10.1029/2007JF000832

Bokhorst, S., Pedersen, S.H., Brucker, L. et al. (2016). Changing Arctic snow cover: a review of recent developments and assessment of future needs for observations, modelling, and impacts. *Ambio* 45: 516–537. https://doi.org/10.1007/s13280-016-0770-0

Bolibar, J., Rabatel, A., Gouttevin, I. et al. (2020. Deep learning applied to glacier evolution modelling. *Cryosphere* 14: 565–584. https://doi.org/10.5194/tc-14-565-2020

Braithwaite, R.J. and Zhang, Y.U. (1999). Modelling changes in glacier mass balance that may occur as a result of climate changes. *Geogr. Ann. Ser. A Phys. Geogr.* 81: 489–496. https://doi.org/10.1111/j.0435-3676.1999.00078.x

Braun, Jand Sambridge, M. (1997). Modelling landscape evolution on geological time scales: a new method based on irregular spatial discretization. *Basin Res.* 9: 27–52. https://doi.org/10.1046/j.1365-2117.1997.00030.x

Braun, J., Zwartz, D., and Tomkin, J.H. (1999). A new surface-processes model combining glacial and fluvial erosion. *Ann. Glaciol.* 28: 282–290. https://doi.org/10.3189/172756499781821797

Briggs, R., Pollard, D., and Tarasov, L. (2013). A glacial systems model configured for large ensemble analysis of Antarctic deglaciation. *Cryosphere* 7: 1949–1970. https://doi.org/10.5194/tc-7-1949-2013

Briggs, R.D. and Tarasov, L. (2013.) How to evaluate model-derived deglaciation chronologies: a case study using Antarctica. *Quat. Sci. Rev.* 63: 109–127. https://doi.org/10.1016/j.quascirev.2012.11.021

Brinkerhoff, D.J. and Johnson, J.V. (2013). Data assimilation and prognostic whole ice sheet modelling with the variationally derived, higher order, open source, and fully parallel ice sheet model VarGlaS. *Cryosphere* 7: 1161–1184. https://doi.org/10.5194/tc-7-1161-2013

Bulthuis, K, Arnst, M, Pattyn, F etal. (2017). Uncertainty quantification of Antarctic contribution to sea-level rise using the fast Elementary Thermomechanical Ice Sheet (f. ETISh) model. *EGU General Assembly* 2017. 19: 14880https://meetingorganizer.copernicus.org/EGU2017/EGU2017-14880-5.pdf

Callegari, M., Marin, C., and Notarnicola, C. (2017). Multi-temporal and multi-source alpine glacier cover classification. *2017 9th Int. Work. Anal. Multitemporal Remote Sens. Images, MultiTemp* 2017: 1–3. https://doi.org/10.1109/Multi-Temp.2017.8035233

Campforts, B., Schwanghart, W., and Govers, G. (2017). Accurate simulation of transient landscape evolution by eliminating numerical diffusion: the TTLEM 1.0 model. *Earth Surf. Dyn.* 5: 47–66. https://doi.org/10.5194/esurf-5-47-2017

Carretier, S. and Lucazeaut, F. (2005>). How does alluvial sedimentation at range fronts modify the erosional dynamics of mountain catchments? *Basin Res.* 17: 61–381. https://doi.org/10.1111/j.1365-2117.2005.00270.x

Chen, A., Darbon, J.Ô., and Morel, J.M. (2014). Landscape evolution models: a review of their fundamental equations. *Geomorphology* 219: 68–86. https://doi.org/10.1016/j.geomorph.2014.04.037

Codilean, A.T., Bishop, P., and Hoey, T.B. (2006). Surface process models and the links between tectonics and topography. *Prog. Phys. Geogr.* 30: 307–333. https://doi.org/10.1191/0309133306pp480ra

Coléou, C., Giraud, G., Danielou, Y. et al. (2009). Use of the models Safran-Crocus-Mepra in operational avalanche forecasting. *ISSW 09 – Int. Snow Sci. Work. Proc.* 341–345.

Coulthard, T.J. (2001). Landscape evolution models: a software review. *Hydrol. Process.* 15: 165–173.

Coulthard, T.J., Neal, J.C., Bates, P.D. et al. (2013). Integrating the LISFLOOD-FP 2D hydrodynamic model with the CAESAR model: implications for modelling landscape evolution. *Earth Surf. Process. Landforms* 38: 1897–1906. https://doi.org/10.1002/esp.3478

Crave, A. and Davy, P. (2001). A stochastic "precipiton" model for simulating erosion/sedimentation dynamics. *Comput. Geosci.* 27: 815–827. https://doi.org/10.1016/S0098-3004(00)00167-9

Davis, W.M. (1899). The geographical cycle. *Geogr. J.* 14: 481–504. doi:10.2307/1774538.

Densmore, A.L., Ellis, M.A., and Anderson, R.S. (1998). Landsliding and the evolution of normal-fault-bounded mountains. *J. Geophys. Res. Solid Earth* 103: 15203–15219. https://doi.org/10.1029/98jb00510

Dornes, P.F., Pomeroy, J.W., Pietroniro, A. et al. (2008). Influence of landscape aggregation in modelling snow-cover ablation and snowmelt runoff in a sub-arctic mountainous environment. *Hydrol. Sci. J.* 53: 725–740. https://doi.org/10.1623/hysj.53.4.725

Egholm, D.L., Andersen, J.L., Knudsen, M.F. et al. (2015). The periglacial engine of mountain erosion. Part 2: Modelling large-scale landscape evolution. *Earth Surf. Dyn.* 3: 463–482. https://doi.org/10.5194/esurf-3-463-2015

Egholm, D.L., Knudsen, M.F., Clark, C.D. et al. (2011). Modeling the flow of glaciers in steep terrains: the integrated second-order shallow ice approximation (iSOSIA). *J. Geophys. Res. Earth Surf.* 116: 1–16. https://doi.org/10.1029/2010JF001900

Egholm, D.L., Pedersen, V.K., Knudsen, M.F. et al. (2012). Coupling the flow of ice, water, and sediment in a glacial landscape evolution model. *Geomorphology* 141–142: 47–66. https://doi.org/10.1016/j.geomorph.2011.12.019

Ellis, M.A., Densmore, A.L., and Anderson, R.S. (1999). Development of mountainous topography in the Basin Ranges, USA. *Basin Res.* 11: 21–41. https://doi.org/10.1046/j.1365-2117.1999.00087.x

Emetc, V., Tregoning, P., Morlighem, M. et al. (2018). A statistical fracture model for Antarctic ice shelves and glaciers. *Cryosph.* 12: 3187–3213. doi: 10.5194/tc-12-3187-2018.

Etzelmüller, B. (2000). On the quantification of surface changes using grid-based Digital Elevation Models (DEMs). *Trans. GIS* 4: 129–143. doi: 10.1111/1467-9671.00043.

Evans, I.S. (2012). Geomorphometry and landform mapping: what is a landform? *Geomorphology* 137: 94–106. doi: 10.1016/j.geomorph.2010.09.029.

Farinotti, D., King, E.C., Albrecht, A. et al. (2014). The bedrock topography of Starbuck Glacier, Antarctic Peninsula, as determined by radio-echo soundings and flow modeling. *Ann. Glaciol.* 55: 22–28. https://doi.org/10.3189/2014AoG67A025

Favier, L., Jourdain, N.C., Jenkins, A. et al. (2019). Assessment of sub-shelf melting parameterisations using the ocean-ice sheet coupled model NEMO(v3.6)-Elmer/Ice(v8.3). *Geosci. Model Dev. Discuss.* 12: 2255–2283. doi:10.5194/gmd-12-2255-2019.

Forte, A.M. and Whipple, K.X. (2019). Short communication: the Topographic Analysis Kit (TAK) for TopoToolbox. *Earth Surf. Dyn. Discuss.* 7: 87–95. doi:10.5194/esurf-7-87-2019.

Gabbi, J., Carenzo, M., Pellicciotti, F. et al. (2014). A comparison of empirical and physically based glacier surface melt models for long-term simulations of glacier response. *J. Glaciol.* 60: 1199–1207. https://doi.org/10.3189/2014JoG14J011

Gagliardini, O., Zwinger, T., Gillet-Chaulet, F. et al. (2013). Capabilities and performance of Elmer/Ice, a new-generation ice sheet model. *Geosci. Model Dev.* 6: 1299–1318. https://doi.org/10.5194/gmd-6-1299-2013

Gao, H., Li, H., Duan, Z. et al. (2018). Modelling glacier variation and its impact on water resource in the Urumqi Glacier No. 1 in Central Asia. *Sci. Total Environ.* 644: 1160–1170. https://doi.org/10.1016/j.scitotenv.2018.07.004

Garcia-Castellanos, D., Fernández, M., and Torne, M. (2002). Modeling the evolution of the Guadalquivir foreland basin (southern Spain). *Tectonics* 21: 9-1-9–17. https://doi.org/10.1029/2001TC001339

Garcia-Castellanos, D., Fernàndez, M., and Torne, M. (1997). Numerical modeling of foreland basin formation: a program relating thrusting, flexure, sediment geometry and lithospehre rheology. *Comput. Geosci.* 23: 993–1003. https://doi.org/10.1016/S0098-3004(97)00057-5

Gilbert, G.K. (1909). The convexity of hilltops. *J. Geol.* 17: 344–350. https://doi.org/10.1086/621620

Gladstone, R., Galton-Fenzi, B., Gwyther, D. et al. (2021). The framework for Ice Sheet–Ocean Coupling (FISOC) V1.1. *Geosci. Model Dev.* 14: 889–905. doi: 10.5194/gmd-14-889-2021.

Greve, R. (2007). Ice-sheet model SICOPOLIS.

Greve, R. (2020). SICOPOLIS V5.1: Quick Start Manual 2019, 1–14.

Greve, R. and Blatter, H. (2016). Comparison of thermodynamics solvers in the polythermal ice sheet model SICOPOLIS. *Polar Sci.* 10: 11–23. https://doi.org/10.1016/j.polar.2015.12.004

Greve, R., Chambers, C., Calov, R. et al. (2021). Long-term future projections for the Greenland and Antarctic ice sheets with the model SICOPOLIS. *Cryosphere* 14: 3071–3096. https://doi.org/10.5194/tc-14-3071-2020

Greve, R., Saito, F., and Abe-Ouchi, A. (2011). Initial results of the SeaRISE numerical experiments with the models SICOPOLIS and IcIES for the Greenland ice sheet. *Ann. Glaciol.* 52: 23–30. https://doi.org/10.3189/172756411797252068

Hack, J.T. (1960). Interpretation of erosional topography in humid temperate regions. *Am. J. Sci.* 258-A: 80–97

Hancock, G.R., Lowry, J.B.C., Coulthard, T.J. et al. (2010). A catchment scale evaluation of the SIBERIA and CAESAR landscape evolution models. *Earth Surf. Process. Landforms* 35: 863–875. https://doi.org/10.1002/esp.1863

Hansen, L.U., Piotrowski, J.A., Benn, D.I. et al. (2020). A cross-validated three-dimensional model of an englacial and subglacial drainage system in a High-Arctic glacier. *J. Glaciol.* 66: 278–290. https://doi.org/10.1017/jog.2020.1

Harbor, J.M. (1992). Numerical modeling of the development of U-shaped valleys by glacial erosion. *Geol. Soc. Am. Bull.* 104: 1364–1375. https://doi.org/10.1130/0016-7606(1992)104<1364:NMOTDO>2.3.CO;2

Harmon, B.A., Mitasova, H., Petrasova, A. et al. (2019). r.sim.terrain 1.0: a landscape evolution model with dynamic hydrology. *Geosci. Model Dev.* 12: 2837–2854. doi: 10.5194/gmd-12-2837-2019.

Herman, F. and Braun, J. (2006). Fluvial response to horizontal shortening and glaciations: a study in the Southern Alps of New Zealand. *J. Geophys. Res. Earth Surf.* 111: 1–23. https://doi.org/10.1029/2004JF000248

Herman, F. and Braun, J. (2008). Evolution of the glacial landscape of the Southern Alps of New Zealand: insights from a glacial erosion model. *J. Geophys. Res. Earth Surf.* 113: F02009. https://doi.org/10.1029/2007JF000807

Hewitt, I.J. and Schoof, C. (2017). Models for polythermal ice sheets and glaciers. *Cryosphere* 11: 541–551. https://doi.org/10.5194/tc-11-541-2017

Hirabayashi, Y., Döll, P., and Kanae, S. (2010). Global-scale modeling of glacier mass balances for water resources assessments: glacier mass changes between 1948 and 2006. *J. Hydrol.* 390: 245–256. https://doi.org/10.1016/j.jhydrol.2010.07.001

Hirabayashi, Y., Zang, Y., Watanabe, S. et al. (2013). Projection of glacier mass changes under a high-emission climate scenario using the global glacier model HYOGA2. *Hydrol. Res. Lett.* 7: 6–11. https://doi.org/10.3178/hrl.7.6

Hock, R. (2005). Glacier melt: a review of processes and their modelling. *Prog. Phys. Geogr.* 29: 362–391. https://doi.org/10.1191/0309133305pp453ra

Hock, R., Bliss, A., Marzeion, B.E.N. et al. (2019). GlacierMIP-A model intercomparison of global-scale glacier mass-balance models and projections. *J. Glaciol.* 65: 453–467. https://doi.org/10.1017/jog.2019.22

Hoffman, M.J., Perego, M., Price, S.F. et al. (2018). MPAS-Albany Land Ice (MALI): a variable-resolution ice sheet model for Earth system modeling using Voronoi grids. *Geosci. Model Dev.* 11: 3747–3780. https://doi.org/10.5194/gmd-11-3747-2018

Howard, A.D. (1994). A detachment-limited model of drainage basin evolution. *Water Resour. Res.* 30: 2261–2285. https://doi.org/10.1029/94WR00757

Huggel, C., Kääb, A., Haeberli, W. et al. (2003). Regional-scale GIS-models for assessment of hazards from glacier lake outbursts: evaluation and application in the Swiss Alps. *Nat. Hazards Earth Syst. Sci.* 3: 647–662.

Hunke, E.C., Lipscomb, W.H., Turner, A.K. et al. (2015). CICE: the Los Alamos Sea Ice model documentation and Software User's Manual version 5.1 LA-CC-06-012. Los Alamos Natl. Lab. Tech. Rep. LA-CC-06-012 76.

Huss, M. and Hock, R. (2015). A new model for global glacier change and sea-level rise. *Front. Earth Sci.* 3: 1–22. https://doi.org/10.3389/feart.2015.00054

Huybrechts, P. (1990). A 3-D model for the Antarctic ice sheet: a sensitivity study on the glacial-interglacial contrast. *Clim. Dyn.* 5: 79–92. doi: 10.1007/BF00207423

Jamieson, S.S.R., Hulton, N.R.J., and Hagdorn, M. (2008). Modelling landscape evolution under ice sheets. *Geomorphology* 97: 91–108. https://doi.org/10.1016/j.geomorph.2007.02.047

Jong, L., Gladstone, R., and Galton-Fenzi, B. (2017). *Coupled ice sheet-ocean modelling to investigate ocean driven melting of marine ice sheets in Antarctica*. in *19th EGU General Assembly, EGU2017, Proceedings From the Conference Held 23–28 April, 2017 in Vienna, Austria*, 1573.

Kay, J.E., Deser, C., Phillips, A. et al. (2015). The Community Earth System Model (CESM) large ensemble project: a community resource for studying climate change in the presence of internal climate variability. *Bull. Am. Meteorol. Soc.* 96: 283–296.

Kessler, M.A., Anderson, R.S., and Stock, G.M. (2006). Modeling topographic and climatic control of east-west asymmetry in Sierra Nevada glacier length during the Last Glacial Maximum. *J. Geophys. Res. Earth Surf.* 111: 1–15. https://doi.org/10.1029/2005JF000365

Kochanski, K., Defazio, G.-C., Green, E. et al. (2019). Rescal-snow: simulating snow dunes with cellular automata. *J. Open Source Softw.* 4: 1699. https://doi.org/10.21105/joss.01699

Kowalewski, D.E., Marchant, D.R., Head, J.W. et al. (2012). A 2D model for characterising first-order variability in sublimation of buried glacier ice, Antarctica: assessing the influence of polygon troughs, desert pavements and shallow subsurface salts. *Permafr. Periglac. Process.* 23: 1–14. https://doi.org/10.1002/ppp.731

Lancaster, S.T (1998). A nonlinear river meandering model and its incorporation in a landscape evolution model. MIT PhD *Thesis*. Availableathttp://hdl.handle.net/1721.1/47672

Larour, E., Seroussi, H., Morlighem, M. et al. (2012). Continental scale, high order, high spatial resolution, ice sheet modeling using the Ice Sheet System Model (ISSM). *J. Geophys. Res. Earth Surf.* 117: F01022. doi: 10.1029/2011JF002140.

Le Brocq, A.M., Payne, A.J., Siegert, M.J. et al. (2009). A subglacial water-flow model for west Antarctica. *J. Glaciol.* 55: 879–888. https://doi.org/10.3189/002214309790152564

Le Clec'h, S., Quiquet, A., Charbit, S. et al. (2019). A rapidly converging initialisation method to simulate the present-day Greenland ice sheet using the GRISLI ice sheet model (version 1.3). *Geosci. Model Dev.* 12: 2481–2499. https://doi.org/10.5194/gmd-12-2481-2019

Lehning, M., Bartelt, P., Brown, B. et al. (1999). SNOWPACK model calculations for avalanche warning based upon a new network of weather and snow stations. *Cold Reg. Sci. Technol.* 30: 145–157. https://doi.org/10.1016/S0165-232X(99)00022-1

Levermann, A., Winkelmann, R., Nowicki, S. et al. (2014). Projecting Antarctic ice discharge using response functions from SeaRISE ice-sheet models. *Earth Syst. Dyn.* 5: 271–293. https://doi.org/10.5194/esd-5-271-2014

Li, H., Beldring, S., Xu, C.Y. et al. (2015). Integrating a glacier retreat model into a hydrological model: case studies of three glacierised catchments in Norway and Himalayan region. *J. Hydrol.* 527: 656–667. https://doi.org/10.1016/j.jhydrol.2015.05.017

Lingle, C.S. and Clark, J.A. (1985). A numerical model of interactions between a marine ice sheet and the solid Earth: application to a West Antarctic ice stream. *J. Geophys. Res.* 90: 1100–1114. https://doi.org/10.1029/JC090iC01p01100

Linsbauer, A., Paul, F., and Haeberli, W. (2012). Modeling glacier thickness distribution and bed topography over entire mountain ranges with glabtop: application of a fast and robust approach. *J. Geophys. Res. Earth Surf.* 117: 1–17. https://doi.org/10.1029/2011JF002313

Linsbauer, A., Paul, F., Machguth, H. et al. (2013). *Comparing three different methods to model scenarios of future glacier change in the Swiss Alps. Ann. Glaciol.* 54: 241–253. https://doi.org/10.3189/2013AoG63A400

Lipscomb, W.H., Price, S.F., Hoffman, M.J. et al. (2019). Description and evaluation of the Community Ice Sheet Model (CISM) v2.1. *Geosci. Model Dev.* 12: 387–424. https://doi.org/10.5194/gmd-12-387-2019

Logan, L.C., Narayanan, S.H.K., Greve, R. et al. (2020.) SICOPOLIS-AD v1: an open-source adjoint modeling framework for ice sheet simulation enabled by the algorithmic differentiation tool OpenAD. *Geosci. Model Dev.* 13: 1845–1864. https://doi.org/10.5194/gmd-13-1845-2020

Lythe, M.Band Vaughan, D.G. (2001). BEDMAP: a new ice thickness and subglacial topographic model of Antarctica. *J. Geophys. Res. Solid Earth* 106: 11335–11351. https://doi.org/10.1029/2000jb900449

Macayeal, D.R., Rommelaere, V., Huybrechts, P. et al. (1996). An ice-shelf model test based on the Ross Ice Shelf, Antarctica. *Ann. Glaciol.* 23: 46–51. https://doi.org/10.1017/s0260305500013240

MacDougall, A.H., Wheler, B.A., and Flowers, G.E. (2011). A preliminary assessment of glacier melt-model parameter sensitivity and transferability in a dry subarctic environment. *Cryosphere* 5: 1011–1028. https://doi.org/10.5194/tc-5-1011-2011

MacGregor, K.R., Anderson, R.S., Anderson, S.P. et al. (2000). Numerical simulations of glacial-valley longitudinal profile evolution. *Geology* 28: 1031–1034. https://doi.org/10.1130/0091-7613(2000)028<1031:NSOGVL>2.3.CO;2

Malamud, B.D. and Baas, A.C.W. (2013). Nine considerations for constructing and running geomorphological models. *Treatise on Geomorphology*. Elsevier Ltd. https://doi.org/10.1016/B978-0-12-374739-6.00024-5

Maris, M.N.A., van Wessem, J.M., van de Berg, W.J. et al. (2015). A model study of the effect of climate and sea-level change on the evolution of the Antarctic Ice Sheet from the Last Glacial Maximum to 2100. *Clim. Dyn.* 45: 837–851. https://doi.org/10.1007/s00382-014-2317-z

Martin, M.A., Winkelmann, R., Haseloff, M. et al. (2011). The Potsdam Parallel Ice Sheet Model (PISM-PIK). Part 2: Dynamic equilibrium simulation of the Antarctic ice sheet. *Cryosphere* 5: 727–740. https://doi.org/10.5194/tc-5-727-2011

Martin, Y. and Church, M. (2004). Numerical modelling of landscape evolution: geomorphological perspectives. *Prog. Phys. Geogr.* 28: 317–339. https://doi.org/10.1191/0309133304pp412ra

Matell, N., Anderson, R.S., Overeem, I. et al. (2013). Modeling the subsurface thermal impact of Arctic thaw lakes in a warming climate. *Comput. Geosci.* 53: 69–79. https://doi.org/10.1016/j.cageo.2011.08.028

Matero, I.S.O., Gregoire, L.J., and Ivanovic, R.F. (2020). Simulating the Early Holocene demise of the Laurentide Ice Sheet with BISICLES (public trunk revision 3298). *Geosci. Model Dev.* 13: 4555–4577. https://doi.org/10.5194/gmd-13-4555-2020

Maussion, F., Butenko, A., Champollion, N. et al. (2019). The Open Global Glacier Model (OGGM) v1.1. *Geosci. Model Dev.* 12: 909–931. https://doi.org/10.5194/gmd-12-909-2019

Meneghel, M., Bondesan, A., Salvatore, M.C. et al. (1999). A model of the glacial retreat of upper Rennick Glacier, Victoria Land, Antarctica. *Ann. Glaciol.* 29: 225–230. https://doi.org/10.3189/172756499781821463

Mokarram, M., Pourghasemi, H.R., Tiefenbacher, J.P. et al. (2022). Prediction of drainage morphometry using a genetic landscape evolution algorithm. *Geocarto Int.* 37: 1364–1377. https://doi.org/10.1080/10106049.2020.1762766

Mugford, R.I. and Dowdeswell, J.A. (2010). Modeling iceberg-rafted sedimentation in high-latitude fjord environments. *J. Geophys. Res. Earth Surf.* 115: 1–21. https://doi.org/10.1029/2009JF001564

Nick, F.M., Van Der Veen, C.J., Vieli, A. et al. (2010). A physically based calving model applied to marine outlet glaciers and implications for the glacier dynamics. *J. Glaciol.* 56: 781–794. https://doi.org/10.3189/002214310794457344

Oerlemans, J. (1984). Numerical experiments on glacial erosion. *Zeitschrift für Gletscherkunde und Glazialgeologie* 20: 107–126. Available at:

Oerlemans, J. (1982). A model of the Antarctic ice sheet. *Nature* 297: 550–553. doi:10.1038/297550a0.

Oerlemans, J. (1997). Climate sensitivity of Franz Josef Glacier, New Zealand, as revealed by numerical modeling. *Arct. Alp. Res.* 29: 233–239. https://doi.org/10.2307/1552052

Oerlemans, J., Anderson, B., Hubbard, A. et al. (1998). Modelling the response of glaciers to climate warming. *Clim. Dyn.* 14: 267–274. https://doi.org/10.1007/s003820050222

Oerlemans, J., Jania, J., and Kolondra, L. (2011). Application of a minimal glacier model to Hansbreen, Svalbard. *Cryosphere* 5: 1–11. https://doi.org/10.5194/tc-5-1-2011

Otero, J., Navarro, F.J., Martin, C. et al. (2010). A three-dimensional calving model: numerical experiments on Johnsons Glacier, Livingston Island, Antarctica. *J. Glaciol.* 56: 200–214. https://doi.org/10.3189/002214310791968539

Paik, K. (2012). Simulation of landscape evolution using a global flow path search method. *Environ. Model. Softw.* 33: 35–47. https://doi.org/10.1016/j.envsoft.2012.01.005

Parizek, B.R., Alley, R.B., and MacAyeal, D.R. (2005). The PSU/UofC finite-element thermomechanical flowline model of ice-sheet evolution. *Cold Reg. Sci. Technol.* 42: 145–168. https://doi.org/10.1016/j.coldregions.2004.12.006

Pattyn, F. (2006). GRANTISM: an Excel$^{TM}$ model for Greenland and Antarctic ice-sheet response to climate changes. *Comput. Geosci.* 32: 316–325. https://doi.org/10.1016/j.cageo.2005.06.020

Pattyn, F. (2010). Antarctic subglacial conditions inferred from a hybrid ice sheet/ice stream model. *Earth Planet. Sci. Lett.* 295: 451–461. https://doi.org/10.1016/j.epsl.2010.04.025

Pattyn, F., de Smedt, B., and Souchez, R. (2004). Influence of subglacial Vostok lake on the regional ice dynamics of the Antarctic ice sheet: a model study. *J. Glaciol.* 50: 583–589. https://doi.org/10.3189/172756504781829765

Pattyn, F., Favier, L., Sun, S. et al. (2017). Progress in numerical modeling of Antarctic ice-sheet dynamics. *Curr. Clim. Chang. Reports* 3: 174–184. https://doi.org/10.1007/s40641-017-0069-7

Payne, A.J., Sugden, D.E., and Clapperton, C.M. (1989). Modeling the growth and decay of the Antarctic Peninsula Ice Sheet. *Quat. Res.* 31: 119–134. https://doi.org/10.1016/0033-5894(89)90002-1

Pazzaglia, F.J. (2003). Landscape evolution models. *Dev. Quat. Sci.* 1: 247–274. https://doi.org/10.1016/S1571-0866(03)01012-1

Pellicciotti, F., Brock, B., Strasser, U. et al. (2005). An enhanced temperature-index glacier melt model including the shortwave radiation balance: development and testing for Haut Glacier d'Arolla, Switzerland. *J. Glaciol.* 51: 573–587. https://doi.org/10.3189/172756505781829124

Penck (1953). *Morphological Analysis of Land Forms: A Contribution to Physical Geography*, 1st ed. London: Macmillan & Co., Ltd. Available at: http://geomorphology.sese.asu.edu/Papers/Penck.pdf

Pimentel, S., Flowers, G.E., Sharp, M.J. et al. (2017). Modelling intra-annual dynamics of a major marine-terminating Arctic glacier. *Ann. Glaciol.* 58: 118–130. https://doi.org/10.1017/aog.2017.23

Plummer, M.A. and Phillips, F.M. (2003). A 2-D numerical model of snow/ice energy balance and ice flow for paleoclimatic interpretation of glacial geomorphic features. *Quat. Sci. Rev.* 22: 1389–1406. https://doi.org/10.1016/S0277-3791(03)00081-7

Pollard, D. and Deconto, R.M. (2012). Description of a hybrid ice sheet-shelf model, and application to Antarctica. *Geosci. Model Dev.* 5: 1273–1295. https://doi.org/10.5194/gmd-5-1273-2012

Putkonen, J., Connolly, J., and Orloff, T. (2008). Landscape evolution degrades the geologic signature of past glaciations. *Geomorphology* 97: 208–217. https://doi.org/10.1016/j.geomorph.2007.02.043

Quiquet, A., Dumas, C., Ritz, C. et al. (2018). The GRISLI ice sheet model (version 2.0): calibration and validation for multi-millennial changes of the Antarctic ice sheet. *Geosci. Model Dev.* 11: 5003–5025. https://doi.org/10.5194/gmd-11-5003-2018

Rabatel, A., Dedieu, J.P., and Vincent, C. (2005). Using remote-sensing data to determine equilibrium-line altitude and mass-balance time series: validation on three French glaciers, 1994-2002. *J. Glaciol.* 51: 539–546. https://doi.org/10.3189/172756505781829106

Ramsankaran, R.A.A.J., Pandit, A., and Azam, M.F. (2018). Spatially distributed ice-thickness modelling for Chhota Shigri Glacier in western Himalayas, India. *Int. J. Remote Sens.* 39: 3320–3343. https://doi.org/10.1080/01431161.2018.1441563

Rawlins, M.A., Nicolsky, D.J., McDonald, K.C. et al. (2013). Simulating soil freeze/thaw dynamics with an improved pan-Arctic water balance model. *J. Adv. Model. Earth Syst.* 5: 659–675. https://doi.org/10.1002/jame.20045

Raymond, C. and Harrison, W.D. (1987). Fit of ice motion models to observations from Variegated Glacier, Alaska. *In: The Physical Basis of Ice Sheet Modeling* (ed. E.D. Waddington and J.S. Walder), 153–166. IAHS Publ., 170.

Reese, R., Albrecht, T., Mengel, M. et al. (2018). Antarctic sub-shelf melt rates via PICO. *Cryosphere* 12: 1969–1985. https://doi.org/10.5194/tc-12-1969-2018

Refice, A., Giachetta, E., and Capolongo, D. (2012). SIGNUM: a Matlab, TIN-based landscape evolution model. *Comput. Geosci.* 45: 293–303. https://doi.org/10.1016/j.cageo.2011.11.013

Renwick, W.H. (1992). Equilibrium, disequilibrium, and nonequilibrium landforms in the landscape. *Geomorphology* 5: 265–276. doi:10.1016/0169-555X(92)90008-C.

Réveillet, M., Six, D., Vincent, C. et al. (2018). Relative performance of empirical and physical models in assessing the seasonal and annual glacier surface mass balance of Saint-Sorlin Glacier (French Alps). *Cryosphere* 12: 1367–1386. https://doi.org/10.5194/tc-12-1367-2018

Ritz, C., Rommelaere, V., and Dumas, C. (2001). Modeling the evolution of Antarctic ice sheet over the last 420,000 years: implications for altitude changes in the Vostok region. *J. Geophys. Res. Atmos.* 106: 31943–31964. https://doi.org/10.1029/2001JD900232

Rounce, D.R., Hock, R., and Shean, D.E. (2020a). Glacier mass change in High Mountain Asia through 2100 using the open-source Python Glacier Evolution Model (PyGEM). *Front. Earth Sci.* 7: 1–20. https://doi.org/10.3389/feart.2019.00331

Rounce, D.R., Khurana, T., Short, M.B. et al. (2020b). Quantifying parameter uncertainty in a large-scale glacier evolution model using Bayesian inference: application to High Mountain Asia. *J. Glaciol.* 66: 175–187. https://doi.org/10.1017/jog.2019.91

Rozier, O., and Narteau, C. (2014). A real-space cellular automaton laboratory. *Earth Surf. Process. Landforms* 39: 98–109. https://doi.org/10.1002/esp.3479

Rückamp, M., Greve, R., and Humbert, A. (2019). Comparative simulations of the evolution of the Greenland ice sheet under simplified Paris Agreement scenarios with the models SICOPOLIS and ISSM. *Polar Sci.* 21: 14–25. https://doi.org/10.1016/j.polar.2018.12.003

Rutt, I., Hagdorn, M., Johnson, J. et al. (2010). Glimmer-CISM: developments in community ice sheet modelling. *EGU Gen. Assem. Conf. Abstr.* 12: 9543.

Rye, C.J., Arnold, N.S., Willis, I.C. et al. (2010). Modeling the surface mass balance of a high Arctic glacier using the ERA-40 reanalysis. *J. Geophys. Res. Earth Surf.* 115: 1–18. https://doi.org/10.1029/2009jf001364

Salles, T. (2019). ESCAPE: regional to global scale landscape evolution model v2.0. *Geosci. Model Dev.* 12: 4165–4184. https://doi.org/10.5194/gmd-12-4165-2019

Schmeltz, M., Rignot, E., Dupont, T.K. et al. (2002). Sensitivity of Pine Island Glacier, West Antarctica, to changes in ice-shelf and basal conditions: a model study. *J. Glaciol.* 48: 552–558. https://doi.org/10.3189/172756502781831061

Shapero, D., Badgeley, J., Hoffmann, A. et al. (2021). icepack: a new glacier flow modeling package in Python, version 1.0. *Geosci. Model Dev. Discuss.* 14:4593–4616

Temme, A.J.A.M., Claessens, L., Veldkamp, A. et al. (2011). Evaluating choices in multi-process landscape evolution models. *Geomorphology* 125: 271–281. https://doi.org/10.1016/j.geomorph.2010.10.007

Temme, A.J.A.M., Schoorl, J.M., and Veldkamp, A. (2006). Algorithm for dealing with depressions in dynamic landscape evolution models. *Comput. Geosci.* 32: 452–461. https://doi.org/10.1016/j.cageo.2005.08.001

Thoma, M., Grosfeld, K., Barbi, D. et al. (2014). RIMBAY: a multi-approximation 3D ice-dynamics model for comprehensive applications: model description and examples. *Geosci. Model Dev.* 7: 1–21. https://doi.org/10.5194/gmd-7-1-2014

Tomkin, J.H. and Braun, J. (2002). The influence of alpine glaciation on the relief of tectonically active mountain belts. *Am. J. Sci.* 302: 169–190. https://doi.org/10.2475/ajs.302.3.169

Tucker, G., Lancaster, S., Gasparini, N. et al. (2001). The Channel-Hillslope Integrated Landscape Development Model (CHILD). *In: Landscape Erosion and Evolution Modeling* (ed. W.W. Harmon and R.S. Doe), 349–388. https://doi.org/10.1007/978-1-4615-0575-4_12

Tucker, G.E. and Hancock, G.R. (2010). Modelling landscape evolution. *Earth Surf. Process. Landforms* 35: 28–50. https://doi.org/10.1002/esp.1952

Tucker, G.E. and Slingerland, R.L. (1994). Erosional dynamics, flexural isostasy, and long-lived escarpments: a numerical modeling study. *J. Geophys. Res.* 99: 12229–12243. https://doi.org/10.1029/94jb00320

Ugelvig, S.V., Egholm, D.L., and Iverson, N.R. (2016). Glacial landscape evolution by subglacial quarrying: a multiscale computational approach. *J. Geophys. Res. Earth Surf.* 121: 2042–2068. doi: 10.1002/2016JF003960

Van Der Beek, P. (2013.) *Modelling landscape evolution*. In*Environmental Modelling*, 309–331. Chichester, UK John Wiley & Sons, Ltd doi:10.1002/9781118351475.ch19

Waddington, E.D. and Clarke, G.K.C. (1988). Stable-isotope pattern predicted in surge-type glaciers. *Can. J. Earth Sci.* 25: 657–668. https://doi.org/10.1139/e88-063

Wadham, J.L., Kohler, J., Hubbard, A. et al. (2006). Superimposed ice regime of a high Arctic glacier inferred using ground-penetrating radar, flow modeling, and ice cores. *J. Geophys. Res. Earth Surf.* 111: 1–11. https://doi.org/10.1029/2004JF000144

Wang, K., Jafarov, E., and Overeem, I. (2020). Sensitivity evaluation of the Kudryavtsev permafrost model. *Sci. Total Environ.* 720: 137538. https://doi.org/10.1016/j.scitotenv.2020.137538

Westermann, S., Langer, M., Boike, J. et al. (2016). Simulating the thermal regime and thaw processes of ice-rich permafrost ground with the land-surface model CryoGrid

3. *Geosci. Model Dev.* 9: 523–546. https://doi.org/10.5194/gmd-9-523-2016

Whitehouse, P.L., Bentley, M.J., and Le Brocq, A.M. (2012a). A deglacial model for Antarctica: geological constraints and glaciological modelling as a basis for a new model of Antarctic glacial isostatic adjustment. *Quat. Sci. Rev.* 32: 1–24. https://doi.org/10.1016/j.quascirev.2011.11.016

Wickert, A. (2014). IceFlow[WWW Document]. URL https://csdms.colorado.edu/wiki/Model:IceFlow accessed 12 November 2021.

Willgoose, G. (2005). Mathematical modeling of whole landscape evolution. *Annu. Rev. Earth Planet. Sci.* 33: 443–459. https://doi.org/10.1146/annurev.earth.33.092203.122610

Willgoose, G., Bras, R.L., and Rodriguez-Iturbe, I. (1991). Results from a new model of river basin evolution. *Earth Surf. Process. Landforms* 16: 237–254. https://doi.org/10.1002/esp.3290160305

Winkelmann, R., Martin, M.A., Haseloff, M. et al. (2011). The Potsdam Parallel Ice Sheet Model (PISM-PIK). Part 1: Model description. *Cryosphere* 5: 715–726. https://doi.org/10.5194/tc-5-715-2011

Wright, A., Wadham, J., Siegert, M. et al. (2005). Modelling the impact of superimposed ice on the mass balance of an Arctic glacier under scenarios of future climate change. *Ann. Glaciol.* 42: 277–283. https://doi.org/10.3189/172756405781813104

Wright, A.P., Wadham, J.L., Siegert, M.J. et al. (2007). Modeling the refreezing of meltwater as superimposed ice on a high Arctic glacier: a comparison of approaches. *J. Geophys. Res. Earth Surf.* 112: 1–14. https://doi.org/10.1029/2007JF000818

Yao, G., Ke, C., Zhou, X. et al. (2020). Identification of Alpine glaciers in the Central Himalayas using fully polarimetric L-band SAR data. *IEEE Trans. Geosci. Remote Sens.* 58: 691–703.

Yu, H., Rignot, E., Seroussi, H. et al. (2018). Retreat of Thwaites Glacier, West Antarctica, over the next 100 years using various ice flow models, ice shelf melt scenarios and basal friction laws. *Cryosphere* 12: 3861–3876. https://doi.org/10.5194/tc-12-3861-2018

Zekollari, H., Huss, M., and Farinotti, D. (2019). Modelling the future evolution of glaciers in the European Alps under the EURO-CORDEX RCM ensemble. *Cryosphere* 13: 1125–1146. https://doi.org/10.5194/tc-13-1125-2019

# 6

# Spectral Indices Across Remote Sensing Platforms and Sensors Relating to the Three Poles

## An Overview of Applications, Challenges, and Future Prospects

*Mallikarjun Mishra[1], Kiran Kumari Singh[2,13], Prem C. Pandey[3], Rahul Devrani[4], Avinash Kumar Pandey[5], KN Prudhvi Raju[1], Prabhat Ranjan[6], Aman Arora[7], Romulus Costache[8], Saeid Janizadeh[9], Nguyen Thuy Linh[10], and Manish Pandey[11,12,*]*

[1] Department of Geography, Ravenshaw University, Cuttack, Odisha, India
[2] Department of Geography, Central University of Punjab, Bathinda, Punjab, India
[3] School of Natural Sciences, Shiv Nadar Institution of Eminence, Greater Noida, Uttar Pradesh, 201314, India
[4] Delhi School of Climate Change & Sustainability, Institution of Eminence, University of Delhi, Delhi - 110007, India
[5] Department of Chemistry, GLA University, Chaumuhan, Mathura, Uttar Pradesh, India
[6] Central Pollution Control Board, Ministry of Environment, Forest, and Climate Change, Parivesh Bhawan, East Arjun Nagar, Shahdara, Delhi, India
[7] Bihar Mausam Seva Kendra, Planning and Development Department, Government of Bihar, Patna, Bihar, India
[8] Department of Civil Engineering, Transilvania University of Brasov, Brasov, Romania
[9] Department of Watershed Management Engineering and Sciences, Faculty in Natural Resources and Marine Science, Tarbiat Modares University, Tehran, Iran
[10] Institute of Applied Technology, Thu Dau Mot University, Binh Duong province, Vietnam
[11] University Center for Research & Development (UCRD), Chandigarh University, Mohali, Punjab, India
[12] Department of Civil Engineering, University Institute of Engineering, Chandigarh University, Mohali, Punjab, India
[13] Department of Geography, Central University of South Bihar, Gaya-824236, Bihar, India
* Corresponding author

## 6.1 Introduction

In recent years, remote-sensing technology has advanced our knowledge in polar research. Satellite-based observations such as optical (Landsat series), synthetic aperture radar (SAR) satellite imagery, and very-high-resolution images have been applied through various analytical techniques to understand the dynamic processes operating in the lithosphere (Özkan et al., 2018), atmosphere (Yang et al., 2020), biosphere (Yang et al., 2020), hydrosphere (Cermakova, et al., 2019), in general, the cryospheric component of the hydrosphere (How et al., 2021), and anthroposphere (DaCamara et al., 2019; Agyeman et al., 2021). Figure 6.1 illustrates the spheres of our planet, and various objects, also called "Elements of Identification (IoE)," that advances in remote-sensing technology has helped, is helping, and will help plan future missions by space agencies of different countries and groups of countries aiming to monitor and manage the betterment of our society. Remote sensing has been widely used to understand geology, geomorphology, ice shelfs, glaciers, coastlines, and vegetation cover. The study of the polar regions is of utmost importance to understand the atmospheric–oceanic process of the globe. This review paper has compiled a comprehensive database on remote-sensing application of polar research through various spectral indices. Several important challenges are discussed that occur when different spectral bands are employed, keeping in view the spectral response to the ground features.

Remote sensing helps to acquire spatial and temporal data and to increase our understanding of polar dynamics. It is a system that measures electromagnetic energy emanating from an object. However, the remotely-sensed images are the result of various interrelated processes and can provide information about the characteristics of the surface materials based on characteristics of electromagnetic radiation, atmospheric interaction, and sensor characteristics (Campbell and Wynne, 2011). Remote-sensing technology has advanced tremendously over the past few decades in terms of its spectral (ranging from visible to microwave), spatial (ranging from kilometer to sub-meter), radiometric, and temporal (from years to months to hours to minutes) characteristics, which have facilitated the research community to experiment with varieties of remotely-sensed data (Liang and Wang, 2020). Fortunately, various satellite platforms have made multispectral data freely available to research community, such as the Landsat series (30 m spatial resolution) and Sentinel 2 (10 m and 20 m spatial resolution of different spectral bands). The commercially available data are available at a very fine spatial and higher spectral resolution. In recent times, the use of UAVs (Unmanned Aerial Vehicles) and LiDAR have surged forward to acquire spatial information with higher spatial and temporal resolution.

*Advances in Remote Sensing Technology and the Three Poles,* First Edition. Edited by Manish Pandey, Prem C. Pandey, Yogesh Ray, Aman Arora, Shridhar D. Jawak, and Uma K. Shukla.

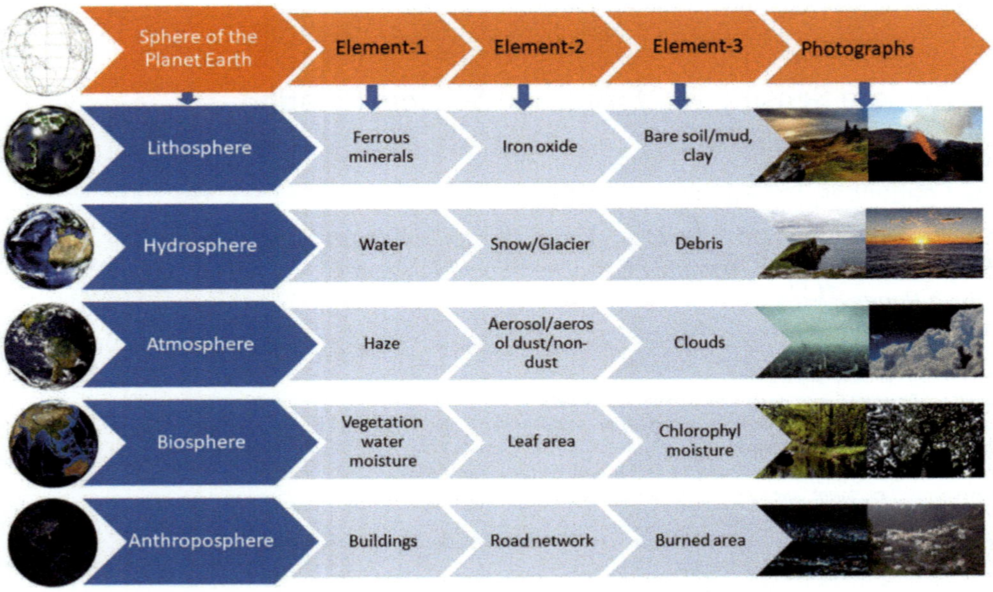


**Figure 6.1** Different spheres of the planet Earth and the Elements of Detection (EoD) that various spectral indices segregate from all other objects. *Source*: AlexAntropov86 / Pixabay, Caniceus / Pixabay, Pravin / Adobe Stock, mary1826 / Pixabay, StevenGoddard / Pixabay, kravtsova maria kravtsova maria / EyeEm / Adobe Stock, AllegroSympatico / Pixabay, Joe Straker / Unsplash.con, dimitrisvetsikas1969 / Pixabay, Peggychoucair / Pixabay, leavealegacy / Pixabay, Игорь Бершауэр / Adobe Stock, Bjonsson / Pixabay.

## 6.2 Database and Methodology

To track the evolution and progress of spectral indices derived from remotely-sensed images, we have used the set of search keywords shown in Figure 6.2. The period of database tracking spans 23 years (from 1999 to 2022) archived in the Web of Science (WoS) scientific literature collection. Only English language documents were filtered that contain any of the search keywords (Figure 6.1) in their titles. While downloading the database, we made sure that references of all the documents were included in the database. For scientometric analysis of all the documents existing in the downloaded WoS database, the "bibliometrix" package in RStudio software was used. Using the "biblioshiny" function of the bibliometrix package, data overview analyses, e.g., "Main Information" and "Three-Fields Plot;" "Sources"-related analyses, e.g., "Most Relevant Sources," has been performed. "WordClouds" was run to give an idea of the most important words in the database over the study period. Under "Conceptual Structure" and "Intellectual Structure," "Thematic Map," with their corresponding "Networks" in each time period, and "Thematic Evolution," analyses have been run.

By examining the strategic diagrams, graphs showing two-dimensional (2D) visualizations of themes as their centrality and density rank values on *x*- and *y*-axes (centrality on *x*-axis and density on *y*-axis) (Viedma-Del-Jesus et al., 2011), for all the four time periods split and randomly chosen by the examiner (authors) to track the theme evolution, helped to understand how the spectral index using remote-sensing theme evolved. The strategic graphs are also termed as "thematic maps" for this study. The centrality index, in Figure 6.1, is the measure of how intense the link of one cluster with other mapped clusters is, and the density represents the tie strength between the clusters. This thematic mapping exercise has helped to classify the clusters into four main categories, such as the following: i) motor themes; ii) niche themes (specialized themes); iii) emerging/declining themes; and iv) basic themes.

These four themes of the thematic evolution map (produced under the conceptual evolution option of the Bibliometrix Tool in RStudio) represent four quadrants having specific interpretation, which according to Pacheco et al. (2020) and López-Robles et al. (2019), are as follows: i) Quadrant Q1: considered motor themes, important for the development of the research field; ii) Quadrant Q2 (Niche themes): themes well developed; iii) Quadrant Q3: represents emergent or declining themes; and v) Quadrant Q4 (basic themes): characterized by basic and transversal themes, nevertheless not well developed.

Motor themes have the properties of high centrality and high density; niche themes (or specialized themes) show low centrality and high density among clusters; emerging/declining themes are low centrality and low-density categories; and basic themes clusters are characterized by high centrality and low density as their salient feature. All the published "KeyWords Plus" during each time period has been thematically mapped for network analysis for each time period.

*"Clay Mineral Ratio" OR "Ferrous Minerals Ratio" OR "Iron Oxide Ratio" OR "WorldView New Iron Index" OR "WorldView Soil Index" OR "Bare Soil Index" OR "Normalized Difference Snow Index" OR "Normalized-difference debris Index" OR "Normalized Difference Glacier Index" OR "Normalized Difference Water Index" OR "Modified Normalized Difference Water Index" OR "WorldView Water Index" OR "Normalized Difference Vegetation Index" OR "Modified Red Edge NDVI" OR "Green Normalized Difference Vegetation Index" OR "Enhanced Vegetation Index" OR "Advanced Vegetation Index" OR "Soil Adjusted Vegetation Index" OR "Normalized Difference Moisture Index" OR "Moisture Stress Index" OR "Green Coverage Index" OR "Difference Vegetation Index" OR "Enhanced Vegetation Index" OR "Global Environmental Monitoring Index" OR "Green Atmospherically Resistant Index" OR "Green Coverage Index" OR "Green Difference Vegetation Index" OR "Green Leaf Index" OR "Green Normalized Difference Vegetation Index" OR "Green Optimized Soil Adjusted Vegetation Index" OR "Green Ratio Vegetation Index" OR "Green Soil Adjusted Vegetation Index" OR "Green Vegetation Index" OR "Infrared Percentage Vegetation Index" OR "Leaf Area Index" OR "Modified Non-Linear Index" OR "Modified Simple Ratio" OR "Modified Soil Adjusted Vegetation Index" OR "Non-Linear Index" OR "Normalized Difference Vegetation Index" OR "Optimized Soil Adjusted Vegetation Index" OR "Renormalized Difference Vegetation Index" OR "Soil Adjusted Vegetation Index" OR "Simple Ratio" OR "Transformed Difference Vegetation Index" OR "Triangular Greenness Index" OR "Visible Atmospherically Resistant Index" OR "Wide Dynamic Range Vegetation Index" OR "WorldView Improved Vegetation Index" OR "Atmospherically Resistant Vegetation Index" OR "Modified Chlorophyll Absorption Ratio Index" OR "Modified Chlorophyll Absorption Ratio Index – Improved" OR "Modified Red Edge Normalized Difference Vegetation Index" OR "Modified Red Edge Simple Ratio" OR "Modified Triangular Vegetation Index" OR "Modified Triangular Vegetation Index – Improved" OR "Red Edge Normalized Difference Vegetation Index" OR "Red Edge Position Index" OR "Transformed Chlorophyll Absorption Reflectance Index" OR "Triangular Vegetation Index" OR "Vogelmann Red Edge Index" OR "Normalized Difference Nitrogen Index" OR "Moisture Stress Index" OR "Normalized Difference Infrared Index" OR "Normalized Difference Water Index" OR "Normalized Multi-band Drought Index" OR "Water Band Index" OR "Cellulose Absorption Index" OR "Lignin Cellulose Absorption Index" OR "Normalized Difference Lignin Index" OR "Plant Senescence Reflectance Index" OR "Anthocyanin Reflectance Index" OR "Anthocyanin Reflectance Index" OR "Carotenoid Reflectance Index" OR "Photochemical Reflectance Index" OR "Structure Insensitive Pigment Index" OR "Red Green Ratio Index" OR "Atmospherically Resistant Vegetation Index" OR "Modified Chlorophyll Absorption Ratio Index" OR "Modified Chlorophyll Absorption Ratio Index – Improved" OR "Modified Red Edge Normalized Difference Vegetation Index" OR "Modified Red Edge Simple Ratio" OR "Modified Triangular Vegetation Index" OR "Modified Triangular Vegetation Index – Improved" OR "Red Edge Normalized Difference Vegetation Index" OR "Transformed Chlorophyll Absorption Reflectance Index" OR "Triangular Vegetation Index" OR "Vogelmann Red Edge Index" OR "Normalized Difference Cloud Index" OR "Cloud Index" OR "Normalized Difference Built-Up Index" OR "WorldView Built-Up Index" OR "Normalized Burned Ratio Index" OR "Burn Area Index" OR "Normalized Multi-band Drought Index" OR "Normalized Burn Ratio" OR "Normalized Burn Ratio Thermal" OR "WorldView Non-Homogeneous Feature Difference"*

**Figure 6.2** Search KeyWords used in the Web of Science Database creation for this study.

## 6.3 Rationale of Different Spectral Indices Across RS Sensors and Platforms

Remote sensing acquires information about the surface features based on their reflectance characteristics since they have unique spectral features (reflectance or emission regions) (Xie and Yu, 2008). The reflectance in various electromagnetic spectra depends on the biophysical, chemical, and morphological properties of the object. In recent times, the higher spectral resolution instruments have made a higher number of bands available. Detailed information about the importance of spectral dimension in remote sensing (spatial and temporal dimension) is critically reviewed by Pandey et al. (2021). Optical remote sensing utilizes the visible (VIS), near (NIR), and mid-infrared (MIR) spectrums of the wavelength, which may be written as VIS-NIR-MIR. Features are recorded differently in VIS-NIR-MIR regions according to their reflectance characteristics, and thus each feature has a unique spectral signature. The spectral signature helps to discriminate targets and even features among targets (plant species). The spectral signature changes with the variation in the composition or changes in its internal properties. In the case of vegetation, reflectance depends upon the pigment concentration, leaf structure, and water content. Therefore, combination of bands or ratioing of bands helps to identify a peculiar characteristic of the target features more precisely. The spectral band ratio is performed to enhance the differences between bands/among bands. Furthermore, band ratioing helps to reduce topographic effects, such as shadow or sunlit differences in the same image. Dividing one spectral band by another produces an index image that provides relative band intensities. The index image enhances the spectral differences among spectral bands. Taking advantage of spectral band ratioing, the remote-sensing community utilizes a combination of bands to generate several indices to identify and discriminate target features such as vegetation, soil, water, ice, cloud, snow, and urban regions. Spectral indices utilize different bands to generate results with varying accuracies for delineation of particular features. The spectral index has been defined as an algorithm that separates different reflectivities using a combination of spectral bands (Lyzenga, 1979). The index is used to classify the pixels into different reflectivity classes. A multitude of spectral indices is devised to generate data on various surface features using different data. Over the years, several spectral indices have emerged to identify and monitor features. The inadequacy or limitations observed in one index encouraged researchers to modify and devise a new index by either adopting constant or correction factors, and inclusion of a new band or replacement with other spectral bands as required. According to Matsushita et al. (2007), indices may have errors due to atmospheric and

**Table 6.1** Salient features of the dataset included for analysing the evolution of this subtheme and spectral index, extracted based on the query set in the Web of Science database for the period 1999–2022.

| Metadata Description | Search Results |
|---|---|
| **MAIN INFORMATION ABOUT DATA** | |
| Timespan | 1999–2022 |
| Sources (journals, books, etc.) | 666 |
| Documents | 2227 |
| Average years from publication | 7.74 |
| Average citations per document | 29.8 |
| Average citations per year per doc | 3.047 |
| References | 48,804 |

topographic effects, which can be removed through modification of the indices or by using factor values.

Landsat series data, due to its open accessibility, is the most widely used satellite data to generate indices. Vegetation indices are used more frequently in the red and infrared bands, as these bands are sensitive to the photosynthesis capacity of leaves and plant canopies or the chlorophyll contents in the plant leaves. The absorption of red light by plant leaves is higher than that of infrared light. One of the most commonly used vegetation indices calculated from multispectral data is the Normalized Difference Vegetation Index (NDVI), which utilizes the red and near-infrared bands (Rouse et al., 1974; Karnieli et al., 2010). However, due to the amplifying effect in the end result of the vegetation indices, several correction factors were added, and the spectral indices were modified such as Atmospherically Resistant Vegetation Index (ARVI), Soil-Adjusted Vegetation Index (SAVI), Enhanced Vegetation Index (EVI), and many others. The Sentinel-1 SAR imagery and Sentinel-2 imagery with improved spatial and spectral resolution capacity led to numerous band ratios and vegetation indices.

The pervious and impervious surfaces in urban areas can be derived and classified using spectral indices such as NDBI, using a ratio combination of NIR and SWIR bands (Zha et al., 2003). The ShortWave Infrared (SWIR) band provides information about forest structure and density and is, therefore, commonly used in vegetation studies (Schroeder et al., 2011). Similarly, the TIR band is also employed to derive various built-up indices due to its sensitivity to albedo (Zhao and Chen, 2005). The combination of NIR, MIR, and TIR bands has been found to produce better results for built-up areas and can separate them from a bare surface (As-syakur et al., 2012). Several spectral indices for the extraction of built-up surfaces are devised due to the challenge of separating built-up surfaces from barren land. The land uses in urban areas require higher resolution data to obtain results with higher accuracy. A few built-up features, such as road extraction from remote-sensing data are a challenging task and require improved versions of spectral indices. Shahi et al. (2015) used a WorldView-2 multispectral high-resolution image and identified the significant bands using a field radio-spectrometer to devise automated techniques based on a spectral index (asphalt road extraction index, REI). Similarly, various indices have been developed to monitor atmospheric dynamics, snow cover, mining areas, etc. Indices are easily calculated and adopted and find specific uses and applications with some limitations such as additive noise effects among others. Hence, they are constantly being revised in the light of limitations and made more robust toward the specific requirement of the user. As advances in sensors and an increasing number of data products become available, new indices are being developed to serve the needs of the scientific community.

## 6.4 RS Sensors and Platforms: Characteristics (Spatial, Temporal, Spectral, and Radiometric Resolutions)

There are a large number of platforms on which several types of sensors (optical-hyperspectral-radar) are mounted for scanning and observing land features. These are known as space-borne satellites, airborne platforms (UAVs), and ground-based platforms (terrestrial laser scanners). Spectral indices use different spectral bands of remote-sensing data to quantify different aspects of elements of detection of any of the spheres of our planet, based on their optical, hyperspectral, and thermal properties. With the advancement in Earth Observation (EO) technology, mostly via space-borne images, research capacity has spread its wings in all directions of ice, snow, and glacier mapping, monitoring, and assessment at different scales. EO datasets have been harnessed to implement the VIR imagery, thermal, and microwave backscattering coefficients at the small local scale, intermediate regional scale, and global scale. With EO technology, researchers are widely implementing studies on lake ice, glaciers, freshwater ice, sea ice, icebergs, and snow in mountainous regions such as the Himalaya. With the advancement of EO datasets at different scales with multi-source, multi sensors, these datasets are helping in the analysis of the following parameters more accurately and more easily, such as:

- Spatial extents of the glaciers/icebergs/sea ice and change monitoring.
- Ice thickness/glacier thickness

- Glacier mass balance and assessment
- Topography as a whole/bedrock topography
- Elevation variation and Digital Elevation Models of glaciers, icebergs
- Ice temperature assessment
- Ice/ glaciers movement
- Accumulation rate of the ice/snow
- Disaster monitoring and assessment for ice sheets/ snow avalanche, glacier outburst and
- Several other applications based on the extent and types of the ice in question.

Further to EO assessment, the images acquired from different sensors (multi-sensors are employed for assessment and monitoring) use spectral reflectance to understand the variation in the different types of ice (new ice, thin ice, fast ice, iceberg), snow deposition, and even with the variation of grain size of snow at VIR ranges (400–2500 nm). Ice does not have very high reflectivity in the thermal range of 8 to 14 um (full spectrum 2–14 um), but can be estimated and differentiated from snow more easily. Also, thermal emissivity varies with grain size (ranging from 10 to 400 um). The emissivity of ice is up to 0.9 μm in the thermal range and exhibits total high absorbance at 10 μm in the region of the electromagnetic spectrum. With the fine grain size of snow and maximum absorbance at 10 μm, this ice-snow acts like a perfect blackbody, and thus is mostly used to perform research to differentiate features. Detailed published information regarding the VIR-thermal properties can be consulted in Kuittinen (1997).

## 6.5 Most Widely and Popularly Used Spectral Indices

### 6.5.1 Spectral Indices and Lithosphere

Surface and subsurface remote sensing along with the satellite-based Digital Elevation Model (DEM) has continuously supported the understanding of lithospheric structure and processes. Topographic complexities, geomorphic structures (fold, fault, fractures), and mineral compositions in rocks/geochemical characters provide signatures and patterns to be captured and recognized through satellite images. Remote sensing has been particularly used in mineral exploration, along with other traditional and modern methods of geosciences. Various EO systems such as MSS Landsat, Hyperspectral, and radar systems have been used to identify mineral compositions. Hyperspectral images, due to their high spectral resolution capacity, are used and processed to identify varieties of minerals with subtle differences in their spectral signatures. On the other hand, radar images provide detailed topographic structure impressions (Sabins, 1999). Most popularly use spectral indices used for quantifying lithospheric Elements of Detection (EoD) and their applications are provided in Table 6.2.

#### i) Mineral Indices

Band ratioing or indexing are the common methods in mineral deposit identification and require a thorough and careful understanding of the reflectance and absorption properties of various mineral ions in spectral regions of different EO data. Attempts were made to identify the sulfide deposits in the Arabian shield using Landsat and ASTER (Advanced Space-borne Thermal Emission and Reflection Radiometer) images through band ratioing (Velosky et al., 2003). Gad and Kusky (2007) used ASTER data for band ratioing (by using SWIR bands and TIR bands) in lithological mapping. Massironi et al. (2008) performed VNIR/ SWIR Band Ratios on ASTER imaging to identify Mg–OH bond stretching, Al–OH absorption of clay minerals, etc., and found hydrothermal alteration around mineralized veins. High spatial and spectral resolutions of ASTER data allow for mapping iron-oxide minerals in Visible Near Infrared (VNIR) bands, carbonate minerals in VNIR and SWIR bands, and silicate and granitoid rocks in TIR bands (Yamaguchi et al., 1998; Yamaguchi and Naito, 2003). In the study by Khalifa et al. (2020), the ASTER data was used for band ratioing of VNIR and SWIR bands in the lithological mapping of various formations and geological offsets. High-resolution World view images were used to extract the ion alteration anomalies using several mineral indices such as hydroxyl index, carbonate index, FeOH Index, Iron-Stained Index, etc. Similarly, the Landsat 8 OLI (Operational Land Imager) data, particularly its MIR bands, has been proved to be effective in the identification of the spatial distribution of hydrothermal alteration zones for gold mineralization potential areas through band ratioing techniques (Osinowo et al., 2021). Furthermore, there are various studies available on the potential of freely available Sentinel-2 images in geological mapping, particularly for the detection of iron-bearing minerals (Ge et al., 2020).

Several different indices are created from band ratioing to map geo-chemical elements after careful observation of the spectral response of various mineral deposits and calculation of the Optimum Index Factor (OIF). For example, various combinations of ASTER band ratios were created to identify clay and carbonate minerals by Fatima et al. (2017). The Iron Oxide Ratio uses Red and Blue bands due to absorption of iron oxide in the Blue band and reflectance in the Red band; the Ferrous Minerals Ratio uses SWIR and NIR bands; the Clay Ratio uses two bands of the SWIR region (Segal, 1982). The high-resolution World view images are very effective in identifying the iron oxide through the World-View New Iron Index (WV-II) which uses Green, Yellow, and Blue bands (Wolf, 2012).

**Table 6.2** List of spectral Indices related to the lithosphere.

| Serial No. | Index (Abbreviation) | Formulae | Satellite/Sensor | Applications | References |
|---|---|---|---|---|---|
| 1. | Clay Mineral Ratio (CMR) | $CMR = \frac{\lambda_{SWIR-I}}{\lambda_{SWIR-II}}$ | Landsat-8/OLI | Separate mine and non-mine water bodies; geological index for identifying mineral features containing clay and alunite | (Drury, 1987; Mukherjee et al., 2019) |
| 2. | Ferrous Minerals Ratio (FMR) | $FMR = \frac{(RED + SWR)}{NIR}$ | Landsat-8/OLI | Highlight Iron Bearing Zones | (Drury, 1987; Ducart et al., 2016) |
| 3. | Iron Oxide Ratio (IOR) | $IOR = \frac{Red}{Blue}$ | Landsat-8/OLI | Highlight Iron Bearing Zones | (Drury, 1987; Ducart et al., 2016) |
| 4. | World-View New Iron Index (WV-II) | $WV - II = \frac{(Green * Yellow)}{(Blue * 1000)}$ | WorldView 2 Vis-NIR MSI | To identify pixels rich in iron oxide | (Wolf, 2012) |
| 5. | World-View Soil Index (WV-SI) | $WV - SI = \frac{(Green - Yellow)}{(Green + Yellow)}$ | WorldView 2 Vis-NIR MSI | To identify pixels that primarily consist of soil | (Wolf, 2012) |
| 6. | Bare Soil Index (BSI) | $BSI = \frac{(B6 + B4) - (B5 + B2)}{(B6 + B4)} + (B5 + B2)$ | Landsat-8/OLI | To capture soil variations | (Bluemarblegeo, 2019; Gabri, 2019) |
| | | $BSI = \frac{(B5 + B3) - (B4 + B1)}{(B5 + B3)} + (B4 + B1)$ | Landsat 4–7/ TM-ETM+ | | |
| | | $BSI = \frac{(B11 + B4) - (B8 + B2)}{(B11 + B4)} + (B8 + B2)$ | Sentinel 2/MSI | | |
| 7. | Normalized Difference Mud Index (NDMI) | $NDMI = \frac{(\rho_{795} - \rho_{990})}{(\rho_{795} + \rho_{990})}$ | TWIN Otter/ AVIRIS | Highlights muddy or shallow water pixels | (Bernstein, 2012) |

### 6.5.2 Spectral Indices and Hydrosphere

Spectral indices also found their use in the assessment and monitoring of the hydrosphere. These are several indices developed to delineate water from other land-use/land-cover classes, to monitor flooded regions, and to monitor ice sheets and snow-ice extent in glaciated regions. This section deals with the spectral indices associated with the hydrosphere. Spectral indices used for quantifying hydrospheric component related Elements of Detection (EoD), e.g., snow, ice, water, etc., and their applications are provided in Table 6.3.

#### i) Snow Indices

The Snow Index generation depends upon the properties of the snow to reflect at different wavelengths in the region of the electromagnetic spectrum of 400–2500 nm. Considering the earliest simple ratio to identify the snow using Landsat™ or ETM+ sensors, Band 2 and Band 5 have been employed to generate a simple ratio (Band 2:Band 5). This is based on the fact that the reflectance of snow is very high in the wavelength range (500–600 nm) for Band 2 TM/ETM+ as compared to the lowest reflectance in the wavelength ranges 1.5–1.75 um (Band 5 of TM/ETM+). This gives the outcome of Band2:Band5 ratio outcome as relatively very high for snow-covered regions. The ratio is the most common band transformation information extraction technique using two-band values, when the reflectance is contrasting in both of the bands, as either low or high; pixel values assist in the diagnosis of the physical quantity of features and may sometimes not be very reliable due to erroneous outcome.

Normalized Difference Snow Index (NDSI) is more reliable, as it will generate the outcome ranging from −1 to +1.

$$SnowRatio = Band2 / Band5 \tag{6.1}$$

$$NDSI = (Band2 - Band5)/(Band2 + Band5) \tag{6.2}$$

The range of NDSI is from −1 to +1, while the ratio outcome may generate any values between 0 to infinity and

**Table 6.3** List of spectral Indices related to the hydrosphere.

| Serial No. | Index (Abbreviation) | Formulae | Satellite/ Sensor | Application | Reference |
|---|---|---|---|---|---|
| 1. | Normalized Difference Snow Index (NDSI) | $NDSI = \frac{(Green - SWIR)}{(Green + SWIR)}$ | EOS (TERRA and AQUA)/MODIS | Mapping, monitoring, and differentiating between snow-ice covered areas and non-snow-and-ice area | (Hall et al., 1995; Salomonson and Appel, 2004) |
| | | $NDSI = \frac{Green_{(ast\_b1)} - SWIR_{(ast\_b5)}}{Green_{(ast\_b1)} + SWIR_{(ast\_b5)}}$ | Landsat/TM | | (Dozier, 1989) |
| | | $NDSI(Landsat8) = \frac{(B3 - B6)}{(B3 + B6)}$ | Landsat 8/OLI | | (Bluemarblegeo, 2019) |
| | | $NDSI(Landsat4-7) = \frac{(B2 - B5)}{(B2 + B5)}$ | Landsat 4–7/TM | | (Erdenetuya et al., 2006) |
| | | $NDSI(Sentinel2) = \frac{(B3 - B11)}{(B3 + B11)}$ | Sentinel 2/MSI | | (Bluemarblegeo, 2019) |
| 2. | Normalized-Difference Debris Index (NDDI) | $NDDI = \frac{SWIR_{(ast\_b6)} - TIR_{(ast\_b14)}}{SWIR_{(ast\_b6)} + TIR_{(ast\_b14)}}$ | TERRA/ASTER | Mapping and differentiating between supraglacial debris and that of the terrain | (Shukla and Ali, 2016) |
| 3. | Normalized Difference Glacier Index (NDGI) | $NDGI = \frac{(NIR - Green)}{(NIR + Green)}$ OR | Landsat-8/OLI | Detect and monitor glaciers and mapping and differentiating between snow ice and ice-mixed debris | (Bluemarblegeo, 2019) |
| | | $NDGI(Landsat8) = \frac{(B3 - B4)}{(B3 + B4)}$ | | | |
| | | $NDGI(Landsat4-7) = \frac{(B2 - B3)}{(B2 + B3)}$ | Landsat 4–7/TM | | |
| | | $NDGI(Sentinel2) = \frac{(B3 - B4)}{(B3 + B4)}$ | Sentinel 2/MSI | | |
| 4. | Normalized Difference Water Index (NDWI) | $NDWI(Landsat8) = \frac{(B3 - B5)}{(B3 + B5)}$ | Landsat 8/OLI | Used for mapping analysis of surface water bodies | (Gabri, 2019) |
| | | $NDWI(Landsat4-7) = \frac{(B2 - B4)}{(B2 + B4)}$ | Landsat 4–7/TM | | (McFeeters, 1996) |
| | | $NDWI(Sentinel2) = \frac{(B3 - B8)}{(B3 + B8)}$ | Sentinel 2/MSI | | (Gabri, 2019) |
| | | $NDWI(Gao) = \frac{(NIR - MIR)}{(NIR + MIR)}$ | Landsat/TM | | (Gao, 1996) |
| 5. | Modified Normalized Difference Water Index (MNDWI) | $MNDWI = \frac{Green - SWIR}{Green + SWIR}$ | Landsat/ETM+ | Enhancing and extracting water from a background that is dominated by built-up land areas | (Xu, 2007) |
| 6. | World-View Water Index (WV-WI) | $WV - WI = \frac{(Coastal - NIR2)}{(Coastal + NIR2)}$ | WorldView 2/ Pan-Sharpened MSI | To highlight areas of standing water greater than one pixel in size | (Wolf, 2012) |

sometimes may cause erroneous results when band 5 values reach near to 0.

NDSI values help in the detection of snow at the study site. It is demonstrated that snow is mostly present when the NDSI value is greater than 0.4 (Hall et al., 1995), but recently it has been illustrated that the optimum range for snow detection using the NDSI index has a threshold value greater than 0.48 which has also been confirmed with field validations results (Vogel, 2002). The band implemented in the NDSI varies from sensor to sensor, but the wavelength range stays similar to the originally proposed index. For example, for Landsat TM/ETM+, Band 2 and Band 5 will work while for MODIS data, Band 4 and Band 6 are employed to generate the NDSI.

AVHRR (Advanced Very High-Resolution Radiometer), a multi-functional instrument, is also used to monitor cloud cover, vegetation, ice, snow coverage, and their characteristics at the global level. But this AVHRR instrument lacks spectral coverage in the range of 500–600 nm (equivalent to Band 2 of Landsat TM/ETM+). Similar to AVHRR, there is a VEGETATION SPOT in the SPOT series of satellites launched in 1998, which is utilized for snow monitoring, and VEGETATION SPOT also lacks the spectral coverage equivalent to Band 2 of Landsat.

Therefore, to monitor snow coverage, an alternate Normalized Difference ratio was proposed by Xiao et al. (2001) for VEGETATION SPOT, and is known as NDSII (Normalized Difference Snow and Ice Index). NDSII is represented as:

$$NDSII = \left(Band3 - Band5\right) / \left(Band3 + Band5\right) \quad (6.3)$$

where Band 3 corresponds to the red region of the electromagnetic wavelength spectrum (600–700 nm) and Band 5 corresponds to MIR (mid infra-red regions: 1.55–1.75 um) of the electromagnetic wavelength spectrum. VEGETATION SPOT generated NDSII can provide snow monitoring at a kilometer scale. Since 2000, MODIS (Moderate Resolution Imaging Spectroradiometer) images acquired from the Terra satellite have been utilized for snow mapping and monitoring at a finer scale (250 and 500 m). Snow coverage and monitoring from MODIS data utilize Band 2, Band 4, and Band 6. The NDSI for MODIS products are formulated as:

$$NDSI_{MODIS} = \left(Band4 - Band6\right) / \left(Band4 + Band6\right) \quad (6.4)$$

NDSI values are used together with NDVI values to improve the detectability of snow from densely vegetated regions. The methods implemented to monitor snow cover in an area not densely forested, are given by the equation of comparative values:

$$NDSI \geq 0.48 \text{ AND } R_2 \geq 0.11 \text{ } AND \text{ } R_4 \geq 0.10 \quad (6.5)$$

For the NDSI value, 0.4/0.48 refers to the previous discussion on NDSI, where $R_2$ corresponds to Band 2 and $R_4$ corresponds to Band 4.

**ii) Water Indices**

Over the past decades, remote-sensing communities has developed several indices to differentiate water from other land-use/land-cover classes. These efforts also include the differentiation of turbid water from clear water bodies as well as extracting the algae-laden water from other types of water bodies or within the extent of similar types of aquatic ecosystem.

**iii) Water Quality Indices**

### 6.5.3 Spectral Indices and Atmosphere

Over the past decades, remote-sensing techniques have improved our understanding of atmospheric processes and their dynamics. A wide range of satellite-based products from Geostationary and Polar orbiting satellite systems are available to capture information on radiation, energy exchange among the planetary systems, water vapor, different cloud properties, precipitation, land surface temperature, sea surface temperature, and chemical composition of the atmosphere, aerosols, and dust particles. Efforts are being made to enhance the technical capabilities to observe cloud and their properties (cloud height and optical thickness, etc.) through active RADAR and LIDAR sensors with high spatial and temporal resolution. Frequent coverage of weather elements by satellite sensors makes weather forecasting possible. EoDs constituting atmosphere; e.g., clouds, aerosols, etc., are tabularly presented in Table 6.4.

**i) Cloud Indices**

Clouds are important weather elements and play an important role in radiant energy flow in the Earth-atmosphere system, the precipitation mechanisms, and more. The identification of clouds is essential not only for understanding weather systems but also for retrieving surface properties in satellite data products. Clouds have the distinctive physical appearance of satellite images and affect image quality by hiding the features beneath them. Clouds are identified and monitored through their spectral signatures stored in different bands of images. Distinctive features of clouds are high reflectivity which gives them a brighter appearance in the visible region but a low luminosity/brightness temperature (BT), which gives them a darker appearance in the thermal infrared regions (Platnick et al., 2003; Zhu et al., 2015). However, the major problem encountered during cloud studies based on their spectral signature is their similarity with other features such as ice cover. The most commonly used methods to study cloud properties are thresholding and multi-spectral approaches (Bendix et al., 2004). Studies on the texture characteristics (e.g., smoothness or roughness) of clouds have increased the potential for cloud classification in more efficient ways, because the texture properties of clouds are less sensitive to atmospheric effects (Kaur and Ganju, 2008).

To date, various cloud-based indices are developed to study cloud optical depth and identify clouds and shadows. Marshak et al. (2000) studied the cloud optical depth of broken cloud using ground zenith radiation measurements by employing the bands from visible and near-infrared regions and proposed the Normalized Difference Cloud Index (NDCI). The method measures the spectral response of cloud and green vegetation and obtains information about the optical depth of a cloud by removing the radiation effects of the 3D cloud structure.

Zhai et al. (2018) developed an algorithm based on spectral indices (CSD-SI) for cloud/shadow detection using the visible and infrared spectral regions. The CSD-SI is based on two different indices: first, the Cloud Index (CI) developed by them to detect similarities in reflectance in RGB (red, green, and blue bands) and three infrared bands (1 NIR and 2 SWIR bands). It detected the brightness properties of the cloud in comparison to other features and furthermore, using threshold values, the clouds were the brighter objects/features. The second index is the Cloud Shadow Index (CSI) which is based on the NIR and SWIR bands to capture the reflective properties of shadows in the longer wavelength region and the blue bands to remove water from the region. In another method, Karthik et al. (2020) detected cloud pixels through the cloud-index dynamic-threshold (CIDT) algorithm that is based on the spectral and spatial characteristics of clouds. The result could also identify cloud contamination with higher accuracy. The index used top-atmospheric reflection in conjunction with dynamic threshold values to detect cloud pixels.

### 6.5.4 Spectral Indices and Biosphere

Remote sensing has provided a great deal of support to map and monitor the ecological functions of the biosphere, thereby providing insights to safeguard the ecosystems. Right from the beginning of remote-sensing techniques, the world has obtained capabilities to monitor different components of the biosphere, particularly the vegetation cover. Red (Visible) and NIR wavelength ranges has been found suitable for spatio-temporal observation of vegetation. A great deal of literature is available on the spectral response of plant characteristics such as plant types, structure, physiology, water content, and interaction with chemicals (Chang et al., 2016). Various techniques have been employed to map vegetation covers such as image processing, image interpretation, use of indices, image classification, image fusion, segmentation, and others. Biospheric EoDs quantifiable using spectral indices are given in Table 6.5.

#### i) Vegetation Indices

The spectral signatures recorded in the red and infrared regions have been widely used to generate vegetation indices. The chlorophyll content of a plant absorbs the visible radiation at wavelengths around 700 nm, while the response changes abruptly around wavelengths of 760–900 nm when the leaves and the cellular structures of a plant strongly reflect the radiation in the near-infrared region (Major et al., 1990). It is this variable response in the two different regions that gives an indication of the presence of green vegetation (Bannari et al., 1995) and can be captured using indices. Indices are the combinations of various bands which represent the quantitative characteristics of the vegetation vigor (Campbell, 1987). The sensitivity to the vegetation characteristics is maximized while the characteristics of other surface features are minimized through the indices, thereby allowing to extract vegetation information efficiently (Fang and Liang, 2014). The first vegetation index, called the Ratio Vegetation Index (RVI), was proposed by Pearson and Miller (1972). Later, Rouse et al. (1974) proposed NDVI which became popular among the scientific community (Karnieli et al., 2010) and has been widely applied in local, regional, and global monitoring and assessment of forest cover. However, soon after the realization of the sensitivities of these indices to soil brightness and atmospheric effects, a new vegetation index was proposed by Kaufman and Tanre (1992) in the form of the Atmospherically Resistant Vegetation Index (ARVI) wherein they used the blue band to modify the spectral value of the red band so that the effect of the atmosphere could be removed. Similarly, various scholars tried to reduce the effect of soil brightness on the end product through separate indices (Richardson and Wiegang, 1977). For example, Huete (1988) devised the Soil-Adjusted Vegetation Index (SAVI) and introduced the L factor (soil conditioning factor) to reduce the soil effect.

Other vegetation indices were developed to monitor the chlorophyll response in the narrow range spectrum to provide information about the photosynthetic material and the overall health of the vegetation. The Chlorophyll Absorption Ratio Index (CARI) was developed by Kim et al. (1994) by using the visible region along with the red edge (it is based on the principle that the absorption by chlorophyll occurs at 670 nm, while reflection occurs at 550 nm and 700 nm). Daughtry et al. (2000) measured the dynamics of chlorophyll concentrations as an indicator of

**Table 6.4** List of spectral Indices related to the biosphere.

| Serial No. | Index (Abbreviation) | Formulae | Satellite/ Sensor | Application | Reference |
|---|---|---|---|---|---|
| 1. | Normalized Difference Vegetation Index (NDVI) | $NDVI = \frac{(NIR - Red)}{(NIR + Red)}$ | ERTS(Landsat-1)/MSS | Vegetation content | (Rouse et al., 1973) |
| | | $NDVI(Landsat8) = \frac{(B5 - B4)}{(B5 + B4)}$ | Landsat-8/OLI | | (Gabri, 2019) |
| | | $NDVI(Landsat5) = \frac{(B4 - B3)}{(B4 + B3)}$ | Landsat-4-7/TM | | |
| | | $NDVI(Sentinel2) = \frac{(B8 - B4)}{(B8 + B4)}$ | Sentinel-2/MSI | | |
| 2. | Modified Red Edge NDVI (*MRENDVI*) | $MRENDVI = \frac{\rho_{750} - \rho_{705}}{\rho_{750} + \rho_{705} - 2 * \rho_{445}}$ | Unispec, PP Systems /field portable spectrometer | Precision agriculture, forest monitoring, and vegetation stress detection | (Datt, 1999; Sims and Gamon, 2002) |
| 3. | Green Normalized Difference Vegetation Index (GNDVI) | $GNDVI = \frac{(NIR - Green)}{(NIR + Green)}$ | EOS/MODIS | To be more sensitive to the variation of chlorophyll content in the crop | (Gitelson et al., 1996; Gabri, 2019) |
| | | $NDSI(Landsat8) = \frac{(B3 - B6)}{(B3 + B6)}$ | Landsat 8/OLI | | |
| | | $GNDVI(Landsat4-7) = \frac{(B4 - B2)}{(B4 + B2)}$ | Landsat 4-7/TM | | |
| | | $GNDVI(Sentinel2) = \frac{(B8 - B3)}{(B8 + B3)}$ | Sentinel2/MSI | | |
| 4. | Enhanced Vegetation Index (EVI) | $EVI = 2.5^{*}\left(\frac{(NIR - Red)}{(NIR + 6^{*} Red - 7.5^{*} Blue) + 1}\right)$ | Landsat/ETM+ | To quantify vegetation greenness | (Huete et al., 2002; Mokarram et al., 2015) |
| | | $EVI(Landsat8) = 2.5^{*}\left(\frac{(B5 - B4)}{(B5 + 6^{*} B4 - 7.5^{*} B2) + 1}\right)$ | Landsat8/OLI | | (USGS, 2019c) |
| | | $EVI(Landsat4-7) = 2.5^{*}\left(\frac{(B4 - B3)}{(B4 + 6^{*} B3 - 7.5^{*} B1) + 1}\right)$ | Landsat 4–7/TM | | |
| | | $EVI(Sentinel2) = 2.5^{*}\left(\frac{(B8 - B4)}{(B8 + 6^{*} B4 - 7.5^{*} B2) + 1}\right)$ | Sentinel/MSI | | |

| | | | | | |
|---|---|---|---|---|---|
| 5. | Advanced Vegetation Index (AVI) | $AVI = \left[IR^{*}\left(255 - R\right)^{*}\left(IR - R\right)\right]^{\frac{1}{3}}$ | Landsat/TM | Forest density and physiognomic vegetation classes | (Roy et al., 1996) |
| | | $AVI = \left[NIR^{*}\left(1 - Red\right)^{*}\left(NIR - Red\right)\right]^{\frac{1}{3}}$ | — | | (Gabri, 2019) |
| | | $AVI = \left[B5^{*}\left(1 - B4\right)^{*}\left(B5 - B4\right)\right]^{\frac{1}{3}}$ | Landsat8-/OLI | | |
| | | $AVI = \left[B4^{*}\left(1 - B3\right)^{*}\left(B4 - B3\right)\right]^{\frac{1}{3}}$ | Landsat-4-7/TM | | |
| | | $AVI = \left[B8^{*}\left(1 - B4\right)^{*}\left(B8 - B4\right)\right]^{\frac{1}{3}}$ | Sentinel-2/MSI | | |
| 6. | Soil-Adjusted Vegetation Index (SAVI) | $\left[\frac{\left(NIR - Red\right)}{\left(NIR + Red + 0.5\right)}\right]^{*}\left(1.5\right)$ | Landsat-8/OLI | In areas with relatively sparse vegetation where soil is visible through the canopy | (Huete, 1988; USGS, 2019a, 2019b) |
| | | $SAVI\left(Landsat8\right) = \left[\frac{\left(B5 - B4\right)}{\left(B5 + B4 + 0.5\right)}\right]^{*}\left(1.5\right)$ | | | |
| | | $SAVI\left(Landsat4-7\right) = \left[\frac{\left(B4 - B3\right)}{\left(B4 + B3 + 0.5\right)}\right]^{*}\left(1.5\right)$ | Landsat 4–7/TM-ETM+ | | |
| | | $SAVI\left(Sentinel2\right) = \left[\frac{\left(B08 - B04\right)}{\left(B08 + B04 + 0.428\right)}\right]^{*}\left(1.428\right)$ | Sentinel-2/MSI | | |
| 8. | Normalized Difference Moisture Index (NDMI) | $NDMI\left(Landsat8\right) = \frac{\left(B5 - B6\right)}{\left(B5 + B6\right)}$ | Landsat 8/OLI | Determine vegetation water content | (Welikhe et al., 2017) |
| | | $NDMI\left(Landsat4-7\right) = \frac{\left(B4 - B5\right)}{\left(B4 + B5\right)}$ | Landsat 4–7/TM-ETM+ | | |
| | | $NDMI\left(Sentinel2\right) = \frac{\left(B8 - B11\right)}{\left(B8 + B11\right)}$ | Sentinel 2/MSI | | |
| 9. | Moisture Stress Index (MSI) | $MSI\left(Landsat8\right) = \frac{B6}{B5}$ | Landsat 8/OLI | Canopy stress analysis, productivity prediction, and biophysical modeling | (Welikhe et al., 2017; USGS, 2019a) |
| | | $MSI\left(Landsat4-7\right) = \frac{B5}{B4}$ | Landsat 4–7/TM-ETM+ | | |
| | | $MSI\left(Sentinel2\right) = \frac{B11}{B08}$ | Sentinel 2/MSI | | |

*(Continued)*

**Table 6.4** (Continued)

| Serial No. | Index (Abbreviation) | Formulae | Satellite/ Sensor | Application | Reference |
|---|---|---|---|---|---|
| **Broadband Greenness** | | | | | |
| 10. | Green Coverage Index (GCI) | $GCI = \left(\frac{\rho NIR}{\rho Green}\right) - 1$ | | used to estimate the content of leaf chlorophyll in various species of plants | (Gitelson et al., 2003) |
| 11. | Difference Vegetation Index (DVI) | $DVI = NIR - Red$ OR $DVI = B5 - B4$ | Landsat/MSS | Distinguishes between soil and vegetation | (Tucker, 1979) |
| | | | Landsat/ETM+ | | (Mokarram et al., 2015) |
| | | $DVI = B8 - B4$ | Sentinel-2/MSI | | (Jordan, 1969) |
| 12. | Enhanced Vegetation Index (EVI) | $EVI = 2.5 * (NIR - Red) / (NIR + 6 * Red - 7.5 * Blue + 1)$ | Landsat-8/OLI | It is most useful in high LAI regions where NDVI may saturate | (Huete et al., 2002) |
| | | $EVI(Sentinel - 2) = 2.5 * ((B8 - B4) / (B8 + 6 * B4 - (7.5 * B2) + 1))$ | Sentinel-2/MSI | | |
| 13. | Global Environmental Monitoring Index (GEMI) | $GEMI = eta(1 - 0.25 * eta) - \frac{Red - 0.125}{1 - Red}$ *Where,* $eta = \frac{2(NIR^2 - Red^2) + 1.5 * NIR + 0.5 * Red}{NIR + Red + 0.5}$ | POES&MetOp/AVHRR | It is similar to NDVI but is less sensitive to atmospheric effects | (Pinty et al., 2011) |
| 14. | Green Atmospherically Resistant Index (GARI) | $GARI = \frac{NIR - [Green - \gamma(Blue - Red)}{NIR + [Green - \gamma(Blue - Red)}$ | EOS/MODIS+ Hitachi-150-20 | Chlorophyll concentrations | (Gitelson et al., 1996) |
| 15. | Green Coverage Index (GCI) | $GCI(Landsat8) = \left(\frac{B5}{B3}\right) - 1$ | Landsat 8/OLI | To estimate leaf chlorophyll content across a wide range of plant species | (Gabri, 2019) |
| | | $GCI(Landsat - 7) = \left(\frac{B4}{B2}\right) - 1$ | Landsat 4–7/TM | | |
| | | $GCI(Sentinel2) = \left(\frac{B9}{B3}\right) - 1$ | Sentinel 2/MSI | | |

| | | | | | |
|---|---|---|---|---|---|
| 16. | Green Difference Vegetation Index (GDVI) | $GDVI = NIR - Green$ | Landsat-8/OLI<br>Trimble AG 132, 35-mm Canon AE-1 camera & Kodak Ektachrome professional Infrared EIR 135–36 film along with a TIFFEN 52-mm Yellow No. 12 filter | To predict nitrogren requirements for corn | (Tucker, 1979; Sripada, 2005; Mokarram et al., 2015) |
| 17. | Green Leaf Index (GLI) | $GLI = \frac{(Green - Red) + (Green - Blue)}{(2 * Green) + Red + Blue}$ | Kodak® Royal Gold® ISO 400 film in a 35-mm camera fitted with a 28-mm wide-angle lens, mounted on a single-engine fixed-wing aircraft | To measure wheat cover | (Louhaichi et al., 2001) |
| 18. | Green Normalized Difference Vegetation Index (GNDVI) | $GNDVI = \frac{(NIR - Green)}{(NIR + Green)}$ | Hitachi 150-20 spectrophotometer, UV-2001 PC, Shimadzu, LICOR LI-1800 spectral radiometer | More sensitive to chlorophyll concentration than NDVI | (Gitelson and Merzlyak, 1998) |
| 19. | Green Optimized Soil Adjusted Vegetation Index (GOSAVI) | $GOSAVI = \frac{NIR - Green}{NIR + Green + 0.16}$ | Trimble AG 132, 35-mm Canon AE-1 camera and Kodak Ektachrome professional Infrared EIR 135–36 film along with a TIFFEN 52-mm Yellow No. 12 filter | To predict nitrogen requirements for corn | (Sripada, 2005) |
| 20. | Green Ratio Vegetation Index (GRVI) | $GRVI = \frac{NIR}{Green}$ | Trimble AG 132, 35-mm Canon AE-1 camera and Kodak Ektachrome professional Infrared EIR 135–36 film along with a TIFFEN 52-mm Yellow No. 12 filter | Sensitive to photosynthetic rates in forest canopies | (Sripada et al., 2006) |
| 21. | Green Soil Adjusted Vegetation Index (GSAVI) | $GSAVI = 1.5 * \frac{(NIR - Green)}{(NIR + Green + 0.5)}$ | Trimble AG 132, 35-mm Canon AE-1 camera and Kodak Ektachrome professional Infrared EIR 135–36 film along with a TIFFEN 52-mm Yellow No. 12 filter | To predict nitrogen requirements for corn | (Sripada, 2005) |

*(Continued)*

**Table 6.4** (Continued)

| Serial No. | Index (Abbreviation) | Formulae | Satellite/ Sensor | Application | Reference |
|---|---|---|---|---|---|
| 22. | Green Vegetation Index (GVI) | $GVI = (-0.2848 * TM_1) + (-0.2435 * TM_2) + (-0.5436 * TM_3) + (0.7243 * TM_4) + (0.0840 * TM_5) + (-0.1800 * TM_7)$ | Landsat/TM and ETM+ | Landsat TM Tasselled Cap green vegetation index | (Kauth, 1976) |
| 23. | Infrared Percentage Vegetation Index (IPVI) | $IPVI = \frac{NIR}{(NIR + R)}$ | Landsat 8 ETM+ | Functionally the same as NDVI, but computationally faster | (Crippen, 1990; Mokarram et al., 2015) |
| 24. | Leaf Area Index (LAI) | $LAI = (3.618 * EVI - 0.118)$ | CASI | To estimate foliage cover and to forecast crop growth and yield | (Boegh et al., 2002) |
| 25. | Modified Non-Linear Index (MNLI) | $MNLI = \frac{(NIR^2 - Red) * (1 + L)}{NIR^2 + Red + L}$ | SPOT | Simpler version of the MSAVI index | (Qi et al., 1994; Yang et al., 2008) |
| 26. | Modified Simple Ratio (MSR) | $MSR = \frac{\left(\frac{NIR}{Red}\right) - 1}{\left(\sqrt{\frac{NIR}{Red}}\right) - 1}$ | Landsat-5/TM | Retrieving biophysical parameters of boreal forests | (Chen, 1996) |
| 27. | Modified Soil Adjusted Vegetation Index 2 (MSAVI2) | $MSAVI2 = \frac{2 * NIR + 1 - \sqrt{(2 * NIR + 1)^2 - 8(NIR - Red)}}{2}$ | MMR and SPOT | Simpler version of the MSAVI index | (Qi et al., 1994) |
| 28. | Non-Linear Index (NLI) | $NLI = \frac{NIR^2 - Red}{NIR^2 + Red}$ | ——— | Relationship between many vegetation indices and surface biophysical parameters is non-linear | (Goel and Qin, 1994) |
| 29. | Normalized Difference Vegetation Index (NDVI) | $NDVI = \frac{(NIR - Red)}{(NIR + Red)}$ | ERTS-1/MSS | Measure of healthy, green vegetation | (Rouse et al., 1973) |
| 30. | Optimized Soil Adjusted Vegetation Index (OSAVI) | $OSAVI = \frac{(NIR - Red)}{(NIR + Red + 0.16)}$ | Landsat/MSS | Best used in areas with relatively sparse vegetation where soil is visible through the canopy | (Rondeaux et al., 1996) |
| 31. | Renormalized Difference Vegetation Index (RDVI) | $RDVI = \frac{(NIR - Red)}{\sqrt{(NIR + Red)}}$ | NOAA/AVHRR | Highlights healthy vegetation | (Roujean and Breon, 1995) |

| | | | | | |
|---|---|---|---|---|---|
| 32. | Soil Adjusted Vegetation Index (SAVI) | $SAVI = \frac{1.5*(NIR - Red)}{(NIR + Red + 0.5)}$ | — | Best used in areas with relatively sparse vegetation where soil is visible through the canopy | (Huete, 1988) |
| | | $SAVI = \frac{(\rho NIR - \rho RED)}{\rho NIR + \rho RED + L}(1 + L)$ | Modular Multispectral Radiometer(MMR) | | (Qi et al., 1994) |
| 33. | Simple Ratio (SR) | $SR = \frac{NIR}{Red}$ | A single-beam spectrophotometer | Indicates and differentiates between healthy vegetation, soil, water, or ice | (Birth and McVey, 1968) |
| 34. | Transformed Difference Vegetation Index (TDVI) | $TDVI = 1.5\left[\frac{(NIR - Red)}{\sqrt{NIR^2 + Red + 0.5}}\right]$ | Portable GER-3700 spectroradiometer (350 to 2500 nm) | For monitoring vegetation cover in urban environments | (Bannari, et al., 2002) |
| 35. | Triangular Greenness Index (TGI) | $TGI = \frac{(\lambda_{Red} - \lambda_{Blue})(\rho_{Red} - \rho_{Green}) - (\lambda_{Red} - \lambda_{Green})(\rho_{Red} - \rho_{Blue})}{2}$ | A narrow-band imaging spectrometer, the Landsat TM multispectral sensor, and a commercial digital camera and ASD FieldSpec Pro FR and LI-COR LI 1800–12 | Leaf chlorophyll content | (Raymond Hunt et al., 2011) |
| 36. | Visible Atmospherically Resistant Index (VARI) | $VARI = \frac{Green - Red}{Green + Red - Blue}$ | Kodak DC-40 digital camera, LI-1800 spectroradiometer, EOS/MODIS, MERIS and Spectron Engineering SE-590 portable spectroradiometer; Landsat/ETM+ | To estimate the fraction of vegetation in a scene with low sensitivity to atmospheric effects | (Gitelson et al., 2002a; Mokarram et al., 2015) |
| 37. | *Wide Dynamic Range Vegetation Index* (WDRVI) | $WDRVI = \frac{(a * NIR - Red)}{(a * NIR + Red)}$ | Licor LI-1800 spectroradiometer, Spectron Engineering SE-590 portable spectroradiometer and Kodak DC-40 camera | Similar to NDVI, but it uses a weighting coefficient ($a$) to reduce the disparity between the contributions of the near-infrared and red signals to the NDVI | (Gitelson, 2004; Henebry et al., 2004) |
| 38. | WorldView Improved Vegetation Index (WV-VI) | $WV - VI = \frac{(NIR2 - Red)}{(NIR2 + Red)}$ | WorldView/MSI | To Support Land Mapping and Feature Extraction | (Wolf, 2012) |

*(Continued)*

Table 6.4 (Continued)

| Serial No. | Index (Abbreviation) | Formulae | Satellite/ Sensor | Application | Reference |
|---|---|---|---|---|---|
| **Narrowband Greenness** | | | | | |
| 39. | Atmospherically Resistant Vegetation Index (ARVI) | $ARVI = \frac{\rho_{800} - [\rho_{680} - \gamma(\rho_{450} - \rho_{680})]}{\rho_{800} + [\rho_{680} - \gamma(\rho_{450} - \rho_{680})]}$ or $(NIR - (Red - 1*(Blue - \mathrm{Red}))) / (\mathrm{NIR} + (\mathrm{Red} - 1*(\mathrm{Blue} - \mathrm{Red})))$ | EOS/MODIS, Landsat/TM and EOS/HIRIS | Enhanced version of NDVI (Vegetation) | (Kaufman and Tanre, 1992) |
| 40. | Modified Chlorophyll Absorption Ratio Index (MCARI) | $MCARI = [(\rho_{700} - \rho_{670}) - 0.2(\rho_{700} - \rho_{550})] * \left(\frac{\rho_{700}}{\rho_{670}}\right)$ | Model 20C Exotech spectroradiometer and Landsat/TM | Relative abundance of chlorophyll and measure of chlorophyll concentration | (Daughtry et al., 2000; Haboudane et al., 2004) |
| 41. | Modified Chlorophyll Absorption Ratio Index – Improved (MCARI2) | $MCARI2 = \frac{1.5[2.5(\rho_{800} - \rho_{670}) - 1.3(\rho_{800} - \rho_{500})]}{\sqrt{(2*\rho_{800} + 1)^2 - (6*\rho_{800} - 5*\sqrt{\rho_{670}}) - 0.5}}$ | CASI/Li-Cor<br>CASI – Compact Airborne Spectrographic Imager | To consider a better predictor of green Leaf Area Index | (Haboudane et al., 2004) |
| 42. | Modified Red Edge Normalized Difference Vegetation Index (MRENDVI) | $MRENDVI = \frac{\rho_{750} - \rho_{705}}{\rho_{750} + \rho_{705} - 2*\rho_{445}}$ | Unispec, PP Systems /field portable spectrometer | Precision agriculture, forest monitoring, and vegetation stress detection | (Datt, 1999; Sims and Gamon, 2002) |
| 43. | Modified Red Edge Simple Ratio (MRESR) | $MRESR = \frac{\rho_{750} - \rho_{445}}{\rho_{705} - \rho_{445}}$ | Unispec, PP Systems /field portable spectrometer | Precision agriculture, forest monitoring, and vegetation stress detection | (Datt, 1999; Sims and Gamon, 2002) |
| 44. | Modified Triangular Vegetation Index (MTVI) | $MTVI = 1.2[1.2(\rho_{800} - \rho_{550}) - 2.5(\rho_{670} - \rho_{550})]$ | Plant Canopy Analyzer (LAI-2000; Li-Cor) and Area Meter (LI-3100,Li-Cor)<br>CASI – Compact Airborne Spectrographic Imager | Leaf Area Index estimations | (Haboudane et al., 2004) |
| 45. | Modified Triangular Vegetation Index – Improved (MTVI2) | $MTVI2 = \frac{1.5[1.2(\rho_{800} - \rho_{550}) - 2.5(\rho_{670} - \rho_{550})]}{\sqrt{(2*\rho_{800} + 1)^2 - (6*\rho_{800} - 5*\sqrt{\rho_{670}}) - 0.5}}$ | Plant Canopy Analyzer (LAI-2000; Li-Cor) and Area Meter (LI-3100, Li-Cor)<br>CASI – Compact Airborne Spectrographic Imager | Better predictor of green LAI | (Haboudane et al., 2004) |

| | | | | | |
|---|---|---|---|---|---|
| 46. | (RENDVI) | $RENDVI = \frac{\rho_{750} - \rho_{705}}{\rho_{750} + \rho_{705}}$ | Hitachi 150-20 spectrophotometer and Unispec, PP Systems/field portable spectrometer | Modified traditional broadband NDVI applied in precision agriculture, forest monitoring, and vegetation stress detection | (Gitelson and Merzlyak, 1994; Sims and Gamon, 2002) |
| 47. | Red Edge Position Index (REPI) | $REPI = 700 + 40\left[\frac{R_{rededge} - R_{700}}{R_{740} - R_{700}}\right]$ | Spectroradiometer | Crop monitoring and yield prediction, ecosystem disturbance detection, photosynthesis modeling, and canopy stress caused by climate and other factors | (Curran et al., 1995) |
| 48. | Transformed Chlorophyll Absorption Reflectance Index (TCARI) | $TCARI = 3\left[\left(\rho_{700} - \rho_{670}\right) - 0.2\left(\rho_{700} - \rho_{550}\right)\left(\frac{\rho_{700}}{\rho_{670}}\right)\right]$ | CASI/Li-Cor/ Compact Airborne Spectrographic Imager | Indicates the relative abundance of chlorophyll | (Haboudane et al., 2004) |
| 49. | Triangular Vegetation Index (TVI) | $TVI = \frac{120\left(\rho_{750} - \rho_{550}\right) - 200\left(\rho_{670} - \rho_{550}\right)}{2}$ | Landsat-5/TM/ CASI – Compact Airborne Spectrographic Imager | Good for estimating green LAI, but its sensitivity to chlorophyll increases with an increase in canopy density | (Broge and Leblanc, 2001) |
| 50. | Vogelmann Red Edge Index 1 (VREI1) | $VREI1 = \frac{\rho_{740}}{\rho_{720}}$ | VIRIS | Vegetation phenology (growth) studies, precision agriculture, and vegetation productivity modeling | (Vogelmann et al., 1993) |
| 51. | Vogelmann Red Edge Index 2 (VREI2) | $VREI2 = \frac{\rho_{734} - \rho_{747}}{\rho_{715} + \rho_{726}}$ | VIRIS | Vegetation phenology (growth) studies, precision agriculture, and vegetation productivity modeling | (Vogelmann et al., 1993) |
| **Narrowband Greenness (Canopy Nitrogen)** | | | | | |
| 52. | Normalized Difference Nitrogen Index (NDNI) | $NDNI = \frac{loglog\left(\frac{1}{\rho_{1510}}\right) - loglog\left(\frac{1}{\rho_{1680}}\right)}{loglog\left(\frac{1}{\rho_{1510}}\right) + loglog\left(\frac{1}{\rho_{1680}}\right)}$ | AVIRIS | To estimate the relative amounts of nitrogen contained in vegetation canopies | (Fourty et al., 1996; Serrano et al., 2002) |
| **Narrowband Greenness (Canopy Water Content)** | | | | | |
| 53. | Moisture Stress Index (MSI) | $MSI = \frac{\rho_{1599}}{\rho_{819}}$ | Landsat/TM | Reflectance measurement that is sensitive to increasing leaf water content | (Hunt and Rock, 1989; Ceccato et al., 2001) |

*(Continued)*

Table 6.4 (Continued)

| Serial No. | Index (Abbreviation) | Formulae | Satellite/ Sensor | Application | Reference |
|---|---|---|---|---|---|
| 54. | Normalized Difference Infrared Index (NDII) | $NDII = \frac{(\rho_{819} - \rho_{1649})}{(\rho_{819} - \rho_{1649})}$ | Prototype to the Goddard Space Flight Center Mark II three band radiometer and Landsat-D Thematic Mapper | Crop agricultural management, forest canopy monitoring vegetation stress detection and sensitive to changes in water content of plant canopies | (Hardisky et al., 1983) |
| 55. | Normalized Difference Water Index (NDWI) | $NDWI = \frac{(\rho_{857} - \rho_{1241})}{(\rho_{857} - \rho_{1241})}$ | Beckman 5270 spectrometer and ER-2 aircraft/AVIRIS | Used for mapping analysis of surface water bodies | (Gao B C, 1996; Jackson et al., 2004) |
| 56. | Normalized Multi-band Drought Index (NMDI) | $NMDI = \frac{\rho_{860} - (\rho_{1640} - \rho_{2130})}{\rho_{860} + (\rho_{1640} - \rho_{2130})}$ | Terra/MODIS | Monitoring soil and vegetation moisture | (Wang and Qu, 2007; Wang et al., 2008) |
| 57. | Water Band Index (WBI) | $WBI = \frac{\rho_{970}}{\rho_{900}}$ | LI-cor/ LAI-2000, LI 1800-12s and GER1500 Portable spectroradiometer | Canopy stress analysis, productivity prediction and modeling, fire hazard condition analysis, cropland management, and studies of ecosystem physiology | (Penuelas et al., 1993; Champagne et al., 2001) |
| **Narrowband Greenness (Dry or Senescent Carbon)** | | | | | |
| 58. | Cellulose Absorption Index (CAI) | $CAI = 0.5(\rho_{2000} - \rho_{2200}) - \rho_{2100}$ | Field Spec Pro spectroradiometer | Crop residue monitoring, plant canopy senescence, fire fuel conditions in ecosystems, and grazing management | (Daughtry, 2001; Daughtry et al., 2004) |
| 59. | Lignin Cellulose Absorption Index (LCAI) | $LCAI = 100 \times [(\rho_{2185to2225} - \rho_{2145to2185}) + (\rho_{2185to2225} - \rho_{2295to2365})]$ | ASD FieldSpec Pro spectroradiometer and AVIRIS | To monitor crop residue | (Daughtry et al., 2005) |
| 60. | Normalized Difference Lignin Index (NDLI) | $NDLI = \frac{loglog\left(\frac{1}{\rho_{1754}}\right) - loglog\left(\frac{1}{\rho_{1680}}\right)}{loglog\left(\frac{1}{\rho_{1754}}\right) + loglog\left(\frac{1}{\rho_{1680}}\right)}$ | AVIRIS | Relative amounts of lignin contained in vegetation canopies | (Melillo et al., 1982; Fourty et al., 1996; Serrano et al., 2002) |
| 61. | Plant Senescence Reflectance Index (PSRI) | $PSRI = \frac{\rho_{680} - \rho_{500}}{\rho_{750}}$ | Hitachi 150-20 Spectrophotometer | Vegetation health monitoring, plant physiological stress detection, and crop production and yield analysis | (Merzlyak et al., 1999) |

| | | | | | |
|---|---|---|---|---|---|
| **Narrowband Greenness (Leaf Pigments)** | | | | | |
| 62. | Anthocyanin Reflectance Index 1 (ARI1) | $ARI1 = \frac{1}{\rho_{550}} - \frac{1}{\rho_{700}}$ | 150-20 Hitachi spectrophotometer and Licor LI-1800 spectroradiometer | To measure stressed vegetation | (Gitelson et al., 2001) |
| 63. | Anthocyanin Reflectance Index 2 (ARI2) | $ARI2 = \rho_{800}\left[\frac{1}{\rho_{550}} - \frac{1}{\rho_{700}}\right]$ | 150-20 Hitachi spectrophotometer and Licor LI-1800 spectroradiometer | Detects higher concentrations of anthocyanins in vegetation | (Gitelson et al., 2001) |
| 64. | Carotenoid Reflectance Index 1 (CRI1) | $CRI1 = \frac{1}{\rho_{510}} - \frac{1}{\rho_{550}}$ | Hitachi 150-20 spectrophotometer and Shimatzu 2101 PC spectrophotometer | Measure of stressed vegetation | (Gitelson et al., 2002b) |
| 65. | Carotenoid Reflectance Index 2 (CRI2) | $CRI2 = \frac{1}{\rho_{510}} - \frac{1}{\rho_{700}}$ | Hitachi 150-20 spectrophotometer & Shimatzu 2101 PC spectrophotometer | Modification to CRI1 that provides better results in areas of high carotenoid concentration | (Gitelson et al., 2002b) |
| **Narrowband Greenness (Light Use Efficiency)** | | | | | |
| 66. | Photochemical Reflectance Index (PRI) | $PRI = \frac{\rho_{531} - \rho_{570}}{\rho_{531} + \rho_{570}}$ | Pulse-amplitude-modulated Fluorometer PAM-2000 and LI-6200, Li-Cor, model LC-10AS HPLC | Used in studies of vegetation productivity and stress and applicable to monitor vegetation health in evergreen shrublands, forests, and agricultural crops prior to senescence | (Penuelas et al., 1995; Gamon et al., 1997) |
| 67. | Structure Insensitive Pigment Index (SIPI) | $SIPI = \frac{\rho_{800} - \rho_{445}}{\rho_{800} - \rho_{680}}$ OR $SIPI = \frac{(NIR - Bue)}{(NIR - Red)}$ | Varian Cary 17.DI spectrophotometer and GER IRIS spectrophotometer | Vegetation health monitoring, plant physiological stress detection, and crop production and yield analysis | (Penuelas, Iolanda Eilella and Gamon, 1995) |
| 68. | Red-Green Ratio Index (RGRI) | $RGRI = \frac{\sum_{i=600}^{699} Ri}{\sum_{j=500}^{599} Rj}$ | ResourceSat-1 AWIFS | An indicator of leaf production and stress, and may also indicate flowering in some canopies | (Gamon and Surfus, 1999; Yang et al., 2008) |
| **Narrowband Greenness** | | | | | |
| 69. | Atmospherically Resistant Vegetation Index (ARVI) | $ARVI = \frac{\rho_{800} - \left[\rho_{680} - \gamma(\rho_{450} - \rho_{680})\right]}{\rho_{800} + \left[\rho_{680} - \gamma(\rho_{450} - \rho_{680})\right]}$ | EOS/MODIS, Landsat/TM and EOS/HIRIS | Enhanced version of NDVI (Vegetation) | (Kaufman and Tanre, 1992) |

*(Continued)*

**Table 6.4** (Continued)

| Serial No. | Index (Abbreviation) | Formulae | Satellite/ Sensor | Application | Reference |
|---|---|---|---|---|---|
| 70. | Modified Chlorophyll Absorption Ratio Index (MCARI) | $MCARI = [(\rho_{700} - \rho_{670}) - 0.2(\rho_{700} - \rho_{550})] \times \left(\frac{\rho_{700}}{\rho_{670}}\right)$ | Model 20C Exotech spectroradiometer and Landsat/TM | | (Daughtry et al., 2000) |
| 71. | Modified Chlorophyll Absorption Ratio Index – Improved (MCARI2) | $MCARI2 = \frac{1.5[2.5(\rho_{800} - \rho_{670}) - 1.3(\rho_{800} - \rho_{550})]}{\sqrt{(2 \times \rho_{800} + 1)^2 - (6 \times \rho_{800} - 5 \times \sqrt{\rho_{670}}) - 0.5}}$ | CASI – Compact Airborne Spectrographic Imager | | (Haboudane et al., 2004) |
| 72. | Modified Red Edge Normalized Difference Vegetation Index (MRENDVI) | $MRENDVI = \frac{\rho_{750} - \rho_{705}}{\rho_{750} + \rho_{705} + 2 \times \rho_{445}}$ | Unispec, PP Systems /field portable spectrometer | | (Datt, 1999; Sims and Gamon, 2002) |
| 73. | Modified Red Edge Simple Ratio (MRESR) | $MRESR = \frac{\rho_{750} - \rho_{445}}{\rho_{705} + \rho_{445}}$ | Unispec, PP Systems /field portable spectrometer | | (Datt, 1999; Sims and Gamon, 2002) |
| 74. | Modified Triangular Vegetation Index (MTVI) | $MTVI = 1.2[1.2(\rho_{800} - \rho_{550}) - 2.5(\rho_{670} - \rho_{550})]$ | CASI – Compact Airborne Spectrographic Imag | | (Haboudane et al., 2004) |
| 75. | Modified Triangular Vegetation Index – Improved (MTVI2) | $MTVI2 = \frac{1.5[1.2(\rho_{800} - \rho_{550}) - 2.5(\rho_{670} - \rho_{550})]}{\sqrt{(2 \times \rho_{800} + 1)^2 - (6 \times \rho_{800} - 5 \times \sqrt{\rho_{670}}) - 0.5}}$ | CASI – Compact Airborne Spectrographic Imager | | (Haboudane et al., 2004) |
| 76. | Red Edge Normalized Difference Vegetation Index (RENDVI) | $RENDVI = \frac{\rho_{750} - \rho_{705}}{\rho_{750} + \rho_{705}}$ | Hitachi 150-20 spectrophotometer and Unispec, PP Systems/field portable spectrometer | | (Gitelson and Merzlyak, 1994; Sims and Gamon, 2002) |
| 77. | Transformed Chlorophyll Absorption Reflectance Index (TCARI) | $TCARI = 3\left[(\rho_{700} - \rho_{670}) - 0.2(\rho_{700} - \rho_{550})\left(\frac{\rho_{700}}{\rho_{670}}\right)\right]$ | CASI – Compact Airborne Spectrographic Imager | | (Haboudane et al., 2004) |
| 78. | Triangular Vegetation Index (TVI) | $TVI = \frac{120(\rho_{750} - \rho_{550}) - 200(\rho_{670} - \rho_{550})}{2}$ | CASI – Compact Airborne Spectrographic Imager | | (Broge and Leblanc, 2001) |
| 79. | Vogelmann Red Edge Index 1 (VREI1) | $VREI1 = \frac{\rho_{740}}{\rho_{720}}$ | VIRIS | | (Vogelmann et al., 1993) |
| 80. | Vogelmann Red Edge Index 2 (VREI2) | $VREI2 = \frac{\rho_{734} - \rho_{747}}{\rho_{715} + \rho_{726}}$ | VIRIS | | (Vogelmann et al., 1993) |

**Table 6.5** List of spectral Indices related to atmosphere.

| Serial No. | Index (Abbreviation) | Formulae | Satellite/ Sensor | Application | Reference |
|---|---|---|---|---|---|
| 1. | Normalized Difference Cloud Index (NDCI) | $NDCI = \frac{r^{\downarrow}(\lambda_1) - r^{\downarrow}(\lambda_1)}{r^{\downarrow}(\lambda_1) + r^{\downarrow}(\lambda_1)}$ | | | (Marshak et al., 2000) |
| 2. | Cloud Index (CI) | $CI_{thin} = \frac{\rho_{Green} / (\rho_{NIR} - \rho_{Red})}{\rho_{Green} / (\rho_{NIR} + \rho_{Red})}$ <br> $CI_{thick.broken} = \frac{\rho_{NIR} / (\rho_{Blue} - \rho_{Red})}{\rho_{NIR} / (\rho_{Blue} + \rho_{Red})}$ | | | (Karthick and Shanmugam, 2020) |

the plant's nitrogen status through the Modified Chlorophyll Absorption Ratio Index (MCARI) using the wavelength sensitive to chlorophyll. Other vegetation indices developed in narrowband spectra are: The Red Edge Normalized Difference Vegetation Index (RENDVI), which utilizes the red edge band (705 nm) and NIR band (750 nm) (Gitelson and Merzlyak, 1994); the Modified Red Edge Normalized Difference Vegetation Index (MRENDVI) developed by Datt (1999) used to predict chlorophyll content in vegetation by using 445 nm, 705 nm, and 750 nm wavelengths; and the Modified Red-Edge Simple Ratio Index (MRESR) was developed to make corrections in the leaf specular using 445 nm, 705 nm, and 750 nm wavelengths (Sims and Gamon, 2002). Broge and Leblanc (2000) developed the Triangular Vegetation Index (TVI) to measure the difference between the red and NIR reflectance and the green reflection magnitude using Hyperspectral data. Since the chlorophyll concentration affects green reflectance, Haboudane et al. (2004) introduced modification in MCARI (in the form of MCARI1) and TVI (in the form of MTVI1) and replaced the 750 nm of wavelength with 800 nm (NIR band) in the respective algorithm to predict the Leaf Area Index (LAI) of crop canopies. They further modified CARI and introduced the Transformed Chlorophyll Absorption Reflectance Index (TCARI) to further reduce the effect of confounding variables, particularly soil reflectance, in LAI estimation.

### 6.5.5 Spectral Indices and Anthroposphere

Advances in remote-sensing technology have improved our knowledge of the dynamics of the anthroposphere and its linkages with other regions of the planet. Humans have left distinct marks upon the Earth and have changed the land use/cover through an increase in agricultural activities and settlements all over the world. The EO satellites capture these impressions, monitor changes, and allow for greater understanding. The current generation of EO satellites is playing a vital role in urban planning (Masek et al. 2000), governance, and policy support, by providing high-capacity data and information relating to urban areas. This section outlines some of the selected spectral indices used for the scientific investigation of the anthroposphere. Readers are directed to Table 6.6 for a formula of the spectral indices used for quantification of the EoDs of the anthroposphere.

#### i) Built-Up Indices

The urban areas have special characteristics due to denser human agglomeration and infrastructural growth in a relatively smaller spatial unit, thereby requiring the spectral bands to reflect such characters. In addition, it represents highly heterogeneous land-use characteristics mixed up with other land uses/covers such as water bodies and vegetation. Extraction of built-up surfaces from the urban areas with higher accuracies relies on better spatial and spectral resolution of images. The freely available Landsat data with 30 m spatial resolution are widely used data in the field of urban studies with limited accuracies, and the Sentinel data with the better spatial and spectral resolution has opened up avenues of experiments in this field. Urban remote sensing continues to rely on high-resolution data to achieve high accuracies, which can be achieved through commercial data platforms.

The popularity of NDVI for monitoring vegetation surfaces opened up possibilities for experiments with spectral bands relevant to other land-use classes, such as in built-up areas. The first index used to study a built-up surface is the Urban Index (UI) (Kawamura et al., 1997), which uses pixel values in the NIR and SWIR bands to extract urban information. The built-up related information is derived from the reflective properties of the surface features, especially the impermeable, stored in the different bands. Therefore, the built-up index mainly focuses on the use of short-wave infrared (SWIR) and NIR spectral regions, as SWIR regions have more reflectance than NIR for built-up

**Table 6.6** List of spectral Indices related to anthroposphere.

| Serial No. | Index (Abbreviation) | Formulae | Satellite/Sensor | Application | Reference |
|---|---|---|---|---|---|
| 1. | Normalized Difference Built-Up Index (NDBI) | $NDBI = \frac{(SWIR - NIR)}{(SWIR + NIR)}$ | Landsat/TM | Highlights urban areas | (Zha et al., 2003) |
| 2. | World-View Built-Up Index (WV-BI) | $WV - BI = \frac{(Coastal - RedEdge)}{(Coastal + RedEdge)}$ | WorldView-II/ MSS | To extract built-up land | (Kumar et al., 2012) |
| 3. | Normalized Burned Ratio Index (NBRI) | $NBR = \frac{(NIR - SWIR)}{(NIR + SWIR)}$ | Landsat-8/OLI | To identify burned areas | (Gabri, 2019) |
| | | $NBRI(Landsat8) = \frac{(B5 - B7)}{(B5 + B7)}$ | Landsat/TM | To identify burned areas | (Key and Benson, 2006) |
| | | $NBRI(Landsat4 - 7) = \frac{(B4 - B7)}{(B4 + B7)}$ | *Sentinel2 / MSI* | To identify burned areas | (Key and Benson, 2006) |
| | | $NBRI(Sentinel2) = \frac{(B8 - B12)}{(B8 + B12)}$ | | | |
| 4. | Burn Area Index (BAI) | $BAI = \frac{1}{(0.1 - Red)^2 + (0.1 - Red)^2}$ | (Landsat/TM) +NOAA/AVHRR | Highlights burned land | (Chuvieco et al., 2010) |
| 5. | Normalized Multi-band Drought Index (NMDI) | $NMDI(Landsat8) = \frac{(B_5 - (B_6 - B_7)}{(B_5 + (B_6 - B_7)}$ | Landsat-8/OLI | Monitoring soil and vegetation moisture | (Wang and Qu, 2007) |
| | | $NMDI = (B_{8a} - (B_{11} - B_{12})$ | Sentinel-2/MSI | | (Wang and Qu, 2007) |
| 6. | Normalized Burn Ratio (NBR) | $NBR = \frac{(NIR - SWIR)}{(NIR + SWIR)}$ | Landsat 5/ Thematic Mapper (TM) | Highlights burned areas in large fire zones greater than 500 acres | (García and Caselles, 1991; Key and Benson, 2006) |
| 7. | Normalized Burn Ratio Thermal 1 (NBRT1) | $NBRT = \frac{\left(NIR - SWIR\left(\frac{Thermal}{1000}\right)\right)}{\left(NIR + SWIR\left(\frac{Thermal}{1000}\right)\right)}$ | Landsat/ TM-ETM+ | To enhance the NBR | (Holden et al., 2005) |
| **Other Indices** | | | | | |
| 1. | WorldView Non-Homogeneous Feature Difference (WV-NHFD) | $WV - NHFD = \frac{(RedEdge - Coastal)}{(RedEdge + Coastal)}$ | Worldview-2/ MSI | Classifying areas which contrast against the background, which can be identified as man-made | (Wolf, 2012) |

regions. As spectral instrumentation evolved into remote sensing-sensors, built-up indices evolved over time.

Zha et al. (2003) developed the Normalized Difference Built-up Index (NDBI) using SWIR and NIR bands to create binary images. Soon it became a popular method in urban remote sensing. However, this method failed to separate built-up areas from barren and bare land. The earlier methods were unable to differentiate impervious surfaces from bare soil due to complexities in urban settings and similarities in spectral response pattern of bare soil; hence, the latter methods focused on resolving the issues by using other bands and applying correction factors. In pursuit of development with better accuracies in built-up indices, several other indices such as Normalized Built-up Area Index (NBAI), Band Ratio for the Built-up Area (BRBA), Index-based Built-up Index (IBI), and Enhanced Built-up and Bareness Index (EBBI) were devised. Xu (2008) tried to enhance the built-up features by creating three thematic index-derived bands such as SAVI (Soil Adjusted Vegetation Index), MNDWI (Modified Normalized Difference Water Index), and NDBI. The idea behind using these three index-derived bands was to enhance the built-up features and suppress other features of urban areas. However, the challenge to separate built-up surfaces from bare surfaces continued which led to further experimentation with thermal bands. As-syakur et al. (2012) devised a new urban index, i.e., EBBI, which utilized NIR, SWIR, and thermal bands to extract urban surfaces. However, the thermal data posed limitations of spatial resolution and the effects of seasonality.

**ii) Burn Indices (NBI)**

Fire management has been an important disaster management approach and EO satellites have provided opportunities to monitor active fire areas, estimate burned areas, and plan strategies for fire management. Spectral indices are proposed and used to identify post-fire conditions/burn areas through spectral bands like NIR and SWIR. The Normalized Burn Ratio (NBR) uses these bands to account for their spectral response changes after the occurance of a fire as the fire causes changes in the biophysical components of the landscape (Key and Benson, 2005). Vegetation gives brighter reflections in the NIR band, which is reduced due to fire activities, allowing the detection of burned areas. On the other hand, the SWIR band gives an idea about changes in soil dryness and fire traces. In addition, for separating the burned regions from the unburned regions, Normalized Burn Ratio Thermal 1 (NBRT1) was employed (Holden et al., 2005).

## 6.6 Thematic Evolution and Trends

Tracking the evolutionary history of spectral indices derived from different spectral bands of remotely-sensed data reveals that these indices have been used for quantifying the elements (or objects) of detection belonging to respective spheres of our planet. The elements of detection (EoD) of the lithosphere for which spectral indices have been devised include clay, mud, ferrous minerals, soil, etc. The evolution of the satellite data derived spectral indices from 1999 to 2021 has been tracked using the Web of Science database.

Evolution of the field of spectral indices, presented in Figure 6.3, derived from satellite and aerial imagery products tracked through Figure 6.3A "KeyWords Plus" (devised by Clarivate) and Figure 6.3B "Author KeyWords" suggest that LAI (vegetation, and model) and LAI (remote sensing and leaf area index), have been the most frequently used and most significant words in the two categories of keywords during the studied period (1999–2022). NDVI, forest, and growth theory are the next significant words in the "KeyWords Plus" category, whereas in the "Author KeyWords" category, MODIS, NDVI, and vegetation indices have frequently been in use throughout the study period.

"Trend Topics" analysis, visualized by Figure 6.4, in bibliometrix package in RStudio reveals that during 1999–2000, "licorlai 2000" and "pine stands" were the trending words in the domain of spectral indices using remote-sensing technology. In 2010, the most important words with highest frequency in the published literature were "transpiration," "reflectance model," "fluxes," "$CO_2$," and "scale." Published scientific literature in the document category indicated in Table 6.1 in 2015, registered trending topics to containing KeyWords Plus "LAI," "NDVI," "Validation," "forest," and "MODIS," meaning that 2015 has witnessed the most significant trending research work in the field of "spectral indices" using remote-sensing technology focusing mostly on devising and validating new vegetation indices such as leaf area index (LAI), and normalized difference vegetation index (NDVI), using MODIS satellite products. During the last phase of the study period; i.e., 2020–2022, the trending topics were related to "Airborne LiDAR," "yield estimation," "Sentinel-2," "wheat yield," "urban heat island," "agricultural drought," "Landsat-8," and "EVI" or enhanced vegetation index.

To analyze the relevance of the published documents on the topic, total number of citations per annum, means total citation per article, and total number of citations were tracked and the results are shown in Figure 6.5.

The Three-Fields Plot in Figure 6.6 visualizes the most significant countries that have published the greatest number of articles in the current field of interest to this chapter. Also, the authors and their main topics of publication are presented in this figure. China followed by the USA are the two most important countries for publication in the current field of interest and their most important topic of research falls in the leaf area index using MODIS and other remote-sensing datasets.

Figure 6.7 shows that most of the articles published during 1999–2022 show local impact of the source in terms of H-Index. *Remote Sensing of Environment*, *Agricultural and Forest Meteorology*, and *International Journal of Remote Sensing* are the highest three journals publishing the articles included in the dataset.

### 6.6.1 Thematic and Network Maps

Theme clusters of "KeyWords Plus" from the WoS dataset for the 1999–2022 period into four randomly selected time slots and their network maps are shown in Figure 6.8. A look at the "motor themes," with high centrality and high density, for time periods 1999–2009, 2010–2014, 2015–2018, and 2019–2022, shown in Figures 6.7A–D, reveal the shifting focus of the remote-sensing scientific community's thrust into the research on leaf area index "LAI" and "vegetation." We can observe that during the 2010–2014 time period, the "vegetation" theme has slightly slipped into the basic theme quadrant, which later, in the 2015–2018 time

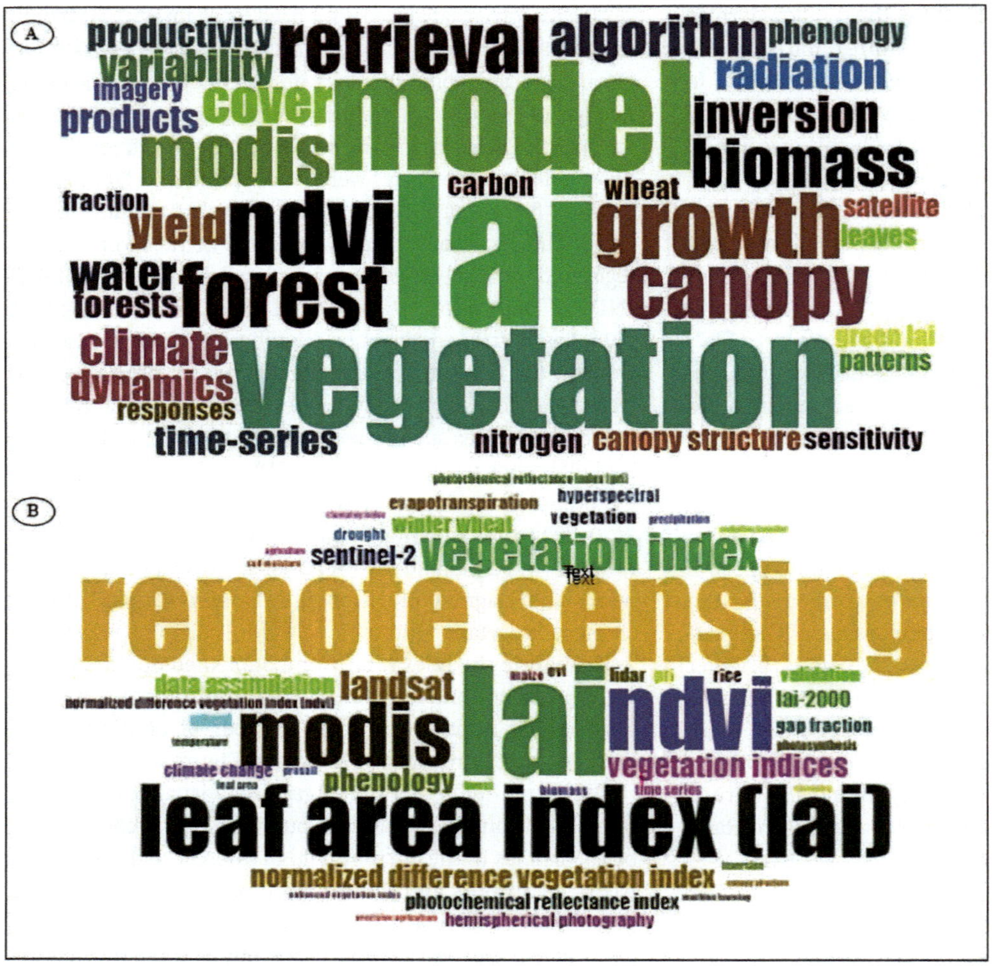


**Figure 6.3** Significance of the most important themes in the published domain related to spectral indices using remote-sensing technology and datasets displayed through WordCloud method. Two types of keywords, Clarivate devised "KeyWords Plus" and "Author KeyWords" are presented in (A) and (B), respectively. The definition of the phrase "KeyWords Plus," according to Clarivate dictates that, "The data in KeyWords Plus are words or phrases that frequently appear in the titles of an article's references, but do not appear in the title of the article itself. Based upon a special algorithm that is unique to Clarivate Analytics databases, KeyWords Plus enhances the power of cited-reference searching by searching across disciplines for all the articles that have cited references in common."

period, moved into the specialized or niche theme quadrant and finally vanished from the scene in 2018–2022. The last time period also shows that though "vegetation" studies in general are no longer in the motor theme quadrant, a proportion of all the published dataset on vegetation indexes during this time period is in this part and has its major portion in basic themes. So, during the last time period (2019–2022), the studies focusing on vegetation indexes, leaf area index (LAI), and NDVI have become basic themes meaning that all the clusters; e.g., vegetation indexes, LAI, and NDVI, are intensely linked but the tie strength among these clusters is weak. Specialization of research themes presented in the niche theme quadrant shows a sense of research focus shift. During 1999–2009, "spectral reflectance"-related research themes were focused on the RS research community (Figure 6.8A), whereas the focus of research shifted to "canopy"-related RS studies employing spectral indices during 2010–2014. Model, vegetation, LAI, NDVI, and vegetation indexes are the keywords (from KeyWords Plus category) in the basic theme group of research topics, out of which NDVI, LAI, and vegetation indexes has turned into topics of research with a high level of inter-cluster linkage intensity but with low tie strengths.

Radiation, growth, NDVI, and canopy are the KeyWords Plus that has been in the emerging/declining themes during the study period but a look in the emerging/declining theme quadrant in Figure 6.8 demonstrates that in the last time period (2019–2022), this particular quadrant became blank, but NDVI and canopy, the two KeyWords Plus terms are reentering this quadrant indicating emergence of these two fields in the era of a plethora of remote-sensing sensor produced satellite products made freely available to the scientific community. In the face of

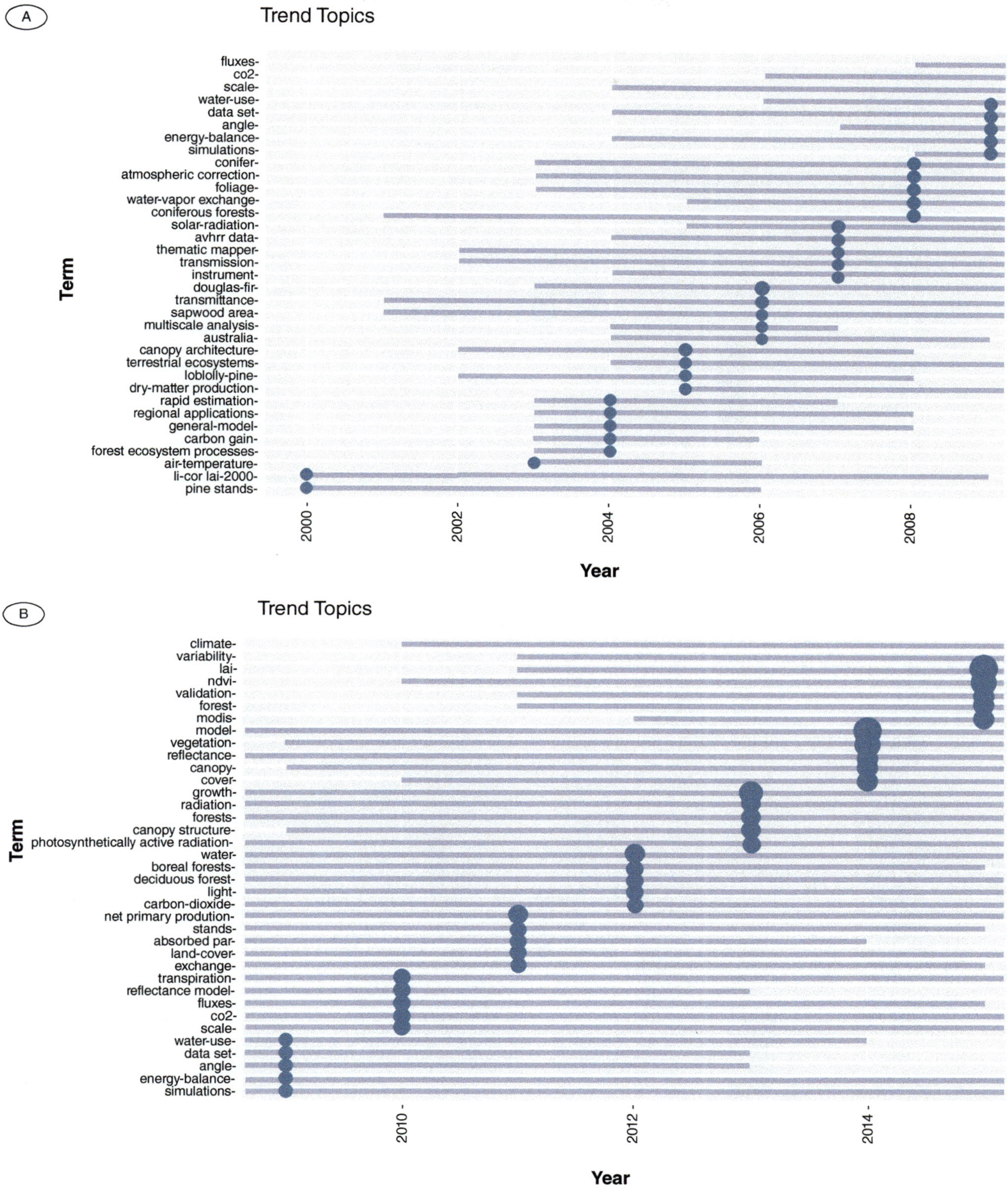


**Figure 6.4** Bibliometrix generated graph showing trending topics in the field of spectral indices using remote sensing during the period 1999–2022.

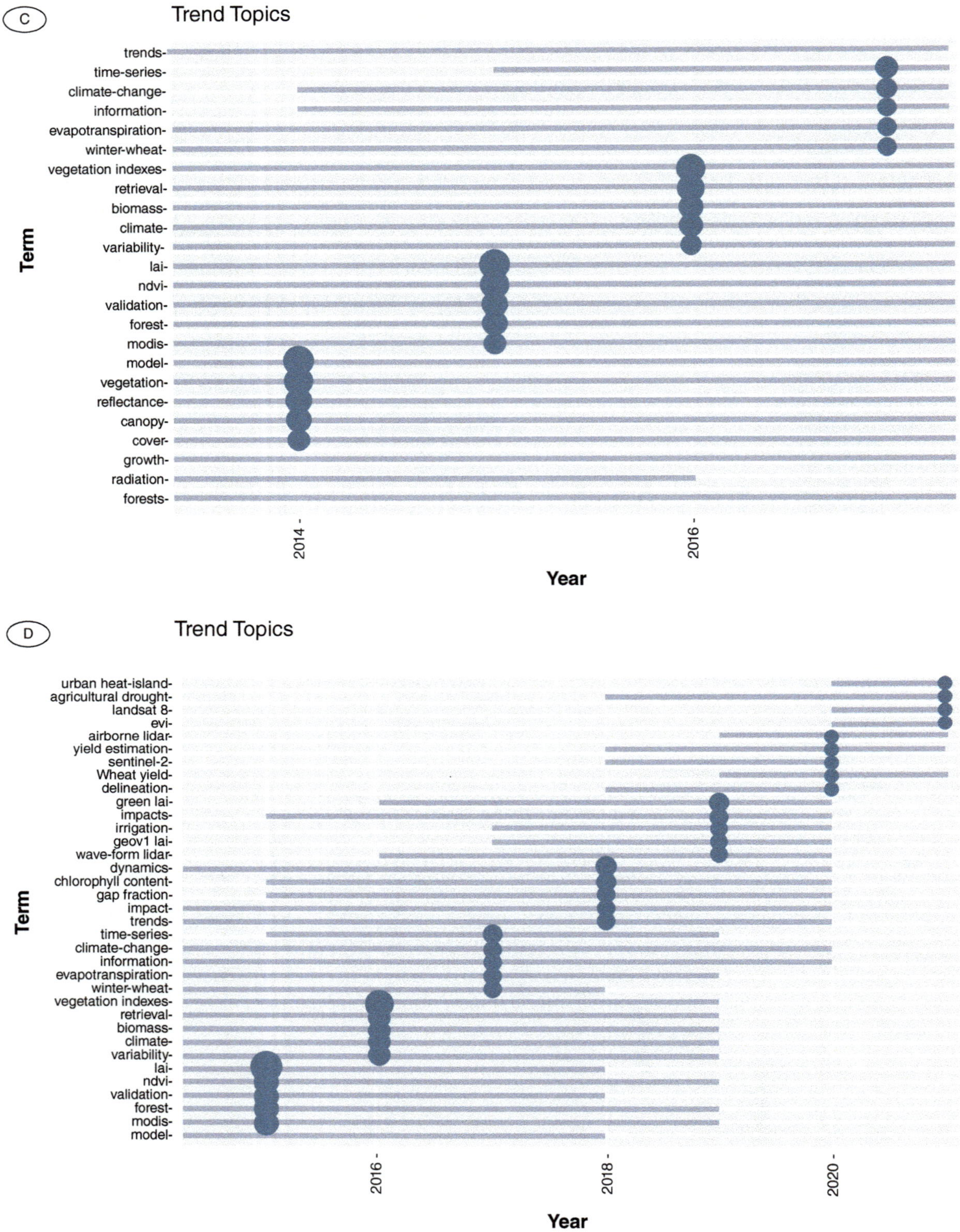


**Figure 6.4** (Continued)

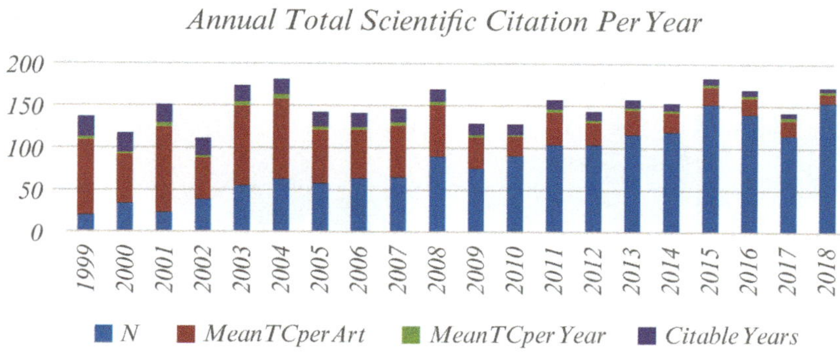


**Figure 6.5** The histogram shows annual total citatione for the articles published on the spectral index using a remote-sensing theme during the period 1999–2018. Mean total citation per article was highest during 2001 followed by 2004, and 2003. Total number of citations for the articles can be seen for the year 2018 (154), followed by 2015 (152) and 2016 (140).

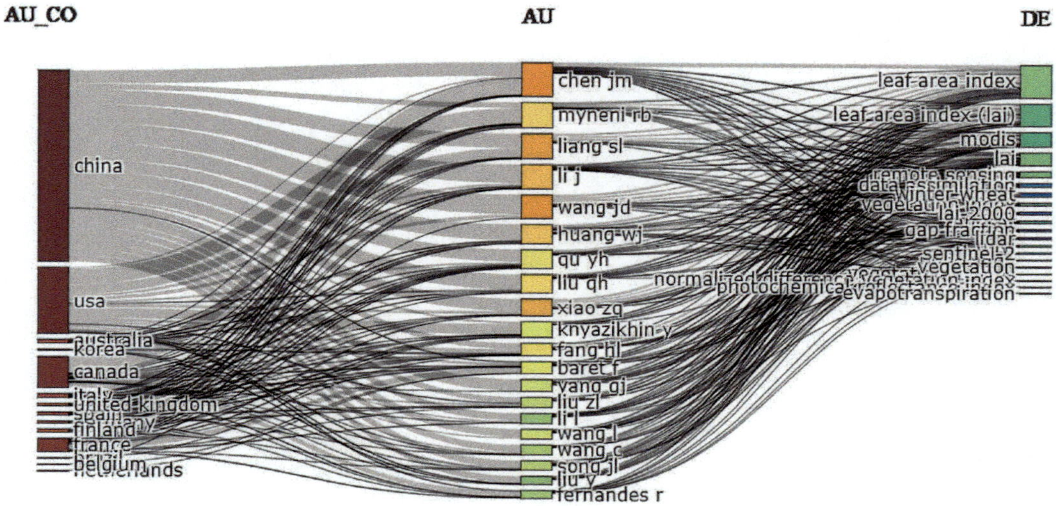


**Figure 6.6** Three-Fields Plot showing correlative presentation of most significant author countries, authors, and topics in the presented dataset for the study period.

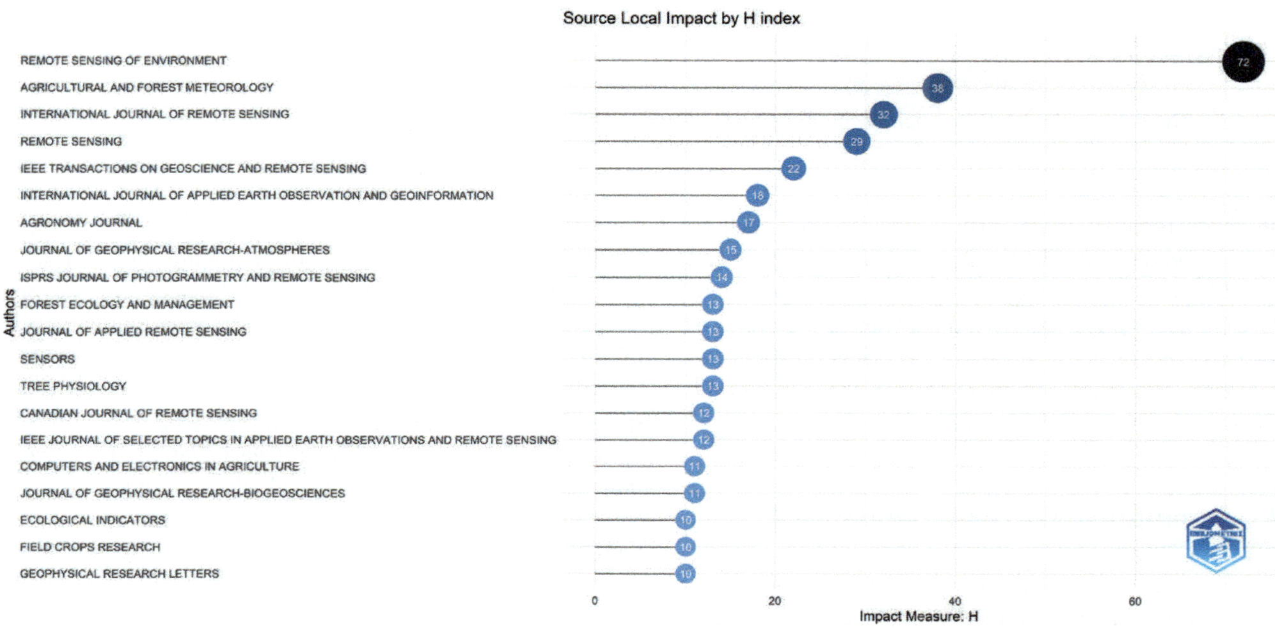


**Figure 6.7** Source impact by the quality of articles published on the topic of interest to this chapter in the journals.

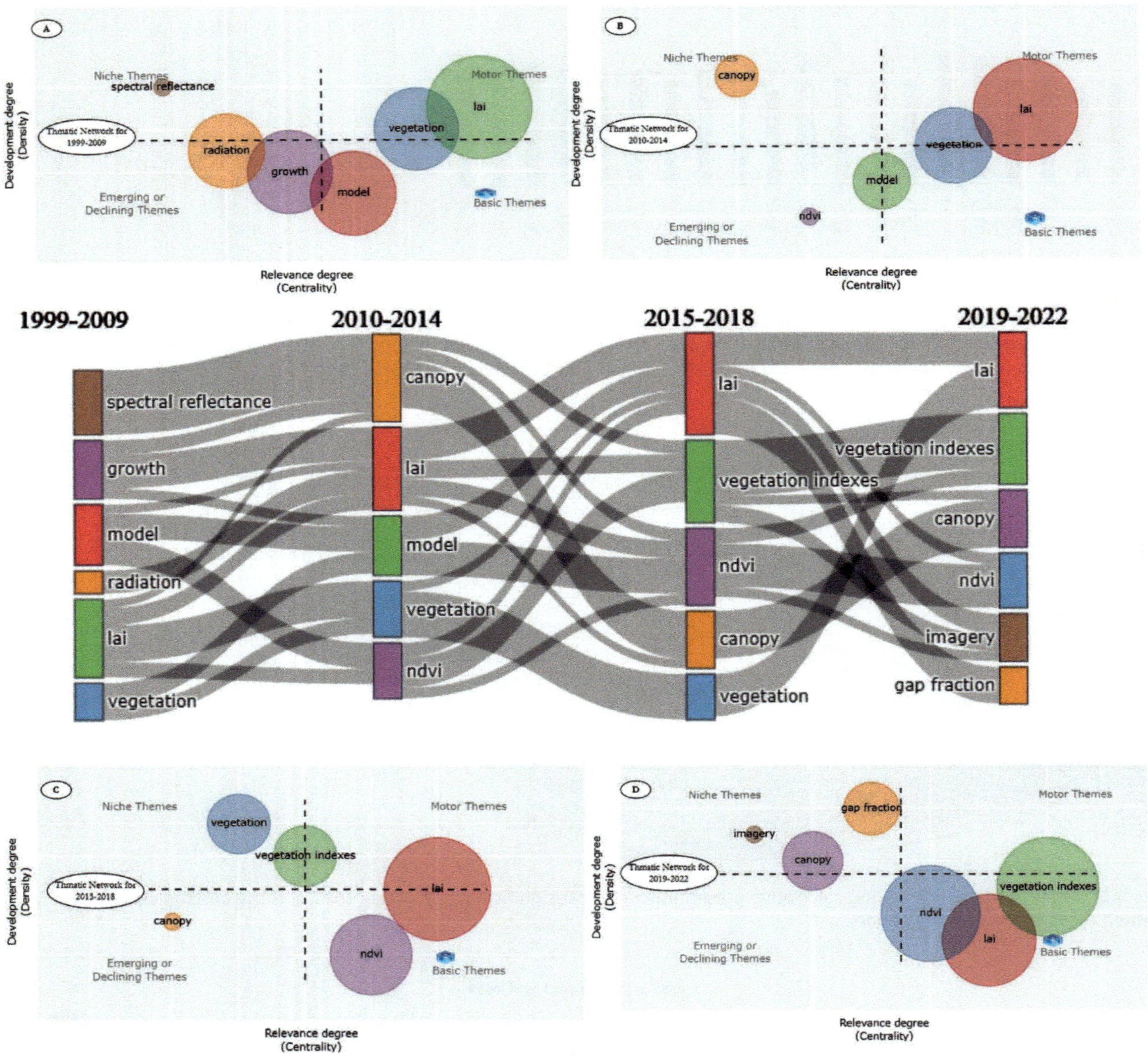


**Figure 6.8** Thematic evolution of the topic in four time periods 1999–2009; 2010–2014; 2015–2018; and 2019–2022.

some of very important future earth observation (EO) missions planned to produce satellite products in different bands and bandwidths of the electromagnetic spectrum, the significance of developing, testing, training, and applying spectral indices for various EoDs of all the spheres of our planet is expected to remain one of the most important tools in the hands of the remote-sensing scientific community.

## 6.7 Summary and Recommendations

Spectral indices regularly in use and still being developed for the remote-sensing products from ongoing missions signify the importance and usefulness of them to spatial data scientists and engineers constantly engaged in monitoring, assessing, quantifying, and modeling earth surface forms and related processes, who feel the need to improve uncertainties relating to quantified EoDs of various spheres of our planet and suggest remedial steps to improve the quantification through remote-sensing product-derived spectral indices in the form of: a) improvement of empirical relationships among different variables in the equations listed in Tables 6.2–6.6 by calculating more accurate constant values; b) enhancement in the spectral, and radiometric, resolutions in future satellite missions; c) more frequent and wide use of unmanned aerial vehicles (UAVs) using a variety of sensors including infrared, optical, LiDAR, and other sensors; and d) improvement in the method of thresholding of the output image of spectral indices with automation of thresholding processes as was done by Firozjaei et al. (2019) for binary class image segmentation and for multiclass segmentation thresholding. From the observation of formulae of spectral indices, it appears that earlier spectral indices were developed using simple relationships and were single sourced, but over time, with the advances in remote-sensing technology, the bandwidth of different remote-sensing products have achieved higher accuracy in monitoring smaller energy fluxes at-sensor, resulting in capturing finer objects with more spatial accuracy, improving signal-to-noise ratios, high temporal repeatability (temporal resolution), etc. Sensitivity of spectral indices will significantly improve the quantifiability and reliability spectral indices applicable in any of the spheres of the planet for detection of any EoDs.

## Acknowledgments

The authors are thankful to their respective institutions for allowing them to utilize their library facilities. MP is thankful to Chandigarh University for subscribing to WoS database from where all the documents for scientometric analysis have been downloaded and analyzed.

## References

As-syakur, A.R., Adnyana, I.W.S., Arthana, I.W. et al. (2012). Enhanced built-up and bareness index (EBBI) for mapping built-up and bare land in an urban area. *Remote Sensing* 4(10): 2957–2970.

Bannari, A., Asalhi, H., and Teillet, P.M. (2002). Transformed difference vegetation index (TDVI) for vegetation cover mapping. *International Geoscience and Remote Sensing Symposium (IGARSS)* 5(C): 3053–3055. doi:10.1109/igarss.2002.1026867.

Bendix, J., Rollenbeck, R., and Palacios, W.E. (2004). Cloud detection in the Tropics: a suitable tool for climate-ecological studies in the high mountains of Ecuador. *International Journal of Remote Sensing* 25: 4521–4540.

Bernstein, L.S. (2012). Quick atmospheric correction code: algorithm description and recent upgrades. *Optical Engineering* 51(11): 111719. doi:10.1117/1.oe.51.11.111719.

Birth, G.S. and McVey, G.R. (1968). Measuring the color of growing turf with a reflectance spectrophotometer 1. *Agronomy Journal* 60(6): 640–643. doi:10.2134/agronj1968.00021962006000060016x.

Bluemarblegeo (2019). Raster Calculator. Available at: https://www.bluemarblegeo.com/knowledgebase/global-mapper-19/Raster_Calculator.htm. (accessed 21 May 2022).

Boegh, E., Soegaard, H., Broge, N., et al. (2002). Airborne multispectral data for quantifying leaf area index, nitrogen concentration, and photosynthetic efficiency in agriculture. *Remote Sensing of Environment* 81(2–3): 179–193. doi:10.1016/S0034-4257(01)00342-X.

Broge, N.H. and Leblanc, E. (2000). Comparing prediction power and stability of broadband and hyperspectral vegetation indices for estimation of green leaf area index and canopy chlorophyll density. *Remote Sensing of Environment* 76: 156–172.

Broge, N.H. and Leblanc, E. (2001). Comparing prediction power and stability of broadband and hyperspectral vegetation indices for estimation of green leaf area index and canopy chlorophyll density. *Remote Sensing of Environment* 76(2): 156–172. doi:10.1016/S0034-4257(00)00197-8.

Campbell, J.B. (1987). *Introduction to Remote Sensing*, 1st ed. New York: Guilford Press. ISBN: 0898627761

Campbell, J.B, and Wynne, R.H. (2011). *Introduction to Remote Sensing*, 1st edn, 670. New York: Guilford Press. ISBN: 1609181778.

Ceccato, P., Flasse, S., Tarantola, S. et al. (2001). Detecting vegetation leaf water content using reflectance in the optical domain. *Remote Sensing of Environment* 77(1): 22–33. doi:10.1016/S0034-4257(01)00191-2.

Cermakova, I., Komarkova, J., and Sedlak, P. (2019). Calculation of visible spectral indices from UAV-based data: small water bodies monitoring. In: *14th Iberian Conference on Information Systems and Technologies (CISTI)*, 1–5.

Champagne, C., Pattey, E., Abderrazak, B.A. et al. (2001). Mapping crop water status: issues of scale in the detection of crop water stress using hyperspectral indices. *Proceedings of the 8th International Symposium on Physical Measurements and Signatures in Remote Sensing* (January), 79–84.

Chang, L., Peng-Sen, S., and Shi-Rong, L. (2016). A review of plant spectral reflectance response to water physiological changes. *Chinese Journal of Plant Ecology* 40(1): 80–91.

Chen, J.M. (1996) Evaluation of vegetation indices and a modified simple ratio for boreal applications. *Canadian Journal of Remote Sensing* 22(3): 229–242. doi:10.1080/07038992.1996.10855178.

Chuvieco, E., Martín, M.P., and Palacios, A. (2010). Assessment of different spectral indices in the red-near-infrared spectral domain for burned land discrimination. *International Journal of Remote Sensing* 23(23): 5103–5110. doi:10.1080/01431160210153129.

Crippen, R.E. (1990). Calculating the vegetation index faster. *Remote Sensing of Environment* 34(1): 71–73. doi:10.1016/0034-4257(90)90085-Z.

Curran, P.J., Windham, W.R., and Gholz, H.L. (1995). Exploring the relationship between reflectance red edge and chlorophyll concentration in slash pine leaves. *Tree Physiology* 15(3): 203–206. doi:10.1093/treephys/15.3.203.

DaCamara, C.C., Libonati, R., Pinto, M. et al. (2019). Near- and middle-infrared monitoring of burned areas from space. In: *Satellite Information Classification and Interpretation* (ed. R.B. Rustamov), (IntechOpen), 1–18. doi:10.5772/intechopen.82444.

Datt, B. (1999). A new reflectance index for remote sensing of chlorophyll content in higher plants: tests using *Eucalyptus* leaves. *Journal of Plant Physiology* 154(1): 30–36. doi:10.1016/S0176-1617(99)80314-9.

Daughtry, C.S.T. (2001). Discriminating crop residues from soil by shortwave infrared reflectance. *Agronomy Journal* 93(1): 125–131. doi:10.2134/agronj2001.931125x.

Daughtry, C.S.T., Walthall, C.L., Kim, M.S. et al. (2000) Estimating corn leaf chlorophyll concentration from leaf and canopy reflectance. *Remote Sensing of Environment* 74(2): 229–239. doi:10.1016/S0034-4257(00)00113-9.

Daughtry, C.S.T., Hunt, E.R., and McMurtrey, J.E. (2004). Assessing crop residue cover using shortwave infrared reflectance. *Remote Sensing of Environment* 90(1): 126–134. doi:10.1016/j.rse.2003.10.023.

Daughtry, C.S.T., Hunt, E.R., and McMurtrey, J.E. et al. (2005). Remote sensing the spatial distribution of crop residues. *Agronomy Journal* 97(3): 864–871. doi:10.2134/agronj2003.0291.

Dozier, J. (1989). Spectral signature of alpine snow cover from the Landsat thematic mapper. *Remote Sensing of Environment* 28(C): 9–22. doi:10.1016/0034-4257(89)90101-6.

Drury, S.A. (1987). *Image Interpretation in Geology*. London: Allen & Unwin.

Ducart, D.F., Silva, A.M., Toledo, C.L.B. et al. (2016). Mapping iron oxides with Landsat-8/OLI and EO-1/Hyperion imagery from the Serra Norte iron deposits in the Carajás Mineral Province, Brazil. *Brazilian Journal of Geology* 46(3): 331–349. doi:10.1590/2317-4889201620160023.

Erdenetuya, M., Khishigsuren, P., Davaa, G. et al. (2006). Glacier change estimation using Landsat TM data. *Remote Sensing and Spatial Information Science*, XXXVI (International Archives of the Photogrammetry), 240–243. Available at: http://www.isprs.org/proceedings/XXXVI/part6/240_XXXVI-part6.pdf (accessed 21 July 2022).

Fang, H. and Liang, S. (2014). Leaf area index models. *Reference Module in Earth Systems and Environmental Sciences*. Amsterdam: Elsevier.

Fatima, K., Muhammad Khattak, M.U.K., Kausar, A.B. et al. (2017). Minerals identification and mapping using ASTER satellite image. *Journal of Applied Remote Sensing* 11(4): 046006. doi:10.1117/1.JRS.11.046006.

Firozjaei, M., Fathololuom, S., Alavipanah, S.K. et al. (2019). Modeling the impact of surface characteristics on the near surface temperature lapse rate. In: *The International Archives of the Photogrammetry, Remote Sensing and Spatial Information Sciences*, vol, XLII-4/W18, 2019 GeoSpatial Conference 2019 – Joint Conferences of SMPR and GI Research.

Fourty, T., Baret, F., Jacquemoud, S. et al. (1996). Leaf optical properties with explicit description of its biochemical composition: direct and inverse problems. *Remote Sensing of Environment* 56(2): 104–117. doi:10.1016/0034-4257(95)00234-0.

Gabri (2019). *List of Spectral Indices for Sentinel and Landsat.* Available at: https://giscrack.com/list-of-spectral-indices-for-sentinel-and-landsat (accessed 21 May 2022).

Gad, S. and Kusky, T. (2007). ASTER spectral ratioing for lithological mapping in the Arabiane Nubian shield, the Neoproterozoic Wadi Kid area, Sinai, Egypt. *Gondwana Research* 11:: 326–335.

Gamon, J.A., Serrano, L., and Surfus, J.S. (1997). The photochemical reflectance index: an optical indicator of photosynthetic radiation use efficiency across species, functional types, and nutrient levels. *Oecologia*. 112(4): 492–501. doi:10.1007/s004420050337.

Gamon, J.A. and Surfus, J.S. (1999). Assessing leaf pigment content and activity with a reflectometer. *New Phytologist* 143(1): 105–117. doi:10.1046/j.1469-8137.1999.00424.x.

Gao, B.C. (1996). A normalized difference water index for remote sensing of vegetation liquid water from space Bo-Cai Gao Joint Center for Earth System Sciences Code 913, NASA Goddard Space Flight Center, Greenbelt, MD 20771. *Remote Sensing of Environment* 58(3): 257–266.

García, M.J.L. and Caselles, V. (1991). Mapping burns and natural reforestation using thematic mapper data. *Geocarto International* 6(1): 31–37. doi:10.1080/10106049109354290.

Ge, W., Cheng, Q., Jing, L. et al. (2020). Assessment of the capability of Sentinel-2 imagery for iron-bearing minerals mapping: a case study in the Cuprite Area, Nevada. *Remote Sensing* 12: 3028.

Gitelson, A.A. (2004). Wide dynamic range vegetation index for remote quantification of biophysical characteristics of vegetation. *Journal of Plant Physiology* 161(2): 165–173. doi:10.1078/0176-1617-01176.

Gitelson, A. and Merzlyak, M.N. (1994). Spectral reflectance changes associated with autumn senescence of *Aesculus hippocastanum* L. and *Acer platanoides* L. leaves: spectral features and relation to chlorophyll estimation. *Journal of Plant Physiology* 143(3): 286–292. doi:10.1016/S0176-1617(11)81633-0.

Gitelson, A.A., Stark, R., Grits, U. et al. (2002a). Vegetation and soil lines in visible spectral space: a concept and technique for remote estimation of vegetation fraction. *International Journal of Remote Sensing* 23(13): 2537–2562. doi:10.1080/01431160110107806.

Gitelson, A.A., Anatoly, A., Zur, Y. et al. (2002b). Assessing carotenoid content in plant leaves with reflectance spectroscopy. *Photochemistry and Photobiology* 75(3): 272. doi:10.1562/0031-8655(2002)075<0272:accipl>2.0.co;2.

Gitelson, A.A., Gritz, Y., and Merzlyak, M.N. (2003). Relationships between leaf chlorophyll content and spectral reflectance and algorithms for non-destructive chlorophyll assessment in higher plant leaves. *Journal of Plant Physiology* 160(3): 271–282. doi:10.1078/0176-1617-00887.

Gitelson, A.A., Kaufman, Y.J., and Merzlyak, M.N. (1996). Use of a green channel in remote sensing of global vegetation from EOS- MODIS. *Remote Sensing of Environment* 58(3): 289–298. doi:10.1016/S0034-4257(96)00072-7.

Gitelson, A.A. and Merzlyak, M.N. (1998). Remote sensing of chlorophyll concentration in higher plant leaves. *Advances in Space Research* 22(5): 689–692. doi:10.1016/S0273-1177(97)01133-2.

Gitelson, A.A., Merzlyak, M.N., and Chivkunova, O.B. (2001). Optical properties and nondestructive estimation of anthocyanin content in plant leaves. *Photochemistry & Photobiology* 74(1): 38. doi:10.1562/0031-8655(2001)074<0038:opaneo>2.0.co;2.

Goel, N.S. and Qin, W. (1994). Influences of canopy architecture on relationships between various vegetation indices and LAI and FPAR: a computer simulation. *Remote Sensing Reviews* 10(4): 309–347. doi:10.1080/02757259409532252.

Haboudane, D., Miller, J.R., Pattey, E. et al. (2004). Hyperspectral vegetation indices and novel algorithms for predicting green LAI of crop canopies: modeling and validation in the context of precision agriculture. *Remote Sensing of Environment* 90(3): 337–352. doi:10.1016/j.rse.2003.12.013.

Hall, D.K., Riggs, G.A., and Salomonson, V.V. (1995). Development of methods for mapping global snow cover using moderate resolution imaging spectroradiometer data. *Remote Sensing of Environment*. 54: 127–140.

Hardisky, M.A., Klemas, V., and Smart, R.M. (1983). The influence of soil salinity, growth form, and leaf moisture on the spectral radiance of *Spartina alterniflora* canopies. *Photogrammetric Engineering and Remote Sensing* 49(1): 77–83.

Henebry, G., Viña, A., and Gitelson, A. (2004). The wide dynamic range vegetation index and its potential utility for gap analysis. *Papers in Natural Resources* 262: 50–56. Available at: http://digitalcommons.unl.edu/natrespapers/262.

Holden, Z.A., Smith, A.M.S., Morgan, P. et al. (2005). Evaluation of novel thermally enhanced spectral indices for mapping fire perimeters and comparisons with fire atlas data. *International Journal of Remote Sensing* 26(21): 4801–4808. doi:10.1080/01431160500239008.

How, P., Messerli, A., Mätzler, E. et al. (2021). Greenland-wide inventory of ice marginal lakes using a multi-method approach. *Sci. Rep.* 11: 4481. doi:10.1038/s41598-021-83509-1.

Huete, A. Didan, K., Miura, T. et al. (2002). Overview of the radiometric and biophysical performance of the MODIS vegetation indices. *Remote Sensing of Environment* 83(1–2): 195–213. doi:10.1016/S0034-4257(02)00096-2.

Huete, A.R. (1988). A soil-adjusted vegetation index (SAVI). *Remote Sensing of Environment* 25(3): 295–309. doi:10.1016/0034-4257(88)90106-X.

Hunt, E.R. and Rock, B.N. (1989). Detection of changes in leaf water content using Near- and Middle-Infrared reflectances. *Remote Sensing of Environment* 30(1): 43–54. doi:10.1016/0034-4257(89)90046-1.

Jackson, T.J., Chen, D., Cosh, M. et al. (2004). Vegetation water content mapping using Landsat data derived normalized difference water index for corn and soybeans. In: *Remote Sensing of Environment* 92: 475–482. doi:10.1016/j.rse.2003.10.021.

Jordan, C.F. (1969). Derivation of leaf-area index from quality of light on the forest floor. *Ecology* 50(4): 663–666. doi:10.2307/1936256.

Karnieli, A., Agam, N., Pinker, R.T. et al. (2010). Use of NDVI and land surface temperature for drought assessment: merits and limitations. *Journal of Climate* 23(3): 618–633. doi:10.1175/2009JCLI2900.1.

Karthick, M. and Shanmugam, P. (2020). Spectral index-based dynamic threshold technique for detecting cloud contamination in ocean colour data. *International Journal of Remote Sensing* 41(5): 1839–1866. doi:10.1080/01431161.2019.1675330.

Kaufman, Y. and Tanre, D. (1992). Atmospherically resistant vegetation index. *IEEE Transactions on Geoscience and Remote Sensing* 30(2): 260–271.

Kaur, R. and Ganju, A. (2008). Cloud classification in NOAA AVHRR imageries using spectral and textural features. *Journal of the Indian Society of Remote Sensing* 36: 167–174.

Kauth, R.J. and Thomas, G.S. (1976). The Tasselled Cap: a graphic description of the spectral-temporal development of agricultural crops as seen by Landsat. *Proceedings, Symposium on Machine Processing of Remotely Sensed Data, Purdue University*, West Lafayette, IN, 29 June–1 July 1976, 41–51.

Kawamura, M., Jayamanna, S., and Tsujiko, Y. (1997). Quantitative evaluation of urbanization in developing countries using satellite data. *Doboku Gakkai Ronbunshu* 580: 45–54).

Key, C. and Benson, N. (2005). Landscape assessment: remote sensing of severity, the normalized burn ratio; and ground measure of severity, the composite burn index. FIREMON: Fire Effects Monitoring and Inventory System, RMRS-GTR, Ogden, UT: USDA Forest Service, Rocky Mountain Research Station. (ESRI, 2018).

Key, C.H. and Benson, N.C. (2006). Landscape Assessment (LA) sampling and analysis methods. USDA Forest Service – General Technical Report RMRS-GTR [Preprint]. (164 RMRS-GTR).

Khalifa, A., Çakır, Z., Kaya, S. et al. (2020). ASTER spectral band ratios for lithological mapping: a case study for measuring geological offset along the Erkenek Segment of the East Anatolian Fault Zone, Turkey, *Arabian Journal of Geosciences* 13: 832.

Kim, M., Daughtry, C., Chappelle, E., McMurtrey, J., III et al. (1994). The use of high spectral resolution bands for estimating absorbed photosynthetically active radiation (Apar). In: *Proceedings of the 6th Symposium on Physical Measurements and Signatures in Remote Sensing*, 299–306. Val D'Isere, France

Kuittinen, R. (1997). Optical and thermal sensors in snow cover monitoring. In: *Proceedings of the EARSeL Workshop on Remote Sensing of Land Ice and Snow*, 7–20.

Kumar, A., Pandey, A.C., and Jeyaseelan, A.T. (2012). Built-up and vegetation extraction and density mapping using WorldView-II. *Geocarto International* 27(7): 557–568. doi:10.1080/10106049.2012.657695.

Liang, S. and Wang, J. (2020). *Advanced Remote Sensing Terrestrial Information Extraction and Applications*, 2nd edn. Amsterdam: Elsevier. doi: 10.1016/C2017-0-03489-4.

López-Robles, J.R., Otegi-Olaso, J.R., Porto Gómez, I. et al. (2019). 30 years of intelligence models in management and business: a bibliometric review. *Internatiional Journal of Information Management* 48: 22–38. https://doi.org/10.1016/j.ijinfomgt.2019.01.013.

Louhaichi, M., Borman, M.M., and Johnson, D.E. (2001). Spatially located platform and aerial photography for documentation of grazing impacts on wheat. *Geocarto International* 16(1): 65–70. doi:10.1080/10106040108542184.

Lyzenga, D.R. (1979). Shallow-water reflectance modeling with applications to Penginderaan Jauh of ocean floor. *Proceedings of the 13th International Symposium on Penginderaan Jauh of Environment*, 583–602.

Major, D.J., Baret, F., and Guyot, G. (1990). A ratio vegetation index adjusted for soil brightness. *International Journal of Remote Sensing* 11(5): 727–740.

Marshak, A., Kntzikhin, Y., Davis, A.B. et al. (2000). Cloud-vegetation interaction: use of normalized difference cloud index for estimation of cloud optical thickness. *Geophysical Research Letters* 27(12): 1695–1698. doi:10.1029/1999GL010993.

Masek, J.G., Lindsay, F.E., and Goward, S.N. (2000). Dynamics of urban growth in the Washington DC metropolitan area, 1973–1996, from Landsat observations. *International Journal of Remote Sensing* 21: 3473–3486. doi: 10.1080/014311600750037507.

Massironi, M.L., Bertoldi, P., Calafa, D. et al. (2008). Interpretation and processing of ASTER data for geological mapping and granitoids detection in the Saghro massif (eastern Anti-Atlas, Morocco). *Geosphere* 4: 736–759.

Matsushita, B., Yang, W., Chen, J. et al. (2007). Sensitivity of the Enhanced Vegetation Index (EVI) and Normalized Difference Vegetation Index (NDVI) to topographic effects: a case study in high-density cypress forest. *Sensors* 7: 2636–2651. https://doi.org/10.3390/s7112636.

McFeeters, S.K. (1996). The use of the Normalized Difference Water Index (NDWI) in the delineation of open water features. *International Journal of Remote Sensing* 17(7): 1425–1432. doi:10.1080/01431169608948714.

Melillo, J.M., Aber, J.D., and Muratore, J.F. (1982). Nitrogen and lignin control of hardwood leaf litter decomposition dynamics. *Ecology* 63(3): 621–626. Available at: http://www.jstor.org/stable/1936780. Linked references are available on JSTOR for this article: (You may need to log in to JSTOR to access the linked reference.)

Merzlyak, M.N., Gitelson, A.A., Chivkunova, G.B. et al. (2015). Non-destructive optical detection of pigment changes during leaf senescence and fruit ripening. *Physiologia Plantarum* 106:135–141. https://doi.org/10.1034/j.1399-3054.1999.106119.x

Mokarram, M. (2015). Modeling the behavior of Vegetation Indices in the salt dome of Korsia in North-East of Darab, Fars, Iran. *Modeling Earth Systems and Environment* 1: 27. doi:10.1007/s40808-015-0029-y.

Mukherjee, J., Mukherjee, J., and Chakravarty, D. (2019). Automated seasonal separation of mine and non-mine water bodies from Landsat 8 oli/tirs using clay mineral and iron oxide ratio. *IEEE Journal of Selected Topics in Applied Earth Observations and Remote Sensing* 12(7): 2550–2556. doi:10.1109/JSTARS.2019.2895385.

Osinowo, O.O., Gomy, A., and Isseini, M. (2021). Mapping hydrothermal alteration mineral deposits from Landsat 8 satellite data in Pala, Mayo Kebbi Region, Southwestern Chad. *Scientific African* 11: e00687. https://doi.org/10.1016/j.sciaf.2020.e00687.

Özkan, M., Çelik, Ö,F., and Özyavas, A. (2018). Lithogical discrimination of accretionary complex (Siva, northern Turkey) using novel hybrid color composites and field data. *Journal of African Earth Sciences* 138: 75–85. doi:10.1016/j.jafrearsci.2017.11.009.

Pacheco, D., Rocha, A.C., Pereira, L. et al. (2020). Microalgae water bioremediation: trends and hot topics. *Applied Science* 10: 1886. https://doi.org/10.3390/app10051886

Pandey, P.C., Koutsias, N., Petropoulos, G.P. et al. (2021). Land use/land cover in view of earth observation: data sources, input dimensions, and classifiers: a review of the state of the art. *Geocarto International* 36(9): 957–988.

Pearson, R.L. and Miller, L.D. (1972). Remote mapping of standing crop biomass for estimation of the productivity of the shortgrass prairie. *Remote Sensing of Environment* 8: 1355.

Penuelas, J., Filell, I., Biel, C. et al. (1993). The reflectance at the 950–970 nm region as an indicator of plant water status. *International Journal of Remote Sensing* 14(10): 1887–1905. doi:10.1080/01431169308954010.

Penuelas, J., Filella, I., and Gamon, J.A. (1995). Assessment of photosynthetic radiation-use efficiency with spectral reflectance. *New Phytologist* 131: 291–296.

Pinty, A.B. and Verstratete, M.M. (1992). GEMI: a non-linear index to monitor global vegetation from satellites. *Vegetation* 101(1): 15–20. https://doi.org/10.1007/BF00031911.

Platnick, S., King, M.D., Ackerman, S. et al. (2003). The MODIS cloud products: algorithms and examples from Terra. *IEEE Transactions on Geoscience and Remote Sensing* 41(2): 459–473.

Qi, J. Chenjhouni A., Huete, A.R., Chenbourni et al. (1994). A modified soil adjusted vegetation index. *Remote Sensing of Environment* 48(2): 119–126. doi:10.1016/0034-4257(94)90134-1.

Raymond Hunt, E., Daugtrey, C.S.T., Eitel, J.U.H. et al. (2011). Remote sensing leaf chlorophyll content using a visible band index. *Agronomy Journal* 103(4): 1090–1099. doi:10.2134/agronj2010.0395.

Richardson, A.J. and Weigand, C. (1977). Distinguishing vegetation from soil background information. *Photogrammetric Engineering and Remote Sensing* 43: 1541–1552. https://www.asprs.org/wp-content/uploads/pers/1977journal/dec/1977_dec_1541-1552.pdf.

Rondeaux, G., Steven, M., and Baret, F. (1996). Optimization of soil-adjusted vegetation indices. *Remote Sensing of Environment* 55(2): 95–107. doi:10.1016/0034-4257(95)00186-7.

Roujean, J.L. and Breon, F.M. (1995). Estimating PAR absorbed by vegetation from bidirectional reflectance measurements. *Remote Sensing of Environment* 51(3): 375–384. doi:10.1016/0034-4257(94)00114-3.

Rouse, J., Haas, R., Schell, J. et al. (1974). Monitoring the vernal advancement and retrogradation (greenwave effect) of natural vegetation. 362 pp, Texas A&M Univ., Remote Sens. Cent., College Station, Texas.

Rouse, J.W., Haas, R.H., Schell, J.A. et al. (1973). Monitoring the vernal advancement and retrogradation (green wave effect) of natural vegetation, Progress Report RSC 1978–1: 112.

Roy, P.S., Sharma, K.P., and Jain, A. (1996). Stratification of density in dry deciduous forest using satellite remote sensing digital data: an approach based on spectral indices. *Journal of Biosciences* 21(5): 723–734. doi:10.1007/BF02703148.

Sabins, F.F. (1999). Remote sensing for mineral exploration. *Ore Geology Reviews* 14: 157–183.

Salomonson, V.V. and Appel, I. (2004). Estimating fractional snow cover from MODIS using the normalized difference snow index. *Remote Sensing of Environment* 89(3): 351–360. doi:10.1016/j.rse.2003.10.016.

Schroeder, T.A., Wulder, M.A., Healey, S.P. et al. (2011). Mapping wildfire and clearcut harvest disturbances in boreal forests with Landsat time series data. *Remote Sensing of Environment* 115: 1421–1433.

Segal, D. (1982). Theoretical basis for differentiation of ferric-iron bearing minerals, using Landsat MSS Data. In: *Proceedings of Symposium for Remote Sensing of Environment, 2nd Thematic Conference on Remote Sensing for Exploratory Geology*. Fort Worth, TX, 949–951 (ESRI, 2018).

Serrano, L., Peñuelas, J., and Ustin, S.L. (2002). Remote sensing of nitrogen and lignin in Mediterranean vegetation from AVIRIS data: decomposing biochemical from structural signals. *Remote Sensing of Environment* 81(2–3): 355–364. doi:10.1016/S0034-4257(02)00011-1.

Shukla, A. and Ali, I. (2016). A hierarchical knowledge-based classification for glacier terrain mapping: a case study from Kolahoi Glacier, Kashmir Himalaya. *Annals of Glaciology* 57(71): 1–10. doi:10.3189/2016AoG71A046.

Sims, D.A. and Gamon, J.A. (2002). Relationships between leaf pigment content and spectral reflectance across a wide range of species, leaf structures and developmental stages. *Remote Sensing of Environment* 81(2–3): 337–354. doi:10.1016/S0034-4257(02)00010-X.

Sripada, R.P. (2005) *Determining In-Season Nitrogen Requirements for Corn Using Aerial Color-Infrared Photography*. Available at: http://www.lib.ncsu.edu/resolver/1840.16/4200 (accessed 20 May 2022).

Sripada, R.P., Heiniger, R.W., White, J.G., Heiniet al. (2006). Aerial color infrared photography for determining early in-season nitrogen requirements in corn. *Agronomy Journal* 98(4): 968–977. doi:10.2134/agronj2005.0200.

Tucker, C.J. (1979). Red and photographic infrared linear combinations for monitoring vegetation. *Remote Sensing of Environment* 8(2): 127–150. doi:10.1016/0034-4257(79)90013-0.

USGS (2019a). Landsat Surface Reflectance-Derived Spectral Indices. Available at: https://www.usgs.gov/core-science-systems/nli/landsat/normalized-difference-moisture-index accessed 20 May 2022).

USGS (2019b). Landsat Surface Reflectance-Derived Spectral Indices, Landsat Soil Adjusted Vegetation Index. Available at: https://www.usgs.gov/core-science-systems/nli/landsat/landsat-soil-adjusted-vegetation-index (accessed 19 March 2022).

USGS (2019c). Landsat Enhanced Vegetation Index. Available at: https://www.usgs.gov/core-science-systems/nli/landsat/landsat-enhanced-vegetation-index?qt-science_support_page_related_con=0#qt-science_support_page_related_con (accessed 20 May 2022).

Velosky, J.C., Stern, R.J., and Johnson, P.R. (2003). Geological control of massive sulfide mineralization in the Neoproterozoic Wadi Bidah shear zone, southwestern Saudi Arabia, inferences from orbital remote sensing and field studies. *Precambrian Research* 123(2–4): 235–247.

Viedma-Del-Jesus, M.I., Perakakis, P., Muñoz, M.Á. et al. (2011). Sketching the first 45 years of the journal *Psychophysiology* (1964–2008): a co-word-based analysis. *Psychophysiology* 48: 1029–1036. https://doi.org/10.1111/j.1469-8986.2011.01171.x

Vogel, S.W. (2002). Usage of high-resolution Landsat-7 band 8 for single-band snow cover classification. *Annals of Glaciology* 34: 53–57.

Vogelmann, J.E., Rock, B.N., and Moss, D.M. (1993). Red edge spectral measurements from sugar maple leaves. *International Journal of Remote Sensing* 14(8): 1563–1575. doi:10.1080/01431169308953986.

Wang, L. and Qu, J.J. (2007). NMDI: a normalized multi-band drought index for monitoring soil and vegetation moisture with satellite remote sensing. *Geophysical Research Letters* 34(20): 1–5. doi:10.1029/2007GL031021.

Wang, L., Qu, J.J., and Hao, X. (2008). Forest fire detection using the normalized multi-band drought index (NMDI) with satellite measurements. *Agricultural and Forest Meteorology*, 148(11): 1767–1776. doi:10.1016/j.agrformet.2008.06.005.

Welikhe, P., Quansah, J.E., Fall, S. et al. (2017). Estimation of soil moisture percentage using Landsat-based moisture stress index. *Journal of Remote Sensing & GIS* 6(2): 1–remote sen5. doi:10.4172/2469-4134.1000200.

Wolf, A.F. (2012). Using WorldView-2 Vis-NIR multispectral imagery to support land mapping and feature extraction using normalized difference index ratios. In: *Algorithms and Technologies for Multispectral, Hyperspectral, and Ultraspectral Imagery XVIII* (ed. S.S. Shen and P.E. Lewis), vol. 8390, 188–195. SPIE. 83900N-83900N–8. doi:10.1117/12.917717.

Xiao, X., Shen, Z,. and Qin, X. (2001). Assessing the potential of VEGETATION sensor data for mapping snow and ice cover: a normalized difference snow and ice index. *International Journal of Remote Sensing* 22(13): 2479–2487.

Xie, Y., Sha, Z., and Yu, M. (2008). Remote sensing imagery in vegetation mapping: a review. *Journal of Plant Ecolology* 1: 9–23. https://doi.org/10.1093/jpe/rtm005

Xu, H. (2007). Modification of normalised difference water index (NDWI) to enhance open water features in remotely sensed imagery modification of normalised difference water index (NDWI) to enhance open water features in remotely sensed imagery. *International Journal of Remote Sensing* 27(14): 3025–3033. doi:10.1080/01431160600589179.

Xu, H. (2008). A new index for delineating built-up land features in satellite imagery. *International Journal of Remote Sensing* 29: 4269–4276.

Yamaguchi, Y., Kahle, A.B., Kawakami, T. et al. (1998). Overview of the advanced spaceborne thermal emission and reflection radiometer (ASTER). *IEEE Transaction on Geoscience and Remote Sensing* 36(4): 1062–1071.

Yamaguchi, Y. and Naito, C. (2003). Spectral indices for lithologic discrimination and mapping by using the ASTER SWIR bands. *International Journal of Remote Sensing* 24(22): 4311–4323.

Yang, Z., Willis, P., and Mueller, R. (2008). Impact of band-ratio enhanced AWIFS image to crop classification accuracy. In: *Pecora 17 – The Future of Land Imaging... Going Operational*, 11.

Yang, L.S., Feng, Q., Adamowski, J.F. et al. (2020). Spatio-temporal variation of reference evapotranspiration in northwest China based on CORDEX-EA. *Atmosphere Research* 238: doi:10.1016/j.atmosres.2020.104868.

Zha, Y., Gao, J., and Ni, S. (2003). Use of normalized difference built-up index in automatically mapping urban areas from TM imagery. *International Journal of Remote Sensing* 24(3): 583–594. doi:10.1080/01431160304987.

Zhai, H., Zhang, H., Zhang, L. et al. (2018). Cloud/shadow detection framework based on spectral indices for multi/hyperspectral optical remote sensing imagery. *ISPRS Journal of Photogrammetry and Remote Sensing* 144: 235–253.

Zhao, H. and Chen, X. (2005). Use of Normalized Difference Bareness Index in quickly mapping bare areas from TM/ETM+. *Proceedings of 2005 IEEE International Geoscience and Remote Sensing Symposium* 3: 1666–1668.

Zhu, Z., Wang, S.X., and Woodcock, C.E. (2015). Improvement and expansion of the F-mask algorithm: cloud, cloud shadow, and snow detection for Landsats 4–7, 8, and Sentinel 2 images. *Remote Sensing of Environment* 159: 269–277.

# Section II

## Antarctica: The Southernmost Continent Having the South Pole Environment and Remote Sensing

# 7

## Glacier Dynamics in East Antarctica

### A Remote Sensing Perspective

*Kiledar Singh Tomar*[1,*], *Sangita Singh Tomar*[2], *Ashutosh Venkatesh Prasad*[1], *and Alvarinho J. Luis*[1]

[1] *Polar Remote Sensing Division, National Centre for Polar and Ocean Research, Ministry of Earth Sciences, Goa, India*
[2] *Department of Civil Engineering, Indian Institute of Technology Bombay, Powai, Maharashtra, India*
* *Corresponding author*

## 7.1 Introduction

The Antarctic cryosphere, which mainly contains ice sheets, store about 87% of the global glacier ice (Frezzotti and Orombelli, 2014). Ice mass loss (on full melting) from the Antarctic Ice Sheet (AIS) has the potential to raise the sea level equivalent by 58.3 m (Vaughan et al., 2013). Particularly for period 1992–2017, the reported average rate of ice mass loss in the overall AIS region was observed to be 109 ± 56 Gt/yr, which corresponds to contributing 0.3 ± 0.15 mm/yr rise in mean sea level (IMBIE, 2018). The majority of the grounded ice has been lost (~80%) to the ocean through processes such as basal melting or calving. In addition to the ice shelves along the Antarctic coast, there are a number of glaciers that contribute to ice loss.

The significance of Antarctica extends beyond the cryosphere. It also stores valuable resources such as minerals and fossil fuels. Monitoring activities by several research facilities have been established over the years. According to the Council of Managers of National Antarctic Programs (COMNAP), Antarctica accommodates 100 research facilities, operated and managed by 30 different countries (COMNAP, 2017). In addition to this, changes in the Antarctic ecosystem has prime environmental influence that affects in a variety of ways and subsequently impacts on sea level, biochemical cycles, and climate that are not yet fully understood.

The dynamic fluctuation in the Antarctic glaciers is considered to be one of the significant processes affecting sea-level rise (Miles, 2012; Foga et al., 2014). With the advancement in satellite technology, remote sensing-based studies have evolved to monitor these changes from space. Studies utilizing satellite remote sensing imagery have reported disintegration of ice shelves and massive icebergs breaking off the coast of Antarctica's coastline. On the contrary, an advance has been reported elsewhere.

Among the dynamic fluctuations, the glacier velocity is one of the crucial parameters that needs regular monitoring and proper understanding. Glacier velocity is sensitive to variations in temperature and precipitation, as any subtle changes in these two meteorological parameters causes variations in ice flow velocity. Glacier velocity measurements give a direct signal of the changing climate, depending upon the increase or decrease of temperature and precipitation. Greenland and Antarctica are of prime concern as these two ice sheets would contribute up to 7.36 m and 58 m respectively to sea-level rise, if they melted completely (Frezzotti and Orombelli, 2014). The glaciers present on these two ice sheets are the source through which they lose their ice mass into the ocean. The Antarctic Peninsula has seen an air temperature rise of 2.5°C since 1950 and as a result it is losing ice mass at a rapid rate (Turner et al., 2005). In such a scenario, the monitoring of glacier velocity becomes important as it acts as an indicator of the changing climate. Increased mass loss from the ice sheets is attributed to the rise in air temperature, subsurface ocean temperature, basal melting, and surface meltwater runoff. The need to monitor the glaciers of the ice sheets regularly has increased but it is difficult to carry out *in-situ* observations under such remote and harsh climatic

*Advances in Remote Sensing Technology and the Three Poles*, First Edition. Edited by Manish Pandey, Prem C. Pandey, Yogesh Ray, Aman Arora, Shridhar D. Jawak, and Uma K. Shukla.

conditions. In order to overcome this problem, we make use of innovative Earth Observation technology, in which, through satellite images, we can timely monitor the dynamics of the glacier.

## 7.2 Satellite Remote Sensing of Glacier Dynamics in East Antarctica

The Antarctic Ice Sheet contains the Antarctic Peninsula Ice Sheet, the West Antarctic Ice Sheet, and the East Antarctic Ice Sheet (EAIS) (Rémy and Frezzotti, 2006). With the efforts being made to investigate the dynamics of Antarctica pertaining to ice loss/gain, the status of EAIS remains uncertain (Hanna et al., 2013). The large outlet glaciers and massive ice streams primarily facilitate ice loss through ice conveyance from the inner region of East Antarctica to the ocean (Rignot, 2002; Stearns, 2011; Rignot et al., 2013). The outlet glaciers, which undergo a change in their speed over time, demonstrate their dynamic response to the changes in climate (Scambos et al., 2004; Stearns et al., 2008). Variations in the dynamics of these Antarctic outlet glaciers are crucial to understand the mass balance of ice sheets and sea-level rise (Stearns et al., 2008). However, the abrupt changes in the flow velocity of outlet glaciers are complex (Stearns et al., 2008). Many studies in West Antarctica, such as seasonal variations in ice velocity of the Larsen B embayment, show a role in summer melt percolation as well as variation of stress field in glacier dynamics (Scambos et al., 2004). On the other hand, East Antarctica gets less consideration due to its comparative stability with a slight positive mass balance. Moreover, recent observations show an accelerated melting in West Antarctica since 2000, while the data are scarce and less concrete (Noble et al., 2020).

The launch of Landsat has paved the way for large-scale monitoring of glacier dynamics. To this end, manual delineation is the most applied approach for extracting ice fronts and glacier termini or ice fronts from satellite images through geolocated coordinates. Few studies have been carried out for ice-front change detection using either automated or semi-automated techniques. This is also evident from the projects on coastline mapping (Williams et al., 1995; Scambos et al., 2007), where the maps were produced manually.

Chen et al. (2020) investigated the Dalk Glacier using remote sensing in order to study the glacier surface velocity and terminus positions and to analyse the dynamic response of this glacier to the changing climate. The remote sensing data included both optical (Landsat imagery) and SAR (Sentinel-1 imagery) data along with digital elevation model (DEM) generated from Worldview datasets, i.e., Reference Elevation Model of Antarctica (REMA). The glacier flow velocity variations at seasonal level have been studied at the Totten Ice Shelf, which attributed to ice shelf acceleration due to buttressing loss caused by the disintegration of landfast sea ice (Greene et al., 2018).

Stearns and Hamilton (2005) observed a deceleration in the estimated ice flow velocity of the Byrd Glacier using the Advanced Spaceborne Thermal Emission and Reflection Radiometer (ASTER) imagery. Here, a cross-correlation based technique was used to process the 2000–2001 imagery. The study also implied that for East Antarctic outlet glaciers, substantial changes can be expected over short timescales relating to annual scale. Another study by Stearns et al. (2008) argued that the summer melting and surface melt water are not the only reasons for changes observed in the dynamics of outlet glaciers. This signifies that, being one among the major outlet glaciers of EAIS region, the Polar Record Glacier (PRG) needs attention.

The glacier surface velocity and ice-front changes are some of the significant processes involved in understanding glacier dynamics. It also serves as an alternative to other glacier features such as mass balance and change in the glacier area over time, to understand the behavior of glaciers. For better glacier dynamics monitoring and associated climate change related responses, spatial and temporal analysis of glacier surface velocity is of paramount importance. Remote sensing techniques are viable for glaciers like PRG, which is full of crevasses to provide detail and timely observation.

The PRG is one of the major outlet glacier in East Antarctica (Liu et al., 2018; Pandit et al., 2018). So far, only a few studies have been carried out to monitor the PRG in East Antarctica. Earlier studies have addressed the monitoring of dynamics of this glacier utilizing remote sensing techniques. Liang et al. (2019) derived surface velocity of the glacier through the feature-tracking approach based on the Multi Image Multi Chip (MIMC) technique and reported ice flow variations at the PRG from 2005 to 2015. They used Landsat 7 and Landsat 8 data along with Sentinel 1 data as a constraint, which showed an overall glacier speedup. Another study by Liu et al. (2018) introduced rotation invariant feature tracking in order to understand the ice velocity variations of PRG. Using optical imagery, they investigated the PRG velocity variations over the period of 2004 to 2015. They also found no significant trend in velocity. Zhou et al. (2014) reported ice velocity changes at seasonal and interannual scale utilizing Synthetic Aperture Radar (SAR) imagery. Moreover, the study utilized SAR imagery for Differential SAR Interferometry (DInSAR) and reported interannual as well as seasonal variations in the ice flow adjacent to the glacier tongue, and inferred that the overall interannual changes were too small. Using the Permanent Scatterer Interferometry (PS-InSAR) technique Pandit et al. (2018) studied PRG and estimated velocity utilizing Sentinel-1 images in Interferometric Wide (IW) mode in order to investigate the spatial distribution of PRG velocity.

The study of glacier dynamics has advanced slowly as the availability of satellite products and their accessibility has increased. In the future, remote sensing-based observation seems to have great potential to monitor glacier surface dynamics. The following reasons support this further:

*Requirement of homogenized information*: Difficulties in ground-based monitoring arises from several concerns when collaborating data from multiple surveys. The data collected from different sources often do not follow similar formats. This can be overcome by remotely estimating glacier processes on a large scale.

*Requirement of long-term, frequent, and spatially comprehensive information*: The data collected from field surveys are generally sparse (both spatially and temporally). To understand the dynamic nature of glacier processes, it is necessary to have full coverage which can be efficiently obtained through the wide variety of satellite imagery available.

## 7.3 Glacier Velocity Estimation Using Remote Sensing

The remote sensing-based measurements are complementary to *in-situ* observations in spatio-temporal ice flow monitoring of glaciers. Different techniques involved in this estimation using radar and optical imagery are: Offset Tracking and Interferometric SAR (InSAR/DInSAR) (Eldhuset et al., 2003; Huang and Li, 2011; Zhou et al., 2014; Satyabala, 2016; Lemos et al., 2018; Gomez et al., 2019; Yellala et al., 2019). The glacier velocity estimation through optical imagery is limited by sun illumination and cloud cover (Huang and Li, 2011). This is mainly as a result of image distortion occurring in optical images which is generated by attitude variations (Heid and Kääb, 2012; Shukla and Garg, 2020). This subpixel noise generated in Landsat images used in the previous studies (Liu et al., 2018; Liang et al., 2019) on PRG can be eliminated by using SAR imagery with its penetration capability and ability to monitor in all-weather conditions.

### 7.3.1 Glacier Velocity Estimation Using SAR Interferometry

Interferometric SAR techniques are used to determine the glacier's flow using the phase information contained in the radar signal. This signal is gathered by receiving antennas operating in different modes (horizontal and vertical polarization) and parted by either time or space. The phase in the SAR signal can be related to the distance from the target. Thus, a pair of scenes collected though SAR antennas generally contain the displacement information due to target movement or change in the antenna's location. Generating an interferogram, which is essentially subtracting the phase in two scenes, gives an idea of the displacement. Apart from the changes in position, these datasets are sensitive to geometric errors. To accurately measure the motion of the target, both geometric and topographical errors need to be removed. Studies have used SAR interferometry to determine glacier surface flow and ice-sheet movement using a variety of datasets (e.g., ERS and TerraSAR-X). This technique is known to give high accuracy measurements. However, it possesses some limitations such as coherence loss, temporal decorrelation, and phase noise over large displacements. In such cases, offset tracking acts as a viable tool (Strozzi et al., 2002; Kumari et al., 2019). Moreover, the offset tracking approach is more suited for monitoring fast-moving glaciers (Sánchez-Gámez and Navarro, 2017; Nela et al., 2019).

### 7.3.2 Glacier Velocity Estimation Using Offset Tracking

The offset tracking method is one among the widely used methods other than the DInSAR method to estimate the flow rate of any glacier. The offset tracking method employs different approaches to monitor the flow rate of the glaciers. These approaches include speckle tracking, intensity tracking, and feature tracking. The algorithm used in the intensity tracking approach applies pixel-to-pixel correlation in the images which generates the offset fields from image patches of intensity images with normalized cross-correlation (Scambos et al., 1992) algorithm and estimate the relative displacement between image pairs. The feature tracking approach involves tracking of the glacial features such as crevasses and moraines, on co-registered image pairs in order to develop a two-dimensional (2D) velocity field. This approach keeps track of the features following two ways, i.e., manual selection and automatic tracking. The requirement for this approach is that the glacial feature should remain undisturbed over the period of observation. The estimation of displacement can be inferred from the relative movement of the glacial features measured between the two images.

In order to study the ice flow over the glacier surface, we need to have an idea of the glacier type, topography, and its orientation. The SAR data parameters such as coherence, availability, accuracy, and spatial resolution are also crucial for glacier motion estimation. Depending upon this information, we can decide the effective way to determine the glacier ice-flow velocity. The offset tracking approach measures the offsets in both azimuth as well as slant range direction, thus providing a 2D surface displacement. One of the major significant advantages of the offset tracking method is that it can be applied where there is low coherence. It can be employed on SAR images with large intervals between the acquisition time.

The offset tracking approach overcomes these limitations and measures the feature movement between master and slave image using cross-correlation optimization of patch intensity in the case of fast and uneven flow of a glacier (Strozzi et al., 2002). It can be implemented with data obtained from optical as well as SAR satellites. For longer time periods, the sequential satellite imageries can be utilized to track the movement of glacier surface features like crevasses. The temporal separation between the image pair can range from a few weeks to years. Essentially, the correlation of an image pair representing the same location is called image matching. It should be noted that the displacement within the image pair must not overcome the method's accuracy Also, any apparent variations between the times of image pair acquisition must be negligible to perform efficient image matching.

In the case of SAR imagery, the cross-correlation-based image matching is performed in the azimuth as well as the range direction. Pixel oversampling of the image pair can be used to determine the matching peak to a further smaller fraction of a pixel. The successful image matching is governed by the presence of nearly comparable or identical features in the SAR imagery at a given scale. In order to achieve adequate image matching, coherence is a necessary prerequisite for optimal accuracy.

The widely used microwave data used for velocity estimations are: ALOS PALSAR, Envisat ASAR, TerraSAR-X, and Sentinel-1. In the optical domain ASTER, SPOT, Landsat ETM+ pan, and Landsat TM are generally used to implement feature tracking to derive the glacier surface flow.

## 7.4 Remote Sensing Based Dynamics of PRG: A Case Study

In this case study, the glacier surface velocity of PRG for the austral summer is investigated through the remote sensing approach. It is located in the Prydz Bay area near the Larsemann Hills and Ingrid Christensen Coast at 69°45"S and 75°30"E (Figure 7.1) on Princess Elizabeth land, to the east of Amery Ice Shelf (AIS) and surrounded by Dodd Island and Meknattane Nunataks. The Indian Research Base Bharati and Chinese Zhongshan station lie toward the east at about 50 km from PRG. About a 26 × 16 km-sized enormous mass of ice detached from the glacier tongue has been reported during the period 1973–1989 (Zhou et al., 2014). The area occupied by the PRG glacier amounts to ~877 km$^2$, with ~47 km length from top to glacier tongue. The PRG has the narrowest width of ~8 km, while the top and glacier tongue width is 25 km and 17.5 km, respectively. The elevation profile of PRG varies from ~15m to ~700m respectively from glacier tongue to the upper reaches of the glacier.

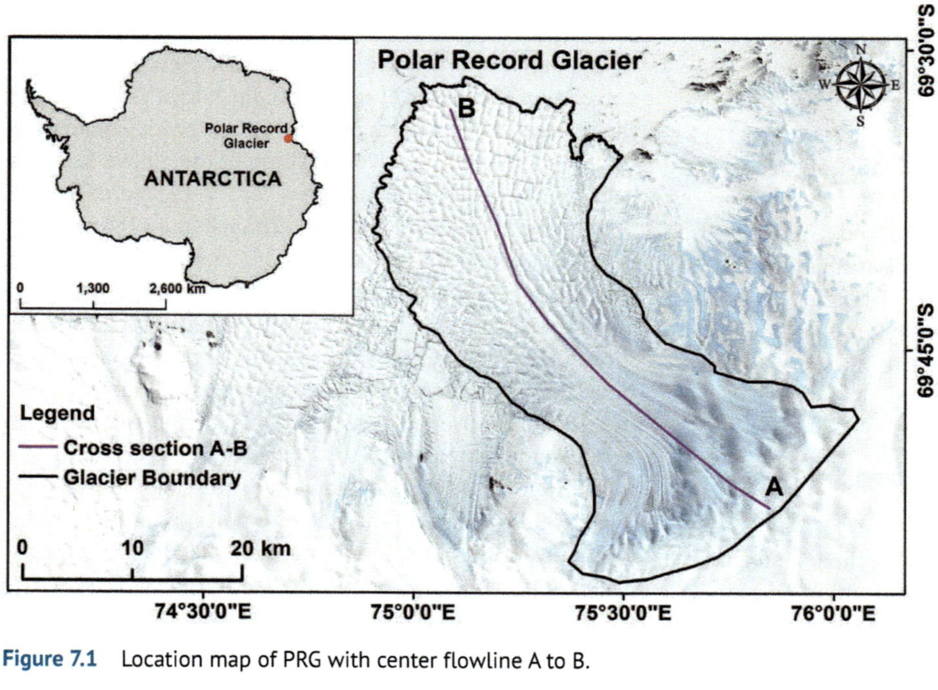


**Figure 7.1** Location map of PRG with center flowline A to B.

### 7.4.1 Data and Methods

In order to estimate the glacier flow rate between two observations, a pair of Sentinel-1 Ground Range Detected (GRD) images were utilized, which were pre-processed before applying the offset tracking algorithm. The images comprise of C-band Single Look Complex (SLC) with 6 days interval. These images were acquired during the 2016–2019 period including ascending and descending track. The polarization of the available scenes were horizontal transmit and horizontal receive (HH) type. Data acquisition information used in the study are mentioned in Table 7.1.

The Reference Elevation Model of Antarctica (REMA), a high resolution Digital Elevation Model (DEM), comprised of stereoscopic DEM extracted from image pairs of 0.32 m to 0.5 m resolution. This DEM has been used as input to the ice-flow velocity estimation approach. The 8-m resolution REMA terrain map may provide corrections for remote sensing based ice-flow modeling. The DEM was vertically registered to Cryosat-2 and ICESat with absolute and relative uncertainties of <1 m and in decimetres, respectively (Howat et al., 2019). The horizontal and vertical offsets of REMA DEM scale down to the 0.2 m of DEM relative accuracy, while the average accuracy of the REMA tiles are ~0.6 m (Howat et al., 2019). In addition to this, optical data of Landsat-8 over the period 2016–2019 have been used to manually demarcate the ice-front position.

The seasonal velocity is derived using SAR offset tracking which has proved to be a reliable technique to determine displacements using the cross-correlation of two images (Strozzi et al., 2002; Pritchard et al., 2005; Luckman et al. 2007; Satyabala, 2016; Singh et al., 2020). This technique measures movement of glacier surface features between two images using cross-correlation optimization of patch intensity (Gray et al., 1998; Kumari et al., 2014). It coregisters SAR images to track offset of distinguishable features to determine glacier flow velocity (Strozzi et al., 2002). The offset tracking approach is not constrained by temporal decorrelation between pairs of SAR datasets (Strozzi et al., 2002) when compared to interferometric techniques. This approach utilizes a selected image pair of Seninel-1 GRD product with a low temporal baseline supplied with precise orbit file for orbit correction. The pre-processing procedure in the velocity tracking approach incorporates the apply orbit file function which provides the SAR image file with the accurate orbital information. Then thermal noise removal is performed which is followed by calibration. Image intensity affected due to additional thermal noise has been reduced by the Thermal Noise Removal module provided under the SNAP tool. The image pair obtained is coregistered using DEM-assisted coregistration with cross-correlation to adjust the alignment from pixel of master and slave image. The approach involves cross-correlation in master and slave images using some common GCPs in these images. Using the master image GCP grid, the corresponding pixel location in the slave image was identified using the peak of normalized cross-correlation. The coefficient of normalized cross-correlation has been estimated following the study conducted by Kumari et al. (2014).

**Table 7.1** Specification of Sentinel-1 SAR data used in the study.

| Parameters | Sentinel-1 |
|---|---|
| Band | C |
| Orbit | Ascending/Descending |
| Product | SLC |
| Mode | IW |
| Wavelength | 5.6 cm |
| Polarization | HH |
| Revisit period | 6 days |
| Acquisition period | 2016–2019 |
| Total scenes | 18 |

### 7.4.2 Results and Discussion

#### 7.4.2.1 Ice Front Location

Ice-front location of PRG has been reported to have changed between 2004 and 2015. Zhou et al. (2014) reported the ice-front locations of the glacier until 2012 and Liu et al. (2018) reported annual frontal margin changes until 2015. Figure 7.2 shows the demarcated ice-front locations of the PRG. The glacier front was manually digitized using freely available Landsat-8 optical imagery. The acquired

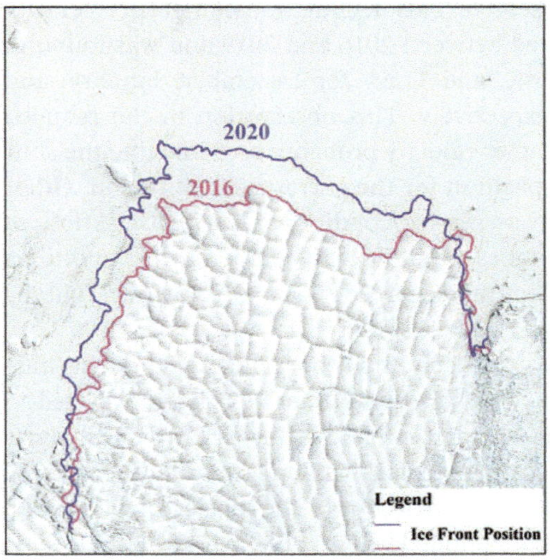


**Figure 7.2** Ice front location of PRG for year 2016 and 2020.

images are from the austral summer when the sea-ice conditions are minimum. The ice-front position of PRG has been reported to vary continuously from 2016 to 2020. The total change in ice-front position from 2016 to 2020 was found to be ~1750 m. It is noted that the rate of ice front position change reported by Liang et al. (2019) is consistent at ~600 $ma^{-1}$ from 2004 to 2012. The modulation of the ice-front area led to a 35 $km^2$ change in glacier area from 877 $km^2$ to 912 $km^2$ over the period 2016–2020.

#### 7.4.2.2 Glacier Velocity Over the Period of 2016–2019

The ice-flow velocity has also been estimated along with the varying ice-front locations of the PRG. Remote sensing-based estimates has been utilized to monitor the ice flow velocity over the period of 2016–2019 for the months of December, January, and February. The ice flow velocity graph on the center flowline A to B is represented in Figure 7.3.

The glacier velocity maps generated using feature tracking have adequate spatial coverage to deliver austral summer ice-flow profiles along the central flow line of PRG (Figure 7.3). The dip at a distance of ~7 km from point A indicates the presence of the grounding line. Overall, the glacier velocity along the central flowline A to B is higher in the terminus region and lower in the upper reaches during 2016–2019. The topographical aspects of PRG have been examined to investigate the probable reasons for the variation in estimated glacier velocity using surface topography from REMA DEM. It is observed that the western part of the glacier showed high velocity as well as higher variations when compared to the eastern part. This trend was also observed in an earlier study by Zhou et al. (2014) that the western part moves faster. The higher slope in the western part appears to be the reason here. An overall decline in the estimated mean glacier velocity has been observed. This decline in mean glacier velocity was calculated between 2016 and 2019 and was found to be 23.9%, 7.4%, and 11.6% for December, January, and February, respectively. This observation in the reduced estimated glacier velocity probably occurred due the shift in ice-front position for the corresponding period. Other factors, such as sea-ice conditions, ice melting, flow of glacier dynamics, and nearby temperature conditions may also affect the observed seasonal velocity variations of PRG (Zhou et al., 2014).

Due to the warm conditions over the glacier during austral summer, the estimated glacier velocity is expected to be high while the present study showed low velocity near the glacier terminus region. This suggests that the temperature conditions do not control the glacier flow entirely. Other factors mentioned below could play a role. Since it is an outlet glacier, the glacier terminus of PRG floats over the sea and the nearby area is surrounded by sea ice. So, the sea-ice variability near the glacier terminus of PRG was analysed from December to February over the period of 2016–2019. From the visual inspection of the optical images, the presence of sea ice during the month of December was observed. This could be a possible reason behind the slow glacier flow near the terminus.

To estimate the underlying uncertainties in the velocity estimates, the velocity over permanent features was assumed to have zero movement. The uncertainty in estimated glacier velocity is 0.025 m/d. It is noted that the seasonal velocity estimated for PRG for 2016–2019 is variable, which concurs with the conclusions presented in earlier studies on PRG (Zhou et al., 2014; Liu et al., 2018; Liang et al., 2019).

This study reports the glacier velocity of >2 m/d for some parts over the period 2016–2019, which was also reported by Zhou et al. (2014). This study focuses on the prime movement of the glacier flow, which is during the austral summer (Liang et al., 2019).

### 7.4.3 Summary and Conclusion

PRG is one of the prominent glaciers located in the East Antarctic region. The glacier is also reported to have calving events. The study of glacier flow is important, since it controls the ice-front position as well as the ablation in the calving region. However, the interaction relationship between these two entities (ice-front change and seasonal variations) in the glacier flow has not been studied comprehensively. In order to explore the underlying mechanism of the glacier dynamics, the austral summer glacier flow speed has been estimated. The study focuses on studied variations in the ice-front location and the austral glacier surface flow. This study does not provide any validation of the estimated velocity against the field measurements. Field visits by some authors of this study have revealed that the PRG contains a huge number of crevasses which made it difficult for field observation. Our remote sensing-based investigation showed a change in the ice-front location and the seasonal variation during austral summer in the glacier surface flow.

For this study, the Sentinel-1 SAR data was chosen since it has an advantage over the datasets of optical sensors in that it provides all weather capability to monitor the glacier dynamics, which is not possible in the case of optical datasets. Moreover, the offset tracking method proves to be advantageous over the DInSAR technique, as offset tracking avoids the cumbersome phase unwrapping, which requires a lot of time and processing power. The other benefit of offset tracking for fast-flowing glaciers such as PRG includes reliable estimation over large displacement. The high-resolution REMA DEM was used in the study to coregister the SAR data pair as well as to analyse the surface topography. Through this investigation, it

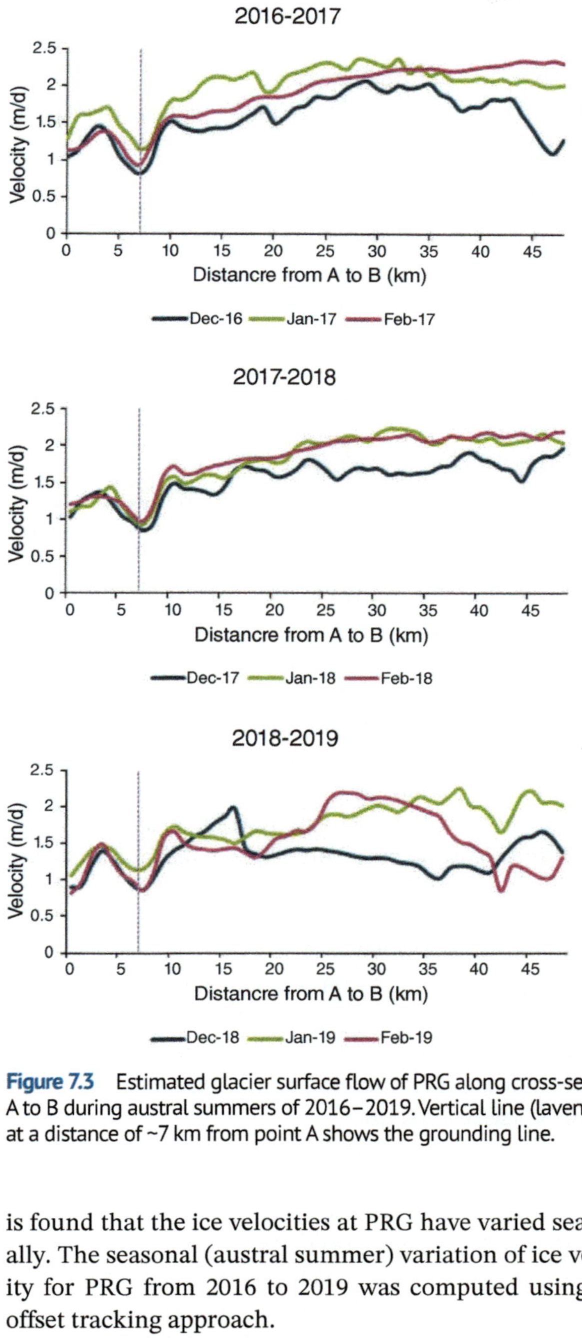


**Figure 7.3** Estimated glacier surface flow of PRG along cross-section A to B during austral summers of 2016–2019. Vertical line (lavender) at a distance of ~7 km from point A shows the grounding line.

is found that the ice velocities at PRG have varied seasonally. The seasonal (austral summer) variation of ice velocity for PRG from 2016 to 2019 was computed using the offset tracking approach.

Significant findings of this study are as follows. An average shift of 1750 m was observed in the glacier-front location during 2016–2020. Particularly, during the 2018–2019 period, a recession in glacier flow near the terminus was observed, which can be due to the thick sea ice conditions during this time. The estimated velocities followed a particular topographical pattern where the velocity in the western part is higher as compared to the eastern part. It shows that the surroundings have a significant impact on the glacier flow and thus its dynamics. However, additional analysis is required to understand the influence of these processes toward an overall spatial distribution of the glacier flow.

## References


Chen, Y., Zhou, C., Ai, S. et al. (2020). Dynamics of dalk glacier in East Antarctica derived from multisource satellite observations since 2000. *Remote Sensing* 12(11): 13–16. doi: 10.3390/rs12111809.

COMNAP (2017). Antarctic Station Catalogue. The Council of Managers of National Antarctic Programs. Available at:www.comnap.aq;https://static1.squarespace.com/static/61073506e9b0073c7eaaf464/t/611497cc1ece1b43f0eeca8a/1628739608968/COMNAP_Antarctic_Station_Catalogue.pdf

Eldhuset, K., Andersen, P.H., Hauge, S. et al. (2003). ERS tandem InSAR processing for DEM generation, glacier motion estimation and coherence analysis on Svalbard. *International Journal of Remote Sensing* 24(7): 1415–1437. doi: 10.1080/01431160210153039.

Foga, S., Stearns, L.A., and Van Der Veen, C.J. (2014). Application of satellite remote sensing techniques to quantify terminus and ice mélange behavior at helheim glacier, East Greenland. *Marine Technology Society Journal* 48(5): 81–91. doi: 10.4031/MTSJ.48.5.3.

Frezzotti, M. and Orombelli, G. (2014). Glaciers and ice sheets: current status and trends. *Rendiconti Lincei* 25(1): 59–70. doi: 10.1007/s12210-013-0255-z.

Gomez, R., Arigony-Neto, J., De Santis, A. et al. (2019). Ice dynamics of union glacier from SAR offset tracking. *Global and Planetary Change* 174: 1–15. doi: 10.1016/j.gloplacha.2018.12.012.

Gray, A.L., Mattar, K.E., Vachon, P.W. et al. (1998). InSAR results from the RADARSAT Antarctic mapping mission data: estimation of glacier motion using a simple registration procedure. *International Geoscience and Remote Sensing Symposium (IGARSS)* 3(1): 1638–1640. doi: 10.1109/igarss.1998.691662.

Greene, C.A., Young, D.A., Gwyther, D.E. et al. (2018). Seasonal dynamics of Totten Ice Shelf controlled by sea ice buttressing. *Cryosphere* 12(9): 2869–2882. doi: 10.5194/tc-12-2869-2018.

Hanna, E., Navarro, F.J., Pattyn, F. et al. (2013). Ice-sheet mass balance and climate change. *Nature.* 498(7452): 51–59. doi: 10.1038/nature12238.

Heid, T. and Kääb, A. (2012). Evaluation of existing image matching methods for deriving glacier surface displacements globally from optical satellite imagery. *Remote Sensing of Environment* 118: 339–355. doi: 10.1016/j.rse.2011.11.024.

Howat, I.M., Porter, C., Smith, B.E. et al. (2019). The reference elevation model of Antarctica. *Cryosphere* 13(2): 665–674. doi: 10.5194/tc-13-665-2019.

Huang, L. and Li, Z. (2011). Comparison of SAR and optical data in deriving glacier velocity with feature tracking. *International Journal of Remote Sensing* 32(10): 2681–2698. doi: 10.1080/01431161003720395.

IMBIE (2018). Mass balance of the Antarctic ice sheet from 1992 to 2017. *Nature* 558: 219–222. doi: 10.1098/rsta.2006.1792.

Kumari, S., Ghosh, S.K., and Buchroithner, M.F. (2014). Measurement of glacier velocity at Pik Lenin, Tajikistan, by feature tracking. *International Archives of the Photogrammetry, Remote Sensing and Spatial Information Sciences – ISPRS Archives* 40(8): 531–536. doi: 10.5194/isprsarchives-XL-8-531-2014.

Kumari, S., Ramsankaran, R., and Walker, J.P. (2019). Impact of window size in remote sensing based glacier feature tracking: a study on Chhota Shigri Glacier, Western Himalayas, India. *International Geoscience and Remote Sensing Symposium (IGARSS)* (December 2020), 4175–4178. doi: 10.1109/IGARSS.2019.8899325.

Lemos, A., Shepherd, A., McMillan, M. et al. (2018). Seasonal variations in the flow of land-terminating glaciers in central-west Greenland using Sentinel-1 imagery. *Remote Sensing* 10(12): 1–12. doi: 10.3390/rs10121878.

Liang, Q.I., Zhou, C., Howat,I.A. et al. (2019). Ice flow variations at Polar Record Glacier, East Antarctica. *Journal of Glaciology* 65(250): 279–287. doi: 10.1017/jog.2019.6.

Liu, T., Niu, M., and Yang, Y. (2018). Ice velocity variations of the Polar Record Glacier (East Antarctica) using a rotation-invariant feature-tracking approach. *Remote Sensing* 10(1): doi: 10.3390/rs10010042.

Luckman, A., Quincey, D., and Bevan, S. (2007). The potential of satellite radar interferometry and feature tracking for monitoring flow rates of Himalayan glaciers. *Remote Sensing of Environment* 111(2): 172–181. doi: 10.1016/j.rse.2007.05.019.

Miles, B. (2012). Synchronous Terminus Change of East Antarctic Outlet Glaciers Linked to Climatic Forcing. Durham, UK: Durham University. http://etheses.dur.ac.uk/7333/

Nela, B.R., Bandyopadhyay, D., Singh, G. et al. (2019). Glacier flow dynamics of the Severnaya Zemlya archipelago in Russian high arctic using the differential sar interferometry (DInSAR) technique. *Water (Switzerland)* 11(12): doi: 10.3390/w11122466.

Noble, T.L., Rohling, E J., Aitken, A.R.A. et al. (2020). The sensitivity of the Antarctic ice sheet to a changing climate: past, present, and future. *Reviews of Geophysics* 58(4): 1–89. doi: 10.1029/2019RG000663.

Pandit, P.H., Jawak, S.D., and Luis, A.J. (2018). Estimation of velocity of the Polar Record Glacier, Antarctica using Synthetic Aperture Radar (SAR). *Proceedings* 2(7): 332. doi: 10.3390/ecrs-2-05145.

Pritchard, H., Murray, T., Luckman, A. et al. (2005). Glacier surge dynamics of Sortebræ, east Greenland, from synthetic aperture radar feature tracking. *Journal of Geophysical Research: Earth Surface* 110(3): 1–13. doi: 10.1029/2004JF000233.

Rémy, F. and Frezzotti, M. (2006). Antarctica ice sheet mass balance. *Comptes Rendus – Geoscience* 338(14–15): 1084–1097. doi: 10.1016/j.crte.2006.05.009.

Rignot, E. (2002). Mass balance of East Antarctic glaciers and ice shelves from satellite data. *Annals of Glaciology* 34: 217–227. doi: 10.3189/172756402781817419.

Rignot, E., Jacobs, S., Mouginot, J. et al. (2013). Ice-shelf melting around Antarctica. *Science* 341(6143): 266–270. doi: 10.1126/science.1235798.

Sánchez-Gámez, P. and Navarro, F.J. (2017). Glacier surface velocity retrieval using D-InSAR and offset tracking techniques applied to ascending and descending passes of sentinel-1 data for Southern Ellesmere ice caps, Canadian Arctic. *Remote Sensing* 9(5): 1–17. doi: 10.3390/rs9050442.

Satyabala, S.P. (2016). Spatiotemporal variations in surface velocity of the Gangotri Glacier, Garhwal Himalaya, India: study using synthetic aperture radar data. *Remote Sensing of Environment* 181: 151–161. doi: 10.1016/j.rse.2016.03.042.

Scambos, T.A., Bohlander, J.A., Shuman, C.A. et al. (2004). Glacier acceleration and thinning after ice shelf collapse in the Larsen B embayment, Antarctica. *Geophysical Research Letters* 31(18): 2001–2004. doi: 10.1029/2004GL020670.

Scambos, T.A., Dutkiewicz, M.J., Wilson, J.C. et al. (1992). Application of image cross-correlation to the measurement of glacier velocity using satellite image data. *Remote Sensing of Environment* 42(3): 177–186. doi: 10.1016/0034-4257(92)90101-O.

Scambos, T.A., Haran, T.M., Fahnestock, M.A. et al. (2007). MODIS-based Mosaic of Antarctica (MOA) datasets: continent-wide surface morphology and snow grain size. *Remote Sensing of Environment* 111(2): 242–257. doi: 10.1016/j.rse.2006.12.020.

Shukla, A. and Garg, P.K. (2020). Spatio-temporal trends in the surface ice velocities of the central Himalayan glaciers, India. *Global and Planetary Change* 190: 103187. doi: 10.1016/j.gloplacha.2020.103187.

Singh, G., Nela, B.R., Bandyopadhyay, D. et al. (2020). Discovering anomalous dynamics and disintegrating behaviour in glaciers of Chandra-Bhaga sub-basins, part of Western Himalaya using DInSAR. *Remote Sensing of Environment* 246: 111885. doi: 10.1016/j.rse.2020.111885.

Stearns, L.A. (2011). Dynamics and mass balance of four large East Antarctic outlet glaciers. *Annals of Glaciology* 52(59): 116–126. doi: 10.3189/172756411799096187.

Stearns, L.A. and Hamilton, G.S. (2005). A new velocity map for Byrd Glacier, East Antarctica, from sequential ASTER satellite imagery. *Annals of Glaciology* 41:71–76. doi: 10.3189/172756405781813393.

Stearns, L.A., Smith, B.E., and Hamilton, G.S. (2008). Increased flow speed on a large east Antarctic outlet glacier caused by subglacial floods. *Nature Geoscience* 1(12): 827–831. doi: 10.1038/ngeo356.

Strozzi, T., Luckman, A., Murray, T. et al. (2002). Glacier motion estimation using SAR offset-tracking procedures. *IEEE Transactions on Geoscience and Remote Sensing* 40(11): 2384–2391. doi: 10.1109/TGRS.2002.805079.

Turner, J., Colwell, S.R., Marshall, G.J. et al. (2005). Antarctic climate change during the last 50 years. *International Journal of Climatology* 25(3): 279–294. doi: 10.1002/joc.1130.

Vaughan, D G., Comiso, J.C., Allison, I. et al. (2013). *Observations: Cryosphere, Climate Change 2013: The Physical Science Basis*. Cambridge, UK and New York: Cambridge University Press. doi: 10.1007/978-1-4020-4411-3_53.

Williams, R.S., Ferrigno, J.G., Swithinbank, C. et al. (1995). Coastal-change and glaciological maps of Antarctica. *Annals of Glaciology* 21: 284–290. doi: 10.3189/s0260305500015950.

Yellala, A., Kumar, V., and Høgda, K.A. (2019). Bara Shigri and Chhota Shigri glacier velocity estimation in western Himalaya using Sentinel-1 SAR data. *International Journal of Remote Sensing* 40(15): 5861–5874. doi: 10.1080/01431161.2019.1584685.

Zhou, C., Zhou, Y., Deng, F. et al. (2014). Seasonal and interannual ice velocity changes of Polar Record Glacier, East Antarctica. *Annals of Glaciology* 55(66): 45–51. doi: 10.3189/2014AoG66A185.

# 8

# Terrestrial Deglaciation Signatures in East Antarctica

*Uday Sharma[1], Yogesh Ray[1,*], and Manish Pandey[2,3]*

[1] National Centre for Polar and Ocean Research, (Ministry of Earth Sciences, Govt. of India) Headland Sada, Vasco Da Gama 403804, Goa, India.
[2] University Centre for Research and Development (UCRD), Chandigarh University, Mohali 140413, Punjab, India
[3] Department of Civil Engineering, Chandigarh University, Mohali 140413, Punjab, India
* Corresponding author

## 8.1 Introduction

Since the earliest expeditions to the Antarctic region at the beginning of the 20th century, humans have established many research stations that facilitate extensive scientific studies on the isolated icy continent. The surface of this cold, rugged continent is 97% covered with ice. The inception of glaciation in Antarctica has been long dated back to 34 Ma after its occupation as an isolated continent in the southernmost corner of the globe and global cooling trends across the Eocene–Oligocene transition (Coxall et al., 2005; Lear et al., 2008). In the present day, terrestrial, marine, and ice core signatures of past ice-sheet fluctuations in Antarctic regions provide a holistic window to understand the dynamics of global climatic variation across the planet.

Along with the current decadal-scale changes, understanding the past deglaciation trends in Antarctica is also essential to predict future climatic behavior of ice sheets in the Antarctic region and subsequent global sea-level rise (Suganuma et al., 2014). As deglaciation and warming across Antarctica have been relatively enhanced in comparison with global temperatures (Cuffey et al., 2016), it is needed that the resolution of deglaciation data from the Antarctic region is increased, in order to better understand the climatic dynamics that can cause subsequent changes in the future. Though deglaciation studies have been sparse in the whole Antarctic region, data available from vast East Antarctica are relatively few compared with West Antarctica. This chapter reviews the terrestrial glacial and ice-sheet fluctuation studies that have been conducted in East Antarctica and tries to provide insights into the palaeoclimatic dynamics across the Late Quaternary.

## 8.2 Geomorphology

Antarctica is the southernmost continent on our planet, separated by the Southern Ocean from Africa and Australia, and the Drake Passage separating it from South America. It is an enormous continent occupying 14.2 million km$^2$ of area. (Watt, 2022). Trans-Antarctic mountains exclusively divide the continent into two major divisions: East Antarctica and West Antarctica (Figure 8.1). Though three ice sheets cover both regions, two of them being the most extensive ice sheets in the world, the East and West Antarctic ice sheets, the geomorphology and ice sheet dynamics across these two regions vary significantly. The West Antarctic part shows more rapid warming trends than the East Antarctic. These Antarctic ice sheets and the Antarctic Peninsula ice sheet contain about 27 million km$^3$ of ice, which, if melted, would raise the global sea level as high as 58 m (Fretwell et al., 2013). The basal topography of the Antarctic continent beneath these two ice sheets varies distinctively and governs the ice-sheet flow above. The East Antarctic Ice Sheet (EAIS) consists of an underlying landmass, whereas the West Antarctic Ice sheet (WAIS) is over the sea. A detailed discussion about the geomorphology of the major regions of East Antarctica is as follows.

*Advances in Remote Sensing Technology and the Three Poles,* First Edition. Edited by Manish Pandey, Prem C. Pandey, Yogesh Ray, Aman Arora, Shridhar D. Jawak, and Uma K. Shukla.

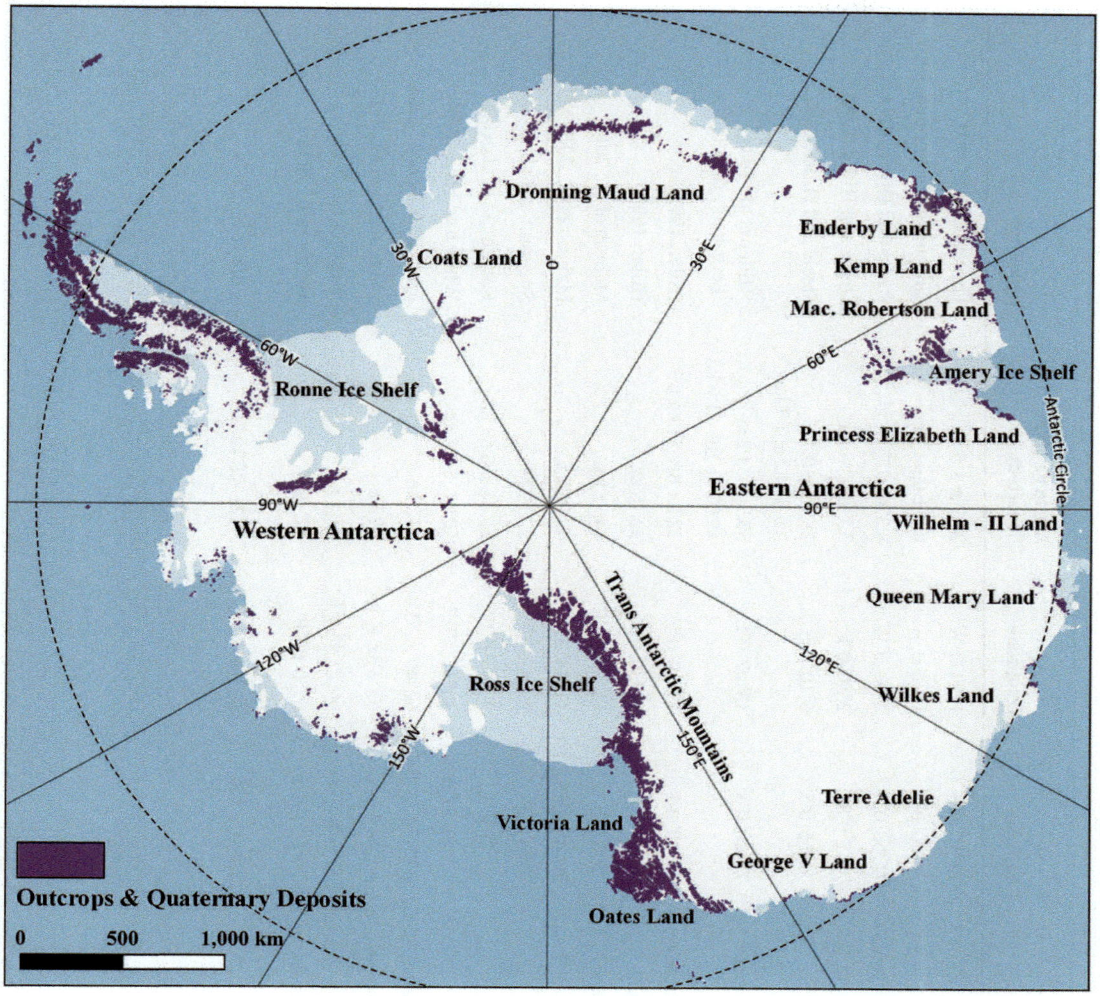


**Figure 8.1** Rock outcrops and moraine deposits located in different regions of Antarctica along with demarcated regions of Dronning Maud Land, Enderby Land, Kemp Land, Mac. Robertson Land, Princess Elizabeth Land, Wilhelm-II Land Queen Marry Land, Wilkes Land, Terre Adelie, George V Land, Oates Land, Victoria Land of Antarctica, and Trans-Antarctic Mountains, along with the location of major three ice shelves, Ronne Ice Shelf, Amery Ice Shelf, and Ross Ice Shelf. Map modified using Quantarctica 3 (Adapted from Matsuoka et al., 2021).

### 8.2.1 East Antarctica

Extending from 45° west to 168° east, East Antarctica consists of the largest ice sheets on our planet. It is classified into the following divisions from west to east: Dronning Maud Land, Enderby Land, Kemp Land, Mac. Robertson Land, Princess Elizabeth Land, Wilhelm II Land, Queen Mary Land, Wilkes Land, Terre Adelie or Adelie Land, George V Land, and Oates Land (Table 8.1). There exist multiple ice shelves along the coastal margins of the regions mentioned above. These ice shelves are the floating extent of ice sheets from the landmass toward the sea. Most of East Antarctica is a large plateau with various outcrops of rocks known as Nunataks and scattered mountain chains mostly following an E–W direction parallel to the coast, other than a few mountains along the Amery Ice Shelf such as Prince Charles Mountains, Mawson Escarpment, and Grove Mountains following a NW–SE direction somewhat parallel to the Lambert Graben. The central part of the EAIS consists of several domes, along the slopes of which ice flows toward the coast in all directions, except one side restricted by the Trans-Antarctic Mountains. The central part of the EAIS consists of several highlands. The highlands represent the semi-continuous ice caps that cover the highest part of the mountains and partly reflect the perturbations of the underlying land surface. The major three highlands are polar gap sub-glacial highlands toward the western part of EAIS, Recovery sub-glacial highlands in the Central part of EAIS, and Gamburtsev sub-glacial mountains toward the eastern part of EAIS underneath Dome A. Along with these highlands, two major sub-glacial basins exist, Pensacola-Pole Basin and South Pole Sub-basin (Paxman et al., 2019).

**Table 8.1** Geographical locations and oldest chronological data signifying the initiation of deglaciation in different regions of East Antarctica.

| Name of the Region | Extent | Location Name | Latitude | Longitude | Age in Ka (Oldest Reported) | Reference |
|---|---|---|---|---|---|---|
| Dronning Maud Land | 20°W–45°E | Vestfjella | 73°10′ 0″S | 14°0′ 0″W | 38.7 ± 1.5 ka | Thor and Low, 2011 |
| | | Heimefrontfjella | 74°40′ 0″S | 11°0′ 0″W | 8.7 ± 0.04 ka | Lintinen and Nenonen, 1997 |
| | | Ahlmannryggen | 71°50′ 0″S | 2°30′ 0″W | 8.3 ± 0.07 ka | Steele and Hiller, 1997 |
| | | Gjelsvikfjella | 72°5′ 0″S | 2°50′ 0″E | 3.7 ± 0.08 ka | Steele and Hiller, 1997 |
| | | Svarthamaren Nunatak, Muhlig-Hafmannfjella | 71°55′ 0″S | 5°10′ 0″E | 28.15 ± 0.42 ka–33.82 ± 1.7 ka | Steele and Hiller, 1997 |
| | | Lake Untersee, Wohlthat Massif | 71°20′ 0″S | 13°27′ 0″E | 33.9 ± 3.02 ka[1], 45.07 ka BP–21.2 ka[2] | [1]Hiller et al., 1988, [2]Berg *et al.*, 2019 |
| | | Petermann Range, Wohlthat Massif | 71°40′ 0″S | 12°20′ 0″E | 22 ka BP–58 ka BP | Hiller, 1995 |
| | | Insel Range, Wohlthat Massif | 71°26′ 0″S | 11°36′ 0″E | 32.48 ka | Hiller, 1995 |
| | | Long Lake, Schirmacher Oasis | 70°45′ 27″S | 11°39′ 20″E | 44.85 ka | Mahesh et al., 2015 |
| | | Zub Lake | 70°45′ 49″S | 11°44′ 25″E | 42.98 ka | Mahesh et al., 2019 |
| | | Sor Rondane Mountains | 72°0′ 0″S | 25°0′ 0″E | 2.319 Ma | Suganuma et al., 2014 |
| | | Soya Coast | 69°28′ 0″S | 39°40′ 0″E | 21.24 ± 0.63 ka | Kawamata et al., 2020 |
| Enderby Land | 45°E–60°E | Condon Hills | 67°52′ 59″S | 48°37′ 59″E | 103.1 ± 9.5 ka | White and Fink, 2014 |
| Mac. Robertson Land | 60°E–73°E | Mt Ward, Framnes Mountains | 67°47′17.9″S | 62°49′5.2″E | ~12 ka | Mackintosh et al., 2014 |
| | | Mt Henderson, Central Mawson | 67°42′7.6″S | 63°3′32″E | ~9 ka | Mackintosh et al., 2014 |
| | | Mt. Steiner | 73°4′30.7″S | 66°21′44.6″E | 21.3 ka | White et al., 2011 |
| | | Aramis Range, Prince Charles Mountains | 70°37′ 0.1″S | 67°0′ 00″E | 17. ka | White et al., 2011 |
| | | Radock Lake | 70°51′ 35″S | 67°59′ 47″E | ~121–28 ka BP | Fink et al., 2006 |

| | | | | | | |
|---|---|---|---|---|---|---|
| Princess Elizabeth Land | 73°E–87°E | Grove Mountains | 72°45′ 0″S | 75°0′ 0″E | 900 ka–50 ka | Lilly et al., 2010 |
| | | Lake Nella, Broknes Peninsula, Larsemann Hills | 69°23′ 4″S | 76°22′ 23″E | 24.95 ka | Hodgson et al., 2001; Burgess et al., 2004 |
| | | Stornes Peninsula, Larsemann Hills | 69°25′ 42″S | 76°8′ 13″E | 140.2 ± 14.2 ka | Kiernan et al., 2009 |
| | | Mirror Peninsula, Larsemann Hills | 69°20′ 00″S | 75°55′ 00″ E | 70.7 ± 4. ka | Liang et al., 2020 |
| | | Mistichelli Hills, Larsemann Hills | 70°1′ 59.9″S | 72°49′ 59.9″E | 38.9 ± 7.2 ka | Liang et al., 2020 |
| | | Long Peninsula | 68°28′ 20″S | 78°14′ 40″E | 15.6 ka BP | Gao et al., 2018 |
| | | Lake Abrax, Vestfold Hills | 68°29′ 21.3″S | 78°17′ 08.9″E | 18,110 ± 440 ka | Gibson et al., 2009 |
| Wilhelm-II Land | 87°E–92°E | No Terrestrial Deglaciation Chronological Data Available | | | | |
| Queen Mary Land | 92°E–100°E | | | | | |
| Wilkes Land | 100°E–136°E | Bunger Hills | 66°10′ 0.1″S | 100°52′ 59.9″E | 41.4 ± 11.8 ka | Gore et al., 2001 |
| | | Holl Island, Windmill Islands | 66°24′ 38.5″S | 110°24′ 32″ E | 7460–7710 ka | Goodwin, 1993 |
| | | Bailey's Peninsula, Windmill Islands | 66°17′ 8.9″S | 110°31′ 59.9″E | 5230–5480 ka | Goodwin, 1993 |
| | | Beall Island,Windmill Islands | 66°17′ 51″S | 110°28′ 58.2″E | 8435–8640 ka | Roberts et al., 2006 |
| Terre Adélie | 136°E–142°E | No Terrestrial Deglaciation Chronological Data Available | | | | |
| George V Land | 142°E–154°E | | | | | |
| Oates Land | 154°E–160°E | | | | | |

## 8.3 Landform Variation Concerning Various Sectors and Elevation

### 8.3.1 Dronning Maud Land

Dronning Maud Land exists in the western part of East Antarctica, extending from Princess Martha Coast (20°W–05°E) to Prince Olav Coast (40°E–45°E). Dronning Maud Land consists of several Nunataks and mountain ranges, mostly running parallel to the coast in a SW–NE trend (Figure 8.2). The subglacial escarpments in the region result in complex bed relief and protruding Nunataks, which restrict the movement of ice flow from the polar plateau into the coastal areas (Newall et al., 2020).

In the extreme western part of Dronning Maud Land, the Heimefrontfjella and Vestfjella mountain range extend ~150 km and ~120 km, respectively. The maximum elevation goes as high as 2600–2700 m a.s.l. The presence of striae and till on Nunataks in Vestfjella gives evidence of overriding of the ice sheet in the past (Hättestrand and Johansen, 2005; Mackintosh et al., 2014; Newall et al., 2020).

To its east, the Ahlmannryggen and Gjelsvikfjella exist south of Princess Martha coast. Most glaciers in this area drain toward the north in the Fimbul Ice Shelf along the coast (Neethling, 1969). The Ahlmannryggen consists of an ice-covered ridge 200 km away from the ice-sheet margin with maximum elevation going up to 1843 m a.s.l. Glacial cirques often restrict Nunataks edges, and erratic boulders, till, and supraglacial moraines are found at various elevations from 1307 m a.s.l. to 1123 m a.s.l., suggesting that the ice sheet exceeded beyond ~1120 m a.s.l. in the past (Mackintosh et al., 2014; Dymova, 2018). In the Gjelsvikfjella area, many rugged peaks are exposed above the ice sheet with a maximum elevation of ~2700 m a.s.l.

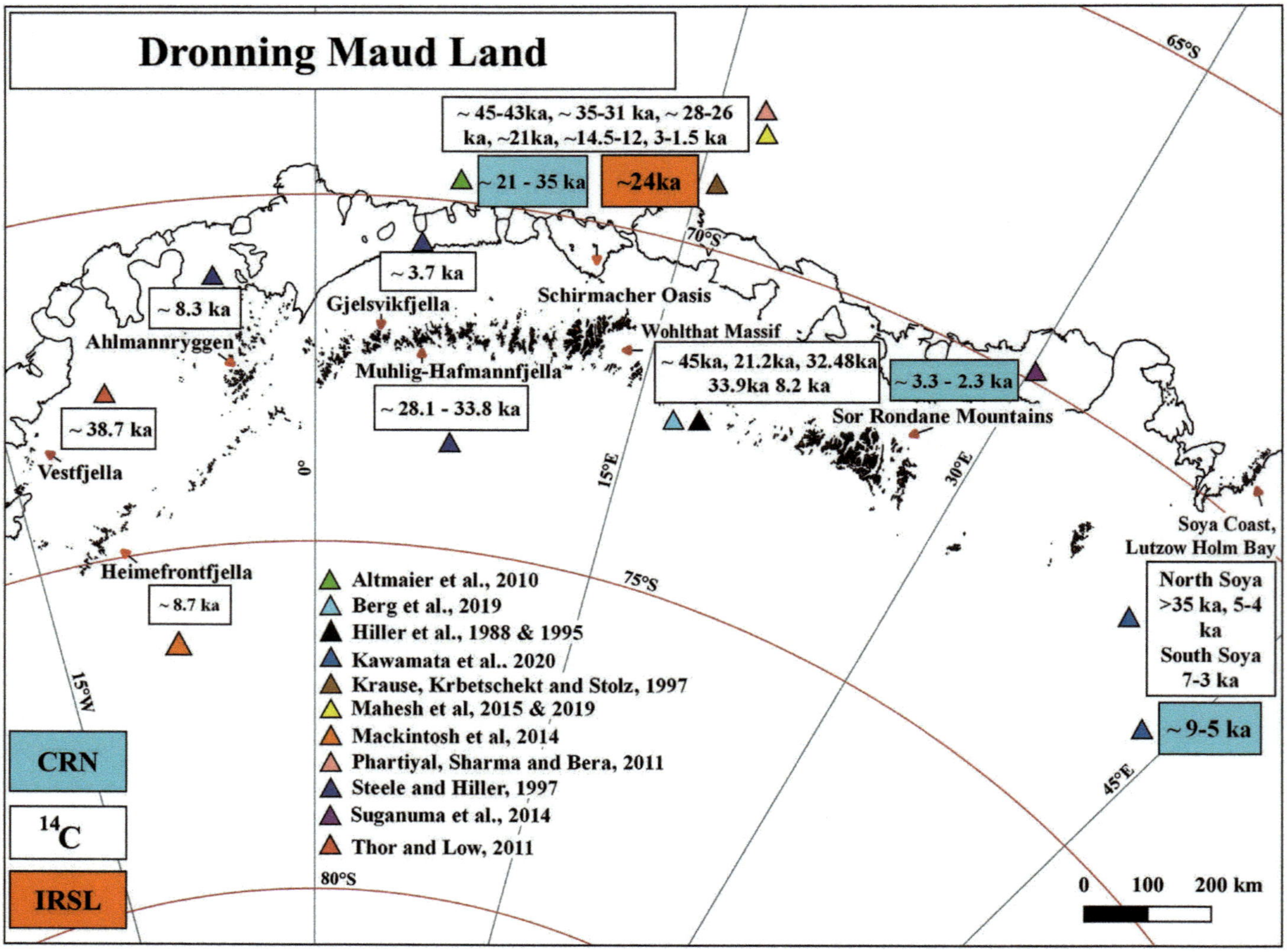


**Figure 8.2** Chronological sequences from the Dronning Maud Land region derived from Cosmogenic radionuclide dating (blue), radiocarbon C-14 dating (white), and IRSL dating (orange). Ages from Vestfjella, Muhlig-Haffmannfjella, Schirmacher Oasis, Wohlthat Massif, and North Soya Coast predates LGM. Map modified using Quantarctica 3 (Adapted from Matsuoka et al., 2021).

The loose sediment covers the western slopes of the peaks. The two significant glaciers, Slithallet and Fjellimellom, drain the area from south to north and exist on the eastern and western sides of Jutulsessen Mountain. The exposed outcrops in the region comprise various aeolian and glacial erosional features such as horns, arêtes, caves, and moraines (Shrivastava et al., 2014).

The Schirmacher Oasis, another ice-free area in the Coast Central Dronning Maud Land, Eastern Antarctica, exists along the Princess Astrid Coast, bounded by the ice sheet in the south and Nivlisen ice shelf in the north. This area covers 34 $km^2$ and trends 20 km from east to west. The elevation rises toward the south with the highest peak, Mt Rebestiya, also called Trishul Peak, at 228 m a.s.l. existing in the south-central part. The area is undulatory in morphology and consists of many lakes, ponds, and other periglacial features. The area consists of developed Rochee mountains, U-shaped deglaciated valleys, escarpments and moraines, and till deposits over the top of hills that give evidence of topographic modification in the past by the ice sheet (Paech and Stackebrandt, 1995; Richter and Borman, 1995; Ravindra, 2001; Ray et al., 2021). The Wolthat Massif, existing ~100 km south of Schirmacher Oasis, trends NE–SW and comprises multiple mountains and Nunataks such as Humboldt Mountains, Gruber Mountains, Petermann Range, and Untersee Oasis (Naslund, 2007). The elevation in the region exceeds close to ~2800 m a.s.l.–3000 m a.s.l. The moraine deposits and glacial benches in the Gruber Mountains and Petermann range indicate the maximum glaciation extent to be around 2000 m in the past (Srivastava et al., 1988).

In middle-eastern Dronning Maud Land, the Sor Rondane Mountains comprise many Nunataks trending in an E-W direction, which are widely scattered across the 2000 $km^2$ area and lie at about 200 km from the coast (Pattyn et al., 2010). The Nunataks consist of various local glaciers and ice-free slopes. The ice-sheet elevation in the region ranges from 1000 m in the north to 2500 m in the south. Most of the area in the central part of the mountains consists of flat-top ridges and thin till covered flanks. The existence of glacial striations at an elevation of ~2500 m in the Walnumfjellet area and ~2000 m at Mt Tekubi, along with several moraine ridges and moraine fields, gives evidence of the extent of the overriding ice sheet in the past in this region (Suganuma et al., 2014).

In the easternmost part of Dronning Maud Land, along with the ice-free areas of the Soya coast, Lutzow Holm Bay, substantial geomorphological evidence is available about the ice sheet's past extent. Skarvsnes, the largest ice-free area in the region spanning over 63 $km^2$, protrudes above the ice sheet with a maximum elevation of up to 400 m a.s.l. The area comprises numerous lakes, low hills. and shallow linear depressions. Raised beaches containing many marine shells are found along the shoreline at various places, rising to 23 m. The ice-free area of Skallen, with an expanse of ~14 $km^2$, has a gently undulating surface with broad E-W trending valleys, with a maximum elevation rising high up to 186 m a.s.l. Numerous lateral moraine ridges, rising to 20–40 m and erosional surfaces with striations and grooves and lack of marine features distinctively separate it from the Skarvsnes area (Kawamata et al., 2020).

### 8.3.2 Enderby Land

Enderby Land lies between Lutzow Holm Bay and Shirase Glacier in the west to Wilma Glacier and Edward VIII Bay in the east (Figure 8.3). There is a lack of geomorphological and palaeo-glacial studies in Enderby Land. The data is sparse except for a few cosmogenic nuclide studies carried out along Rayner Glacier, Enderby Land. The flow of the Rayner Glacier is confined in nature across the narrow bedrock ridge called the Condon Hills, located at 560 m a.s.l. and aligned perpendicular to the glacier. The bedrock is highly weathered and eroded into small hillocks oriented E–W, with visible iron staining and crumbling. The glacial sediments are sparse in the upper Condon Hills area, except for a few boulders and cobbles erratics.

The lower Condon Hills, at an elevation of 240 m a.s.l., are highly variable compared to the upper Condon Hills, with glacial erratics covering the whole area. The bedrock in the area is relatively less weathered or unweathered than upper Condon Hills, with distinctive striation in an almost true north direction. The areal observational studies marked the two distinctive trim lines at ~300 and 450 m (White and Fink, 2014).

### 8.3.3 Mac. Robertson Land, Amery Ice Shelf, and Prince Elizabeth Land

Framnes Mountains, in Mac. Robertson Land, comprises various Nunataks extending from 10–50 km inland of the coast and rising up to 400 m above the present-day ice-sheet surface (Figure 8.3). The Nunataks rises as high as ~1500 m a.s.l. The widely-spread granitic gneiss erratics over the charnockite bedrock give evidence of the past overriding by the ice sheet in this area. The presence of tafoni and weathering pits on the erratics proves the extended surface exposure (Mackintosh et al., 2014).

To the east of Framnes Mountains lies one of the largest and fast-flowing catchments of the Lambert Glacier and Amery Ice Shelf system, with Prince Charles Mountains bordering the western margin. The Prince Charles Mountains are home to numerous ice-free areas with some glacial deposits dating even further back to the late Pliocene

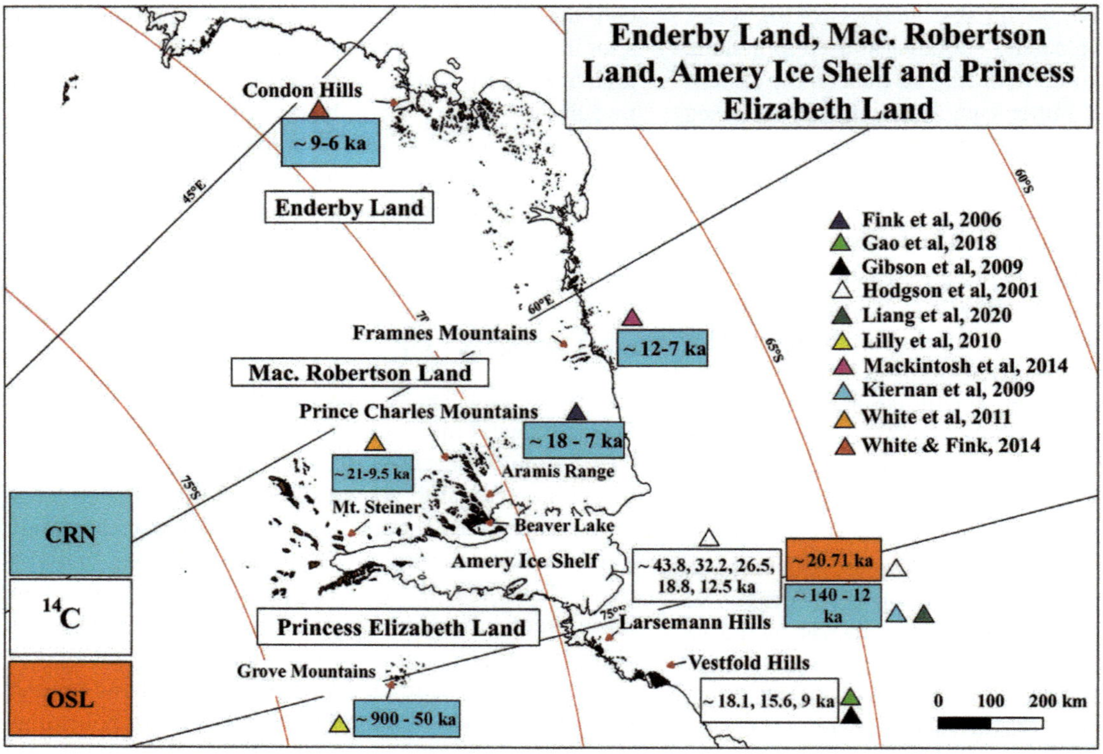


**Figure 8.3** Chronological sequences from Enderby Land, Mac. Robertson Land, Princess Elizabeth Land, and the area surrounding the Amery Ice shelf derived from Cosmogenic radionuclide dating (blue), radiocarbon C-14 dating (white), and OSL dating (orange). Ages from Larsemann Hills, Grove Mountains, predate LGM. Map modified using Quantarctica 3 (Adapted from Matsuoka et al., 2021).

to early Pleistocene. The presence of mesas and eroded slopes and ventifacts along the flanks of various lakes suggest the dominant aeolian and fluvial action modifying the landscape. In the northern Prince Charles Mountains, fresh moraine deposit occurs up to 100–120 m a.s.l. in height to the east of Fisher Massif and Mount Meredith, linked with deposition during the LGM. The two major lakes, Beaver Lake in the Beaver Basin and Raddock Lake in the Raddock Basin, exist in the Amery Oasis. Beaver Lake is a horseshoe-shaped basin shielded from the surrounding ice on three sides by a 300 m ridge and opened toward the north. Raddock Lake exists 7 m above Beaver Lake, spans over 2 $km^2$ and is fed by the 14 km long Battaye Glacier. In summer, the overflowing Raddock Lake drains the fluvially cut V-shaped valley of the Pagodroma Gorge. Based on the amount of weathering, the hummocky debris, scattered ice-cored moraines and erratic boulders that occupy the glaciers' flanks are distinguishable from the older moraines. The modern LGM deposits show evidence of previously existing low-angle ice streams (Adamson et al., 1997; Mackintosh et al., 2014).

On the opposite side, the Grove Mountains (Figure 8.3) comprise of many secluded Nunataks on the eastern side of the Amery Ice Shelf, 500 km south of the Vestfold Hills. The Nunataks are scattered over a vast area of ~1600 $km^2$. The ice sheet elevation in the region varies from ~1900 m a.s.l. to 2150 m a.s.l. The location of the Grove Mountains, far from any outlet glacier or ice stream, creates a unique environment in the region that makes it ideal for palaeoglaciation studies. The distinctive glacial geomorphological features such as ventifacts, ice-core pyramids, ice-cored moraines and Roches moutonnées are reported here (Lilly et al., 2010).

The Larsemann Hills (Figure 8.3), comprising the peninsula and various islands, lie as a coastal oasis along the Prince Ingrid Christensen Coast in Princess Elizabeth Land. It is the second-largest oasis in Eastern Antarctica, home to many periglacial lakes, glacial braised rounded hilltops, and highly weathered bedrock. The maximum elevation of hills ranges up to ~180 m a.s.l., which are highly dissected in nature. The area's geomorphology is attributed to the local lithology and geological structures, supplemented by physical glacial, glacio-fluvial. and aeolian weathering. The lower sites near the coast are also impacted by salt weathering (Hodgson et al., 2001).

The Vestfold Hills lie east of the Larsemann Hills, separated by Sorsdal Glacier and Rauer Island (Figure 8.3). The hills span over ~200 km$^2$ and have low relief, with ridges rising to ~157 m a.s.l. The area is a low elevation plateau comprising majorly bare rock terrain, with sporadic patches of unconsolidated sediment. The isolated patches of scattered erratics, occasional moraines, and thin debris over the bare rock terrain is the rare surface exposure of sediments. The glacial, fluvial, and marine sediments accumulate in valleys and marine inlets (Fitzsimons, 1996). The thickness of deposits is high in the western part, attributed to salt weathering of erratics and bedrock. The coast area is marked by various islands and fjords, out of which some have risen as isolated lakes attributed to isostatic rebound (Gore et al., 2003; Mackintosh et al., 2014).

### 8.3.4 Wilkes Land

Wilkes Land has the same orientation concerning glacial features as the Shackelton Ice Shelf, which bounds it in the west and north, EAIS in the east, and an outlet Apfel glacier in the south, and salt being a principal geomorphic and sedimentary agent (Figure 8.3) (Mackintosh et al., 2014; Gore and Leishman, 2020). The ice-free area of Bunger Hills in Wilkes Land covering up to 950 km$^2$, is similar in topography to the Vestfold Hills. Both sites have low relief with summits of less than ~200 m a.s.l., the presence of both fresh water and saline lakes, and various marine inlets across the whole area, although they vary in glacial history.

Further east, Windmill Islands are a group of islands spanning ~75 km$^2$, which lie along Vincennes Bay in Wilkes Land. The relief is relatively rugged in the southern isles because of igneous terrain, whereas the northern islands are mostly metamorphic. The islands exist 15 km offshore from the Law Dome rising to 1300 m a.s.l. on the mainland, separated from EAIS by a saddle at 1000 m a.s.l. The southern margin of the island is marked by an outlet glacier draining ice from EAIS and the Law Dome known as the Vanderford Glacier. The low depressions in the bedrock are occupied by various lakes that exist throughout the area. The low rounded hills with asymmetric Roches moutonnées profile, glacial striae, chatter marks, and moderately weathered bedrock give evidence of recent glacial activity in the region (Goodwin, 1993).

## 8.4 Chronology

In the Quaternary, EAIS has been considered more or less stable. However, the story seems slightly different with recent exploration and detailed deglaciation studies across Eastern Antarctica. Multiple studies have been carried out concerning the LGM to define the deglaciation trend across the deglaciated regions of EAIS using three absolute dating techniques: i) terrestrial cosmogenic nuclide exposure dating; ii) carbon dating; iii) and luminescence dating. These techniques provide specific age ranges about when an event has occurred.

The radiocarbon technique is based on the measurement of decaying radioactive isotope of carbon – $^{14}C$, which is found in any organic material. Once the organism/plant stops exchanging carbon with the environment, i.e., after death, the concentration of $^{14}C$ starts decreasing, as $^{14}C$ will decay as an exponential function. The half life of $^{14}C$ is 5730 years. If we know the present and initial concentration of $^{14}C$, we can determine the age of the organic material. The radiocarbon dating technique has two major limitations; it can only be used to date the material up to 50–60 ka. Second is the issue of dead carbon contamination, which may happen when older carbon mixes into younger sediment. Similarly, if younger material is carried and flows into the older deposit, it will underestimate the age (Fowler and Gillespie, 1986).

The other commonly used absolute dating method in Antarctica is the Cosmogenic Radionuclide (CRN) dating method. It involves measuring *in-situ* produced CRNs ($^{10}Be$, $^{26}Al$) formed due to spallation reactions when cosmic rays interact with the rock surface. The concentration of CRNs on the rock surface provides the timeline for the rock's surface exposure and serves as an excellent tool to determine the deglaciation history of glacial landforms, boulders, and moraines (Nishiizumi et al., 1993).

The limitation of CRN dating is the inheritance of cosmogenic isotopes from previous exposure, which may result in older ages. The other issue is repeated exhumation and burial of the boulder, which results in a complex exposure history and a wide range of ages (Darvill, 2013).

The third absolute dating technique is the Luminescence dating technique. This technique is used on the landforms, which primarily contain sand (Wintle and Huntley, 1982; Ray and Srivastava, 2010; Rhodes, 2011), though, in recent years, depositional ages from boulders have also been obtained (Rades et al., 2018). The OSL signal in sediment is reset when exposed to sunlight while sediment is undergoing transportation. Once the sediment is buried and no longer exposed to sunlight, the OSL signal starts accumulating when quartz or feldspar is exposed to ionizing radiation emitted by uranium, thorium, and potassium isotopes present in the sediment. We can determine the age of the sediment by measuring the total radiation dose and determining the dose rate in the grain (Rhodes, 2011).

The challenge with OSL dating common in the glacial environment is partial bleaching (resetting), which occurs when the sand grain is not exposed to sufficient light before deposition, which may result in overestimating the age (Fuchs and Owen, 2008; Rhodes, 2011).

In obscure areas, where terrestrial studies have been difficult to carry out, the sedimentology and dating of offshore marine sedimentary cores (Bentley et al., 2014; Mackintosh et al., 2014) have been carried out to determine the chronological sequence of ice sheet retreat across the different regions in Eastern Antarctica. The chronological findings up to the present are discussed in the following sections

### 8.4.1 Dronning Maud Land

In Western Dronning Maud Land (Figure 8.2), in the region of Vestfjella and Heimefrontfjella, no exact chronological order has yet been established. Few radiocarbon studies, which have been carried out in the area, give the minimum age of deglaciation. Radiocarbon dating of Mumiyo deposits from Vestfjella at an elevation of 1100 m a.s.l. has been reported at 38.7 ± 1.5 ka (Thor and Low, 2011), which suggests that Snow petrel colonies have persisted in the regions atleast before ~38 ka. In Heimefrontfjella, which exists ~150 km inland compared to Vestfjella, the basal layer of Snow petrel nests existing at 30–230 m above the present-day ice surface was dated earlier. The radiocarbon ages from these deposits are found at 8.7 ± 0.04 ka, suggesting that ice has been lowered at least 30 m from the ice level around ~8.7 ka BP (Lintinen and Nenonen, 1997).

Further to the east, in the west-central part of Dronning Maud Land, in Ahlmannryggen (Figure 8.2), the oldest radiocarbon age of 8.3 ± 0.07 ka is reported from various samples of Mumiyo deposit collected from the base to the summits of Nunataks found in this region, again suggesting that the ice sheet from summits retreated at or before ~8.3 ka. In Gjelsvikfjella (Figure 8.2), which exists ~200 km from the ice-sheet margin, the age of the Mumiyo deposit has been reported at 3.7 ± 0.08 ka, whereas in Svarthamaren Nunatak, lying in another area close to western Muhlig-Hafmannfjella, ages of 28.15 ± 0.42 ka–33.82 ± 1.7 ka BP have been reported in deposits found at 1610 m a.s.l. (Steele and Hiller, 1997). This suggests that ice-sheet levels in the area have remained close to present levels. even before the LGM (Mackintosh et al., 2014).

Carbon dating of Snow petrel and Mumiyo deposits in Central Dronning Maud Land (CDML) (Figure 8.2) has been extensively studied and offers a good record of colonization history by Snow petrels, which provides insights into local deglaciation trends. In the earlier works, in Lake Untersee, age of 33.9 ± 3.02 ka BP was reported from Mumiyo deposits. Another age of 8.2 ka ± 0.18 from the same region suggests that this area's climate has not fluctuated much with the LGM (Hiller et al., 1988). Even the recent radiocarbon dating of Mumiyo deposits (Berg et al., 2019) near Lake Untersee also report between 45.07 ka BP and 21.2 ka BP, suggesting the area to be ice free well before ~45 ka. In the Petermann Range, the radiocarbon ages of Mumiyo range from 22 ka BP to >58 ka BP, which suggest the area was occupied by Snow petrels long before ~58 ka, indicating hardly any changes in ice level during the LGM. In the Insel Range, Wohlthat Massif (Figure 8.2), the radiocarbon age of 32.48 ka is reported 50 m above the present-day ice sheet at an elevation of 1470 m a.s.l. (Hiller, 1995), again suggesting that the higher reaches in the area have been deglaciated before the LGM.

In the ice-free region of Schirmacher Oasis (Figure 8.2), north of Wohlthat Massif and the Humboldt Mountains, Central Dronning Maud Land, various researchers have worked extensively regarding palaeo-glaciation studies. Regarding the Lacustrine deposit, basal sediments of the proglacial Long Lake, Schirmacher Oasis, were reported at 12.9 ka and upper layers at 3 ka (Phartiyal et al., 2011). The proglacial lake became much more prominent after 3 ka, because of excess ablation and recession rates in this region (Phartiyal et al., 2011). Later, from the other cores from the same lake, the oldest radiocarbon age of 44.855 ka BP was reported from the bottom of the core (Mahesh et al., 2015). In Lake Dlinnoye, Schirmacher Oasis, the IRSL ages are reported at 24 ka at the base, which though considerably older, suggests that the area was ice free before the LGM (Krause et al., 1997). In Zub Lake, the radiocarbon dating of the bottom of the lake sediment core reported the date of 44, 42.98 ka BP. The chronological studies as discussed above suggest various phases of cold and warm environments repeatedly over the last ~45 ka BP in the Schirmacher Oasis region. The older age of sediments may likely be because of glacially eroded till deposited in the lake (Mahesh et al., 2019). Even the radionuclide exposure ages from Schirmacher Oasis region are between ~21 ka and 35 ka (Altmaier et al., 2010), suggesting that the older radiocarbon dates may be because of dead carbon accumulated in the younger sediment. However, this suggests that some parts of the Schirmacher Oasis areas were already ice free even during the LGM.

In Eastern Dronning Maud Land, the Sor Rondane Mountains (Figure 8.2), lying 150 km inland, comprises a vast network of Nunataks, comprising various signatures that provide evidence of the past existence of EAIS overriding these Nunataks. The cosmogenic radionuclide dates derived from erratic boulders from the region lie between 3.39 ka and 2.319 Ma, within 1132–2470 m a.s.l. elevation and deglaciation taking place over four stages (Suganuma et al., 2014). Over time, since the Early Pleistocene, the ice sheet lost its thickness to more than 500 m, but no significant thinning of EAIS in this sector of the Dronning Maud Land sector has been reported post-LGM in the Late-Pleistocene–Holocene (Suganuma et al., 2014).

In the Lutzow Holm Bay, Eastern Dronning Maud Land region, the ice-free area of the Soya Coast extends from north to south (Figure 8.2). The radiocarbon ages obtained from fossil shells from raised beaches of North Soya Coast gives two different groups of ages, one dating pre-35 ka BP, and another 5–4 ka BP. In contrast, in the southern part of the Soya coast, the radiocarbon ages are limited to only Holocene from 7–3 ka BP, and surface exposure ages are limited from 9–5 ka BP. Based on geomorphological studies carried out in this region along with surface exposure ages and radiocarbon ages, it can be deciphered that the northern part of the Soya coast had remained ice free during the LGM. In the southern part of the Soya coast, abrupt thinning of the ice sheet has taken place in the Early–Middle Holocene. The deglaciation pattern in this region has been attributed to the surrounding bedrock topography (Kawamata et al., 2020).

### 8.4.2 Enderby Land

In Enderby Land (Figure 8.3), the exposure ages from two boulders dated at 8.7 ka and 9.2 ka in the Upper Condon Hills suggest deglaciation occurring around ~9 ka BP, except for a boulder which was dated back to 103.1 ± 9.5 ka. This suggests that the last glaciation was not strong enough to remove old boulders and thus be cold based. In the Lower Condon Hills, the deglaciation stage seems close to ~6 ka BP, as evidenced by three dated erratics at 5.8 ka, 6.1 ka, and 6.6 ka BP (White and Fink, 2014). The 300 m lowering of ice from Upper Condon Hills to Lower Condon Hills between 9 and 6 ka BP in the Early Holocene gives the approximate idea about the timing of the final phase of thinning of the ice sheet, close to the present ice margin in this area (White and Fink, 2014).

### 8.4.3 Mac. Robertson Land, Amery Ice Shelf's and Princess Elizabeth Land

Concerning the terrestrial records of deglaciation in the Mac. Robertson Land region, the surface exposure ages of the erratic boulders from the Framnes Mountains (Figure 8.3) suggests the sharp thinning of EAIS, about 350 m from the coast in Mt Ward, around 12 ka–7 ka, after which the current ice sheet level has remained more or less stable to date. The exposure ages of boulders from North Mawson are reported between 12 and 10 ka, suggesting a lowering of about 200 m, whereas, from Mt Henderson, Central Mawson, David ranges, the exposure age lies between 9 and 7 ka, again suggesting about 200m of ice thinning. The Mumiyo deposit carbon age from North Mawson is also reported at 7.103 ± 0.048 ka BP, which relates to exposure ages from the same region and suggests the bedrock became ice free by 7 ka. (Mackintosh et al., 2014).

Further to the southwest, Prince Charles Mountains (Figure 8.3) in Mac. Robertson Land contain many ice-free Nunataks and oases, lying east of the Amery Ice Shelf. The different exposure TCN ages from the Aramis Range of Prince Charles Mountains were reported between 17.4 ka (329 m a.s.l.) and 11.8 ka (211 m a.s.l.) in the Loewe Massif region and between 21.3 ka (993 m a.s.l.) and 9.5 ka (310 m a.s.l.) in the Mt Steiner region, which suggest post-LGM retreat of ice after 18 ka (White et al., 2011), and attaining modern ice margins at ~11 ka in the Loewe Massif and ~9.5 ka near the Mt Steiner area. The $^{10}$Be TCN dating in the Radock Lake ranges from ~121–28 ka BP up to an elevation of 220 m, suggesting that the surrounding Battye Glacier might have advanced into the Radock basin at 120 ka or 28 ka BP, and the last retreat occurred between 20 and 11 ka BP (Fink et al., 2006).

Across the Amery Ice Shelf, in the Grove Mountains, southwest of the Larsemann Hills, the TCN evidence suggests a lower ice level than present in the area during the LGM (Figure 8.3). The exposure ages ranged from 900 ka–50 ka BP, indicating the present or lower levels of the EAIS were attained far before ~50 ka (Lilly et al., 2010; Mackintosh et al., 2014).

In the Broknes Peninsula, Larsemann Hills (Figure 8.3), the OSL studies carried out on glacio-fluvial sands gave the age of 20.71 ka, and the moss from Lake Nella has been dated back to 24.95 ka, suggesting the area was ice free during the LGM. The $^{14}$C dating of Lacustrine sediment from Lake Reid and Progress Lake dates back to 43 ka BP; however, no intelligible constraint is available to determine the deglaciation initiation (Hodgson et al., 2001; Burgess et al., 2004). The $^{10}$Be exposure ages from the Stornes Peninsula, Mirror Peninsula, and Mistichelli Hills helped in this cause, recording ages between ~140 ka and 12 ka (Kiernan et al., 2009) and ~70 ka and 13.5 ka (Liang et al., 2020). The single age of 140.2 ± 14.2 ka (Kiernan et al., 2009) suggests the possibility of deglaciation in the Larsemann Hills region, even before ~100 ka, probably back in MIS-6. However, this age may result due to several factors such as sample conditions, issue because of older $^{10}$Be inheritance, or other unknown information (Liang et al., 2020); hence, the deglaciation initiation of Larsemann Hills is generally considered during the MIS-4 (Figure 8.5).

In the Vestfold Hills (Figure 8.3), lying east of the Larsemann Hills separated by Sorsdal Glacier and Rauer Island, radiocarbon ages determined from remains of penguin colonies and mosses give evidence of a coastal area and the Watts Lake region being ice free at around ~9 ka (Pickard and Seppelt, 1984; Huang et al., 2009). This suggests that this area became deglaciated after the LGM. In Lake Abraxas, the radiocarbon dating on lake sediments gave the age of 18,110 ± 440 ka, suggesting that the area might have been ice free during the LGM (Gibson et al., 2009). Moreover, another study to determine initial penguin colonization from the ornithogenic sedimentary core, from Long Peninsula in the Vestfold Hills, suggests the local deglaciation was initiated by ~15.6 ka BP (Gao et al., 2018).

### 8.4.4 Wilkes Land

In Wilkes Land, extending from 100°E–140°E, deglaciation studies are specific to Bunger Hills and Windmill Islands only (Figure 8.4). In the central part of the Bunger Hills, the OSL studies on glacio-fluvial and lacustrine sands have constrained the deglaciation commencing at around 40 ka BP and the southern hills becoming completely ice free by 20 ka. In contrast, radiocarbon studies from the glacial lake shoreline have been dated at ~13.6 ± 0.14 ka BP (Gore et al., 2001). The radiocarbon studies on the epi-shelf lakes from coastal margins have suggested another deglaciation stage starting at around 9.6 ka BP and continuing intensively until 7.9 ka BP.

In the southern part of Windmill Island (Figure 8.4), ice retreat has been constrained at around ~8.1 ka through the evidence provided by radiocarbon dating of basal biogenic sediment in Holl Pond on Holl Island, southern Windmill Island. In the northern part of Windmill Island at Bailey Peninsula, the radiocarbon age of 5.930 ± 0.12 ka has been dated (Goodwin, 1993). In Beall Lake, Beall Island, a radiocarbon age of 7.84 ka ± 0.04 has been reported (Roberts et al., 2006). The extrapolation of sediment rate to radiocarbon ages suggests the inception of deglaciation at ~10.7 ka BP to ~10.5 ka BP in the southern Windmill Island and ~8.5 ka BP in the northern Windmill Island, extending up to ~5.5 ka BP in the Bailey Peninsula, northern Windmill Island (Mackintosh et al., 2014).

## 8.5 Discussion

The terrestrial deglaciation chronological sequence suggests variable retreat of the East Antarctic Ice Sheet across East Antarctica. The EAIS behavior across the Late Quaternary has fluctuated in different regions; however, the documentation and evidences to verify the extent of paleo ice-sheet extent and vertical thickness reduction of existing ice are sparse.

Other than the cosmogenic radionuclide studies, radiocarbon-based lake core studies and OSL studies of terrestrial Quaternary samples, ice core data from Antarctica also provide direct evidence of the climatic fluctuations and ice-sheet surface lowering, but sparse studies of these are available for the period around the LGM (Lorius et al., 1984; Jouzel et al., 1989; Delmonte et al., 2007; Siddall et al., 2012). The ice-core data may also contain mixed signals of local, regional, and global climatic fluctuations for the site chosen and proxies used (Thamban, 2020). The different ice-sheet models derived from distinct ice-core data suggest >100 m lowering of the surface ice sheet after the LGM, irrespective of disparities between the methods (Mackintosh et al., 2014). The ice cores beyond East Antarctica's interior region suggest the ice to be comparatively thinner during the LGM because of less accumulation (Jouzel et al., 1989; Siddall et al., 2012).

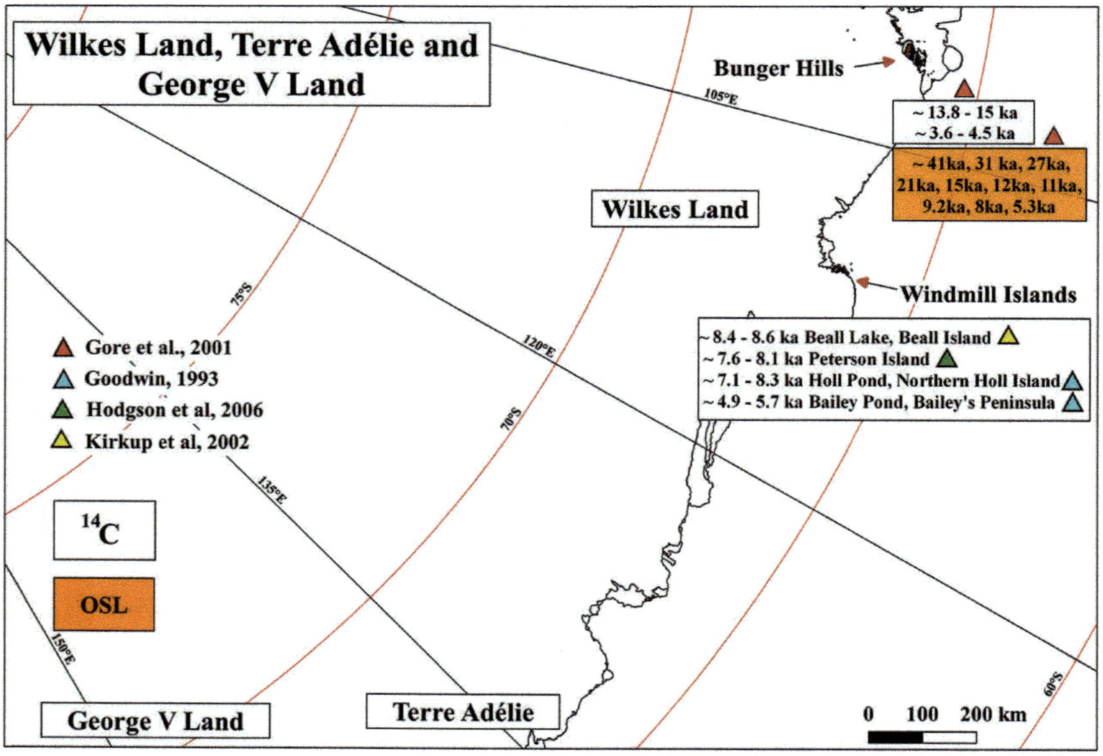


Figure 8.4 Chronological sequence from Wilkes Land region derived from radiocarbon C-14 (white) and OSL dating (orange). OSL Ages from Bunger Hills predates LGM. Map modified using Quantarctica 3 (Adapted from Matsuoka et al., 2021).

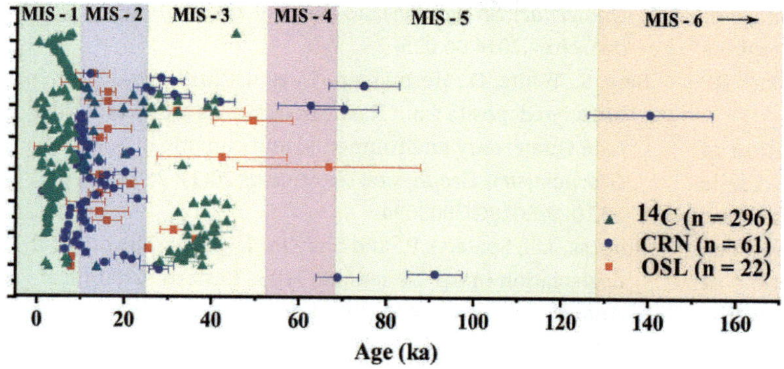


**Figure 8.5** Available chronological data from East Antarctica with respect to Marine Isotopic Stages [MIS −1 (0–14 ka), MIS−2 (14–29 ka) (LGM), MIS−3 (29–57 ka), MIS−4 (57–71 ka), MIS−5 (71– 29 ka)]. CRN ages depict initiation of deglaciation at around ~70 ka.

As per terrestrial chronological studies carried out in East Antarctica, it has been observed that two major deglaciation stages affected the area in the last 50 ka (Figure 8.6). The first was initiated before 45–40 ka during MIS-3 and the next post-LGM. Several hills and coastal areas of Dronning Maud Land, Larsemann Hills, and Bunger Hills (Figures 8.2 and 8.4) were already ice-free before the LGM, and no significant change to EAIS thickness has been observed post-LGM surrounding these regions (Kiernan et al., 2009; Suganuma et al., 2014; Liang et al., 2020). With respect to the studies from other areas, the earliest signs of significant deglaciation around the LGM are observed from regions of Mt Steiner, Loewe Massif, and Raddock Basin of Prince Charles Mountains, which suggest the lowering of ice beginning around ~21 ka and continuing up to ~9.5 ka (Fink et al., 2006; White et al., 2011).

The changes in ice thickness have varied from 100 m to even 700 m in some regions, except for areas such as the Grove Mountains, where ice levels have remained more or less the same over the Late Quaternary (Lilly et al., 2010; Mackintosh et al., 2014).

## 8.6 Conclusion

1) The available age constraints based on $^{14}C$, OSL, and CRN suggest two major phases of deglaciation, latest one at MIS-1, post LGM and another during MIS-3 at around ~ 45–40 ka (Figures 8.5 and 8.6). The available data also suggests that some regions in Eastern Antarctica had little impact of the LGM, and ice margins were relatively stable with respect to the present day.

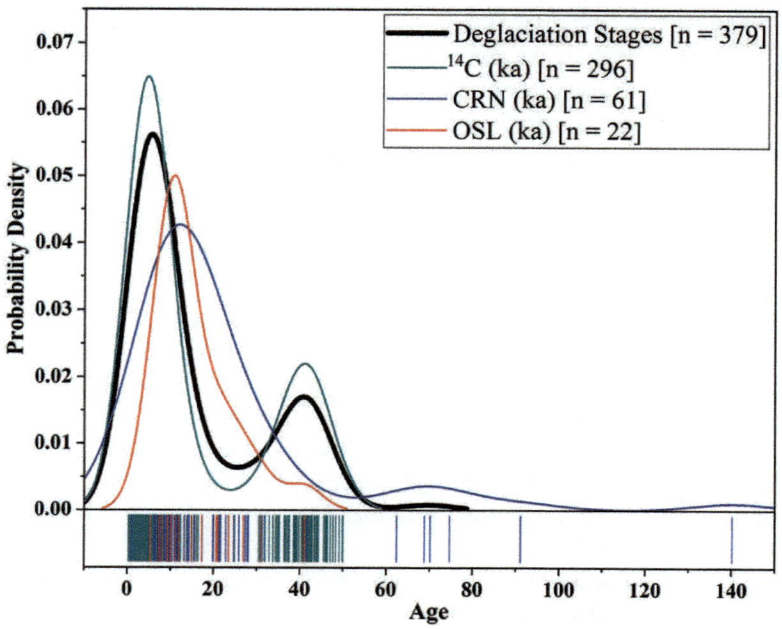


**Figure 8.6** Probability Plot for the available chronologies depicting two major deglaciations based on three absolute dating techniques OSL, CRN, and $^{14}C$, one before LGM at 45–40 ka BP and the other post-LGM.

2) In addition to this, CRN data suggest another phase of deglaciation during MIS-4, particularly affecting regions of the Larsemann Hills around ~80–100 ka BP (Figure 8.5).
3) Several studies concerning the geological evolution of exposed rocks in Kemp Land, Wilhelm II Land, Adelie Land, and George V Land (Stüwe and Oliver, 1989; Halpin et al., 2007; Mikhalsky et al., 2015) have been carried out, but studies focusing on deglaciation or exposure of the Nunataks in these regions is unavailable. This is maybe due to poorly-exposed, obscure terrain and lack of sediments in these regions. The dating from offshore sediment cores at George V Land and the Terre Adélie basin does complement the terrestrial deglaciation data to enhance the resolution of chronological sequence (Mackintosh et al., 2014); however, terrestrial deglaciation study is necessary to decipher the local and regional deglaciation trend in these regions.
4) In other areas, such as Enderby Land, Mac. Robertson Land, and Wilkies Land, data is sparse to decipher the definite trend of deglaciation around these regions of East Antarctica.

## Acknowledgments

US and YR are thankful to The Director, National Centre for Polar and Ocean Research, Goa, for his support and encouragement to carry out this work. YR acknowledges the Indian Antarctic project Ant/2017/ESG-08. We are also thankful to the Editors of this book for providing the opportunity to publish the chapter in this book. This is NCPOR contribution no. B-1/2022-23.

## References

Adamson, D.A., Mabin, M C.G., and Luly, J.G. (1997). Holocene isostasy and late Cenozoic development of landforms including Beaver and Radok Lake basins in the Amery Oasis, Prince Charles Mountains, Antarctica. *Antarctic Science* 9(3): 299–306. doi: 10.1017/s0954102097000382.

Altmaier, M., Herpers, U., Delisle, G. et al. (2010). Glaciation history of Queen Maud Land (Antarctica) reconstructed from *in-situ* produced cosmogenic 10Be, 26Al and 21Ne. *Polar Science* 4(1): 42–61. doi: 10.1016/j.polar.2010.01.001.

Bentley, M.J., Ocofaigh, C., Anderson, J. et al. (2014). A community-based geological reconstruction of Antarctic Ice Sheet deglaciation since the Last Glacial Maximum. *Quaternary Science Reviews* 100: 1–9. doi: 10.1016/j.quascirev.2014.06.025.

Berg, S., White, D., Hermichen, W. et al. (2019). Evaluation of Mumiyo deposits from East Antarctica as archives for the Late Quaternary environmental and climatic history. *Geochemistry, Geophysics, Geosystems* 20(1): 260–276. doi: 10.1029/2018GC008054.

Burgess, J.S., Spate, A.P., and Shevlin, J. (2004). The onset of deglaciation in the Larsemann Hills, Eastern Antarctica. *Antarctic Science* 6: 491–495. doi: 10.1017/S095410209400074X.

Coxall, H.K., Wilson, P., Pälike, H. et al. (2005). Rapid stepwise onset of Antarctic glaciation and deeper calcite compensation in the Pacific Ocean. *Nature* 433(7021): 53–57. doi: 10.1038/nature03135.

Cuffey, K.M., Clow, G., Steig, E. et al. (2016). Deglacial temperature history of West Antarctica. *Proceedings of the National Academy of Sciences of the United States of America* 113(50): 14249–14254. doi: 10.1073/pnas.1609132113.

Darvill, C.M. (2013). Cosmogenic nuclide analysis. In: *Geomorphological Techniques (Online Edition)*, 1–25, London: British Society of Geomorphology.

Delmonte, B., Petit, J.B., Doelsch, I.B. et al. (2007). Late Quaternary interglacials in East Antarctica from ice-core dust records. *Developments in Quaternary Science* 7(C): 53–73. doi: 10.1016/S1571-0866(07)80031-5.

Dymova, T. (2018). *Paleoglaciological Study of the Ahlmannryggen, Borgmassivet and Kirwanveggen Nunatak Ranges, Dronning Maud Land, East Antarctica Using WorldView Imagery*. Stockholm University.Available at: https://www.diva-portal.org/smash/record.jsf?pid=diva2%3A1423429&dswid=-8723.

Fink, D., McKelvey B., Hambrey, M. et al. (2006). Pleistocene deglaciation chronology of the Amery Oasis and Radok Lake, northern Prince Charles Mountains, Antarctica. *Earth and Planetary Science Letters* 243(1–2): 229–243. doi: 10.1016/j.epsl.2005.12.006.

Fitzsimons, S.J. (1996). Paraglacial redistribution of glacial sediments in the Vestfold Hills, East Antarctica. *Geomorphology* 15(2): 93–108. doi: 10.1016/0169-555X(95)00122-L.

Fowler, A.J. and Gillespie, R. (1986). Radiocarbon dating of sediments. *Radiocarbon* 28(2): 441–450. doi:10.1017/S0033822200007578.

Fretwell, P., Pritchard, H., Vaughan, D. et al. (2013). Bedmap2: improved ice bed, surface and thickness datasets for Antarctica. *Cryosphere* 7(1): 375–393. doi: 10.5194/tc-7-375-2013.

Fuchs, M. and Owen, L.A. (2008). Luminescence dating of glacial and associated sediments: review, recommendations and future directions. *Boreas* 37(4): 636–659. doi: 10.1111/j.1502-3885.2008.00052.x.

Gao, Y., Yang, L., Wang, J. et al. (2018). Penguin colonization following the last glacial-interglacial transition in the Vestfold Hills, East Antarctica. *Palaeogeography, Palaeoclimatology, Palaeoecology* 490: 629–639. doi: 10.1016/j.palaeo.2017.11.053.

Gibson, J.A.E., Paterson, K.S., White C.A. et al. (2009). Evidence for the continued existence of Abraxas Lake, Vestfold Hills, east Antarctica during the last glacial maximum. *Antarctic Science* 21(3): 269–278. doi: 10.1017/S0954102009001801.

Goodwin, I.D. (1993). Holocene deglaciation, sea-level change, and the emergence of the Windmill Islands, Budd Coast, Antarctica. *Quaternary Research* 55–69. doi: 10.1006/qres.1993.1057.

Gore, D.B. and Leishman, M.R. (2020). Salt, sediments and weathering environments in Bunger Hills. *Antarctic Science* 32(2): 138–152. doi: 10.1017/S0954102020000073.

Gore, D.B., Rhodes, E J., Augustinus, P.C. et al. (2001). Bunger Hills, East Antarctica: ice free at the last glacial maximum. *Geology* 29(12): 1103–1106. doi: 10.1130/0091-7613(2001)029<103:BHEAIF>2.0.CO;2.

Gore, D. B., Snape, I., and Leishman, M.R. (2003). Glacial sediment provenance, dispersal and deposition, Vestfold Hills, East Antarctica. *Antarctic Science* 15(2): 259–269. doi: 10.1017/S0954102003001263.

Halpin, J.A., White, R.W., Clarke G.L. et al. (2007). The Proterozoic P-T-t evolution of the Kemp Land Coast, East Antarctica: constraints from Si-saturated and Si-undersaturated metapelites. *Journal of Petrology* 48(7): 1321–1349. doi: 10.1093/petrology/egm020.

Hättestrand, C. and Johansen, N. (2005). Supraglacial moraines in Scharffenbergbotnen, Heimefrontfjella, Dronnig Maud Land, Antarctica: significance for reconstructing former blue ice areas. *Antarctic Science* 17(2): 225–236. doi: 10.1017/S0954102005002634.

Hiller, A., Wand, U., Kämpf, H. et al. (1988). Occupation of the Antarctic continent by petrels during the past 35,000 years: inferences from a 14C study of stomach oil deposits. *Polar Biology* 9(2): 69–77. doi: 10.1007/BF00442032.

Hiller, A. (1995). Radiocarbon-dated subfossil stomach oil depsoits from petrel nesting sites: novel paleoenvironmental records from continental Antarctica. *Radiocarbon* 37(2): 171–180.

Hodgson, D.A., Noon P.E., Vyverman, W. et al. (2001). Were the Larsemann Hills ice-free through the Last Glacial Maximum? *Antarctic Science* 13(4): 440–454. doi: 10.1017/S0954102001000608.

Huang, T., Sun, L., Wang, Y. et al. (2009). Penguin occupation in the Vestfold Hills. *Antarctic Science* 21(2): 131–134. doi: 10.1017/S095410200800165X.

Jouzel, J., Raisbeck G., Benoist, J. et al. (1989). A comparison of deep Antarctic ice cores and their implications for climate between 65,000 and 15,000 years ago. *Quaternary Research* 31(2): 135–150. doi: 10.1016/0033-5894(89)90003-3.

Kawamata, M., Suganuma, Y., Doi, K. et al. (2020). Abrupt Holocene ice-sheet thinning along the southern Soya Coast, Lützow-Holm Bay, East Antarctica, revealed by glacial geomorphology and surface exposure dating. *Quaternary Science Reviews* 247: 106540. doi: 10.1016/j.quascirev.2020.106540.

Kiernan, K., Gore D.B., Fink, D. et al. (2009). Deglaciation and weathering of Larsemann Hills, East Antarctica. *Antarctic Science* 21(4): 373–382. doi: 10.1017/S0954102009002028.

Krause, W.E., Krbetschekt, M.R., and Stolz, W. (1997). Dating of Quaternary lake sediments from the Schirmacher Oasis (East Antarctica) by infra-red stimulated luminescence (IRSL) detected at the wavelength of 560 NM. *Quaternary Science Reviews* 16(96): 387–392.

Lear, C. H., Bailey, T.R., Pearson, P.N. et al. (2008). Cooling and ice growth across the Eocene-Oligocene transition. *Geology* 36(3): 251–254. doi: 10.1130/G24584A.1.

Liang, X., Huang, F., Yan, J. et al. (2020). Last exposure process of the Larsemann Hills and adjacent area, East Antarctica, based on bedrock exposure ages. *Quaternary International* 568:116–121. doi: 10.1016/j.quaint.2020.10.031.

Lilly, K., Fink, D., Fabel, D. et al. (2010). Pleistocene dynamics of the interior East Antarctic ice sheet. *Geology* 38(8): 703–706. doi: 10.1130/G31172x.1.

Lintinen, P. and Nenonen, J. (1997). Glacial history of the Vestfjella and Heimefrontfjella Nunatak ranges in western Dronning Maud Land, Antarctica. In: *The Antarctic Region Geological Evolution & Processes* (ed. C.A. Ricci), 845–852. Sienna: Terra Antarctica Publications.

Lorius, C., Raynaud, D., Petit, J.R. et al. (1984). Late-glacial maximum: Holocene atmospheric and ice-thickness changes from Antarctic ice-core studies. *Annals of Glaciology* 5: 88–94. doi: 10.3189/1984aog5-1-88-94.

Mackintosh, A.N., Verleyen, E., O'Brien, P.E. et al. (2014). Retreat history of the East Antarctic Ice Sheet since the Last Glacial Maximum. *Quaternary Science Reviews* 100: 10–30. doi: 10.1016/j.quascirev.2013.07.024.

Mahesh, B.S., Kumar, A., Mohan, R. et al. (2015). Response of Long Lake sediments to Antarctic climate: a perspective gained from sedimentary organic geochemistry and particle size analysis. *Polar Science* 9(4): 359–367. doi: 10.1016/j.polar.2015.09.004.

Mahesh, B.S., Warrier, A.K., Mohan, R. et al. (2019). Impact of Antarctic climate during the Late Quaternary: records from Zub Lake sedimentary archives from Schirmacher Hills, East Antarctica. *Palaeogeography, Palaeoclimatology, Palaeoecology* 514: 398–406. doi: 10.1016/j.palaeo.2018.10.029.

Matsuoka, K., Skoglund, A., Roth, G. et al. (2021). Quantarctica, an integrated mapping environment for Antarctica, the Southern Ocean, and sub-Antarctic islands. *Environment Modelling and Software* 140: 1–14.

Mikhalsky, E.V., Belyatsky, B.V., Presnyakov, S.L. et al. (2015). The geological composition of the hidden Wilhelm II Land in East Antarctica: SHRIMP zircon, Nd isotopic and geochemical studies with implications for Proterozoic supercontinent reconstructions. *Precambrian Research* 258: 171–185. doi: 10.1016/j.precamres.2014.12.011.

Naslund, J. (2007). Landscape development in western and central Dronning Maud. *Antarctic Science* 13(3): 302–311.

Neethling, D.C. (1969). Geology of the Ahlmann Ridge, Western Queen Maud Land. In: *Antarctic Map Folio Series – 12*. New York: American Geographical Society.

Newall, J.C.H, Dymova, T., Serra, E. et al. (2020). The glacial geomorphology of western Dronning Maud Land, Antarctica. *Journal of Maps* 16(2): 468–478. doi: 10.1080/17445647.2020.1761464.

Nishiizumi, K., Kohl, C.P., Arnold, J.R. et al. (1993). Role of *in situ* cosmogenic nuclides $^{10}$be and $^{26}$al in the study of diverse geomorphic processes. *Earth Surface Processes and Landforms* 18(5): 407–425.

Paech, H.-J. and Stackebrandt, W. (1995). Geology. In: *The Schirmacher Oasis, Queen Maud Land, East Antarctica, and Its Surroundings* (ed. P. Bormann and F. Diedrich), 59–169. Gotha, Germany: Justus Perthes Verlag.

Pattyn, F., Matsuoka, K., and Berte, J. (2010). Glacio-meteorological conditions in the vicinity of the Belgian Princess Elizabeth Station, Antarctica. *Antarctic Science* 22(1): 79–85. doi: 10.1017/S0954102009990344.

Paxman, G.J.G., Jamieson, S.S., Ferraccioli, F. et al. (2019). Subglacial geology and geomorphology of the Pensacola-Pole Basin, East Antarctica. *Geochemistry, Geophysics, Geosystems* 20(6): 2786–2807. doi: 10.1029/2018GC008126.

Phartiyal, B., Sharma, A., and Bera, S.K. (2011). Glacial lakes and geomorphological evolution of Schirmacher Oasis, East Antarctica, during Late Quaternary. *Quaternary International* 235(1–2): 128–136. doi: 10.1016/j.quaint.2010.11.025.

Pickard, J. and Seppelt, R.D. (1984). Holocene occurrence of the moss *Bryum algens* Card. in the Vestfold Hills, Antarctica. *Journal of Bryology* 13(2): 209–217. doi: 10.1179/jbr.1984.13.2.209.

Rades, E.F., Shobati, R., Lüthgens, C. et al. (2018). First luminescence-depth profiles from boulders from moraine deposits: insights into glaciation chronology and transport dynamics in Malta valley, Austria. *Radiation Measurements* 120: 281–289. doi: 10.1016/j.radmeas.2018.08.011.

Ravindra, R. (2001). Geomorphology of Schirmacher Oasis, East Antarctica. In: *Proceedings of the Symposium on Snow, Ice and Glacier*. Geological survey of India Special Publication 53: 379–390.

Ray, Y., Sen, S., Sen, K. et al. (2021). Quantifying the past glacial movements in Schirmacher Oasis, East Antarctica. *Polar Science* 30: 1–7. doi: 10.1016/j.polar.2021.100733.

Ray, Y. and Srivastava, P. (2010). Widespread aggradation in the mountainous catchment of the Alaknanda-Ganga River System: timescales and implications to hinterland–foreland relationships. *Quaternary Science Reviews* 29: 2238–2260. doi: 10.1016/j.quascirev.2010.05.023.

Rhodes, E J. (2011). Optically stimulated luminescence dating of sediments over the past 200,000 years. *Annual Review of Earth and Planetary Sciences* 39(1): 461–488. doi: 10.1146/annurev-earth-040610-133425.

Richter, W. and Borman, P. (1995). Geomorphology. In: *The Schirmacher Oasis, Queen Maud Land, East Antarctica, and Its Surroundings* (ed. P. Bormann and F. Diedrich), 171–206. Gotha, Germany: Justus Perthes Verlag.

Roberts, D., Hodgson, D.A., McMinn, A. et al. (2006). Recent rapid salinity rise in three East Antarctic lakes. *Journal of Paleolimnology* 36(4): 385–406. doi: 10.1007/s10933-006-9010-0.

Shrivastava, P.K., Dharwadkar, A., and Asthana, R. (2014). The sediment properties of glacial diamicts from the Jutulsessen area of Gjelsvikfjella, East Antarctica: a reflection of source materials and regional climate. *Polar Science* 8(3): 264–282. doi: 10.1016/j.polar.2014.03.001.

Siddall, M., Milne, G.A., and Masson-Delmotte, V. (2012). Uncertainties in elevation changes and their impact on Antarctic temperature records since the end of the last glacial period. *Earth and Planetary Science Letters* 315–316: 12–23. doi: 10.1016/j.epsl.2011.04.032.

Srivastava, D., Kaul, M.K., Singh, R.K. et al. (1988). *Some Observations on the Glacial Geomorphological Features of Wohlthat Mountains, Central Queen Maud Land, Antarctica Geological Survey of India, Fifth Indian Expedition to Antarctica, Scientific Report.*

Steele, W.K. and Hiller, A. (1997). Radiocarbon dates of snow petrel (*Pagodroma nivea*) nest sites in central Dronning Maud Land, Antarctica. *Polar Record* 33(184): 29–38. doi: 10.1017/S0032247400014145.

Stüwe, K. and Oliver, R. (1989). Geological history of Adélie Land and King George V Land, Antarctica: evidence for a polycyclic metamorphic evolution. *Precambrian Research* 43(4): 317–334. doi: 10.1016/0301-9268(89)90063-6.

Suganuma, Y., Miura, H., Zondervan, A. et al. (2014). East Antarctic deglaciation and the link to global cooling during the Quaternary: evidence from glacial geomorphology and 10Be surface exposure dating of the Sør Rondane Mountains, Dronning Maud Land.

*Quaternary Science Reviews* 97: 102–120. doi: 10.1016/j.quascirev.2014.05.007.

Thamban, M. (2020). Palaeoclimatic records from Antarctica and Southern Ocean: a review of Indian contributions. *Episodes* 43(1): 575–585. doi: 10.18814/EPIIUGS/2020/020038.

Thor, G. and Low, M. (2011). The persistence of the snow petrel (*Pagodroma nivea*) in Dronning Maud Land (Antarctica) for over 37,000 years. *Polar Biology* 34(4): 609–613. doi: 10.1007/s00300-010-0912-y.

Watt, L.-M. van der. (2022). *Antarctica, Britannica Online Encyclopedia*.

White, D.A. and Fink, D. (2014). Late Quaternary glacial history constrains glacio-isostatic rebound in Enderby Land, East Antarctica. *Journal of Geophysical Research: Earth Surface* 119(3): 401–413. doi: 10.1002/2013JF002870.

White, D.A., Fink, D., and Gore, D.B. (2011). Cosmogenic nuclide evidence for enhanced sensitivity of an East Antarctic ice stream to change during the last deglaciation. *Geology* 39(1): 23–26. doi: 10.1130/G31591.1.

Wintle, A G. and Huntley, D.J. (1982). Thermoluminescence dating of sediments. *Quaternary Science Reviews* 1(1): 31–53. doi: 10.1016/0277-3791(82)90018-X.

# 9

# Geospatial Tools for Monitoring Vertebrate Populations in Antarctica With a Note on the Ecological Component of the Indian Antarctic Program

*Anant Pande[1,2,*], Ankita Anand[1], Shailendra Saini[3], and Kuppusamy Sivakumar[4]*

[1] Department of Endangered Species Management, Wildlife Institute of India, Chandrabani, Dehradun 248001, Uttarakhand
[2] Marine Program, Wildlife Conservation Society -India, Kodigehalli, Bengaluru, Karnataka 560097
[3] Antarctic Logistics Division, National Centre for Polar and Ocean Research, Vasco-da-Gama, Goa 403804
[4] Department of Ecology and Environmental Sciences, Pondicherry University, Puducherry 605014, India
* Corresponding author

## 9.1 Introduction

Antarctica has attracted scientists to study its gigantic glaciers, icebergs, pristine lakes, and alluring biodiversity, despite its extreme environmental conditions. It is also the prime focus of the global community in the wake of climate change related impacts such as rising sea levels, accelerated glacial melting, and global warming (Convey and Peck, 2019). Since the signing of the Antarctica Treaty in 1959, several countries have set up permanent or temporary research stations to conduct seasonal (austral summer mostly) or year-long studies in Antarctica, to decode its impact on global weather patterns and effects of global activities on its environment. Initially signed and adopted by 12 countries, this treaty now has 54 countries as official signatories steering "considerable scientific research activity" (as per Article IX.2) following The Environmental (Madrid) Protocol (Article 2), which entitles Antarctica as a "natural reserve, devoted to peace and science" (https://www.ats.aq/e/science.html). However, despite the Antarctic Treaty regulating activities to minimize human impact on Antarctica, the human footprint is ever increasing (Kariminia et al., 2013; Coetzee and Chown, 2016; Pertierra et al., 2017), putting pressure on its unique and endemic terrestrial as well as marine biodiversity.

The Southern Ocean, surrounding the Antarctic continent, supports myriad life forms, serving as a home for a variety of species of marine mammals (seals, whales), seabirds (penguins, skua, petrels), fish, plankton, and diverse benthic communities. Many of these species, inter-dependent on each other, consume Antarctic krill, a super-abundant (zoo) planktonic inhabitant of the Southern Ocean. Krill serves as a keystone species of the Southern Ocean food chain, foraged by many pelagic seabirds (i.e., procellarid petrels, skua, etc.), penguins (Adelie, Gentoo, etc.), seals (Weddell, Crabeater, etc.), and whales (Antarctic blue whale, Fin whale, Humpback whale). Due to their critical role in maintaining the health of the Southern Ocean ecosystem, Antarctic wild populations are monitored to generate long-term data on their population dynamics, to build conservation strategies, and understand the variations in the ecosystem (Fraser et al., 1999; Duley et al., 1999; Ratcliffe et al., 2015). Multiple studies have been pursued assessing species distribution and population abundance as predictors of current and future threats to Antarctic wildlife. For example, determining the distribution of the Weddell seal *Leptonychotes weddellii* contributed to a comprehensive analysis of the species with their neighboring populations to predict the influence of environmental change (LaRue et al., 2011, 2019). A report on another charismatic endemic fauna of Antarctica, the Emperor penguin *Aptenodytes forsteri*, found that populations were declining in the wake of sea-ice melting, which is predicted to substantially decrease in the near future. The other factors that influence Emperor penguin breeding colonies are disturbance of the marine food chain and more frequent storms (Fretwell et al., 2012). Seasonal changes in Leopard seal *Hydrurga leptonyx* (a large seal which typically predates on penguins and other pinnipeds among other smaller prey) populations was observed where their density increased during the summer due to

*Advances in Remote Sensing Technology and the Three Poles,* First Edition. Edited by Manish Pandey, Prem C. Pandey, Yogesh Ray, Aman Arora, Shridhar D. Jawak and Uma K. Shukla.

sea-ice melting (Rogers and Bryden, 1997). These examples show that changes in the Antarctic environment affect species at population scales with varying consequences. Furthermore, population counts of several species vary drastically across the continent (Fretwell et al., 2012) necessitating robust data collection and monitoring (Hyun et al., 2020). Hence, it becomes crucial to devise methods which help monitor their populations and decipher the effect of climate change on animal communities (Allan, et al., 2019).

Several methods have been utilized for surveying and monitoring wildlife populations in Antarctica. These methods include time-consuming on-foot or ground truthing surveys (Pemberton and Kirkwood, 1994; Rogers and Bryden, 1997; Southwell et al., 2011) and cost-intensive ship-based (Bengtson et al., 2011; Bester et al., 2019) or manned aircraft-based (Southwell et al., 2008) surveys. Vessel surveys have been frequently used to count populations of ice-breeding seals (e.g., onboard the ship MV RSA to conduct census of Crabeater, Leopard, Ross, and Weddell seals in the King Haakon VII Sea; Condy, 1977) or determine their distribution (e.g., onboard the ship MV SA Agulhas II in pack ice of the Lazarev Sea and Weddell Sea; Bester et al., 2002, 2019). Antarctic seal populations and seabird colonies have also been scrutinized using intensive aerial surveys to count numbers, determine distribution, or carry out ecological experiments over single or multiple seasons (but mostly during austral summers). Surveys on board lightweight helicopters, such as Squirrel AS350B or Puma, to determine seal haul-out areas (Rogers and Bryden, 1997) or in a combination of vessel and manned aircraft surveys, for example, in the (APIS) program to estimate the numbers of Antarctic phocid seals (Gurarie et al., 2017) were conducted. Similarly, aerial surveys were recommended as a tool to census seals and penguins while experimenting with their response toward aircraft noise (Southwell, 2005). These conventional surveying methods are often time-consuming, expensive, and require considerable human effort. Many times these methods are insufficiently employed over spatial or temporal scales, resulting in under- or overestimation of populations and frequently disturb the resting or breeding animals (Ratcliffe et al., 2015). The biggest hurdle in conducting ground-based or aerial surveys is the challenging weather conditions further exacerbated by the lack of mobility to access the remote locations across the continent for extensive data collection of Antarctic biodiversity (Wauchope et al., 2019). These limitations hinder accurate data collection (Fretwell et al., 2012), especially in the context of the Antarctic continent demanding the use of modern tools of GIS and remote sensing.

In recent years, to reduce costs and leverage maximum effort during the short windows in the extreme weather, use of advanced tools such as kite-based aerial photography (Fraser et al., 1999), Unmanned Aerial Vehicle (Goebel et al., 2015), and geospatial technology (Schwaller et al., 2013) have gained momentum. Many researchers have utilized spatially acquired data from aerial surveys and space technology for seabird population estimation (Kerr and Ostrovsky, 2003; Ratcliffe et al., 2015). Photogrammetry is a more reliable method for Antarctic ecology as marine mammals can be easily detected because they rest and breed on land (LaRue et al., 2011). Image-based methods have also been used to evaluate the body condition of seals by detecting their body shape and color (Allan et al., 2019; Hyun et al., 2020). Over the last few decades, UAVs, especially vertical take-off landing (VTOL) drones, have been found to be highly efficient tools for examining seabird colonies (Goebel et al., 2015; Rümmler et al., 2021). These tools are further aiding spatial data collection in tandem with high-resolution remote-sensed imagery such as SPOT, QuickBird-2, WorldView, IKONOS, DigitalGlobe, and GeoEye to study seabird and marine mammal distribution (LaRue et al., 2014). Remote-sensed imagery contains fine-scale information and can be used to cover large spatial scales for analysis (Turner et al., 2003). Researchers have used remote sensing techniques for direct species mapping, habitat mapping, and species detection using the, spectral contrast method. Satellite-based imagery coupled with advanced aerial (UAV) and ground surveys have further eased precise estimation of species numbers, identification, and biodiversity mapping (Ratcliffe et al., 2015; Borowicz et al., 2018).

## 9.2 Novel Geospatial Tools for Biodiversity Monitoring in Antarctica

### 9.2.1 Unmanned Aerial Vehicles

Population counts of seals and penguins by Unmanned Aerial Vehicles (UAVs) have outperformed ground-based census methods (Table 9.1). In a study using drones to determine breeding colonies of Elephant seals at King George Island, orthophotos were generated using Pix4DMapper. Furthermore, the counting of individual seals was performed on QGIS (Fudala and Bialik, 2020). In another, penguin populations were estimated by flying a UAV over penguin colonies (Ratcliffe et al., 2015). The drone photographs were processed as a single image using an Image Composite Editor and the head count of penguins was determined using iTag software (Viquerat, 2015). The count obtained from UAV imagery was further compared to the direct count conducted during fieldwork. It was observed that the count obtained from UAV images resulted in higher accuracy when compared to the direct count conducted during fieldwork. This is because the ground survey led to double counting of penguins, whereas the iTag

**Table 9.1** Studies over the past decade (2010–2020) using Unmanned Aerial Vehicles to monitor Antarctic vertebrate species.

| Year | Species | Study Area/ Data Collected | Methods | Software | UAV Used | Reference |
|---|---|---|---|---|---|---|
| 2015 | Leopard seal | Cape Shirreff (62°27′30″ S, 60°47′17″ W) | Using VTOLs to estimate size of Leopard seals | Adobe Photoshop CS5 | VTOL aircrafts: two quadcopters (MD4-1000, Micro-drones; APQ-18 quadrocopter and APH-22 hexacopter) | (Goebel et al., 2015) |
| 2015 | Multiple penguin species (Adelie, Chinstrap, Gentoo, Emperor, King) | Falkland Island (UK) | UAV acquired aerial imagery of penguin colonies; Open-Source software for image processing | Image Composite Editor, GIMP, iTag | Multi-rotor UAVs; DJI Naza F550 hexacopter | (Ratcliffe et al., 2015) |
| 2017 | Leopard seal | Cape Shirreff (62°28′12″S 60°46′16″W) | To test aerial photographs for manual measurements of body size and mass of Leopard seals | ImageJ (Java-based open access software package) | APH-22 (battery-powered VTOL UAS system) | (Krause et al., 2017) |
| 2018 | Penguins and seals | Penguin Island (62°06′00″S 57°55′41″W) | Environmental data collection; estimate penguin and pinniped breeding population numbers; map vegetation cover | Flight routes were prepared in HORIZON software<br>Agisoft Photoscan Professional software<br>ArcGIS | Fixed-wing UAV BVLOS<br>PW-ZOOM | (Zmarz et al., 2018) |
| 2019 | Adelie penguin, Chinstrap, Southern Giant petrel, Antarctic shag, Southern Elephant seal and Weddell seal | Penguin Island (62°06′00″S 57°55′41″W), Turret Point Oasis | The UAV images were classified as Adelie penguin, Chinstrap, Southern Giant petrel, Antarctic shag, Southern Elephant seal and Weddell seal | Image Composite Editor, ArcGIS | PW-ZOOM UAV | (Korczak-Abshire et al., 2019) |
| 2019 | Antarctic seals (Antarctic fur seal, Crabeater seal, Leopard seal, Southern Elephant seal, Weddell seal), seabirds (Antarctic shag, Antarctic Tern, Kelp Gull, Skua, Snowy Sheathbill, Southern Giant petrel) | Fildes Peninsula (King George Island, South Shetland Islands) (62°02′S 58°21′W) | To evaluate flight height for Antarctic flying seabird and seal species; body size measurements | ArcGIS | octocopter UAV "Mikrokopter MK" with a "Samsung MX500" and a "EX-W20NB" lens, a quadrocopter UAV "DJI Phantom 4 pro" and "Bormatec Ninox" fixed-wing UAV with a "MAPIR Survey-2 RGB" | (Mustafa et al., 2019) |

**Table 9.1** (Continued)

| Year | Species | Study Area/ Data Collected | Methods | Software | UAV Used | Reference |
|---|---|---|---|---|---|---|
| 2020 | Southern Elephant seals | King George Island (62°02'S 58°21'W) | Detecting Southern Elephant seals and shape property extraction using thermal-imaging sensor | Metashape 1.6.2 software, ArcGIS | RPAS is equipped with a thermal camera; Phantom 4 | (Hyun et al., 2020) |
| | | Patelnia Point, King George Island (62°02'S 58°21'W) | ody length and surface area; individual age and sex | PIX4D Mapper, QGIS | quadcopter Inspire 2 | (Fudala and Bialik, 2020) |
| 2020 | Antarctic shag | Harmony Point, Nelson Island, South Shetland Islands (62°18'S 59°03'W) | Population colony count | Agisoft PhotoScan Professional, ArcGIS | DJI Phantom 4 Advanced | (Oosthuizen et al., 2020) |
| 2021 | Southern Elephant seal, Adelie and Chinstrap penguins | South Georgia-Sub Antarctica (UK) | Comparative ground and aerial count | DotDotGoose counting software | quadcopter, DJI Mavic 2 Pro | (Dickens et al., 2021) |
| 2021 | Emperor penguins | Atka Bay Emperor penguin Colony (70.6125°S, 8.1236°W) | To analyse the utility of UAVs for penguin data collection | | Phantom UAV, Magpy UAV | (Rümmler et al., 2021) |

software uniquely assigns different colored dots to adults, chicks, and incubating eggs. (Korczak-Abshire et al., 2019) used a PW-ZOOM fixed-wing UAV to detect penguins and seals. The UAV images were classified as Adelie penguin, Chinstrap, Southern Giant petrel, Antarctic shag, Southern Elephant seal, and Weddell seal. Several factors like flight height, type of UAV, and weather conditions were examined to analyse the effect of UAVs for penguin data collection. The flight height of 70 m or higher using a quadcopter did not distract the adult penguins, whereas the height was unknown for chicks and adults when fixed-wing UAV was operated. The penguins were found to be more sensitive to Phantom UAV than Magpy UAV. Therefore, fixed-wing UAVs were used with Agisoft PhotoScan Professional to process the overlapping UAV images and were further geo-referenced using ArcGIS. The digital elevation model was produced to understand the features of Antarctic shag breeding colonies. In total, 69 breeding colonies were detected at the height of 10–20 m above sea level around South Shetland Islands. With the increasing use of UAVs, vertical take-off and landing (VTOL) has proved to be the advanced method for surveying to understand the population distribution in Antarctica. The other unmanned aerial system (UAS) or fixed wing UAVs are noisy due to their fuel-burning engines, thus are capable of disturbing the target species. These are also limited in capturing high-resolution images and take-off or landing on rocky terrain. The battery powered VTOL has overcome the limitations of other platforms (Goebel et al., 2015). Apart from population surveys, body size (e.g., length and mass of the target animal) can be determined using thermal images acquired from remotely piloted aircraft system (RPAS). A case study conducted on King George Island used DJI Inspire V1 with thermal imaging sensors to detect Southern Elephant seals. The seals were distinctly observable in the co-registered mosaiced thermal image and further temperature contours were drawn to determine their size. The thermal photography obtained using RPAS technique is suggested to determine the body size of seals and is applicable to differentiate sub-adults from adults. Moreover, sex of the seals can also be determined using the thermal approach (Hyun et al., 2020), recommended over multi-rotor UAVs as the behavioural response by penguins was minimal (Rümmler et al., 2021; Oosthuizen et al., 2020).

### 9.2.2 Satellite Imagery

Satellite imagery over the years has improved in resolution and spatial scale covering large parts of Antarctica (Table 9.2). Previously, studies have used high-resolution satellite imagery such as WorldView-1 and QuickBird-2 and further compared the results with ground surveys taken during the same time period (e.g., for Weddell seals at Erebus bay; LaRue et al., 2011). Additionally, incorporating various ecological factors

**Table 9.2** Some relevant past studies monitoring of Antarctic vertebrate species using satellite imagery.

| Year | Species | Study Area/ Data | Tools/Algorithm | Software Used | Satellite Imagery Used | Reference |
|---|---|---|---|---|---|---|
| 1989 | Adelie Penguin | Ross Islands (77°30'S 168°00'E) and Beaufort Islands (76°57'S 166°57'E) | Linear discriminant analysis | SAS VMS version 5 | Landsat TM-30 m | (Schwaller, et al., 1989) |
| 1992 | Adelie penguin | Vestfold Hills and Rauer Islands (68°51'S 77°50'E), East Antarctica | minimum distance classifier | microBRIAN image processing system | SPOT HRV | (Bhikharidas et al., 1992) |
| 1995 | King penguin | Sub-Antarctic (southern Indian Ocean) lIe aux Cochons, Crozet Archipelago | Estimation of the surface area | Didactim software (AES image 1989) | Spot – 10 m (panchromatic and multispectral mode) | (Guinet et al., 1995) |
| 2011 | Weddell seals | Erebus Bay (77°45'S 166°33'E) | Manual counting | ArcGIS | QuickBird-2, WorldView-1 | (LaRue et al., 2011) |
| 2012 | Emperor penguin | Coastline of Antarctica | Multivariate supervised classification, Linear Regression algorithm | ArcGIS | QuickBird, WorldView-2, Ikonos | (Fretwell et al., 2012) |
| 2013 | Adelie penguin | Coastline of Antarctica | Affine transformation classification, spatial pixel clustering algorithm | Not mentioned | LANDSAT ETM+ | (Schwaller et al., 2013) |
| 2014 | Adelie, Chinstrap, Gentoo Penguin | Sub-Antarctic Signy Island (60°43'S 45°36'W), South Orkneys (62°S 60°N) field-based digital mapping system (DMS) | Maximum likelihood multivariate classification | DMS- ESRI ArcPad Application builder in ArcPad6 GIS- ArcGIS | QuickBird2- 0.6 m (Panchromatic) 2.4 m (multispectral) | (Waluda et al., 2014) |
| 2014 | Adelie penguin | Victoria Land in Ross Sea (70°30'S to 78°00'S) | Maximum likelihood classification | ArcGIS | VHR, DigitalGlobe | (LaRue et al., 2014) |
| 2015 | Multiple Seabirds (Adelie penguin, Antarctic shag, Dominican gull, Skua, Snow petrel, Southern Fulmar, Southern giant petrel) | Marguerite Bay, Antarctic peninsula (68°30'S 068°30'W) | Spectral angle mapper in ENVI | ERDAS Imagine | Landsat ETM - 30 m | (Fretwell et al., 2015) |

**Table 9.2** (Continued)

| Year | Species | Study Area/ Data | Tools/Algorithm | Software Used | Satellite Imagery Used | Reference |
|---|---|---|---|---|---|---|
| 2018 | Antarctic petrel | Coastline of Antarctica | Affine transformation classification, spatial pixel clustering algorithm | Not mentioned | Landsat 8 OLI | (Schwaller et al., 2018) |
| 2019 | Weddell seals | Ross sea (75°S 175°W) | CrowdRank | Crowd-sourcing platform Tomnod | DigitalGlobe | (LaRue et al., 2019) |
| 2020 | Weddell seals | Ross Sea (75°S 175°W), Amundsen Sea (70°S 105°W) and Bellingshausen Sea (57°18'W 102°20'W), Weddell Sea, East Antarctica | CrowdRank | Crowd-sourcing platform Tomnod | DigitalGlobe | (LaRue et al., 2020) |

to determine the animal populations (e.g., for Weddell seals in the Ross Sea; LaRue et al., 2019), Tomnod's CrowdRank (CR) algorithm on DigitalGlobe images with high spatial resolution (~60 cm) was used. The model predicted the potential habitat of seals considering fast-ice cracks, distance from the sea, and penguin colonies. The spectral bands of remote sensing data have distinctive features to distinguish penguins and guano from their surroundings (Schwaller et al., 2013). Guano and Emperor penguin colonies can further be differentiated based on spectral contrast using multispectral band of QuickBird imagery (Fretwell et al., 2012). Following a similar methodology, Adelie penguin colonies were identified using maximum likelihood (MLC) classification on very high-resolution DigitalGlobe of 0.6 m resolution. Spectral signatures were used to classify new guano stains and old guano stains to identify penguin breeding colonies in the Cape Royds colony on Ross Island (LaRue et al., 2014). Moderate resolution satellite Landsat-7 ETM imageries were also found suitable to detect Adelie penguin colonies along the Antarctica coastline. The proposed algorithm eliminated radiometric noise from the satellite image. Using the corrected Landsat-7 ETM+ remote sensing data, six new penguin colonies were mapped (Schwaller et al., 2013).

## 9.3 Spatial Mapping of Seabirds Under the Indian Antarctic Program

Indian Antarctic Expeditions have a strong component of monitoring wildlife under the aegis of the Antarctic Wildlife Monitoring Program (Pande et al., 2017, 2018, 2020). In the early 1990s, ecological studies on vertebrates were undertaken by Indian Antarctic expeditioners, i.e., researchers from Central Marine Fisheries Research Institute (CMFRI), Kochi; National Institute of Oceanography (CSIR- NIO), Goa; and Wildlife Institute of India, Dehradun. These scientific investigations generated baseline information on marine mammals and seabirds near Indian stations in Antarctica (Chattopadhyay, 1995; Sathyakumar, 1998; Venkataraman, 1998; Bhatnagar, 1999; Bhatnagar and Sathyakumar, 1999; Hussain and Saxena, 2008).

Later, WII, Dehradun initiated a project to develop a long-term monitoring program for wildlife occurring near the Indian area of operation which was designed to ascertain the possibility of conducting long-term research on wildlife species in Antarctica. The subsequent Phase II and III generated large spatial scale data on seals and penguins in the Indian sector of operation in Antarctica (Sivakumar and Sathyakumar, 2012; Kumar and Johnson, 2014). With knowledge on existing species around the Indian research station and logistical capabilities, the Phase III of the program was launched in 2013/14 (33rd Indian Scientific Expeditions to Antarctica) to undertake detailed long-term monitoring work on selected indicator species of the polar ecosystem. The Phase III of the program was conducted during three successive expeditions (33rd, 34th, and 35th) and later in the 39th Indian Scientific Expedition to Antarctica, and resulted in key understanding of the distribution and breeding biology of species such as pelagic seabirds, penguins, and seals (Pande et al., 2017, 2018, 2020).

With seabirds being some of the most threatened groups of birds in the world, it becomes imperative to understand their breeding distribution to implement successful monitoring programs. With their vulnerability to slight fluctuations in the environment, they serve as an important indicator species for monitoring the health of the Antarctic ecosystem. Under Phase III of the Antarctic Wildlife Monitoring Program, seabird colonies were accessed to understand their spatial distribution in the Indian sector of operation. These colonies were mapped by ground surveys and later uploaded on to ArcGIS and Google Earth Engine to visualize their distribution (Figure 9.1).

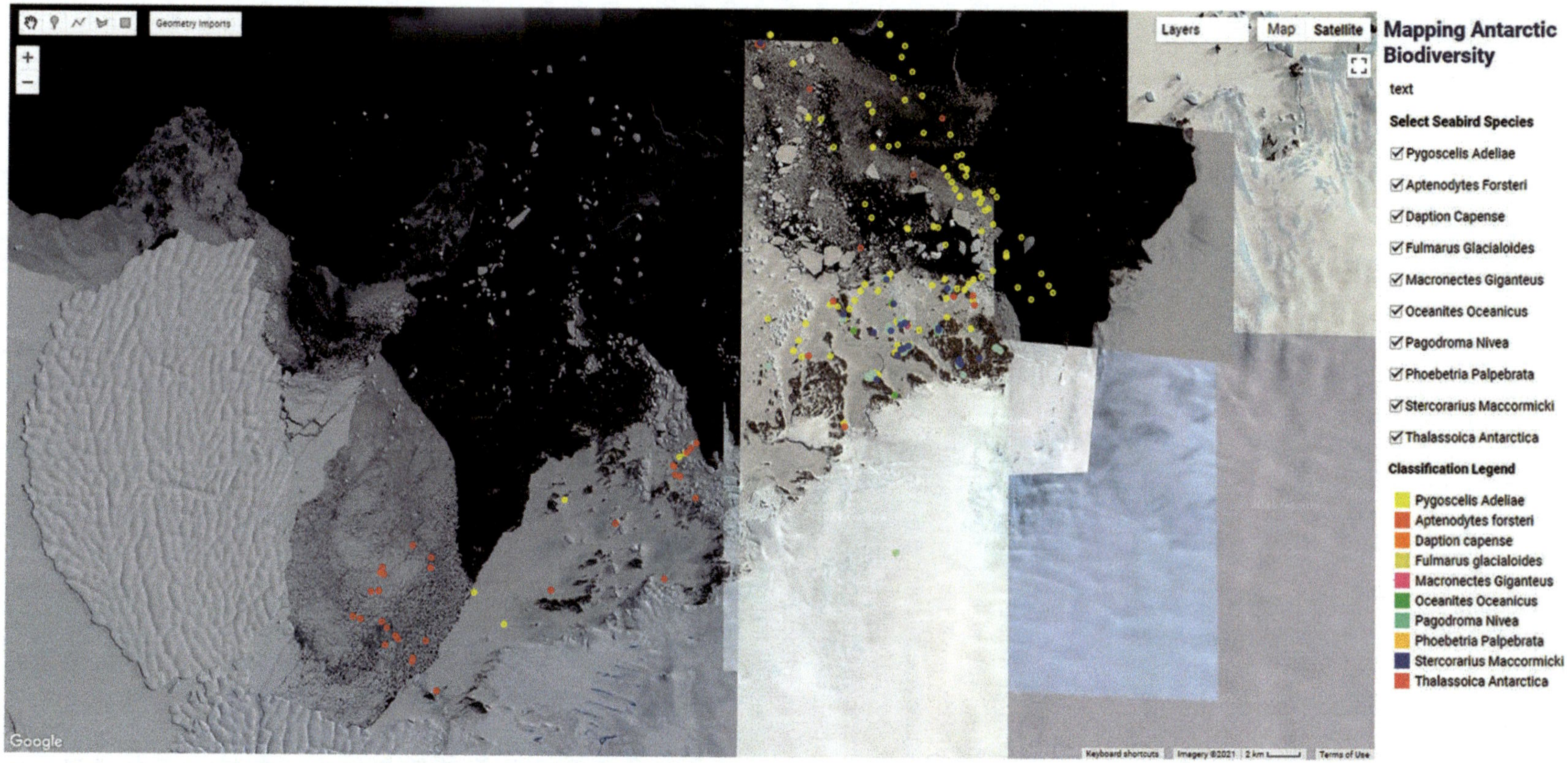


**Figure 9.1** Spatial distribution of Antarctic seabirds near Bharati station, Larsemann Hills, East Antarctica (the data can also be requested at the Google Earth Engine app link – https://code.earthengine.google.co.in/47fa8341777b792b09ba85e043b29094).

The goal of this exercise is to make the data available to a large user group in Antarctica for further research and monitoring. Further work involves uploading this data along with accessory information such as numbers, nest status, and age-class of seabirds detected onto global data portals like Antarctic Biodiversity Portal (https://www.biodiversity.aq) and Ocean Biodiversity Information System (OBIS – https://obis.org) or OBIS-SEAMAP (https://seamap.env.duke.edu).

## 9.4 Recommendations to Incorporate New Tools for Antarctic Wildlife Monitoring Program

The current methods being used to collect data on seabirds and marine mammals under the Indian Antarctic Program have great scope for improvement. Previously mapped seabird colonies can be identified on satellite imagery (LANDSAT, SENTINEL, etc.) for exploring new areas for seabird presence. For cavity-nesters like snow petrels (Figure 9.2) or Wilson's storm petrels, topographic features can aid in determining potential new sites to visit through species distribution modeling methods. For open nesters, like penguins and skua (Figures 9.3 and 9.4), direct observations need to be upgraded by using UAVs. High-resolution images collected by the UAVs could be used to create 3D models for studying habitat use, quality, and monitoring changes with respect to changing land and sea-ice cover, etc. Adelie penguins nest very close to the Indian Antarctic station Bharati, but on rocky outcrops which are inaccessible for most of the year. Utilizing UAVs to map the colony would aid population census and monitoring breeding

Figure 9.2 Snow petrels (here a snow petrel chick) nest in rock cavities in Antarctica. Satellite imagery aided with ground data would help in developing species distribution models to understand distribution in the Antarctic landscape.

Figure 9.3 South Polar skua is an open nesting species which nests around both Indian Antarctic stations of Maitri and Bharati.

Figure 9.4 Adelie penguins use multiple rocky outcrops around Bharati station for breeding and molting purpose. Use of Unmanned Aerial Vehicles would aid their population monitoring. Photo Credit: Shashank Arya.

Figure 9.5 Author visiting inland mountains in East Antarctica to search for seabird breeding sites. Photo Credit: C.P Singh.

patterns over the years. Furthermore, use of high-resolution imagery is needed to monitor their colony movement in addition to the emperor penguin colony at nearby Amanda Bay. As the Indian Antarctic program expands west of the Larsemann Hills, toward the Amery ice shelf and beyond, VHR imagery is required to plan and execute short expeditions and classify ice-free areas for potential animal presence.

## 9.5 Conclusion

Wildlife monitoring in Antarctica has moved from rigorous on-ground surveys to use of advanced UAVs and high-resolution satellite imagery. However, there are some parts of the continent for which little is known in terms of habitat quality and animal distribution. In the coming decades, the technology will certainly outclass on-field survey efforts but would require ground truthing to substantiate the datasets. Long-ranging UAVs would be needed to minimize costs or replace large spatial scale aerial surveys, whereas Very High Resolution (VHR) imagery would be needed at a lower cost to maximize effort for monitoring critical species (such as emperor penguins) in areas that are inaccessible to most researchers. In the Indian sector of operations in Antarctica, seabird nest counts and spatial distribution data would benefit from geospatial tools such as spectral analyses or VHR imagery (Pande et al., 2020), though with limitations toward birds breeding in low numbers and cavity-nesting species (La Rue et al., 2014; Fretwell et al., 2015). Lastly, the Indian Antarctic Program would need to upscale the tools and technology used in wildlife monitoring to strengthen the country's obligations toward climate change research in the polar regions.

## Acknowledgments

The authors would like to thank the National Centre for Polar and Ocean Research, Goa for providing logistics support during the Indian Antarctic Expeditions. Thanks to the Director, Wildlife Institute of India, Director, National Centre for Polar and Ocean Research, Rahul Mohan, and M. Javed Beg for their constant encouragement. Authors are grateful for valuable field support provided by voyage leaders, stations leaders, and members of 33rd, 34th, 35th, and 39th Indian Antarctic Expeditions.

ORCIDs:
Anant Pande: https://orcid.org/0000-0002-2835-1481

## References

Allan, B.M., Ierodiaconou, D., Hoskins, A.J. et al. (2019). A rapid UAV method for assessing body condition in fur seals. *Drones* 3(1): 1–7. doi: 10.3390/drones3010024.

Bengtson, J.L., Laake, J.L., Boveng, P.L. et al. (2011). Distribution, density, and abundance of pack-ice seals in the Amundsen and Ross Seas, Antarctica. *Deep Sea Research Part II: Topical Studies in Oceanography* 58(9–10): 1261–1276.

Bester, M.N., Wege, M., Lübcker, N. et al. (2019). Opportunistic ship-based census of pack ice seals in eastern Weddell Sea, Antarctica. *Polar Biology* 42(1): 225–229. doi: 10.1007/s00300-018-2401-7.

Bester, M.N., Ferguson, J.W.H., and Jonker, F.C. (2002). Population densities of pack ice seals in the Lazarev Sea, Antarctica. *Antarctic Science* 14(2): 123–127. doi: 10.1017/S0954102002000676.

Bhatnagar, Y.V. (1999). Daily monitoring and aerial census of penguins and seals in Antarctica. *Fifteenth Indian Expedition to Antarctica, Scientific Report, Department of Ocean Development, Technical Publication* 13: 165–182.

Bhatnagar, Y.V. and Sathyakumar, S. (1999). Developing a long-term monitoring programme for birds and mammals in the Indian Ocean and Antarctica. *Fifteenth Indian Expedition to Antarctica, Scientific Report, Department of Ocean Development, Technical Publication* 13: 131–164.

Bhikharidas, A.K., Whitehead, M.D., and Peterson, J.A. (1992). Mapping Adélie penguin rookeries in the Vestfold Hills and Rauer Islands, East Antarctica, using SPOT HRV data. *International Journal of Remote Sensing* 13(8): 1577–1583. doi: 10.1080/01431169208904211.

Borowicz, A., McDowall, P., Youngflesh, C. et al. (2018). Multi-modal survey of Adélie penguin mega-colonies reveals the Danger Islands as a seabird hotspot. *Scientific Reports* 8(1): 1–9.

Chattopadhyay, S. (1995). On the avian forms encountered during the Eleventh Indian Scientific Expedition to Antarctica. *Eleventh Indian Expedition to Antarctica, Scientific Report, Department of Ocean Development, Technical Publication* 9: 163–197.

Coetzee, B.W. and Chown, S.L. (2016). A meta-analysis of human disturbance impacts on Antarctic wildlife. *Biological Reviews* 91(3): 578–596.

Condy, P.R. (1977). Results of the fourth seal survey in the King Haakon VII Sea, Antarctica. *South African Journal on Antarctic Research* 7: 2–8. Available at: http://alp.lib.sun.ac.za/handle/123456789/7552.

Convey, P. and Peck, L.S. (2019). Antarctic environmental change and biological responses. *Science Advances* 5(11): 16. doi: 10.1126/sciadv.aaz0888.

Dickens, J., Hollyman, P.R., Hart, T. et al. (2021). Developing UAV monitoring of South Georgia and the South Sandwich Islands' iconic land-based marine predators. *Frontiers in Marine Science* 8: 1–16. doi: 10.3389/fmars.2021.654215.

Fraser, W.R., Carlson, J.C., Duley, P.A. et al. (1999). Using kite-based aerial photography for conducting Adelie penguin censuses in Antarctica. *Waterbirds: The International Journal of Waterbird Biology* 22(3): 435–440. doi: 10.2307/1522120.

Fretwell, P.T., LaRue, M.A., Morin, P. et al. (2012). An Emperor penguin population estimate: the first global, synoptic survey of a species from space. *PLoS ONE* 7(4): doi: 10.1371/journal.pone.0033751.

Fretwell, P.T., Phillips, R.A., Brooke, M.D.L. et al. (2015). Using the unique spectral signature of guano to identify unknown seabird colonies. *Remote Sensing of Environment* 156: 448–456. doi: 10.1016/j.rse.2014.10.011.

Fudala, K. and Bialik, R.J. (2020). Breeding colony dynamics of Southern Elephant seals at Patelnia Point, King George Island, Antarctica. *Remote Sensing* 12(2964): 1–18. doi: 10.3390/rs12182964.

Goebel, M.E., Perryman, W.L., Hinke, J.T. et al. (2015). A small unmanned aerial system for estimating abundance and size of Antarctic predators. *Polar Biology* 38(5): 619–630. doi: 10.1007/s00300-014-1625-4.

Guinet, C., Jouventin, P., and Malacamp, J. (1995). Satellite remote sensing in monitoring change of seabirds: use of Spot Image in King Penguin population increase at Ile aux Cochons, Crozet Archipelago. *Polar Biology* 15(7): 511–515. doi: 10.1007/BF00237465.

Gurarie, E., Bengtson, J.L., Bester, M.N. et al. (2017). Distribution, density and abundance of Antarctic ice seals off Queen Maud Land and the eastern Weddell Sea. *Polar Biology* 40(5): 1149–1165.

Hussain, S.A. and Saxena, A. (2008). Distribution and status of Antarctic seals and penguins along the Princess Astrid Coast, East Antarctica. *Indian Journal of Marine Science* 37: 7.

Hyun, C.U., Park, M., and Lee, W.Y. (2020). Remotely piloted aircraft system (Rpas)-based wildlife detection: a review and case studies in maritime Antarctica. *Animals* 10(12): 1–17. doi: 10.3390/ani10122387.

Kariminia, S., Ahmad, S.S., Hashim, R. et al. (2013). Environmental consequences of Antarctic tourism from a global perspective. *Procedia-Social and Behavioral Sciences* 105: 781–791.

Kerr, J.T. and Ostrovsky, M. (2003). From space to species: ecological applications for remote sensing. *Trends in Ecology and Evolution* 18(6): 299–305. doi: 10.1016/S0169-5347(03)00071-5.

Korczak-Abshire, M., Zmarz, A., Rodzewicz, M. et al. (2019). Study of fauna population changes on Penguin Island and Turret Point Oasis (King George Island, Antarctica) using an unmanned aerial vehicle. *Polar Biology* 42: 217–224. doi: 10.1007/s00300-018-2379-1.

Krause, D.J., Hinke, J.T., Perryman, W.L. et al. (2017). An accurate and adaptable photogrammetric approach for estimating the mass and body condition of pinnipeds using an unmanned aerial system. *PLoS ONE* 12(11): 1–20. doi: 10.1371/journal.pone.0187465.

Kumar, R.S. and Johnson, J.A. (2014). Aerial surveys for pack-ice seals along the Ingrid Christensen and Princess Astrid Coasts, East Antarctica. *Journal of Threatened Taxa* 6: 6230–6238. doi: 10.11609/JoTT.o3817.6230-8

LaRue, M.A., Rotella, J.J., Garrott, R A. et al. (2011). Satellite imagery can be used to detect variation in abundance of Weddell seals (*Leptonychotes weddellii*) in Erebus Bay, Antarctica. *Polar Biology* 34(11): 1727–1737. doi: 10.1007/s00300-011-1023-0.

LaRue, M.A., Lynch, H.J., Lyver, P.O.B. et al. (2014). A method for estimating colony sizes of Adélie penguins using remote sensing imagery. *Polar Biology* 37(4): 507–517. doi: 10.1007/s00300-014-1451-8.

LaRue, M.A., Salas, L., Nur, N. et al. (2019). Physical and ecological factors explain the distribution of Ross Sea Weddell seals during the breeding season. *Marine Ecology Progress Series* 612: 193–208. doi: 10.3354/meps12877.

LaRue,M.A., Ainley, D.G., Pennycook, J. et al. (2020). Engaging "the crowd" in remote sensing to learn about habitat affinity of the Weddell seal in Antarctica. *Remote Sensing in Ecology and Conservation* 6(1): 70–78. doi: 10.1002/rse2.124.

Mustafa, O., Braun, C., Esefeld, J. et al. (2019). Detecting Antarctic seals and flying seabirds by UAV. *ISPRS Annals of the Photogrammetry, Remote Sensing and Spatial Information Sciences* 4(2/W5): 141–148. doi: 10.5194/isprs-annals-IV-2-W5-141-2019.

Oosthuizen, W.C., Krüger, L., Jouanneau, W. et al. (2020). Unmanned aerial vehicle (UAV) survey of the Antarctic shag (*Leucocarbo bransfieldensis*) breeding colony at Harmony Point, Nelson Island, South Shetland Islands. *Polar Biology* 43(2): 187–191. doi: 10.1007/s00300-019-02616-y.

Pande, A., Sivakumar, K., Sathyakumar, S. et al. (2017). Monitoring wildlife and their habitats in the Southern Ocean *and around Indian research stations in Antarctica. Proceedings of the Indian National Science Academy* 83(2): 483–496.

Pande, A., Rawat, N., Sivakumar, K. et al. (2018). Cross-species screening of microsatellite markers for individual identification of Snow Petrel *Pagodroma nivea* and

Wilson's Storm Petrel *Oceanites oceanicus* in Antarctica. *Peer Journal* 6: e5243.

Pande, A., Mondol, S., Sathyakumar, S. et al. (2020). Past records and current distribution of seabirds at Larsemann Hills and Schirmacher Oasis, East Antarctica. *Polar Record*, 56.

Pemberton, D. and Kirkwood, R.J. (1994). Pup production and distribution of the Australian fur seal, *Arctocephalus pusillus doriferus*, in Tasmania. *Wildlife Research* 21(3): 341–351.

Pertierra, L.R., Hughes, K.A., Vega, G.C. et al. (2017). High resolution spatial mapping of human footprint across Antarctica and its implications for the strategic conservation of avifauna. *PloS One* 12(1): e0168280.

Ratcliffe, N., Guihen, D., Robst, J. et al. (2015). A protocol for the aerial survey of penguin colonies using UAVs. *Journal of Unmanned Vehicle Systems* January: 1–16. doi: 10.1139/juvs-2015-0006.

Rogers, T.L. and Bryden, M.M. (1997). Density and haul-out behavior of leopard seals (*Hydrurga leptonyx*) in Prydz Bay, Antarctica. *Marine Mammal Science* 13(2): 293–302. doi: 10.1111/j.1748-7692.1997.tb00632.x.

Rümmler, M.C., Esefeld, J., Pfeifer, C. et al. (2021). Effects of UAV overflight height, UAV type, and season on the behaviour of Emperor Penguin adults and chicks. *Remote Sensing Applications: Society and Environment* 23(May): 10. doi: 10.1016/j.rsase.2021.100558.

Sathyakumar, S. (1998). Developing a long-term monitoring programme for birds and mammals in the Indian Ocean and Antarctica using GPS and GIS technologies. *Fourteenth Indian Expedition to Antarctica, Scientific Report, 1998 Department of Ocean Development, Technical Publication* 12: 207–219.

Schwaller, M.R., Olson Jr, C.E., Ma, Z. et al. (1989). A remote sensing analysis of Adélie Penguin rookeries. *Remote Sensing of Environment* 28(C): 199–206. doi: 10.1016/0034-4257(89)90113-2.

Schwaller, M.R., Lynch, H.J., Tarroux, A. et al. (2018). A continent-wide search for Antarctic petrel breeding sites with satellite remote sensing. *Remote Sensing of Environment*. 210: 444–451. doi: 10.1016/j.rse.2018.02.071.

Schwaller, M.R., Southwell, C.J., and Emmerson, L.M. (2013). Continental-scale mapping of Adélie penguin colonies from Landsat imagery. *Remote Sensing of Environment* 139: 353–364. doi: 10.1016/j.rse.2013.08.009.

Sivakumar, K. and Sathyakumar, S. (2012). *Climate change and its potential impacts on the distribution of birds in Southern Indian Ocean and Antarctica*. New Delhi: Daya Publishers.

Southwell, C. (2005). Response behavior of seals and penguins to helicopter surveys over the pack ice off East Antarctica. *Antarctic Science* 17(3): 328–334.

Southwell, C., Paxton, C.G.M., and Borchers, D.L. (2008). Detectability of penguins in aerial surveys over the pack-ice off Antarctica. *Wildlife Research* 35(4): 349–357. doi: 10.1071/WR07093.

Southwell, D.M., Einoder, L.D., Emmerson, L M. et al. (2011). Using the double-observer method to estimate detection probability of two cavity-nesting seabirds in Antarctica: the snow petrel (*Pagodroma nivea*) and the Wilson's storm petrel (*Oceanites oceanicus*). *Polar Biology* 34(10): 1467–1474.

Turner, W., Spector, S., Gardiner, N. et al. (2003). Remote sensing for biodiversity science and conservation. *Trends in Ecology and Evolution* 18(6): 306–314. doi: 10.1016/S0169-5347(03)00070-3.

Venkataraman, K. (1998). Studies on phylum Tardigrada and other associated fauna, South Polar Skua and bird and mammal logging during 1994–1995 expedition. *Fourteenth Indian Expedition to Antarctica,Scientiflc Report, 1998 Department of Ocean Development, Technical Publication* 12: 221–243.

Viquerat, S. (2015). iTAG: an Open-Source software facilitating the analysis of count data from still images. doi: 10.13140/RG.2.2.32344.29441.

Waluda, C.M.. Dunn, M.J., Curtis, M.L. et al. (2014). Assessing penguin colony size and distribution using digital mapping and satellite remote sensing. *Polar Biology* 37(12): 1849–1855. doi: 10.1007/s00300-014-1566-y.

Wauchope, H.S., Shaw, J.D., and Terauds, A. (2019). A snapshot of biodiversity protection in Antarctica. *Nature Communications* 10(946): 1–6. doi: 10.1038/s41467-019-08915-6.

Zmarz, A., Rodzewicz, M., Dąbski et al. (2018). Application of UAV BVLOS remote sensing data for multi-faceted analysis of Antarctic ecosystem. *Remote Sensing of Environment* 217: 375–388. doi: 10.1016/j.rse.2018.08.031.

# 10

# Bryophytes of Larsemann Hills, East Antarctica and Future Prospects

*Devendra Singh**

*Botanical Survey of India, AJC Bose Indian Botanic Garden, Howrah 711103, West Bengal, India*
** Corresponding authors*

## 10.1 Introduction

The Antarctica is the harshest region in the world and only about 2% ice free regions of the continent are able to sustain any type of terrestrial plant life. The Antarctic flora comprises only Bacteria, Algae, Fungi, Lichens, and Bryophytes. Bryophytes dominate in sheltered and moister habitats in Antarctic regions. These plants have responded to the adverse conditions with different strategies of survival which become extreme and sophisticated with increasing altitude and proximity to the pole. The Antarctic environmental conditions, such as temperature and moisture availability, favor the growth of bryophytes.

Bryophytes, because of their relative lack of cuticle, absorb water and nutrients through the surface of their plant body. In some mosses, where the conducting tissue is fairly evolved, and liverworts where rhizoids are abundantly present, they normally depend on surface absorption of water and nutrients. Therefore, they are growing near water bodies and thus the various smaller and larger water bodies of the Larsemann Hills support the growth of bryophytes. Bryophytes are poikilo-hydric in nature and having an alternative strategy of adaptation, they are able to grow in Antarctica. To properly understand the ecosystem functionality of the harsh Antarctic environment and contribution of bryophytes, it is imperative to know the diversity, distribution, and association of these plants.

The terrestrial habitat of the Larsemann Hills is mostly colonized by the thallophytes and dominated by mosses. To thrive in this harshest of environments they develop various biochemical, physiological, and morphological mechanisms. Bryophytes are of particular signifiance, in that they play an important role in habitat modification, nutrient cycling, primary production, and providing shelter and security for the associated micro-fauna and microflora such as Algae and Lichens, etc. They not only serve as a model organism for understanding cold tolerance mechanisms but also for studying biological responses to climate change.

Therefore, repeated and comprehensive biological collections encompassing the entire range of variability is important for proper identification of bryophytes. This proposed study provides baseline information on bryophytes, their taxonomic diversity, associates, and distribution in the Larsemann Hills area. It is hoped that the study adds to our existing knowledge of these important components of the Antarctic diversity of bryophytes and helps in other environmental studies.

In Antarctica, bryophytes are represented by 27 species of liverworts (Bednarek-Ochyra et al., 2000) and 134 species of mosses (Ochyra et al., 2008; Sollman, 2015). One species, i.e., *Cephaloziella varians* (Gottsche) Steph. is a liverwort and six species of mosses, i.e., *Bryum archangelicum* Bruch & Schimp., *Bryum pseudotriquetrum* (Hedw.) P. Gaertn., B. Mey & Scherb., *Bryoerythrophyllum antarcticum* (L.I. Saviz & Smirnova P. Sollman, *Ceratodon purpureus* (Hedw.) Brid., *Coscinodon lawianus* (J.H. Willis) Ochyra, and *Syntrichia sarconeurum* Ochyra & R.H. Zanderwas, were reported earlier from the Larsemann Hills (Kurbatova and Andreev, 2015)). Recently Das and Singh (2020) reported six species of bryophytes during the study of epiphytic algae on bryophytes

A total of 12 species of bryophytes belonging to a single genus and species, i.e. *C. varians* (Gottsche) Steph. of Marchantiophyta (liverworts) and 11 species belonging

*Advances in Remote Sensing Technology and the Three Poles,* First Edition. Edited by Manish Pandey, Prem C. Pandey, Yogesh Ray, Aman Arora, Shridhar D. Jawak, and Uma K. Shukla.

to 3 families and 6 genera of Bryophyta (mosses), i.e. *Bryum archangelicum*, *Bryum argenetum* var. *argenetum*, *Bryum argenetum* var. *muticum*, *Bryum bharatiensis*, *Bryum pseudotriquetrum*, *Bryoerythrophyllum antarcticum*, *Bryoerythrophyllum recurvirostrum* (Hedw.) P.C. Chen, *Ceratodon purpureus*, *Coscinodon lawanius*, *Guembelia longirostris* (Hook.) Ochrya & Zarnowiec, and *Syntrichia sarconeurum* Ochrya & R.H. Zander, have been recorded based on the author's own collection during XXXV ISEA (summer) and based on earlier reports. *Guembelia longirostris* (Hook.) Ochrya & Zarnowiec is reported here for the first time in detail from East Antarctica, earlier known from West Antarctica, and *Bryum argenetum* Hedw. var. *muticum* is newly reported from the Larsemann Hills.

## 10.2 Study Area

The Larsemann Hills (69°20′–69°30′S and 76°55′–76°30′E) is bordered by two main peninsulas, Broknes Peninsula and Stornes Peninsula, and several small and medium sized islands of 40 km$^2$. The exposed rock and low rolling hills containing hundreds of freshwater lakes supports rich growths of Bryophytes. Because of their relative lack of cuticle, they absorb water and nutrients through the general surface of their plant body (Figure 10.1).

The Larsemann Hills have persistent and strong katabatic winds that blow from the north-east on most summer days. During the summer, the air temperatures from December to February are in the range of 4° to 10°C, with a mean monthly temperature of about 0°C, whereas in winter the mean monthly temperature ranges between −15°C and −18°C. Precipitation accounts for approximately 250 mm water equivalent annually.

## 10.3 Materials and Methods

The author participated in the XXXV Indian Scientific Expedition to Antarctica to carry out bryological studies in the Larsemann Hills during the austral summer of 2015/2016. The survey and collection were done from Brokness and Stornes Penansulas and various small and medium sized islands (Figure 10.2). He was accompanied by scientific officials of NCPOR, GSI, WII, BSIP, IITM, NHO, SAC, and SOI between January and February 2016 and collected approximately 110 specimens of Bryophytes. The specimens were preserved in drying paper packets and field data on their habit, habitat, color, and abundance were recorded. After shade-drying, the specimens have been preserved in herbarium packets of 15 × 10 cm size, in accordance with internationally accepted herbarium methodology, and have been deposited in the herbarium of Botanical Survey of India, Central National Herbarium, Howrah (CAL). The terrestrial mosses, abundant in the Larsemann Hills, colonize a range of habitats and grow in sheltered sites in exposed areas, around the south polar Skua nests or sometimes on calcareous bones, etc., but they usually grows along the banks of water bodies and meltwater streams (Figure 10.3).

For microscopic study, the dried specimens were first soaked in normal or warm water to allow them to expand and regain their original shape. The external morphology was studied under Stereozoom Dissecting Microscope, model Leica S8APO. The anatomical details and microstructures were studied with the help of Biological Research Microscope model Nikon Eclipse Ni-1. The observations were made by preparing temporary slides of different dissected/sectioned materials in 30% aqueous Glycerin. All parts were measured along their maximum and minimum dimensions.

The photomicrographs of the habit and dissected plant parts and sections were taken with the help of Olympus Camedia C-7070 digital camera, and Leica DFC550 and Nikon DS-Ri1-U3 digital cameras with the help of image analysing software Montage multifocous and NIS-D and EDF, respectively. For the study of leaf surface under the Scanning Electron Microscope, the specimens were mounted on double adhesive tape affixed to metal stubs and coated with gold. The specimens were then stereoscanned at an accelerating potential of 20–30 KV under suitable magnification with FEI Quanta 200 Scanning Electron Microscope at Central National Herbarium, Botanical Survey of India, Howrah. Phytosociological studies were done by squeezing the samples from wet moss patches and during the anatomical studies of bryophytes algal species were found to grow in the intercellular regions of the bryophytes. Bryophytes and the algae were identified using standard monographs flora and research articles.

## 10.4 Taxonomic Treatment

Key to the species

1a. Leaves in 2–3-rows, non-spiral, midrib absent, rhizoids unicellular, smooth-walled Marchantiophyta (*C. varians*) ............................................................1

1b. Leaves in 3-rows, spirally arranged, midrib present, rhizoids multicellular, septate Bryophyta (Mosses) ............................................................2

2a. Leaf surface papillose ............................................3

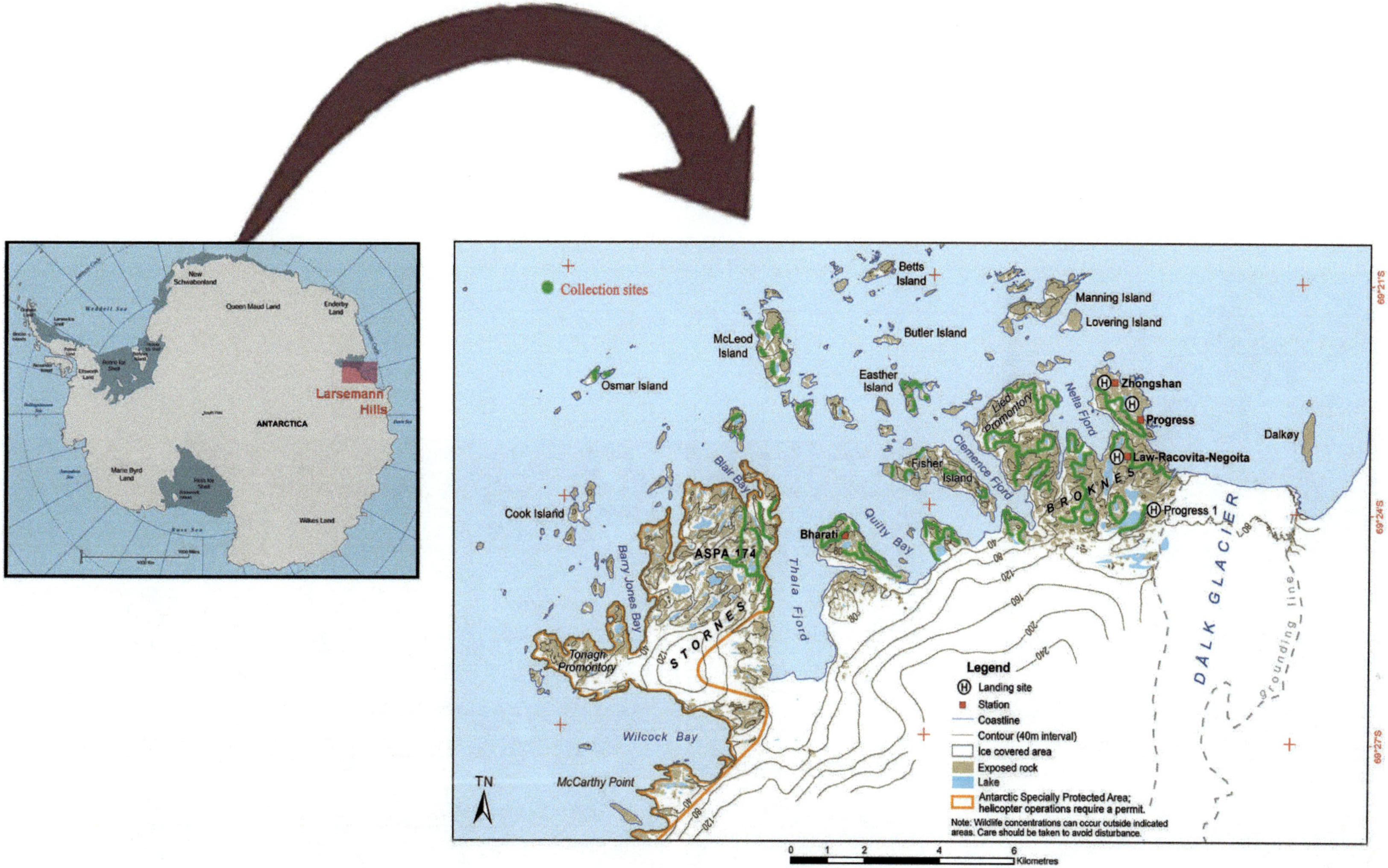


**Figure 10.1** Study area: Larsemann Hills, East Antarctica *Source*: Das & Singh, 2020 / Springer Nature.

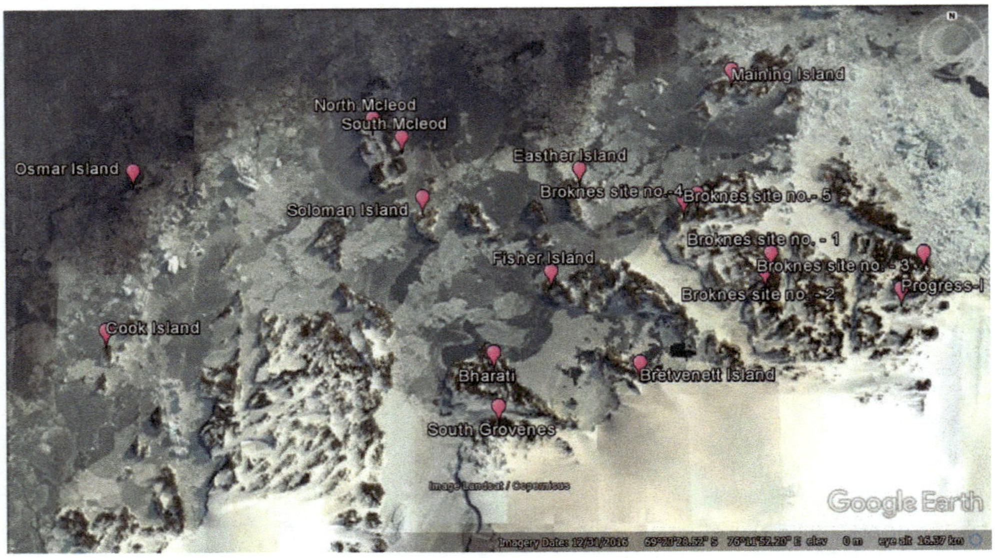


**Figure 10.2** Major collection sites.

**Figure 10.3** Bryophyte community on calcareous soil, near seepage area or growing in association with Lichens.

2b. Leaf surface smooth ....................5

3a. Leaves slightly constricted at or below the middle; costa with one stereid bands in transverse section *S. sarconeurum* ....................12

3a. Leaves usually not constricted at or below the middle; costa with two stereid bands in transverse section ....................4

4a. Leaves ovate, entire at apex; rhizoids scattered along the whole stem *B. antarcticum* ....................2

4b. Leaves lanceolate, 1–2 (–3) teeth at apex; rhizoids present only on the basal parts of stem *B. recurvirostrum* ....................3

5a. Leaves with long, hyaline hairs ....................6

5b. Leaves without long, hyaline hairs ....................7

6a. Leaves distinctly plicate; laminal cells nodulose *C. lawaniuos* ....................10

6b. Leaves not plicate; laminal cells straight *G. longirostris* ....................11

7a. Plants usually reddish brown; laminar cells subqudrate–short rectangular *C. purpureus* ....................9

7b. Plants light green–whitish green or yellowish green; laminar cells rombic, rhomboidal or oblong-hexagonal or sometimes linear-rhomboidal ....................8

8a. Upper parts of leaves usually echlorophyllose, hyaline; costa weak ending below the apex ....................9

8b. Upper parts of leaves usually chlorophyllose; costa relatively strong, usually excurrent or sometimes ending a few cells below the apex ....................10

9a. Leaf mostly apiculate; apiculus typically more than 12.5 μm long, usually much larger *B. argenteum* var. *argenteum* ....................5

9b. Leaf apex rounded; apiculus absent or rarely present, very short, less than 5 μm long *B. argenteum* var. *muticum* ....................6

10a. Leaves usually oblong-lanceolate, apex long acuminate; costa long-excurrent *B. archangelicum* ....................4

10b. Leaves usually ovate-lanceolate or sometimes oblong-lanceolate; costa precurrent or sometimes excurrent ....................11

11a. Leaf apex acute to short acuminate; costa precurrent to short excurrent. *B. pseudotriquetrum* ....................8

11b. Leaf apex long acuminate; costa always precurrent *B. bharatiensis* ....................7

1) ***Cephaloziella varians*** (Gottsche) Steph., Wiss. Ergebn. Schwed. Südpolar-Exped. [1901–1903] 4 (1): 4. 1905; Bednarek-Ochrya, Váňa, R. Ochrya & R.I.L. Smith in *Liverw. Fl. Antarctica* 81. 2000. *Jungermannia varians* (Gottsche) Steph., *Jungermannia varians* Gottsche, Int. Polarforsch., Deutsch. Exped. 2: 452. 1890. (figures 10.4:1; 10.5, 10.6).

Plants prostrate, growing in compact patches, green-dark green when fresh, blackish-brown in herbarium; shoots small, 5–9 mm long, 0.20–0.25 mm wide, delicate, sparsely branched; branches lateral intercalary. Stems oval, in outline in transverse section, 100–140 × 75–100 μm, 7–8 cells across diameter, undifferentiated; outer cells subquadrate-polygonal, 10.0–20.0 × 10.0–15.0 μm, slightly thick-walled; inner cells more or less polygonal, 7.5–20.0 × 7.5–15.0 μm, thin-walled, light yellowish to hyaline. Leaves transversely-subtransversely inserted, usually distant, rarely contiguous toward apical portion, quadrate–subquadrate, 0.17–0.25 mm long, 0.18–0.23 mm wide; bilobed to 1/5–1/6 of their length, sinus acute–subacute, margin entire; ventral surface of leaf surface smooth, lobes ovate–ovate-triangular, 8–11 cells long, 6–8 (–10) cells wide at base, apex acute–subacute or sometimes obtuse, dorsal, and ventral leaf base more or less equally decurrent; apical leaf cells more or less subquadrate–polygonal, 10.0–17.5 × 7.5–17.5 μm; median leaf cells polygonal, 10.0–22.5 × 10.0–20.0 μm, thin-walled with minute trigones; basal leaf cells polygonal, 12.5–27.5 × 12.5–20.0 μm, cells thin-walled with indistinct trigones; cuticle striolate-verrucose. Oil bodies not seen. Underleaves distant, small, ligulate–lanceolate or bilobed, quadrate–rectangulate, 50–110 μm long, 40–90 μm wide, lobes irregular, triangulate, margin entire. Gemmae in masses at the apex of leaves, brown, bicelled, subglobose, 17.5–22.5 × 7.5–12.5 μm. Rhizoids present at the base of leaves and underleaves, hyaline. Fertile plants not seen.

*Habitat:* Terricolous, growing in moist habitats near the edge of lakes or commonly as a component of communities dominated by turf and carpet forming moss of *Bryum argenetum* var. *argenetum* and *B. pseudotriquartum*.

*Distribution:* Bipolar: Russia, Antarctica – widely distributed in Continental and Maritime Antarctic but scattered in Victoria Land, coastal sites in Wilkes Land and Princess Elizabeth Lands, Europe, New Zealand, North America (Bednarek-Ochyra et al., 2000; Kurbatova and Andreev, 2015).

*Cephaloziella varians* is characterized by quadrate–subquadrate leaves, bilobed to 1/5–1/6 of leaf length with entire, 6–8 (–10) cells wide leaf lobe at base, striolate-verrucose cuticle and small, distant, ligulate–lanceolate or bilobed underleaves.

*Specimens examined:* East Antarctica, Larsemann Hills, Mining Island, 69°21'407"S, 76°18'902"E, 6 m, 08.01.2016, D. Singh 70811 (CAL); South Mcleod, 69°22'078" S, 76°09'32.59"E, 2 m, 09.01.2016, D. Singh 70813, 70820 (CAL); Fisher Island, 69°23'34.5"S, 76°13'24.7"E, 13 m, 11.01.2016, D. Singh 70836 (CAL).

2) ***Bryoerythrophyllum antarcticum*** (L.I. Savicz & Smirnova) P. Sollman, *Polish Bot. J.* 60(1): 20. 2015. *Bryoerythrophyllum recurvirostrum* (Hedw.) P.C. Chen var. *antarcticum* L.I. Savicz & Smirnova, *Bot. Zhurn.* (Moscow & Leningrad) 48: 356. 1963.

Description: Sollman (2015); illustration Savicz-Lyubitskaya and Samirnova (1963: 356, figures 1 and 2); Ochyra et al., (2008; figures 148: 2, 3, 7–13 and 15–27).

*Distribution:* Widely distributed in Antarctica (Maritime Antarctic, South Orkney Islands to Alexander Island); East coast of the Antarctic Peninsula; continental, Larsemann Hills, Vestfold Hills, and Bunger Hills (Kurbatova and Andreev, 2015; Sollman, 2015).

*Bryoerythrophyllum antarcticum* is characterized by dark reddish-brown plants; large central strand with smaller cells of stem in transverse section and without epidermis; rhizoids in clusters, scattered throughout the stem surface; oblong-ovate leaves with blunt apices, margin

**Figure 10.4** 1. *Cephaloziella varians* (Gottsche) Steph.; 2. *Bryoerythrophyllum recurvirostrum* (Hedw.) P.C. Chen; 3. *Bryum argenetum* Hedw. var. *argenetum*; 4. *Bryum pseudotriquetrem* (Hedw.) Gaertn., B. Mey & Schreb.; 5. *Bryum argenetum* Hedw. var. *muticum* Brid.; 6. *Coscinodon lawanius* (J.H. Willis) Ochrya; 7. *Guembelia longirostris* (Hook.) Ochyra & Zamowiec; 8. *Syntrichia sarconeurum* Ochyra & R.H. Zander.

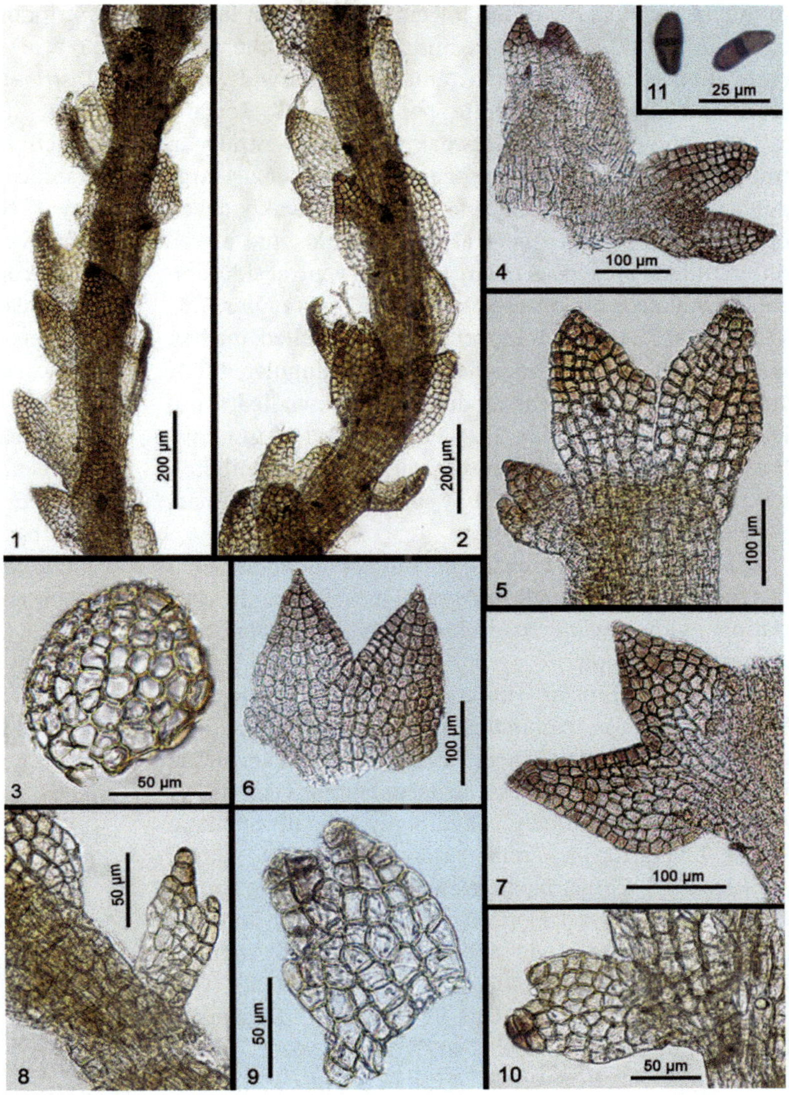


**Figure 10.5** *Cephaloziella varians* (Gottsche) Steph. 1, 2. Habit; 3. Transverse section of stem; 4, 5. Leaves and underleaves; 6, 7. Leaves; 8–10. Underleaves; 11. Gemmae.

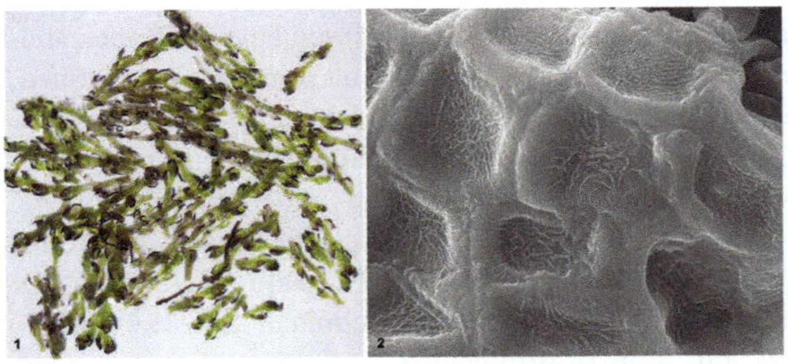


**Figure 10.6** *Cephaloziella varians* (Gottsche) Steph. 1. Habit; 2. Leaf surface under SEM.

entire, not or weakly recurved below, upward leaves apiculate with some papillae; subpercurrent to percurrent costa, dorsally strongly convex, ventrally flat with median row of 3–4 (–5) large, one row of hyaline guide cells and weakly developed or often reduced ventral stereid band, dorsal side more distinct epidermal cells in compared to ventral side (Sollman, 2015).

3) ***Bryoerythrophyllum recurvirostrum*** (Hedw.) P.C. Chen, *Hedwigia* 80: 255. 1941; Ochyra, Smith & Bednarek-Ochyra in *Illustratd. Moss Fl.* Antarctica 329. 2008. *Weissia recurvirostra* Hedw., Sp. Muse. Frond.: 71. 1801. (figures 10.4: 2; 10.7, 10.8)

Plants yellowish-green or reddish-brown, rusty or brick-red tinge when fresh and brownish in herbarium 8–14 mm long, 1.5–1.8 mm wide including leaves. Stems, sparingly branched, circular in transverse section, consisting of a small central strand, 3–4 layers of large, hyaline, thin-walled medullary cells, 1–2-layered cortex of relatively large cells with brown and moderately thickened walls. Rhizoids scattered throughout the stem surface, reddish brown. Leaves erect with sheathing base, when dry usually twisted to tightly appressed and in wet condition erect-spreading, ovate or oblong-lanceolate, linear-lanceolate, (1–) 1.8–2.8 mm long, (0.25–) 0.3–0.45 mm wide, apex acute or bluntly acute to apiculate, margins recurved to the apex or usually in the lower 3/4 portion of leaves, entire or with 1–2 irregular, minute teeth at the apex; costa subpercurrent to percurrent, 65–83 µm wide, covered ventrally with quadrate papillose cells, dorsally strongly convex and flat to weakly convex ventrally in transverse section, consisting of a median row of 2–4 large guide cells, reniform in outline dorsal and weakly developed or often reduced ventral stereid bands and a well differentiated epidermis of large cells on ventral side; apical laminal cells quadrate–subquadrate or irregular, 7.5–15.0 × 7.5–12.5 µm, cells moderately thick-walled; median laminal cells subquadrate–narrowly rectangular, 10.0–20.0 × 7.5–15.0 µm, cells moderately thick-walled; basal laminal cells rectangular, 25.0–37.5 × 10.0–15.0 µm, hyaline, thin-walled; cuticle densely multipapillose with hollow, bifid papillae, usually 4 or sometimes 6 per lumen. Fertile plants not seen.

*Habitat:* Terrestrial, growing in both dry and moist habitats frequently on soil, silt, and gravel in wet places in association with *Bryum argenetum* var. *argenetum* and *B. pseudotriquetrum*.

*Distribution:* Bipolar [India (Western Himalaya and Western Ghats), Antarctica (East Antarctica: East Antarctica Peninsula, Princes Elizabeth Land, Larsemann Hills, Vestfold Hills), James Ross Island group Vega I; West Antarctica: South Orkney Island Signy I, South Shetland Island Livingston I, Loubet Cost, George VI Sound, Alexander I)], China, Japan, New Guinea, Hawaii, Africa, Europe, New Zealand, North America, and South America (Lal, 2005; Ochyra et al., 2008; Kurbatova and Andreev, 2015; Singh and Nayaka, 2017; Das and Singh, 2020).

*Specimens examined:* East Antarctica, Larsemann Hills, Mining Island, 69°21'407"S, 76°18'902"E, 6 m, 08.01.2016, D. Singh 70819 (CAL); Fisher Island, 69°23'34.5"S, 76°13'24.7"E, 13 m, 11.01.2016, D. Singh 70838 (CAL); West Broknes, 69°24'46.2"S, 76°21'40.1"E, 26 m, 16.01.2016, D. Singh 70862 (CAL); North Grovenes, 69°24'755"S, 76°12'340"E, 65 m, 12.02.2016, D. Singh 70898 (CAL).

*Bryum recurvirostrum* is easily recognized by the rusty or brick-red tinge of the older parts of the plants, lanceolate to linear-lanceolate leaves with strongly recurved margin, densely papillose leaf cuticle toward apex and basal leaf cells are rectangular. *B. recurvirostrum* is somewhat similar to *Didymodon brachyphyllus* in the basal leaf cells and highly papillose apical leaf cells. However, the latter differs from the former in the absence of pellucid apiculus and brown basal laminal cells (Ochyra et al., 2008).

4) ***Bryum archangelicum*** bruch & Schimp in Bruch, Schimp. & W. Gümbel, *Bryol. Eur.* 4: 153., figure 333. 1846; Bednarek-Ochyra in *Illustratd. Moss Fl.* Antarctica 482. 2008.

Description and illustration: Ochyra et al. (2008; figure 233)

*Distribution*: Bipolar: Afghanistan, Iran, Japan, Turkey, Russia, Africa, Europe, North America, Antarctica (East Antarctica; Larsemann Hills, Schirmacher Oasis); West Antarctica (West Antarctic Peninsula, South Orkney Island Signy I, South Shetland Islands) (Ochyra et al., 2008; Kurbatova and Andreev, 2015; Singh and Nayaka, 2017).

*Bryum archangelicum* is characterized by yellowish green to brownish green plants, stem with distinct central strand, 3-4-layered, thin-walled medullary cells and 1–2 layered thick-walled, smaller cortical cells; erect, appressed, oblong-lanceolate leaves with acuminate to long acuminate apices and a broad unistratose border of elongated and narrow cells, recurved to revolute margin throughout; stout costa, long-excurrent, in transverse section consisting of central stereid band covered with dorsal epidermis and with a row of guide cells immediately below the ventral epidermis and unistratose laminal cells (Ochyra et al., 2008).

5) ***Bryum argenteum*** Hedw., Sp. Muse. Frond. 181. 1801; Ochyra, Smith & Bednarek-Ochyra in *Illustratd. Moss Fl. Antarctica* 454. 2008. var. **argenteum** (figures 10.4: 3; 10.9, 10.10)

Plants pale green to green or bright green when fresh, whitish, silvery and shining in herbarium; shoots 5–8 mm long, 0.4–0.7 mm wide including leaves. Stems subcircular in outline in transverse section with a distinct central strand, 3–4-layered large, medullary cells, thin-walled, 1–2-layered of thick-walled cortical cells. Rhizoids light brown at basal portion of stem. Leaves imbricate or somewhat distant at older portion of shoot, slightly concave, broadly ovate, 0.4–0.9 mm long, 0.30–0.60 mm wide, base non-decurrent, apex acute–apiculate, lower half slightly chlorophyllose, upper half hyaline, margins entire, plane, unbordered; costa single, ending well below the apex and extending 3/4 of leaf length, but sometimes ill-defined above and extending into the apiculus, 35–45 µm wide at the base, gradually narrowed upwards, pale brownish, with a row of thin-walled, large epidermal cells on the ventral and dorsal sides and a central band of stereid or substereid cells; laminal cells pellucid, unistratose throughout; apical leaf cells rhomboidal, 22.5–25.0 × 12.5–20.0 µm, thin to firm-walled, hyaline; median leaf cells slightly elongated–rhomboidal 30.0–55 × 15.0–22.5 µm, becoming slightly narrower toward the margins but not forming distinct borders; basal leaf cells short-rectangular to subquadrate, 22.5–62.5 × 15.0–25.0 µm µm, thin-walled. Fertile plants not seen.

*Habitat*: Terricolous or saxicolous, growing in both dry and wet habitats at the moist margins of drainage channels, beneath rocks and on sand, clay, or gravel. Also grows in dry, open, or sheltered sites in rock cervices in association

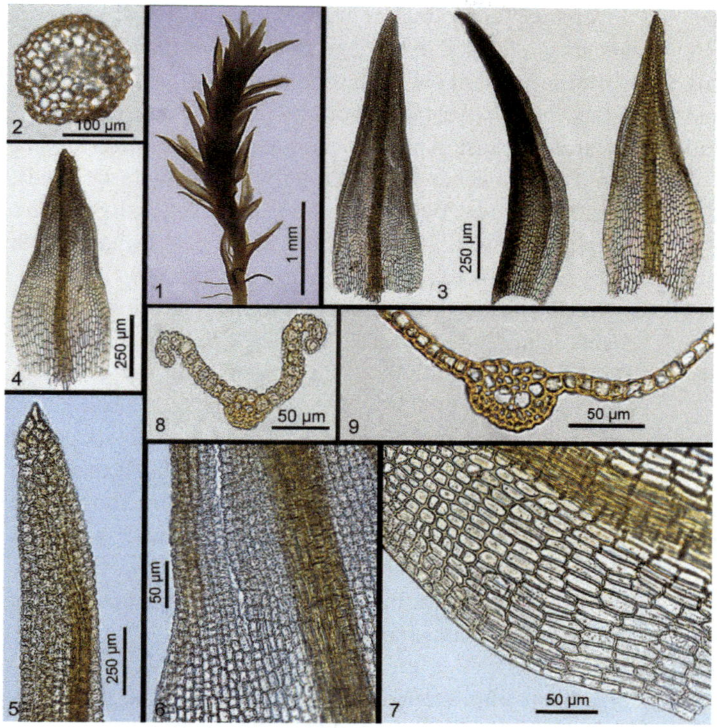


**Figure 10.7** *Bryoerythrophyllum recurvirostrum* (Hedw.) P.C. Chen; 1. A portion of the plants; 2. Transverse section of stem; 3, 4. Leaves; 5. Apical leaf lobe cells; 6. Median leaf cells; 7, Basal leaf cells; 8, 9. Transverse section of leaf.

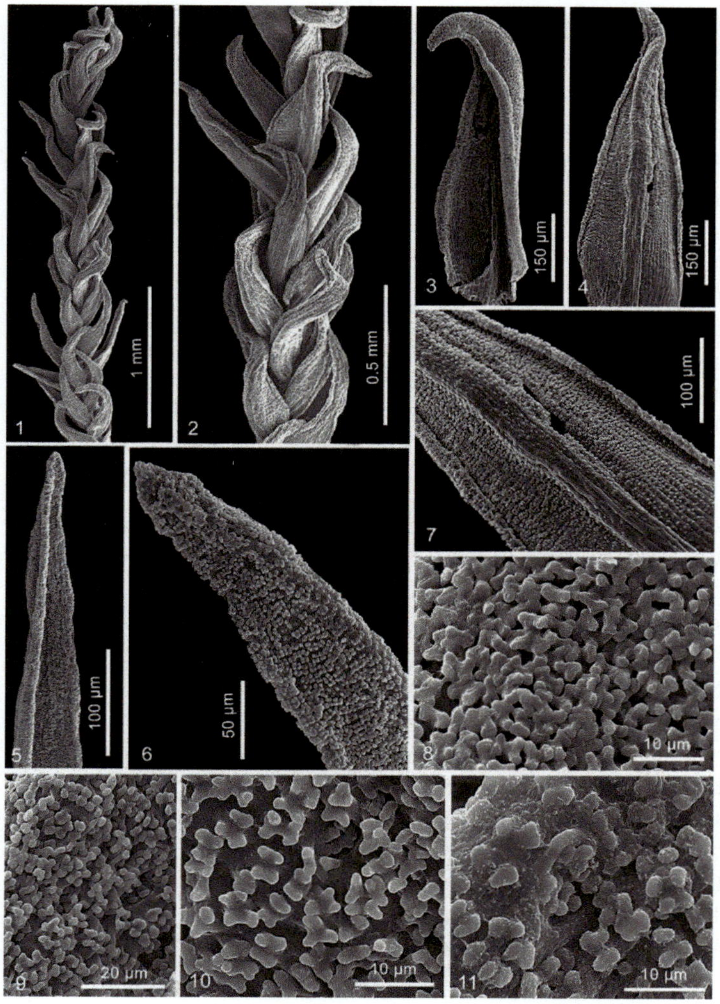


**Figure 10.8** *Bryoerythrophyllum recurvirostrum* (Hedw.) P.C. Chen 1, 2. A portion of the plants; 3, 4. Leaves; 5, 6. Apical leaf cells showing cuticle; 7. Median leaf cells showing cuticle; 8–11. The same enlarged.

with *Bryoerythrophyllum recurvirostrum, Bryum pseudotriquetrum Cepaloziella varians* and various algal species.

*Distribution*: Cosmopolitan moss species occurring throughout the glove. In Antarctica it is widely distributed in East and West Antarctica (Ochyra et al., 2008; Singh and Nayaka, 2017).

*Specimens examined*: East Antarctica, Larsemann Hills, Broknes Peninsula (Progress-1), 69°23'47.2"S, 76°23'40.14"E, 65 m, 07.01.2016, D. Singh 70801, 70803, 70806 (CAL); Mining Island, 69°21'407"S, 76°18'902"E, 6 m, 08.01.2016, D. Singh 70816, 70820 (CAL); Fisher Island, 69°23'34.5"S, 76°13'24.7"E, 13 m, 11.01.2016, D. Singh 70835, 70839, 70842, 70846 (CAL); West Broknes, 69°24'46.2"S, 76°21'40.1" E, 26 m, 16.01.2016, D. Singh 70864, 70866, 70874 (CAL); Stornes Penansula, 69°24'703"S, 76°08'25.1"E, 45 m, 25.01.2016, D. Singh 70877 (CAL); Easther Island, 69°22'30.9"S, 76°16'084"E, 23 m, 28.01.2016, D. Singh 70855, 70856, 70859, 70866 (CAL).

*Bryum argenteum* var. *argenteum* is characterized by imbricate, broadly ovate leaves, acute to acuminate apices, hyaline in the distal half and green or clorophyllose near the base with a slender costa extending to 3/4 of leaf length.

6) ***Bryum argenteum*** Hedw. var. **muticum** Brid., Bryol. Univ. 1: 846. 1826; Ochyra, Smith & Bednarek-Ochyra in *Illustratd. Moss Fl. Antarctica* 458. 2008 (figures 10.4:5; 10.11).

Plants yellowish green or reddish brown at older portion of shoots when fresh, brown in herbarium, shoots 4–8 mm long, 0.3–0.4 mm wide including leaves. Stem subcircular in outline in transverse section with a small indistinct central strand; medullary cells 2–3-layered large, thin-walled, hyaline; cortical cells 1 (–2)-layered, somewhat smaller thick-walled. Rhizoids scattered throughout the stem surface, reddish brown, branched. Leaves closely imbricate, slightly concave, broadly ovate, suborbicular, 0.3–0.6 mm long, 0.3–0.5 mm wide, base non-decurrent, apex usually obtuse or sometimes rounded or bluntly acute, usually hyaline in the distal half, light green, yellowish green to pale yellowish brown below, often reddish at base; margins entire, plane, unbordered; costa single, ending well below 3/4 of leaf length, subpercurrent, very weak in distal part or obsolete, 27.5–37.5 μm wide at base, gradually narrowed upwards, yellowish-brown to pale brownish, terete, slightly convex on the dorsal side and ventral sides, consisting of large dorsal and ventral epidermal cells and a small highly reduced central group of a few substereid cells; apical leaf cells rhomboidal, 22.5–35.0 × 12.5–17.5 μm, thin to firm-walled, hyaline in the distal half; middle leaf cells subquadrate to short rectangular, 22.5–37.5 × 15.0–20.0 μm; basal leaf cells rectangular, 25.0–50.0 × 15.0–22.5 μm; moderately thick-walled. Androecial and gynoecial branches not seen.

*Habitat*: Terricolous or saxicolous, usually growing in wet areas, silty soil in seepage areas on slopes or at margins of drainage channels or on sand, clay, or gravel.

*Distribution*: Australia, South America, Antarctica (East Antarctica: East Antarctic Peninsula, Dronning Moaud Land, Schirmacher Oasis, Enderby Land, Victoria Land); West Antarctica (West Antarctic Peninsula South Shetland Island, Sandwich Island) (Ochyra and S.M. Singh, 2008; Ochyra et al., 2008; S.M. Singh and Nayaka, 2017).

*Specimens examined*: East Antarctica, Larsemann Hills, Fisher Island, 69°23'34.5"S, 76°13'24.7"E, 13 m, 11.01.2016, D. Singh 70828 (CAL); Maining Island, 69°21'40.7"S, 76°18'90.2"E, 24 m, 07.02.2016, D. Singh 70817 (CAL).

*Bryum argenteum* var. *muticum* is characterized by closely imbricate, broadly ovate–suborbicular leaves with usually obtuse or sometimes rounded to very short acute apices, leaf cells hyaline in the distal half with a slender costa extending to 3/4 of leaf length.

*Note*: This species has been recorded here for the first time from the Larsemann Hills.

7) ***Bryum bharatiensis*** Rehman, K. Gupta & F. Bast, J. *Asia-Pacific Biodiversity* 14 (3): 288. 2021.

Description and illustration: Rehman et al. (2021; figures 4 and 5)

*Distribution*: Endemic to Antarctica, Princes Elizabeth Land (Larsemann Hills) (Rehman et al., 2021).

*Bryum bharatiensis* is characterized by unbranched shoots with long-acuminate, serrulate apex and decurrent base, strong, percurrent coasta, narrow, elongated leaf laminal cells and rhomboidal–rectangular cells toward the coasta (Rehman et al., 2021).

8) ***Bryum pseudotriquetrum*** (Hedw.) P. Gaertn., B. Mey & Scherb., *Oekon. Fl. Wetterau* 3: 102. 1802; Ochyra, Smith & Bednarek-Ochyra in *Illustratd. Moss Fl. Antarctica* 473. 2008. *Mnium pseudotriquetrum* Hedw., *Sp. Musc. Frond.* 190. 1801. (figures 10.4: 4; 10.12, 10.13)

Plants growing in dense, often very compact and tightly coherent tufts, yellowish green or brownish-green when fresh, blackish-brown in herbarium; shoots 10–14 mm long, 0.5–0.9 mm wide including leaves; simple or sparingly branched. Stems erect or ascending from a decumbent base, subcircular in transverse section, with large central strand, 3–5-stratose, large, medullary cells, hyaline, thin-walled; cortical cells l–2-layered, thick-walled, red to reddish-brown. Rhizoids long, densely branched, pale brown to reddish-brown. Leaves closely imbricate or distant toward basal portion of stem, erect to erecto-patent, shrunken and flexuose when dry, erect-spreading when wet, oblong-ovate or sometimes broadly ovate at older portion of shoots, suborbicular, (0.7–) 1.1–2.1 mm

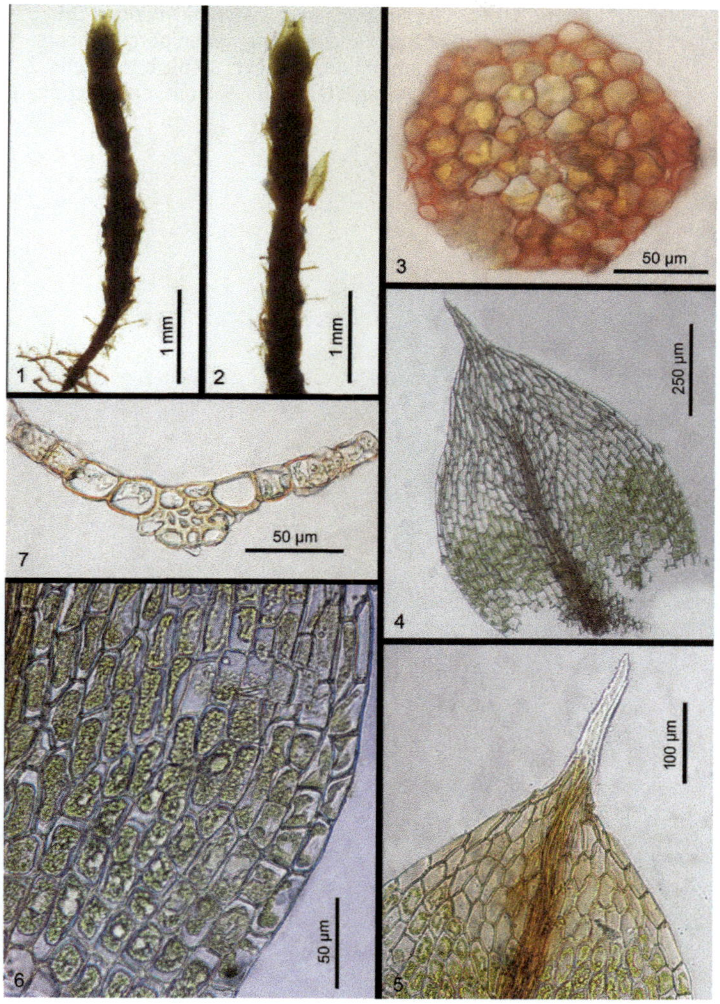


**Figure 10.9** *Bryum argenetum* Hedw. var. *argenetum* 1, 2. A portion of plant; 3. Transverse section of stem; 4. A leaf; 5. Apical leaf cells; 6. Median and basal leaf cells; 7. Transverse section of leaf.

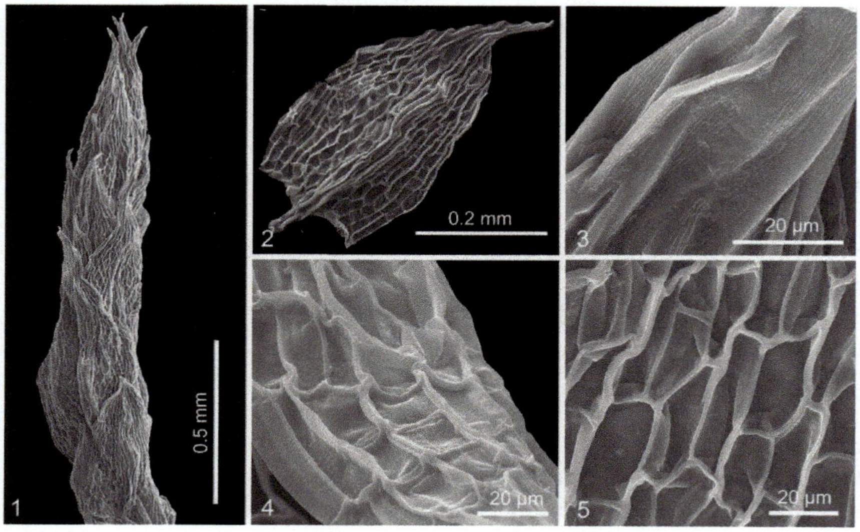


**Figure 10.10** *Bryum argenetum* Hedw. var. *argenetum* 1. A portion of plant under SEM; 2. 4. A leaf; 3. Apex of leaf lobe; 4. Median leaf cells; 5. Basal leaf cells.

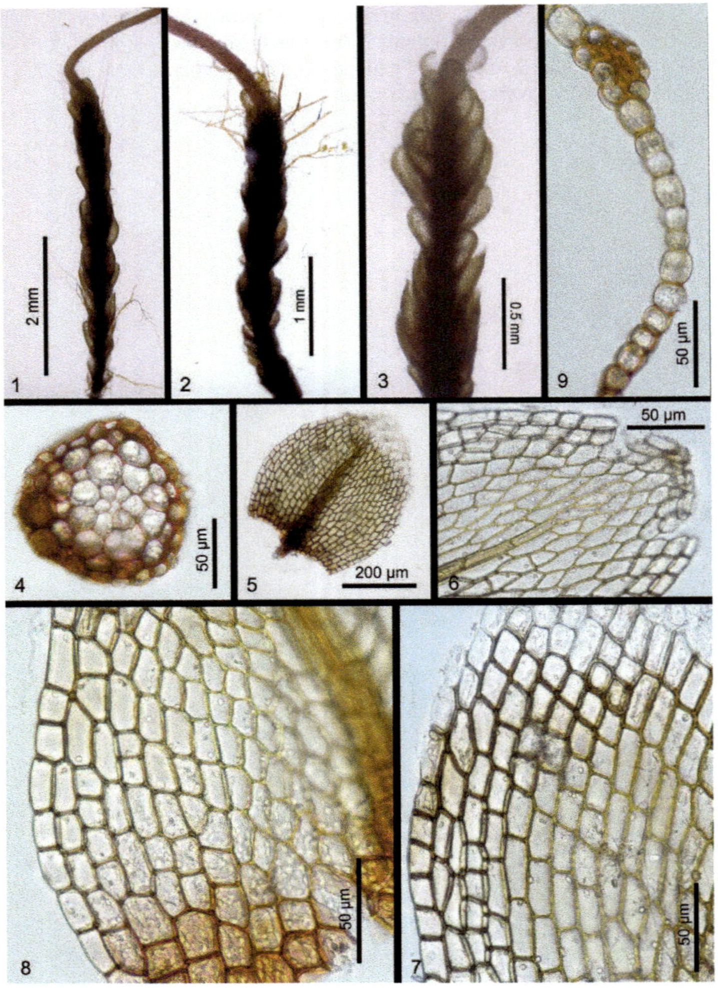


**Figure 10.11** *Bryum argenetum* Hedw. var. *muticum* Brid. 1–3. A portion of the plants bearing gemmiperous shoots; 4. Transverse section of leaf; 5. A leaf; 6. Apical leaf cells; 7. Median leaf cells; 8. Basal leaf cells; 9. Transverse section of leaf.

long, (0.30–) 0.5–0.8 mm wide, apex acuminate or acute to broadly acute or short acuminate, moderately to strongly concave; margins entire or weakly crenulate at the apex, narrowly recurved or plane, distinctly bordered with 2–3 rows of elongate, unistratose cells or unbordered; costa percurrent to short-excurrent, seldom ending well below the apex, 45.0–87.5 µm wide at the base, terete, consisting of a large central stereid band with a single layer of central and dorsal large epidermal cells and a row of large guide cells immediately below the ventral epidermis in transverse section; laminal cells unistratose throughout, transparent, reddish at the base; apical leaf cells hexagonal to rhomboidal-hexagonal to rhomboidal, 27.5–50.0 × 15.0–20.0 µm, thin-walled; median leaf cells rhomboidal, 25.0–42.5 × 12.5–20.0 µm; becoming narrowed and elongate, firm-walled at the margin and forming a distinct border or only somewhat longer than the inner cells and forming indistinct uniseriate border; basal leaf cells rectangular, 30.0–55.0 × 15.0–25.0 µm, cells thin-walled, reddish-brown. Fertile plants not seen.

*Habitat*: Terrestrial or saxicolous, growing in both dry and wet habitats in exposed or sheltered places, margins of lakes, margins of melt-water channels in association with *Bryum argenteum var. argenteum, Bryoerythrophyllum recurvirostrum* and various algal species.

*Distribution*: Bipolar [India (Western and Eastern Himalaya), Antarctica: widely distributed in East Antarctica: Dronning Maud Land (Schirmacher Osias), Enderby Land, Princes Elizabeth Land (Larsemann Hills, Westfold Hills), Queen Mary Land, Wilkes Land; West Antarctica: South Sandwich Island, South Orkney Island, South Shetland, West Antarctic Peninsula], Tropical Asia, China, Japan, Iran, Taiwan, Australia, Africa, Europe, New Zealand, North America, and South America (Singh and Semwal, 2000; Lal, 2005; Ochyra et al., 2008; Kurbatova and Andreev, 2015; Singh and Nayaka, 2017).

*Specimens examined*: East Antarctica, Larsemann Hills, Broknes Peninsula (Progress-1), 69°23'47.2"S, 76°23'40.14"E, 65 m, 07.01.2016, D. Singh 70808 (CAL); Mining Island, 69°21'407"S, 76°18'902"E, 6 m, 08.01.2016,

D. Singh 70814 (CAL); Fisher Island, 69°23'34.5"S, 76°13'24.7"E, 13 m, 11.01.2016, D. Singh 70824, 70827 (CAL); West Broknes, 69°24'46.2"S, 76°21'40.1"E, 26 m, 16.01.2016, D. Singh 70879 (CAL); Stornes Penansula, 69°24'703"S, 76°08'25.1"E, 45 m, 25.01.2016, D. Singh 70882 (CAL); Easther Island, 69°22'30.9"S, 76°16'084"E, 23 m, 28.01.2016, D. Singh 70848, 70849 (CAL); North Mcleod, 69°22'25.7"S, 76°08'19.08"E, 34 m, 11.02.2016, D. Singh 70894, 70896 (CAL).

*Bryum pseudotriquetrum* is characterized by usually ovate-lanceolate or sometimes oblong-lanceolate leaves with acute to short-acuminate leaf apices, short excurent costa, shorter distal laminal cells and withb reddish leaf base.

9) ***Ceratodon purpureus*** (Hedw.) Brid., Bryol. Univ. 1: 480. 1826. *Dicranum purpureum* Hedw., *Spec. Muse. Frond.* 136: 36. 1801; Ochyra, Smith & Bednarek-Ochyra in *Illustratd. Moss Fl. Antarctica* 155. 2008 (Figure 10.14)

Plants yellowish to brownish-green, reddish brown above. Stems simple or sparsely branched, circular in outline in transverse section, composed of 1–2 layers of small, thick-walled, reddish brown cortical cells; 2–4 layers of large, thin-walled, hyaline medullary cells; small, distinct central strand. Rhizoids light brown at basal portion of stem. Leaves contiguous to distant, loosely appressed, straight or slightly twisted and flexuose when dry, erecto-patent to patent when wet, oblong-lanceolate, (0.5–) 0.8–1.7 mm long, 0.3–0.5 mm wide, carinate to concave, straight. or somewhat falcate toward apex, apex short-acuminate to acute, concave, and curved when dry; margins recurved to revolute toward apical portion of leaves, unistratose, entire or slightly denticulate at extreme apex; costa single, stout, (37–) 45–66 μm wide at the base, reddish-brown, straight or sometimes flexuose, percurrent to short-excurrent, occasionally ending shortly below the apex, composed of a strong dorsal stereid band and weaker, sometimes reduced to a few cells or lacking ventral stereid group in transverse section and a row of large guide cells; laminal cells unistratose, pellucid; apical leaf cells quadrate, 10.0–22.5 × 7.5–20.0 μm wide, slightly thick-walled; median leaf cells subquadrate to short rectangular, 12.5–25.0 × 10.0–20.0 μm, thin to slightly thick-walled; basal leaf cells short rectangular, 22.5–35.0 × 15–20.0 μm, moderately thick to thin-walled; cuticle smooth. Fertile plants not seen.

*Habitat*: Terricolous, growing in both dry and wet habitats in exposed or sheltered places, margins of lakes, seepage areas.

*Distribution*: Cosmopolitan India (Eastern Himalaya), widely distributed in Antarctica: East Antarctica: Dronning Maud Land (Schirmacher Oasis), Enderby Land, Princes Elizabeth Land (Larsemann Hills, Westfold Hills), Queen

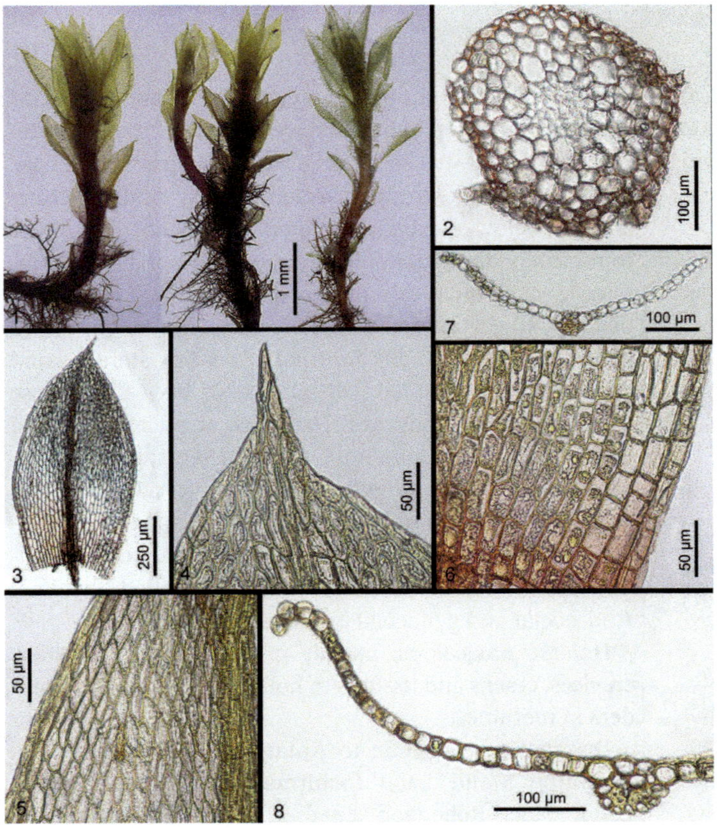


**Figure 10.12** *Bryum pseudotriquetrem* (Hedw.) Gaertn., B. Mey & Schreb. 1. A portion of plants; 2. Transverse section of stem; 3. A leaf; 4. Apical leaf cells; 5. Median leaf cells; 6. Basal leaf cells; 7, 8. Transverse section of leaf.

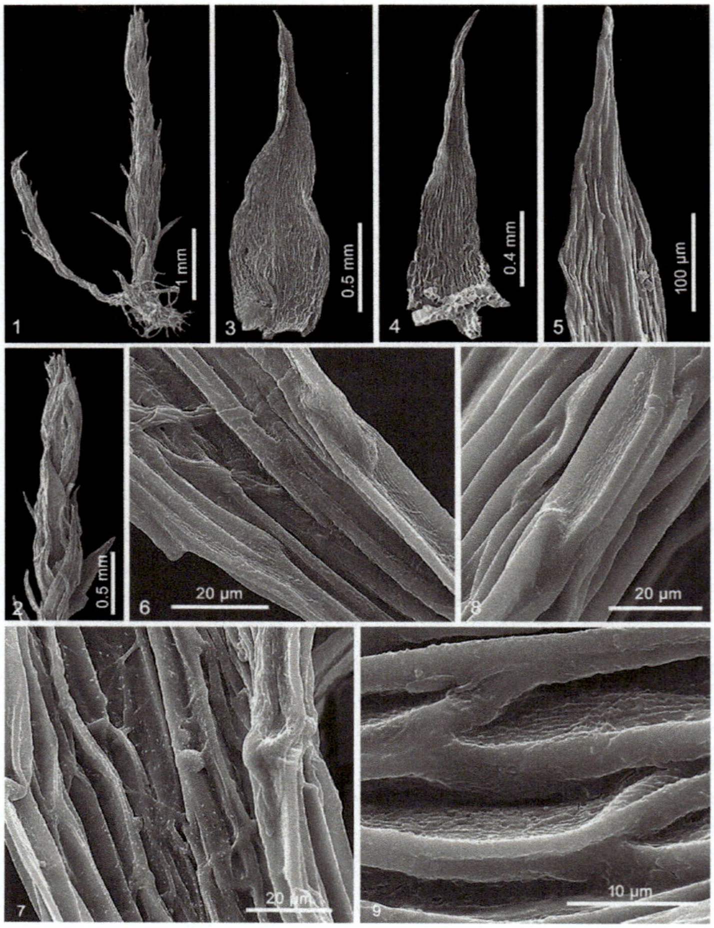


**Figure 10.13** *Bryum pseudotriquetrem* (Hedw.) Gaertn., B. Mey & Schreb. 1, 2 A portion of the plants under SEM; 3, 4. Leaves; 5. Apical leaf lobe cells; 6–9. A portion of the same enlarged showing cuticle.

Mary Land, Wilkes Land, Victoria Land; West Antarctica: South Sandwich Island, Ellsworth Land, West Antarctic Peninsula] (Singh and Semwal, 2000; Lal, 2005; Ochyra et al., 2008; Kurbatova and Andreev, 2015; Singh and Nayaka, 2017).

*Specimens examined:* East Antarctica, Larsemann Hills, Broknes Peninsula, 69°23'47.2"S, 76°23'40.14"E, 65 m, 07.02.2016, D. Singh 72834 (CAL).

*Ceratodon purpureus* is characterized by oblong-lanceolate leaves with apex short-acuminate apices, strongly recurved to revolute leaf margins and smooth cuticle.

10) ***Coscinodon lawianus*** (J.H. Willis) Ochyra, *Polish Pol. Res.* 25: 112. 2004; Ochyra, Smith & Bednarek-Ochyra in *Illustratd. Moss Fl. Antarctica* 224. 2008. *Grimmia lawiana* J.H. Willis in Filson, *Anare Sci.* Rep. Ser. B (II) Bot. 82: 148. 1966. (figures 10.4: 6; 10.15, 10.16)

Plants pale olive-green, 6–12 mm long, 0.5–0.8 mm wide. Stem circular in in transverse section with a small, distinct central strand surrounded by 3–4-layered, large, thin-walled medullary cells; cortex 1–2-layers, cells slightly smaller, thick-walled. Rhizoids usually at older portion of stem, light brown. Leaves imbricate, appressed when dry, erecto-patent when wet, ovate to broadly ovate-lanceolate, 0.8–1.2 mm long, 0.3–0.5 mm wide, apex acute–apiculate, strongly keeled distally and biplicate, concave below, mostly terminated with a stout, flat, entire to minutely denticulate, hyaline; margins plane, unistratose, erect to obviously incurved in the distal half; costa distinct, 40–55 μm wide, dorsally strongly convex, with 2 ventral epidermal cells internally, central stereid band distinct; laminal cells of distal part of leaves bistratose throughout or sometimes tristratose at plication; apical leaf cells rounded-quadrate to short-rectangular, often oblate along the margins, 5.0–12.5 × 5.0–10.0 μm, thick-walled; median leaf cells short rectangular, 15.0–30.0 × 8.0–12.5 μm, thick-walled; basal leaf cells rectangular, 25.0–37.5 × 10–15.0 μm, thin-walled, hyaline. Androecial and gynoecial branches not seen.

*Habitat:* Saxicolous, usually growing in dry habitats, crevices, cracks and fissures in hollows among larger boulders in moraines.

*Distribution:* Endemic to Antarctica [(East Antarctica: Dronning Maud Land (Schirmacher Oasis)], Enderby Land, Mac. Robertson Land, Princes Elizabeth Land

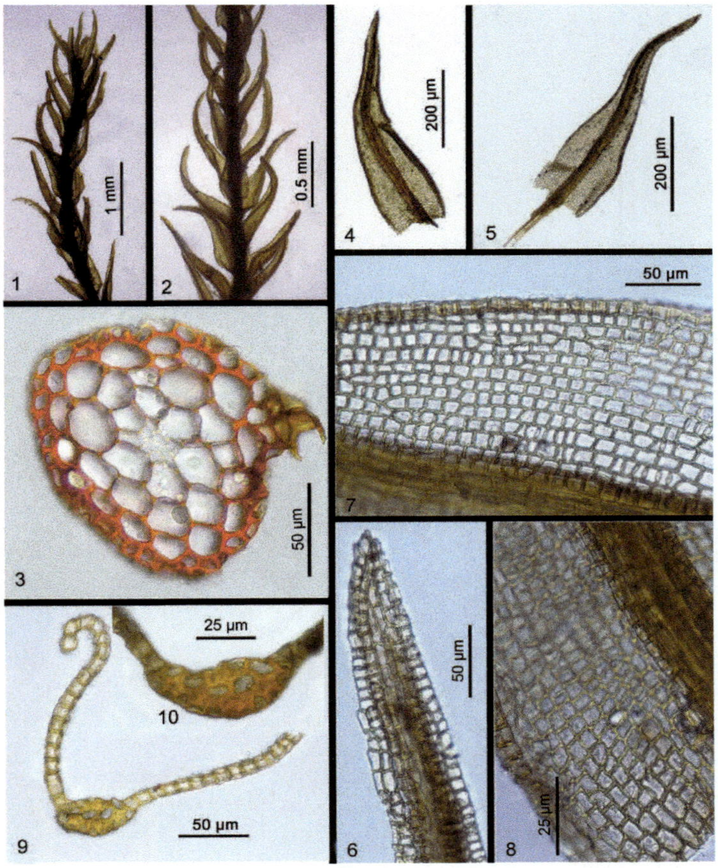


**Figure 10.14** *Ceratodon purpureus* (Hedw.) Brid. 1, 2. A portion of plant; 3. Transverse section of stem; 4, 5. Leaves; 6. Apical leaf cells; 7. Median leaf cells; 8. Basal leaf cells; 9, 10. Transverse section of leaf.

(Larsemann Hills) (Ochyra et al., 2008; Kurbatova and Andreev, 2015).

*Specimens examined:* East Antarctica, Larsemann Hills, Mining Island, 69°21'407"S, 76°18'902"E, 6 m, 08.01.2016, D. Singh 70812 (CAL); Stornes Peninsula, 69°24'703"S, 76°08'25.1"E, 45 m, 25.01.2016, D. Singh 70881 (CAL); North Grovenes, 69°24'755"S, 76°12'340"E, 65 m, 12.02.2016, D. Singh 70889, 70890, 70900 (CAL).

*Coscinodon lawianus* is characterized by imbricate, distinctly biplicate leaves with strongly pilose long hyaline hair-points.

11) ***Guembelia longirostris*** (Hook.) Ochyra & Zamowiec in Ochyra, Zamowiec & Bednarek-Ochyra, *Cens. Cat. Polish Mosses*: 127. 2003; Ochyra, Smith & Bednarek-Ochyra in *Illustratd. Moss Fl. Antarctica* 276. 2008. *Grimmia longirostris* Hook., *Musci Exot.* 1: tab. 62. 1818. (figures 10.4:7; 10.17, 10.18)

Plants small, densely caespitose, 5–8 mm long, 0.8–1.4 mm wide including hair-point, olive-brown to blackish-brown below, light yellowish green above in fresh, blackish brown in herbarium; sparingly branched. Stem circular to subcircular in outline in transverse section, 100–120 × 90–110 µm with a small central strand, cells 5.0–7.5 × 3.7–5.0 µm, thin-walled; 1 (–2) rows of brownish, thick-walled cortical cells, 7.5–12.5 × 5.0–10.0 µm; 3-4-layered hyaline large medullary cells, 7.5–20.0 × 7.5–17.5 µm, thin-walled. Rhizoids few at older portion of stem, brownish. Leaves erect, closely appressed, not or very slightly changed on drying, erect spreading, rigid in wet condition, ovate-lanceolate, 1.0–1.4 mm long (including hair-point), 0.22–0.30 mm wide, widest at middle, not plicate, apex acute, canaliculate, weakly keeled distally, usually epilose below, margins entire, recurved on one or both sides to 1/2–2/3 of leaf length, hyaline-awned longer toward stem apex, hair-point terete, erect, rigid, smooth to minutely distantly denticulate; costa percurrent, 35–40 µm wide at the base, clearly delimited from the laminal cells at basal portion, weakly differentiated at apical portion, semi-elliptic to reniform in transverse section, U-shaped, bistratose at apical portion, tristratose toward basal portion, with 2–4 large ventral epidermal cells, a median layer of substereid cells and a row of larger dorsal epidermal cells; laminal cells bistratose in the distal half and unistratose toward the costa; apical leaf cells isodiametric to short-rectangular, 5.0–12.5 × 5.0–10.0 µm, thick-walled without distinct trigones, walls often sinuate; median leaf cells subquadrate–rectangular,

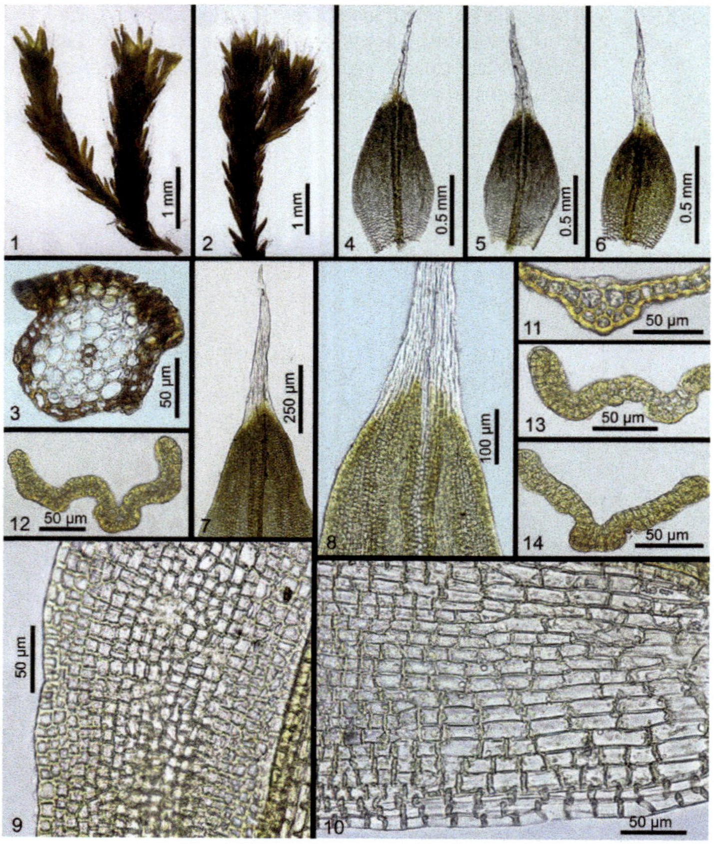


**Figure 10.15** *Coscinodon lawanius* (J.H. Willis) Ochrya 1, 2. A portion of plant; 3. Transverse section of stem; 4–6. Leaves. 7, 8. Apical leaf cells; 9. Median leaf cells; 10, Basal leaf cells; 11–14. Transverse section of leaf from base toward apex.

10.0–20.0 × 7.5–12.5 μm, thick-walled with incrassate, nodulose vertical walls; basal cells short to long rectangular or subquadrate, 15.0–45.0 × 10.0–12.5 μm, slightly thin-walled with thickened nodulose longitudinal walls, transverse walls markedly thicker than the longitudinal walls at the basal marginal region; leaf surface minutely verrucose. Fertile plants not found.

*Habitat*: Saxicolous, growing in dry conditions in association with Cyanobacteria (*Stigonema minutum* Hassall ex Bornet & Flahault) on fine to coarse sandstones in sheltered places near the edge of lake.

*Distribution*: Antarctica (East Antarctica – present study; West Antarctica), Sri Lanka, Philippines, Indonesia, New Guinea, Africa, North and South America (Ochyra et al., 2008).

*Guembelia longirostris* is characterized by canaliculated, U-shaped leaves with recurved margins in the distal part and semi-elliptic–reniform and curved costa; costa with 2–4 ventral epidermal, markedly larger than the dorsal ones and the basal cells with strongly thickened nodulose longitudinal walls.

*Specimens examined*: East Antarctica, Larsemann Hills, Broknes Peninsula (Progress – 1), 69°23'47.2"S, 76°23'40.14"E, 65 m, 07.01.2016, D. Singh 70802 (CAL).

*Note*: *G. longirostris* is recorded for the first time from the Larsemann Hills of East Antarctica. The plants of the Larsemann Hills are slightly atypical in having smaller size of the plants and leaves described by (Ochyra et al., 2008) from West Antarctica, but these differences are very minor and could be because of different ecological conditions at the two locations and it is covered well within the range of variations in the species. *G. longirostris* is easily mistaken in the field for the widely-distributed *Coscinodon lawianus* in the Larsemann Hills in having general appearance of the plants, transverse section of stem, leaf shapes and hyaline hair point of the leaves, but it can be readily distinguished in having non-plicate, not or weakly keeled leaves distally, median and basal leaf cells with distinctly thickened, sinuate-nodulose longitudinal walls as compared to strongly plicate leaves on both sides of the costa in the distal portion of the leaves, strongly keeled distally and leaf cell with straight longitudinal walls in

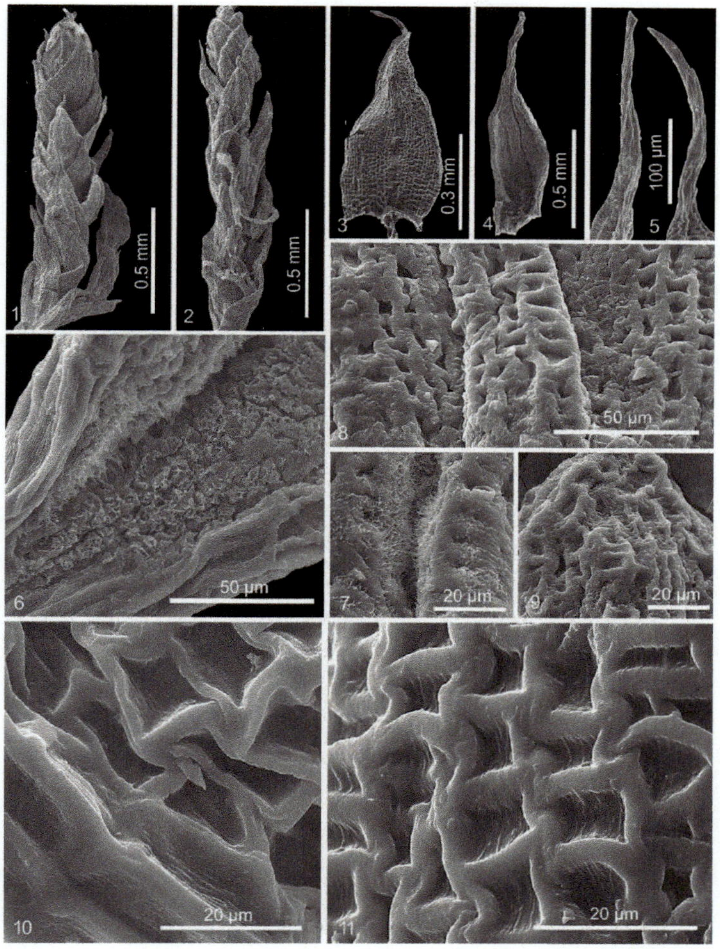


**Figure 10.16** *Coscinodon lawanius* (J.H. Willis) Ochrya 1, 2. A portion of the plants under SEM; 3, 4. Leaves; 5. Apical portion of leaf; 6–9. Apical potion of leaf showing cuticle in dorsal surface; 10, 11. Median leaf cells showing cuticle.

the latter (Ochyra et al., 2008). *G. longirostris* is somewhat similar to *Grimmia plagiopodia* Hedw. in having general appearance of the plants, transverse section of stem and hyaline hair point of the leaves, but it can be distinguished in having non-plicate leaves and leaf cell with straight longitudinal walls in the latter (Ochyra et al., 2008)

12) ***Syntrichia sarconeurum*** Ochyra & R.H. Zander, *Fragm. Florist. Geobot. Polonica* 14: 210. 2007; Ochyra, Smith & Bednarek-Ochyra *in Illustratd. Moss Fl. Antarctica* 377. 2008. *Sarconeurum glaciale* (Miill.Hal.) Cardot & Bryhn, *Natl. Antarct. Exped. Nat. Hist.* 3 Musci: 3. 1907. (Figures 10.4: 8; 10.19)

Plants small, 2.5–3.5 mm long, 0.3–0.4 mm wide, growing in loose patches, bright green or yellowish reddish-brown. Stems erect, simple or sparingly branched, mostly slender and fragile, circular in outline in transverse section without or with central strand and a uniform parenchymatous tissue of large, almost uniform, thin-walled, somewhat collenchymatous, hyaline or yellowish-hyaline cells, cortex and hyalodermis lacking or with 1–2 peripheral layers smaller and dark-colored cells present. Rhizoids long, vigorous, reddish, branched. Leaves crowded at apex, somewhat smaller below, (0.5–) 1.2–1.7 mm long, (0.20–) 0.3–0.4 mm wide, non-decurrent, erect, somewhat shrunken and tightly or loosely appressed with incurved apices to erecto-patent when dry, spreading–recurved when moist, narrowly oblong–lingulate-lanceolate or lanceolate, sometimes ovate-oblong, gradually tapering or quite rapidly contracted near or above mid-length into a long subula, terete in the upper quarter, forming a cylindrical, flexuose, solid, fragile, multistratose, caducous, snout-like propagulum, acute or sharply apiculate with 1–2 smooth, pellucid cells; margins plane, entire, sometimes slightly undulate; costa single, prominent, (22.5–) 27.5–42.5 µm wide at base, light brown, ending in the solid deciduous apex, convex dorsally, covered on both ventral and dorsal surfaces with quadrate to short-rectangular

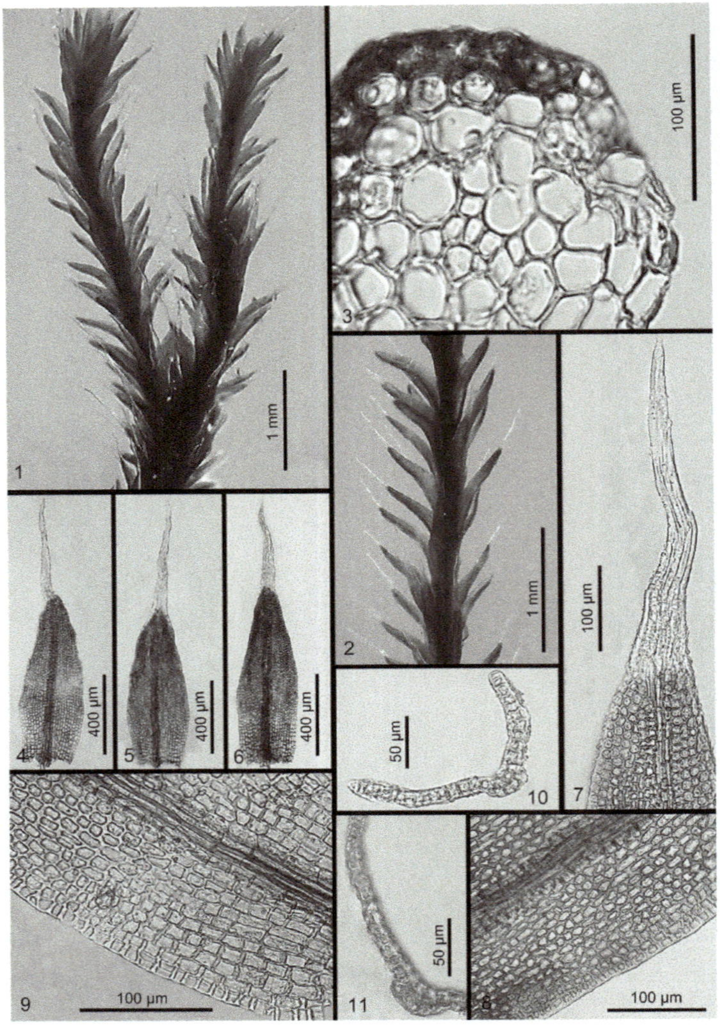


**Figure 10.17** *Guembelia longirostris* (Hook.) Ochyra & Zamowiec 1, 2. A portion of plant. 3. A portion of transverse section of stem; 4–6. Leaves; 7. Apical leaf cells; 8. Median leaf cells; 9, Basal leaf cells; 10, 11. Transverse section of leaf from base toward apex.

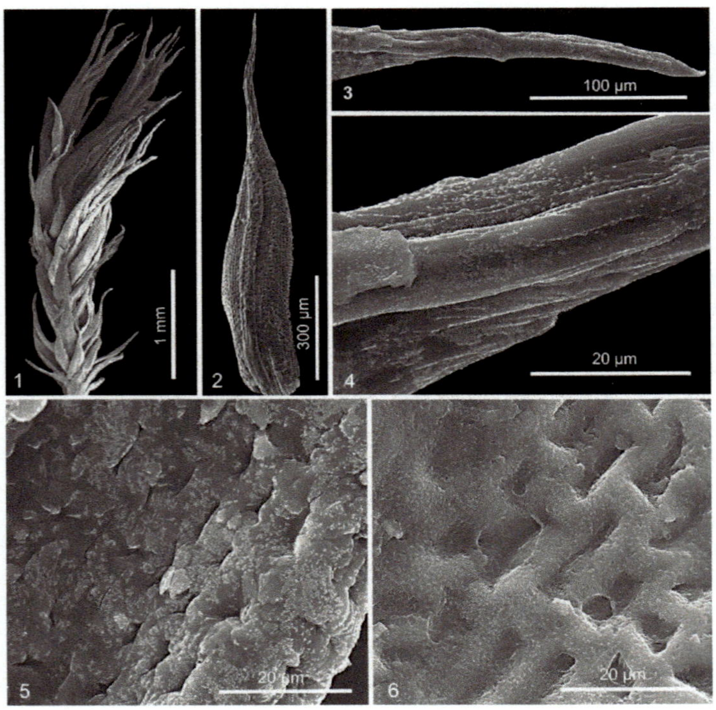


**Figure 10.18** *Guembelia longirostris* (Hook.) Ochrya & Zarnowiec 1. A portion of the plant under SEM; 2. A leaf; 3. Apical portion of leaf; apex; 4. The same a portiin enlarged showing cuticle; 5, 6. Median leaf cells showing cuticle.

cells in 3–4 rows of cells across the costa, one stereid or substereid band dorsally, 2–3 large guide cells in 1 or very seldom 2 layers, without ventral stereid band or occasionally with small ventral stereid group near the leaf base on large leaves with distinct ventral epidermis of large, papillose cells and lacking a dorsal epidermis in transverse section; apical leaf cells quadrate short-rectangular, 15.0–27.5 × 10.0–22.5 μm, evenly thick-walled; median leaf cells short-rectangular, 15.0–27.5 × 10.0–22.5 μm, thick-walled; basal leaf cells narrowly rectangular, 17.5–37.5 × 15.0–22.5 μm, thin-walled, smooth, pellucid, hyaline; cuticle densely papillose with low, small, solid, conic or bifid papillae, several per lumen. Androecial and gynoecial branches not seen.

*Habitat*: Terrestrial or saxicolous, usually growing in dry or moist habitats, exposed and sheltered places on rock faces in association with *Bryum argenetum* var. *Argenetum* and *B. pseudotriquatrum*.

*Distribution*: Endemic to Antarctica (West Antarctica: West Antarctic Peninsula, South Shetland Island, South Orkney Island; East Antarctica: East Antarctic Peninsula, Costa Land, Dronning Maud Land (Schirmacher Oasis), Princess Elizabeth Land (Larsemann Hills, Westfold Hills), Queen Mary Land, Wilkes Land, Victoria Land, Marie Byrd Land) (Ochyra et al., 2008; Kurbatova and Andreev, 2015; Singh and Nayaka, 2017).

*Specimens examined*: East Antarctica, Larsemann Hills, Broknes Island, 69°23'47.2"S, 76°23'40.14"E, 65 m, 07.02.2016, D. Singh 70844 (CAL); South Grovenes, 69°23'40.75"S, 76°12'48.8"E, 45 m, 12.02.2016, D. Singh 72835, 72836 (CAL).

*Syntrichia sarconeurum* is characterized by fleshy and swollen deciduous leaf apices and slightly constricted at or below the middle, papillose leaf cuticle and costa with one stereid bands in transverse section.

**Figure 10.19** *Syntrichia sarconeurum* Ochyra & R.H.Zander 1, 2. A portion of plants; Habit; 3. Transverse section of stem; 4–7. Leaves; 8, 9. Apical leaf lobe cells; 10, 11. Median leaf cells; 12. Basal leaf cells; 13. 14. Transverse section of leaf.

## 10.5 Phytosociological Studies

In the harsh climatic conditions of both the Arctic and Antarctica, cyanobacteria and algae are the important primary producers, which survive climatic adversity in the refuge of the bryophytes and lichen, both extra and intracellularly. This phytosocialism among them is mutualistic, where the bryophytes provide suitable ambience, and the cyanobacteria contribute toward nitrogen fixation. In the course of survivability, the spores of cyanobacteria and algae, swayed by wind, colonize mostly near the rhizoidal and axial areas of the moss patches, where the water retention is at its maximum. But during more unfavorable climatic conditions they also colonize the shoot apical area to mitigate the poor light availability (Broady, 1979). Likewise, several cyanobacterial taxa also thrive intracellularly, within the pores present in the hyaline leaf cells of mosses, to withstand unsuitable pH (Granhall and Hofsten, 1976).

Our recent study, to understand to the diversity of these photosynthetic microbes in the vegetational consortium (Das and Singh, 2020), has identified a total of 16 taxa of cyanobacteria and algae from different sites of the Larsemann Hills, as presented in Table 10.1. The host bryophytic patches include *Cephaloziella varians*, which is the only known species of liverwort from the Larsemann Hills, along with six moss species of the genera *Bryum*, *Bryoerythrophyllum*, *Syntrichia*, *Coscinodon*, and *Guembelia*. The rhizoidal regions of these bryophytes represent a consortium of the most diverse cyanobacterial and algal species, whereas colonization of only heterocystous cyanobacteria were detected, growing both in the exterior and interior pores of the leaf blades, as recorded earlier from other regions of East Antactica (Nakatsubo and Ino, 1986). The unbranched nostocalean members of the genera *Nostoc* and *Gloeocapsopsis* preferred the exteriors of leaf lamina, and the branched cyanophyta *Stigonema minutum* symbiosed in the internal pores. This uniqueness of cyanobacterial

**Table 10.1** Epiphytic algae growing with bryophytes in Larsemann Hills (Das & Singh, 2020 / Springer Nature).

| Epiphytic Algae | Family | Bryophyte Species |
|---|---|---|
| **Cyanophyta** | | |
| *Gloeocapsopsis magma* | *Chroococcaceae* | *Bryoerythrophyllum recurvirostrum*, *Syntrichia sarconeurum* |
| *Nostoc commune* | *Nostocaceae* | *Bryoerythrophyllum recurvirostrum*, *Bryum argenteum*, *B. pseudotriquetrum*, *Syntrichia sarconeurum* |
| *N. fuscescens* | *Nostocaceae* | *Bryum pseudotriquetrum* |
| *N. lichenoides* | *Nostocaceae* | *Bryum pseudotriquetrum* |
| *N. punctiforme* | *Nostocaceae* | *Bryum argenteum*, *B. pseudotriquetrum* |
| *N. sphaericum* | *Nostocaceae* | *Bryoerythrophyllum recurvirostrum*, *Bryum argenteum*, *B. pseudotriquetrum*, *Syntrichia sarconeurum* |
| *Nostoc sp.* | *Nostocaceae* | *Bryoerythrophyllum recurvirostrum* |
| *Nostoc sp.* | *Nostocaceae* | *Bryoerythrophyllum recurvirostrum* |
| *Oscillatoria sancta* | *Oscillatoriaceae* | *Bryum pseudotriquetrum* |
| *Stigonema minutum* | *Stigonemataceae* | *Coscinodon lawianus*, *Guembelia longirostris* |
| **Charophyta** | | |
| *Coleochaete scutata* | *Coleochaetaceae* | *Bryum argenteum* |
| *Coleochaete* sp. | *Coleochaetaceae* | *Bryum argenteum* |
| **Chlorophyta** | | |
| *Hormidiopsis crenulata* | *Ulotrichaceae* | *Bryum pseudotriquetrum* |
| **Xanthophyta** | | |
| *Xanthonema antarcticum* | *Xanthonemataceae* | *Bryum argenteum* |
| **Bacillariophyta** | | |
| *Luticola muticopsis* | *Diadesmidaceae* | *Bryum argenteum*, *B. pseudotriquetrum*, *Cephaloziella varians* |
| *Pinnularia borealis* | *Pinnulariaceae* | *Bryum argenteum*, *B. pseudotriquetrum*, *Cephaloziella varians* |

population was common in the patches of *Coscindon* and *Guembelia*, found growing on sandy soils in the lake slopes or gravel pits near Bharati station and Broknes Peninsula. Cyanobacteria are more tolerant to hostile environments than the other groups of eukaryotic algae. The latter are rather opportunistic, as revealed from our observation of their occurrence in the rhizoidal area of bryophytes. They could have migrated from the adjacent terrestrial or aquatic habitats. The only two diatom species recorded from the epibryic habitats of the Larseman hills, i.e., *Pinnularia borealis* and *Luticola muticopsis*, were found in the moss patches in close vicinity to a stream in Solomon Island. In general, the moss species supporting the most diversified cyanobacteria and algae are *Bryum argenteum* var. *argenteum* and *B. pseudotriquetrum*, which are also the most commonly occurring bryophytes in Larsemann Hills.

Prior to our study (Das and Singh, 2020), exploration and documentation of epibryic cyanobacteria and algae by any Indian biologist was confined to the Schirmacher Oasis. Singh et al. (2013) reported 14 taxa of algae from the moss patches of *Bryum* and *Pottia* in and around the oasis. Our study ellucidated a clear picture of the cyanobacterial and algal distribution and their seasonal preferences. The occurrence of several cyanobacteria on the leaves of *Bryum pseudotriquetrum* and *Bryoerythrophyllum recurvirostrum*, while combating the adverse conditions during the winter, may interpret their host specificity. Furthermore, the geological and climatic uniqueness of the Larsemann Hills, supported by adequate influential aquatic habitats with nearly 130 shore islands and hundreds of lakes, are the primary forces behind this phytosociology and inherent diversity.

## 10.6 Results and Discussion

The present study revealed the occurance of twelve taxa of Bryophytes belonging to seven genera and five families based on own collection and an earlier report from the Larsemann Hills, East Antarctica (Figure 10.20); (Table 10.2). *Guembelia longirostris* (Hook.) Ochrya & Zarnowiec is reported here for the first time in detail from East Antarctica, earlier reported from West Antarctica. *Bryum argenetum* Hedw. var. *muticum* Brid. is recorded for the first time from the Larsemann Hills and other species also extend their range of distribution on various islands of the Larsemann Hills (Figure 10.21).

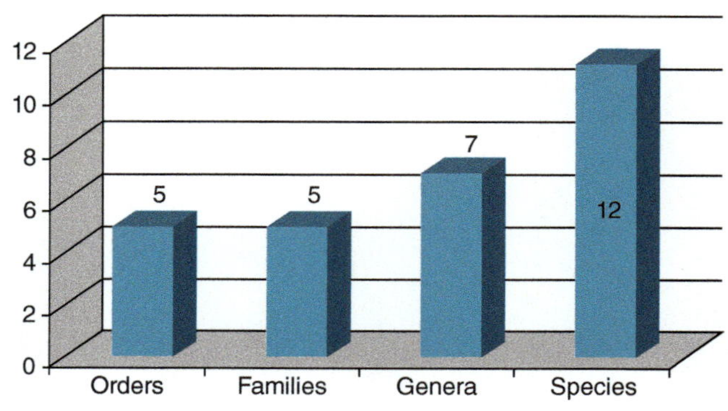


**Figure 10.20** Order, families, genera, and species of Bryophytes in the Larsemann Hills.

The maximum diversity of Bryophytes (08 spp.) was recorded from the Broknes Peninsula, followed by Fisher, Manning, Bharati islands and Stones Peninsula with 05 spp. each, and Soloman and Bettnobett with 03 spp. each, and Easther and Osmar islands with 02 species each. The Cook Islands is represented by a single species. i.e., *Bryum argenteum* var. *argenteum* (Figure 10.22).

The present study has been compared with bryoflora of Schirmacher Oasis and the Indian mainland, especially the Indian Himalaya and found that out of 12 taxa from the Larsemann Hills, 5 are restricted to the Larsemann Hills, whereas in the Schirmacher Oasis 17 taxa were reported, of which 10 are restricted to the oasis and 7 taxa are common between the Larsemann Hills and the Schirmacher Oasis and 4 species are common between the Larsemann Hills, Indian mainland and Schirmacher Oasis, respectively, whereas 4 species are common in all three regions. As a result, more or less 70% taxa from the Larsemann Hills are similar to the Schirmacher Oasis and only one liverwort *Cephaloziella varians* and three mosses, i.e., *Bryoerythrophyllum antarcticum*, *Coscinodon lawianus*, and *Guembelia longirostris*, are restricted to the Larsemann Hills (Table 10.3).

In the present study, *Guembelia longirostris* and *Bryum argenteum* var. *muticum* are rare, as known only by a single collection, while *Bryoerythrophyllum antarcticum, Cephaloziella varians, Ceratodon purpureus*, and *Syntrichia sarconeurum* are frequent and *Bryoerythrophyllum recurvirostrum, Bryum argenteum* var. *argenteum*, and *B. pseudotriquetrum* are the most common species in the Larsemann hills.

## 10.7 Future Prospects

Bryophytes usually provide substratum for the growth of various Lichens in the Larsemann Hills, and usually prefer to grow on a thick bed of *Bryum* species like *Xanthoria elegans, Candelariella flava*, and *Lecanora epibryon*. As the similar type of nutrients and microclimatic requirements may result in such an association, study is required to draw any further conclusions. *Candelariella flava* is the most common association associated with the Bryophytes in the Larsemann Hills.

Bryophytes grow in harsh environments and the scope is required for the experimental bryologist to study their responses to climate change as they are potential

**Table 10.2** Conspectus of Bryophytes in the Larsemann Hills.

| Orders | Families | Genera | Species (Including Infraspecific Taxa) |
|---|---|---|---|
| **Marchantiophyta (Liverworts)** | | | |
| Jungermanniales | Cephaloziellaceae | *Cephaloziella* | 01 |
| **Bryophyta (mosses)** | | | |
| Dicranales | Ditrichaceae | *Ceratodon* | 1 |
| Grimmiales | Grimmiaceae | *Coscinodon* | 1 |
| | | *Guembelia* | 1 |
| Phascales | Pottiaceae | *Bryoerythrophyllum* | 02 |
| | | *Syntrichia* | 01 |
| Bryales | Bryaceae | *Bryum* | 4/1 |
| **5** | **5** | **7** | **12** |

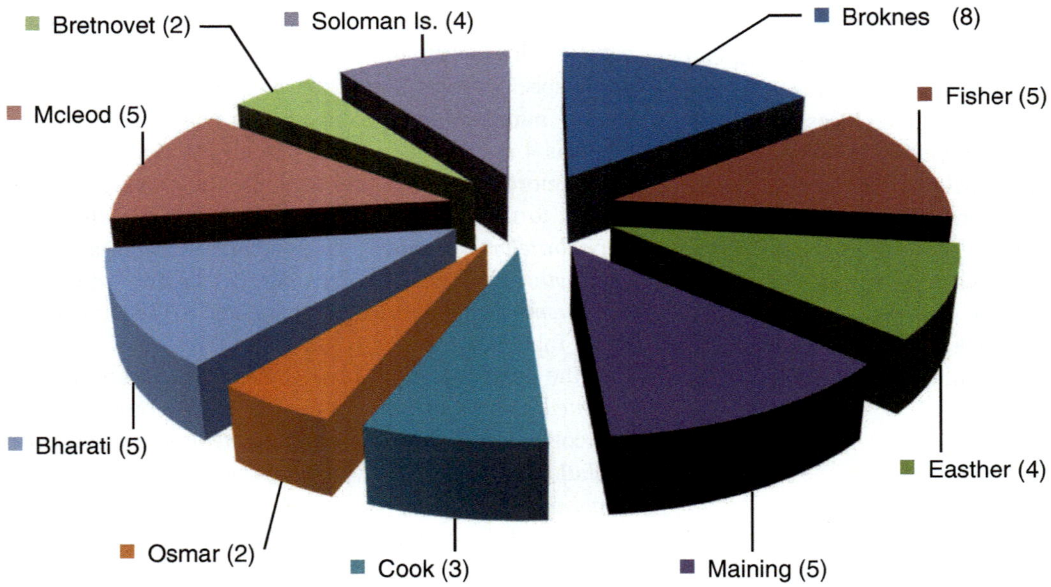


**Figure 10.21** Island-wise diversity of Bryophytes of the Larsemann Hills.

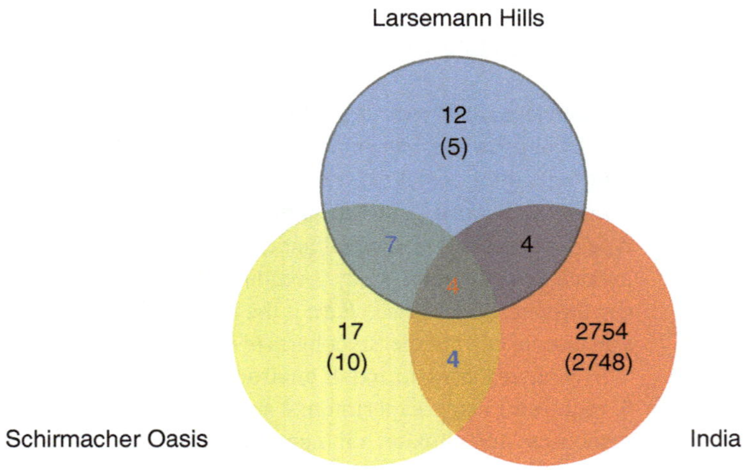


**Figure 10.22** Comparative diversity of Bryophytes.

**Table 10.3** Comparative analysis of Bryophyte diversity at LH, SO, and Indian mainland.

| S. No. | Name of taxa | Larsemann Hills | Schirmacher Oasis | India |
|---|---|---|---|---|
| 1. | *Cephaloziella varians* | + | – | – |
| 2. | *Bryoerythrophyllum antarcticum* | + | – | – |
| 3. | *Bryoerythrophyllum recurvirostrum* | + | + | + |
| 4. | *Bryum archangelicum* | + | + | – |
| 5. | *Bryum argenteum* var. *argenteum* | + | + | + |
| 6. | *Bryum argenteum* var. *muticum* | + | + | – |
| 7. | *Bryum bharatiensis* | + | – | – |
| 8. | *Bryum pseudotriquetrum* | + | + | + |
| 9. | *Ceratodon purpureus* | + | + | + |
| 10. | *Coscinodon lawianus* | + | – | – |
| 11. | *Guembelia longirostris* | + | – | – |
| 12. | *Syntrichia sarconeurum* | + | + | – |

indicators of damaging UV-B levels and other stress factors, and they may be in synthesis with both cryo- and photo-protectant organic compounds. The present findings will help to gain a better understanding of plant–microbe interactions as well as the possibility of any host preferences by the algal members in future studies. The present baseline information may be of help in the study of the various other applied aspects such as biochemical, cytological, physiological, molecular, and response to climate change.

## Acknowledgments

I thank the former and present Directors of Botanical Survey of India, Kolkata for providing the facilities and encouragement; the Director National Centre for Polar and Ocean Research (NCPOR), Goa for providing the opportunity to participate in the XXXV ISEA; Dr D.K. Singh, formerly scientist-G, Botanical Survey of India, Kolkata for valuable suggestion; team leader Dr Shailendra Saini (NCPOR) and team members of the expedition for help and cooperation; Dr S.K. Das, ex PDF of National Mission on Himalayan Studies (NMHS) project for identification of algal species.

## References

Bednarek-Ochyra, H., Váňa, J., Ochyra, R., Smith, R.I.L. (2000). *The Liverwort Flora of Antarctica*, 236. Cracow: Polish Academy of Sciences, Institute of Botany.

Borady (1979). Wind dispersal of terrestrial algae at Signy Island, South Orkney Island. *Br. Antarct. Surv. Bull.* 48: 99–102.

Das, S.K. and Singh, D. (2020). Epiphytic algae on the bryophytes of Larsemann Hills, East Antarctica. *Natl. Acad. Sci. Lett.* 44: 161–165. https://doi.org/10.1007/s40009-020-00947-7.

Granhall, U. and Hofsten, A.V. (1976). Nitrogenase activity in relation to intracellular organisms in Sphagnum mosses. *Physiol. Plant* 36: 88–94.

Kurbatova, L.E. and Andreev, M.P. (2015). Bryophytes of the Larsemann Hills (Princess Elizabeth Land, Antarctica). *Novosti Sist. Nizsh. Rast.* 49: 360–368.

Lal, J. (2005). *A Checklist of Indian Mosses*, 164. Dehradun: Bishen Singh Mahendra Pal Singh.

Nakatsubo, T. and Ino, Y. (1986). Nitrogen cycling in an Antarctic ecosystem. 1: Biological nitrogen fixation in the vicinity of Syowa Station. *Mem. Natl. Inst. Polar Res. Ser.* 37: 1–10.

Ochyra, R. and Singh, S.M. (2008). Three remarkable moss records from Dronning Maud Land, continental Antarctica. *Nova Hedwigia* 86: 497–506.

Ochyra, R., Smith, R.I.L., and Bednarek-Ochyra, H. (2008). *The Illustrated Moss Flora of Antarctica*, 685. Cambridge UK: Cambridge University Press.

Rehman, W.U., Gupta, K., and Bast, F. (2021). Morphophylogenetic assessment of a new moss species *Bryum bharatiense* sp. nov. (Bryaceae) from Larsemann Hills, Eastern Antarctica. *J. Asia-Pacific Biodiversity* 14(3): 288–289. https://doi.org/10.1016/j.japb.2021.07.001.

Savicz-Lyubitskaya, L.L. and Samirnova, Z.N. (1963). A contribution to the biology and geography of

*Bryoerythrophyllum recurvirostre* (Hedw.) Chen. a new species in the bryoflora of the Antarctica. *Bot. J.* 48: 350–361.
Singh, D.K. and Semwal, R.C. (2000). Bryoflora of Schirmacher Oasis, East Antarctica: a preliminary Study. *Scientific Report: Sixteeth Indian Expedition to Antarctica. Department of Ocean Development, Tech. Pub.* 14: 173–182.
Singh, D.K., Sharma, J.R., Gupta, R.K., Palnisamy, M. (2013). Plant diversity in Schirmacher Oasis, East Antarctica with special reference to bryophytes, fungi and diatoms. In: *Studies in Biological Sciences and Human Physiology. Three Decades of Indian Scientific Activities in Antarctica* (ed. S.C. Tripathy, R.K. Mishra, R. Mohan, et al.), 117–138. New Delhi: CSIR-NISCAIR
Singh, S.M. and Nayaka, S (2017). Contributions to the floral diversity of Schirmacher Oasis and Larsemann Hills, Antarctica. *Proc. Indian Natl. Sci. Acad.* 83(2): 469–481.
Sollman, P. (2015). The genus *Bryoerythrophyllum* (Musci, Pottiaceae) in Antarctica. *Polish Bot. J.* 60(1): 19–25.

# 11

## Antarctic Sea Ice Variability and Trends Over the Last Four Decades

*Swathi M., Juhi Yadav, Avinash Kumar*, and Rahul Mohan*

*National Centre for Polar and Ocean Research, Ministry of Earth Sciences, Govt. of India, Goa, India*
** Corresponding author*

### 11.1 Introduction

In the Earth's climate system, polar sea ice is a critical component as it modifies heat, momentum, and moisture exchanges within the ocean-ice-atmosphere system. Additionally, it provides early indication of climate change related to variations in solar energy absorption (Comiso and Nishio, 2008) and influences climate change of regional and global scales via ice-albedo feedback (Parkinson, 2004; Bintanja et al., 2013).

Antarctic sea ice expansion in a warming world remains a puzzle. Some data and model results indicate that sea ice in Antarctica will decline in a warmer climate, although slower than the anticipated rate of decline in the Arctic (Kumar et al, 2020, 2021; Bromwich et al., 2013; Siegert et al., 2019). The relationship between regional Antarctic sea-ice changes and coastal air temperatures is inverse (Massom and Stammerjohn, 2010). However, a study using a coupled model of ice-ocean-atmosphere shows that sea-ice cover will grow with global warming (Zhang, 2007; Bintanja et al., 2013) due to increased snow precipitation on the sea ice. This would reduce the salinity of the near-surface ocean layers, resulting in an increased mixed layer stability and decreased surface heat flux, leading to more sea ice.

Several studies have examined interannual variations in Antarctic sea ice and their climate implications using satellite remote-sensing techniques (Cavalieri et al., 1997; Parkinson, 2002, 2019; Zwally, 2002; Cavalieri and Parkinson, 2008; Parkinson and Cavalieri, 2012; Parkinson and DiGirolamo, 2021). Prior to the 1970s, the sea-ice record was incomplete due to the continent's remoteness and harsh weather, which made obtaining consistent science-quality data difficult. Since the late 1970s, passive and active microwave observations from space have proven to be highly effective in mapping the distribution and Antarctic sea-ice extent and its variabilities (Zwally et al.,1983; Stammerjohn and Smith, 1997; Zwally, 2002; Parkinson, 2019). The Antarctic Sea ice cover has changed dramatically in the last few decades; nevertheless, decadal-scale sea-ice changes imply minor variations that are difficult to statistically quantify (Zwally, 2002; Goosse and Zunz, 2014; Arndt and Haas, 2019; Yu et al., 2021). Additionally, previous research concluded that the positive seasonal trend in Antarctic sea-ice cover is driven by summer and autumn, with little change in the winter. On decadal and regional timescales, the annual growth and decline of Antarctic sea ice are governed by very complex physical processes involving the ocean-atmosphere system. On the basis of visible and infrared satellite images acquired during the period 1973–1977, the Antarctic SIE demonstrated a sharp decline followed by a gradual increase (Parkinson, 2019). There has been a minor variation in the trend of the total Antarctic sea-ice extent across the period (Tables 11.1 and 11.2). These variations were also observed across the Antarctic's several sectors, implying contrasting changes in sector trends.

This study addresses the sea-ice variability on monthly, interannual, and seasonal basis in detail for the total Southern Ocean (Figure 11.1). Additionally, we conduct a comprehensive analysis of decadal, monthly, seasonal, and annual sea-ice changes and compare them to previous studies (Cavalieri et al., 1999; Parkinson, 2002, 2019; Zwally, 2002; Cavalieri and Parkinson, 2008). Finally, we analyzed long-term annual and seasonal ocean-atmospheric temperatures such as SST and 2 m AT to understand their linkages with sea ice.

*Advances in Remote Sensing Technology and the Three Poles*, First Edition. Edited by Manish Pandey, Prem C. Pandey, Yogesh Ray, Aman Arora, Shridhar D. Jawak, and Uma K. Shukla.

**Table 11.1** For the Southern Ocean, yearly and seasonal SIE trends and standard deviations were computed from 1979 to 2020. *R* denotes the ratio of the trend's magnitude to its standard deviation. For example, assuming a null hypothesis of zero trend and degrees of freedom is 40, *R* values in bold are significant at 95% and higher, while those in bold and italics are significant at 99% and above.

| Year/Season | Slope, $10^3$ km$^2$ yr$^{-1}$ | Slope %Decade$^{-1}$ | *R* |
|---|---|---|---|
| Yearly (1979–2020) | 6.78 ± 5.16 | 0.56 ± 0.42 | 1.3 |
| Summer (JFM) | 6.3 ± 7.2 | 1.48 ± 1.71 | 0.9 |
| Autumn (AMJ) | 9.06 ± 7.7 | 0.84 ± 0.72 | 1.2 |
| Winter (JAS) | 7.8 ± 4.4 | 0.43 ± 0.25 | **1.7** |
| Spring (OND) | 0.5 ± 7.1 | 0.034 ± 0.6 | 0.1 |

**Table 11.2** Comparison of the yearly trend of SIE from the cited literature available for six different periods along with the present study of the 42-year period. Assuming a null hypothesis of zero trends and 40 degrees of freedom, *R* values in bold are significant at 95% and above, and those in italics and bold are significant at 99% and above.

| Periods | Slope $10^3$km$^2$yr$^{-1}$ | Slope %Decade$^{-1}$ | *R* | Source |
|---|---|---|---|---|
| 1979–1998 (20 years) | 10.9 ± 6.9 | 0.96 ± 0.61 | 1.58 | (Zwally, 2002) |
| 1979–2006 (28 years) | 11.5 ± 4.6 | 1.0 ± 0.4 | **2.5** | (Cavalieri and Parkinson, 2008) |
| 1979–2010 (32 years) | 17.5 ± 4.1 | 1.5 ± 0.4 | ***4.28*** | (Parkinson and Cavalieri, 2012) |
| 1979–2014 (36 years) | 22.4 ± 4.3 | 2.0 ± 0.4 | ***5.25*** | (Parkinson, 2019) |
| 1979–2018 (40 years) | 11.3 ± 5.3 | 1.0 ± 0.5 | 2.12 | (Parkinson, 2019) |
| 1979–2020 (42 years) | 6.8 ± 5.2 | 0.56 ± 0.4 | 1.3 | The present study (Swathi et al., 2022) |

## 11.2 Datasets and Methods

### 11.2.1 Sea Ice Extent Analysis

We used a 42-year record (1979–2020) of satellite passive-microwave measurements derived from the Scanning Multichannel Microwave Radiometer (SMMR) on board NASA's Nimbus 7, and the Special Sensor Microwave Imager (SSM/I) and Special Sensor Microwave Imager/Sounder (SSMIS) on board the US Defense Meteorological Satellites Program (DMSP) satellites: F8, F11, F13, F17, and F18. The daily sea-ice concentration (SIC) measurements from SMMR (sensor frequencies: 6, 10, 18, 21, 37 GHz) span October 1978 to August 1987, SSMI (sensor frequencies: 19, 22, 37, 85 GHz) span September 1987 to December 2008, and SSMIS (sensor frequencies: 19, 22, 37, 91 GHz) span January 2000 to December 2020. The area covered by ice represents the amount of ice present; it is calculated by multiplying the entire surface area of a pixel by the ice concentration in that pixel. Ice persistence is the percentage of months during which ice was present at a place during the data collection period. The ice extent determines whether ice is present or not in a pixel; in this case, ice is deemed to exist in a pixel if the sea-ice concentration in that pixel is more than 15%. The ice extent is derived by summing the areas of all grid cells (25 × 25 km grids with polar stereographic projection, in the region of interest, having at least 15% sea-ice concentration (Cavalieri et al., 1999; Zwally, 2002; Cavalieri and Parkinson, 2008; Parkinson, 2019). The monthly SIE derived from the NASA team (Gloersen, 1992; Cavalieri et al., 1999; Parkinson, 2019) is accessible from the National Snow and Ice Data Center (NSIDC), which provides daily data from 1979 to 2020 for the Southern Ocean.

Figures 11.1b and 11.1c show 1979–2020 mean SICs for the austral summer (February) and austral winter (September). The 42-year long-term linear trends were computed using the lines of least-squares fit to monthly, interannual, and seasonal mean data, in accordance with the literature (Zwally, 2002; Cavalieri and Parkinson, 2008; Parkinson and Cavalieri, 2012; Parkinson, 2019). The seasonal cycle was eliminated, and the anomaly was calculated by subtracting the 30-year mean (1981–2010) from each individual monthly mean (Figures 11.2b, 11.3, 11.5b, and 11.6). Over the 42 years, trends for monthly, seasonal, and yearly averages were also computed (Figure 11.2c and Table 11.1). The seasonal cycle was removed in the yearly averages of the monthly mean, and seasonal means were calculated by taking the average of four months, i.e., January–March (JFM) for austral summer; April–June

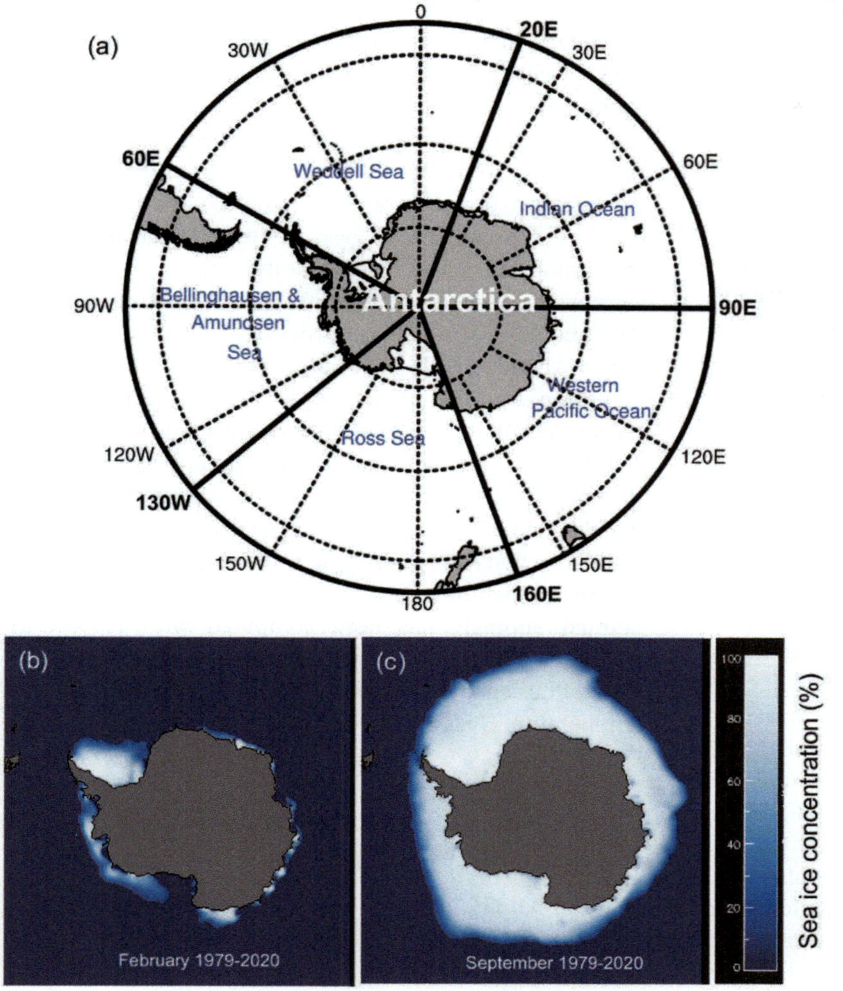


**Figure 11.1** (a) Map showing the five sectors of Antarctica – Weddell Sea, Indian Ocean, Western Pacific Ocean, Ross Sea, and Bellingshausen-Amundsen Sea; (b) minimum sea-ice concentration (SIC) derived from long-term February mean (1979–2020) retrieved from SMMR, SSMI, and SSMIS satellite data for the total Southern Ocean; (c) same as in Figure 11.1b, but for September (austral winter maximum).

(AMJ) for austral autumn; July–September (JAS) for austral winter; and October–December (OND) for austral spring (Zwally, 2002). The standard deviations/errors of the trend were calculated as described in Taylor (1997). The trends were tested for statistical significance using Student's $t$-test assuming the null hypothesis of a zero trend and 40 degrees of freedom (42 – 2 = 40).

We establish a statistical metric. R denotes the relationship between the trend and its standard deviation (Santer et al., 2000). Trends are considered significant at the 95% and 99% confidence levels if $R$ exceeds 2.02 and 2.70, respectively (Santer et al., 2000). According to Cavalieri and Parkinson (2008), the statistical significance test is criticized for its use of the null hypothesis and the arbitrary levels of significance and issues related to the autocorrelation of the data (Santer et al., 2000). These arbitrary levels of statistical significance were utilized in this study to provide only a relative measure of the robustness of those trends with lower values of $R$.

### 11.2.2 Analysis of Physical Parameters

To understand the sea-ice dynamics in the Southern Ocean, we analyzed the relation between sea ice and ocean-atmospheric temperatures (SST and 2 m AT) from ERA-5 monthly means provided by the European Centre for Medium-Range Weather Forecasts (ECMWF; https://www.ecmwf.int/en/forecasts/datasets/reanalysis-datasets/era5). ERA5 model reanalysis data on a single level with a spatial resolution of 25 × 25 km. ERA5 is the 5th-generation ECMWF reanalysis for the global climate and weather for the past 4 to 7 decades. SST and 2 m AT data were analyzed over the poleward domain of 55°S to 85°S latitude (with the land areas masked).

## 11.3 Results and Discussion

### 11.3.1 Sea Ice Variability in the Southern Ocean

The Southern Ocean SIE anomaly and trend vary substantially on monthly, yearly, and seasonal scales from January 1979 to December 2020 (Figure 11.2). Over the 42 years, the minimum and maximum ice extents occurred in February ($3.1 \times 10^6$ km$^2$) and September ($18.5 \times 10^6$ km$^2$), respectively (inset in Figure 11.2a). This inference is consistent with previous work (Zwally et al., 1983; Parkinson, 2002, 2019; Zwally, 2002; Cavalieri and Parkinson, 2008; Parkinson and Cavalieri, 2012), which revealed the existence of a significant seasonal cycle (Figure 11.2a). The SIE in February ranged from a minimum of $2.3 \times 10^6$ km$^2$ in 2018 to a maximum of $4 \times 10^6$ km$^2$ in 2008, whereas in September, the ice extent varied from a minimum of $18 \times 10^6$ km$^2$ in 1986 to a maximum of $20 \times 10^6$ km$^2$ in 2014 (Figure 11.2a).

The Antarctic SIE trends vary substantially over seasonal and interannual cycles. The lines of least-square fit exhibit positive trends in the yearly and seasonal ice extent (Figure 11.2c, Table 11.1). The yearly SIE trend is similar to the monthly deviations. The $R$ values suggest that the yearly trend is significant at the 99% level for the 28-, 32-, and 36-year periods; however, the 42 years annual trend observed is not statistically significant (Table 11.2). Seasonal trend analysis shows positive trends for all seasons (Figure 11.2c). The Antarctic SIE trends are positive for each season and on a yearly average, with autumn having the highest magnitude seasonal trend at $9060 \pm 7700$ km$^2$ yr$^{-1}$ (Table 11.1). However, the seasonal trend for spring ice extent ($500 \pm 7100$ km$^2$ yr$^{-1}$) shows the lowest magnitude compared to that of summer ice extent ($6300 \pm 7200$ km$^2$ yr$^{-1}$) (Table 11.1).

The Southern Ocean as a whole has had a positive trend over the last 42 years; however, it is significantly less than previously recorded trends for various periods (Zwally, 2002; Cavalieri and Parkinson, 2008; Parkinson and Cavalieri, 2012; Parkinson, 2019). These previously reported SIE trends show that Antarctic sea ice is very dynamic in nature (Table 11.2). From 1979 to 2014 (over 36 years), it had the highest positive trend ($22{,}400 \pm 4300$ km$^2$ yr$^{-1}$); however, the ice extent trends declined dramatically for 40 years (1979–2018) and 42 years (1979–2020) (Table 11.2). The computed trends for 40 and 42 years recorded declines of nearly 50% and 70%, respectively, compared to the 36-year trends (Table 11.2).

The variability in sea-ice extent is considered to be influenced by large-scale atmospheric forcing in the Southern Ocean (Cavalieri and Parkinson, 2008; Kusahara et al., 2017; Schroeter et al., 2017; Stuecker et al., 2017; Scott et al., 2019), which results in a weakening of both the positive (Weddell Sea, western Pacific Ocean, and Ross Sea sectors) and negative (Bellingshausen-Amundsen Sea sector) trends, as well as a trend reversal in the Indian Ocean sector. There are several atmospheric forcing mechanisms making a significant contribution to these trends. Local (increases in air temperature, enhanced wind stress due to pressure changes induced by cyclones and katabatic winds, precipitation, ocean temperature, etc.) (Parish and Bromwich, 1998; Holland et al., 2005; Uotila et al., 2011; Holland and Kwok, 2012; Turner et al., 2013; Clem et al., 2017; Alkama et al., 2020) and remote forcing (ENSO), Southern Annular Mode (SAM), Antarctic Circumpolar Wave (ACW), etc.), are well-known mechanisms that affect Antarctic sea ice via the atmosphere and ocean (Yuan and Martinson, 2001; Cobb et al., 2003; Bertler et al., 2004; Lefebvre and Goosse, 2005; Yadav et al., 2022).

In this study, we investigated the role of ocean-atmospheric temperatures on sea-ice variability. The long-term annual anomaly has been computed with respect to 30 years of climatology (1981–2010) to understand the anomalous changes that occurred in sea-ice variations and temperature (Figures 11.3–11.7).

### 11.3.2 Sea Ice Distribution With Respect to Ocean-Atmospheric Temperature

The annual SIC anomaly reveals an overall positive trend over the last four decades (Figure 11.3a). The SST trend in the region is slightly negative, whereas the AT anomaly trend is positive. After 2015, the largest deviation was observed when there was a continual decline in the SIC anomaly. In this period, the SST and AT anomalies had a notable positive trend (Figure 11.3a). The spatial trend of the SIC anomaly indicated an overall positive trend (Figure 11.3b), but the SST anomaly indicated warming in East Antarctica (Figure 11.3c). The AT anomaly shows overall cooling in the Southern Ocean, whereas an increase in AT has been observed in the Indian Ocean and Ross Sea sectors (Figure 11.3d). Several studies have been conducted to investigate the relationship between Antarctic sea-ice variability and seasonal air temperatures (Jacobs and Comiso, 1993; Zwally, 2002) and found that sea-ice deviations are negatively correlated with air and sea surface temperature, and their associations are stronger on regional and seasonal scales than over the whole Antarctica.

The Southern Ocean SIC and SST in summer show a slight negative anomaly trend, whereas the AT trend is nearly constant over the study period. In 2016, SIC showed a clear decreasing trend, which is consistent with the positive SST and AT anomalies observed during the same

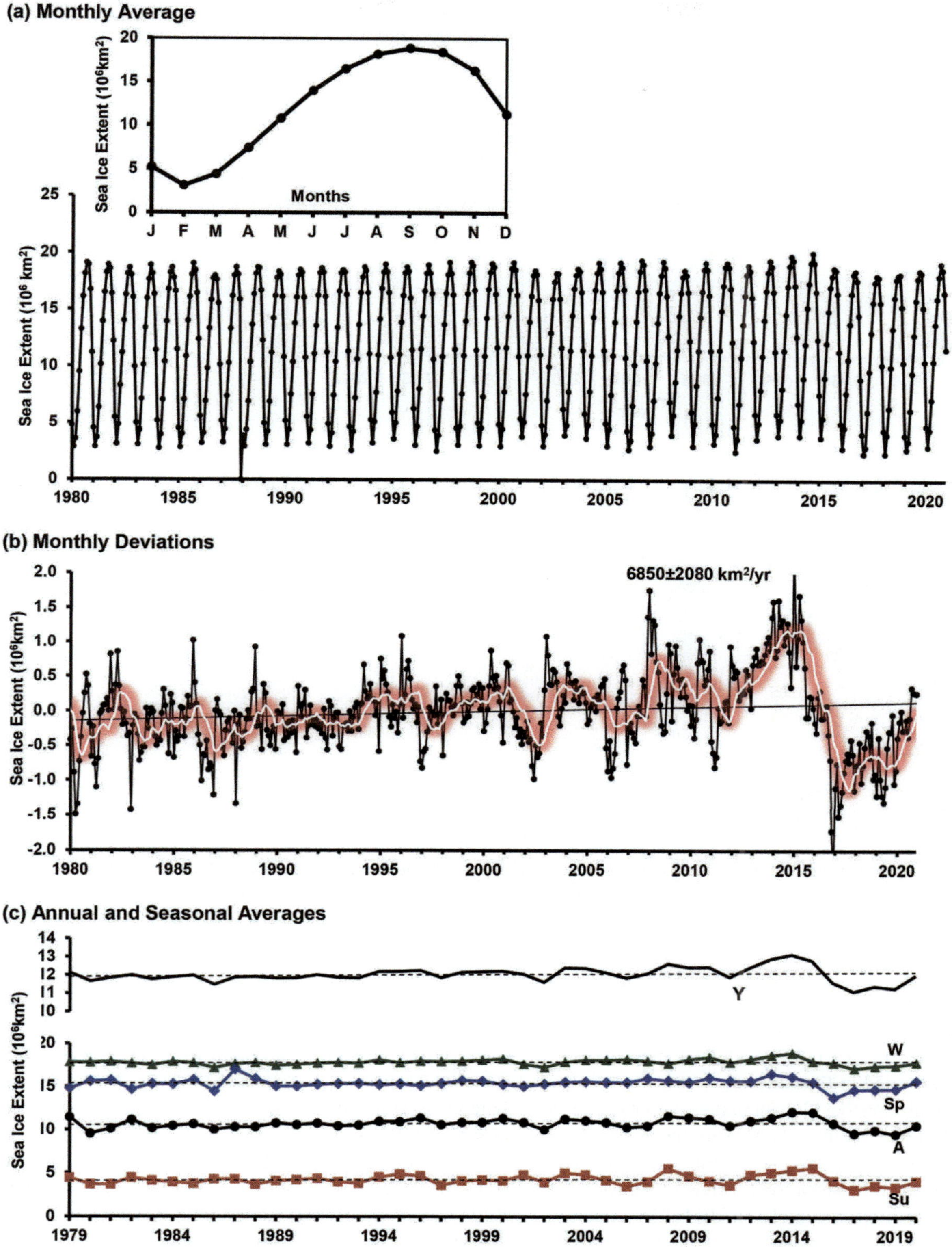


**Figure 11.2** Timeseries of SIE derived from SMMR, SSMI, and SSMIS satellite data. (a) Southern Ocean SIE monthly averages from January 1979 to December 2012, with an inset illustrating the average annual cycle; (b) monthly deviations for the SIE in (a), along with the slope and standard deviation of the line of least squares fit through the data points and 12 month moving average; and (c) yearly (Y) and seasonal averages of sea-ice extents, with the corresponding lines of least-square fit. Summer (Su), autumn (A), winter (W), and spring (Sp) values correspond to the months of JFM, AMJ, JAS, and OND, respectively.

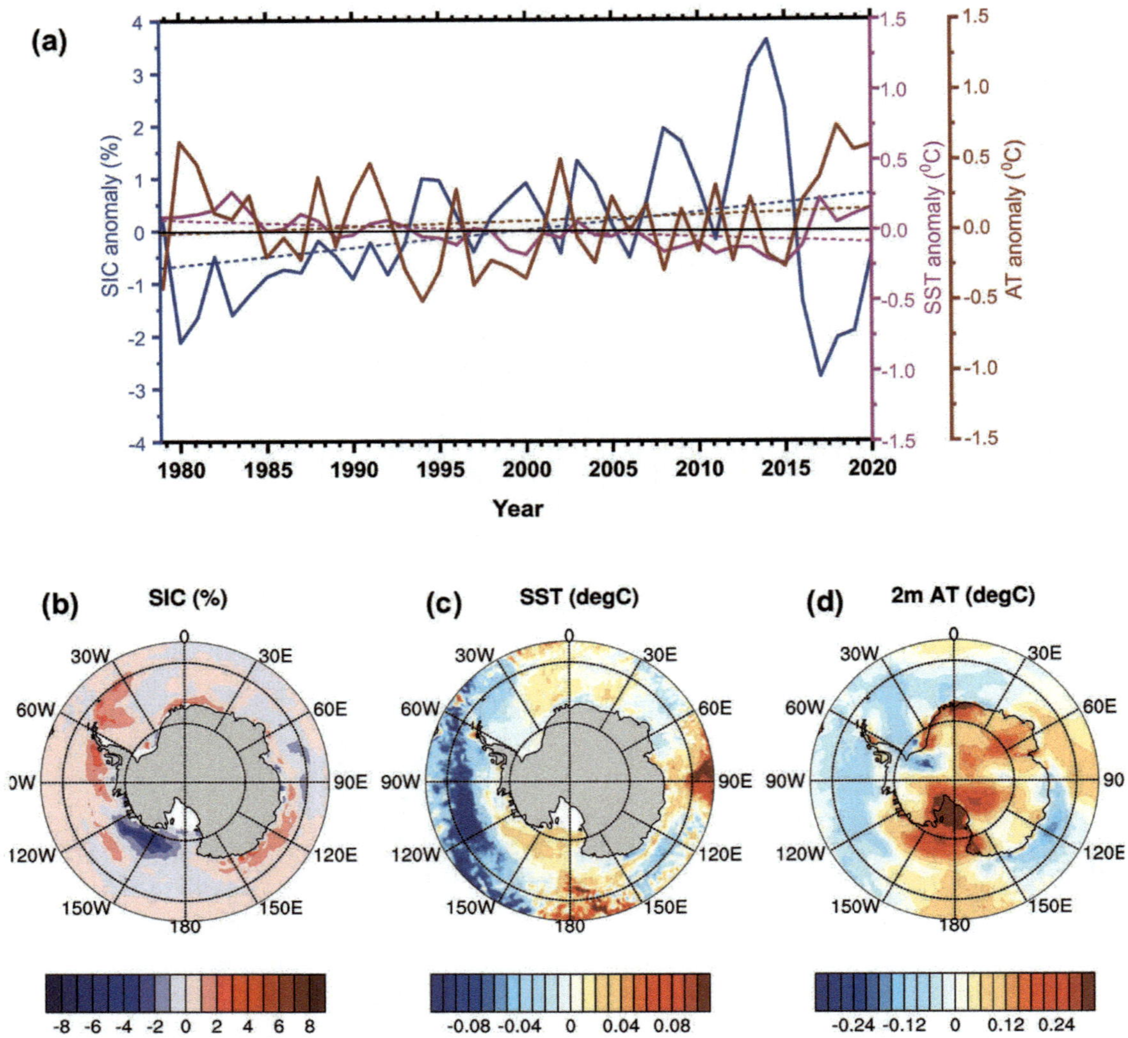


**Figure 11.3** (a) Timeseries plot showing annual anomalies and trends (dashed line) for SIC, SST, and 2 m AT during 1979–2020. Spatial plots represent the 42-year (1979–2020) annual anomalies of (b) SIC (%); (c) SST (°C); and (d) 2 m AT (°C). Anomalies are computed based on the climatology years 1981–2010.

period (Figure 11.4a). The highest sea-ice deduction is caused by an increased surface heat exchange between the atmosphere and the ocean. Considerable negative SIC anomalies are located in the Amundsen Sea, closer to the coast, while high positive SIC anomalies are observed in the Weddell Sea sector (Figure 11.4b). Summer SIC anomalies in the Weddell sector have been found to be positively linked with spring intensification (decreased sea ice loss), which persists until autumn (Holland, 2014). The enhanced wind patterns result in intensification of the Weddell Gyre, which results in the westward advection of sea ice (Kumar et al., 2021). In contrast, the Amundsen Sea has been observed previously to undergo large spring sea-ice loss, which results in a negative summer sea ice concentration. The SST anomaly in comparison to the 30 years climatology increases, with a concentration along the coast (Figure 11.4c). The warming in this region has been attributed to the advection of warm air from the tropics to the pole, which warms the atmosphere over the Antarctica landmass (Holland and Kwok, 2012), and can also be observed in Figure 11.4d.

For autumn, the SIC anomaly shows an overall positive trend, while SST and AT show a negative trend. After 2015, the AT anomaly increased substantially, which is associated with a negative SIC anomaly (Figure 11.5a). For the long-term anomaly with respect to the 30 years (1981–2010) climatology, the spatial analysis for autumn observed similar results with an overall positive SIC anomaly in the Southern Ocean, except the negative anomaly was observed in the Bellingshausen-Amundsen Sea sector (Figure 11.5b).

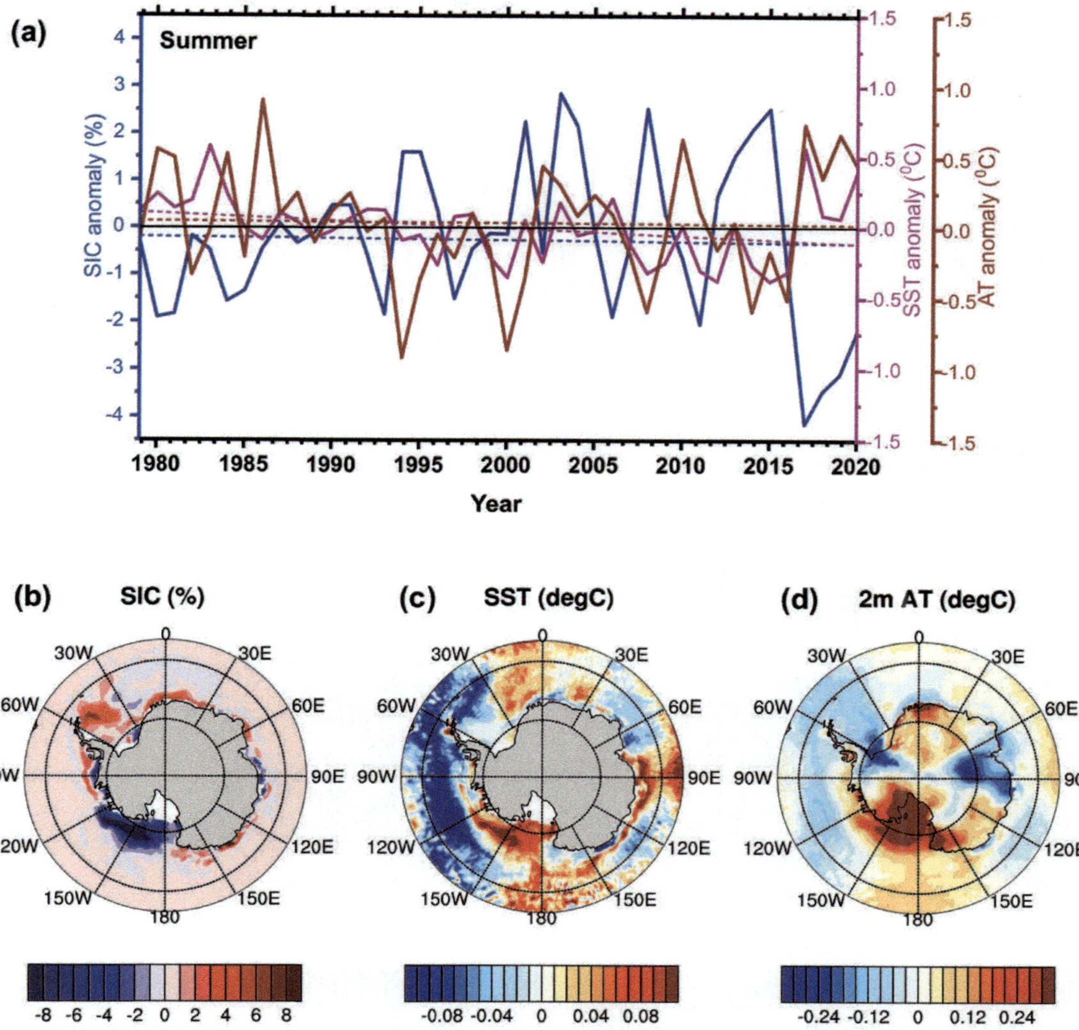


**Figure 11.4** Same as Figure 11.3, but for summer season (JFM).

The SST anomaly trend shows a negative pattern over the last four decades (Figure 11.5c), whereas the AT shows a prominent positive anomaly in West Antarctica, especially closer to the coasts (Figure 11.5d). The decrease in Bellingshausen-Amundsen Sea SIC has been attributed to the advection of a warm and moist air mass from lower latitudes to the Bellingshausen sea region, resulting in melting (Stammerjohn et al., 2012). A major SIC increase can be observed west of the Weddell Sea, primarily attributable to the advancement of the ice edge following the westerlies during the autumn season (Holland, 2014).

The winter SIC anomaly shows a significant positive trend, whereas SST and AT trends have remained constant over the last 42 years. The AT shows inconsistency in its anomaly throughout the years, whereas after 2016, an increase in SST can be observed (Figure 11.6a). In winter, the computed spatial anomaly of SIC shows an overall positive and maximum in the Western Pacific Ocean and Bellingshausen-Amundsen Sea sectors, while it is negative in the Indian Ocean and Ross Sea sectors (Figure 11.6b). This can be attributed to the significant positive AT anomaly observed in these regions (Figure 11.6d). The SST anomaly exhibits a general negative trend in winter (Figure 11.6c). During cold seasons, the surface heat flux is directed from the ocean to the atmosphere, which results in the cooling of the sea surface, thus creating an ideal environment for sea-ice formation. According to (Hobbs et al., 2020), strong cooling during the winter season resulted in higher negative SIC-AT relationships.

In spring, the SIC anomaly shows a positive trend, and the SST anomaly is almost constant with the lowest trend. The positive AT anomaly can be attributed to the substantial positive trend, especially in recent decades (Figure 11.7a). The spatial computed SIC anomaly exhibits an

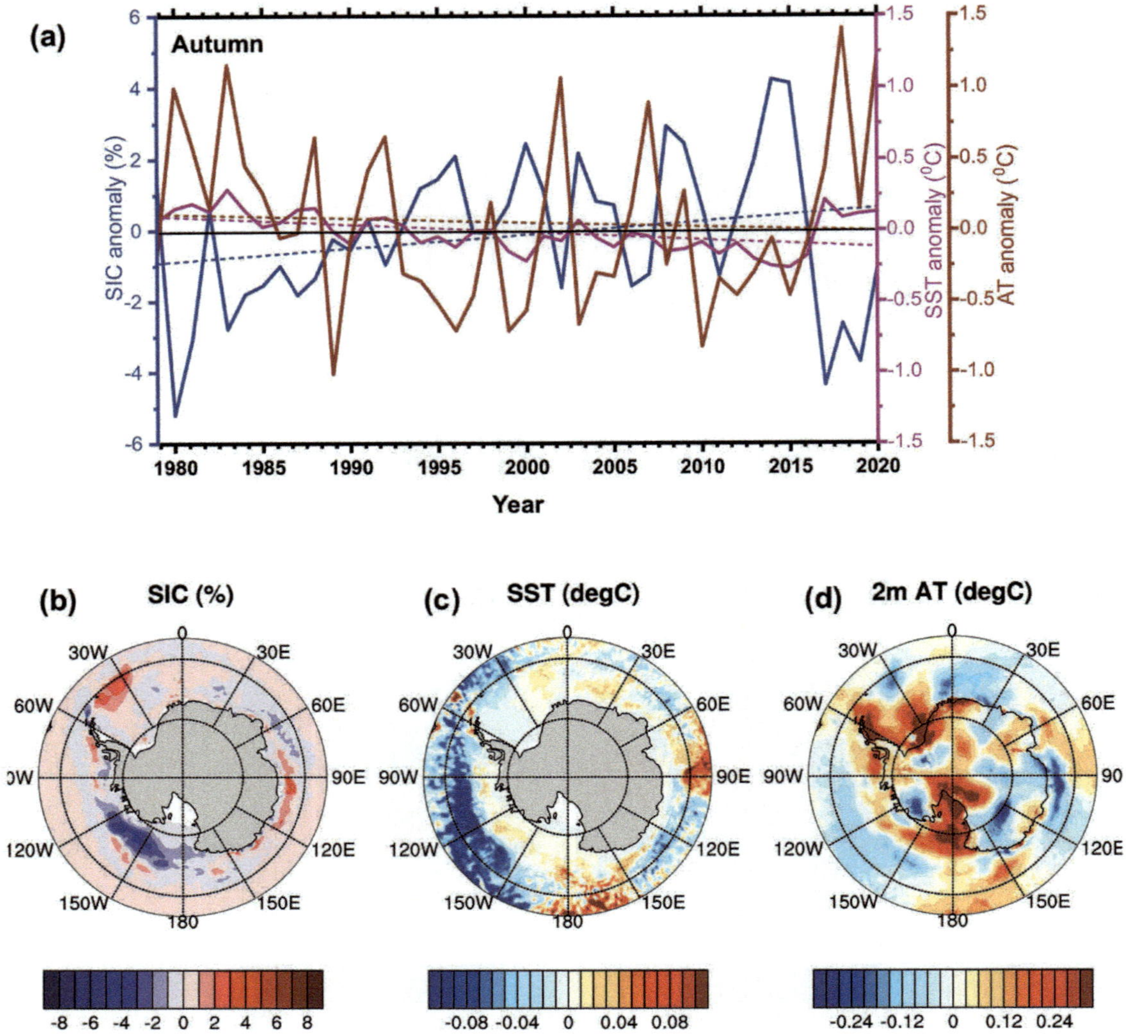


**Figure 11.5** Same as Figure 11.3, but for autumn season (AMJ).

overall positive trend, especially along the Antarctic coastlines (Figure 11.7b), but a decrease in the magnitude of the positive anomaly can be observed compared to previous seasons (Figure 11.6b) indicating the onset of the melt season or spring retreat. The SST anomaly shows a dipole in the Southern Ocean, with a positive anomaly in the east and a negative anomaly in the west (Figure 11.7c). However, the AT anomaly is mostly positive except in a few regions such as the Bellingshausen Sea, the Western Weddell Sea, and the Western Pacific Ocean sectors (Figure 11.7d). Sea-ice changes in spring are substantially associated with zonal wind trends (Holland, 2014). Furthermore, earlier studies reveal that the SAM is strongest in spring and winter, resulting in significant sea-ice retreat in various sectors (Fogt et al., 2012; Raphael and Hobbs, 2014).

Sea-ice variability can be connected with the interaction of ocean and atmospheric drivers on regional and seasonal scales. Gordon, (1981) used the Ekman divergence model to explain the mechanism of seasonal sea-ice growth and decline. Cyclonic wind in the Southern Ocean causes Ekman divergence of sea ice, thus creating open water commonly known as "leads." During the winter months, an ascending heat flux toward the atmosphere fills these leads to the filling of freshly formed ice. This enables the sea ice to form in the interiors rather than on the periphery of the ice edge. As the ice advances northwards, melting is induced by local heat balances, and sea ice gradually retreats. During warm months, these "leads" also play an important role in melting sea ice. Even with higher insolation during summer, the high albedo of sea ice counteracts radiation, resulting in little melt over the sea ice cover. However, open spaces between sea ice such as leads absorb incoming solar radiation and warm the ocean, thus melting the sea ice from below (Stammerjohn et al., 2012). Several studies indicate that sea ice can advance

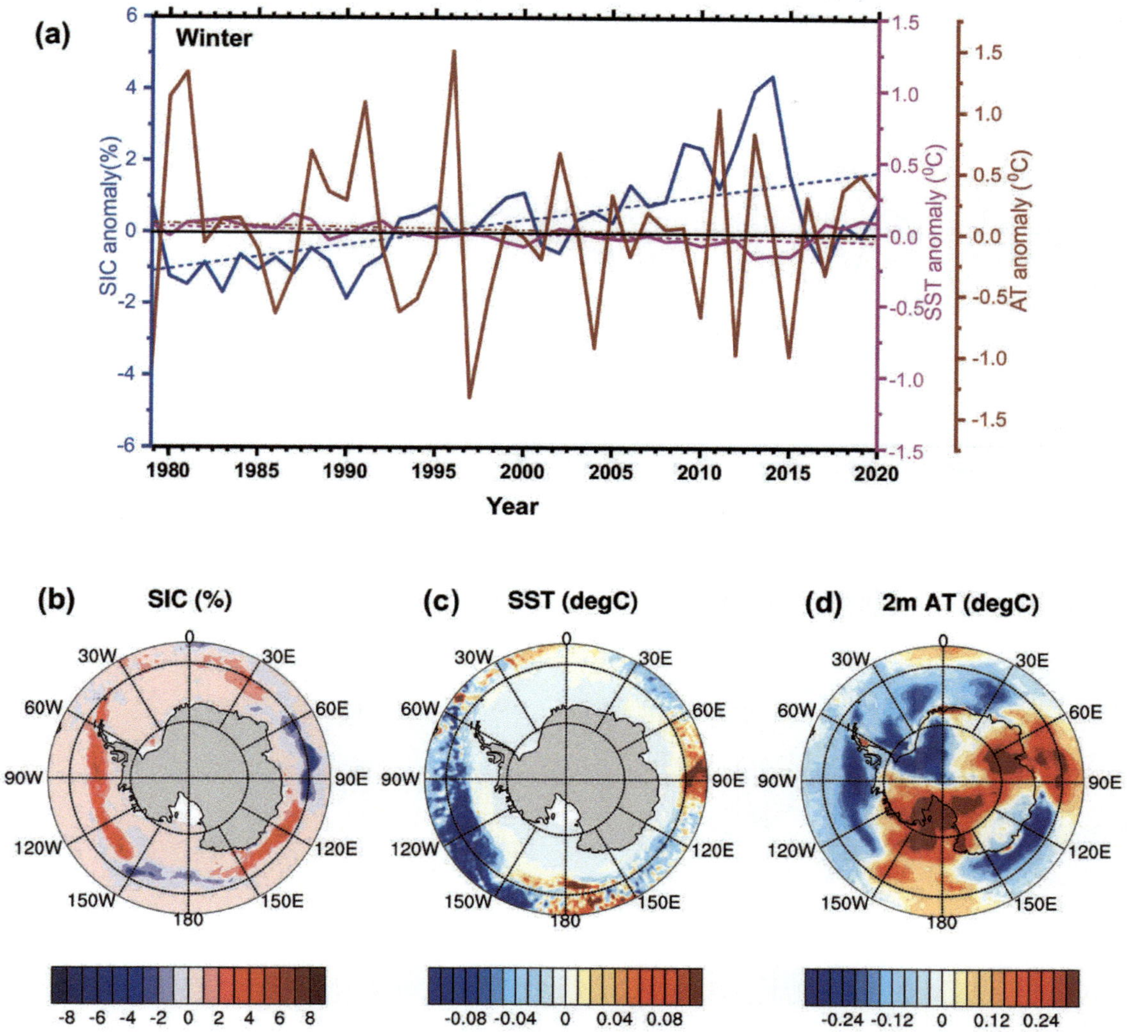


**Figure 11.6** Same as Figure 11.3, but for winter season (JAS).

under warm atmospheric conditions (Zhang, 2007; Bintanja et al., 2013). They argue that when AT increases, downward longwave radiation increases, causing the sea surface and increasing sea-ice melt. This results in reduced brine rejection leading to a decrease in ocean salinity and density. This, in turn, results in ocean stratification, which restricts upward heat transport in the water, leading to sea-ice melting. (Hobbs et al., 2016).

## 11.4 Summary and Conclusions

Sea-ice variability in Antarctica is associated with intricate interactions between the ice, ocean, and atmosphere. This study has addressed some of the key factors that have an important role in influencing the sea-ice variability of the Southern Ocean on a monthly, interannual, and seasonal scale from 1979 to 2020. After the unprecedented decline in 2016, the Southern Ocean can be assumed to be in a "restoration mode," showing a trend of 0.56% per decade. Even though the Southern Ocean has had a positive trend over the last 42 years, the trend value has been observed to vary for different periods. It has been calculated that the 42-year trend is nearly 70% less than what was obtained from 1979 to 2014. After 2015, the dramatic reduction in SIC was attributed to multiple interconnected ocean–atmosphere interactions. Strong northerly warm airflow into the higher latitudes, and the negative phase of SAM and El Niño conditions lingered over the summer of 2016.

Sea ice shows a distinct seasonal cycle, with the minimum summer extent occurring in February and the maximum winter extent typically occurring in September. It has been estimated that approximately 90% of Antarctic sea ice melts during the summer each year. We investigated the

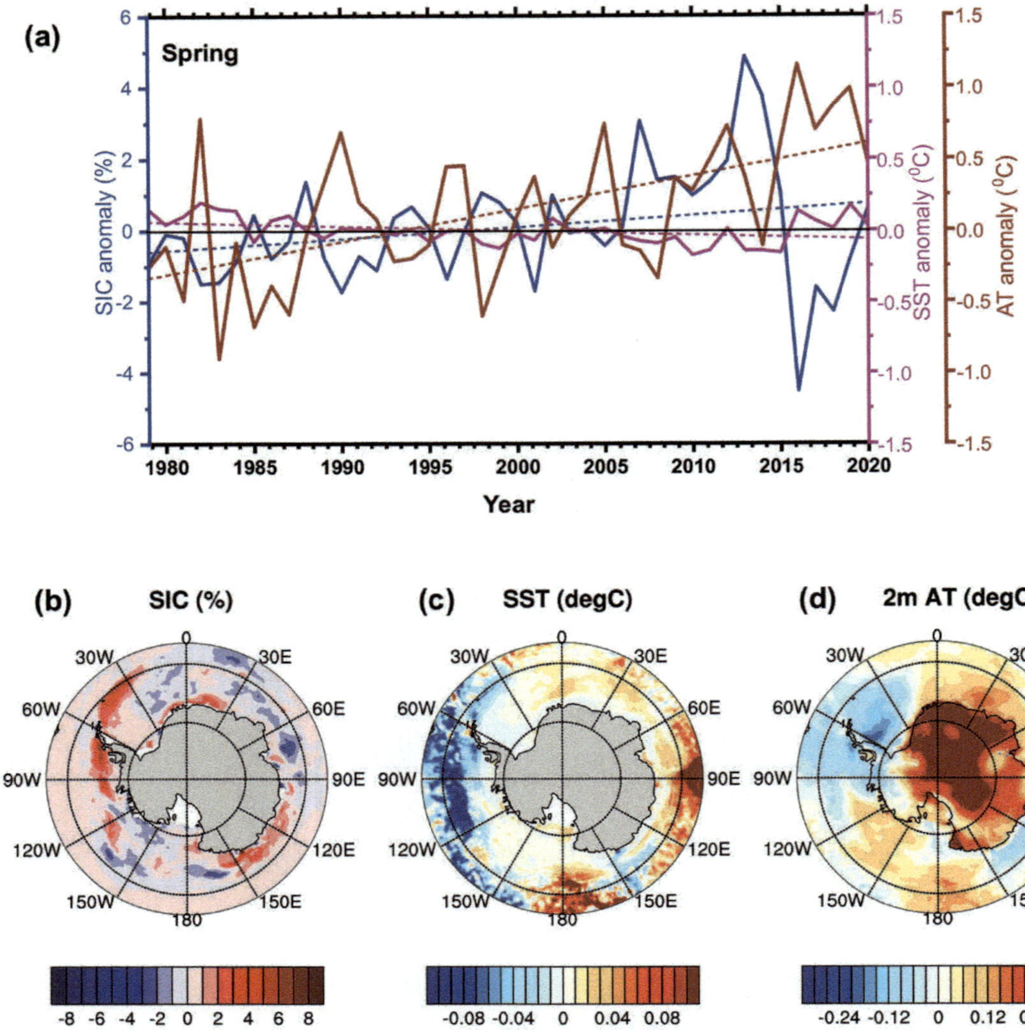


**Figure 11.7** Same as Figure 11.3, but for the spring season (OND).

seasonal character of sea ice in the Southern Ocean and observed a strong negative relationship with ocean–atmosphere temperature. The seasonal characteristics of sea ice and their relationship to air-sea temperature are more precise and understood at the regional level than at the global level, as each sector exhibits distinct spatial patterns. It has been observed that Southern Ocean Sea ice melting begins with the onset of spring (OND), known as the spring retreat, reaches a seasonal minimum in summer (JFM), and then begins advancement in autumn (AMJ), eventually reaching a maximum in winter (JAS). The study reveals that with ongoing data modeling and reanalysis techniques, we may be able to accurately monitor ocean-atmosphere-anthropogenic drivers and their relationship to the Antarctic climate, providing a tool for monitoring and forecasting future sea-ice variations.

## Acknowledgments

We gratefully acknowledge Director, National Centre for Polar and Ocean Research (NCPOR) and Ministry of Earth Sciences, for the continuous support and encouragement. Swathi M thanks the Council of Scientific & Industrial Research (CSIR), India, for the award of the Junior Research Fellowship. Juhi Yadav thanks the University Grants Commission (UGC), India, for the award of the Senior Research Fellowship. We greatly acknowledge various organizations such as the National Snow and Ice Data Center (NSIDC), National Oceanic and Atmospheric Administration (NOAA), National Centre for Atmospheric Research (NCAR), and European Centre for Medium Range Weather Forecast (ECMWF) for making various datasets available in their portals. The authors thank the Editor/s of the book, for their insightful comments and suggestions. This is NCPOR contribution no B-2/2022–23.

## References

Alkama, R., Koffi, E.N., Vavrus, S.J. et al. (2020). Wind amplifies the polar sea ice retreat. *Environmental Research Letters* 15(12): 1–16. doi: 10.1088/1748-9326/abc379.

Arndt, S. and Haas, C. (2019). Spatiotemporal variability and decadal trends of snowmelt processes on Antarctic sea ice observed by satellite scatterometers. *Cryosphere* 13(7): 1943–1958. doi: 10.5194/tc-13-1943-2019.

Bertler, N.A.N., Barrett, P.J., Mayewski, P.A. et al. (2004). El Niño suppresses Antarctic warming. *Geophysical Research Letters* 31(15): L15207. doi: 10.1029/2004GL020749

Bintanja, R., Van Olderborgh, G.J., Drijfhout, S.S. et al. (2013). Important role for ocean warming and increased ice-shelf melt in Antarctic sea-ice expansion. *Nature Geoscience* 6(5): 376–379. doi: 10.1038/ngeo1767.

Bromwich, D.H., Nicholas, J.P., Monaghan, A.J. et al. (2013). Central West Antarctica among the most rapidly warming regions on Earth. *Nature Geoscience* 6(2): 139–145. doi: 10.1038/ngeo1671.

Cavalieri, D.J., Gloersen, P., Parkinson, C.L. et al. (1997). Observed hemispheric asymmetry in global sea ice changes. *Science* 278(5340): 1104–1106. doi: 10.1126/science.278.5340.1104.

Cavalieri, D.J. (1999). Deriving long-term time series of sea ice cover from satellite passive-microwave multisensor data sets. *Journal of Geophysical Research* 104(C7): 15803. doi: 10.1029/1999JC900081.

Cavalieri, D.J. and Parkinson, C.L. (2008). Antarctic sea ice variability and trends, 1979–2006. *Journal of Geophysical Research* 113(C7): 1–19. doi: 10.1029/2007jc004564.

Clem, K.R., Renwick, J.A., and Mcgregor, J. (2017). Large-scale forcing of the Amundsen Sea low and its influence on sea ice and west Antarctic temperature. *Journal of Climate* 30(20): 8405–8424. doi: 10.1175/JCLI-D-16-0891.1.

Cobb, K M., Charles, C.D., Cheng, H. et al. (2003). El Niño/ Southern Oscillation and tropical Pacific climate during the last millennium. *Nature* 424(6946): 271–276. doi: 10.1038/nature01779.

Comiso, J.C. and Nishio, F. (2008). Trends in the sea ice cover using enhanced and compatible AMSR-E, SSM/ I,and SMMR data. *Journal of Geophysical Research* 113(C2): C02S07. doi: 10.1029/2007JC004257.

Fogt, R.L., Jones, J.M., and Renwick, J. (2012). Seasonal zonal asymmetries in the southern annular mode and their impact on regional temperature anomalies. *Journal of Climate*. 25(18): 6253–6270. doi: 10.1175/JCLI-D-11-00474.1.

Gloersen, P., Campbell, W.J., Cavalieri, D.J. et al. (1992). *Arctic and Antarctic Sea Ice, 1978–1987: Satellite Passive-Microwave Observations and Analysis*. Spec. Publ., 511, 290. Washington, DC: NASA, 1992

Goosse, H. and Zunz, V. (2014). Decadal trends in the Antarctic sea ice extent ultimately controlled by ice-ocean feedback. *Cryosphere* 8(2): 453–470. doi: 10.5194/tc-8-453-2014.

Gordon, A.L. (1981). Seasonality of Southern Ocean sea ice. *Journal of Geophysical Research* 86(C5): 4193–4197. doi: 10.1029/JC086iC05p04193.

Hobbs, W.R., Massom, R., Stammerjohn, S. et al. (2016). A review of recent changes in Southern Ocean sea ice, their drivers and forcings. *Global and Planetary Change* 143: 228–250. doi: 10.1016/j.gloplacha.2016.06.008.

Hobbs, W.R., Klekociuk, A.R., and Pan, Y. (2020). Validation of reanalysis Southern Ocean atmosphere trends using sea ice data. *Atmospheric Chemistry & Physics* 20(23): 14757–14768. doi: 10.5194/acp-20-14757-2020.

Holland, M.M., Bitz, C.M., and Hunke, E.C. (2005). Mechanisms forcing an Antarctic dipole in simulated sea ice and surface ocean conditions. *Journal of Climate* 18(12): 2052–2066. doi: 10.1175/JCLI3396.1.

Holland, P.R. (2014) The seasonality of Antarctic sea ice trends. *Geophysical Research Letters* 41(12): 4230–4237. doi: 10.1002/2014GL060172.

Holland, P.R. and Kwok, R. (2012). Wind-driven trends in Antarctic sea-ice drift. *Nature Geoscience* 5(12): 872–875. doi: 10.1038/ngeo1627.

Jacobs, S.S. and Comiso, J.C. (1993). A recent sea-ice retreat west of the Antarctic Peninsula. *Geophysical Research Letters* 20(12): 1171–1174. doi: 10.1029/93GL01200.

Kumar, A., Yadav J., and Mohan R. (2020). Global Warming Leading to Alarming Recession of the Arctic Sea-Ice Cover: Insights from Remote Sensing Observations and Model Reanalysis. *Heliyon* 6(7):e04355. doi: 10.1016/J.HELIYON.2020.E04355.

Kumar, A., Yadav J., and Mohan R. (2021). Spatio-Temporal Change and Variability of Barents-Kara Sea Ice, in the Arctic: Ocean and Atmospheric Implications. *Science of the Total Environment* 753:142046. doi: 10.1016/j.scitotenv.2020.142046.

Kumar, A., Yadav, J., and Mohan, R. (2021). Seasonal sea-ice variability and its trend in the Weddell Sea sector of West Antarctica. *Environmental Research Letters* 16(2): 024046. doi: 10.1088/1748-9326/ABDC88.

Kusahara, K., Williams, G.D., Massom, R. et al. (2017). Roles of wind stress and thermodynamic forcing in recent trends in Antarctic sea ice and Southern Ocean SST: an ocean-sea ice model study. *Global and Planetary Change* 158: 103–118. doi: 10.1016/j.gloplacha.2017.09.012.

Lefebvre, W. and Goosse, H. (2005). Influence of the Southern Annular Mode on the Sea Ice-ocean System: the role of the thermal and mechanical forcing, European Geosciences

Union. *Ocean Science* 1(3): 145–157. https://doi.org/10.5194/os-1-145-2005

Massom, R.A. and Stammerjohn, S.E. (2010). Antarctic sea ice change and variability: physical and ecological implications. *Polar Science* 4(2): 149–186. doi: 10.1016/j.polar.2010.05.001.

Meehl, G.A., Arblaster, J.M., Chung, C.T.Y. et al. (2019). Sustained ocean changes contributed to sudden Antarctic sea ice retreat in late 2016. *Nature Communications* 10(1): 1–9. https://doi.org/10.1038/s41467-018-07865-9

Parish, T.R. and Bromwich, D.H. (1998). A case study of Antarctic katabatic wind interaction with large-scale forcing. *Monthly Weather Review* 126(1): 199–209. doi: 10.1175/1520-0493(1998)126<0199:ACSOAK>2.0.CO;2.

Parkinson, C.L. (2002). Trends in the length of the Southern Ocean sea-ice season, 1979–99. *Annals of Glaciology* 34: 435–440. doi: 10.3189/172756402781817482.

Parkinson, C.L. (2004). Southern Ocean sea ice and its wider linkages: insights revealed from models and observations. *Antarctic Science* 16(4): 387–400. doi: 10.1017/S0954102004002214.

Parkinson, C.L. (2019). A 40-y record reveals gradual Antarctic sea ice increases followed by decreases at rates far exceeding the rates seen in the Arctic. *Proceedings of the National Academy of Sciences of the United States of America* 116(29): 14414–14423. doi: 10.1073/pnas.1906556116.

Parkinson, C.L. and Cavalieri, D.J. (2012). Antarctic sea ice variability and trends, 1979-2010. *Cryosphere* 6(4): 871–880. doi: 10.5194/tc-6-871-2012.

Parkinson, C.L. and DiGirolamo, N.E. (2021). Sea ice extents continue to set new records: Arctic, Antarctic, and global results. *Remote Sensing of Environment* 267: 112753. doi: 10.1016/j.rse.2021.112753.

Raphael, M.N. and Hobbs, W. (2014). The influence of the large-scale atmospheric circulation on Antarctic sea ice during ice advance and retreat seasons. *Geophysical Research Letters* 41(14): 5037–5045. doi: 10.1002/2014GL060365.

Santer, B.D. (2000). Statistical significance of trends and trend differences in layer-average atmospheric temperature time series. *Journal of Geophysical Research Atmospheres* 105(D6): 7337–7356. doi: 10.1029/1999JD901105.

Schroeter, S., Hobbs, W., and Bindoff, N.L. (2017). Interactions between Antarctic sea ice and large-scale atmospheric modes in CMIP5 models. *The Cryosphere* 11(2): 789–803. doi: 10.5194/tc-11-789-2017.

Scott, R.C., Nicolas, J.P., Bromwich, D.H. et al. (2019). Meteorological drivers and large-scale climate forcing of West Antarctic surface melt. *Journal of Climate* 32(3): 665–684. doi: 10.1175/JCLI-D-18-0233.1.

Siegert, M., Atkinson, A., Banwell, A. et al. (2019). The Antarctic Peninsula under a 1.5°C global warming scenario. *Frontiers in Environmental Science* 7: 1–7. doi: 10.3389/fenvs.2019.00102.

Stammerjohn, S., Massom, R., Rind, D. et al. (2012). Regions of rapid sea ice change: an inter-hemispheric seasonal comparison. *Geophysical Research Letters* 39(6): 1–8. doi: 10.1029/2012GL050874.

Stammerjohn, S.E. and Smith, R.C. (1997). Opposing Southern Ocean climate patterns as revealed by trends in regional sea ice coverage. *Climatic Change 37(4): 617–639.* doi: 10.1023/A:1005331731034.

Stuecker, M.F., Bitz, C.M., and Armour, K.C. (2017). Conditions leading to the unprecedented low Antarctic sea ice extent during the 2016 austral spring season. *Geophysical Research Letters* 44(17): 9008–9019. doi: 10.1002/2017GL074691.

Turner, J., Phillips, T., Hosking, J.S. et al. (2013). The Amundsen Sea low. *International Journal of Climatology* 33(7): 1818–1829. doi: 10.1002/joc.3558.

Turner, J., Phillips, T., Marshall, G.J. et al. (2017). Unprecedented springtime retreat of Antarctic sea ice in 2016. *Geophysical Research Letters* 44(13): 6868–6875. https://doi.org/10.1002/2017GL073656

Uotila, P., Vihma, T., Pezza, A.B. et al. (2011). Relationships between Antarctic cyclones and surface conditions as derived from high-resolution numerical weather prediction data. *Journal of Geophysical Research Atmospheres* 116(7): 1–14. doi: 10.1029/2010JD015358.

Wang, Z., Turner, J., Wu, Y. et al. (2019). Rapid decline of total Antarctic sea ice extent during 2014–16 controlled by wind-driven sea ice drift. *Journal of Climate* 32(17): 5381–5395. https://doi.org/10.1175/JCLI-D-18-0635.1

Wille, J.D., Favier, V., Dufour, A. et al. (2019). West Antarctic surface melt triggered by atmospheric rivers. 12(11): 911–916. https://doi.org/10.1038/s41561-019-0460-1

Yadav, J., Kumar A., Srivastava, A. et al. (2022). Sea Ice Variability and Trends in the Indian Ocean Sector of Antarctica: Interaction with ENSO and SAM. *Environmental Research* 212(PD):113481. doi: 10.1016/j.envres.2022.113481.

Yu, L.J., Zhong, S.Y., Sui, C.J. et al. (2021). Synoptic mode of Antarctic summer sea ice superimposed on interannual and decadal variability. *Advances in Climate Change Research* 12(2): 147–161. doi: 10.1016/j.accre.2021.03.008.

Yuan, X. and Martinson, D.G. (2001). The Antarctic dipole and its predictability. *Geophysical Research Letters* 28(18): 3609–3612. doi: 10.1029/2001GL012969.

Zhang, J. (2007). Increasing Antarctic sea ice under warming atmospheric and oceanic conditions. *Journal of Climate* 20(11): 2515–2529. doi: 10.1175/JCLI4136.1.

Zwally, H.J. (2002). Variability of Antarctic sea ice 1979–1998. *Journal of Geophysical Research* 107(C5): 9–1. doi: 10.1029/2000jc000733.

Zwally, H.J., Parkinson, C.L., and Comiso, J.C. (1983). Variability of Antarctic sea ice and changes in carbon dioxide. *Science* 220(4601): 1005–1012. doi: 10.1126/science.220.4601.1005.

# Section III

## Himalayas: The Third Pole Environment and Remote Sensing

# 12

# Some Unresolved Problems in the Himalaya

## A Synoptic View

*Om N. Bhargava**

*Centre of Advanced Studies in Geology, Panjab University, Chandigarh 160014*
** Corresponding authors*

## 12.1 Introduction

The first account of the regional geology of the Himalaya between the rivers "Ravee" and "Ganges" was by Medlicott (1864). Later, he was followed by many illustrious geologists, notable among them McMahon (1882), Oldham (1888, 1918), Hayden (1904), Middlemiss (1910), Pilgrim and West (1928), Wadia (1931), Heim and Gansser (1939); Auden (1934), West (1939), and Gansser (1964). Their contributions mainly pertained to the western Himalaya. During the Second World War, there was lull in geological activities. The mapping was half-heartedly resumed in the early 1950s, gaining fresh impetus after the reorganization of the Geological Survey of India (GSI) in 1961. Several universities also participated in research activities in the Himalaya, which so far have been monopolized by the GSI. Since then, enough water has flown down the Ganga and Ravi.

Acharyya (1978), Valdiya (1996), Bhargava (1995), Kumar (1997, 2005), Srikantia and Bhargava (1998), Bhargava and Bassi (1998), Dhital (2014), and Jain et al. (2020) summarize the latest state-of-art pertaining to the Himalaya. The geology, like any other science due to continuous addition of new data and technological advancement, needs constant revision of old concepts and so no word can be final. Thus, despite extensive contributions, there is enormous scope of further research as several aspects of this mountain chain have remained untackled.

This chapter catalogues some of the important topics awaiting resolution. It is hoped that once the right questions have been framed, sooner or later, these will provoke new thinking, open new vistas for future research, and finally right answers will emerge with new questions.

The unresolved problems are classified under: i) stratigraphy and palaeontology; ii) sedimentology; iii) tectonics and structure; iv) magmatism and geochronology; v) metamorphism; vi) mineral deposits; vii) paleomagnetic studies; viii) glaciological studies; and ix) geomorphological studies.

## 12.2 Stratigraphic Ages, Basin Configuration, and Palaeontology

The stratigraphic age of unfossiliferous sequences until recently has been conjectural. In recent decades, dating of detrital zircons (DZ) has been widely used to determine the age of these sediments. DZ as a chronometer provides the maximum age of the sediments, but the question of how young the sediment enclosing the dated DZ are remains unanswered. The method has limitation; for example, if the Bundelkhand granite is exclusive provenance then the sediments regardless of their age will yield DZ of the Archaean age! The Kashala Formation (Alpurai Group, Northern Pakistan) has yielded just one zircon of 229 Ma (Lutfi et al., 2021), based on which the Kashala Formation is assigned a late Triassic or younger age. Fortuitously, if even one 229 Ma zircon was not detected, the age assignment to the Kashala Formation could be different! Also vast and/or coeval basins may have provenances of different ages and as a result these will have DZ of different ages, yet the sediments in the basin may be of the same age. In this context it is necessary to systematically sample the sequences and date the DZ of underlying and overlying sediments in different sections along the strike.

*Advances in Remote Sensing Technology and the Three Poles,* First Edition. Edited by Manish Pandey, Prem C. Pandey, Yogesh Ray, Aman Arora, Shridhar D. Jawak, and Uma K. Shukla.

Even in fossiliferous sequences in the Tethyan Himalaya, the finer stratigraphic resolution is lacking, which hampers precise event and sequence stratigraphic reconstruction. Precise age limits of the Palaeozoic Thango, Takche, and Ganmachidam formations and the Mesozoic Kioto Group of the Spiti Valley (Bhargava and Bassi, 1998; Bhargava et al., 2004) are debatable. The Thango and the Ganmachidam formations, in general, lacking body fossils, are variously assigned lower or upper Ordovician age and late Carboniferous or early Permian, respectively, and the upper age limit of the Takche Formation is assigned Llandovery or Wenlock age. Based on correlation of the Ralam Conglomerate with the Thango and discovery of *Redlichia* from the underlying Milam Formation (Kacker et al., 1993), the Garbyang Formation resting over the Ralam was assigned to Ordovician; however, based on geochemical attributes, the Garbyang is reassigned a Cambrian age (Ansari et al., 2019). Similar controversies exist for contemporary sequences in other parts of the Tethyan sector. Save, the Permian/ Triassic (Nakazawa et al., 1975; Ghosh et al., 2016) and Jurassic/ Cretaceous (Pandey and Pathak, 2017; Pathak, 2007) demarcation of other stratigraphic boundaries i.e., Ordovician-Silurian, Devonian-Carboniferous, and Triassic-Jurassic, has not received due attention. The following sections are ideal for the delineation of various boundaries: i) the Koti Dhaman section in the Nigali Dhar Syncline (Sirmaur district) for the Precambrian/Cambrian boundary (Bhargava et al., 2021); ii) Manchap (Tidong Valley, Kinnaur) and the Pin Valley (Spiti) sections for the Ordovician/Silurian boundary (Bhargava and Bassi, 1998); iii) Takche section (Spiti) for the Devonian/Carboniferous (Bhargava and Bassi, 1998); iv) Ganmachidam Hill (Spiti) for the Carboniferous/Permian boundary (Bhargava and Bassi, 1998); and v) lower Pin Valley for the Triassic/Jurassic boundary.

Likewise, controversy persists regarding the Eocene-Oligocene boundary. The Subathu Formation, based on foraminiferal remains, is assigned a Middle Lutetian age. The overlying Dagshai Formation was assigned a Miocene age on the basis of vertebrate fauna recovered from the Murree, considered to be the equivalent of the Dagshai (Pilgrim, 1910). This fauna has now been re-dated as Oligocene (Marivaux et al., 1997). Bhatia and Bhargava (2006) regarded a gradual passage from the Subathu to the Dagshai Formation. Based on detrital mica (Najman et al., 1997) and fission track (Jain et al., 2009) studies, a 10 My break was postulated between these two formations. Samples of Najman et al., (1997) are from Morni and Jamuna Valley, at both these localities the rock exposed is the Nahan Formation. Both localities are, therefore, dubious. The Dagshai rocks have been heated to temperatures higher than the annealing temperature of zircon; hence, Jain et al.'s (2009) fission track data are not valid (Bhargava, 2020). Retallack et al. (2018) reported a paleosol along the Subathu-Dagshai contact. It is worthwhile to have detailed integrated palaeontological and sedimentological investigations of the above-mentioned boundary sequences to arrive at definite conclusions.

Major and regional anomalies exist in the Palaeozoics and Mesozoics of the Tethyan part in the Himalaya, which so far have not attracted attention of any worker. These pertain to the Ordovician and Triassic periods. The Ordovician conglomerate (Thango/Ralam Formation) with identical characteristics is developed from Zanskar-Spiti to Kinnaur-Painkhanda (Garhwal). It is absent in the Byans (Kumaon) basin, but reappears in Nepal (Damgad Formation). Similarly, the Triassic sequence (Lilang Supergroup) has similar lithostratigraphy from Zanskar to Painkhanda; it is different in Byans, but in the Nepal Himalaya it is similar to that of Zanskar-Spiti-Kinnaur-Painkhanda (Figure 12.1). More puzzling is the faunal resemblance of the Zanskar-Painkhanda-Nepal with the Oman section and of Byans with that of the Timor section. Why is the Byans Basin, sandwiched between the identical Zanskar-Spiti-Kinnaur-Painkhanda and Nepal, so different?

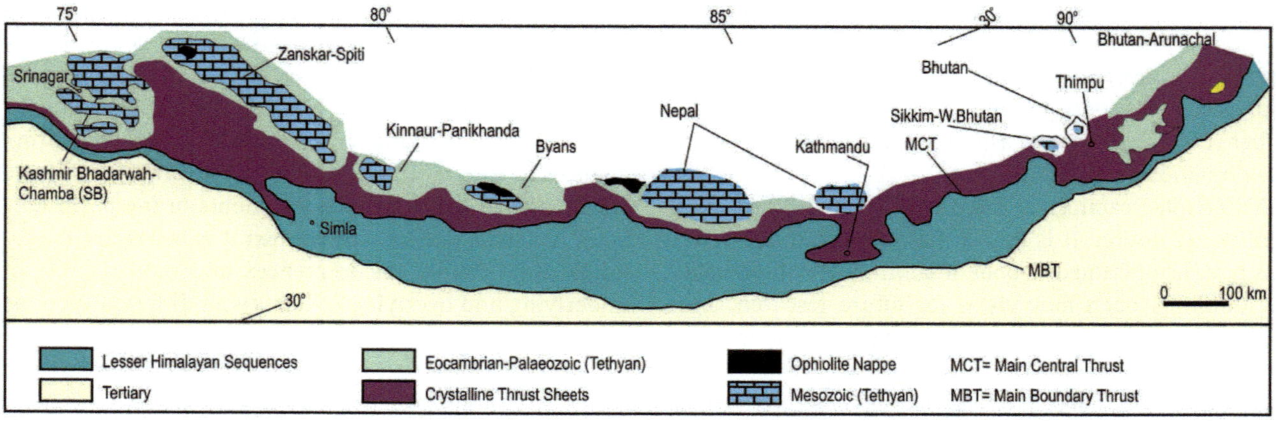


**Figure 12.1** Map of the Himalaya showing positions of various Tethyan basins (modified after Bhargava, O.N. and Singh, B.P. (2020). Geological evolution of the Tethys Himalaya. *Episode* 43(1): 404–416.

Similarly, in the Triassic of the Byans basin, the Carnian and Norian fossils occur in the same horizon. The anomalous mingling is known to exist for more than 100 years (von Kraft in Diener, 1912), yet remains unexplained. There are many possibilities: i) Are the Carnian fossils reworked? ii) Is there an unusual condensation of the stratigraphic section? or (iii) Does a disconformity exist?

Several branches of palaeontology have not been tackled. There is a need for systematic study of palynomorphs, conodonts, chitonozoa, sponge, stromatoporoid, formaminifera, ostracoda, and radiolarian, particularly in the Tethyan sequences. Even old classic palaeontologic data need to be restudied and revaluated in the light of modern concepts and revisions. The Triassic of the Byans has a sizable amount of shale, which may yield rich palynomorph remains. An integrated palyno- and conodont/ammonoid biostratigraphy of the Byans will help in standardizing the Triassic palynostratigraphy, which can be used in the Gondwana sequence where marine fossil control is lacking.

Very little has been done regarding the paleoecology, palaeoenvironment, spatial, and temporal dispersion and evolution of the India fauna in relation to other continents. In some sequences, guide fossils are not preserved; as a result, precise age of the sediments cannot be determined. Methods are now available to find the absolute age of the fossils (McArthur et al., 2012; Meinhold et al., 2020, and references therein). These studies will immensely help in palaeogeographic reconstruction.

## 12.3 Sedimentology

In most cases, the sedimentological studies have been carried out for a particular formation that is also in one or two sections. What is necessary is to take up study of the entire sequence, sandwiched between two unconformities along and across the strike in various sections. This kind of study, associated with lithofacies and isopach maps, will help in working out the evolution of the basin and its depth variation for proper basin analysis studies. Combined with the study of DZ, it can be worked out if the entire basin had one single provenance or different parts had different sources. Except for the Krol Basin (Jiang et al., 2002), such studies are lacking in all the sequences in the Himalaya.

In the Lesser Himalaya, the Rampur-Berinag Group, the Shali Group (and equivalents), the Simla/Jaunsar Group, and the Tal Group are ideal for such studies.

Besides these studies, certain formations whose environment of deposition is highly controversial should be critically examined. These are: i) the Blaini Formation is variously interpreted as glacial, glaciomarine, flysch, mud/debris flows, turbidite, and fanglomerate; ii) the Subathu Formation is regarded as a shallow marine deposit (Singh, 1978) and also as calciturbidite (Bera et al., 2008); iii) the Thango Formation is regarded as fresh water (Bagati et al., 1991) and also shallow marine (Bhargava and Bassi, 1998); iv) the Muth Formation as mid-shelf (Shanker et al., 1993), beach (Bhargava and Bassi, 1998), and sand barrier island (Draganits et al., 2001, 2003); and v) the Giumal Formation is considered a proximal flysch (Kumar et al., 1977) and shallow marine (Bhargava and Bassi, 1998).

These studies may be combined for a holistic basin analysis, event, and sequence stratigraphy for global correlation of sea-level curves and events.

## 12.4 Tectonics and Structure

Until recently, only the Himalayan orogeny, deformation, and metamorphism were recognized, and the earlier tectonic phases mentioned by a few (Srikantia, 1977; Bhargava et al., 2011) were ignored and not taken seriously. The Late Cambrian Kurgiakh orogenic episode (Srikantia, 1977; DeCelles et al., 2000; Wiesmayr and Grasemann, 2002; Gehrels et al., 2003; Bhargava et al., 2011; Cawood, 2017) is now well established. There are, though maybe minor, events that need a thorough probe. For example, the Blaini Formation rests over all three of the formations of the Simla Group and over the Nagthat and the Mandhali formations. It clearly demonstrates uplift of the Simla and Jaunsar sequences and their erosion to expose different formations to receive the deposition of the Blaini Formation. Did this event involve a pre-Blaini folding too?

The Paleoproterozoic Ramgarh (=Munsiari=Baragaon) gneiss is overlain by the Nathuakhan sequence of the Tonian age (Mandal et al., 2015). This disposition implies that the Ramgarh gneiss was already exhumed and exposed in the Tonian time. Does the Ramgarh gneiss exposure indicate a Precambrian thrusting event (Bhargava, 2000) over which the Nathuakhan Formation was deposited? Presence of the Tonian and Cambrian DZs in the Garhwal metasediments and the Vaikrita sequence also point to their exposure during those periods. Indirectly, such distribution suggests that the thrusting of the Ramgarh Gneiss and Almora thrust sheet was pre- or syn-Tonian, and Almora thrusting was possibly related to the Kurgiakh Orogeny. These aspects need critical evaluation.

Each thrust sheet is characterized by granite of a particular age. The Kulu/Munsiari, the lower-most thrust sheet containing only ~1860 Ma gneisses, and the Jutogh Group of the Himachal Himalaya, unfortunately not differentiated in other sectors and clubbed with the Vaikrita Thrust Sheet, is characterized by Tonian granites, and doubtful Lower Palaeozoic granites, while the upper-most Vaikrita Thrust Sheet encompasses exclusively Lower Palaeozoic granites (Bhargava et al., 2021). Why the granites are so compartmentalized is a moot question?

So far, except in Himachal (Bhargava et al., 2011; Singh et al., 2019) and Nepal (Cawood et al., 2007), the pre-Himalayan deformation has not been categorically identified in other sectors of the Himalaya.

In the current models, the highest thrust sheet (MCT/Vaikrita) is regarded to emerge first, followed by the lower thrust sheet as out of sequence thrust (Stübner et al., 2018). If it were so, the Lower Siwalik would have contained high-grade metamorphic heavy minerals; on contrary, such minerals are present in the Upper Siwalik, indicating reverse order of thrusting, i.e., the low-grade metamorphic thrust sheet to move out first, followed by high-grade crystalline thrust sheets. Is it that these thrust sheets were initiated in the Precambrian (Bhargava et al., 2021), and the present dating represents reactivation in response to the Himalayan orogeny. This aspect requires a critical review.

Neotectonic activity has been recorded in the foothills (Thakur et al., 2020 and references therein), Lesser Himalaya (Bhargava et al., 2010 and references therein), and Tethyan Himalaya (Bhargava, 1990). Except for the foothills, most of the studies are confined to isolated sections. There is a need to carry out studies systematically in all the sectors to weave a cogent neotectonic history. These studies with dating will aid in the related field of paleoseismology, help in delineating areas of potential earthquakes, and work out the geomorphological evolution of the Himalaya.

Balanced cross-sections, except for a few (Dubey et al., 2001; Mukhopadhyay and Mishra, 2004; Baruah and Joshi, 2011; Webb et al., 2011) are almost totally lacking.

## 12.5 Magmatism and Geochronology

Granites and basic rocks of different ages are known in the Himalaya. Except for granites of the Ladakh, adjoining the Indus Suture to some extent, there has been no systematic dating of the magmatic rocks in the Lesser Himalaya, though quite a few granitic rocks have been dated (Le Fort et al., 1983; Singh, 2020, and references therein). In many instances, there are more than one type of granite in a complex (e.g., Chaur Granite; Bhargava and Srikantia, 2014, Jeori-Wangtu Complex), which may represent altogether different events. Most studies, however, do not specify the type of granites that have been dated. Moreover, most of the dates are based on Rb/Sr isotopes, which are no longer recognized. The granites in the Vaikrita Group, though basically of Lower Palaeozoic age, do show variation. Whole-rock petrochemistry, U-Pb, and Lu-Hf isotope studies of these granites are required to find their source together with determination of the time of metamorphism and tectonic discrimination. The isotopic dates of the Lower Palaeozoic granites, by more acceptable modern and robust methods, will help in working out the stages of the Kurgiakh Orogeny.

The situation in the case of basic rocks is worse. So far, dates only for the Rampur (Miller et al., 2001), Bhowali, and Panjal volcanics are available. The age of the Darla volcanics, dated earlier by K/Ar method, is not valid. Besides, there are numerous enclaves of basic/ultrabasic rocks and dikes which have not been dated. The tectonic affinities of the basic rocks should be determined, in conjunction with data on granites, sedimentology, and palaeontology to enable reconstruction of the tectonic history of the Himalaya.

## 12.6 Metamorphism

Conventionally, metamorphism is related to the Himalayan orogeny, though many authors have suggested pre-Himalayan metamorphism (Acharyya, 1979; Thakur and Patel, 2012; Chakrabarti, 2016). Precise age in most of cases is lacking and precise age based on isotopic dates are useful in deciphering the tectonic history of the Himalaya. For example, isotopic date for Jutogh metasediments indicates an Early Palaeozoic metamorphism (Bhargava et al., 2016), and a similar age is also available in Nepal (Gehrels et al., 2006), which adds a new dimension to the Kurgiakh Orogeny. Indirect evidence is available to suggest Paleoproterozoic metamorphism (Bhargava et al., 2016). Careful petrographic study combined with dating of minerals is needed to unravel the geological history of the Himalaya. Sedimentary cycles, particularly the Event stratigraphy, magmatic rocks, and metamorphism form a compact package, which aids in unraveling the geological history; hence, it is necessary to have a precise age of all the component.

Also, so far, no totally acceptable model to explain the inverted grade of metamorphism is available.

## 12.7 Mineral Deposits

All studies should find an applied use, geological studies being no exception. Mineral deposits according to the concept of the current Plate Tectonic Model are formed at a specific sites/locale. It is, thus, necessary to identify favorable locales for mineral modeling, which can be identified only after the various studies mentioned in the aforementioned sections, in addition to isotopic mineral dating. It is due to inadequate knowledge of various components, except for a few preliminary exercises (Bhargava et al., 1991), that no serious mineral modeling has been attempted.

## 12.8 Palaeomagnetic Studies

The comparative studies of palaeomagnetism of the rocks in the Himalaya that have counterparts in the peninsula can help in determining the horizontal translation that the rocks in the Himalaya have undergone, which can also aid in working out the crustal shortening. These shall also help in palaeogeographic reconstruction. The most important will be the study of *Eurydesma*-bearing beds in the Himalaya, as such a horizon also exists in the peninsula.

## 12.9 Glaciological Studies

Inventory of the glaciers in the Himalaya is more or less complete (Vohra, 1978; Dobhal and Kumar, 1996). Selectively, some detailed and modern investigations have been carried out mainly in the western Himalaya (Ali et al., 2013; Bali et al., 2013; Dobhal et al., 2013; Mehta et al., 2013, 2014; Bisht et al., 2015, 2017; Pratap et al., 2015; Sharma et al., 2016; Ganju et al., 2018; Shukla et al., 2018; Kumar et al., 2020, 2021), and a few in the eastern Himalaya (Debnath et al., 2019). It is only appropriate to undertake modern systematic studies of the glaciers all over the Himalaya to find their secular temporal and spatial behavior. Future studies should be carried sequentially at specified time intervals to monitor the rate of recession, and to find if the recession was uniform in all the sectors or did it vary? In the event of variation, it would be interesting to find the causative factor. To find the past rate of recession and climatic changes, it is essential to date old moraines. Only a few glacial moraines have been dated (Owen et al., 1996).

Equally important is to systematically trace the footprints of previous glaciations, some which are believed to have descended to altitudes as low as 1000–1500 m in the Kangra Valley. It may also be necessary to examine the core logs of the Indo-Gangetic plains drilled by the ONGC.

Systematic dating of the earliest moraine in different sectors shall enable correlation of various stages of Himalayan glaciation with those of the Alps and elsewhere. It could be different, just as Palaeolithic, Neolithic, or Bronze dates are different. Glacial dates, if variable, can form another thrust area for research, as glaciation could have affected Human migration.

## 12.10 Geomorphological Studies

Very few geomorphological studies have been carried out according to modern concepts.

Studies of some river basins, which have anthropological and archaeological implications, have been carried out (Ghosh et al., 2019). Only in the Ganga valley has a holistic approach been adopted (Singh, 1996). Similar studies are required in all the river basins. Even in the legendry lost river of the Saraswati, an integrated approach has been missing. The Ghaghar river is supposed to be remnant of the Saraswati, which originates in the Siwalik Hills near Markanda. The concept is faulty, as the river originating in the Siwalik Hills can at best be seasonal and not turbulent as described in the Rig Veda. The terraces of the Ghaghar river contain pebbles of metamorphic rocks (Puri and Verma, 1998). No metamorphic rock is exposed in the present catchment area of the Ghaghar. Obviously, in the past, the river must have coursed through a metamorphic terrain. The nearest country rocks constituted of the metamorphic rocks are in the Tons valley; thus, originally the river must have extended to the Higher Himalaya. The present Tons could have been the original Saraswati, which would have been a glacially-fed turbulent river. Later, the Tons got deflected and joined the Yamuna. The concept could be tested by detailed chemical and isotopic studies of DZ and heavy minerals in the terraces throughout the interpreted course of the Saraswati. The point I wish to make is that lack of an integrated approach led to a wrong interpretation. While reconstructing the history of a river, studies should include detailed lithology of the terraces, their biota, OSL dates, paleocurrent directions (which will be different in the case of drainage reversal/disorganization), and neotectonic features (Shukla et al., 2012).

Another controversial aspect that needs critical appraisal is cause of river aggradation. Was it tectonically or climatically controlled (Srivastava and Misra, 2008; Srivastava et al., 2008; Ray and Srivastava, 2010; Sharma et al., 2016; Kumar and Srivastava, 2017; Kumar et al., 2017; Chahal et al., 2019)?

Lakes in the past fall into two categories: glacial and those formed due to huge slides blocking the rivers. There are very few studies reconstructing past flood histories in the Himalaya (Sharma et al., 2021); also were the burstings triggered by tectonic activities or a normal wear and tear of the dams. Several lacustrine sediments of erstwhile such lakes have been studied. According to OSL dating, these lakes lasted for several thousand years, some even >69 Ka. At the outset, the dates look improbable on two counts: i) in an active mountain, it is unlikely that the dams that bounded the lakes remained intact for such a long duration, and ii) the glaciers of last glacial maxima at 12 Ka would have bulldozed the existing lacustrine/fluvial deposits/terraces; hence, there is little chance for lake beds being older than 12 Ka in the Higher and Tethyan Himalayan domains. There is a need to perfect the dating techniques.

One aspect which has fascinated me is the pre-thrusting geomorphological setup. The Lesser Himalaya in particular and the Higher Himalaya remained positive areas between the Ordovician and the Early Permian, between the Early Permian and the Cretaceous, and between the Cretaceous and the Eocene; presumably, the terrain would have had a well-developed geomorphology. What happened to it when the thrusting was initiated? Was the pre-thrusting topography on the hanging wall translated as piggy back? How was it modified? Did the physiography on the footwall side exercise control on the disposition of the thrust sheets? Some illustrious worker may be able find the answer to these fascinating aspects.

## 12.11 Conclusion

1) We are over-relying on instrumentation. Unless exact location of a sample is known in the regional detailed map and lithocolumn, its analysis may lead to conflicting results, as enumerated in this chapter.
2) In 1976, when Prof A. Gansser had visited Chandigarh, some faculty members of the Panjab University asked him, "How to tackle the problems in the Himalaya?" His reply was, "Map, map and map." This advice is still valid.A single swallow does not make a summer. No interpretation should be based on a single observation/result.
3) With over-specialization in each branch, there is a need to have a multidisciplinary integrated approach for complicated problems.

## Acknowledgments

Thanks are due to Prof U.K. Shukla for inviting me for this contribution. A very painstaking and detailed review by Prof P. Srivastava has greatly improved the manuscript.

## References

Acharyya, S.K. (1978). Stratigraphy and tectonic features of the eastern Himalaya. In: *Tectonic Geology of the Himalaya* (ed. P.S. Salani), 243–269. New Delhi: Today and Tomorrow's Publishers.

Acharyya, S.K. (1979). Pre-Tertiary fabric and metamorphism in eastern Himalaya. In: *Metamorphic Rock Sequences of the Eastern Himalaya* (ed. P.K. Verma), 67–82. New Delhi: Today's and Tomorrow's Publishers.

Ali, S.N., Biswas, R.H., Shukla, A.D. et al. (2013). Chronology and climatic implications of Late Quaternary glaciations in the Goriganga valley, central Himalaya, India. *Quaternary Science Reviews* 73: 59–76.

Ansari, A.H., Singh, I.B., Bhattacharya, S.K. et al. (2019). Note on C and O stratigraphy of the Garbyang Formation (Malla JoharArea), Tethyan Himalaya, India. *Journal of Palaeontological Society of India* 64(2): 266–275.

Ansari, Z., Ahmad, S., and Khan, M.A. (2019). Seasonal variations of streams hydrochemistry and relationships with morphometric/landcover parameters in the Bhagirathi watersheds, Garhwal Himalaya, India. *Journal Geological Society of India* 94: 493–500

Auden, J.B. (1934). Geology of the Krol Belt. *Records Geological Survey of India* 67: 357–454.

Bagati, T.N., Kumar, R., and Ghosh, S.K. (1991). Regressive-transgressive sedimentation in the Ordovician sequence of the Spiti (Tethyan) Basin, Himachal Pradesh, India. *Sedimentary Geology* 73: 171–184.

Bali, R., Ali, S.N., Agarwal, K.K. et al. (2013). Chronology of late Quaternary glaciation in the Pindar Valley, Alaknanda basin, Central Himalaya (India). *Journal of Asian Earth Sciences* 66: 224–233.

Baruah, M., and Joshi, G. (2011). Section balancing in the area south of Solan, H.P., India for reinforcing subsurfacing model related to ongoing hydrocarbon exploration in the area. *Journal of Indian Association of Sedimentologists* 30(1): 1–10.

Bera, M.K., Sarkar, A., Chakraborty, P.P. et al. (2008). Marine to continental transition in Himalayan foreland. *Bulletin Geological Society of America* 120(9/10): 1214–1232. https://doi.org/10.1130/B26265.

Bhargava, O.N. (1990). Holocene tectonics, south of Indus Suture: a consequence of Indian Plate motion. *Tectonophysics* 174: 315–320.

Bhargava, O.N. (1995). The Bhutan Himalaya: a geological account. *Special Publication Geological Survey of India* 39: 1–245.

Bhargava, O.N. (2000). The Precambrian sequences in the western Himalaya. *Special Publication Geological Survey of India* 55: 69–84.

Bhargava, O.N., Krystyn, L., Balini, M. et al. (2004). Revised litho- and sequence stratigraphy of the Spiti Triassic. *Albertiana* 30: 21–38.

Bhargava, O.N. (2020). Grey areas in the Palaeogene of the Lesser Himalaya: a review of various controversies. *Indian Journal of Geosciences* 74(3): 197–204.

Bhargava, O.N. and Bassi, U.K. (1998). Geology of Spiti-Kinnaur, Himachal Himalaya. *Memoirs Geological Survey of India* 124: 1–210.

Bhargava, O.N., Bassi, U.K., and Sharma, R.K. (1991). Crystalline thrust sheets, age of metamorphism, evolution and mineralisation of Himachal Himalaya. *Indian Minerals* 45(1 and 2): 1–18.

Bhargava, O.N., Frank, W., and Bertle, R. (2011). Late Cambrian deformation in the Lesser Himalaya. *Journal of Asian Earth Sciences* 40: 201–212.

Bhargava, O.N., Kumbkarni, S., and Ahluwalia, A.D. (2010). *Geomorphology and Landforms: Illustrations from the*

*Himachal Himalaya*, 212. Dehradun: Technology Publications.

Bhargava, O.N. and Singh, B.P. (2020). Geological evolution of the Tethys Himalaya. *Episode* 43(1): 404–416.

Bhargava, O.N., Singh, B.P., Frank, W. et al. (2021). Evolution of the Lesser Himalaya in space and time. *Himalayan Geology* 42(2): 273–289.

Bhargava, O.N. and Srikantia, S.V. (2014). Geology and age of metamorphism of the Jutogh and Vaikrita Thrust Sheets, Himachal Himalaya. *Himalayan Geology* 35(1): 1–15.

Bhargava, O.N., Thoni, M., and Miller, C. (2016). Isotopic evidence of Early Palaeozoic metamorphism in the Lesser Himalaya (Jutogh Group), Himachal Pradesh, India: its implication. *Himalayan Geology* 37(2): 73–84.

Bhatia, S.B. and Bhargava, O.N. (2006). Biochronological continuity of the Paleogene sediments of the Himalayan foreland basin: paleontological and other evidences. *Journal of Asian Earth Sciences* 26: 477–487.

Bisht, P., Ali, S.N., Rana, N. et al. (2017). Pattern of Holocene glaciation in the monsoon-dominated Kosa Valley, central Himalaya, Uttarakhand, India. *Geomorphology* 284: 130–141.

Bisht, P., Ali, S.N., Shukla, A.D. et al. (2015). Chronology of late Quaternary glaciation and landform evolution in the upper Dhauliganga valley (Trans Himalaya), Uttarakhand, India. *Quaternary Science Reviews* 129: 147–162.

Cawood, P.A., Johnson, M.R.W., and Nemchin, A.A. (2007). Early Palaeozoic orogenesis along the Indian margin of Gondwana: tectonic response to Gondwana assembly. *Earth and Planetary Science Letters* 255: 70–84.

Chahal, P., Kumar, A., Sharma, C.P. et al. (2019). Late Pleistocene history of aggradation and incision, provenance and channel connectivity of the Zanskar River, NW Himalaya. *Global and Planetary Change* 178: 110–128.

Chakrabarti, B.K. (2016). *Geology of the Himalayan Belt Deformation, Metamorphism, Stratigraphy*, 248. Amsterdam: Elsevier.

Debnath, M., Sharma, M.C., and Syiemlieh, H.J. (2019). Glacier dynamics in Changme Khangpu Basin, Sikkim Himalaya, India, between 1975 and 2016. *Geosciences* 9(6): 1–21. https://doi.org/10.3390/geosciences9060259.

DeCelles, P.G., Gehrels, G.E., Quade, J. et al. (2000). Tectonic implications of U-Pb Zircon ages of the Himalayan orogenic belt in Nepal. *Science* 288(5465): 497–499. https://doi.org/10.1126/science.288.5465.497.

Dhital, M.R. (2015). *Geology of the Nepal Himalaya, Regional Perspective of the Classic Collided Orogen*, 498. London: Springer.

Diener, C. (1912). Trias of the Himalaya. *Memoirs Geological Survey of India* 36(2): 202–358.

Dobhal., D.P. and Kumar, S. (1996). Inventory of glacier basins in Himachal Himalaya. *Journal of Geological Society of India* 48: 671–683.

Dobhal, D.P., Mehta, M., and Srivastava, D. (2013). Influence of debris cover on terminus retreat and mass changes of Chorabari Glacier, Garhwal region, Central Himalaya, India. *Journal of Glaciology* 59(217): 961–971. doi: 10.3189/2013JoG12J180.

Draganits, E., Brady, S.J., and Briggs, D.E.G. (2001). A Gondwana coastal arthropod ichnofauna from the Muth Formation (Lower Devonian), Northern India: paleoenvironment and tracemaker behaviour. *Palaios* 16: 126–147.

Draganits, E., Grassemann, B., and Schmid, H.P. (2003). Fluidization of pipes and spring pits in a Gondwana barrier-island environment: groundwater phenomenon, paleo-seisimicity or a combination of both. In: *Subsurface Mobilization* (ed. A.J. Maltman and C.K. Morley). London: Geological Society of London, Special Publication, 216: 109–121.

Dubey, A.K., Misra, R., and Bhakuni, S.S. (2001). Erratic shortening from balanced cross-sections: causes and implication for basin evolution. *Journal of Asian Earth Sciences* 19(6): 765–775

Ganju, A., Nagar, Y.C., Sharma, L.N. et al. (2018). Luminescence chronology and climatic implication of the late quaternary glaciation in the Nubra valley, Karakoram Himalaya, India. *Palaeogeography, Palaeoclimatology, Palaeoecology* 502: 52–62.

Gansser, A. (1964). *Geology of the Himalayas*, 289. London: Wiley-Interscience.

Gehrels, C.E., DeCelles, P.G., Ojha, T.P. et al. (2006.) Geologic and UPb geochronological evidences for early Palaeozoic tectonism in the Dadeldhura Thrust sheet, far-west Nepal Himalaya. *Journal of Asian Earth Sciences* 28: 385–408.

Gehrels, G.E., DeCelles, P.G., Martin, A. et al. (2003). Initiation of the Himalayan orogen as an early Paleozoic thin-skinned thrust belt. *Geological Society of America Today* 13: 4–9.

Ghosh, N.P., Basu, A.R., Bhargava, O.N. et al. (2016). Catastrophic environmental transition at the Permian-Triassic Neo-Tethyan margin of Gondwanaland: geochemical, isotopic and sedimentological evidence in the Spiti Valley, India. *Gondwana Research* 34: 324–345.

Ghosh, R., Srivastava, P., Shukla, U.K. et al. (2019). 100 kyr sedimentary record of Marginal Gangetic Plain: implications for forebulge tectonics. *Palaeogeography, Palaeoclimatology, Palaeoecology* 520: 78–95.

Hayden, H.H. (1904). The geology of Spiti with parts of Bashahr and Rupshu. *Memoir Geological Survey India* 36: 1–121.

Heim, A. and Gansser, A. (1939). Central Himalaya geological observations of the Swiss expedition 1936. *Me'moires de la Socie'te' Helve'tique des Sciences Naturelles* 73: 1–246.

Jain, A.K., Lal, N., Sulemani, B. et al. (2009). Detrital-zircon fission-track ages from the Lower Cenozoic sediments, NW Himalayan foreland basin: clues for exhumation and denudation of the Himalaya during the India–Asia collision. *Bulletin Geological Society of America* 121: 519–535.

Jain, A.K., Mukherjee, P.K., and Singhal, S. (2020). Terrane characterization in the Himalaya since Paleoproterozoic. *Episodes* 43(1): 346–357.

Jiang, G., Christie-Blick, N., Kaufman, A.J. et al. (2002). Sequence stratigraphy of the Neoproterozoic Infra Krol Formation and Krol Group, Lesser Himalaya, India. *Journal of Sedimentary Research* 72(4): 524–542.

Kacker, A.K. and Srivastava, M.C. (1993). Redlichid trilobite find in Milam Formation: a new dimension in Precambrian–Cambrian boundary. *Geological Survey of India, News* 23/24: 14.

Kumar, A. and Srivastava, P. (2017). The role of climate and tectonics in aggradation and incision of the Indus River in the Ladakh Himalaya during the late Quaternary. *Quaternary Research* 87(3): 363–385.

Kumar, G. (1997). *Geology of Arunachal Pradesh*, 405. Bangalore: Geological Society of India.

Kumar, G. (2005). *Geology of Uttar Pradesh and Uttaranchal*, 384. Bangalore: Geological Society of India.

Kumar, S., Singh, I.B., and Singh, S.K. (1977). Lithostratigraphy, structure, depositional environment, palaeocurrent and trace fossils the Tethyan sediments of Malla Johar area, Pithoragarh-Chamoli districts, Uttar Pradesh, India. *Journal of Palaeontological Society of India* 20: 396–435.

Kumar, V., Shukla, T., Mehta, M. et al. (2021). Glacier changes and associated climate drivers for the last three decades, Nanda Devi region, Central Himalaya, India. *Quaternary International* 575: 213–226.

Kumar, V., Shukla, T., Mishra, A. et al. (2020). Chronology and climate sensitivity of the post-LGM glaciation in the Dunagiri valley, Dhauliganga basin, Central Himalaya, India. *Boreas* 49(3): 594–614.

Le Fort, P., Debon, F., and Sonet, J. (1983). The Lower Paleozoic "Lesser Himalayan" granitic belt: emphasis on the Simchar pluton of Central Nepal. In: *Granites of Himalaya, Karakoram, and Hindukush* (ed. F.A. Shams), 235–255. Lahore, Pakistan: Institute of Geology, Punjab University.

Lutfi, W., Sheikh, L., Zhao, Z.et al. (2021). The detrital zircon U-Pb-Hf isotopes of the Triassic sediments in northern Pakistan: implications for crustal evolution of the NW Indian continent. *Precambrian Research* 357: 1–13. https://doi.org/10.1016/j.precamres.2021.106146.

Mandal, S., Robinson, D.M., Khanal, S. et al. (2015). Redefining the tectonostratigraphic and structural architecture of the Almora klippe and the Ramgarh–Munsiari thrust sheet in NW India. In: *Tectonics of the Himalaya* (ed. S. Mukherjee, R. Carosi, P.A. van der Beek, et al.). Geological Society, London, Special Publications, 412: 247–269.

Marivaux, L., Vianey-Liaud, M., and Welcomme, J.L. (1999). First discovery of Oligocene Cricetidae (Rodentia, Mammalia) in South Gandoin Syncline, Bugti Hill, Balochistan, Pakistan. Paris: Compte Rendu Academy Science. *Earth and Planetary Science Letters* 329: 839–844.

McArthur, J.M., Howarth, R.J., and Shields, G.A. (2012). Strontium isotope stratigraphy. *The Geologic Time Scale* 127–144.

McMahon, C.A. (1882). The Geology of Dalhousie, Northwest Himalaya. *Records Geological Survey of India* 15: 34–51

Medlicott, H.B. (1864). On the geological structure and relations of the southern portion of the Himalayan ranges between the rivers Ganges and Ravee. *Memoirs Geological Survey of India* 3(2): 1–212.

Mehta, M., Dobhal, D.P., Pratap, B. et al. (2014). Late Quaternary glacial advances in the Tons River valley, Garhwal Himalaya, India and regional synchronicity. *The Holocene* 24(10): 1336–1350.

Mehta, M., Majeed, Z., Dobhal, D.P. et al. (2012). Geomorphological evidences of post-LGM glacial advancements in the Himalaya: a study from Chorabari Glacier, Garhwal Himalaya, India. *Journal of Earth System Science* 121(1): 149–163.

Meinhold, G., Roberts, N.K.W., Arslan, A. et al. (2020). U–Pb dating of calcite in ancient carbonates for age estimates of syn- to post-depositional processes: a case study from the upper Ediacaran strata of Finnmark, Arctic Norway. Geological Magazine 328: 99–110. https://doi.org/10.1017/S0016756820000564.

Middlemiss, C.S. (1910). The Kangra earthquake of 4th April 1905. *Memoirs Geological Survey of India* 37: 1–409.

Miller, C., Klötzli, U., Frank, W. et al. (2000). Proterozoic crustal evolution in the NW Himalaya (India) as recorded by ca. 1.80 Ga mafic and 1.84 Ga granitic magmatism. *Precambrian Research* 103: 191–206.

Mukhopadhyay, D.K. and Mishra, P. (2004). The Main Frontal Thrust (MFT), Northwestern Himalayas: thrust trajectory and hanging wall fold geometry from balanced cross-sections. *Journal of Geological Society of India* 64: 739–746.

Najman, Y., Pringle, M.S., Johnson, M.R.W. et al. (1997). Laser 40Ar/39Ar dating of single detrital muscovite grains from early foreland-basin sedimentary deposits in India: implications for early Himalayan evolution. *Geology* 25(6): 535–538.

Nakazawa, K., Kapoor, H.M., Ishi, K. et al. (1975). The upper Permian and the lower Triassic in Kashmir, India. *Memoir Faculty of Science, Kyoto University, Series Geology and Mineralogy* 42(1): 1–106.

Oldham, R.D., (1888). Notes on the geology of the Northwest Himalaya. *Records Geological Survey of India* 21: 1–149.

Oldham, R.D. (1918). The structure of the Himalaya and of the Gangetic plain as elucidated by geodetic observations in India. *Memoirs Geological Survey of India* 42(2): 149–301

Owen, L.A., Derbyshire, E., and Richardson, S. (1996). The Quaternary glacial history of the Lahul Himalaya, northern India. *Journal of Quaternary Science* 11: 25–42. https://doi.org/10.1002/(SICI)1099-1417(199601/02)11:1<25:AID-JQS209>3.0.CO;2-K.

Pathak, D.B. (2007). Jurassic/Cretaceous boundary in the Spiti Himalaya, India. *Journal of Palaeontological Society of India* 52(1): 51–57.

Pilgrim, G.E. (1910). Notices of new mammalian genera and species from the Tertiaries of India. *Records Geological Survey of India* 40(1): 63–71.

Pilgrim, G.E. and West, W.D. (1928). The structure and correlation of the Simla rocks. *Memoirs Geological Survey of India* 53: 1–140.

Puri, V.M.K., and Verma, B.C. (1998). Glaciological and geological source of Vedic Saraswati in the Himalayas. *Itihas Darpan* 4(2): 7–36.

Pratap, B., Dobhal, D., Mehta, M. et al. (2015). Influence of debris cover and altitude on glacier surface melting: a case study on Dokriani Glacier, Central Himalaya, India. *Annals of Glaciology* 56(70): 9–16. doi:10.3189/2015Ao G70A971

Ray, Y., and Srivastava, P. (2010). Widespread aggradation in the mountainous catchment of the Alaknanda–Ganga River System: timescales and implications to hinterland–foreland relationships. *Quaternary Science Reviews* 29(17–18): 2238–2260.

Retallack, G.J., Bajpai, S., Liu, X. et al. (2018). Advent of strong South Asian monsoon by 20 million years ago. *Journal of Geology* 126: 1–24.

Shanker, R., Bhargava, O.N., Bassi, U.K. et al. (1993). Biostratigraphy controversy: an evaluation in Lahaul-Spiti, Himachal Pradesh. *Indian Minerals* 47(4): 1–60.

Sharma, C.P., Chahal, P., Kumar, A. et al. (2021). Late Pleistocene–Holocene flood history, flood-sediment provenance and human imprints from the upper Indus River catchment, Ladakh Himalaya. *Geological Society of America Bulletin.*

Sharma, S., Chand, P., Bisht, P. et al. (2016). Factors responsible for driving the glaciation in the Sarchu Plain, eastern Zanskar Himalaya, during the late Quaternary. *Journal of Quaternary Science* 31(5): 495–511.

Shukla, T., Mehta, M., Jaiswal, M.K. et al. (2018). Late Quaternary glaciation history of monsoon-dominated Dingad Basin, central Himalaya, India. *Quaternary Science Reviews* 181: 43–64.

Shukla, U.K., Srivastava, P., and Singh, I.B. (2012). Migration of the Ganga River and development of cliffs in the Varanasi region, India during the late Quaternary: role of active tectonics. *Geomorphology* 171: 101–113.

Singh, B.P., Bhargava, O.N., Mikuláš, R. et al. (2019). Discovery of Ordovician trace fossils from the Lesser Himalaya, India: its stratigraphic, tectonic and palaeogeographic implications. *Journal Palaeontological Society of India* 64(2): 283–303.

Singh, I.B. (1978). On some sedimentological and paleoecological aspects of Subathu–Dagshai–Kasauli succession of Simla Hills. *Journal Palaeontological Society of India* 21–22: 19–28.

Singh, I.B. (1996). Geological evolution of Ganga Plain: an overview. *Journal Palaeontological Society of India* 41: 99–137.

Singh, S. (2020). Himalayan magmatism through space and time. *Episodes* 43:358–368. https://doi.org/10.18814/epiiugs/2020/020021.

Srikantia, S.V. (1977). Sedimentary cycles in the Himalaya and their significance on the Orogenic evolution of the mountain belt. *International Colloquium, CNRS* 268: 395–407.

Srikantia, S.V. and Bhargava, O.N. (1998). *Geology of Himachal Pradesh*, 406. Bengaluru: Geological Society of India.

Srivastava, P. and Misra, D.K. (2008). Morpho-sedimentary records of active tectonics at the Kameng River exit, NE Himalaya. *Geomorphology* 96(1–2): 187–198.

Srivastava, P., Tripathi, J.K., Islam, R. et al. (2008). Fashion and phases of late Pleistocene aggradation and incision in the Alaknanda River Valley, western Himalaya, India. *Quaternary Research* 70(1): 68–80.

Stübner, K., Djordje, G., István, D. et al. (2018). Pliocene episodic exhumation and the significance of the Munsiari thrust in the northwestern Himalaya. *Earth and Planetary Science Letters* 481: 273–283.

Thakur, S.S. and Patel, S.C. (2012). Mafic and pelitic xenoliths in the Kinnaur Kailash Granite, Baspa River valley, NW Himalaya: evidence of pre-Himalayan granulite metamorphism followed by cooling event. *Journal of Asian Earth Sciences* 56: 105–117.

Thakur, V., Joshi, M.R., and Jayangondaperumal, R. (2020). Active tectonics of Himalayan Frontal Fault Zone in the Sub-Himalaya. In: *Geodynamics of the Indian Plate* (ed. N. Gupta, and S.K. Tandon), 439–466. Springer Nature Switzerland. https://doi.org/10.1007/978-3-030-15989-4_12.

Valdiya, K.S (1996). River Piracy: Saraswati that disappeared. *"Resonance" Journal of Academy of Sciences, Bangalore* 1(5): 19–28.

Valdiya, K.S. (1980). *Geology of the Kumaun Lesser Himalaya*, 291. Wadia Institute of Himalayan Geology.

Vohra,C.P. (1978).Glacier resources of the Himalaya and their importance to environment studies. In: *Proceedings*

*National Seminar on Resources Development and Environment in Himalaya Region*, 441–460. D.S.T.

Wadia, D.N. (1931). The Syntaxis of the northwest Himalaya: its rocks, tectonics and orogeny. *Records Geological Survey of India* 65(2): 189–220.

Webb, A.A.G., Yin, A., Harrison, T.M. et al. (2011). Cenozoic tectonic history of the Himachal Himalaya (north-western India) and its constraints on the formation mechanism of the Himalayan orogen. *Geosphere* 7(4): 1013–1061.

West, W.D. (1939). Structure of the Shali Window near Simla. *Records Geological Survey of India* 74: 133–163.

Wiesmayr, G. and Grasemann, B. (2002). Eohimalayan fold and thrust belt: implication for the geodynamic evolution of the NW Himalaya (India). *Tectonics* 21: 1–18.

# 13

## Fluctuations of Kolahoi Glacier, Kashmir Valley, Its Assessment With Tree-Rings of *Pinus wallichiana* and Comparable Satellite Imageries and Field Survey Records

*Uttam Pandey[1,2,*], Santosh K. Shah[1,*], and Nivedita Mehrotra[1]*

[1] *Birbal Sahni Institute of Palaeosciences, 53 University Road, Lucknow 226 007, India*
[2] *Graduate School of Environmental Studies, Nagoya University, Japan*
* *Corresponding author*

### 13.1 Introduction

The change in climatic conditions in the higher elevated mountainous regions affects various environmental conditions. Among these, the fluctuation of glaciers is of growing concern for glaciologists, climate policy-makers, and of interest to the general public. These concerns arise from the fact that glaciers control the hydrological condition of downstream inhabitants and various other ecosystems. The glacier fluctuation is mostly associated with the change of area, ice volume, and mass and length of the glacier, and is considered as direct evidence of climate change and its impact on the environment. In the twentieth century, worldwide glaciers are showing higher retreat and loss in glacier mass (Zemp et al., 2015). It is reported that the global temperature may rise by 1.5°C in the next two to three decades (Shukla et al. 2019), which may trigger the melting of the glacial ice in the polar areas and the glaciers of high-elevation Himalaya regions. In the last few decades, temperatures increased at a higher rate and accelerated the rate of glacier retreat in the Himalaya region (Maurer et al., 2019; Prakash, 2020; Romshoo et al., 2020). The Indian part of the Himalaya has around 9575 glaciers extending from northwestern to southeastern, among which few have been studied for their fluctuations (Raina et al., 2015). Several studies showed that most of the glaciers of the Western Himalaya region showed a higher retreat rate and might be associated with change in climate; however, the rate of retreat is variable (Table 13.1). It was found that the glaciers of the Karakoram region are growing and coined the term as "Karakorum anomaly" (Forsythe et al., 2017). Some studies also reported the advancement of glaciers in the western Himalayas (Bhambri and Bolch, 2009; Bolch et al., 2012).

The fluctuation in snout position of the glaciers in the Himalaya has been analyzed using various low-resolution proxy records that provided the dynamics of the glaciers in thousands to million year scales (McAndrews, 1984; Owen et al., 2001; Zhang et al., 2012). To understand the response of glaciers to the current climate change scenario, high-resolution long-term glacier fluctuation records are a prerequisite. The remotely-sensed satellite data provides the high-resolution real-time position of the glacier snout, but its availability is only for a short timespan and has only been available since the 1960s. Such studies are carried out from various parts of the globe such as Southeast Tibet (Loibl et al., 2014), Austria (Fey et al., 2017), Kazakhstan (Bolch, 2007), Russia (Kutuzov and Shahgedanova, 2009), and Switzerland (Kos et al., 2016), to identify the glacial fluctuations during the present and geological past. In the Himalaya region, similar studies (Bahuguna et al., 2007; Kulkarni et al., 2011; Vashisht et al. 2017) are present where satellite images have been analyzed to study similar glacial fluctuations. In most of the studies, satellite images such as LANDSAT MSS, LANDSAT TM, ASTER GDEM, Tera ASTER, CARTOSAT PAN, IRS P6-LISS IV, and SRTM are used (Latief et al., 2016; Vashisht et al., 2017; Murtaza and Romshoo, 2017; Rashid et al., 2017; Shukla et al., 2017; Shukla and Yusuf, 2017; Tawde et al., 2017; Mir, 2018; Rashid et al., 2020). The resolutions of the remote sensing data for these studies ranges between 30 and 60 m and analyzed the glaciers fluctuation back to the mid-twentieth century. This limits the remote sensed data to restrict the study of glacial fluctuation for long-term records. Thus, various proxy records and geological features can be utilized to study the long-term glacial history. Among the proxy records, a biological proxy, tree-rings have an annual-scale resolution and provides a precise past

*Advances in Remote Sensing Technology and the Three Poles,* First Edition. Edited by Manish Pandey, Prem C. Pandey, Yogesh Ray, Aman Arora, Shridhar D. Jawak, and Uma K. Shukla.

**Table 13.1** Differential glacial retreat rates of Himalayas.

| Name of Glacier | ~Length (Km) | Period of Study (CE) | Retreat Rate (m $y^{-1}$) | Reference |
|---|---|---|---|---|
| **Jammu and Kashmir** | | | | |
| Drang Drung | 23 | 1975–2008 | 9.4 | (Kamp et al., 2011) |
| Kange | 15.4 | 1992–2002 | 2 | (Kamp et al., 2011) |
| Parkachik | 14.5 | 1979–1990 | 13 | (Kamp et al., 2011) |
| Haskira | 14.2 | 1990–2003 | 4 | (Kamp et al., 2011) |
| Lalun | 13.75 | 1975–2006 | 30.96 | (Kamp et al., 2011) |
| Zanskar glacier 6 | 10.5 | 1990–203 | 22.3 | (Kamp et al., 2011) |
| Shafat | 9.8 | 1990–2003 | 16.9 | (Kamp et al., 2011) |
| Dalung | 6 | 1975–2003 | 28.29 | (Kamp et al., 2011) |
| Tanak Nala | 3.5 | 1999–2004 | 16 | (Kamp et al., 2011) |
| Kolahoi | 5 | 1962–2014 | 19.5 | (Shukla et al., 2017) |
| **Himachal Pradesh** | | | | |
| Bara Shigri | 26 | 1906–1995 | 33.7 | (Sangewar and Kulkarni, 2010) |
| Samudra Tapu | 16.78 | 1980–2010 | 19.9 | (Pandey and Venkataraman, 2013) |
| Chhota Shigri | 9 | 1988–2003 | 18.5 | (Kulkarni et al., 2007) |
| Shaune Garang | 6 | 1962–1997 | 29.7 | (Kulkarni and Bahuguna, 2002) |
| Panchi Nala | 5.5 | 1971–2011 | 9.25 | (Negi et al., 2013) |
| Zing-zing bar | 3.1 | 1971–2011 | 22.5 | (Negi et al., 2013) |
| Baralacha La | 2.54 | 1971–2011 | 10 | (Negi et al., 2013) |

environmental history (Fritts, 1976). There are some studies from high elevation mountainous region, which showed the application of tree-rings toward evaluation of glacial fluctuation such as from Rio Frias Valley, Argentina (Villalba et al., 1990), Canadian Rockies (Luckman, 1993; Wood and Smith, 2004), British Columbian Coast Mountains (Larocque and Smith, 2004), and from various parts of the Himalaya region (Bhattacharyya and Yadav, 1996; Singh and Yadav, 2000; Bhattacharyya et al., 2001, 2006; Bräuning, 2006). Therefore, the comparison between satellite image-based studies, field survey records, and tree-ring-based studies for glacier fluctuation might be useful to assess the long-term glacier fluctuation records in the Himalaya region.

Within this study, we aimed to compare already published glacier fluctuation records based on remote sensing data, field survey data, and tree-ring chronology developed using ring-widths of Himalayan Blue Pine (*Pinus wallichiana*). The main aim of this study is to analyze the relationship and utility of tree-rings to understand the fluctuation in the Kolahoi glacier in the Kashmir Valley from a long-term prospective. The tree taxa *P. wallichiana* is growing throughout the Himalayan region from Pakistan to Arunachal Pradesh of India along with Nepal and Bhutan (Sahni, 1990; Dhakal et al., 2016) on deep moist soils ranging from height 1800 m to more than 4200 m, on mountain scree, and on glacier forelands. This tree species has showed its affinity with glacial fluctuation in earlier studies carried out from other glaciers of the western Himalayan region (Bhattacharyya and Yadav, 1996; Singh and Yadav, 2000).

## 13.2 Tree-Ring Sampling Site and Data Acquisition

For the present study, tree-ring samples of *P. wallichiana* were collected from the sites located in the end moraine of the Kolahoi glacier, Lidder Valley region of Kashmir. The region has a temperate climate with long and cold winters and short and mild summers. In the Lidder Valley, Pahalgam has the only continuous operational meteorological station, which has climate records for the period from 1979 to 2013 CE. The trend analysis for Pahalgam climate during 1980–2010 CE indicated significant positive trends in the annual and seasonal mean, maximum, and minimum temperatures (Rashid et al., 2015). The maximum mean temperature of 25.5°C in July and a minimum of −6.8°C in January was recorded from the Lidder river basin. The highest (lowest) rainfall recorded in March (October and November) is 201 mm (45 mm and 49 mm), respectively. The average annual temperature and precipitation are 9.8°C and 1288 mm, respectively. It was also observed that the minimum temperature in the region is increasing faster as compared to the maximum temperature (Rashid et al., 2015).

Tree cores of *P. wallichiana* were collected from two sites, Aru Valley (ARU) and Green Top of Aru (GTA) (Figure 13.1). The samples from ARU were collected in the field during 2014 and from GTA were collected during 2017. For sample collection, undamaged trees were selected and from each tree, two tree cores were collected from opposite sides of the tree at breast height using a Swedish increment borer. The details of sites and samples are given in Table 13.2.

All the tree cores were air-dried at room temperature and processed further using standard methodology for

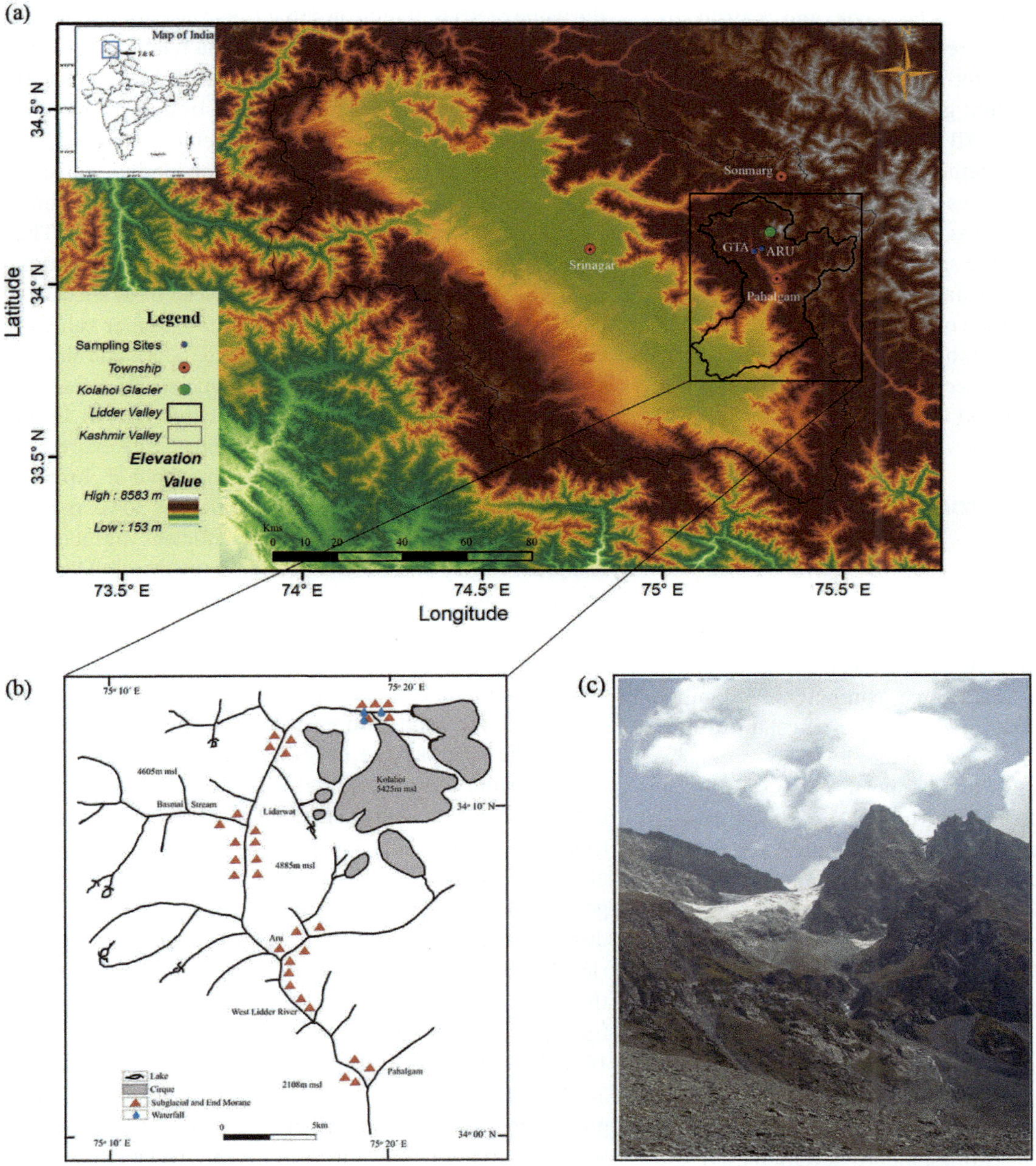


**Figure 13.1** Location map of tree-ring sampling sites and Kolahoi glacier: (a) Location of the Lidder river basin in the Kashmir Valley, showing the location of tree-ring sampling sites and the Kolahoi glacier; (b) a sketchmatic route map of the Lidder Valley showing the location of the Kolahoi glacier (modified after Ahmad and Hashimi, 1974); (c) present location of the Kolahoi glacier snout, photograph during a field visit in 2017 by lead author of this chapter.

**Table 13.2** Sampling site details of *P. wallichiana* from the Lidder river basin, Kashmir Valley.

| Site | Code | LAT | LONG | ELEV | NT | NC | AS | SLR |
|---|---|---|---|---|---|---|---|---|
| Aru Valley | ARU | 34°10′ | 75° 28′ | 2777 | 22 | 45 | SW | 5–10 |
| Green Top Aru | GTA | 34°27′ | 75° 36′ | 2800 | 25 | 50 | W | 10–20 |
| Regional Site | – | – | – | – | 47 | 95 | – | – |

LAT = latitude in degree north; LONG = longitude in degree east; ELEV = elevation a msl; NT = number of trees; NC = number of cores; AS = aspect; SLR = slope range in degree

tree-rings samples (Fritts, 1976; Cook and Kairiukstis, 1990; Speer, 2010). The cores were mounted on wooden blocks with the help of water-soluble glue, dried overnight, and polished with different grades of sand paper to make the ring boundaries clear. All the samples were cross-dated using the Skeleton plot method (Stokes and Smiley, 1968) to assign a calendar year to each ring. The Velmax tree-ring measurement machine was used to measure the ring widths of each cross-dated sample with 0.001 mm precision. The accuracy of cross-dating and measurements were further checked using the computer program COFECHA (Holmes, 1983) and by combining two sites. A total of 95 cores from 47 trees were successfully cross-dated and considered as regional datasets (Table 13.2).

## 13.3 Tree-Ring Chronology and Its Assessments

A regional tree-ring chronology of *P. wallichiana* was developed by standardizing successfully cross-dated tree ring series using a signal-free standardization technique (Melvin and Briffa, 2008) in RCSigFree computer Programme (http://www.ldeo.columbia.edu/tree-ring-laboratory/resources/software). This technique is used to remove growth-related biological trends and to maximize the climate signal related to radial growth. The age-dependent smoothing spline (Melvin et al., 2007) was used to standardize each raw ring-width series. The standardized series were followed by a bi-weight robust estimation of mean to develop signal-free tree-ring chronology (Cook, 1985; Cook and Kairiukstis, 1990). With this we developed a 177-year (1840–2016 CE) long regional tree-ring chronology of *P. wallichiana* from the study region (Figure 13.2). The different descriptive statistics (Fritts, 1976; Speer, 2010) for the full (1840–2016 CE) and for common periods (1923–2006 CE) was calculated. The full period statistics (Fritts, 1976), such as mean sensitivity (MS), standard deviation (SD), and first-order autocorrelation (AC-1) are 0.126, 0.244, and 0.778, respectively. The higher value of MS indicates a presence of high-frequency variation in chronology. The higher the value of SD is, it indicates that tree-ring chronology has a high-degree potentiality to analyze the relationship with the glacial fluctuation. The high value of AC-1 indicates that the climate of the current year is strongly affected by the previous year's climate, which was removed by applying Auto-regressive modeling. The common period statistics were calculated to evaluate the common strength of chronology and to express the degree of common signal. The common period statistics such as mean series correlation between all trees, within trees, and between trees are 0.310, 0.338, and 0.282, respectively. In addition, higher value of variance in first principal component, i.e., 33.3% and signal-to-noise ration with a value of 29.16, indicates the presence of a common environmental signal of tree growth in all trees, which contributed in the chronology. The expressed population signal, that represents the threshold of chronology for further analysis, is calculated with the 50 years running window and depends on a sufficient number of samples (Wigley et al., 1984). Based on the threshold value of 0.85, the present regional tree-ring chronology is reliable since 1855 CE for further analysis.

The regional tree-ring chronology of *P. wallichiana* shows several high and low growth periods, which is associated with the temperature variation of the region. This is due to the fact that the previous studies of tree-rings of *P. wallichiana* from the Lidder Valley showed that growth of this tree is directly associated with winter temperatures (Shah et al., 2019). The low growth during 1855–1868 CE and 1901–1980 CE indicates a cool phase that blocked the growth of trees. The higher growth rates were observed during the last decades of twentieth and twenty-first centuries. In addition, single-year low growth was observed in the years 1864, 1885, 1911, 1934, and 1958 CE, whereas high growth was observed in the years 1840, 1843, 1862, 1871, 1879, 1880, 1970, 1971, 1974, 1981, 1985, 1988, 2000, 2005, 2008, 2010, 2014, and 2015 CE. After 1980, tree-ring growth increased abruptly, which is associated with the warming conditions of the region. This recent increased growth in *P. wallichiana* tree-ring records may be because of significant photosynthesis due to an increase in environmental temperature (Ram and Borgaonkar, 2013).

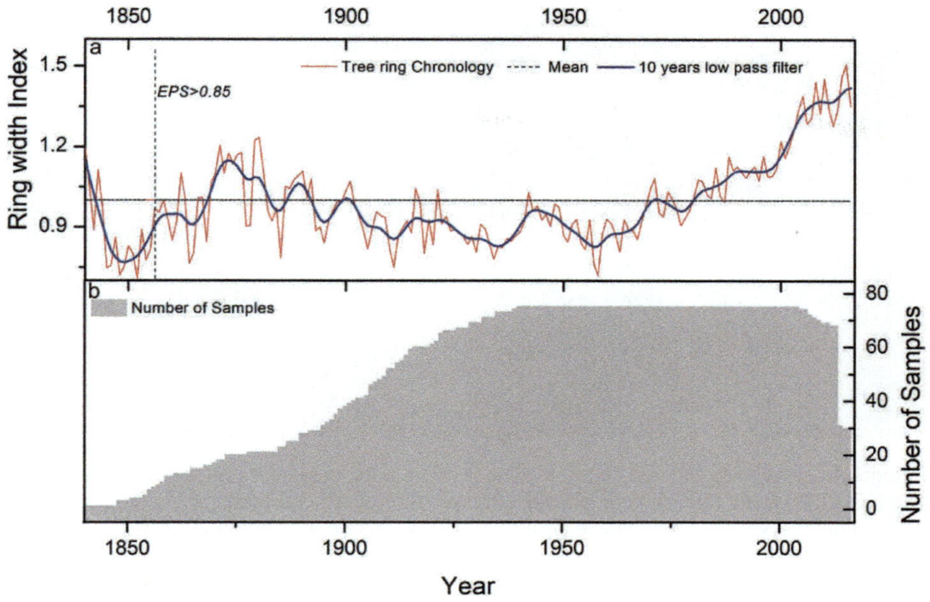


**Figure 13.2** (a) A 177-year (1840 to 2016 CE) long tree-ring chronology of *P. wallichiana* from the Liddar river basin, Kashmir Valley. Dash line in upper panel of the graph is EPS threshold of >0.85 (Adapted from Wigley et al., 1984); (b) number of tree cores used to generate tree-ring chronology.

## 13.4 Fluctuations of Kolahoi Glacier: Existing Records and Its Assessment With Tree-Rings

The Kolahoi glacier fluctuation was first studied based on the trigonometrically surveyed map of 1857 by Neve (1910) and its fluctuation was analyzed for the period 1857 to 1910 CE. Later, Odell (1963) had studied and analyzed the retreat rate of the glacier for the period 1912 to 1962 CE and Kaul (1990) had observed the receding snout of the Kolahoi glacier from 1961 to 1984 CE. All these studies, available for the nineteenth and the first half of twentieth century, were based on field survey records. The actual position of the snout based on the photographs taken by the geologist in different timeframes is given in Figure 13.3. The longest glacier fluctuation record of the Kolahoi glacier was synthesized for fluctuation of the glacier snout for the past 157 years (1857–2013 CE) based on remote sensed data of different sources such as LANDSAT (5, 7, and 8), TERR ASTER, CARTOSAT 1, IRS P6, and ASTER GDEM (Shukla et al., 2017). The positions of the snout of the Kolahoi glacier marked in satellite images for 1857 to 2014 CE are shown in Figure 13.4. In addition, other remote sensing-based studies (Latief et al., 2016; Rashid et al., 2017; Vashisht et al., 2017; Mir, 2018) are also carried out to understand the changes in retreat rate, area, and volume of the Kolahoi glacier from a long-term prospective. These satellite image-based studies of the region, which have been available since 1965, provide continuous glacier dynamics of the Kolahoi glacier in the Lidder basin of the Kashmir Valley. These studies suggest that the glaciated area of the upper Lidder Basin has decreased at a very fast rate in recent decades, with a threat to survival of livelihoods and sustainable development of the basin.

Based on records of satellite images and field observations, it was observed that the Kolahoi glacier fluctuated at various rates in the past and has been assessed in terms of retreat-advancement rate, area change, and volume change. The fluctuation of the Kolahoi glacier has been further assessed with tree-ring chronology *P. wallichiana* developed in the present study and on the basis of existing satellite images and field observations for the glacier. For this, tree-ring chronology of *P. wallichiana* has been converted to anomaly to identify the departure of low and high growth in the time series from the mean. This anomaly is compared with different parameters of the glacier receding, such as retreat rate, area, and volume change from the various earlier studies discussed above (Figure 13.5).

The retreat rate of the Kolahoi glacier for the time period of 1857 to 2013 was synthesized by Shukla et al. (2017) and Rashid et al. (2017) based on earlier field survey records and satellite images. They suggested that from 1857 to 2013, the snout of the Kolahoi glacier had retreated by about 2.83 km. The study based on a trigonometrically surveyed map of 1857 (Neve, 1910) suggested that the Kolahoi glacier receded around 1600 m with an

Figure 13.3 The yellow arrow in the photographs represents the position of the snout of the Kolahoi glacier captured by various researchers during their fieldwork: (a) Neve (1910); (b) Odell (1963); Kolahoi peak obscured by clouds; (c) Gilbert (1977), Kolahoi peak visible; (d) Shah and Kanth (2012), part of Kolahoi peak visible; (e) Shah and Kanth (2012); (f) position of snout in 2017 (current study) Modified after Shukla, A et al. [2017] / Springer Nature.

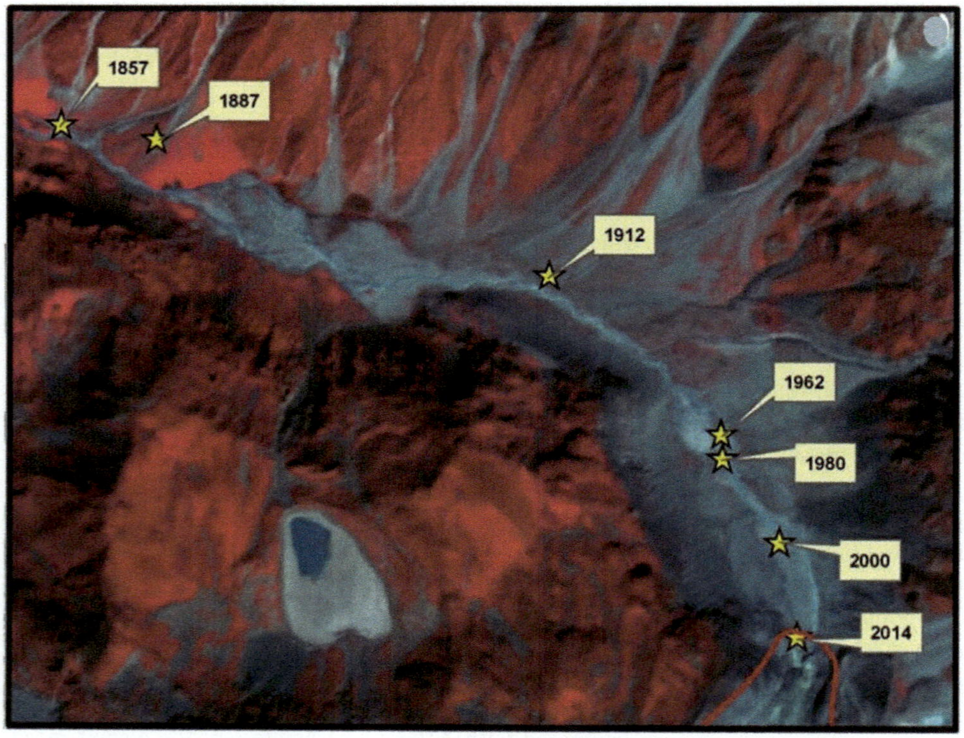


Figure 13.4 Retreat of Kolahoi glacier from 1857 to 2014. Yellow star presents the snout position in a specific year. Modified after Shukla, A et al. [2017] / Springer Nature.

average of 30.94 m $y^{-1}$ retreat rate from 1857 to 1910. This higher retreat rate is analogous to the higher growth rate of the *P. wallichiana* for the same time period. The retreat rate (16.3 m $y^{-1}$) (Odell, 1963) decreases for the period of 1912–1961 CE, which is associated with lowering of tree growth for the same period. Along with retreat rate, an area and volume change of the Kolahoi glacier analyzed by Latief et al. (2016) and Vashisht et al. (2017) respectively, is also synchronized with the growth of *P. wallichiana* (Figure 13.5). The area loss and area change of glacier synthesized in these two studies indicates that the temperature is a driving factor to glacier dynamics and the loss of area coincides with the tree growth of *P. wallichiana* in the Lidder river basin.

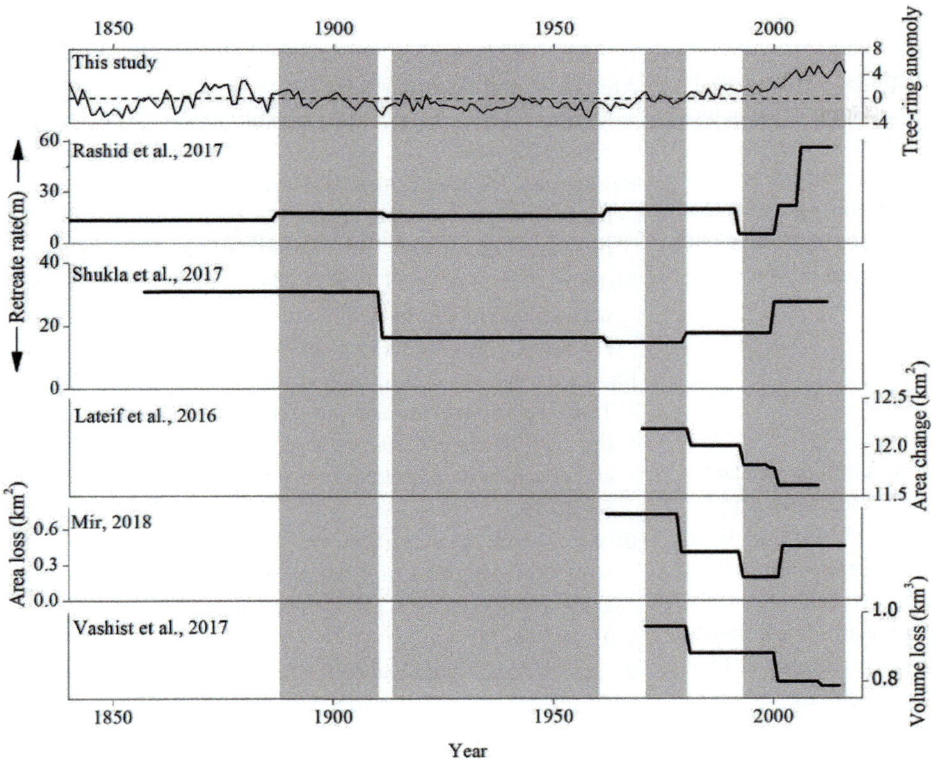


**Figure 13.5** Tree-ring chronology of *P. wallichiana* from the Lidder river basin, Kashmir Valley, along with the glacier retreat rate of the Kolahoi glacier given by earlier studies based on remote sensing.

Tree-ring chronology has the highest growth rate in the last decade of the twentieth century to present. This growth is analogous with higher winter temperatures in the Kashmir Himalaya (Fowler and Archer, 2005; Bhutiyani et al., 2007; Singh et al., 2008; Khattak et al., 2011; Rashid et al., 2015; Shafiq et al., 2019; Shah et al., 2019) and Western Himalaya (Bhutiyani et al., 2007; Dimri and Dash, 2012; Dimra et al. 2015; Ren et al., 2017). Both higher growth rates and warming in winter coincide with the higher retreat rate of the Kolahoi glacier observed in studies carried out by Shukla et al. (2017) and Rashid et al. (2017) and the area changed from 12.02 to 11.82 $km^2$ for the period of 1980–1998 (Latief et al., 2016). Tree growth was relatively slow during 1915 to 1960 that indicates a cooler environment that stops the growth of the trees by reduced photosynthesis. This cooling is reflected by the least retreat rate (16 m $y^{-1}$) of the Kolahoi glacier (Rashid et al., 2017). The glacier showed highest retreat rates during 1890s to 1910 and for this period trees were showing higher growth rates. The volume change in the Kolahoi glacier from 1980 to 2015 (Vashisht et al. 2017) coincides with the growth of *P. wallichiana*. The volume loss of the glacier due to temperature rise has a strong negative relationship with the growth of *P. wallichiana* in the Lidder river basin.

The studies on the Kolahoi glacier dynamics suggest a ~2.83 km decline of the glacier snout in the last 158 years (Rashid et al., 2017; Shukla et al., 2017). This decline in snout position of the glacier since 1857 indicates the region experienced warming at a temporal extent. The Kolahoi glacier had the lowest retreat rate during 1962–1980 and during that period, low radial growth of *P. wallichiana* is observed. Furthermore, higher melting of glacial ice was observed during 1980–2010 and during this period, correspondingly, we observed high radial growth of *P. wallichiana*. This abrupt increase of tree growth after 1980 is associated with the warming conditions of the region. The impact of global warming reflects in the tree-ring chronology of *P. wallichiana*, showing higher growth rates in recent decades. In the Lidder river basin, the growth of *P. wallichiana* shows a strong significant positive response to the winter temperature (Shah et al., 2019). The reconstructed winter temperature from the Lidder river basin showed abrupt warming in the last five decades at regional scale (Shah et al., 2019). During the winter or dormant period, photosynthesis is enhanced by higher temperatures in the pine trees and is responsible for their higher radial growth (Hepting, 1945; Kramer, 1958). The increase in warming conditions is responsible for the enhancement of melting of the Kolahoi glacier ice of the Lidder river basin and is reflected in the radial growth of *P. wallichiana* (Figure 13.5). A similar response to the growth of *P. wallichiana* was observed for the Gangotri glacier, western Himalaya (Singh and Yadav, 2000), where retreat of the glacier is comparable with recent increases in growth of the same tree taxa. These assessments showed that the glaciers in the Lidder river basin are highly sensitive to climatic variability and especially for increased temperature warming rates.

## 13.5 Conclusions

This is the first attempt to assess the satellite imagery-based records of the Kolahoi glacier's fluctuations supplemented with tree-ring chronology of the Himalayan Blue Pine (*Pinus wallichiana*) from the Kashmir Valley. The growth of *P. wallichiana* is found to be suitable to study glacier fluctuations in the Kashmir Valley and it is consistent with the earlier records of the Western Himalaya. This study provides the future possibility to assess remotely sensed glaciers fluctuation records in the Himalayan region supplemented with tree-rings. Furthermore, to understand the glacier dynamics of the Himalaya region from a longer perspective, ring-width data from trees growing at high elevations, in end moraine and from the tree line, can be assessed in future studies.

## Acknowledgements

The authors would like to express their gratitude to Director, Birbal Sahni, Institute of Palaeosciences, Lucknow for support and permission to publish this work (BSIP Contribution No. 08/2021-22) under Institute In-house Project 6. The authors (SKS and UP) also wish to thank the officials of the Kashmir Forest Department, for giving permission to collect tree core samples; Shakil Ahmad Romshoo, Rakesh Chandra, and Irfan Rashid from the Department of Earth Sciences, Kashmir University, for their kind help and support that they provided in Kashmir. We would like to thank Asif Lone, Showket, and Mushber for helping in fieldwork. We are grateful to Shiva for tree-ring sample processing in the lab. The author (NM) is highly grateful to DST WOS-A for the support given to her. The author (UP) would like to thank Prof Takeshi Nakatsuka for continuing his research in Graduate School of Environmental studies, Nagoya University, Japan. The tree-ring samples were collected with support provided by the Department of Science and Technology, New Delhi (Project No: SR/S4/ES-621/2012).

## References

Ahmad, N. and Hashimi, N.H. (1974). Glacial history of Kolahoi Glacier, India. *International Journal of Glaciology* 13(68): 279–283.

Bahuguna, I.M., Kulkarni, A.V., Nayak, S. et al. (2007). Himalayan glacier retreat using IRS 1C PAN stereo data. *International Journal of Remote Sensing* 28(2): 437–442.

Bhambri, R. and Bolch, T. (2009). Glacier mapping: a review with special reference to the Indian Himalayas. *Progress in Physical Geography* 33(5): 672–704.

Bhattacharyya, A., Chaudhary, V., and Gergan, J.T. (2001). Tree ring analysis of *Abies pindrow* around Dokriani Bamak (Glacier), Western Himalayas, in relation to climate and glacial behavior: preliminary results. *Palaeobotanist* 50: 71–75.

Bhattacharyya, A., Shah, S.K., and Chaudhary, V. (2006). Would tree ring data of *Betula utilis* be potential for the analysis of Himalayan glacial fluctuations? *Current Science* 91(6): 754–761.

Bhattacharyya, A. and Yadav, R.R. (1996). Dendrochronological reconnaissance of *P. wallichiana* to study glacier behavior in the Western Himalaya. *Current Science* 70(8): 739–743.

Bhutiyani, M.R., Kale, V.S., and Pawar, N.J. (2007). Long-term trends in maximum, minimum and mean annual air temperatures across the Northwestern Himalaya during the twentieth century. *Climatic Change* 85(1): 159–177.

Bolch, T. (2007). Climate change and glacier retreat in northern Tien Shan (Kazakhstan/Kyrgyzstan) using remote sensing data. *Global and Planetary Change* 56(1–2): 1–12.

Bolch, T., Kulkarni, A., Kääb, A. et al. (2012), The state and fate of Himalayan glaciers. *Science* 336(6079): 310–314.

Bräuning, A. (2006). Tree-ring evidence of "Little Ice Age" glacier advances in southern Tibet. *The Holocene* 16(3): 369–380.

Cook, E.R. (1985). A time series analysis approach to tree ring standardization. PhD Thesis, University of Arizona.

Cook, E.R. and Kairiukstis, L.A. (1990). *Methods of Dendrochronology*. Dordrecht, The Netherlands: Kluwer Academic Press.

Dhakal, Y.R., Gaire, N.P., Aryal, S. et al. (2016.) Treeline shift in central Nepal Himalaya and climate reconstruction of past millennia. In: *Building Knowledge for Climate Resilience in Nepal* (ed. D.R. Bhuju, K. McLaughlin, and J. Sijapati), 41–44. Khumaltar, Lalitpur: Nepal Academy of Science and Technology.

Dimri, A.P. and Dash, S.K. (2012). Wintertime climatic trends in the western Himalayas. *Climatic Change* 111(3): 775–800.

Dimri, A.P., Niyogi, D., Barros, A.P. et al. (2015). Western disturbances: a review. *Reviews of Geophysics* 53(2): 225–246.

Fey, C., Wichmann, V., and Zangerl, C. (2017). Reconstructing the evolution of a deep seated rockslide (Marzell) and its response to glacial retreat based on historic and remote sensing data. *Geomorphology* 298: 72–85.

Forsythe, N., Fowler, H.J., Li, X.F. et al. (2017). Karakoram temperature and glacial melt driven by regional atmospheric circulation variability. *Nature Climate Change* 7(9): 664–670.

Fowler, H.J. and Archer, D.R. (2005). Hydro-climatological variability in the Upper Indus Basin and implications for water resources. In: *Regional Hydrological Impacts of Climatic Change Impact Assessment and Decision Making*. Proceedings of symposium S6 held during the Seventh IAHS Scientific Assembly at Foz do Iguaçu, Brazil. IAHS Publication, 131–138.

Fritts, H.C. (1976). *Tree Rings and Climate*. New York: Academic Press.

Gilbert, R. (1977). Schoolboys on Kolahoi. *Alpine Journal* 83: 174–178.

Hepting, G.H. (1945). Reserve food storage in shortleaf pine in relation to little-leaf disease. *Phytopathology* 35: 106–119.

Holmes, R.L. (1983). Computer-assisted quality control in tree-ring dating and measurement. *Tree-Ring Bulletin* 43: 69–78.

Kamp, U., Byrne, M., and Bolch, T. (2011). Glacier fluctuations between 1975 and 2008 in the Greater Himalaya Range of Zanskar, Southern Ladakh. *International Journal of Mountain Sciences* 8: 374–389.

Kaul, M.N. (1990). *Glacial and Fluvial Geomorphology of Western Himalaya (Liddar Valley)*. New Delhi: Concept Publishing Company.

Khattak, M.S., Babel, M.S., and Sharif, M. (2011). Hydro-meteorological trends in the upper Indus River basin in Pakistan. *Climate Research* 46(2): 103–119.

Kos, A., Amann, F., Strozzi, T. et al. (2016). Contemporary glacier retreat triggers a rapid landslide response, Great Aletsch Glacier, Switzerland. *Geophysical Research Letters* 43(24): 12466–12474.

Kramer, P.J. (1958). *The Physiology of Forest Trees* (ed. K.V. Thinmann), 156–186. New York: The Ronald Press Co.

Kulkarni, A.V. and Bahuguna, I.M. (2002). Glacial retreat in the Baspa Basin, Himalaya, monitored with satellite stereo data. *Journal of Glaciology* 48(160): 171–172.

Kulkarni, A.V., Bahuguna, I.M., Rathore, B.P. et al. (2007). Glacial retreat in Himalaya using Indian remote sensing satellite data. *Current Science* 92(1): 69–74.

Kulkarni, A.V., Rathore, B.P., Singh, S.K. et al. (2011). Understanding changes in the Himalayan cryosphere using remote sensing techniques. *International Journal of Remote Sensing* 32(3): 601–615.

Kutuzov, S. and Shahgedanova, M. (2009). Glacier retreat and climatic variability in the eastern Terskey–Alatoo, inner Tien Shan between the middle of the 19th century and beginning of the 21st century. *Global and Planetary Change* 69(1–2): 59–70.

Larocque, S.J. and Smith, D.J. (2004). Calibrated *Rhizocarpon* spp. growth curve for the Mount Waddington area, British Columbia coast mountains, Canada. *Arctic, Antarctic, and Alpine Research* 36(4): 407–418.

Latief, S.U., Rashid, S.M., and Singh, R. (2016). Impact analysis of climate change on Kolahoi Glacier in Liddar Valley, north-western Himalayas. *Arabian Journal of Geosciences* 9(18): 1–15.

Loibl, D., Lehmkuhl, F., and Grießinger, J. (2014). Reconstructing glacier retreat since the Little Ice Age in SE Tibet by glacier mapping and equilibrium line altitude calculation. *Geomorphology* 214: 22–39.

Luckman, B.H. (1993). Glacier fluctuation and tree-ring records for the last millennium in the Canadian Rockies. *Quaternary Science Reviews* 12(6): 441–450.

Maurer, J.M., Schaefer, J.M., Rupper, S. et al. (2019). Acceleration of ice loss across the Himalayas over the past 40 years. *Science Advances* 5(6): 1–13.

McAndrews, J.H. (1984). Pollen analysis of the 1973 ice core from Devon Island Glacier, Canada. *Quaternary Research* 22(1): 68–76.

Melvin, T.M. and Briffa, K.R. (2008). A "signal-free" approach to dendroclimatic standardisation. *Dendrochronologia* 26: 71–86.

Melvin, T.M., Briffa, K.R., Nicolussi, K. et al. (2007). Time-varying-response smoothing. *Dendrochronologia* 25: 65–69.

Mir, R.A. (2018). Recent changes of two parts of Kolahoi Glacier and its controlling factors in Kashmir Basin, western Himalaya. *Remote Sensing Applications: Society and Environment* 11: 265–281.

Murtaza, K.O. and Romshoo, S.K. (2017). Recent glacier changes in the Kashmir Alpine Himalayas, India. *Geocarto International* 32(2): 188–205.

Negi, H.S., Saravana, G., Rout, R. et al. (2013). Monitoring of great Himalayan glaciers in Patsio region, India using remote sensing and climatic observations. *Current Science* 105(10): 1383–1392.

Neve, E.F. (1910). Mt Kolahoi and its northern glacier. *Alpine Journal* 25(187): 39–42.

Odell, N.E. (1963). The Kolahoi northern glacier, Kashmir. *Journal of Glaciology* 4(35): 633–635.

Owen, L.A., Gualtieri, L.Y.N., Finkel, R.C. et al. (2001). Cosmogenic radionuclide dating of glacial landforms in the Lahul Himalaya, northern India: defining the timing of Late Quaternary glaciation. *Journal of Quaternary Science: Published for the Quaternary Research Association* 16(6): 555–563.

Pandey, P. and Venkataraman, G. (2013). Changes in the glaciers of Chandra–Bhaga basin, Himachal Himalaya, India, between 1980 and 2010 measured using remote sensing. *International Journal of Remote Sensing* 34(15): 5584–5597.

Prakash, A. (2020). Retreating glaciers and water flows in the Himalayas: implications for governance. *ORF Brief Issue* 400: 1–14.

Raina, V.K. and Sangewar, C.V. (2015). *Glacier Snout Monitoring in the Himalayas*. Bengaluru, India: Geological Society of India.

Ram, S. and Borgaonkar, H.P. (2013). Growth response of conifer trees from high-altitude region of western Himalaya. *Current Science* 105: 225–231.

Rashid, I., Majeed, U., Aneaus, S. et al. (2020). Linking the recent glacier retreat and depleting streamflow patterns with land system changes in Kashmir Himalaya, India. *Water* 12(4): 1168.

Rashid, I., Romshoo, S.A., and Abdullah, T. (2017). The recent deglaciation of Kolahoi valley in Kashmir Himalaya, India in response to the changing climate. *Journal of Asian Earth Sciences* 138: 38–50.

Rashid, I., Romshoo, S.A., Chaturvedi, R.K. et al. (2015). Projected climate change impacts on vegetation distribution over Kashmir Himalayas. *Climate Change* 134: 601–613.

Rashid, A., Sayyad, M.R.G., and Bhat, F.A. (2015) The dynamic response of Kolohai Glacier to climate change. *Proceedings of the International Academy of Ecology and Environmental Sciences* 5(1): 1–6.

Ren, Y.Y., Ren, G.Y., Sun, X.B. et al. (2017). Observed changes in surface air temperature and precipitation in the Hindu Kush Himalayan region over the last 100-plus years. *Advances in Climate Change Research* 8(3): 148–156.

Romshoo, S.A., Fayaz, M., Meraj, G. et al. (2020). Satellite-observed glacier recession in the Kashmir Himalaya, India, from 1980 to 2018. *Environmental Monitoring and Assessment* 192(9): 1–17.

Sahni, K.C. (1990). *Gymnosperms of India and Adjacent Countries*. Dehradun: Bishen Singh and Mahendra Pal Singh Publication.

Sangewar, C.V. and Kulkarni, A. (2010). Observational studies of the recent past, report of the study group on Himalayan glaciers. Prepared for Principal Scientific Advisor Government of India, PSA/2011/2, 25–76.

Shafiq, M.U., Ramzan, S., Ahmed, P. et al. (2019). Assessment of present and future climate change over Kashmir Himalayas, India. *Theoretical and Applied Climatology* 137(3): 3183–3195.

Shah, A.A. and Kanth, T.A. (2012). Impact of glaciers on the hydrology of Kashmir Rivers: A case study of Kolahoi Glacier. An unpublished MPhil dissertation, Department of Geography and Regional Development.

Shah, S.K., Pandey, U., Mehrotra, N. et al. (2019). A winter temperature reconstruction for the Lidder Valley, Kashmir, Northwest Himalaya based on tree-rings of *Pinus wallichiana*. *Climate Dynamics* 53(7): 4059–4075.

Shukla, A., Ali, I., Hasan, N. et al. (2017). Dimensional changes in the Kolahoi glacier from 1857 to 2014. *Environmental Monitoring and Assessment* 189(5): 1–18.

Shukla, A. and Yousuf, B. (2017). Evaluation of multisource data for glacier terrain mapping: a neural net approach. *Geocarto International* 32(5): 569–587.

Shukla, P.R., Skeg, J., Buendia, E.C. et al. (2019). *Climate Change and Land*. An IPCC special report on climate change, desertification, land degradation, sustainable land management, food security, and greenhouse gas fluxes in terrestrial ecosystems, 3–38.

Singh, J. and Yadav, R.R. (2000). Tree-ring indications of recent glacier fluctuations in Gangotri, western Himalaya, India. *Current Science* 79(11): 1598–1601.

Singh, P., Haritashya, U.K., and Kumar, N. (2008). Modelling and estimation of different components of streamflow for Gangotri Glacier basin, Himalayas. *Hydrological Sciences Journal* 53(2): 309–322.

Speer, J.H. (2010). *Fundamentals of Tree-Ring Research*. Tucson, AZ: The University of Arizona Press.

Stokes, M.A. and Smiley, T.L. (1968). *An Introduction to Tree-Ring Dating*. Chicago, IL: University of Chicago Press.

Tawde, S.A., Kulkarni, A.V., and Bala, G. (2017) An estimate of glacier mass balance for the Chandra basin, western Himalaya, for the period 1984–2012. *Annals of Glaciology* 58(75pt2): 99–109.

Vashisht, P., Pandey, M., Ramanathan, A.L. et al. (2017). Comparative assessment of volume change in Kolahoi and Chhota Shigri glaciers, Western Himalayas, using empirical techniques. *Journal of Climate Change* 3(1): 37–48.

Villalba, R., Leiva, J.C., Rubulls, S. et al. (1990). Climate, tree-ring, and glacial fluctuations in the Rio Frias Valley, Rio Negro, Argentina. *Arctic & Alpine Research* 22(3): 215–232.

Wigley, T.M.L., Briffa, K.R., and Jones, P.D. (1984). On the average value of correlated time series, with applications in dendroclimatology and hydrometeorology. *Journal of Applied Meteorology & Climatology* 23: 201–213.

Wood, C. and Smith, D. (2004). Dendroglaciological evidence for a neoglacial advance of the Saskatchewan glacier, Banff National Park, Canadian Rocky Mountains. *Tree-Ring Research* 60(1): 59–65.

Zemp, M., Frey, H., Gärtner-Roer, I. et al. (2015). Historically unprecedented global glacier decline in the early 21st century. *Journal of Glaciology* 61(228): 745–762.

Zhang, B., Ou, X., and Lai, Z. (2012). OSL ages revealing the glacier retreat in the Dangzi valley in the eastern Tibetan Plateau during the Last Glacial Maximum. *Quaternary Geochronology* 10: 244–249.

# 14

# Applications of ICESat-2 Photon Data in the Third Pole Environment

*Giribabu Dandabathula**

*Regional Remote Sensing Centre (West), NRSC, Indian Space Research Organization, Jodhpur, India*
** Corresponding author*

## 14.1 Introduction

Elevation information plays a significant role in scientific inquiry about the Earth and its associated processes by enabling interpretation of the data in a 3D framework, especially when the landscape of the study area contains rugged or irregular topography (National Research Council, 2010). Topography influences the Earth's surface processes that include geology, hydrology, climatology, and human settlement patterns (Wallace, 1881; Xiao et al., 2018; Palazzi et al., 2019; Rocha et al., 2020).

The Himalayas, a large mountain belt, are the result of topographic expression from the interplay of tectonics activities, climatic fluctuation, and Earth surface processes (Lamb and Watts, 2010; Champagnac et al., 2012). The Himalayas host some of the highest mountain peaks on Earth, where snow and icy environmental conditions rival those existing in the polar regions and thus the region is termed as "the third pole," "water tower of Asia," and "the roof of the world" (Bahadur, 1993; Bandhyopadhyay, 2013; Yao et al., 2020). The Himalayan region influences the "ice-water-air-ecosystem-energy-human" interactions on most parts of the South Asian countries as it contains numerous glaciers, peaks, valleys, and is the origin of major river systems (Vemsani, 2015; Yao et al., 2020). Elevation information is a critical need to investigate a range of surface processes if the study area falls within the rugged terrains of the Himalayan region (Liu et al., 2019). The availability of void-filled Global Digital Elevation Model (GDEM) products has paved a better way to integrate remote sensing methods and 3D models. Data from multi-view optical satellites images and Synthetic Aperture Radar (SAR) satellites have been commonly used to generate DEMs. Giribabu et al. (2013), Pinel et al. (2014), Mukul et al. (2017), and Liu et al. (2019) through their research, mentioned the challenges and issues related to the accuracies in the elevation information that are associated with the contemporary processes of generating DEMs and publicly available GDEMs for the Himalayas or high mountainous regions. Giribabu et al. (2013) and Zhang and Pavelsky (2019) have mentioned the limitations of using optical remote sensing data for studying high mountainous regions and these include cloud cover, shadows, and certain times due to out-of-stereo coverage. Data acquired by SAR imaging systems in the high mountainous regions, due to the side-looking nature of the sensor, is often not completely imaged and also phenomena like foreshortening and layover effects may induce inaccuracies in the elevation information (Pinel et al., 2014)

One more branch of technology to retrieve elevation information is the satellite laser altimetry, where the working mechanism involves transmitting the short flashes of laser light toward a target surface, after which part of the surface reflected energy will enter the onboard sensor. The space-borne sensors measure the travel time of the light pulses right from the point of transmission to receiving the returned surface reflection; thus, determination of the distance between the spacecraft and respective surface target can be computed and further used to yield the true height of the topographic feature on the Earth's surface (Falkner and Schulz, 2015). Ice, Cloud, and land Elevation Satellite-2 (ICESat-2) is the latest addition in space-borne laser altimetry missions launched by the National Aeronautics and Space Administration (NASA) and hosts a solo sensor titled Advanced

*Advances in Remote Sensing Technology and the Three Poles,* First Edition. Edited by Manish Pandey, Prem C. Pandey, Yogesh Ray, Aman Arora, Shridhar D. Jawak, and Uma K. Shukla.

Topographic Laser Altimeter System (ATLAS); the major goal of ICESat-2 mission is to monitor the Earth's surface dynamics in the polar regions (Markus et al., 2017). However, as the mission acquisition system acquires the data on all the surface types like glaciers, oceans, inland water-bodies, land, and canopy, the usage of ICESat-2 photon data can also be extended to other landforms (Brown et al., 2016). In this chapter, the usage of ICESat-2 laser altimeter data for various strategic applications in the Himalayan region is discussed. All the applications mentioned in the subsequent sections have harnessed the high-resolution and along-track nature of ICESat-2 in the Himalayan region.

## 14.2 Brief Background About NASA's ICESat-2 Mission

Earlier, NASA's first ICESat mission hosted a sensor, namely Geoscience Laser Altimeter System (GLAS), which was launched in 2003 and operated until 2009. The mission accomplished measuring the dynamics of ice sheets, water surface elevations, canopy heights, and height profiles of aerosols and clouds to a satisfactory precision. Technical specification of the GLAS sensor and its mission objectives were mentioned by Schutz et al. (2005). Surface elevation measurements are computed by combining the observatory position and attitude along with the flight times of the laser pulses. The laser pulses of this first ICESat mission produced ~70 m diameter footprints on the surface of the Earth, which were spaced at ~150 m along track. The relatively high precision of the elevation measurements enabled reliable estimates of glacier changes. Wang et al. (2011) have reviewed the Earth Science application of ICESat/GLAS. However, the results from ICESat have shown limitations due to the limited specifications of the GLAS sensor (Markus et al., 2017).

Following the success of the initial ICESat mission and substantial technological improvements, the ICESat-2 mission was successfully commissioned as a follow-on mission to continue certain proposed scientific objectives (Markus et al., 2017). Collection of data from the ICESat-2 mission started in September 2018 and is being disseminated to the scientific fraternity by the National Snow and Ice Data Centre (NSIDC); mission overview and launch proceedings can be found at NSIDC-ICESat-2 (2018). The ICESat-2's ATLAS instrument is configured with two lasers of which one is primary and the other is a backup. The wavelength of the laser light is 532 nm, which in the visible spectrum takes a slot of bright green. The fast-firing laser from the ICESat-2 platform yields one transmitted laser pulse for every 0.7 m (2.3 ft) along the satellite's ground path, during which photons in the order of trillions leave the ATLAS sensor, are transmitted through the Earth's atmosphere, and finally a few dozen surface reflected photons will be counted back in the ATLAS telescope (Markus et al., 2017). The time difference between the point of transmission and reflection will enable computation of range measurement; furthermore, this range is used to estimate the height of the ground feature above a reference ellipsoid.

The ATLAS sensor on-board ICESat-2 splits the laser into a six-beam configuration containing three strong beams and three weak beams, of which a strong and a weak beam will be paired with a distance of 90 m across-track (Neumann et al., 2019). These three pairs are separated by a distance of 3.3 km in the cross-track direction. Each footprint pattern formed from these six beams is identifiable as a Ground Track (GT). After undergoing numerous correction algorithms during the product generation phase, the Level 2A product, namely ATL03, is obtained which contains a geolocated ellipsoidal height for each geotagged photon (Neumann et al., 2018). The ATL03 product generation algorithm separates the erroneous photon events and actual signal events with a confidence attribute ranging from 0 to 4 (Neumann et al., 2018, 2019). A geolocated photon event is represented by a record of latitude, longitude, ellipsoidal height, and its corresponding confidence value. From a series of photons events, along-track elevation profiles can be generated that represent surface heights of the ground feature. Table 14.1 represents the specifications of ICESat and ICESat-2. Figure 14.1 depicts a representation of the multi-beam data acquisition process of the ICESat-2 laser altimeter.

From the ATL03 product, various specialized geophysical products (with different product nomenclature) for different surface types such as land-ice, sea-ice, land/vegetation, atmosphere, oceans, and inland water-bodies are derived and distributed by NASA'S Distributed Active Archive Centre (DAAC) at NSIDC in .hdf format (ICESat-2 Data Sets, 2018; NSIDC-ICESat-2, 2018; Neumann et al., 2021).

A web portal at https://openaltimetry.org provides a facility for online visualization and downloading various ICESat-2 data products (Khalsa et al., 2020). Experiments, accuracy assessments, and validations that are done by various researchers conclude that surface height from ICESat-2's ATL03 data products are better than 15 cm (Brunt et al., 2019; Dandabathula et al., 2021a; Fricker et al., 2021).

**Table 14.1.** Technical specifications of ICESat and ICESat-2.

| Detail/Specification | ICESat | ICESat-2 |
|---|---|---|
| Launch Date/Operational period | 13 January 2003 to14 August 2010 (operated in campaign mode) | 15 September 2018 and operational |
| Sensor | Geoscience Laser Altimeter System (GLAS) | Advanced Topographic Laser Altimeter System (ATLAS) |
| Orbit/altitude/inclination/repeat cycle | Near polar LEO/600 km/94°/ 91 days | Near polar LEO frozen orbit/496 km/92°/91 days |
| Footprint width | 66 m diameter | 14 m diameter |
| Pulses per second | 40 | ~10,000 |
| Number of beams | Single beam | 6 (3 pairs) |
| Energy of laser | 35 mJ | 120 μJ (strong beam) |
| Wavelength | 532 nm (visible green) | 532 nm (visible green) |
| Reflection entity | Waveform | Photon |
| Along-track spacing | 170 m | ~0.7 m |

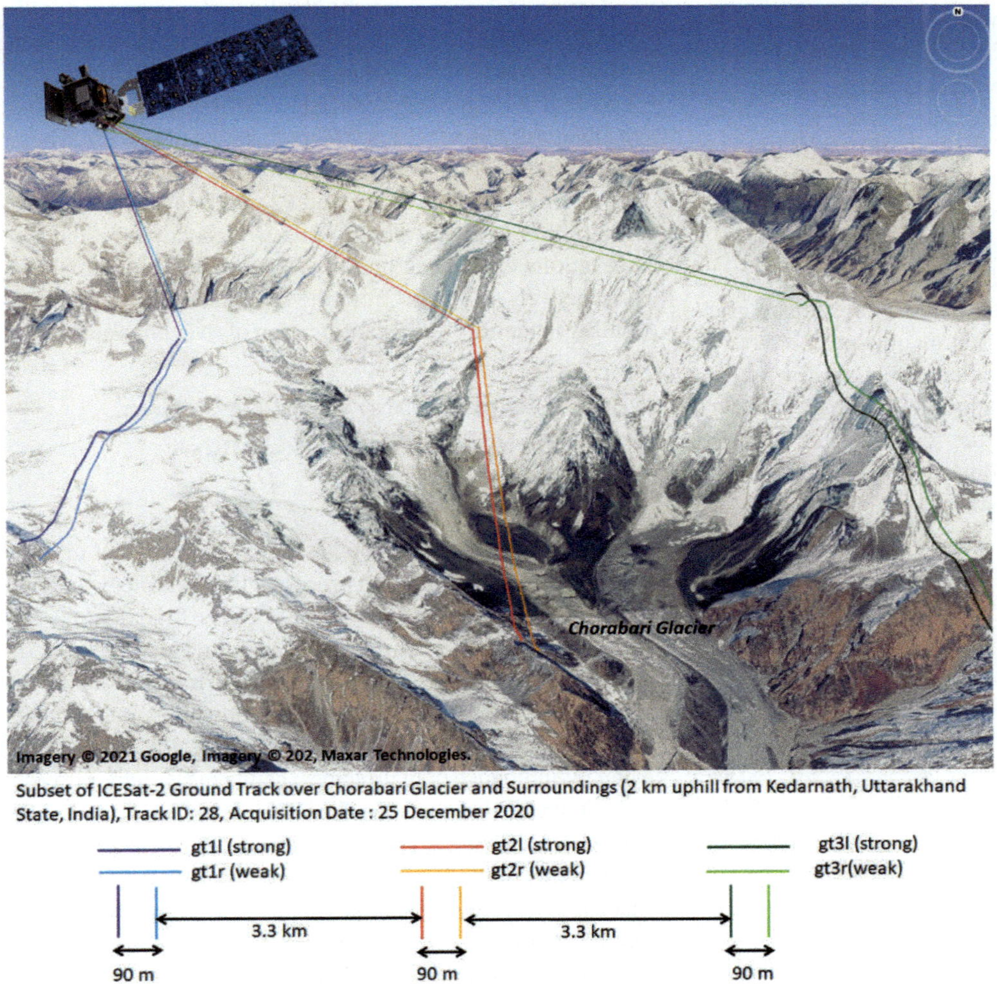


**Figure 14.1** Multi-beam data acquisition process of ICESat-2. ICESat-2's solo sensor ATLAS, which is equipped with photon-counting technology, produces a six-beam laser configuration of three-beam pairs. Each pair contains a strong and a weak beam. Figure 14.1 shows the extent of the Chorabari glacier and surroundings that are uphills to Kedarnath, Uttarakhand state, India. The Level 2A product from ICESat-2, namely ATL03, undergoes numerous corrections and contains geolocated ellipsoidal heights for every successful photon event.

## 14.3 Terrain Profiling From ICESat-2 Photon Elevations Over a Mountainous Region

A terrain profile or an elevation profile over a surface is a two-dimensional (2D) cross-sectional representation of the landscape and is useful to recreate the side view of the terrain relief along a line drawn between two locations. Profiling enables the graphical representation of changes in terrain height along a particular line, which enables us to identify the crests of the ridges, percent slope of the terrain, and lowest points like thalwegs of river channels. The ability to remotely create elevation profiles for inaccessible areas of Himalayan rugged mountains could bring cost benefits, reduce time constraints, and help in understanding the preliminary conditions of the terrain. One possible means is the use of elevation data from the surface-reflected photons of ICESat-2 (preferably Level 2A ATL03 data product) to generate a terrain profile between two points for every 70 cm in between. Earlier, Wang, A. et al. (2021) have utilized elevation profiles generated from ICESat-2 photon data in parts of Amery Ice Shelf in East Antarctica. The authors in their research characterized the fracture features that include surface fractures, surface expressions of basal fractures, and rifts. Dandabathula et al. (2021b) have utilized the elevation profiles from ICESat-2 to study the morphological properties related to various types of sand dunes. Lai and Wang (2021) used ICESat-2 photon data to record the elevation changes in the Greenland Ice Sheet. Wang, Yi et al. (2021) used surface elevation from ICESat-2 photons to estimate glacier mass balance in parts of Asia's high mountains.

Figure 14.2 represents an elevation profile constructed using a subset of ICESat-2's strong beam containing high confidence surface reflected photons over the region of Auli/Joshimath and their surroundings in the Himalayan region. The details of the dataset are given in the legend section of Figure 14.2. The *y*-axis component in this elevation profile contains ellipsoidal height and the *x*-axis component contains the corresponding latitude and is as shown in Figure 14.2b. The subset of the beam data covers an along-track path of ~9 km distance. From the elevation profile, the crest with a maximum elevation of 3400 m (earmarked with blue circle) and the lowest point with an elevation of 1370 m at the Alaknanda River (earmarked with green circle) was identifiable. The difference in elevation between the crest and the lowest point at Alaknanda River is 2030 m in a span of 6.1 km.

## 14.4 Longitudinal Profiling of Rivers in a Mountainous Region

Generating a terrain profile over a surface latitude-wise can be a tailor-made job from ICESat-2 geolocated photon data due to its along-track data acquisition system. However, the multi-beam nature of ICESat-2, which acquires three pairs of beams for a single ground track, enables an understanding of the longitudinal profiling of the surface for a distance of 90 m across the pair of a beam (one strong beam and one weak beam) and 3.3 km across the three-paired beams spread over ~6.6 km. However, using the advantage of multi-date data from ICESat-2, we can extend the measurements beyond this specific limitation. The ICESat-2 mission is commissioned to acquire the data along 1387 Reference Ground Tracks (RGT) and each RGT is targeted once every 91 days to allow elevation changes to be detected (Neumann et al., 2018).

A longitudinal profile for a stretch of mountains enables the study of the variations of elevation across and within the mountain ranges, and gives a hint about the low gradient downstream segment (Hack, 1957; Aiken and Brierley, 2013). Earlier, Prerna et al. (2018) computed longitudinal profiles to analyze the Indus basin of the Himalayan region by using elevation information from two DEMs, namely, CartoDEM and SRTM (Shuttle Radar Topography Mission).

Figure 14.3 shows longitudinal profiles generated with three segments along the stretch of Alaknanda River from Karnaprayag (confluence of the Alaknanda and Pindar rivers) and Rudraprayag (confluence of the Mandakini and Alaknanda rivers) with a stream distance of ~30 km where the stretch of this stream direction is from east to west. The three cross profiles (P1, P2, and P3) were constructed using multi-date ICESat-2 photon data which have yielded the elevations of thalweg points at 710 m, 654 m, and 606 m respectively. A spinoff from the longitudinal profiles is the ability to identify and quantification of river channel width.

## 14.5 Inland Water Level Detection in Mountainous Regions Using ICESat-2 Photon Data

The role of inland water-bodies is highly significant in various cycles of Earth system processes like hydrology and climate change. Water-level monitoring is fundamental for understanding the hydrological phenomenon; conventionally, *in-situ* gauging stations were used to measure the water levels. Due to harsh weather conditions and inaccessible tracts of remote parts in the Himalayan region, installation and maintenance of hydrological gauge measurements pose challenges. Also the records mentioning the quantitative water-level changes are poorly available (Srivastava et al., 2013). One of the proposed applications of the ICESat-2 mission is to quantify the changes in inland water bodies (Brown et al., 2016). Dandabathula et al. (2020), Yuan et al. (2020), and Xiang et al. (2021) experimented to measure the changes in water levels of inland water bodies using ICESat-2 laser altimetry and concluded achieving unprecedented accuracy at the centimeter-level when compared with near real-time gauge data. Similarly, Zhang et al. (2019) and Luo et al.

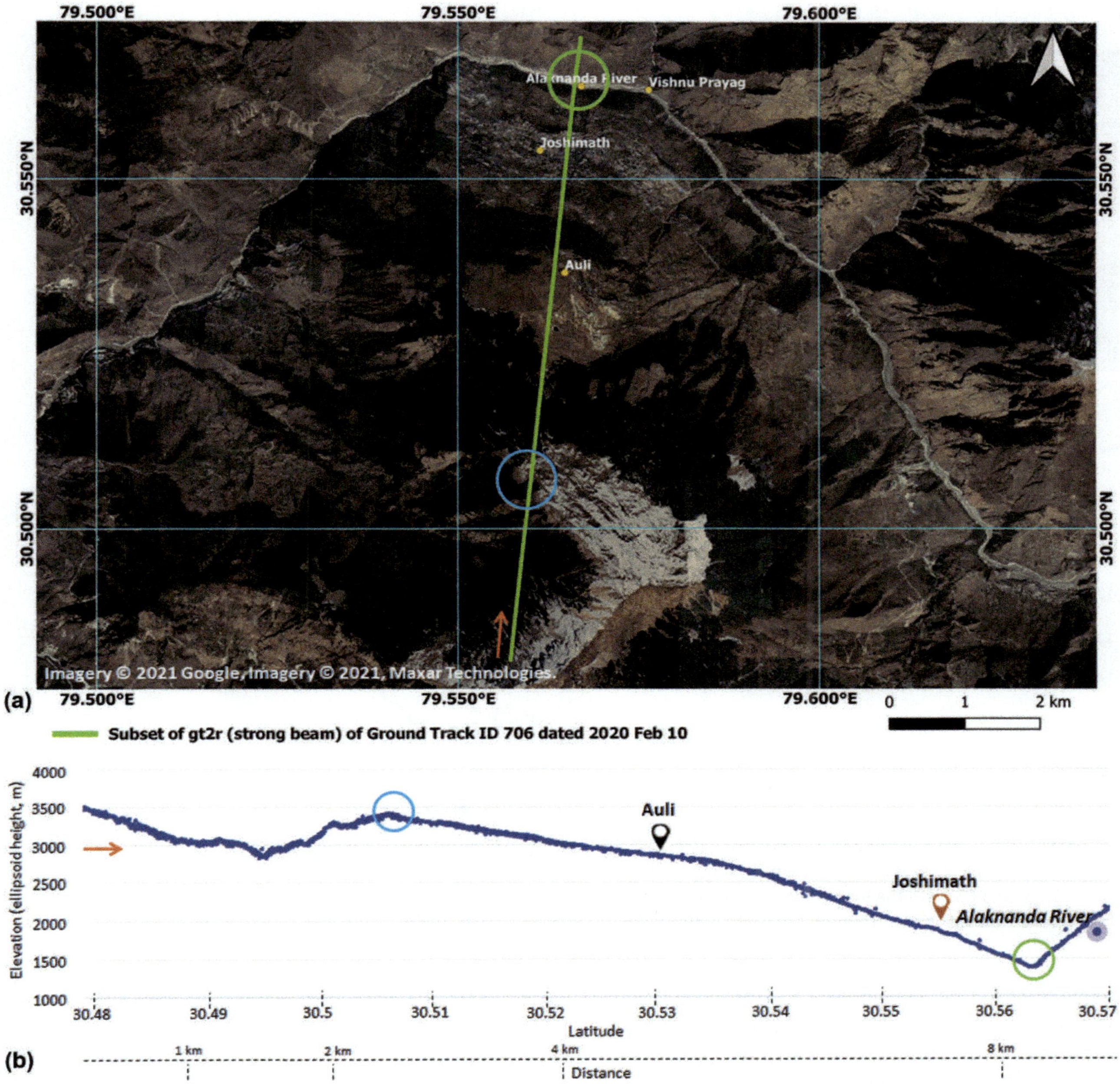


**Figure 14.2** Along-track elevation profile generated from ICESat-2 photon data over a mountainous region. (a) Subset of the strong beam (gt2r) of a ground track acquired at Auli/Joshimath and surroundings dated 10 February 2020 has been overlaid on satellite data; (b) elevation profile generated from the high confidence ICESat-2 photons corresponding to the area shown in (a). *y*-axis of the elevation profile contains ellipsoidal height and the *x*-axis is latitude. The total profile covers an along-track path of ~9 km distance. From the elevation profile, the crest with a maximum elevation of 3400 m (earmarked with blue circle) and the lowest point with an elevation of 1370 m at the Alaknanda River (earmarked with green circle) was identifiable.

(2021) utilized ICESat-2 data to measure water levels in numerous lakes of the Tibetan Plateau and their results concluded that the accuracy is in agreement with the gauge measurements at the centimeter-level.

Figure 14.4 shows a typical example for retrieval of water-level height from the elevation profile using the surface reflected photons of ICESat-2. Here, in this example, a subset from a strong beam belonging to a ground track of ICESat-2's ATL03 data product for the extent of Tsomoriri Lake in the Ladakh region shows the water level as ~4499 m (ellipsoidal height). Gupta and Shukla (2016) reported the details of Tsomoriri Lake while

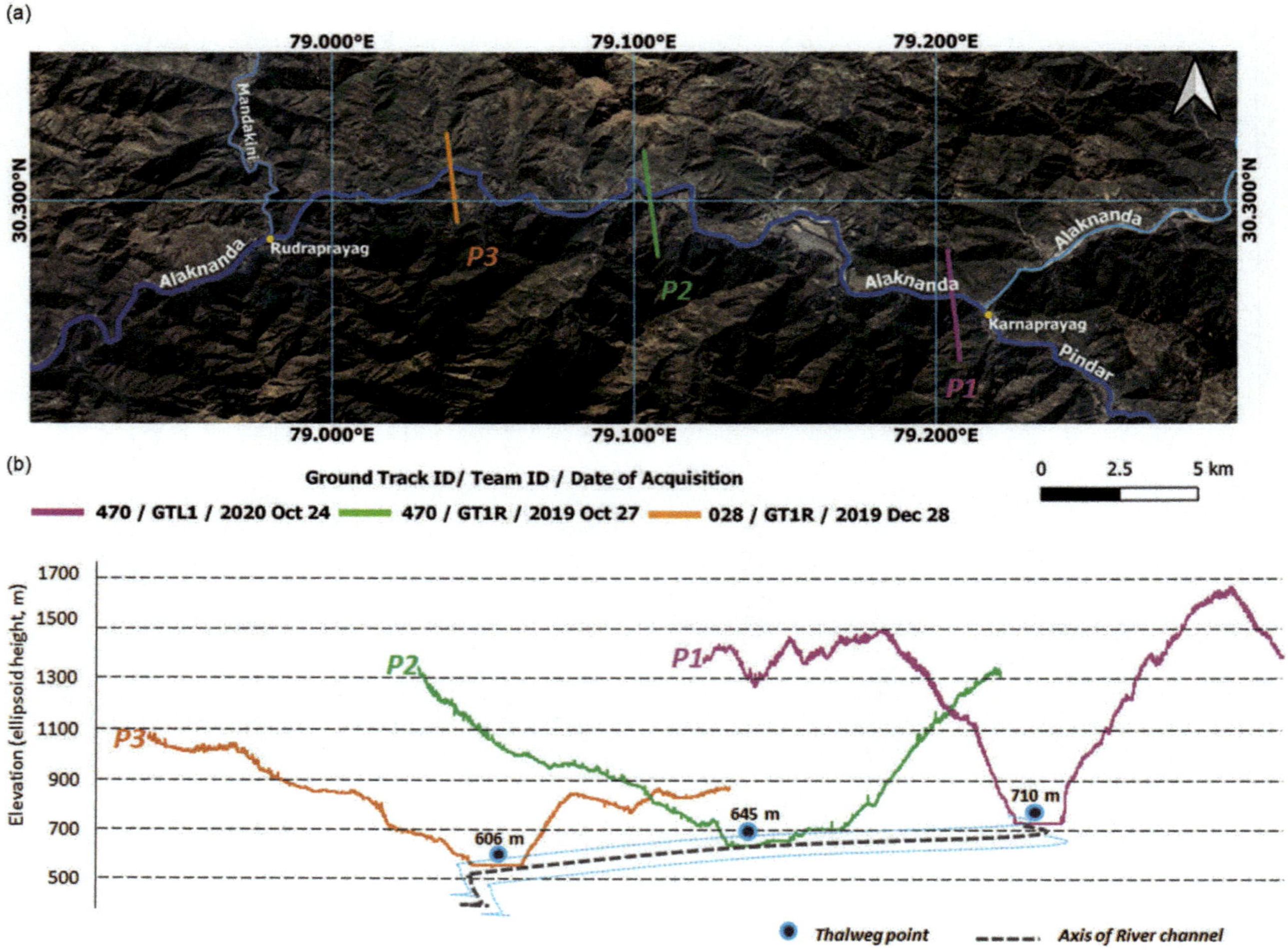


Figure 14.3 Longitudinal profile for a stretch of Alaknanda river. (a) Subsets of ICESat-2 photon beams from various ground tracks acquired from various dates for Alaknanda River between Karnaprayag and Rudraprayag overlaid on satellite data; (b) elevation profiles P1, P2, and P3 yield the longitudinal profiling of the Alaknada river in this part of the study area. Through the profiles P1, P2, and P3, the heights of the river channel's thalweg points were retrieved as 710 m, 645 m, and 606 m respectively.

studying its surrounding land-use and land-cover details. Tsomoriri Lake is stretched north to south with an approximate length of 27 km and the extent in the east-west direction varies between 5 and 7 km.

At certain times, in clear-water conditions, photons from ICESat-2 can penetrate through the water column due to the green (532 nm) energy pulses (Jasinski et al., 2016; Parrish et al., 2019). Thus, a pattern containing double surface returned photons from the water bodies can be witnessed in the plots of along-track heights of the individual ICESat-2 photons. But, the majority of photons can be seen reflected from the water body's surface through which the actual water level can be recorded. As shown in Figure 14.4c, a major chunk of photons were reflected from the water surface of Lake Tsomoriri at a height of 4499 m, but few photons penetrated the sub-surface of the water body up to a depth of nearly 25 m. For applications needing precise water levels, the ICESat-2 team has developed a dedicated geophysical data product, namely ATL13 (Jasinki et al., 2020). The subsequent section elaborates the usage of the ATL13 data product with an example.

## 14.6 Inferring Annual Variations of Water Levels in Mountain Lakes Using ICESat-2's ATL13 Data Product

Alsdorf et al. (2007) have reviewed various space-borne methods for water-level retrieval and concluded that altimetry techniques provide the highest accuracy among others. Dandabathula and Rao (2020), in their research, mentioned that detection of water-level changes helps in addressing the issues of flood hazards, modeling of global water, and energy cycles.

ICESat-2's ATL13 is a thematic data product that represents the along-track surface water height for each beam of ATLAS. Jasinski et al. (2020) have detailed the list of parameters that are available in the ATL13 data product.

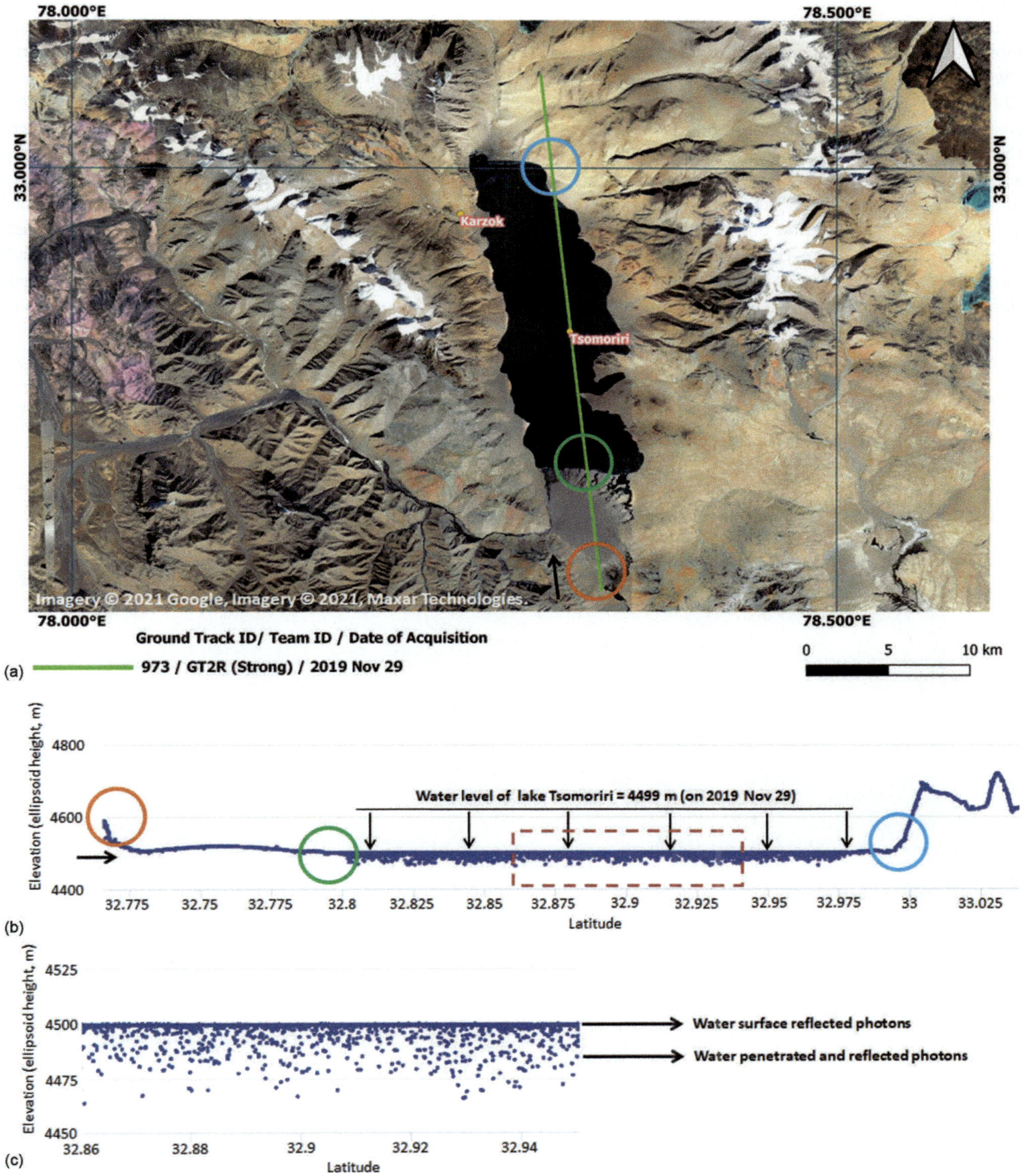


**Figure 14.4** A typical profile for an inland water body in the Himalayan region: Tsomoriri Lake. (a) Subset of ICESat-2 photon beam dated 29 November 2019 overlaid on satellite data. (b) Elevation profile representing surface water level of Tsomoriri Lake. Note the constant water level height of 4499 m (ellipsoidal height) in the entire stretch of the lake during the month of August in 2020; (c) Subsurface penetrated photons as evident in the profile. Depending on the clear-water conditions, photons from ICESat-2 can penetrate down through the water column due to the green (532 nm) energy pulses. Here, in this case, photos have penetrated to a depth of ~25 m.

The product line includes the statistics related to surface water height (like standard deviation, mean, and slope), subsurface attenuation, significant wave height, and shallow bathymetry. The ATL13 data product is available for all the global inland water bodies like lakes and reservoirs greater than ~0.01 km$^2$, rivers greater than an width of 100 m, transitional water bodies including bays and estuaries, and a near-shore 7 km buffer (Jasinski et al., 2020). The Algorithm Theoretical Basis Document (ATBD) document of ATL13 mentions the source of inland water-body shapes from various sources like Hydrolakes, Global lakes and wetlands database, Named Marine water bodies, etc. (Jasinski et al., 2020). Thus, we have to ensure the shape of the water body to maintain the accuracy of water level. One advantage of using the ATL13 data product is that it provides the height of the surface water in orthometric meters with an attribute titled "ht_ortho" for all the six ground tracks. Earlier, Yuan et al. (2020), Dandabathula and Rao (2020), and Xu et al. (2021) have experimented with ATL13 data products and assessed their accuracies, which are in the order of centimeters.

The ATL13 data product can be downloaded at web portal (https://nsidc.org/data/ATL13/version/3) maintained by NSIDC or alternatively we can visualize and download in the OpenAltimetry portal. In this section, Gobind Sagar reservoir (Figure 14.5), which is situated in the Bilaspur

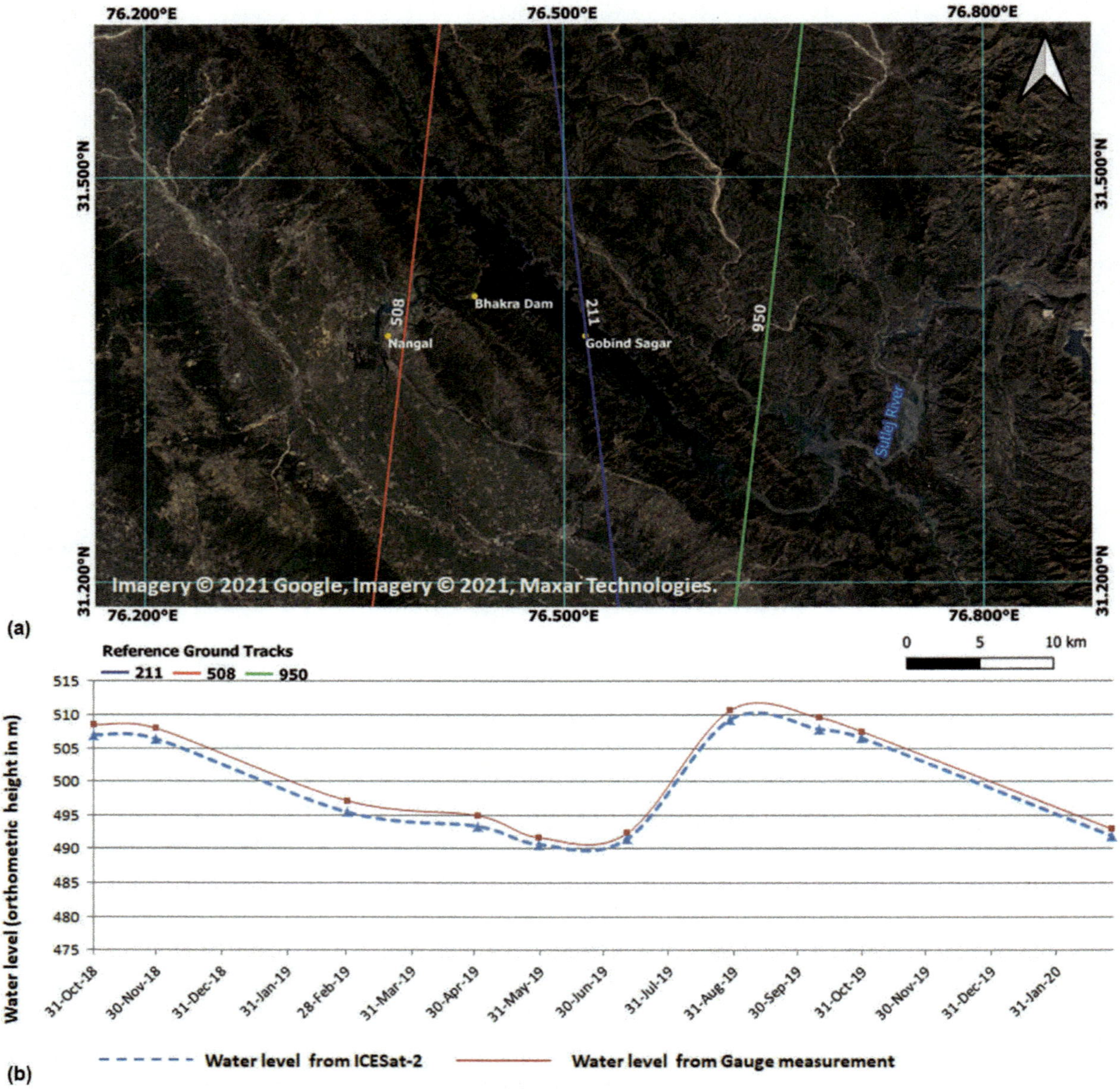


Figure 14.5 Annual variations of water level in the Gobind Sagar reservoir. Water levels from ICESat-2 ATL13 data product for the year 2019. (a) Extent of Gobind Sagar reservoir situated in the Bilaspur districts of Himachal Pradesh state, India and located in the foothills of Himalaya on the northern edge of Indo Gangetic plain, with three reference ground tracks overlaid; (b) interpolated annual water levels in Gobind Sagar (from October 2018 to January 2020) using ICESat-2 ATL13's surface water-level height data product and CWC reported gauge-based water levels.

district, Himachal Pradesh state in India and located at the foothills of the Himalayas on the northern edge of the Indo-Gangetic plain, is used to demonstrate the application of the ATL13 data product toward analyzing the annual variations of water level. Water contribution in the Gobind Sagar reservoir is due to the glacier and snow melt runoff that is carried in the annual flow in the Satluj river at Bhakra dam. Earlier, Singh and Jain (2002) and Singh et al. (2018) carried out extensive studies related to the Gobind Sagar. Table 14.1 shows the water levels retrieved using the ATL13 data product over the Gobind Sagar reservoir from the available ground tracks between the years 2018 and 2020, along with the water levels reported in weekly water bulletins by the Central Water Commission (India) (CWC, 2021). Figure 14.5(b) shows the interpolated profile of water levels in Gobind Sagar using the data shown in Table 14.2 and CWC reported gauged-based water levels. A systematic deviation of 90 cm to 1.2 m is observed between ICESat-2 derived water levels and CWC reported water levels. This deviation can be attributed to height conversion from ellipsoidal height to orthometric height.

Water levels in the reservoirs that are located at the foothills of the Himalayan region are totally controlled by the precipitation dynamics in the form of snow and glaciers. The Gobind Sagar reservoir is part of Sutlej basin and covers the outer Himalayas (Siwalik ranges), middle Himalayas (Dhauladhar range), and greater Himalayas. Water inflow in the reservoirs is due to snowmelt and glacier melt runoff as well as discharge due to monsoon rains. From Figure 14.5b it is understood that the water level in Gobind Sagar reservoirs gradually increases during mid-summer and reaches its peak during July and September. The results shown here strongly agree with the earlier studies done by Singh and Jain (2002) and Singh et al. (2018). Earlier, using similar methodology, Ma et al. (2019) and Xu et al. (2020a) have computed the volumes of various lakes post retrieving the water levels from the ICESat-2 photon data.

## 14.7 Inferring Lake Ice Phenology in Mountainous Regions Using ICESat-2 Photon Data

Lake ice phenology exhibits processes like freezing, the break-up of surface ice, and melting which are sensitive indicators of climate change. Himalayan mountains embrace a large number of glaciers and lakes which are highly influenced by climate change. Most of the lakes in the Himalayas are no stranger to cold and snowy winters during which ice forms naturally on lakes. Many elements drive ice to melt on the lakes, and the sunlight is the most significant factor among them. Especially during summer, direct sunlight heats the lake ice which then melts the ice from beneath. During the freeze-up period, the surface of lake ice will be much flatter than surrounding terrain and ICESat-2 photons can be used to probe the lake-ice phenology. Initial assessment of ICESat-2 derived ice-sheet surface heights in Antarctica was done by Brunt et al. (2019) who concluded that the ATL03 data product is accurate to better than 13 cm of surface measurement precision. Fricker et al. (2021) utilized ICESat-2 data to study the surface melt dynamics of the Amery ice shelf in East Antarctica.

Figure 14.6 shows an example to retrieve lake ice phenology in Spanggur Lake during the January and July months

**Table 14.2** Water levels retrieved from ICESat-2's ATL13 data product (from the available ground tracks between the years 2018 and 2020) and Central Water Commission's (India) weekly bulletins for the Gobind Sagar reservoir.

| Date of Acquisition | Ground Track ID | Water Level (Orthometric Height) m From ICESat-2 | Water Level (Orthometric Height) m From Gauge Measurement |
|---|---|---|---|
| 31 October 2018 | 508 | 506.98 | 508.45 |
| 29 November 2018 | 950 | 506.39 | 507.97 |
| 28 February 2019 | 950 | 495.46 | 497.15 |
| 1 May 2019 | 508 | 493.35 | 494.94 |
| 30 May 2019 | 950 | 490.66 | 491.69 |
| 11 July 2019 | 211 | 491.50 | 492.33 |
| 29 August 2019 | 950 | 509.32 | 510.68 |
| 10 October 2019 | 211 | 507.85 | 509.57 |
| 30 October 2019 | 508 | 506.54 | 507.51 |
| 26 February 2020 | 950 | 491.86 | 492.88 |

of 2020. Spanggur Lake is located to the south of Pangong Lake situated on the Indo-China border and exhibits similar characteristics to that of Pangong Lake's ice phenology (Yan et al., 2019). Figure 14.6a shows the satellite data (Sentinel 2A) acquired over Spanggur Lake dated 30 January 2020 with an overlay of a subset of beams of ICESat-2 ATL03 data acquired on 26 January 2020 during which the season exhibits severe winter conditions and

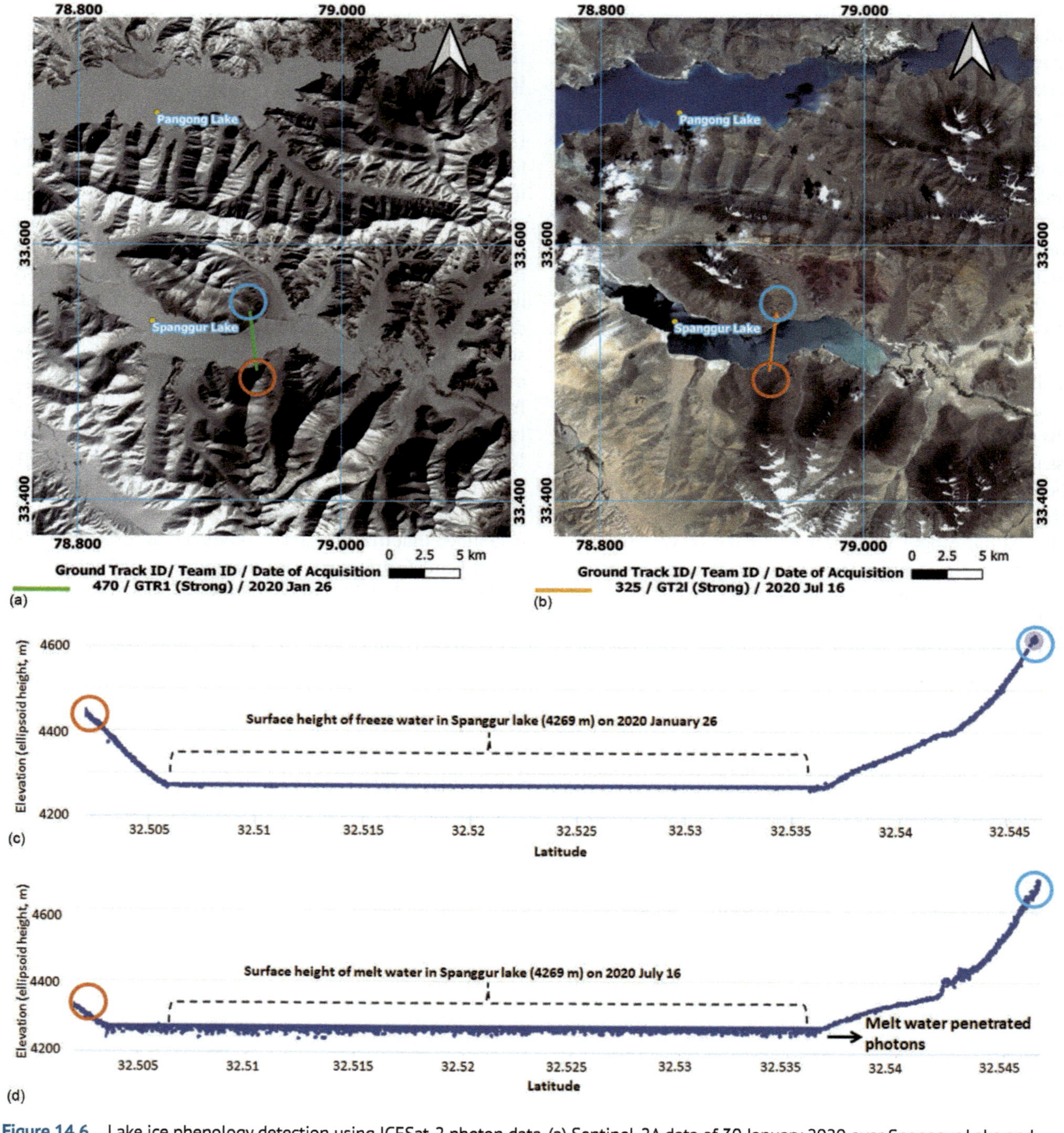


Figure 14.6 Lake ice phenology detection using ICESat-2 photon data. (a) Sentinel-2A data of 30 January 2020 over Spanggur Lake and overlaid with a subset of ICESat-2 photon data dated 26 January 2020. Note the frozen water in Lake Spanngur as the season in January exhibits severe winter conditions; (b) Sentinel-2A data dated 13 July 2020 with an overlay of ICESat-2 data that was acquired on 16 July 2020. During these late summer seasons, water melts; (c) lake surface height profile derived from the subset of a beam of ICESat-2 photons acquired on 26 January 2020 showing smooth surface; (d) lake surface height profiles derived from the subset of a beam of ICESat-2 photons acquired on 16 July 2020 on meltwater showing the photons penetrating up to a depth of ~25 m during non-frozen water conditions.

triggers lake water to freeze. Similarly, Figure 14.6b shows the satellite data acquired on 13 July 2020 with an overlay of ICESat-2 data acquired on 16 July 2020 and during this late summer season, the surface ice on the lake melts. Figures 14.6c and 14.6d show the profile of water surface height retrieved from the ICESat-2 ATL03 data product. It is evident from the profiles that during the freezing time of the lake the profile exhibits smooth nature of surface height, whereas the non-frozen water conditions enable photons to penetrate to a depth of ~25 m.

## 14.8 Estimating Tree Heights in Mountain Regions Using ICESat-2 Photon Data

Biomass monitoring is vital to carbon accounting programs like "Reducing Emissions from Deforestation and forest Degradation" (REDD+), which aim to provide economic incentives toward forest conservation (Virgilio et al., 2010). Researchers like Feldpausch et al. (2012) have appreciated the usage of tree height as a parameter during the computation of forest biomass. Lidar data can quantify the canopy height and to some extent indicates the structure of vegetation (Andersen et al., 2005). Though the primary scientific objective of ICESat-2 is to measure changes in the cryosphere, it provides an opportunity to measure changes in various global surfaces like oceans, seas, inland water bodies, land, and canopy. Investigations done by Neuenschwander and Magruder (2019) provided the first assessment of using ICESat-2 data to map the vegetation over three ecosystems – boreal forest, tropical forest, and semi-arid woodlands. Lin et al. (2020) and Sun et al. (2020) have successfully retrieved canopy height by utilizing ICESat-2 data.

Figure 14.7 shows a typical example of using ICESat-2 photon data to detect the tree heights in an undulated terrain in the surroundings of Chakban Pathiar village located in the Kangra district, Himachal Pradesh state, India. The vicinity is shrouded with Ban Oak (*Quercus leucotrichphora*) species. The Ban Oak in this stretch of the region, which is confined to the elevation range of 900 m to 1200 m, are generally pure, open canopied, and of short bole. From the field assessment, it is observed that the height of this species is in the range of 20 to 22 m (Bharti et al., 2019). Figure 14.7a shows the satellite data for the extent of Chakban Pathiar and surroundings with an overlay of the beam of ICESat-2 photon data (ATL03 data product) and Figure 14.7b shows the corresponding canopy height profile. Figure 14.7c shows the zoomed profile for the area earmarked with the red box in the profile as illustrated in Figure 14.7b. Double surface returned photons are evident from this profile, of which certain photons have been reflected from the canopy and some from the ground surface. From the profile analysis, it is understood that the difference between the ground-hit photons and canopy-hit photons is in the range of 21 m to 22 m and this accuracy is close to the investigation done by Bharti et al. (2019). Evaluation done by Duncanson et al. (2020), Sun et al. (2020), and Zhu et al. (2020) appreciated the application of ICESat-2 photon data for biomass estimation due to its accuracy.

## 14.9 Utilization of ICESat-2 Photon Data to Generate Digital Elevation Models

Earlier, Berry et al. (2010, 2019) have synergistically merged data from multi-mission satellite radar altimetry and the Shuttle Radar Topography Mission (SRTM), and the product was made available to the public under the title of Altimeter Corrected Elevation (ACE). Works done by Brenner et al. (2007) suggest that elevation information from laser altimeters is better in terms of accuracy in comparison to radar altimeters. Wang et al. (2018) proposed a method to generate an accurate DEM for the Antarctica coastal areas by merging ICESat data and ASTER GDEM. On similar lines, Yue et al. (2017) used altimeter data from the ICESat GLAS sensor to enhance ASTER GDEM and further used the resultant product to fill the voids in SRTM.

The unprecedented accuracy of elevation information and the availability of a trillion photons enables ICESat-2 photon data to harness its potential during the process of generating DEMs through photogrammetric techniques or improving the existing open-access global DEMs by using various fusion and interpolation methods. Softcopy photogrammetric solutions constitute bundle block adjustments, which compute a single solution for all the stereo images in the project block and attempts to reduce the total re-projection error between the observed and predicted image points through a mathematical expression as the sum of squares of a large nonlinear and real-valued functions. During this process, there is need of ground control points (GCPs), tie points, and checkpoints (Dandabathula, 2015). Elevation values from the ICESat-2 photons can be utilized as vertical control points to strengthen the bundle block adjustments.

DEM generation in Himalayan regions is highly challenging due to the inherent presence of rugged topography. In the extent of the Himalayan region, processing the stereoscopic imagery to generate DEM and orthoimages using photogrammetric techniques possess certain challenges and requires manual methods to correct the final products (Giribabu et al., 2013). Similarly, publicly-available global

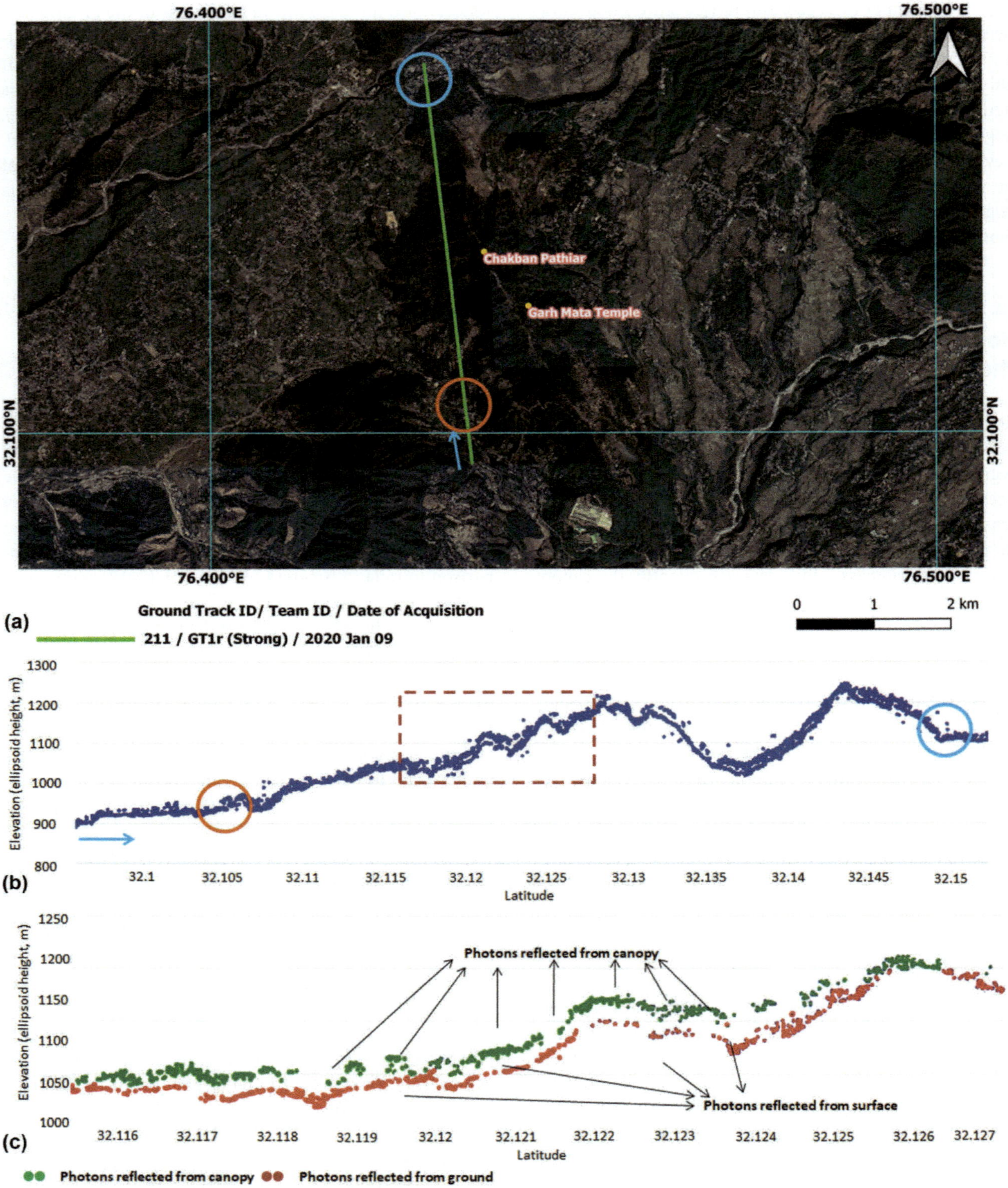


**Figure 14.7** Canopy height detection in mountain regions using ICESat-2 photon data. Satellite data showing the extent of undulating terrain in the vicinity of Chakban Pathiar and its surroundings overlaid with a subset of ICESat-2 photon data dated 09 January 2020 (GT1R – strong beam). The elevation range of this study area falls between 900 m and 1200 m; (b) surface height profile derived from the subset of ICESat-2 photon beam over Chakban Pathiar hills. Note the double reflections in the profile that are from the top of the canopy and surface reflected photons; (c) part of the profile zoomed (earmarked with the red color box in (b)) demonstrating the ability to distinguish the photons reflected from canopy and photons reflected from the ground. The distinguishability of canopy (represented in green) and surface reflected photons (represented with red) enables to compute the tree heights and in this region dominated with Ban Oak trees, the tree heights computed from these profiles stand in the range of 21 m to 22 m.

DEMs like SRTM and ASTER do have large errors in high mountainous regions of the Himalayas and this restricts their applications for practical purposes (Berthier et al., 2006; Gupta et al., 2014; Robinson et al., 2014; Mukul et al., 2017; Liu et al., 2019). Figure 14.8 shows the application of using ICESat-2 photons to improve the accuracy and also a super-resolution reconstruction from existing DEM. Solutions from commercial off-the-shelf software like PCI Geomatica contain functionality like DEMAdjust. This sort of function will adjust the existing DEM (surface) to the accurate elevation points (PCI, 2021). Thus, existing DEMs can readjust or re-interpolate with a large number of high-confidence ICESat-2 photons. Figure 14.8a shows a part of Triangulated Irregular Network (TIN) of SRTM GL1 (ellipsoidal) DEM of 30 m pixel posting (dataset available from SRTM, 2013) and overlaid with the high-confidence photons from the strong beams of ICESat-2 ground tracks of various dates. The study area shown in Figure 14.8 is the surroundings of the Bhagirath Kharak and Satopanth glaciers along with Balakun peaks. SRTM GL1 DEM datasets are provided in WGS84 ellipsoidal

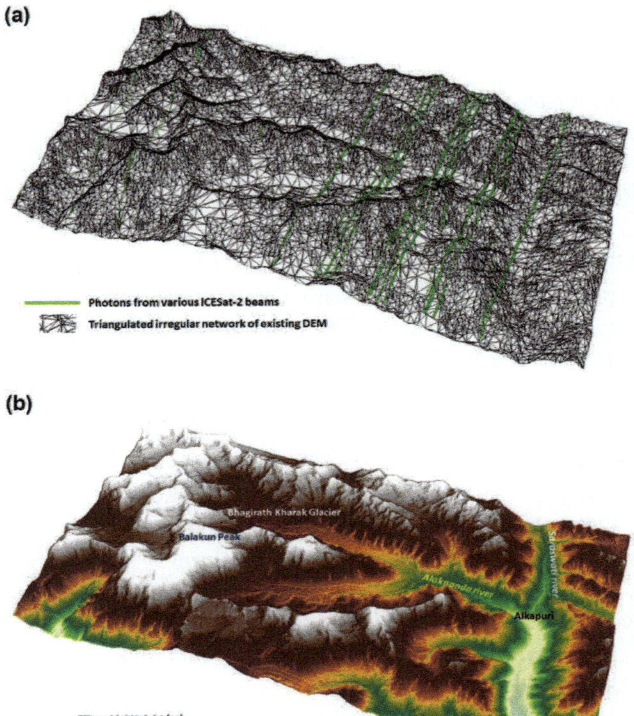


**Figure 14.8** Utilization of ICESat-2 photon data to readjust/refine digital elevation models. (a) Triangulated irregular network (TIN) of SRTM GL1 (ellipsoidal) of 30 m pixel posting overlaid with the high confidence photons from the strong beams of ICESat-2 ground tracks of various dates; (b) the readjusted/refined DEM with super-resolution (10 m) that was generated from existing DEM and ICESat-2 photons. The study area shown here contains Bhagirath Kharak and Satopanth glaciers along with Balakun peak.

vertical datum and thus will be in harmony with the elevation values from ICESat-2 photon data, which also gives surface height relative to the WGS-85 ellipsoid. Figure 14.8b shows the readjusted/refined DEM with super-resolution (10 m) that was generated by readjusting the existing SRTM DEM with ICESat-2 photons. However, during the process of this refinement, we need to take care of the distribution of photon points in the extent of the study area and the accuracy of the final DEM depends on the interpolation mechanism used. On the same lines, water-body smoothing is an important component in the DEM generation process. Toward this, surface water levels that were retrieved from the ICESat-2 data products can be used to smooth the elevation values at water bodies (Xu et al., 2020b)

## 14.10 Conclusion

Launched in September 2018, NASA's ICESat-2, which was commissioned with the intention to quantify the amount of changes in certain Earth system processes, has truly provided key insights into their behavior. Even though ICESat-2 is primarily intended for cryosphere applications, the unique operational mechanism has opened the doors to utilize the surface reflected photons for various domains of Earth system processes. In this chapter, emphasis has been given to the applications of ICESat-2 in the Himalayan region. Approximately 185 RGT of ICESat-2 are available over the extent of the Himalayas (Figure 14.9) and each ground track has a 91-day repetition with 6 separated beams.

In this chapter, a brief introduction to the ICESat-2 working mechanism has been provided along with probable applications in the Himalayan region. The cutting-edge photonics technology has enabled the retrieval of high-resolution and precise elevation profiles over rugged terrains. ICESat-2 data, if used in synchrony with the optical remote sensing data, can provide quantitative information. Level 2A data product of ICESat-2, namely ATL03, which is a finished grade product, can be readily used to study the along-track topographic features of any terrain. Application areas that include studies related to ice phenology, surface water height, and canopy height may pose a challenge for optical remote sensing methods, but these challenges can be overcome using ICESat-2 photon data in synergy with imaging products. Numerous studies have shown the potential of ICESat-2 to detect and quantify the changes in the Antarctica and Arctic glaciers. As the Himalayan region, which is treated as a third pole of Earth, comprises a large number of dynamic glaciers, ICESat-2 can be used to extract the minute features of glaciers in this region. As ICESat-2 photons penetrate surface water in melt ponds that form during summer, studies related to freshwater stores can be done which will provide

**Figure 14.9** Availability of ICESat-2 photon data for the Himalayan region. Reference Ground Tracks (GRT) of ICESat-2 as seen in OpenAltimetry web-portal (https://openaltimetry.org). With each ground track having a repetition rate of 91 days, a trillion laser pulses have been harvested from numerous geolocated photons with surface height information. The vast collection of this geolocated photons archive can be used for a variety of applications in the Himalayan region.

insight into the dynamics of the annual water melt. The archived records of surface heights in the order of a trillion photons for the extent of the Himalayan region can be utilized effectively to refine the existing DEMs or regenerate new DEMs. These DEMs will have a prominent role to perform the hydrological and geomorphological studies in the Himalayan region. ICESat-2 photon data, when appropriately incorporated into the models along with the complementary atmospheric parameters, will help scientists to understand the processes that influence climate change and fill the gaps in climate prediction capability. Similarly, there is the larger scope and potential for ICESat-2 data in the area of hydrology like water surface height retrieval, bathymetric studies, and understanding lake ice phenology.

## Acknowledgments

The author would like to thank Dr Apurba Kumara, Bera, Dr Sitiraju Srinivasa Rao, General Manager, Regional Remote Sensing Centre (West), Jodhpur, Rajasthan, and Dr Chandra Shekhar Jha, Chief General Manager, Regional Centres, NRSC, Hyderabad for providing valuable encouragement toward completing this work. The author would like to express special thanks to Director, National Remote Sensing Centre for facilitating the institutional support and providing basic infrastructure for this work.

## References

Aiken, S.J. and Brierley, G.J. (2013). Analysis of longitudinal profiles along the eastern margin of the Qinghai-Tibetan Plateau. *Journal of Mountain Science* 10(4): 643–657. doi: 10.1007/s11629-013-2814-2.

Alsdorf, D.E., Rodríguez, E., and Lettenmaier, D.P. (2007). Measuring surface water from space. *Reviews of Geophysics* 45(2): RG2002. doi: 10.1029/2006RG000197.

Andersen, H.E., McGaughey, R.J., and Reutebuch, S.E. (2005). Estimating forest canopy fuel parameters using LIDAR data. *Remote Sensing of Environment* 94(4): 441–449. doi: 10.1016/j.rse.2004.10.013.

Bahadur, J. (1993). The Himalayas: a third polar region. In: *Snow and Glacier Hydrology. Proceedings of Kathmandu Symposium – 1992*. 181–190.

Bandyopadhyay, J. (2013). Securing the Himalayas as the water tower of Asia: an environmental perspective. *Asia Policy* 16(1): 45–50.

Berry, P.A.M., Smith, R.G., and Benveniste, J. (2010). ACE2: the new global digital elevation model. In: *Gravity, Geoid and Earth Observation: IAG Commission 2: Gravity Field, Chania, Crete, Greece* (ed. P.S. Mertikas), 231–237. Berlin, Heidelberg: Springer.

Berry, P.A.M., Smith, R.G., and Benveniste, J. (2019). Altimeter Corrected Elevations, Version 2 (ACE2). Palisades, NY: NASA Socioeconomic Data and

Applications Center (SEDAC). Available online: https://doi.org/10.7927/H40G3H78

Berthier, E., Arnaud, Y., Vincent, C. et al. (2006). Biases of SRTM in high-mountain areas: implications for the monitoring of glacier volume changes. *Geophysical Research Letters* 33(8): L08502. doi: 10.1029/2006GL025862.

Bharti, H., Panatu, A., Kiran, L. et al. (2019). *Temporal Change in Tree Species Composition in Palampur Forest Division of Dharamshala Forest Circle, Himachal Pradesh.* Shimla, India: Himachal Pradesh State Centre on Climate Change (HIMCOSTE).

Brenner, A.C., DiMarzio, J.P., and Zwally, H.J. (2007). Precision and accuracy of satellite radar and laser altimeter data over the continental ice sheets. *IEEE Transactions on Geoscience and Remote Sensing* 45(2): 321–331. doi: 10.1109/TGRS.2006.887172.

Brown, M.E., Arias, S.D., Neumann, T. et al. (2016). Applications for ICESat-2 data: from NASA's early adopter program. *IEEE Geoscience and Remote Sensing Magazine* 4(4): 24–37. doi: 10.1109/MGRS.2016.2560759.

Brunt, K.M., Neumann, T.A., and Smith, B.E. (2019). Assessment of ICESat-2 ice sheet surface heights, based on comparisons over the interior of the Antarctic ice sheet. *Geophysical Research Letters* 46(22): 13072–13078. doi: 10.1029/2019GL084886.

Champagnac, J.D., Molnar, P., Sue, C. et al. (2012). Tectonics, climate, and mountain topography. *Journal of Geophysical Research: Solid Earth* 117(B2): B02403. doi: 10.1029/2011JB008348.

CWC (2021). Central Water Commission. Available online: http://www.cwc.gov.in/reservoir-level-storage-bulletin

Dandabathula, G. (2015). Generation of statewide DEMs and orthoimages: guidelines and methodology. *Kartografija i geoinformacije (Cartography and Geoinformation)* 13(22): 4–19.

Dandabathula, G. and Rao, S.S. (2020). Validation of ICESat-2 surface water level product ATL13 with near real time gauge data. *Hydrology* 8(2): 19. doi: 10.11648/j.hyd.20200802.11.

Dandabathula, G., Sitiraju, S.R., and Jha, C.S. (2021a). Retrieval of building heights from ICESat-2 photon data and evaluation with field measurements. *Environmental Research: Infrastructure and Sustainability* 1: 011003. doi: 10.1088/2634-4505/abf820.

Dandabathula, G., Sitiraju, S.R., and Jha, C.S. (2021b). Morphological profiles of sand dunes from ICESat-2 geolocated photons. *Journal of Geoscience and Environment Protection* 9(2): 71–91. doi: 10.4236/gep.2021.92005.

Duncanson, L., Neuenschwander, A., Hancock, S. et al. (2020). Biomass estimation from simulated GEDI, ICESat-2 and NISAR across environmental gradients in Sonoma County, California. *Remote Sensing of Environment* 242: 111779. doi: 10.1016/j.rse.2020.111779.

Dandabathula, G., Verma, M., Satyanarayana, P. et al. (2020). Evaluation of ICESat-2 ATL08 data product: performance assessment in inland water. *European Journal of Environment and Earth Sciences* 1(3): 1–7. https://doi.org/10.24018/ejgeo.2020.1.3.15.

Falkner, P. and Schulz, R. (2015). Instrumentation for planetary exploration missions. In: *Treatise on Geophysics* (ed. Schubert, G.), 719–755. Amsterdam: Elsevier.

Feldpausch, T.R., Lloyd, J., Lewis, S.L. et al. (2012). Tree height integrated into pantropical forest biomass estimates. *Biogeosciences* 9(8): 3381–3403. doi: 10.5194/bg-9-3381-2012.

Fricker, H.A., Arndt, P., Brunt, K.M. et al. (2021). ICESat-2 meltwater depth estimates: application to surface melt on Amery Ice Shelf, East Antarctica. *Geophysical Research Letters* 48(8): e2020GL090550. doi: 10.1029/2020GL090550.

Giribabu, D., Kumar, P., Mathew, J. et al. (2013). DEM generation using Cartosat-1 stereo data: Issues and complexities in Himalayan terrain. *European Journal of Remote Sensing* 46(1): 431–443. doi: 10.5721/EuJRS20134625.

Gupta, R.D., Singh, M.K., Snehmani, S. et al. (2014). Validation of SRTM X band DEM over Himalayan Mountain. *The International Archives of Photogrammetry, Remote Sensing and Spatial Information Sciences* 40(4): 71. doi: 10.5194/isprsarchives-XL-4-71-2014.

Gupta, S.K. and Shukla, D.P. (2016). Assessment of land use/land cover dynamics of Tso Moriri Lake, a Ramsar site in India. *Environmental Monitoring and Assessment* 188(12): 1–3. doi: 10.1007/s10661-016-5707-3.

Hack, J.T. (1957). Studies of longitudinal stream profiles in Virginia and Maryland. US Government Printing Office. Available online: https://pubs.usgs.gov/pp/0294b/report.pdf

ICESat-2 Data Sets. (2018). ICESat-2 Data Sets at NSIDC. Available online: https://nsidc.org/data/icesat-2/data-sets

Jasinski, M.F., Stoll, J.D., Cook, W.B. et al. (2016). Inland and near-shore water profiles derived from the high-altitude Multiple Altimeter Beam Experimental Lidar (MABEL). *Journal of Coastal Research* 76: 44–55. doi: 10.2112/SI76-005.

Jasinski, M.F., Stoll, J.D., Hancock, D. et al. (2020). Algorithm Theoretical Basis Document (ATBD) for Inland Water Data Prodcuts, ATL13, Ver. 3. NASA Goddard Space flight Centre, Greenbelt, MD, 112 pp. Available online: https://nsidc.org/sites/nsidc.org/files/technical-references/ICESat2_ATL13_Known_Issues_v003_Aug2020.pdf

Khalsa, S.J.S., Borsa, A., Nandigam, V. et al. (2020). OpenAltimetry-rapid analysis and visualization of Spaceborne altimeter data. *Earth Science Informatics*. doi: 10.1007/s12145-020-00520-2.

Lai, Y.R. and Wang, L. (2021). Monthly surface elevation changes of the Greenland Ice Sheet from ICESat-1,

CryoSat-2, and ICESat-2 altimetry missions. *IEEE Geoscience and Remote Sensing Letters* 19: 1–5. doi: 10.1109/lgrs.2021.3058956.

Lamb, S. and Watts, A. (2010). The origin of mountains: implications for the behaviour of Earth's lithosphere. *Current Science* 99(12): 1699–1718.

Lin, X., Xu, M., Cao, C., Dang, Y. et al. (2020). Estimates of forest canopy height using a combination of ICESat-2/ATLAS data and stereo-photogrammetry. *Remote Sensing* 12(21): 3649. doi: 10.3390/rs12213649.

Liu, K., Song, C., Ke, L. et al. (2019). Global open-access DEM performances in Earth's most rugged region High Mountain Asia: a multi-level assessment. *Geomorphology* 338: 16–26. doi: 10.1016/j.geomorph.2019.04.012.

Luo, S., Song, C., Zhan, P. et al. (2021). Refined estimation of lake water level and storage changes on the Tibetan Plateau from ICESat/ICESat-2. *CATENA* 200: 105177. doi: 10.1016/j.catena.2021.105177.

Ma, Y., Xu, N., Sun, J., Wang, X.H. et al. (2019). Estimating water levels and volumes of lakes dated back to the 1980s using Landsat imagery and photon-counting lidar datasets. *Remote Sensing of Environment* 232: 111287. doi: 10.1016/j.rse.2019.111287.

Markus, T., Neumann, T., Martino, A. et al. (2017). The Ice, Cloud, and land Elevation Satellite-2 (ICESat-2): science requirements, concept, and implementation. *Remote Sensing of Environment* 190: 260–273. doi: 10.1016/j.rse.2016.12.029.

Mukul, M., Srivastava, V., Jade, S. et al. (2017). Uncertainties in the shuttle radar topography mission (SRTM) heights: insights from the Indian Himalaya and Peninsula. *Scientific Reports* 7(1): 1–10. doi: 10.1038/srep41672.

National Research Council (2010). *Landscapes on the Edge: New Horizons for Research on Earth's Surface*, 180. Washington, DC: The National Academies Press.

Neuenschwander, A.L. and Magruder, L.A. (2019). Canopy and terrain height retrievals with ICESat-2: A first look. *Remote Sensing* 11(14): 1721. doi: 10.3390/rs11141721.

Neumann, T., Brenner, A., Hancock, D. et al. (2018). *ICE, CLOUD, and Land Elevation Satellite-2 (ICESat-2) Project: Algorithm Theoretical Basis Document (ATBD) for Global Geolocated Photons: ATL03*. National Aeronautics and Space Administration, Goddard Space Flight Center, Greenbelt, MD. Available online: https://icesat-2.gsfc.nasa.gov/sites/default/files/files/ATL03_05June2018.pdf.

Neumann, T.A., Brenner, A., Hancock, D. et al. (2021). *ATLAS/ICESat-2 L2A Global Geolocated Photon Data, Version 4*. Boulder, CO: NASA National Snow and Ice Data Center Distributed Active Archive Center. doi: 10.5067/ATLAS/ATL03.004.

Neumann, T.A., Martino, A.J., Markus, T. et al. (2019). The Ice, Cloud, and Land Elevation Satellite – 2 mission: a global geolocated photon product derived from the Advanced Topographic Laser Altimeter System. *Remote Sensing of Environment* 233: 111325. doi: 10.1016/j.rse.2019.111325.

NSIDC-ICESat-2 (2018). National Snow & Ice Data Centre – ICESat-2. Available online: https://nsidc.org/data/icesat-2

Palazzi, E., Mortarini, L., Terzago, S. et al. (2019). Elevation-dependent warming in global climate model simulations at high spatial resolution. *Climate Dynamics* 52(5): 2685–2702. doi: 10.1007/s00382-018-4287-z.

Parrish, C.E., Magruder, L.A., Neuenschwander, A.L. et al. (2019). Validation of ICESat-2 ATLAS bathymetry and analysis of ATLAS's bathymetric mapping performance. *Remote Sensing* 11(14): 1634. doi: 10.3390/rs11141634.

PCI (2021). DEMADJUST – DEM adjustment to elevation points. Available online: https://www.pcigeomatics.com/geomatica-help/references/pciFunction_r/easi/E_demadjust.html

Pinel, V., Poland, M.P., and Hooper, A. (2014). Volcanology: lessons learned from synthetic aperture radar imagery. *Journal of Volcanology and Geothermal Research* 289: 81–113. doi: 10.1016/j.jvolgeores.2014.10.010.

Prerna, R., Pandey, D.K., and Mahender, K. (2018). Longitudinal profiling and elevation-relief analysis of the Indus. *Arabian Journal of Geosciences* 11(13): 1–8. doi: 10.1007/s12517-018-3657-5.

Robinson, N., Regetz, J., and Guralnick, R.P. (2014). EarthEnv-DEM90: a nearly-global, void-free, multi-scale smoothed, 90 m digital elevation model from fused ASTER and SRTM data. *ISPRS Journal of Photogrammetry and Remote Sensing* 87: 57–67. doi: 10.1016/j.isprsjprs.2013.11.002.

Rocha, J., Duarte, A., Silva, M. et al. (2020). The importance of high resolution digital elevation models for improved hydrological simulations of a Mediterranean forested catchment. *Remote Sensing* 12(20): 3287. doi: 10.3390/rs12203287.

Schutz, B.E., Zwally, H.J., Shuman, C.A. et al. (2005). Overview of the ICESat mission. *Geophysical Research Letters* 32: 1–4 (L21S01). doi: 10.1029/2005GL024009.

Singh, P. and Jain, S.K. (2002). Snow and glacier melt in the Satluj River at Bhakra Dam in the western Himalayan region. *Hydrological Sciences Journal* 47(1): 93–106. doi: 10.1080/02626660209492910.

Singh, S., Dhasmana, M.K., Shrivastava, V. et al. (2018). Estimation of revised capacity in Gobind Sagar reservoir using Google Earth Engine and GIS. *International Archives of the Photogrammetry, Remote Sensing and Spatial Information Sciences* 42: 5. doi: 10.5194/isprs-archives-XLII-5-589-2018.

Srivastava, P., Bhambri, R., Kawishwar, P. et al. (2013). Water level changes of high altitude lakes in Himalaya–Karakoram from ICESat altimetry. *Journal of*

*Earth System Science* 122(6): 1533–1543. doi: 10.1007/s12040-013-0364-1.

SRTM (2013). Shuttle Radar Topography Mission (SRTM) global. OpenTopography. doi: 10.5069/G9445JDF

Sun, T., Qi, J., and Huang, H. (2020). Discovering forest height changes based on spaceborne lidar data of ICESat-1 in 2005 and ICESat-2 in 2019: a case study in the Beijing-Tianjin-Hebei region of China. *Forest Ecosystems* 7(1): 1–2. doi: 10.1186/s40663-020-00265-w.

Vemsani, L. (2015). The Himalayan ranges, glaciers, lakes and rivers: an international ecological, economic and military outlook. In: *The Political Economy of Conflict in South Asia* (ed. M. Webb and A. Wijeweera), 171–190. London: Palgrave Macmillan.

Virgilio, N.R., Marshall, S., Zerbock, O. et al. (2010). *Reducing Emissions from Deforestation and Degradation (REDD): A Casebook of On-the-ground Experience*. Arlington, Virginia: The Nature Conservancy, Conservation International and Wildlife Conservation Society.

Wallace, S.J. (1881). Mountain elevation, and changes of temperature, in geology. *Science* os-2(46): 206–206. doi: 10.1126/science.os-2.46.206.

Wang, Q., Yi, S., and Sun, W. (2021). Continuous estimates of glacier mass balance in High Mountain Asia based on ICESat-1, 2 and GRACE/GRACE Follow-On data. *Geophysical Research Letters* 48(2): e2020GL090954. doi: 10.1029/2020GL090954.

Wang, S., Alexander, P., Wu, Q. et al. (2021). Characterization of ice shelf fracture features using ICESat-2: a case study over the Amery Ice Shelf. *Remote Sensing of Environment* 255: 112266. doi: 10.1016/j.rse.2020.112266.

Wang, X., Cheng, X., Gong, P. et al. (2011). Earth science applications of ICESat/GLAS: a review. *International Journal of Remote Sensing* 32(23): 8837–8864. doi: 10.1080/01431161.2010.547533.

Wang, X., Holland, D.M., and Gudmundsson, G.H. (2018). Accurate coastal DEM generation by merging ASTER GDEM and ICESat/GLAS data over Mertz Glacier, Antarctica. *Remote Sensing of Environment* 206: 218–230. doi: 10.1016/j.rse.2017.12.041.

Xiang, J., Li, H., Zhao, J. et al. (2021). Inland water level measurement from spaceborne laser altimetry: validation and comparison of three missions over the Great Lakes and lower Mississippi River. *Journal of Hydrology* 11: 126312. doi: 10.1016/j.jhydrol.2021.126312.

Xiao, C.W., Feng, Z.M., Li, P. et al. (2018). Evaluating the suitability of different terrains for sustaining human settlements according to the local elevation range in China using the ASTER GDEM. *Journal of Mountain Science* 15(12): 2741–2751. doi: 10.1007/s11629-018-5058-3.

Xu, N., Ma, Y., Zhang, W. et al. (2020a). Monitoring annual changes of lake water levels and volumes over 1984–2018 using landsat imagery and ICESat-2 data. *Remote Sensing* 12(23): 4004. doi: 10.3390/rs12234004.

Xu, N., Ma, Y., Zhou, H., Zhang, W. et al. (2020b). A method to derive bathymetry for dynamic water bodies using ICESat-2 and GSWD datasets. *IEEE Geoscience and Remote Sensing Letters* 19: 1–5. 1500305. doi: 10.1109/LGRS.2020.3019396.

Xu, N., Zheng, H., Ma, Y. et al. (2021). Global estimation and assessment of monthly lake/reservoir water level changes using ICESat-2 ATL13 products. *Remote Sensing* 13(14): 2744. doi: 10.3390/rs13142744.

Yan, Y., Xu, H., Liu, G. et al. (2019). Analysis of the variations of the lake ice phenology in the Pangong Lake area from 2013 to 2017: remote sensing survey of the cryosphere in the high altitude and alpine region, West China. *Remote Sensing for Land & Resources* 31(3): 209–125. Available online: http://www.gtzyyg.com/EN/10.6046/gtzyyg.2019.03.26.

Yao, T., Thomsan, L., Chend, D. et al. (2020). Third Pole climate warming and cryosphere system changes. *World Meteorological Organisation Bulletin* 69(1): 35–38. Available online: https://public.wmo.int/en/resources/bulletin/third-pole-climate-warming-and-cryosphere-system-changes.

Yuan, C., Gong, P., and Bai, Y. (2020). Performance assessment of ICESat-2 laser altimeter data for water-level measurement over lakes and reservoirs in China. *Remote Sensing* 12(5): 770. doi: 10.3390/rs12050770.

Yue, L., Shen, H., Zhang, L. et al. (2017). High-quality seamless DEM generation blending SRTM-1, ASTER GDEM v2 and ICESat/GLAS observations. *ISPRS Journal of Photogrammetry and Remote Sensing* 123: 20–34. doi: 10.1016/j.isprsjprs.2016.11.002.

Zhang, G., Chen, W., and Xie, H. (2019). Tibetan Plateau's lake level and volume changes from NASA's ICESat/ICESat-2 and Landsat missions. *Geophysical Research Letters* 46(22): 13107–13118. doi: 10.1029/2019GL085032.

Zhang, S. and Pavelsky, T.M. (2019). Remote sensing of lake ice phenology across a range of lakes sizes, ME, USA. *Remote Sensing* 11(14): 1718.

Zhu, X., Nie, S., Wang, C. et al. (2020). The performance of ICESat-2's strong and weak beams in estimating ground elevation and forest height. In: *IGARSS 2020-2020 IEEE International Geoscience and Remote Sensing Symposium*. IEEE.

# 15

# Extreme Hydrological Event-Induced Temporal Variation in Soil Erosion of the Assiganga River Basin, NW Himalaya

*Rohit Kumar[1], Rahul Devrani[2,*], Astha Dangwal[1], Benidhar Deshmukh[1], and Som Dutt[3]*

[1] *Discipline of Geology, School of Sciences, Indira Gandhi National Open University, New Delhi 110068, India.*
[2] *Delhi School of Climate Change & Sustainability, Institution of Eminence, University of Delhi, Delhi, India*
[3] *Wadia Institute of Himalayan Geology, 33 GMS Road, Dehradun 248001, India*
* *Corresponding author*

## 15.1 Introduction

Soil erosion is a significant aspect of land degradation because of its global extent and effect on agriculture and natural resources (Borrelli et al., 2020). It impedes various natural ecosystems such as aquatic and crop production by eroding the top fertile layer rich in nutrients required by plants (Montanarella et al., 2016; Poesen, 2018; Kumar and Hole, 2021) and identified as a threat to sustainable development and growth (Poesen, 2018). Natural factors such as high precipitation, strong wind, catastrophic storms, flash floods, snow avalanches, and anthropogenic activities such as deforestation, mining, and overgrazing, accelerate soil loss (Pani and Carling, 2013; Poesen, 2018; Borrelli et al., 2020). However, in the last two decades, anthropogenic activity and its corresponding land use-changes outmanoeuvre most of the parameters accelerating soil erosion (Borrelli et al., 2013). On a global scale, water-induced global soil erosion has reached a possible rate of $43 \pm 9.2$ petagrams per year (for the year 2015) (Borrelli et al., 2020). The climatic projections suggest a more vigorous hydrological cycle on the global scale for the coming decades, increasing water-soil erosion up to +30% to +66% (Borrelli et al., 2020). Hence, water-induced soil erosion rates become a critical issue for policy-makers and stakeholders to customize their decision-making.

A global-scale study carried out by Borrelli et al. (2020) showed well distributed high soil erosion patterns associated with concentrated high rainfall events, steep slopes, and high-relief topography across all continents. The Himalayas facilitates such conditions and has distinct rainfall peaks and high mean relief regions (Bookhagen and Burbank, 2006). In comparison to other geographical regions within the Indian subcontinent, the Himalayas face severe soil erosion rates ranging from 20–25 ton/ha/yr to 80 ton/ha/yr that may increase in the future due to changing climate conditions (Garde and Kothyari, 1987; Kumar et al., 2014, 2022; Kumar, Naqvi and Devrani et al. 2022; Mahapatra et al., 2018; Swarnkar et al., 2018). The steep and dissected Himalayan topography, coarse-textured soil, shallow soil depth, agricultural slopes, and active seismicity make it a highly erodible region (Sati et al., 2011); among the different physiographic of around ~50.63 ton/ha/yr, followed by the Middle (34.50 ton/ha/yr) and Greater Himalaya (32.21 ton/ha/yr); and the lowest in Tarai region with a mean value of ~7.45 ton/ha/yr (George et al., 2021). Similarly, there is significant spatial variation in the soil erosion rates across the Himalayas (Garde and Kothyari, 1987; Kumar et al., 2014; Mahapatra et al., 2018; Swarnkar et al., 2018). For instance, the soil erosion rate in the Kangra district, Himachal Pradesh (NW Himalaya) is around 25.63 ton/ha/yr (Kumar et al., 2014). The soil erosion rate in Nepal (Central Himalaya) is around ~25 ton/ha/yr (Koirala et al., 2019). Jaiswal and Amin (2020) calculated soil erosion rate in the alluvial plain areas (near the mountain front) and hilly areas of the Panchnoi River basin (NE Himalaya). The soil erosion rate near alluvial plain areas increased from 0.52 ton/ha/yr in 1990 to 0.94 ton/ha/yr in 2015, whereas it increased from 12.06 ton/ha/yr in the hilly region to 18.30 ton/ha/yr. In the Tirap district of Arunachal Pradesh (NE Himalayas), the soil erosion rate varies from 1.38 ton/ha/yr to 59.05 ton/ha/yr (Das et al., 2018). Another study in the Ri-Bhoi district of Meghalaya (NE Himalaya) by Jena et al. (2018) suggests a soil erosion rate of ~36 ton/ha/yr

*Advances in Remote Sensing Technology and the Three Poles,* First Edition. Edited by Manish Pandey, Prem C. Pandey, Yogesh Ray, Aman Arora, Shridhar D. Jawak, and Uma K. Shukla.

in the Jirang block. Furthermore, Barman et al. (2020) calculated the average annual soil erosion in the Tuirial River basin in Mizoram (NE Himalaya) with the help of the Revised Universal Soil Loss Equation (RUSLE) model. The calculated soil erosion rate is ~59.94 ton/ha/yr, whereas the Morgan-Morgan-Finney (MMF) model estimated soil loss as 55.30 ton/ha/yr. Most soil erosion rate studies are based on the annual average analysis where annual rainfall is included, but it overlooks extreme rainfall events (Martínez-Casasnovas et al., 2002).

The hydrogeological hazards associated with extreme hydrological events are frequent in Himalayan river basins (Bookhagen et al., 2005; Devrani et al., 2015, 2021; Dimri et al., 2017; Sain et al., 2021). Extreme hydrological event induced by flash floods result in the sudden movement of water and sediments (known as a sediment pulse) and create destruction in the downstream regions of the mountain river basin (Devrani et al., 2021). The frequency and intensity of flash floods are rising rapidly due to rapid climatic warming, e.g., changes in short-term rainfall intensity (Pratap et al., 2020) and changes in sedimentation rate (Cendrero et al., 2020). The ideal location for the occurrence of cloudbursts in the hilly regions is the site where clouds are restricted in a closed valley and when the moisture-laden air lifts with sufficient rapidity and forms cumulonimbus clouds shedding a water load with great strength and intensity (Kumar et al., 2012; Dimri et al., 2017). The flash floods linked with cloudbursts frequently occur in the Ganga River basin and are often associated with massive soil erosion. Some of the significant events that have occurred in the Ganga River basin are Chirgaon (Yamuna Valley) in 1998, Malpa, Kali Valley of Kumaun Division on 17 August 1998, Birehi Ganga Valley at Gonain August 2001, BudhaKedar at Balganga Valley in August 2002, Badrinath Shrine, Chamoli district on 24 July 2004, Nachni near Pithoragarh district on August 2009, Assiganga on August 2012 and Kedarnath, Rudraprayag district on 14–15 June 2013 (Bist and Sah, 1999; Paul et al., 2000; Naithani et al., 2002; Sah et al., 2003; Rana et al., 2012; Gupta et al., 2013; Devrani et al., 2015, 2021; Devrani and Kumar, 2021). The high magnitude and erosive forces characterize these flash floods, which can massively erode land and degrade the quality and quantity of soil. The intensity and frequency of the flash flood, which can drastically alter the rate of soil loss, and assessment of multitemporal short-term soil loss rates can provide new information on the fierce natural disaster.

The soil erosion rates can be estimated using field-based approaches employing various techniques, such as field photography (Watson and Evans, 1991), Microfilometer and Erosion pin techniques (Joshi, 2014), as well as integrated remote sensing (RS) and Geographical Information System (GIS) (Pandey et al., 2007; Dabral et al., 2008; Benzer, 2010; Kumar et al., 2014, 2021). In the past few decades, the soil loss estimation models based on remote sensing data; e.g., SWAT (Soil and Water Assessment Tool), WEPP (Water Erosion Prediction Project), USLE (Universal Soil Loss Equation), and its other forms such as RUSLE (Revised Universal Soil Loss Equation), are developed and used in the mountainous regions across the globe (Kumar et al., 2014). Among these, the RUSLE model is widely used to estimate the long-term average annual soil erosion rate in different environments such as agriculture, forest, cropland, mining sites, and construction sites (Stone and Hilborn, 2000; Dais, 2008). In extreme hydrological event affected regions, the pre-post field data are absent. Hence, models like RUSLE are helpful for calculating the initial estimates of soil erosion rates.

On 3 August 2012, an extreme hydrological event produced havoc in the Assiganga River (an upstream tributary of the Ganga River). In this event (hereafter referred to as the 2012 event), a cloudburst induced a Lake Outburst Flow (LOF) in the Assiganga River basin. This event killed ~500 individuals, displaced around 12,000 people, and caused property damage worth INR 6.12 billion (Gupta et al., 2013). The discharge during this event reached 2665 $m^3$/sec, compared to an average discharge of 100–200 $m^3$/sec, and the water level had risen by around 25–30 m, causing considerable havoc in the valley (Gupta et al., 2013; Devrani and Kumar, 2021). Due to this event, several locations were filled with sediments (i.e., sand, cobbles, pebbles, boulders), and noticeable bank erosion occurred (Gupta et al., 2013; Devrani and Kumar, 2021) (Figure 15.1). Hence, a significant amount of eroded and transported sediments has changed the geomorphology of the lower Assiganga River valley (Devrani and Kumar, 2021). Accordingly, such an extreme hydrological event must have changed the land use-land cover (LULC) of the Assiganga River basin and correspondingly, it must have affected the soil erosion rate in the basin. Hence, to understand temporal variation in the post-event soil erosion rate, this present study focuses on LULC changes due to a flash flood and their impact on soil loss rate in the Assiganga River basin using temporal remote sensing data employing the RUSLE model.

## 15.2 Study Area

The present study is conducted in the Assiganga River basin located between 30°45'21"N and 30°56'40"N latitudes and 78°22'52"E, and 78°35'27"E longitudes (Figure 15.2A). The Assiganga River is a 29.54 km long 6th-order river and is a tributary of the Bhagirathi River (an upstream tributary of the Ganga River). It originates in

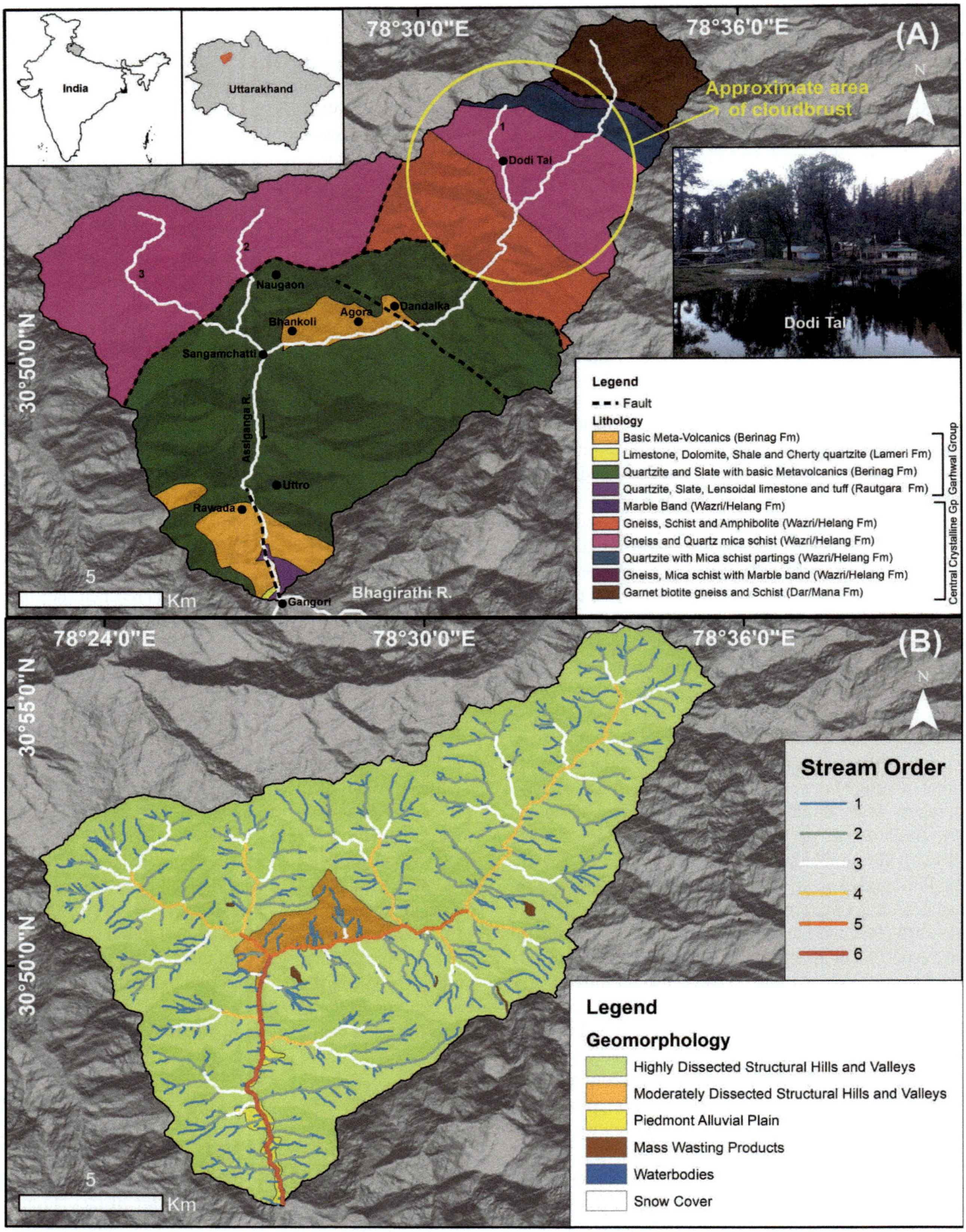


**Figure 15.1** (A) The Geological map of the Assiganga River basin is draped over the hillshade map of the adjoining region. The white color streams represent the Assiganga River and its major tributaries. Note that the yellow circle shows the approximate location of the cloudburst during the 2012 event. In the inset, the red color polygon represents the location of the Assiganga River basin in the state of Uttrakhand. (B) represents regional-scale geomorphology in the Assiganga River basin taken from Geological Survey of India. *Source*: Based on Data from Bhukosh (https://bhukosh.gsi.gov.in/Bhukosh/Public).

the Dar/Mana formation of the central crystalline group with a catchment area of 195.94 $km^2$ (Figures 15.2A and 15.2B). Its higher reach is Dodital Gad and Binsi Gad, the middle reach as Kaldi Gad, and after Sangamchatti, the lower reach is called the Assiganga River (Figures 15.2A and 15.2B). Near Gangori, the Assiganga River merges with the Bhagirathi River at an elevation of 1132 m a.s.l. (Gupta et al., 2013). The NE–SW trending Pandrasu Dhar ridge, located at an altitude between 4000 m a.s.l. and 4800 m a.s.l. at the northern part of the basin serves as a drainage divide between the Bhagirathi and Yamuna rivers (Gupta et al., 2013; Patel et al., 2021). At 3650 m and 4287 m a.s.l., the Devkund Dhar ridge, located south of the Pandrasu Dhar ridge, serves as a drainage divide between the Assiganga and Bhagirathi rivers (Gupta et al., 2013). The Assiganga region receives ~1369 mm annual rainfall, and most of it occurs from July to September of the southwest monsoon season with sub-humid to humid temperate climatic conditions (Gupta et al., 2013). The Assiganga River basin comprises of gneiss, mica schist, schist, quartzite and amphibolite, quartzite, slate, limestone bands, and meta-volcanic rocks belonging to Central Crystalline (Dar/Mana and Wazri/Helang formations) and Garhwal group (Rautgara, Berinag and Lameri formations) (Figure 15.2) (Geological Survey of India: https://bhukosh.gsi.gov.in/Bhukosh/MapViewer.aspx). Highly and moderately dissected structural hills cover the entire basin (Figure 15.2B) (Geological Survey of India: https://bhukosh.gsi.gov.in/Bhukosh/MapViewer.aspx). The fluvial processes have shaped the present landscape, exhibiting highly unstable valley topography, as evidenced by high relief and active erosional processes (Gupta et al., 2013).

## 15.3 Methodology and Dataset

The Assiganga River basin boundary has been delineated from ALOS-PALSAR 12.5 m spatial resolution DEMs (Digital Elevation Models) using hydrology tools in ArcMap 10.6 software (Kumar and Deshmukh, 2018). We have used the RUSLE model integrated with GEE (Google Earth Engine) and GIS (Geographic Information System) in the present study. GEE, which leverages Google servers enormous computing features for high processing power and large storage capacity and self-programming classification algorithms, is a commonly used platform for LULC classification (Stromann et al., 2020; Pan et al., 2021). The RUSLE model is generally used for long-term estimation of soil loss (Kumar et al., 2014; Biswas and Pani, 2015; Das et al., 2018). However, in the present study, we have estimated monthly rates of soil loss during the pre-monsoon months of May (before the flash flood) and post-monsoon months of October (after the flash flood) to estimate temporal uncertainty in soil loss. The RUSLE model requires

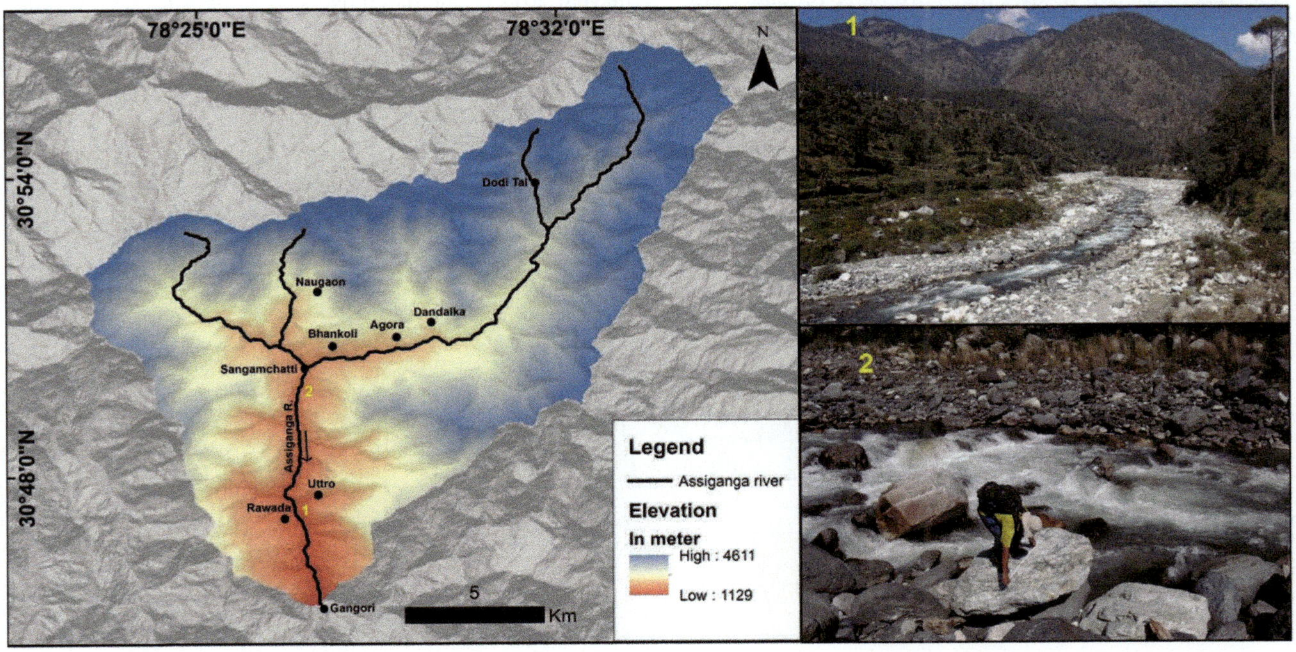


**Figure 15.2** The Elevation map of the Assiganga River basin shows a maximum elevation of 4611 m and minimum elevation of 1129 m with a mean of 2757.62 m. Black color streams represent the Assiganga River and its major tributaries. (1) shows the sandy area created by flashflood, and (2) shows the size of the boulder carried during the event.

derivation of six factors; e.g., factor "K" (soil erodibility in Mg h/MJ/mm), factor "R" (rainfall erosivity in MJ mm/ha/hr/yr) and dimensionless factor "L" (slope length), factor "S" (steepness factor), factor "C" (crop management) and factor "P" (support practices) (Figure 15.3). These factors contribute to an annual rate of soil loss as the RUSLE model is expressed as (Renard et al., 1991, 1997):

$$A = R \times K \times L \times S \times C \times P \tag{15.1}$$

This RUSLE model equation can be converted to a monthly soil erosion equation by including monthly temporal data of factors R and C (Schmidt et al., 2019). It will change the temporal limit for LULC and the spatial distribution of the P factor. Hence, the RUSLE formula can be rewritten as:

$$A_{month} = R_{month} \times K \times L \times S \times C_{month} \times P_{month} \tag{15.2}$$

## 15.3.1 Soil Erodibility (K Factor)

Soil erodibility is the soil's inherent sensitivity to erosion depending upon organic content (Pérez-Rodríguez et al., 2007), texture, and chemical characters of soil (Devatha et al., 2015). The soil texture can be determined by field survey, remote sensing, or both. This study utilized the soil texture maps prepared by the National Bureau of Soil Survey and Land Use Planning (NBSS-LUP), Nagpur, at the scale of 1:50000. The soil maps were scanned in high-resolution (600 dpi) and digitized to extract various soil textures. NBSS-LUP soil texture maps are more reliable than other sources because they combine image interpretation, field surveys, and laboratory investigation to prepare soil texture maps (Shyampura and Sehgal, 1995; Tamgadge, 1996; Singh et al., 2004). The Assiganga River basin comprises loamy, loamy skeletal, and coarse loamy soil texture classes, and K values are assigned from the literature (Figures 15.4A and 15.4B) (Table 15.1).

## 15.3.2 Rainfall Erosivity (R Factor)

The rainfall erosivity factor is a function of rainfall intensity and raindrop size that can erode soil at a higher rate (Pandey et al., 2007) or, in other words, the ability of rain or its power to induce soil erosion (Nearing, 2001). The rate of soil erosion fluctuates in response to changes in rainfall erosive characteristics (Nearing et al., 2004). The rainfall does not occur evenly throughout the year in the Himalayas, and the majority of the rainfall occurs during the monsoon season (July to September) (Bookhagen and Burbank, 2006). Estimating soil erosion requires seasonal or monthly rainfall data to depict the uncertainty in soil erosion throughout the year. Hence, monthly rainfall data is best for calculating soil erosion since it provides a monthly accumulation of eroded soil over time (Diodato and Bellocchi, 2007; Tsitsagi et al., 2017). Month-wise rainfall data of 2012 at 4 km × 4 km spatial resolution are procured from the CHRS and converted into 10 m spatial resolution to match other thematic layers (Figure 15.5).

The rainfall erosivity in the May and October months of the year 2012 is calculated by using Equation 15.3 given by

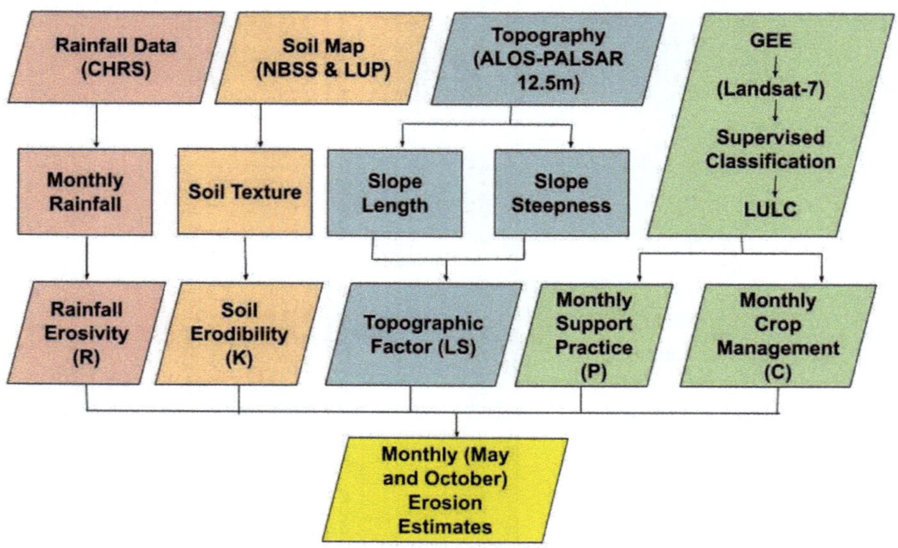


Figure 15.3 The flowchart represents the methodology used in the present study. The main variables used in the RUSLE model are soil erodibility, rainfall erosivity, slope length and steepness factor, crop management, and support practices. These variables are modified and developed using different methods (for more details, please refer to Section 3.1).

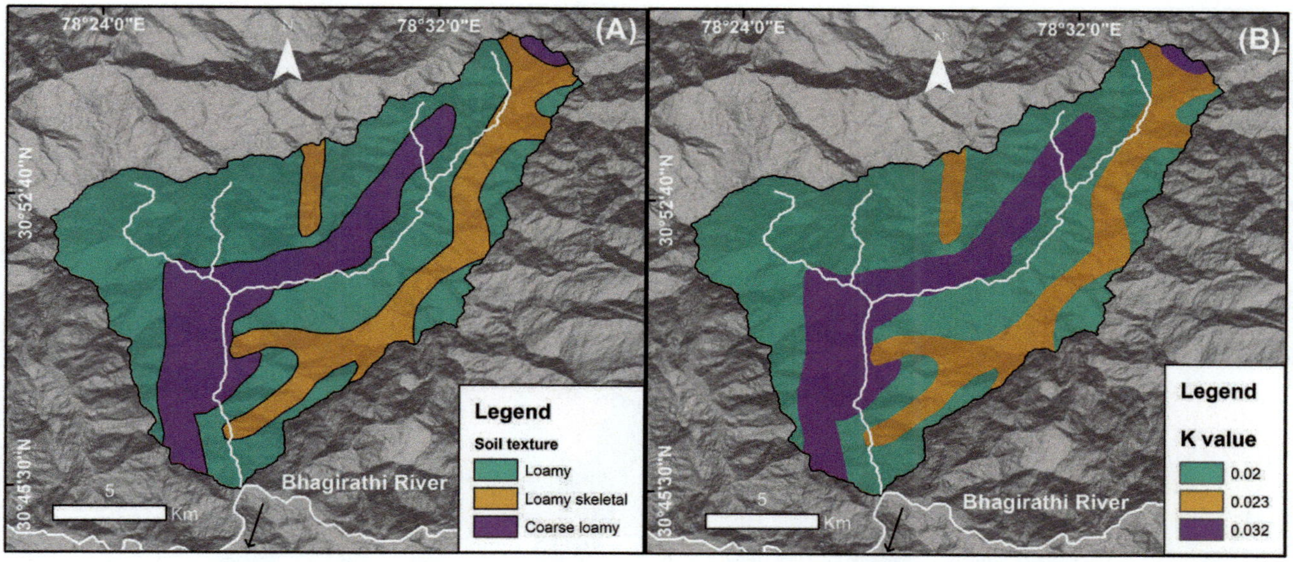


**Figure 15.4** (A) Loamy, loamy skeletal, and coarse loamy soil textures are represented by green, orange, and violet colors in the Assiganga River basin. Note the domination of the loamy soil texture in the study area. (B) The K values for the soil textures are taken from the literature (for more details, please refer to Section 3.1).

**Table 15.1** Soil texture of Assiganga River basin and the K values are assigned from literature.

| S. No | Soil Texture | K-Value | Area (Km$^2$) | Area (%) |
|---|---|---|---|---|
| 1 | Loamy | 0.02 | 117.85 | 60.15 |
| 2 | Loamy skeletal | 0.023 | 36.49 | 18.62 |
| 3 | Coarse loamy | 0.032 | 41.59 | 21.23 |
| **Total** | | | **195.93** | **100** |

Singh et al. (1981) for India (Figure 15.6). The rainfall erosivity (R) equation is expressed as:

$$R = 79 + 0363 \times \text{ Rainfall} \tag{15.3}$$

### 15.3.3 Slope Length and Steepness Factor (LS Factor)

The slope length (factor L) is the distance between the point of an overland stream to the point where the slope angle decreases sufficiently to allow the deposition to begin or the runoff water enters a well-defined channel in a drainage system (Wischmeier and Smith, 1978). The incline of a bed or area from the beginning of an overland stream to the deposition is defined as slope steepness (factor S) (Wischmeier and Smith, 1978). An increase in values of the LS factor will increase the runoff, and it is affected by crop, forest cover, and the surface roughness (Wischmeier and Smith, 1978; Pandey et al., 2007; Das et al., 2018). ALOS-PALSAR 12.5 m spatial resolution DEM procured from Alaska Satellite Facility (ASF) is used to calculate the LS factor through a tool developed in Arc Marco Langauge by Hickey (2001) (Figure 15.7). This tool is based on an equation given by Wischmeier and Smith (1978) and expressed as:

$$LS = \left(\frac{\ell}{72.6}\right)^m \left(65.41\, sin^2\beta + 4.56\, sin\beta + 0.065\right)$$

where $\ell$ *is* cumulative slope length in meter, $m$ dimensionless exponents that rely on slope steepness (slope <5% then m value 0.5 is used, if the slope is 4%, the m value is 0.4, and if the slope is >3%, the m value is 0.3), β is the downhill slope angle, and m is the slope contingent variable.

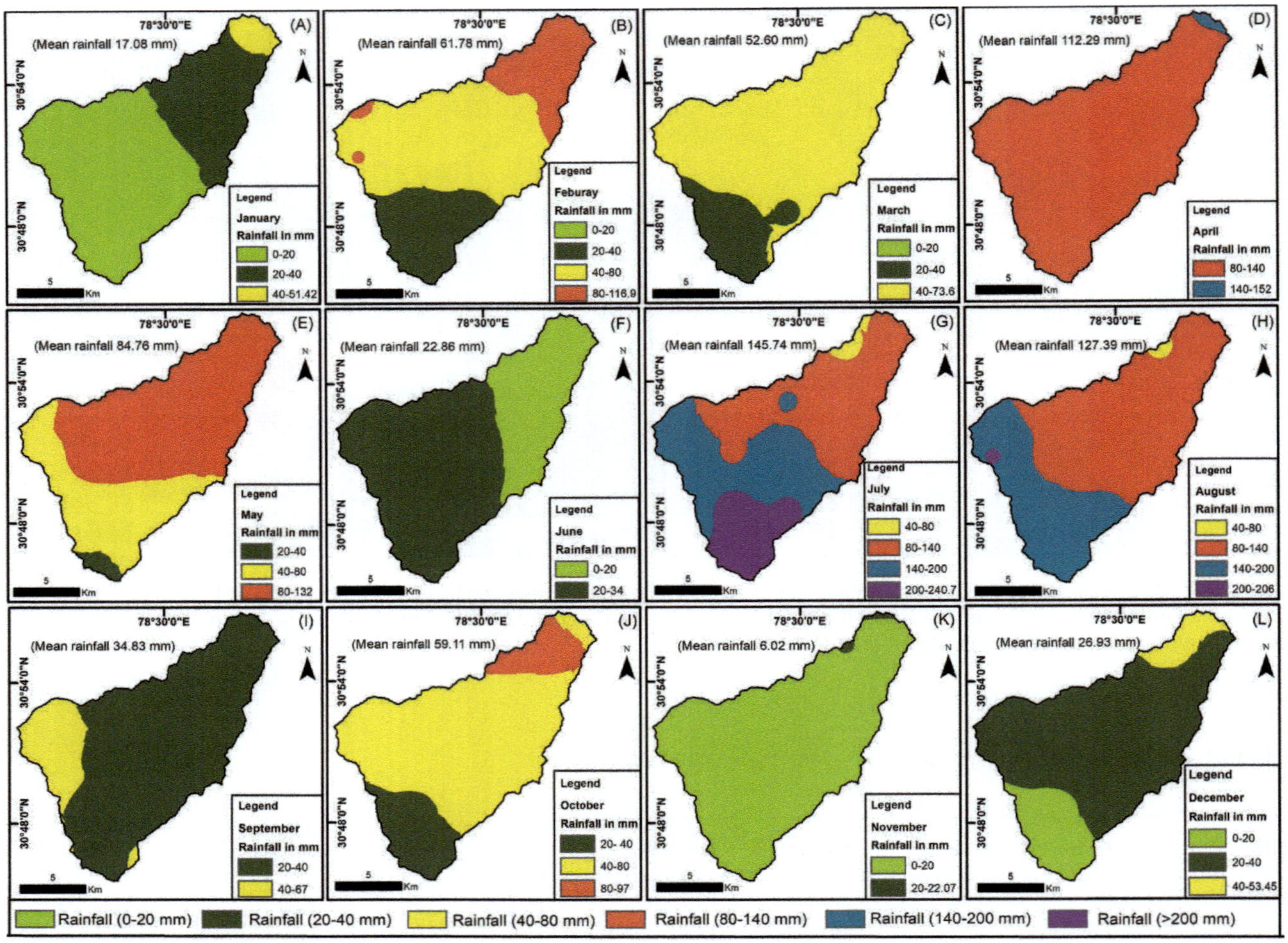


**Figure 15.5** Monthly variation in annual rainfall for the year 2012 in the Assiganga River basin (A: January, B: February, C: March, D: April, E: May, F: June, G: July, H: August, I: September, J: October, K: November, L: December). Please note that July, August, and April received the highest amount of mean monthly rainfall.

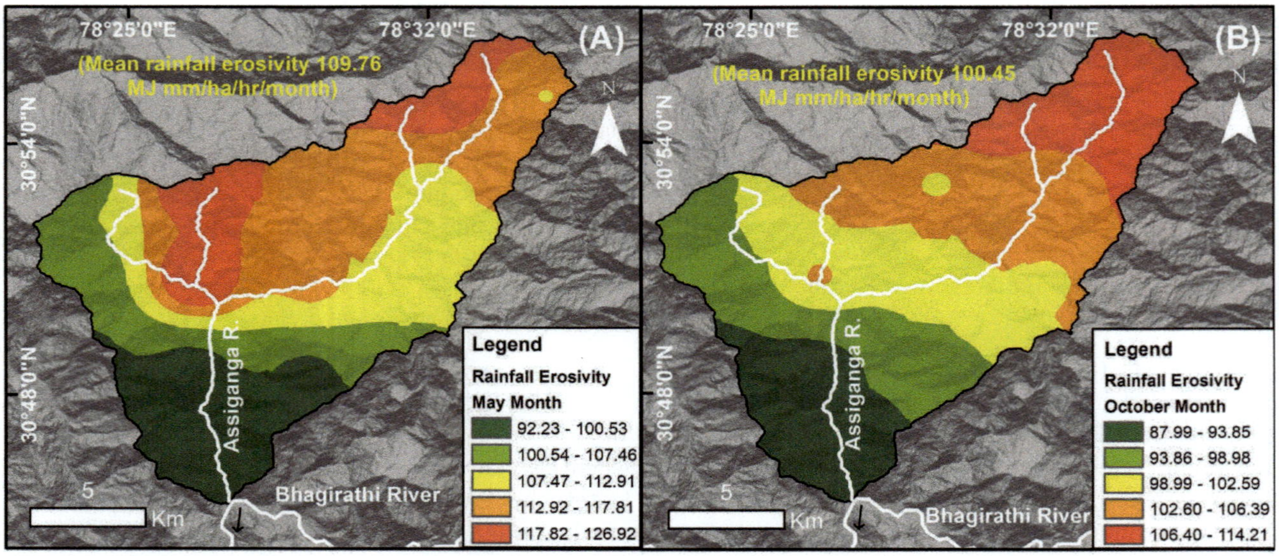


**Figure 15.6** (A) The rainfall erosivity (MJ mm/ha/hr/month) during May 2012 (pre-2012 event) in the Assiganga River basin. The erosivity is higher in the upstream part of Assiganga and its tributaries 1 and 2. (B) The rainfall erosivity during October 2012 (post-2012 event) in the Assiganga River basin. The erosivity is higher in the upstream part of Assiganga and its tributaries 1.

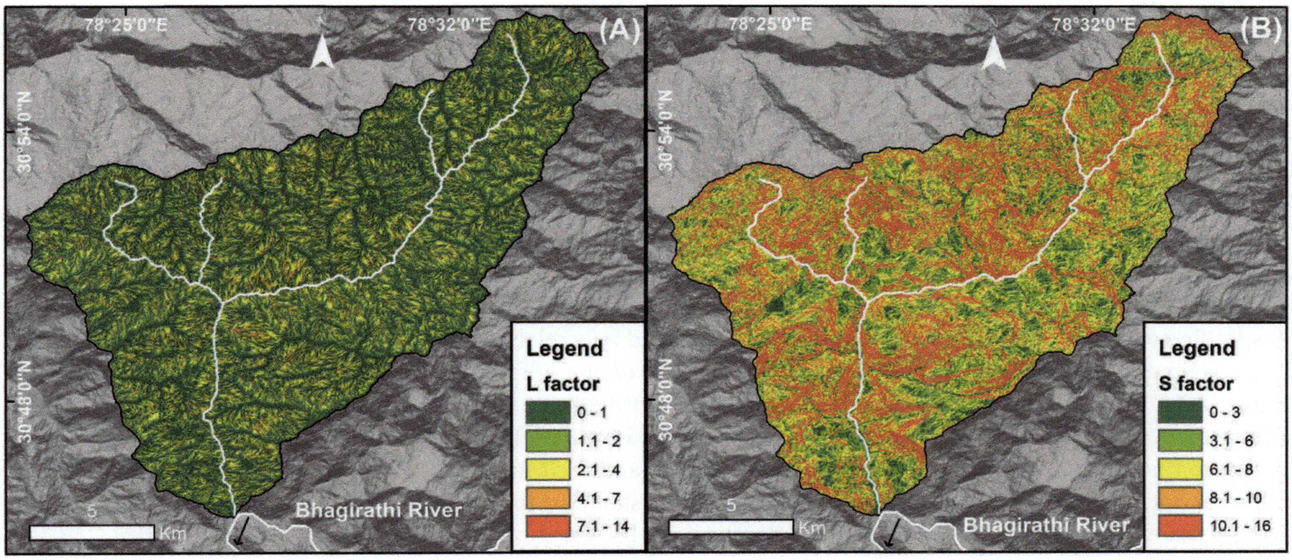


**Figure 15.7** (A) Represents the variation in the slope length (L) of the Assiganga River basin. The mean value of the L variable is around 1.70. (B) The steepness variation factor (S) is well distributed in the Assiganga River basin. The lower S values are prominent around the downstream part of the basin. The mean value of the S variable is around 7.62 of the Assiganga River basin.

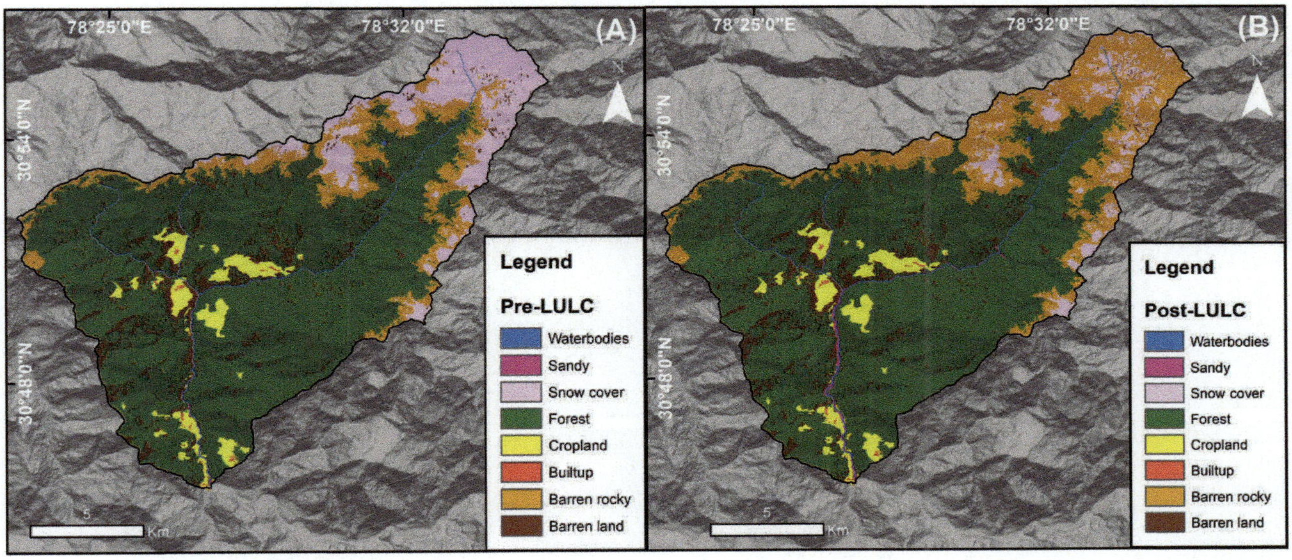


**Figure 15.8** (A) The land use and land cover (LULC) map for the May 2012 (pre-2012 event) in the Assiganga River basin. The study area includes regions with snow cover, forests, terrace farming (cropland), anthropological activities (builtup), barren land, and barren rocky land. For land use and land cover (LULC) map classification details, please refer to methodology Section 3.4 and Table 15.3. (B) The land use and land cover (LULC) map for the October 2012 (post-2012 event) in the Assiganga River basin. Please note the increase in the area of the sandy class post-2012 event. The changes in the pre-post LULC class area details are given in Table 15.3.

### 15.3.4 Crop Management (C Factor) and Support Practices (P Factor)

The crop management (C factor) and support practices (P factor) are based on the landuse and landcover (LULC) classes, and LULC classes can be derived from geospatial data and their accuracy assessed through ground-truthing or analysis of high-resolution satellite or aerial images. The C factors estimate the vegetation canopy and its density, as it serves as a barrier between the direct interaction of soil and runoff to prevent soil erosion (Das et al., 2018). The P factor reflects the soil loss ratio to the specific support practice to reduce runoff (Renard et al., 1997). The C and P factors are calculated by assigning appropriate values to LULC classes derived by supervised classification of Landsat-7 surface reflectance images in the GEE platform (Figure 15.8). A total of eight LULC classes; e.g., builtup, cropland, forest, barren land, snow cover, barren rocky, sandy, and waterbodies, were classified by following the LULC classification of National Natural Resources Management System,

ISRO, 2019 (https://bhuvan-app1.nrsc.gov.in/thematic/thematic/index.php). Temporal LULC of May and October 2012 dated 2 May 2012 and 25 October 2012 with cloud coverage of 0–10% are prepared to estimate temporal uncertainty in soil erosion. Images of the July and August months could not be used as these images had very high ~100% cloud cover. We have used Google Earth May and October images for better bar area estimation and digitized the bar area. The accuracy of bar area computation has been improved by merging these files with pre- and post-LULC.

A specific code was compiled in Google Earth Engine (GEE) to clip and mosaic the cloud-free images and prepare the LULC maps. The Google Earth Engine (GEE) is a platform that combines a host of geospatial datasets and computation power for running user algorithms (Python and JavaScript) for cloud-based data processing making planetary-scale geospatial analysis possible (Gardner, 2010). The data processing is rapid because it does not require any software, works in a cloud computing platform, and is particularly useful in analysis that involves large geospatial datasets. Resampling (downscaling) of Landsat 30 m to panchromatic assisted 15 m was created for better visualization and image interpretation. Thirty signature files of each LULC class were created on the 15 m spatial resolution composite image to train the pixel for supervised LULC classification. The Classification And Regression Tree (CART) classifier was used for supervised classification, creating a set of decision trees from a randomly selected subset of training sets and works on voting among different decision trees (Belgiu and Drăguţ 2016; Pelletier et al., 2016). The extracted LULC thematic maps were exported in 10 m resolution to match other datasets, and their accuracy is increased by Google Earth assisted stamping in GIS software. Appropriate values from the literature were assigned to LULC classes to create C and P factors (Figure 15.9) (Table 15.2). In May (pre-event) and

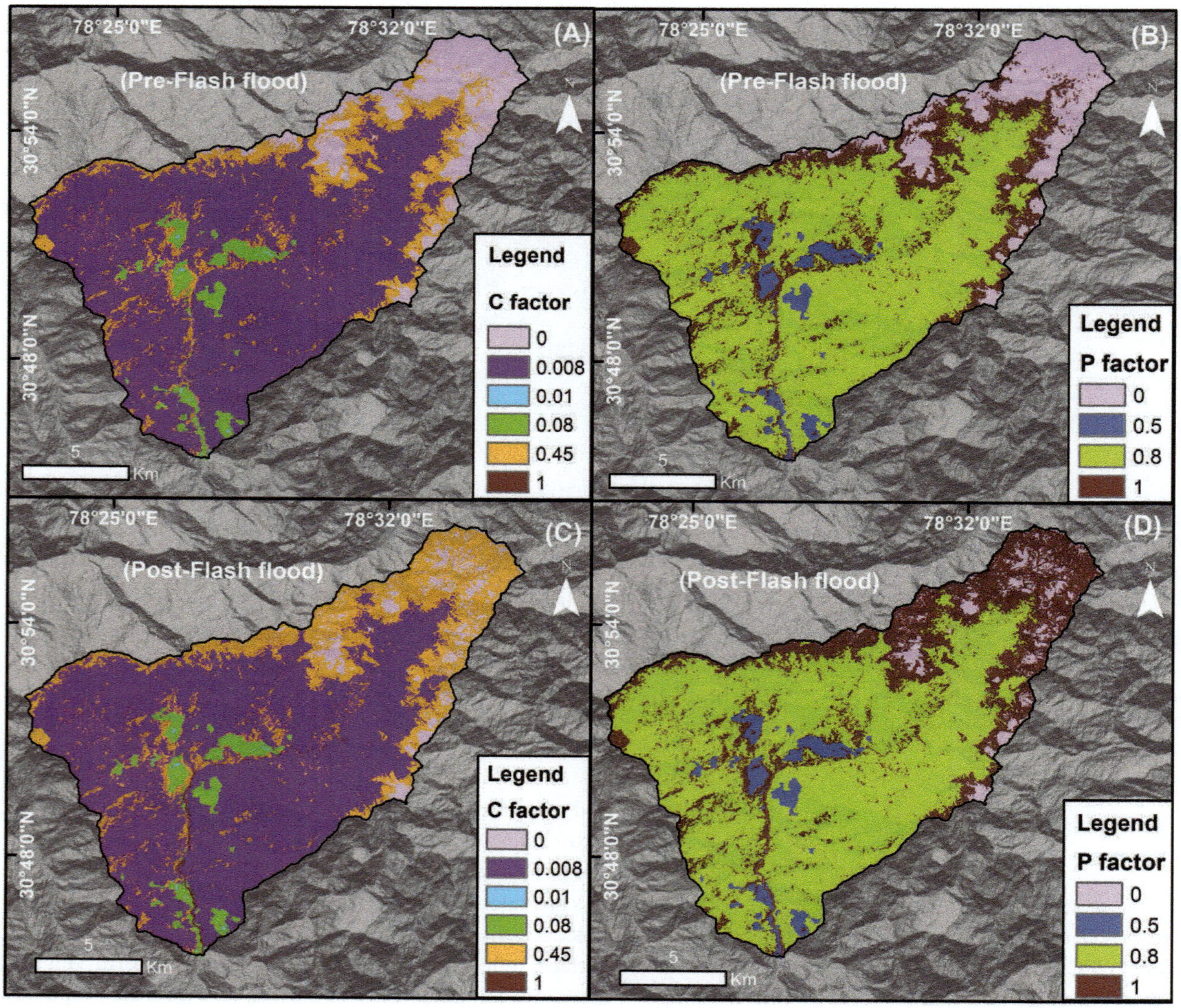


**Figure 15.9** (A) Represents crop management factor during May 2012 (pre-2012 event). (B) represents the support practices factor during May 2012 (pre-2012 event). (C) is the crop management factor for October 2012 (post-2012 event). (D) is support practices factor during October 2012 (post-2012 event).

**Table 15.2** LULC classes present in the Assiganga River basin, C and P values are assigned from the literature.

| LULC class | P-Value | C-Value | Reference |
|---|---|---|---|
| Builtup | 1 | 0.01 | Wanielista and Yousef, 1993 |
| Cropland | 0.5 | 0.08 | Wanielista and Yousef, 1993 |
| Forest | 0.8 | 0.008 | Pandey et al., 2007 |
| Barren land | 1 | 0.45 | Wanielista and Yousef, 1993 |
| Snow cover | 0 | 0 | Wanielista and Yousef, 1993 |
| Barren rocky | 1 | 0.45 | Wanielista and Yousef, 1993 |
| Sandy | 1 | 1 | Wanielista and Yousef, 1993 |
| Waterbodies | 0 | 0 | Jain et al., 2001 |

October (post-event) 2012, LULC maps producer accuracy of 75% and 82% calculated within the GEE platform. Google Earth images are used to assess the accuracy of LULC maps due to availability of the high-resolution satellite images on a regional scale (Jensen, 2015). We randomly selected 100 points to assess accuracy, which achieved user accuracy of 80% and 78% for pre- and post-event LULC maps, respectively.

## 15.4 Results and Discussion

The estimation of soil loss based on temporal data (pre- and post-flash flood periods) for May and October 2012 is utilized for the assessment of extreme event (i.e., cloudburst) induced flash flood changes in soil loss rate in a part of the third pole region (Indian Himalayan region). The geomorphic processes such as snow gliding, snowmelt, and avalanches are not included in the RUSLE model and need to be analyzed separately using other models (Ceaglio et al., 2012; Meusburger et al., 2014; Stanchi et al., 2014). The RUSLE model K, L, and S factors are constant, while combining the effect of R, C, and P factors shows the temporal variability in the soil loss during the different months of a year (Panagos et al., 2016). In the Assiganga River basin, around 60.15% is covered by loamy soil, followed by coarse loamy 21.23% and loamy skeletal 18.62% (Table 15.1). Furthermore, the L factor ranges from 0 to 14 in the study area. Around 76.27% area is covered by the L factor ranging from 0 to 2, and the remaining 23.73% area ranges from 3 to 14, favoring high soil loss. The S factor ranges from 0 to 16, 55.1% of the area is covered from 0 to 8, and the remaining 44.87% area has 8 to 16 values, supporting more soil loss.

### 15.4.1 Pre-Post R, C, and P Variation

The monthly distribution of the R factor provides a spatial distribution of the potential hotspots of rain to cause erosion (Panagos et al., 2016). The mean R factor value from January to March does not show significant variation, followed by a remarkable increase in July and August (21.07% of the total rainfall erosivity), then again a smooth decrease in the value in the last four months has been observed in the Assiganga River basin. As per the LULC pre-post data, May and October 2012 rainfall is used to calculate pre- and post-rainfall erosivity (Figures 15.6A and 15.6B ). As the monsoon mainly arrives in May to June, the mean R factor value in May is 109.76 MJ mm ha/hr/month higher than 100.45 MJ mm ha/hr/in October.

Flood susceptibility can influence changes in LULC (Kafi et al., 2014) and can occur due to natural and anthropogenic factors (Wentz et al., 2014; Yesuph and Dagnew, 2019). The snow melts in the Assiganga River basin from June to September, resulting in a decrease in snow cover on the post-LULC (October) map. As a result, a larger barren rocky area has been exposed in the Assiganga River basin (Table 15.3). Other than these LULC classes, the sandy class increased 86.7% in the study area (from 74,503.2 $m^2$ to 891,080 $m^2$) due to flash floods. In the Assiganga River basin, the bar area has increased between the Sangamchatti and Gangori due to the deposition of 1.5–3.0 m-thick sediments such as pebbles, cobbles, and boulders (Figure 15.2) (Gupta et al., 2013; Devrani and Kumar, 2021). The sediments mainly consist of blackish and greenish gneisses and a few light-colored quartzites (Gupta et al., 2013). An increase in the sediment bar area results in higher runoff and reduced rainfall infiltration, accelerating soil loss.

**Table 15.3** Change detection in LULC due to flashflood. A negative sign indicates a decrease in the area, and a positive sign indicates an area increase. Note the significant change in sandy class.

| LULC class | Area in May 2012 (Km$^2$) | % Area | Area in October 2012 (Km$^2$) | % Area | Change in Km$^2$ | Change in % |
|---|---|---|---|---|---|---|
| Builtup | 0.25 | 0.13 | 0.25 | 0.13 | 0.00 | 0 |
| Cropland | 7.63 | 3.89 | 7.64 | 3.90 | 0.01 | 0.131 |
| Forest | 131.86 | 67.30 | 131.10 | 66.91 | −0.76 | −0.57 |
| Barren land | 12.49 | 6.37 | 11.93 | 6.09 | −0.56 | −4.48 |
| Snow cover | 22.88 | 11.68 | 9.31 | 4.75 | −13.57 | −59.30 |
| Barren rocky | 20.42 | 10.42 | 34.47 | 17.59 | 14.05 | 68.807 |
| Sandy | 0.07 | 0.04 | 0.89 | 0.45 | 0.82 | 1171.427 |
| Waterbodies | 0.33 | 0.17 | 0.34 | 0.17 | 0.01 | 3.03 |
| **Total** | **195.93** | **100** | **195.93** | **100** | | |

## 15.4.2 Soil Loss Spatial Pattern and Extent

The mean rate of soil loss in May and October 2012 is estimated to be 2.64 ton/ha/month and 3.35 ton/ha/month respectively, in the Assiganga River basin (Figure 15.10). The estimated soil loss values are classified into six erosion classes proposed by Singh et al. (1992) in the Indian context with some modifications for better visualization. We observed that 86.59% (in May) and 81.22% (in October) areas are experiencing slight to no erosion due to forest cover and crop cultivation in the Assiganga River basin. On the contrary, only 8.92% (in May) and 11.81% (in October) of the study area are categorized into very high to severe erosion categories. In May 2012, the rainfall of <40 mm resulted in a mean rate of soil loss of 1.83 tons/ha/month, a 40–80 mm rainfall can cause soil loss of 1.54 ton/ha/month, and rainfall of >80 mm can cause 3.28 tons/ha/month. The rainfall in October was less than in May, but the region with more barren exposure and rainfall of >80 mm area reflects 8.77 ton/ha/month mean soil loss. Furthermore, in October, the rainfall with <40 mm and 40–80 mm shows only 1.64 ton/ha/month, and 3.03 tons/ha/month mean soil loss. In May, the coarse loamy soil shows the highest mean rate of soil loss, around ~4.43 ton/ha/month. It is followed by loamy soil (~2.11 ton/ha/month) and loamy skeletal (~2.41 ton/ha/month). Similarly, in October, the highest mean rate of soil loss is observed in the coarse loamy soil (4.75 ton/ha/month), 4.2 tons/ha/month in loamy skeletal soil, and 2.69 tons/ha/month in loamy soil. The L and S factors show significant variation, ranging from 0 to 14 and 0 to 16 (Figures 15.7A and 15.7B). The L value ranges from 7 to 14, with the highest mean soil loss being 12.59 tons/ha/month in May and 12.44 tons/ha/month in October. The S factor value ranges from 10 to 16, with a mean soil loss value of 4.36 tons/ha/month in May and 5.36 in October, indicating that the higher slopes contribute to high erosion. The forest class covers more than 67% of the basin, and the mean rate of soil loss in this class is significantly less due to the high density of vegetation (Table 15.4). The highest mean rates of soil loss are observed in the barren land and barren rocky areas in May and October due to lack of vegetation and higher slope.

Sandy areas (sediment bars) show a significant change in the area (Table 15.3) due to a flash flood (Figures 15.11C and 15.11D). The flash flood carried 2665 m$^3$/sec discharge

**Table 15.4** Mean rate of soil in May and October.

| LULC class | Mean Rate of Soil Loss in May (Pre-Flash Flood) | Mean Rate of Soil Loss in October (Post-Flash FLOOD) | Change | Change in % |
|---|---|---|---|---|
| Builtup | 0.11 | 0.11 | 0 | 0 |
| Cropland | 1.02 | 0.85 | −0.17 | −16.66 |
| Forest | 0.14 | 0.13 | −0.01 | −7.14 |
| Barren land | 17.13 | 15.58 | −1.72 | −10.04 |
| Snow cover | – | – | – | – |
| Barren rocky | 13.51 | 12.61 | −0.9 | −6.66 |
| Sandy | 11.35 | 13.31 | 1.96 | 17–26 |
| Water bodies | – | – | – | – |

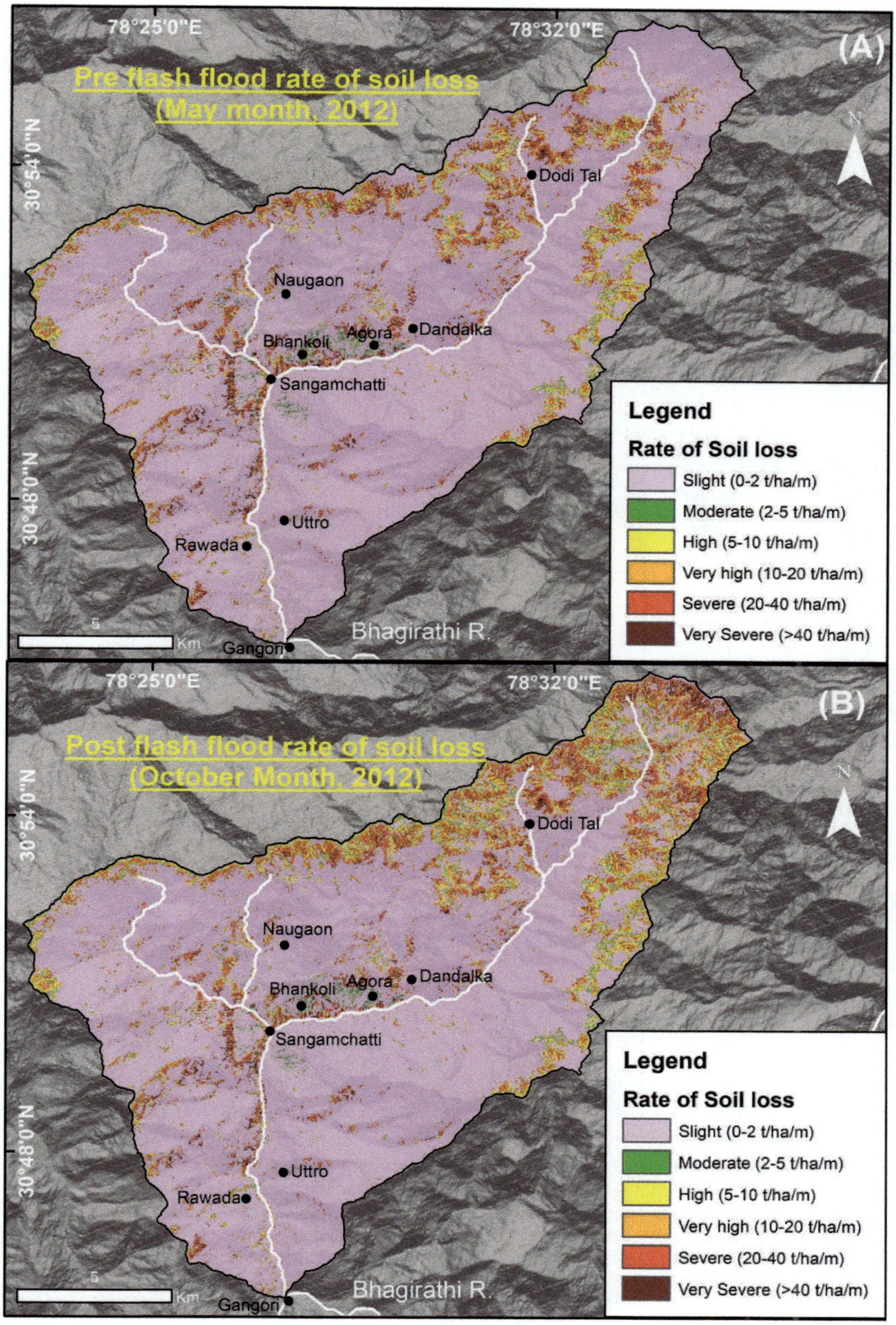


**Figure 15.10** (A) The spatial pattern of soil loss in the Assiganga River basin for May 2012 (pre-2012 event). (B) The spatial pattern of soil loss in the Assiganga River basin for October 2012 (post-2012 event). For more details, please refer to Section 4.2 and Table 15.4.

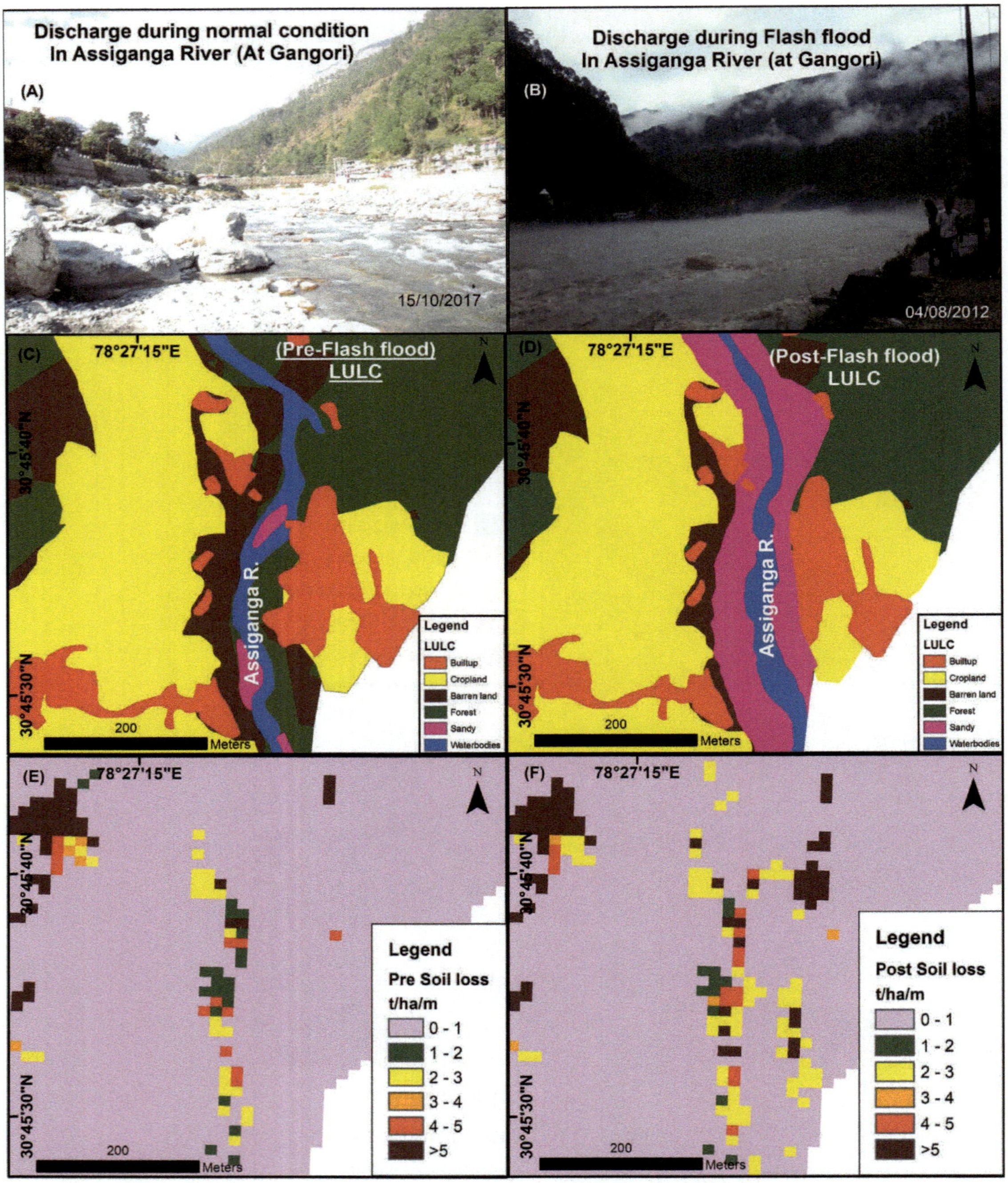


**Figure 15.11** The field photograph showing the confluence of the Assiganga and Bhagirathi river (Gangori location). Note the discharge in normal conditions in the Assiganga River (photograph from the right bank). (B) The field photograph showing the discharge of the Assiganga River during the flash flood (photograph from the left bank). (C) Pre-2012 event LULC data near the mouth of the Assiganga River basin. (D) Post-2012 event LULC data near the mouth of the Assiganga River basin. Note the increase in the sandy area. (E) Pre-2012 event soil loss in the Assiganga river basin. (F) Post-2012 event soil loss in the Assiganga river basin. Note the increase in the post-event soil loss.

instead of 100–200 $m^3$/sec discharge (Gupta et al., 2013; Devrani and Kumar, 2021) under normal conditions and caused significant erosion. Massive amounts of sediments were deposited on the banks of the Assiganga River (Figures 15.11A and 15.11A). These changes in the basin geomorphology accelerate soil loss in sandy areas. For example, the mean rate of soil loss before the flash flood (May) was 11.35 tons/ha/month, which increased after the flash flood to 13.31 tons/ha/month (October) (Figures 15.11E and 15.11F).

## 15.5 Conclusion

Cloudburst-induced flash floods have altered the geomorphology of the Assiganga River basin, particularly after Sangamchatti to confluence, as evidenced by the increased bar area. The pre- and post-flash flood rates of soil loss in May and October 2012 have been estimated to predict how much change in soil erosion can occur due to a single event. The spatiotemporal pattern of soil erosion estimated by the monthly RUSLE model proposed by Wischmeier and Smith (1965) reveals that the mean rate of soil loss has increased from 2.64 ton/ha/month in May to 3.35 ton/ha/month in October month. As a result of the flash flood heavy discharge, sandy areas increased from 0.07 $km^2$ to 0.89 $km^2$, resulting in a distinct increase in the mean rate of soil loss from 11.35 ton/ha/month to 13.31 ton/ha/month in this class. Such studies are essential to understand other extreme hydrological events of different magnitude and determine the sensitivity of the soil erosion rates. Furthermore, the RUSLE model does not include snow in the soil loss calculation; hence, the soil-loss value may fluctuate, but this model can provide soil-erosion hotspots critical for policy-makers, soil scientists, and others. For a more in-depth understanding of climatic variations, high spatial resolution data is recommended, along with monthly soil-loss estimations.

## Acknowledgments

Authors thank GEE for providing a cloud-based platform for LULC, Alaska Satellite Facility for DEM and IGNOU library for providing NBSS-LUP soil texture maps. RD and SD thank the Wadia Institute of Himalayan Geology for providing infrastructure. RK and BD acknowledge the Director, School of Sciences, IGNOU for infrastructure and encouragement.

## References

Barman, B.K., Rao, K.S., Sonowal, K. et al. (2020). Soil erosion assessment using revised universal soil loss equation model and geo-spatial technology: a case study of upper Tuirial River basin, Mizoram, India. *AIMS Geosciences* 6(4): 525–544. doi.org/10.3934/geosci.2020030.

Belgiu, M. and Drăguţ, L. (2016). Random forest in remote sensing: a review of applications and future directions. *ISPRS Journal of Photogrammetry and Remote Sensing* 114: 24–31. doi.org/10.1016/j.isprsjprs.2016.01.011.

Benzer, N. (2010). Using the geographical information system and remote sensing techniques for soil erosion assessment. *Polish Journal of Environmental Studies* 19(5): 881–886.

Bist, K.S. and Sah, M.P. (1999). The devastating landslide of August 1998 in Ukhimath area, Rudraprayag district, Garhwal Himalaya. *Current Science* 76(4): 481–484.

Biswas, S.S. and Pani, P. (2015). Estimation of soil erosion using RUSLE and GIS techniques: a case study of Barakar River basin, Jharkhand, India. *Modeling Earth Systems and Environment* 1(4): 1–13. doi.org/10.1007/s40808-015-0040-3.

Bookhagen, B. and Burbank, D.W. (2006). Topography, relief, and TRMM-derived rainfall variations along the Himalaya. *Geophysical Research Letters* 33(8): 1–5. doi.org/10.1029/2006GL026037.

Bookhagen, B., Thiede, R.C., and Strecker, M.R. (2005). Abnormal monsoon years and their control on erosion and sediment flux in the high, arid northwest Himalaya. *Earth and Planetary Science Letters* 231(1–2): 131–146. doi.org/10.1016/j.epsl.2004.11.014.

Borrelli, P., Robinson, D.A., Panagos, P. et al. (2020). Land use and climate change impacts on global soil erosion by water (2015–2070). *Proceedings of the National Academy of Sciences* 117(36): 21994–22001. doi.org/10.1073/pnas.2001403117.

Borrelli, P., Rondón, L.A.S., and Schütt, B. (2013). The use of Landsat imagery to assess large-scale forest cover changes in space and time, minimising false-positive changes. *Applied Geography* 41: 147–157. doi.org/10.1016/j.apgeog.2013.03.010.

Ceaglio, E., Meusburger, K., Freppaz, M. et al. (2012). Estimation of soil redistribution rates due to snow cover related processes in a mountainous area (Valle d'Aosta, NW Italy). *Hydrology and Earth System Sciences* 16(2): 517–528. doi.org/10.5194/hess-16-517-2012, 2012.

Cendrero, A., Forte, L.M., Remondo, J. et al. (2020). Anthropocene geomorphic change. Climate or human activities? *Earth's Future* 8(5): 1–13. doi.org/10.1029/2019EF001305.

Dabral, P.P., Baithuri, N., and Pandey, A. (2008). Soil erosion assessment in a hilly catchment of north eastern India using USLE, GIS and remote sensing. *Water Resources Management* 22: 1783–1798. doi.org/10.1007/s11269-008-9253-9.

Dais (2008). Directorate Agricultural Information Services. Directorate Agricultural Land and Resources Management. Directorate Agricultural Inforsmation Services, Pretoria. pp. 1–7.

Das, B., Paul, A., Bordoloi, R. et al. (2018). Soil erosion risk assessment of hilly terrain through integrated approach of RUSLE and geospatial technology: a case study of Tirap District, Arunachal Pradesh. *Modeling Earth Systems and Environment* 4(1): 373–381. doi.org/10.1007/s40808-018-0435-z.

Devatha, C.P., Deshpande, V., and Renukaprasad, M.S. (2015). Estimation of soil loss using USLE model for Kulhan Watershed, Chattisgarh: a case study. *Aquatic Procedia* 4: 1429–1436. doi.org/10.1016/j.aqpro.2015.02.185.

Devrani, R. and Kumar, R. (2021). Geomorphic response and paleohydrology of a debris flow event in the upper Ganga River basin, NW Himalaya. In: *EGU General Assembly Conference Abstracts EGU21-14294*. doi.org/10.5194/egusphere-egu21-14294.

Devrani, R., Singh, V., Mehta, M. et al. (2021). Assessing sediment pulse during an extreme hydrological event in the Alaknanda Basin, Northwestern Himalaya, India. *Journal of the Geological Society of India* 97(1): 48–54. doi.org/10.1007/s12594-021-1624-3.

Devrani, R., Singh, V., Mudd, S.M. et al. (2015). Prediction of flash flood hazard impact from Himalayan river profiles. *Geophysical Research Letters* 42(14): 5888–5894. doi.org/10.1002/2015GL063784.

Dimri, A. P., Chevuturi, A., Niyogi, D. et al. (2017). Cloudbursts in Indian Himalayas: a review. *Earth-Science Reviews* 168: 1–23. doi.org/10.1016/j.earscirev.2017.03.006.

Diodato, N. and Bellocchi, G. (2007). Estimating monthly (R) USLE climate input in a Mediterranean region using limited data. *Journal of Hydrology* 345(3–4): 224–236. doi.org/10.1016/j.jhydrol.2007.08.008.

Garde, R J. and Kothyari, U.C. (1987). Flood estimation using regional flood frequency analysis. Hydraul. Eng. Rep. No. HYD-8702, Civ. Eng. Dep., University of Roorkee, Roorkee.

Gardner, T. (2010). Google unveils satellite platform to aid forest efforts. Reuters. Retrieved http://www.reuters.com/article/2010/12/02/us-climate-google-idUSTRE6B13RK20101202.

George, K., Suresh, K., and Rani, M.H. (2021). Geospatial modelling of soil erosion and risk assessment in Indian Himalayan region: a study of Uttarakhand state. *Environmental Advances* 4: 100039. doi.org/10.1016/j.envadv.2021.100039.

Gupta, V., Dobhal, D.P., and Vaideswaran, S.C. (2013). August 2012 cloudburst and subsequent flash flood in the Asi Ganga, a tributary of the Bhagirathi river, Garhwal Himalaya, India. *Current Science* 105(2): 249–253.

Hickey, R. (2001). Slope angle and slope length solutions for GIS. *Cartography* 29(1): 1–8. doi.org/10.1080/00690805.2000.9714334.

Jain, S. K., Kumar, S., and Varghese, J. (2001). Estimation of soil erosion for a Himalayan watershed using GIS technique. *Water Resources Management* 15(1): 41–54. https://doi.org/10.1023/A:1012246029263.

Jaiswal, M.K. and Amin, N. (2020). The impact of land use dynamics on the soil erosion in the Panchnoi river basin, northeast India. Journal of the Geographical Institute, Jovan Cvijic. *SASA*, 70(1): 1–14. doi.org/10.2298/IJGI2001001J.

Jena, R.K., Padua, S., Ray, P. et al. (2018). Assessment of soil erosion in sub-tropical ecosystem of Meghalaya, India using remote sensing, GIS and RUSLE. *Indian Journal of Soil Conservation* 46(3): 273–282.

Jensen, J.R. (2015). *Introductory Digital Image Processing: A Remote Sensing Perspective*. 1–378. USA: Prentice Hall Press, .

Joshi, V.U. (2014). Soil loss estimation by field measurements in the badlands along Pravara river (Western India). *Journal of the Geological Society of India* 83(6): 613–624.

Kafi, K.M., Shafri, H.Z.M., and Shariff, A.B.M. (2014). An analysis of LULC change detection using remotely sensed data: a case study of Bauchi City. *IOP Conference Series: Earth and Environmental Science* 20(1): 1–9. doi.org/10.1088/1755-1315/20/1/012056.

Koirala, P., Thakuri, S., Joshi, S. et al. (2019). Estimation of soil erosion in Nepal using a RUSLE modeling and geospatial tool. *Geosciences* 9(147): 1–19. DOI.org/10.3390/geosciences9040147.

Kumar, S. and Hole, R.M. (2021). Geospatial modelling of soil erosion and risk assessment in Indian Himalayan region: a study of Uttarakhand state. *Environmental Advances* 4: 100039.

Kumar, A., Devi, M., and Deshmukh, B. (2014). Integrated remote sensing and Geographic Information System based RUSLE modelling for estimation of soil loss in western Himalaya, India. *Water Resources Management* 28: 3307–3317. doi.org/10.1007/s11269-014-0680-5.

Kumar, M.S., Shekhar, M.S., Krishna, S.R. et al. (2012). Numerical simulation of cloudburst event on August 05, 2010, over Leh using WRF mesoscale model. *Natural Hazards* 62(3): 1261–1271. doi.org/10.1007/s11069-012-0145-1.

Kumar, R. and Deshmukh, B. (2018). Soil loss assessment using geospatial technology: challenges and opportunities with reference to the Chambal basin, central India. Paper presented at the Environmental and Ecological Sustainability, IGNOU, November 2018.

Kumar, R., Deshmukh, B., and Kumar, A. (2022). Using Google Earth Engine and GIS for basin scale soil erosion risk assessment: A case study of Chambal river basin, central India. *Journal of Earth System Science* 131(4): 1–16. https://doi.org/10.1007/s12040-022-01977-z.

Kumar, R., Deshmukh, B., and Sathunuri, K. (2021). RUSLE model-based assessment of soil erosion in Parbati River basin, Central India using Google Earth Engine and GIS. In: *EGU General Assembly Conference Abstracts* EGU21-1524. doi.org/10.5194/egusphere-egu21-1524.

Kumar, R., Naqvi, H. R., Devrani, R. et al. (2022). Sediment Yield Assessment, Prioritization and Control Practices in Chambal River Basin Employing SYI Model. *Journal of the Geological Society of India* 98(11): 1585–1594. https://doi.org/10.1007/s12594-022-2215-7

Mahapatra, S.K., Obi Reddy, G.P., Nagdev, R. et al. (2018). Assessment of soil erosion in the fragile Himalayan ecosystem of Uttarakhand, India using USLE and GIS for sustainable productivity. *Current Science* 115(1): 108–121. 10.18520/cs/v115/i1/108-12.

Martínez-Casasnovas, J.A., Ramos, M.C., and Ribes-Dasi, M. (2002). Soil erosion caused by extreme rainfall events: mapping and quantification in agricultural plots from very detailed digital elevation models. *Geoderma* 105(1–2): 125–140. doi.org/10.1016/S0016-7061(01)00096-9.

Meusburger, K., Leitinger, G., Mabit, L. et al. (2014). Soil erosion by snow gliding: a first quantification attempt in a subalpine area in Switzerland. *Hydrology and Earth System Sciences* 18(9): 3763–3775. doi.org/10.5194/hess-18-3763-2014.

Montanarella, L., Pennock, D.J., McKenzie, N. et al. (2016). World's soils are under threat. *Soil* 2: 79–82. doi.org/10.5194/soil-2-79-2016.

Naithani, A.K., Joshi, V., and Prasad, C. (2002). Investigation on the impact of cloudburst in Tehri District, Uttaranchal: 31 August 2001. *Journal of Geological Society of India* 60(5): 573–577.

Nearing, M.A. (2001). Potential changes in rainfall erosivity in the US with climate change during the 21st century. *Journal of Soil and Water Conservation* 56(3): 229–232.

Nearing, M.A., Pruski, F.F., and O'Neal, M.R. (2004). Expected climate change impacts on soil erosion rates: a review. *Journal of Soil & Water Conservation* 59(1): 43–50.

Pan, X., Wang, Z., Gao, Y. et al. (2021). Detailed and automated classification of land use/land cover using machine learning algorithms in Google Earth Engine. *Geocarto International* 36(1): 1–18.

Panagos, P., Borrelli, P., Spinoni, J. et al. (2016). Monthly rainfall erosivity: conversion factors for different time resolutions and regional assessments. *Water* 8(4): 119. DOI.org/10.3390/w8040119.

Pandey, A., Chowdary, V.M., and Mal, B.C. (2007). Identification of critical erosion prone areas in the small agricultural watershed using USLE, GIS and remote sensing. *Water Resources Management* 21: 729–746. doi.org/10.1007/s11269-006-9061-z.

Pani, P. and Carling, P. (2013). Land degradation and spatial vulnerabilities: a study of inter-village differences in Chambal Valley, India. *Asian Geographer* 30(1): 65–79. DOI.org/10.1080/10225706.2012.754775.

Patel, M.P., Sharma, N., and Pandey, A. (2021). Research needs for stream power moderation in hilly torrents for disaster mitigation. In: *Hydrological Extremes*, 185–201. Cham: Springer. doi.org/10.1007/978-3-030-59148-9_14.

Paul, S.K., Bartarya, S.K., Rautela, P. et al. (2000). Catastrophic mass movement of 1998 monsoons at Malpa in Kali Valley, Kumaun Himalaya (India). *Geomorphology* 35(3–4): 169–180. DOI.org/10.1016/S0169-555X(00)00032-5.

Pelletier, C., Valero, S., Inglada, J. et al. (2016). Assessing the robustness of Random Forests to map land cover with high resolution satellite image time series over large areas. *Remote Sensing of Environment* 187: 156–168. doi.org/10.1016/j.rse.2016.10.010.

Pérez-Rodríguez, R., Marques, M.J., and Bienes, R. (2007). Spatial variability of the soil erodibility parameters and their relation with the soil map at subgroup level. *Science of the Total Environment* 378(1–2): 166–173. doi.org/10.1016/j.scitotenv.2007.01.044.

Poesen, J. (2018). Soil erosion in the Anthropocene: research needs. *Earth Surface Processes and Landforms* 43(1): 64–84. doi.org/10.1002/esp.4250.

Pratap, S., Srivastava, P.K., Routray, A. et al. (2020). Appraisal of hydro-meteorological factors during extreme precipitation event: case study of Kedarnath cloudburst, Uttarakhand, India. *Natural Hazards* 100(2): 635–654. doi.org/10.1007/s11069-019-03829-4.

Rana, N., Sundriyal, Y.P., and Juyal, N. (2012). Recent cloudburst-induced landslides around Okhimath, Uttarakhand. *Current Science* 103(12): 1389–1390.

Renard, K.G., Foster, G.R., Weesies, G.A. et al. (1997). *Predicting Soil Erosion by Water: A Guide to Conservation Planning with the Revised Universal Soil Loss Equation (RUSLE)*. 703. Washington, DC: United States Department of Agriculture.

Renard, K.G., Foster, G.R., Weesies, G.A. et al. (1991). RUSLE: revised universal soil loss equation. *Journal of Soil and Water Conservation* 46(1): 30–33.

Sah, M.P., Asthana, A.K.L., and Rawat, B.S. (2003). Cloudburst of 10 August, 2002 and related landslides and debris flows around BudhaKedar (ThatiKathur) in Balganga valley, district Tehri. *Himalayan Geology* 24(2): 87–101.

Sain, K., Kumar, A., Mehta, M. et al. (2021). A perspective on Rishiganga-Dhauliganga Flash Flood in the Nanda Devi Biosphere Reserve, Garhwal Himalaya, India. *Journal of the Geological Society of India* 97(4): 335–338. doi.org/10.1007/s12594-021-1691-5.

Sati, S.P., Sundriyal, Y.P., Rana, N. et al. (2011). Recent landslides in Uttarakhand: nature's fury or human folly. *Current Science* 100(11): 1617–1620.

Schmidt, S., Alewell, C., and Meusburger, K. (2019). Monthly RUSLE soil erosion risk of Swiss grasslands. *Journal of Maps* 15(2): 247–256. doi.org/10.1080/17445647.2019.1585980.

Shyampura, R.L. and Sehgal, J. (1995). Soils of Rajasthan for optimising land use. Soils of India Series. NBSS Publication 51(b). ISBN: 81-85460-13-2.

Singh, G., Babu, R., Narain, P. et al. (1992). Soil erosion rates in India. *Journal of Soil and Water Conservation* 47(1): 97–99. Online ISSN: 1941-3300.

Singh, G., Chandra, S., and Babu, R. (1981). Soil loss and prediction research in India, Central Soil and Water Conservation Research Training Institute. Bulletin No T-12 D: 9.

Singh, S.P., Ram, J., Walia, C.S. et al. (2004). Soils of Uttar Pradesh for optimising land use. *NBSS Publication* 68(b): 91.

Stanchi, S., Freppaz, M., Ceaglio, E. et al. (2014). Soil erosion in an avalanche release site (Valle d'Aosta: Italy): towards a winter factor for RUSLE in the Alps. *Natural Hazards and Earth System Sciences* 14(7): 1761–1771. doi.org/10.5194/nhess-14-1761-2014.

Stone, R.P. and Hilborn, D. (2000). Universal Soil Loss Equation (USLE). Ministry of Agriculture, Food and Rural Affairs, pp. 1–12.

Stromann, O., Nascetti, A., Yousif, O. et al. (2020). Dimensionality reduction and feature selection for object-based land cover classification based on Sentinel-1 and Sentinel-2 time series using Google Earth Engine. *Remote Sensing* 12(1): 76.

Swarnkar, S., Malini, A., Tripathi, S. et al. (2018). Assessment of uncertainties in soil erosion and sediment yield estimates at ungauged basins: an application to the Garra River basin, India. *Hydrology and Earth System Sciences* 22(4): 2471–2485. doi.org/10.5194/hess-22-2471-2018.

Tamgadge, D.B. (1996). Soils of Madhya Pradesh for optimising land use. National Bureau of Soil Survey & Land Use Planning Publication 59(b): 1–523. ISBN: 81-85460-32-9.

Tsitsagi, M., Berdzenishvili, A., and Gogidze, K. (2017). Monthly variations of rainfall erosivity (R factor) in Shida Kartli, Georgia. *Earth* 6(5–1): 87–92. DOI.org/10.11648/j.earth.s.2017060501.23.

Wanielista, M.P. and Yousef, Y.A. (1993). Design and analysis of an irrigation pond using urban stormwater runoff. Proceedings Paper, 526–729. https://api.elsevier.com/content/abstract/scopus_id/0027245273.

Watson, A. and Evans, R. (1991). A comparison of estimates of soil erosion made in the field and from photographs. *Soil and Tillage Research* 19(1): 17–27. doi.org/10.1016/0167-1987(91)90106-8.

Wentz, E.A., Anderson, S., Fragkias, M. et al. (2014). Supporting global environmental change research: a review of trends and knowledge gaps in urban remote sensing. *Remote Sensing* 6(5): 3879–3905. doi.org/10.3390/rs6053879.

Wischmeier, W.H. and Smith, D.D. (1965). Predicting rainfall-erosion losses from cropland east of the Rocky Mountains: guide for selection of practices for soil and water conservation (No. 282). Agricultural Research Service, US Department of Agriculture.

Wischmeier, W.H. and Smith, D.D. (1978). Predicting rainfall erosion losses: a guide to conservation planning. USDA, Science and Education Administration, Department of Agriculture, Science and Education Administration. 537: 58.

Yesuph, A.Y. and Dagnew, A.B. (2019). Soil erosion mapping and severity analysis based on RUSLE model and local perception in the Beshillo Catchment of the Blue Nile Basin, Ethiopia. *Environmental Systems Research* 8(1): 1–21. doi.org/10.1186/s40068-019-0145-1.

# 16

# Understanding the Present and Past Climate-Human-Vegetation Dynamics in the Indian Himalaya

## A Comprehensive Review

*Mehta Bulbul[1]*, Yadav Ankit[1], Aljasil Chirakkal[1], Ambili Anoop[1]*, and Praveen K. Mishra[2]*

[1] *Indian Institute of Science Education and Research Mohali, Sector 81, Manauli, Punjab 140306, India*
[2] *Wadia Institute of Himalayan Geology, 33 GMS Road, Dehradun, Uttarakhand 248001, India*
* *Corresponding author*

## 16.1 Introduction

Mountains cover ~24% of the world's land area and impact livelihoods of approximately 40% people worldwide (Singh et al., 2010). However, this ecosystem is largely influenced by climate variability that in turn impacts food security and human settlements, as well as biodiversity (Sharma and Rai, 2012). Previous research concerning vegetation–climate relationship has been conducted at spatio-temporal scales worldwide (D'Arrigo et al., 2004; Quamar and Bera, 2014). Despite many evidences on global environmental change impact, various ecologically sensitive and significant areas of the globe are under-studied (Korner and Paulsen, 2004). Sub-tropical mountain systems in Asia, such as the Himalaya, represent such inadequately studied region. The Himalaya is known to regulate monsoon climate over the Indian subcontinent (Yang, 1996; Bhattacharyya et al., 2006). This region serves as a potential site for understanding the spatio–temporal vegetation climatic interactions due to high biodiversity, higher elevations, and steepness of mountains (Mittermeier et al., 1999). The sensitive landscape of this diverse ecosystem is susceptible to climate variability and impacts of hazards that include droughts and floods, thereby causing serious threats to food security (Cruz et al., 2007; Sharma and Rai, 2012). Besides climate drivers, intense human intervention is substantially impacting larger areas of forests in this ecosystem (Pandit and Grumbine, 2012). Moreover, high anthropogenic influences in the recent past have accelerated the rate of climate change. Therefore, spatio-temporal studies from the Himalayan region are important for understanding vegetational dynamics under climate change as well as human influences. The present study sheds light on the potential link between vegetation dynamics and climatic variability during the Holocene in Indian Himalaya. The Holocene epoch (~11,500 cal. BP to Present) is significant in many respects, as changes in climate during this time have impacted vegetation globally (Quamar and Bera, 2014). This timeframe includes the long-lived (Holocene Climate Optimum: HCO) and short-lived climatic events involving 8.2 kyr BP and 4.2 kyr BP. Therefore, to have a detailed understanding about climate and vegetation relationships, Quaternary lacustrine deposits serve as important archives as they record signatures of fluctuations in the lake basins as well as perturbations induced by climate system.

Vegetation is a crucial part of the biosphere, that regulates the global carbon and biogeochemical cycles and fundamentally reflects changes in the environment (Billings, 1952). Climate change (e.g., rainfall, temperature), human activities (e.g., land-use practices, cultural diversity, urbanization), and disturbance regimes (e.g., fire) are governing factors for vegetation dynamics in the complex Indian Himalaya (Rosenzweig and Neofotis, 2013). Therefore, broad characteristics of past and present climate can largely be unfolded by various terrestrial archives and proxy behavior. The multiproxy approach using palynological records, stable carbon isotope signatures ($\delta^{13}C$), and phytolith assemblages was utilized to deduce past as well as modern climate variability controls on vegetation. As modern flora and vegetation are broadly related to modern climate; hence pollen data can be used to assess knowledge about palaeoenvironment and palaeovegetation on local, regional, and global scales (Misra and Bhattacharyya, 2014). The pollen investigations serve as a useful tool for reconstruction of past climate over timespans ($10^2$–$10^5$ years) and sample

*Advances in Remote Sensing Technology and the Three Poles*, First Edition. Edited by Manish Pandey, Prem C. Pandey, Yogesh Ray, Aman Arora, Shridhar D. Jawak, and Uma K. Shukla.

resolutions of 10–$10^3$ years (Birks, 1981; Quamar et al., 2021). Despite its widespread use, this approach has a few limitations including long-distance transportation, pollen dispersion, reworking and pollen corrosion owing to microbial attack, as well as mechanical forces (Prentice, 1985). Therefore, palynological datasets need to be supplemented by additional proxies so as to provide ample information about vegetation–climate interactions (Guiot et al., 2009). Phytolith assemblages provide a new perspective on interpretation of environment, climate, and vegetation as they tend to be less susceptible to decomposition in comparison to pollen (Prebble et al., 2002). Moreover, they are considered as a reliable proxy in oxidizing conditions such as lacustrine sediments, where pollen tends to get degraded (Rasbold et al., 2020). Furthermore, stable isotope signature ($\delta^{13}C$) is another reliable proxy that is often used in reconstruction of past vegetation as well as climate (Meyers, 1994). $\delta^{13}C$ is a useful tool that helps to distinguish contribution from two principal photosynthetic pathways, C3 and C4 cycles. Carbon isotopic values in C3 and C4 plants are altered due to isotopic fractionation resulting from temperature and precipitation (Amesbury et al., 2015). C3 plants inhabited to humid environmental conditions produce $\delta^{13}C$ values ranging between −20‰ and −34‰, and C4 plants in dry to arid conditions show values between −9‰ and −19‰ (O'Leary, 1981; Agrawal et al., 2012). Hence, the contemporary pollen, $\delta^{13}C$, and phytolith relationship with vegetation, can be employed for understanding modern and palaeoclimate studies (Quamar, 2019). Moreover, an integrated approach based on proxy records will help to delineate long-term climate change impacts on ecosystems as well as provide a baseline study for understanding future vegetation–climate–human interactions (Misra et al., 2020).

The objectives of the chapter are to: i) understand the potential relationship of vegetation response to climate dynamics in Indian Himalaya; ii) assess the multiproxy behavior and observational data in order to delineate long-term vegetation shifts at local and regional levels; and iii) investigate the role of anthropogenic drivers impacting terrestrial ecosystems in Indian Himalaya.

## 16.2 Study Site

The Himalaya account for ~18% of India's geographical area, thereby influencing ecological along with socioeconomic factors (Saxena et al., 2001; Agrawal et al., 2010). The Himalaya has been designated as a "biodiversity hotspot" due to diversified ecological, biogeographical, and evolutionary factors (Mittermeier et al., 1999; Joshi and Pant, 2012). Due to the steepness of Himalaya, variation in climate and biodiversity has been observed in a north-south gradient. Also, rapid warming at higher elevations in comparison to lower elevations has affected vegetation cover in response to climate variability (Shrestha et al., 1999). In order to delineate the future climate variability influence on the ecosystem, the understanding about past events occurring in a few thousand years along with present climatic conditions is crucial.

The Himalaya exhibit great variability in climatic conditions. The foothills, middle hills, and high mountains are characterized by the sub-tropical, temperate, and alpine climatic conditions, respectively (Sanwal et al., 2013). The western Himalaya is marked by the increasing impact of Indian summer monsoon (ISM) and decreasing influence of mid-latitude Westerlies from west to east (Bookhagen and Burbank, 2010; Dimri and Niyogi, 2013). Moreover, west-east variation occurs due to interplay of various weather systems (monsoon vs. Westerlies), which has caused strengthening of the monsoon from west to east. The summer monsoonal precipitation occurs during May to October, while winter precipitation contributed by mid-latitude Westerlies predominantly occurs during November to April (Wulf et al., 2010). The Indian central Himalaya is also influenced by two atmospheric circulation patterns, i.e., ISM in summer and Westerlies in winter (Benn and Owen, 1998). The annual precipitation contributed by the Westerlies in the region varies from 15 to 20% (Sanwal et al., 2013). The eastern Himalaya remarkably differ from western and central parts due to high biological diversity as well as humid climate. Moreover, this part receives high precipitation from the Bay of Bengal branch of ISM and has a floristically vast region (Ghosh et al., 2018). The treeline is higher in this region as compared to the western Himalaya.

## 16.3 Climate Vegetation Interaction in the Indian Himalaya

### 16.3.1 Present-Day Conditions

According to the IPCC (2013), the increasing temperature (0.85°C, during the period between 1880 and 2012), melting glaciers, rising sea levels, and elevating emissions of greenhouse gases (GHGs), are most concerning and notable climate changes observed in the present-day scenario. The available evidence for the climate variations suggested that the effect of global warming or increased temperature has a severe impact on the Himalayan region. The average increase in air temperature (1.6°C) in the Himalaya enhances glacier meltwater and variability in river inflow, which affects health and the socio-economic condition of billions of people in the Indian subcontinent (Bhutiyani et al., 2007).

In recent years, the Himalayan region shows significant change in the vegetation cover from continuous impact of anthropogenic activity and climate variability (Shrestha et al., 2012, 2019; Batar et al., 2017; Ahmed et al., 2021; Sarania et al., 2021). Due to scarcity of datasets, the climate/human induced vegetation biodiversity changes and their extent in the Himalaya is difficult to identify. Therefore, an integration of ground-based data along with remote-sensing derived datasets (maps) is necessary to disentangle the climate–vegetation interactions.

The remote-sensing data (based on normalized difference vegetation index – NDVI) from the huge extension of Himalaya during the period 1982 to 2006, have shown a significant change in both the temperature (average increase of 0.06°C/year) along with precipitation (increased precipitation of ~6.52 mm/year) affecting biodiversity hotspots in the region (Shrestha et al., 2012). Although temperature in the Himalaya shows continuous increasing trends, precipitation trends are spatially variable, where some of the region shows increasing precipitation during the last few decades, while others indicate a decreasing precipitation pattern (Shrestha et al., 2000).

In a study from northeast Himalaya (NEH), based on the Enhanced Vegetation Index, Sarania et al. (2021) derived a relation between vegetation cover with mean annual precipitation in the region. The relation indicates that the vegetation cover in NEH is monotonically increasing with moderate rainfall (i.e., 1000–2000 mm of mean annual precipitation (MAP)) and is characterized by broad-leaved forests, whereas the higher rainfall regions (e.g., >2000 mm of MAP) are characterized by tropical evergreen and semi-evergreen forest (Sarania et al., 2021). Another study based on NDVI from the remote-sensing imagery of NEH has observed that major changes in the vegetation distribution are largely controlled by monsoonal precipitation and anthropogenic activity (based on aerosol cover) in the region (Prasad et al., 2007). This interpretation has been supported by remote-sensing imagery for Assam and Arunachal Pradesh (NEH) showing decrease in forest cover between 1925 and 2009 due to increased human activity in terms of agricultural practice and huge demands for land cultivation (Kushwaha et al., 2018). Furthermore, the major precipitation is largely governed by the spatio-temporal variability in various teleconnections (e.g., El Niño-Southern Oscillation, Indian Ocean Dipole, North Atlantic Oscillation, etc.), over the Indian sub-continent. However, in the NEH, the precipitation variability (and hence the vegetation cover) is largely controlled by the local topography and micro-climate of the region, and no significant role of teleconnection has been identified (Prasad et al., 2007).

Furthermore, geospatial modeling based on the Landsat data (Multispectral Scanner, Thematic Mapper, and Enhanced Thematic Mapper) from Kashmir Himalaya for the years from 1980 to 2009 indicate that the region witnessed significant loss of forest cover due to the continuous impact of human activity in terms of deforestation, illicit felling, and increased grazing activity (Wani et al., 2016). However, a few regions in the Kashmir Himalaya show an increase in the forest cover due to the continuous conservation efforts. Furthermore, the Land-Use/Land-Cover (LULC) change in the Garhwal area in northwest Himalaya shows that increasing anthropogenic activities and natural climate variability negatively impact forest ecosystems (Batar et al., 2017).

In a nutshell, the vegetation cover in the Himalaya manages the local hydrological balance, maintains environmental stability, and prevents land degradation (Negi, 2009). It is important to assess the forest cover and the factors controlling their balance in the Himalaya. However, due to scarcity in the datasets, and inaccessibility in the harsh terrain, it is difficult to identify the factors controlling the vegetation cover in the region. Therefore, multilevel approaches, including remote-sensing applications along with ground-based surveys, are important to understand the extent of the forest cover in any region and the factors, e.g., climate variability and human activity controlling their extension.

### 16.3.2 The Holocene Epoch

During the Holocene epoch, climate has been influenced by different teleconnections, and specifically late Holocene by human intervention (Conroy et al., 2008; Menzel et al., 2014). The climate during this timeframe is well reconstructed globally using a multiproxy approach (Alley et al., 1997). Owing to several factors, e.g., spatial variability in rainfall, different proxy response time, and variable impact of teleconnections, the knowledge about climate variability mainly continental records of precipitation variability is limited from the Indian subcontinent. Moreover, comprehensive study concerning climate and vegetation variability is sparser from the Himalayan region. Therefore, in order to delineate such interactions in the Himalaya, several paleorecords derived from lacustrine sediments are useful (Demske et al., 2009; Mishra et al., 2015). The present work emphasizes on using multiproxy datasets to unravel climate as well as vegetation history of the western, central, and eastern Indian Himalaya from the early Holocene to Present. The pollen records, carbon isotopes, and phytolith assemblages were used to reconstruct changes in vegetation and ISM.

#### 16.3.2.1 Western Himalaya

The multiple records from different sedimentary deposits (e.g., Tso Moriri, Tso Kar, Surinsar Lake, Sangla Valley, Chandra Valley, and Rukti Valley) (Figure 16.1) were utilized to delineate climatic conditions associated with ISM during the early Holocene (11,500 to 8200 yr BP;

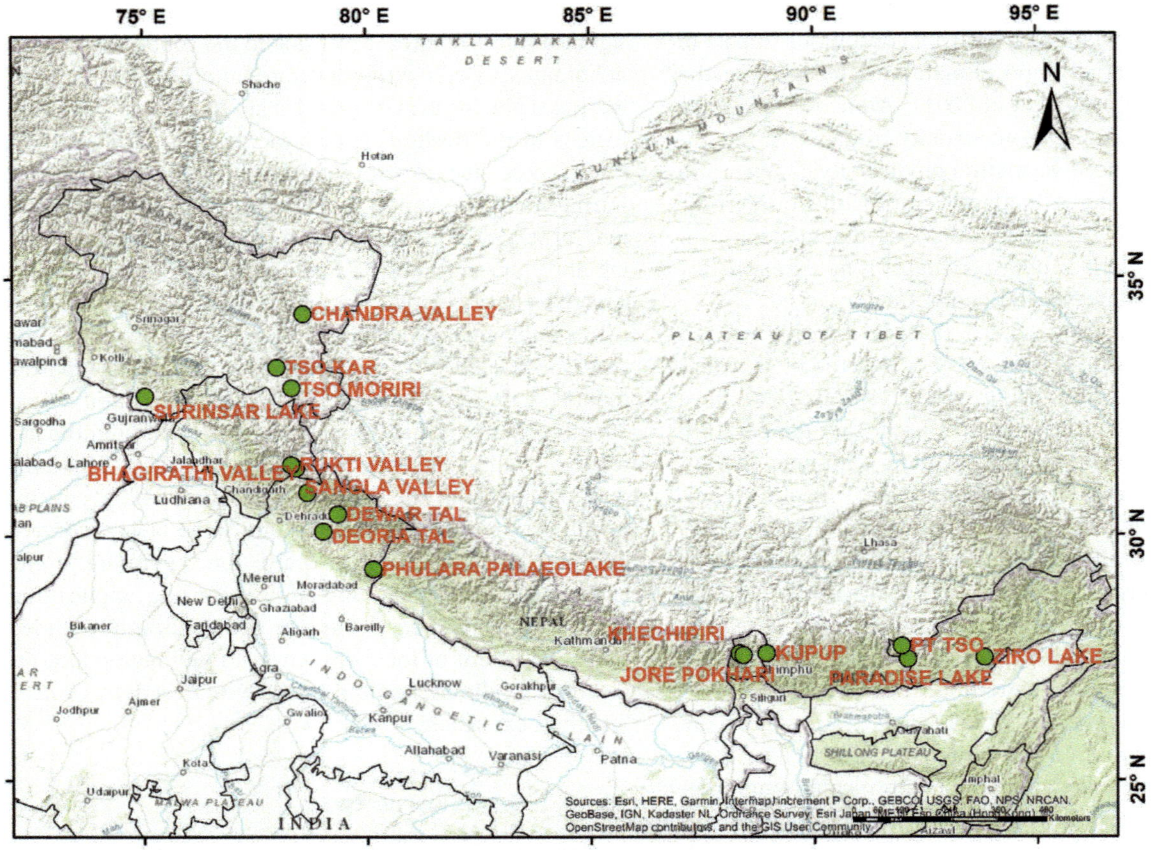


**Figure 16.1** Paleoclimatic sites from Indian Himalaya discussed in the text.

Walker et al., 2012). In general, the lake records from Tibetan Plateau and Indian Himalaya indicate an increase in moisture during the early Holocene (Van Campo et al., 1996; Wei and Gasse, 1999). The investigation from Tso Kar Lake reveals presence of arboreal pollen types (*Carpinus viminea* type and *Pinus roxburghii* type) accompanied by subalpine *Pinus wallichiana* forests, ferns, and alpine Artemisia-dominated steppes that suggest increase in monsoonal intensity between ca. 10,900 and 9200 yr BP (Demske et al., 2009). Likewise, rise in percentage ratio of Artemisia to Chenopodiaceae and reconstructed mean annual precipitation values along with low scores of desert biome and high pollen concentrations from Tso Moriri Lake suggest a phase of maximum moisture conditions from ca. 11,000 to 9600 yr BP (Leipe et al., 2014). Presence of Cyperaceae, Impatiens, fern spores, and aquatic elements (*Typha* and *Potamogeton*) from Sangla Valley in Kinnaur (10,450 to 4310 yrs BP) also indicate a warm and moist climate (Chakraborty et al., 2006). Furthermore, an increase in *Betula*, *Quercus*, and fern spores (monoletes) in Rukti Valley during the time period ~11,500 to ~8500 yr BP indicate high moisture conditions (Ranhotra et al., 2018). Likewise, occurrence of high pollen influx, trends of broad-leaved taxa, and the lowest value of $\delta^{13}C$ (−26.5‰) (Rawat et al., 2015) from a Chandra peat bog (~11,640 to 8810 yr BP) and mixed oak-broad-leaved along with chir-pine forest from Surinsar Lake (9500 and 7700 yr BP) also indicate the commencement of warm and moist conditions (Trivedi and Chauhan, 2009) (Figure 16.2).

During the mid-Holocene (8200 to 4200 yr BP) (Walker et al., 2012), the obtained continental records show various short-term climate events (8200 yr BP, HCO, 4200 yr BP events). The pollen records obtained from Tso Kar, Chandra peat bog, Rukti Valley, and Tso Moriri Lake roughly indicate a global 8.2 ka event (Demske et al., 2009; Leipe et al., 2014; Rawat et al., 2015; Ranhotra et al., 2018). A fair rise in *Potamogeton*, Cyperaceae, and *Quercus* pollen from the Rukti Valley suggest cool/arid conditions (during ~8200 yr BP). The 8200 yr BP climate event also resulted in reduced precipitation in the Asian monsoon region and cooling in the North Atlantic (Morrill et al., 2013). The pollen data from Tso Kar (ca. 6900 to 4800 cal BP) (Demske et al., 2009) and Surinsar Lake (~6125 and 4330 cal BP) (Trivedi and Chauhan, 2009) suggest relatively warm and humid climatic conditions. Similarly, results from palynological analysis as well as $\delta^{13}C$ values from Chandra Valley (6732–3080 cal BP) highlight sufficient moisture supply

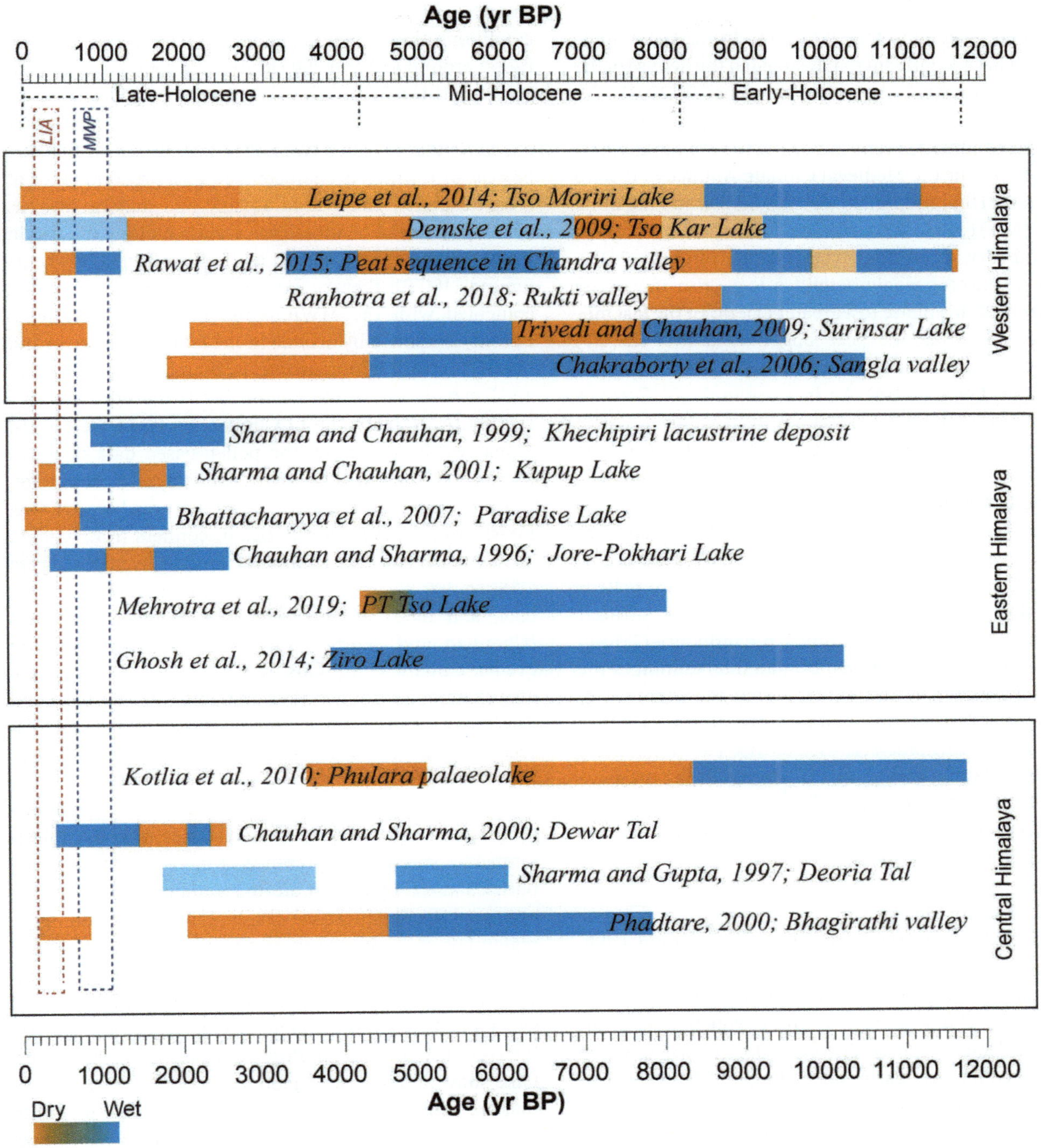


**Figure 16.2** Compilation of paleoclimate records from Indian Himalaya.

(Rawat et al., 2015). The variability in climatic conditions observed in various study sites corresponds to prevalence of active ISM and occurrence of HCO (Benarde, 1996). The latter part of the middle Holocene witnessed a cold and dry spell at Chandra Valley from ~4808 to 4327 cal BP, as shown by decrease in broad-leaved and meadow constituents coinciding with the global 4.2 ka event (Staubwasser et al., 2003). A similar dry pulse was also evident in Tso Kar (~4800 and 3700 cal BP), Triloknath paleolake in Lahaul Himalaya (5.3 to 3.1 kyr BP), and Tso Moriri during ca ~4500 to 4300 cal BP, which indicated the declining phase of monsoonal precipitation but high winter westerly flow (Demske et al., 2009; Leipe et al., 2014; Bali et al., 2017). This arid and cold event is also reported in Dongge Cave, China and Mawmluh

Cave, NEH (Wang et al., 2005; Berkelhammer et al., 2012). This abrupt weakening of the ISM led to drought conditions in the Indian subcontinent that resulted in the demise of the urban Harappan civilization (Staubwasser et al., 2003).

The late Holocene is marked by various short-term centennial-scale climatic events involving the Medieval Warm Period (MWP) or the Medieval Climate Anomaly (1050 to 600 years BP) that resulted in wet conditions, and the Little Ice Age (LIA) (450 to 100 years BP) that is characterized by cold and dry conditions (Dixit and Tandon, 2016). Moreover, anthropogenic activities (e.g., rise in GHGs) also impacted this phase to a greater extent. In the last millennium, the Chandra peat bog (~1158–647 cal BP) recorded a warm and humid phase that corresponds to MWP (Rawat et al., 2015). An increase in the $\delta^{13}C$ values from the Zanskar Valley (~400 cal BP) (Ali et al., 2020) along with decrease in broad-leaved taxa in the Chandra Valley (~650–350 cal BP) (Rawat et al., 2015) closely predates the LIA cooling event. Similar MWP and LIA events have also been reported from Benital and Badanital lakes (Demske et al., 2016; Bhushan et al., 2018). Also, the occurrence of Cerealia pollen along with other culture pollens during the phase 4000–2100 cal BP from the Surinsar region implies intensive anthropogenic influence (Trivedi and Chauhan, 2009). The high human intervention in the western Himalaya is also evident from the presence of charcoal in sediments obtained from the upper reaches of Rukti Valley during phase ca. 4300 to 1800 ka (Chakraborty et al., 2006; Ranhotra et al., 2018). Moreover, the presence of *Cannabis* and *Triticum* in Tso Kar from ca. 4800 to Present emphasize anthropogenic disturbances in the region (Demske et al., 2009).

#### 16.3.2.2 Eastern Himalaya

The multi-proxy records from Khechipiri, Kupup Lake, Paradise Lake, Jore-Pokhari, Pankang Teng Tso (PT Tso) Lake, and Ziro Lake were used to understand vegetation and climatic changes (Figure 16.1). Pollen data from Khechipiri lacustrine deposits reveal dominance of mixed broad-leaved forests (warm moist climate) 2500 years ago followed by increase in *Rhododendron* and *Alnus* (more humid) after 1000 years (Sharma and Chauhan, 1999). Moreover, palynological studies from Kupup Lake (Sharma and Chauhan, 2001) indicate presence of *Pinus*, *Quercus*, *Betula*, and *Rhododendron* along with sedges and grasses from 2000 to 1800 cal BP, decrease in *Betula*, *Alnus*, and *Rhododendron* along with increase in sedges and grasses from 1800 to 1450 cal BP, and increase in broad-leaved taxa from 1450 to 450 cal BP, thereby suggesting cold-moist, drier and cold humid conditions (MWP), respectively (Sharma and Chauhan, 2001) (Figure 16.2). Furthermore, an increase in Cyperaceae, Poaceae, and Ranunculaceae, and decline in arboreal taxa observed in Kupup Lake from 450 to 200 cal BP reveal cold and dry conditions, thus coinciding with LIA. The climate deterioration around 450 and 200 cal BP corresponds to the LIA. Such short-term climatic events based on pollen analysis were also noticed from Paradise Lake (Bhattacharyya et al., 2007) that revealed two pollen zones, i.e., PL-I (1780–684 cal BP), which is characterized by a predominance of arboreal pollen and rise in aquatic and marshy taxa, as well as existence of sub-alpine forests along with rise in *Tsuga*, *Pinus*, *Abies*, *Picea*, and broad-leaved elements around 1100 BP, thereby indicating a warmer climate that coincided with the MWP, while PL-II (684 cal BP to Recent) is marked by decline in the total pollen count and overall increase in non-arboreals, thereby suggesting cool and less moist conditions around 550 BP that corresponded to the LIA. Notably, the LIA impact is less in the eastern Himalaya. Furthermore, the pollen-based data obtained from other sites in the eastern Himalaya, such as Jore-Pokhari, highlight presence of mixed broad-leaved Oak forests around 2500 cal BP, rise in conifers from 1600 to 1000 cal BP, and occurrence of broad-leaved taxa from 1000 to 300 cal BP, reflecting warm-temperate climate, cool, and moist environmental conditions respectively (Chauhan and Sharma, 1996). Along with short-term events, climatic conditions over a longer timescale were also examined from two sites, including PT Tso Lake and Ziro Lake. Palynological and carbon isotope datasets recorded during ~8000–906 cal BP from PT Tso Lake sediments reveal the prevalence of a cold-moist period around 4625 cal BP, which was followed by cold-arid conditions (Mehrotra et al., 2019). The dominance of sub-tropical-temperate species and presence of mixed vegetation, i.e., both C3 and C4 plants, indicate humid conditions during ~4814 cal BP. The presence of sub-alpine vegetation in the region favors a cold-dry phase and abrupt climate change (ACC) around 4200 cal BP. The strength of the SW monsoon reduced in the eastern Himalaya during the ACC event and 4.2 ka event, which affected vegetation–climate interactions. Likewise, multiproxy-based results from Ziro Lake show an increase in globular echinate morphotypes that are considered to be dominant in arboreal phytolith assemblages, hygrophilous elements, particularly taxa including *Pteris*, *Lycopodium*, and *Typha*, coupled with $\delta^{13}C$ values ranging from 29.1‰ to 28.2‰ that suggest SW monsoon intensification during 10,200–3800 yr BP (Ghosh et al., 2014). A rise in heliophytic taxa, absence of globular echinate morphotypes, and increase in C4 species highlight rising trends of dryness as well as thinning of forest cover after 3.8 ka. Furthermore, anthropogenic changes during the late Holocene were also noticed in the eastern Himalaya, mainly from Jore-Pokhari, PT Tso Lake, and Ziro Lake. Human influence from Jore-Pokhari is evidenced from an increase in culture pollen, particularly

Caryophyllaceae, Asteraceae, and Cerealia, while replacement of dry temperate forests by sub-alpine forests and adoption of practices like cultivation and farming in PT Tso Lake. Also, the variation observed in isotopic ratios as well as rise in grass suggest human-induced landscape changes in the Ziro Lake Basin (Ghosh et al., 2014) as well as in Mirik Lake basin since 2000 yr BP (Sharma and Chauhan, 1994).

#### 16.3.2.3 Central Himalaya

The past climate and vegetation changes from the central Himalaya were reconstructed using paleorecords from the Phulara palaeolake, Bhagirathi Valley, Dewar Tal, and Deoria Tal (Figure 16.1). The lowering of carbon isotope values (up to −21.8‰) from Phulara palaeolake indicate abundant C3 vegetation in the region that further suggest high moisture conditions during early Holocene, while sharp increase in $\delta^{13}C$ values (−18.8‰) mark cold excursions between ca. 8300 and 6000 cal BP (Kotlia et al., 2010). Moreover, rising $\delta^{13}C$ (−21‰) values reflect weakening of the SW monsoon and a shift to dry conditions between ca. 5000 and 4000 cal BP and this arid period played a significant role in the draining of Phulara Lake in Kumaun Himalaya. However, the pollen assemblage from Bhagirathi Valley shows abundant pollen of Oak (*Q. semecarpifolia*), sub-aquatic elements, and a mixed evergreen deciduous forest which indicates a cool and moist climate during the time period 7800–7000 cal BP (Phadtare, 2000). An increase in grasses as well as Abies (11%) reflect wetter hydrological conditions from 6000 to 5000 cal BP, while abundant conifers in sub-alpine forest and grasses in alpine meadow hint at a warming climate and increasing monsoon rainfall, therefore reflecting mid-HCO in the Garhwal Himalaya. The gradual decrease in Abies marks cooling of the climate during the time period 4500–4000 cal BP and increase in the Quercus/Pinus ratio suggests prevalence of aridity (800 cal BP) that corresponds to LIA (Figure 16.2). Furthermore, the pollen assemblage obtained from Deoria Tal suggests the presence of Oak-dominated forests as well as fern and spores that highlight warm-temperate humid climate conditions during the time period from 6000 to 4600 years BP (Sharma and Gupta, 1997). Decrease in *Quercus, Carpinus*, and *Engelhardtia*, increase in Poaceae and Cyperaceae, and enhancement of Cerealia-type pollen, indicate cool and humid climate conditions during 3600–1700 years BP (Figure 16.2). Moreover, the presence of abundant culture pollen, specifically Cerealia-type, along with charcoal pieces, demonstrate agricultural activities in the region. After 1700 years BP, an increase of arboreals over nonarboreals and continued dominance of culture pollen indicate regeneration of forests.

The pollen data obtained from Dewar Tal located in Lesser Garhwal Himalaya were utilized to understand palaeovegetation changes during the late Holocene (Chauhan and Sharma, 2000). The pollen sequence observed in the time period 2500–2300 years BP indicates presence of chirpine-dominated open forest, Oak forest, and scarce ground vegetation that comprise grasses linked with sedges, suggesting a cool and dry climate. The expansion of chirpine as well as broad-leaved elements including *Rhododendron* and *Viburnum* demonstrates spread of arboreal taxa from 2300 to 2000 years BP. Moreover, decrease in *Pinus* and *Quercus* frequencies, while enhancement in grasses and sedges during 2000–1400 years BP, suggest cool and dry climate conditions. The development of Oak forest and decrease in mixed chirpine forest mark increased precipitation during 1400–400 years coinciding closely with MWP (Figure 16.2). The pollen data during the last 400 years BP highlight change in the paleovegetational scenario, i.e., decrease in mixed Oak forest and rise in chirpine-dominated forest that further suggest deterioration in the climate. Moreover, presence of culture pollen taxa including grasses, Cheno/Ams, *Artemisia*, Caryophyllaceae, and Urticaceae reflect agricultural practices. The regional comparison of paleovegetation records from western, eastern, and central Himalaya shows spatio-temporal inhomogeneity in vegetational changes. Therefore, asynchronous timing of vegetation development can be attributed to factors including changes in moisture pattern, altitude, and topography, along with human influence.

## 16.4 Conclusions

The present study provides an overview on the vegetation relating to climatic changes from the ecologically sensitive Indian Himalaya during the Holocene. The multiproxy approach involving stable isotope ($\delta^{13}C$), pollen analyses, and phytolith assemblages was utilized to delineate the changes in the western, eastern, and central Himalaya. The study unravels the impact of various Holocene climatic events including HCO, MWP, and LIA, as well as 8.2 ka and 4.2 ka global events on vegetation changes in Indian Himalaya. The observational dataset clearly highlight that the Himalaya is under the impact of the monsoonal precipitation and escalating human influence. The occurrence of culture pollen along with charcoal pieces in various sediment profiles mark intensive anthropogenic pressure during the late Holocene. Moreover, remote-sensing derived datasets also reveal that anthropogenic stressors involving agricultural practices and land cultivation tend to decrease forest cover in various regions. Therefore, ground-based data coupled with remote-sensing data serve as a baseline study for assessment of the future vegetation climate interactions and management of the ecosystem.

## References


Agrawal, D.K., Lodhi, M.S., and Panwar, S. (2010). Are EIA studies sufficient for projected hydropower development in the Indian Himalayan region? *Current Science* 98: 154–161.

Agrawal, S., Sanyal, P., Sarkar, A. et al. (2012). Variability of Indian monsoonal rainfall over the past 100 ka and its implication for C3–C4 vegetational change. *Quaternary Research* 77(1): 159–170.

Ahmed, R., Wani, G.F., Ahmad, S.T. et al. (2021). Spatiotemporal dynamics of glacial lakes (1990–2018) in the Kashmir Himalayas, India using Remote Sensing and GIS. *Discover Water* 1(1): 1–17.

Ali, S.N., Agrawal, S., Sharma, A. et al. (2020). Holocene hydroclimatic variability in the Zanskar Valley, northwestern Himalaya, India. *Quaternary Research* 97: 140–156.

Alley, R.B., Mayewski, P.A., Sowers, T. et al. (1997). Holocene climatic instability: a prominent, widespread event 8200 yr ago. *Geology* 25(6): 483–486.

Amesbury, M.J., Charman, D.J., Newnham, R.M. et al. (2015). Carbon stable isotopes as a palaeoclimate proxy in vascular plant dominated peatlands. *Geochimica et Cosmochimica Acta* 164: 161–174.

Bali, R., Khan, I., Sangode, S.J. et al. (2017). Mid- to late Holocene climate response from the Triloknath palaeolake, Lahaul Himalaya based on multiproxy data. *Geomorphology* 284: 206–219.

Batar, A.K., Watanabe, T., and Kumar, A. (2017) Assessment of land-use/land cover change and forest fragmentation in the Garhwal Himalayan region of India. *Environ – MDPI* 4: 1–16.

Benarde, M.A. (1996). *Global Warming*, 317. New York: John Wiley.

Benn, D.I. and Owen, L.A. (1998). The role of the Indian summer monsoon and the mid-latitude westerlies in Himalayan glaciation: review and speculative discussion. *Journal of the Geological Society* 155(2): 353–363.

Berkelhammer, M., Sinha, A., Stott, L. et al. (2012). An abrupt shift in the Indian Monsoon 4000 years ago. *Geophysical Research Letters* 198: 75–87.

Bhattacharyya, A., Ranhotra, P.S., and Shah, S.K. (2006). Temporal and spatial variations of late Pleistocene–Holocene climate of the western Himalaya based on pollen records and their implications to monsoon dynamics. *Geological Society of India* 68(3): 507.

Bhattacharyya, A., Sharma, J., Shah, S.K. et al. (2007). Climatic changes during the last 1800 yrs BP from paradise lake, Sela Pass, Arunachal Pradesh, Northeast Himalaya. *Current Science* 93: 983–987.

Bhushan, R., Sati, S.P., Rana, N. et al. (2018). High-resolution millennial and centennial scale Holocene monsoon variability in the Higher Central Himalayas. *Palaeogeography Palaeoclimatology Palaeoecology* 489: 95–104.

Bhutiyani, M.R., Kale, V.S., and Pawar, N.J. (2007). Long-term trends in maximum, minimum and mean annual air temperatures across the Northwestern Himalaya during the twentieth century. *Climatic Change* 85(1): 159–177.

Billings, W.D. (1952). The environmental complex in relation to plant growth and distribution. *The Quarterly Review of Biology* 27(3): 251–265.

Birks, H.J.B. (1981). The use of pollen analysis in the reconstruction of past climates: a review. In: *Climate and History: Studies in Past Climates and Their Impact on Man*, 111–138. Cambridge, UK: Cambridge University Press.

Bookhagen, B. and Burbank, D.W. (2010). Toward a complete Himalayan hydrological budget: spatiotemporal distribution of snowmelt and rainfall and their impact on river discharge. *Journal of Geophysical Research: Earth Surface* 115(3): F03019.

Chakraborty, S., Bhattacharya, S.K., Ranhotra, P.S. et al. (2006). Palaeoclimatic scenario during Holocene around Sangla Valley, Kinnaur northwest Himalaya based on multi proxy records. *Current Science* 91(6): 777–782.

Chauhan, M.S. and Sharma, C. (1996). Late Holocene vegetation of Darjeeling (Jore-Pokhari) eastern Himalaya. *Paleobotanist* 45: 125–129.

Chauhan, M.S. and Sharma, C. 2000. Late Holocene vegetation and climate in Dewar Tal area, Inner Lesser Garhwal Himalaya. *Paleobotanist* 49: 509–514.

Conroy, J.L., Overpeck, J.T., Cole, J.E. et al. (2008). Holocene changes in eastern tropical Pacific climate inferred from a Galápagos lake sediment record. *Quaternary Science Reviews* 27(11–12): 1166–1180.

Cruz, R.V., Harasawa, H., Lal, M. et al. (2007). Asia. In: *Impacts Adaptation and Vulnerability. Contribution of Working Group II to the Fourth Assessment Report of the Intergovernmental Panel on Climate Change* (ed. M.L. Parry, O.F. Canziani, J.P. Palutikof, et al.), 469–506. Cambridge, UK: Cambridge University Press.

D'Arrigo, R.D., Kaufmann, R.K., Davi, N. et al. (2004). Thresholds for warming-induced growth decline at elevational tree line in the Yukon Territory, Canada. *Global Biogeochemical Cycles* 18(3): GB3021.

Demske, D., Tarasov, P.E., Leipe, C. et al. (2016). Record of vegetation, climate change, human impact and retting of hemp in Garhwal Himalaya (India) during the past 4600 years. *The Holocene* 26(10): 1661–1675.

Demske, D., Tarasov, P.E., Wünnemann, B. et al. (2009). Late glacial and Holocene vegetation, Indian monsoon and westerly circulation in the Trans-Himalaya recorded in the lacustrine pollen sequence from Tso Kar, Ladakh, NW India. *Palaeogeography, Palaeoclimatology, Palaeoecology* 279(3–4): 172–185.

Dimri, A.P. and Niyogi, D. (2013). Regional climate model application at subgrid scale on Indian winter monsoon over the western Himalayas. *International Journal of Climatology* 33(9): 2185–2205.

Dixit, Y. and Tandon, S.K. (2016). Hydroclimatic variability on the Indian subcontinent in the past millennium: review and assessment. *Earth-Science Reviews* 161: 1–15.

Ghosh, R., Biswas, O., Paruya, D.K. et al. (2018). Hydroclimatic variability and corresponding vegetation response in the Darjeeling Himalaya, India over the past ~2400 years. *Catena* 170: 84–99.

Ghosh, R., Paruya, D.K., Khan, M.A. et al. (2014). Late Quaternary climate variability and vegetation response in Ziro Lake Basin, Eastern Himalaya: a multiproxy approach. *Quaternary International* 325: 13–29.

Guiot, J., Wu, H.B., Garreta, V. et al. (2009). A few prospective ideas on climate reconstruction: from a statistical single proxy approach towards a multi-proxy and dynamical approach. *Climate of the Past* 5(4): 571–583.

IPCC (2013). Summary for policymakers. In: *Climate Change 2013: The Physical Science Basis. Contribution of Working Group I to the Fifth Assessment Report of the Intergovernmental Panel on Climate Change* (ed. T.F. Stocker, D. Qin, G-K. Plattner, et al.), 3–29. Cambridge, UK and New York: Cambridge University Press.

Joshi, B. and Pant, S.C. (2012). Ethnobotanical study of some common plants used among the tribal communities of Kashipur, Uttarakhand. *Indian Journal of Natural Products & Resources* 3: 262–265.

Körner, C. and Paulsen, J. (2004). A world-wide study of high altitude treeline temperatures. *Journal of Biogeography* 31(5): 713–732.

Kotlia, B.S., Sanwal, J., Phartiyal, B. et al. (2010). Late Quaternary climatic changes in the eastern Kumaun Himalaya, India, as deduced from multi-proxy studies. *Quaternary International* 213(1–2): 44–55.

Kushwaha, S.P.S., Nandy, S., Shah, M.A. et al. (2018). Forest cover monitoring and prediction in a Lesser Himalayan elephant landscape. *Current Science* 115: 510–516.

Leipe, C., Demske, D., Tarasov, P.E. et al. (2014). A Holocene pollen record from the northwestern Himalayan lake Tso Moriri: implications for palaeoclimatic and archaeological research. *Quaternary International* 348: 93–112.

Mehrotra, N., Shah, S.K., Basavaiah, N. et al. (2019). Resonance of the "4.2 ka event" and terminations of global civilizations during the Holocene, in the palaeoclimate records around PT Tso Lake, Eastern Himalaya. *Quaternary International* 507: 206–216.

Menzel, P., Gaye, B., Mishra, P.K. et al. (2014.) Linking Holocene drying trends from Lonar Lake in monsoonal central India to North Atlantic cooling events. *Palaeogeography, Palaeoclimatology, Palaeoecology* 410: 164–178.

Meyers, P.A. (1994). Preservation of elemental and isotopic source identification of sedimentary organic matter. *Chemical Geology* 114(3–4): 289–302.

Mishra, P.K., Prasad, S., Ambili, A. et al. (2015). Carbonate isotopes from high altitude Tso Moriri Lake (NW Himalayas) provide clues to late glacial and Holocene moisture source and atmospheric circulation changes. *Palaeogeography, Palaeoclimatology, Palaeoecology* 425: 76–83.

Misra, S., Bhattacharya, S., Mishra, P.K. et al. (2020). Vegetational responses to monsoon variability during Late Holocene: inferences based on carbon isotope and pollen record from the sedimentary sequence in Dzukou Valley, NE India. *Catena* 194: 104697.

Misra, S. and Bhattacharyya, A. (2014). Analysis of the late Holocene climate vis-a-vis vegetation dynamics along the southwest coast of India: Thrissur (Kerala). *Quaternary International* 325: 150–161.

Mittermeier, R.A., Myers, N., Mittermeier, C.G. et al. (1999). Hotspots: Earth's biologically richest and most endangered terrestrial ecoregions. CEMEX, SA, Agrupación Sierra Madre, SC.

Morrill, C., Anderson, D.M., Bauer, B.A. et al. (2013). Proxy benchmarks for intercomparison of 8.2 ka simulations. *Climate of the Past* 9(1): 423–432.

Negi, S.P. (2009). Forests cover in Indian Himalaya states: an overview. *Indian Journal of Forestry* 32: 1–5.

O'Leary, M.H. (1981). Carbon isotope fractionation in plants. *Phytochemistry* 20(4): 553–567.

Pandit, M.K. and Grumbine, R.E. (2012). Potential effects of ongoing and proposed hydropower development on terrestrial biological diversity in the Indian Himalaya. *Conservation Biology* 26(6): 1061–1071.

Phadtare, N.R. (2000). Sharp decrease in summer monsoon strength 4000–3500 cal yr BP in the Central Higher Himalaya of India based on pollen evidence from alpine peat. *Quaternary Research* 53(1): 122–129.

Prasad, A.K., Sarkar, S., Singh, R.P. et al. (2007). Inter-annual variability of vegetation cover and rainfall over India. *Advances in Space Research* 39(1): 79–87.

Prebble, M., Schallenberg, M., Carter, J. et al. (2002). An analysis of phytolith assemblages for the quantitative reconstruction of late Quaternary environments of the Lower Taieri Plain, Otago, South Island, New Zealand. I: Modern assemblages and transferfunctions. *Journal of Paleolimnology* 27(4): 393–413.

Prentice, I.C. (1985). Pollen representation, source area, and basin size: toward a unified theory of pollen analysis. *Quaternary Research* 23(1): 76–86.

Quamar, M.F. (2019). Vegetation dynamics in response to climate change from the wetlands of Western Himalaya, India: Holocene Indian Summer Monsoon variability. *The Holocene* 29(2): 345–362.

Quamar, M.F. and Bera, S.K. (2014). Vegetation and climate change during mid and late Holocene in northern Chhattisgarh (central India) inferred from pollen records. *Quaternary International* 349: 357–366.

Quamar, M.F., Kar, R., and Thakur, B. (2021). Vegetation response to the Indian Summer Monsoon (ISM) variability during the Late-Holocene from the central Indian core monsoon zone. *The Holocene* 31(7): 1197–1211.

Ranhotra, P.S., Sharma, J., Bhattacharyya, A. et al. (2018). Late Pleistocene-Holocene vegetation and climate from the palaeolake sediments, Rukti Valley, Kinnaur, Himachal Himalaya. *Quaternary International* 479: 79–89.

Rasbold, G.G., Stevaux, J.C., Parolin, M. et al. (2020). Phytoliths indicate environmental changes correlated with facies analysis in a paleo island-lake, Upper Paraná River, Brazil. *Journal of South American Earth Sciences* 99: 102513.

Rawat, S., Gupta, A.K., Sangode, S.J. et al. (2015). Late Pleistocene–Holocene vegetation and Indian summer monsoon record from the Lahaul, northwest Himalaya, India. *Quaternary Science Reviews* 114: 167–181.

Rosenzweig, C. and Neofotis, P. (2013). Detection and attribution of anthropogenic climate change impacts. *Wiley Interdisciplinary Reviews: Climate Change* 4(2): 121–150.

Sanwal, J., Kotlia, B.S., Rajendran, C. et al. (2013). Climatic variability in Central Indian Himalaya during the last ~1800 years: evidence from a high resolution speleothem record. *Quaternary International* 304: 183–192.

Sarania, B., Guttal, V., and Tamma, K. (2021). Characterising the vegetation rainfall relationship in the Northeast Himalaya, India. bioRxiv. https://doi.org/10.1101/2021.10.19.464965.

Saxena, K.G., Rao, K.S., Sen, K.K. et al. (2001). Integrated natural resource management: approaches and lessons from the Himalaya. *Conservation Ecology* 5(2): 14.

Sharma, C.and Chauhan, M.S. (1994). Vegetation and climate since Last Glacial Maxima in Darjeeling (Mirik Lake), Eastern Himalaya. In: *Proceedings of 29th International Geological Congress, Japan, Part B*, 279–288. VSP International Science Publishers.

Sharma, C. and Chauhan, M.S. (1999). Inferences from Quaternary palynostratigraphy of the Himalayas. In: *The Himalayan Environment*, 193–207. New Delhi: New Age International

Sharma, C. and Chauhan, M.S. (2001). Late Holocene vegetation and climate of Kupup (Sikkim), Eastern Himalaya, India. *Journal of the Palaeontological Society of India* 46: 51–58.

Sharma, C. and Gupta, A. (1997). Vegetation and climate in Garhwal Himalaya during Early Holocene: Deoria Tal.

Sharma, G. and Rai, L.K. (2012). Climate change and sustainability of agrodiversity in traditional farming of the Sikkim Himalaya. In: *Climate Change in Sikkim: Patterns, Impacts, Initiatives* (ed. M.L. Arawatia and S. Tambe), 193–218. Gangtok, India: Information and Public Relations Department, Government of Sikkim.

Shrestha, A.B., Wake, C.P., Dibb, J.E. et al. (2000). Precipitation fluctuations in the Nepal Himalaya and its vicinity and relationship with some large-scale climatological parameters. *International Journal of Climatology* 20(3): 317–327.

Shrestha, A.B., Wake, C.P., Mayewski, P.A. et al. (1999). Maximum temperature trends in the Himalaya and its vicinity: an analysis based on temperature records from Nepal for the period 1971–94. *Journal of Climate* 12(9): 2775–2786.

Shrestha, S., Yao, T., and Adhikari, T.R. (2019). Analysis of rainfall trends of two complex mountain river basins on the southern slopes of the Central Himalayas. *Atmospheric Research* 215: 99–115.

Shrestha, U.B., Gautam, S., and Bawa, K.S. (2012). Widespread climate change in the Himalayas and associated changes in local ecosystems. *PLoS One* 7: 1–10.

Singh, S.P., Singh, V., and Skutsch, M. (2010). Rapid warming in the Himalayas: ecosystem responses and development options. *Climate and Development* 2(3): 221–232.

Staubwasser, M., Sirocko, F., Grootes, P.M. et al. (2003). Climate change at the 4.2 ka BP termination of the Indus Valley civilization and Holocene Asian monsoon variability. *Geophysical Research Letters* 30(8): 1425–1428

Trivedi, A. and Chauhan, M.S. (2009). Holocene vegetation and climate fluctuations in northwest Himalaya, based on pollen evidence from Surinsar Lake, Jammu region, India. *Journal of the Geological Society of India* 74(3): 402.

Van Campo, E., Cour, P., and Sixuan, H. (1996). Holocene environmental changes in Bangong Co basin (Western Tibet). Part 2: The pollen record. *Palaeogeography, Palaeoclimatology, Palaeoecology* 120(1–2): 49–63.

Walker, M.J.C., Berkelhammer, M., Björck, S. et al. (2012). Formal subdivision of the Holocene Series/Epoch: a discussion paper by a working group of INTIMATE (Integration of ice-core, marine and terrestrial records) and the Subcommission on Quaternary Stratigraphy (International Commission on Stratigraphy). *Journal of Quaternary Science* 27: 649–659.

Wang, Y., Cheng, H., Edwards, R.L. et al. (2005). The Holocene Asian Monsoon: links to solar changes and North Atlantic climate. *Science* 308(5723): 854–857.

Wani, A.A., Joshi, P.K., Singh, O. et al. (2016). Multi-temporal forest cover dynamics in Kashmir Himalayan region for assessing deforestation and forest degradation in the context of REDD+ policy. *Journal of Meteorological Research* 13(8): 1431–1441.

Wei, K. and Gasse, F. (1999). Oxygen isotopes in lacustrine carbonates of West China revisited: implications for post glacial changes in summer monsoon circulation. *Quaternary Science Reviews* 18(12): 1315–1334.

Wulf, H., Bookhagen, B., and Scherler, D. (2010). Seasonal precipitation gradients and their impact on fluvial sediment flux in the Northwest Himalaya. *Geomorphology* 118(1–2): 13–21.

Yang, S. (1996). ENSO-Snow-monsoon associations and seasonal inter-annular prediction. *International Journal of Climatology* 16: 125–134.

# 17

# Flash Flood Susceptibility Mapping of a Himalayan River Basin Using Multi-Criteria Decision-Analysis and GIS

*Pratik Dash*[1,*], *Kasturi Mukherjee*[2], *and Surajit Ghosh*[3]

[1] *Department of Geography, Khejuri College (affiliated with Vidyasagar University), Purba Medinipur 721431, India*
[2] *Department of Geography, Adamas University, Kolkata 700126, India*
[3] *International Water Management Institute, Colombo, Sri Lanka*
[*] *Corresponding author*

## 17.1 Introduction

Among several hazards, the hydrological-meteorological disaster, more specifically flooding, is one of the most commonly occurring and impactful natural calamities worldwide (Shen and Hwang, 2019). The magnitude and frequency of floods, and resulting damage, loss, and casualties are surprisingly high for developing countries, like India, China, Bangladesh, etc. (Nkwunonwo et al., 2020). Floods are primarily caused by various hydro-meteorological factors, i.e. heavy downpours, snow melt, storm surges, etc. Severity of this flooding is increasing day-by-day due to rapid urbanization, land-use dynamics, river flow control, and climate change (Dash and Punia, 2019). Due to such environmental changes, the flash flood situation becomes more common in mountainous topography, especially in the snow-fed river catchments of the Himalaya.

In the Himalayan region, flood disasters are often triggered by intense accumulation of moisture that leads to cloudbursts. The torrential downpour not only causes flash floods, but also landslides that, in turn, trigger further flash floods (Asthana and Asthana, 2014). The shifting pattern of rainfall and unusual rainfall distribution that is primarily found to be caused by climate change, intensifies flooding calamities in the Western Himalaya (Kumar et al., 2018). Recent past events of cloudburst-induced and rainfall-induced flash floods suggests that Himachal Himalaya and Uttarakhand Himalaya are the most prone to flash flooding. The major flash-flood events of the Himachal Himalaya include 29 September 1988 (Soldan Khad, Sutlej Valley), 11 August 1997 (Chirgaon), 30 July 2000 (Sutlej Valley), and 16 July 2003 (Kullu) (Thakur, 2000; Kumar et al., 2018). The major flash floods of the Uttarakhand Himalaya include 5–10 June 2000 (Gangotri Glacier), 15–25 August 2010 and 2011 (Dokriani Glacier), 3 August 2012 (Asiganga), 13–14 September 2012 (Okhimath), 16–17 June 2013 (Kedarnath and Gangotri Glacier), and 1 July 2016 (Bastadi Narula) (Arora et al., 2016; Kumar et al., 2018; Dash and Punia, 2019).

Due to the devastating nature of floods in Uttarakhand Himalaya, as well as in other parts of India, adaptation of suitable strategies for flood risk management is more appropriate, as flooding is unavoidable. Noticeably, the flood warning system is not sufficient to reduce the magnitude of calamities; therefore, in-depth study of flood management is of great concern for researchers worldwide (Ahrens and Rudolph, 2006). Flood susceptibility mapping is one of the fundamental approaches in flood management that could assist in adopting suitable structural and non-structural measures (Dash and Punia, 2019).

Flooding or flash floods are governed and triggered by several environmental factors like topography, geomorphology and geology, climate, and hydrological characteristics, etc., depending upon local and regional settings. In the last two decades, several methodological developments have been made research. Most of them ensemble multiple predictors or controlling factors for developing a flood susceptibility map. The popular approaches for flood susceptibility are: multi-criteria analysis, like the analytical hierarchy process (de Brito and Evers, 2016; Dash and Sar, 2020); statistical models like frequency ratio, Shannon entropy, weights of evidence, multiple linear regression, etc. (Khosravi et al., 2016; Arora et al., 2019; Costache and Bui, 2019); and machine learning models, like artificial neural networks, support vector machine, classification,

*Advances in Remote Sensing Technology and the Three Poles*, First Edition. Edited by Manish Pandey, Prem C. Pandey, Yogesh Ray, Aman Arora, Shridhar D. Jawak, and Uma K. Shukla.

and regression trees, etc. (Costache, 2019; Li et al., 2019). The analytical hierarchy process (AHP)-based multicriteria approach is the most popular for susceptibility mapping because of the simplicity and predictability of the model, despite requiring limited input variables and ground information of past flooding. Besides applicability in data-sparse conditions, another advantage of AHP is the implementation of geospatial techniques by integrating remote sensing and GIS for data preparation and modeling.

The Uttarakhand Himalaya, especially the Bhagirathi River basin, is very sensitive to various natural hazards floods, landslides, earthquake, forest fires, etc., due to tectonic setup, geographical position, hydro-meteorological conditions, topography, and human activities (Kala, 2014). Besides cloudbursts and heavy rainfall, flash floods in Uttarakhand are often triggered by the breaking up of glaciers or glacial lake outburst (Dobhal et al., 2013). The widespread anthropogenic activities like construction of large dams and hydropower projects that restrict natural flow, mock disposal at valley sides, deforestations, encroachment of river beds, mining activities, construction of settlements and hotels near river beds, heavy traffic flow, etc., are found to be the triggering factors to maximize disasters (Kala, 2014; Dash and Punia, 2019).

In the context of the Bhagirathi River basin, few studies have attempted to calculate runoff, flood water volume, and inundation extent, mostly of a single event, through hydrological analysis and hydrodynamic modeling (Rao et al., 2014; Nandargi, Gaur and Mulye, 2016), and the role of anthropogenic activities and governance issues (Kala, 2014; Dash and Punia, 2019). However, there is a gap in research to identify the potential flood hazard zones that could help in decision-making for reducing vulnerability. Hence, the present study is aimed to prepare a comprehensive flood susceptibility map of the Bhagirathi River basin through multicriteria analysis and geospatial modeling. The study may assist decision-makers and administrators in reducing flood impacts.

## 17.2 Study Area

In this chapter, the flash flood susceptibility map was prepared for the Bhagirathi River basin, located in Uttarakhand Himalaya, India (Figure 17.1). The Bhagirathi River is a

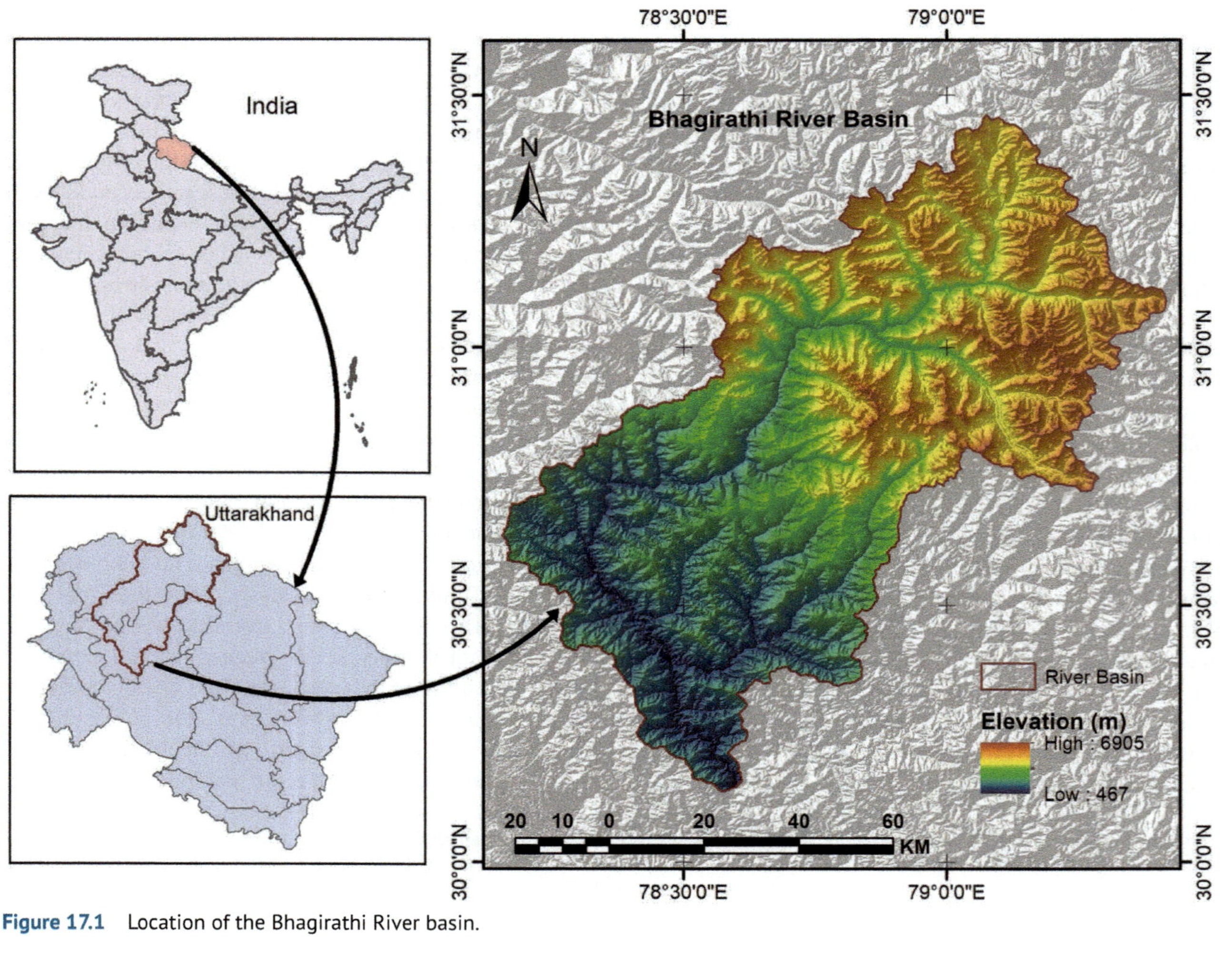


Figure 17.1 Location of the Bhagirathi River basin.

tributary of Ganga River, flowing for 220 km and covering an 8850 $km^2$ area in Uttarakhand Himalaya or Garhwal Himalaya. The basin is characterized by ridge and valley topography along with narrow flood plains, piedmont slopes, and glacial landforms. The basin is characterized by different climatic conditions from sub-tropical humid to sub-arctic climate. Due to geographical location, the area receives a large amount of rainfall (1500–2000 mm), mostly during the monsoon season between June and September. Besides, the upper catchment also receives a significant amount of snowfall in winter. The steep surface gradient and contribution of a large amount of rainfall along with snow melt have caused high runoff potentiality in the basin. Therefore, hydrologically, the basin is prone to flash flooding. Also, climate change and alteration of surface characteristics by human intervention have increased flood vulnerability.

The region is famous for pilgrimage and tourism due to the location of holy shrines and its natural beauty. The continuous increasing population pressure, on the one hand, and hydroelectricity generation potentiality and other economic activities, on the other, have accelerated the alteration of the natural environment of this fragile ecosystem (Kala, 2014; Dash and Punia, 2019).

## 17.3 Data and Methodology

### 17.3.1 Data

This study has applied GIS-based multi-criteria analysis for the mapping of potential areas prone to flash flood hazards in the Bhagirathi River basin. Therefore, several spatial data pertaining to topography, morphometry, geomorphology, and surface characteristics have been utilized. Several topographic and morphometric parameters have been derived from Shuttle Radar Topography Mission (SRTM) Digital Elevation Model (DEM) data of 30 m spatial resolution. The spatial data of surface cover, i.e., land use–land cover (LULC) information, have been extracted from Landsat 8 OLI data for the year 2016. The satellite data (SRTM-DEM and Landsat 8) have been retrieved from https://earthexplorer.usgs.gov. The geomorphological map (250 K scale) published by the Geological Survey of India (GSI), has been downloaded from BHUKOSH (https://bhukosh.gsi.gov.in/Bhukosh/Public). All datasets have been projected in UTM projection Zone 44 N (datum WGS 1984) and clipped for the study domain.

### 17.3.2 Multicriteria Analysis

The multi-criteria decision analysis (MCDA) is one of the suitable approaches to flood hazard analysis, as it is capable of incorporating the inherent complexity and uncertainty of the flood events resulting from the influence of multiple factors (de Brito and Evers, 2016). The MCDA adopts a participatory approach for considering the subjective judgements from a wide variety of stakeholders and experts to enhance the decision-making for a complex phenomenon. Analytical hierarchy process (AHP) is a commonly applied MCDA tool that solves a complex decision choice through pairwise comparison of criteria arranging in a hierarchical structure based on their importance or influence for an event, marked according to a predefined scale (Saaty, 1980). Based on the evaluation of the decision-makers (user, experts, and stakeholders) the pairwise comparisons matrix of the controlling factors is created where a higher weight is assigned to the more important criteria over the other. Finally, influence of each criterion is computed as a global score from a weighted sum of scores assigned in a row for a give criterion against other criteria. The overall procedure of AHP can be summarized in six steps: establishing the unstructured problem, designing the AHP hierarchy, assigning a score for a pairwise comparison matrix, assessment of the relative scores, consistency check, and performance evaluation of the model (Papaioannou et al., 2015). In this study, seven experts from the field of geomorphology, hydrology, and disaster management have provided the pairwise comparison score matrix. The experts were also asked to iterate scores in pairwise comparison until the threshold of acceptable consistency ratio limit (CR <10%) was achieved. Finally, the roundoff of average score of each pairwise comparison was taken to prepare the final score matrix.

The subjective judgement of the score in the comparison matrix is compulsorily checked by the consistency ratio (*CR*), calculated as the ratio between consistency index (*CI*) and random index (*RI*). The value of *RI*, as provided by Saaty (1980), varies with the total number of criteria (*N*) used for the pairwise comparison. For this study, the value of *RI* is 1.45 as nine criteria were selected. *CI* is estimated from the maximum eigenvalue ($\lambda_{max}$) of the score matrix and total number of criteria (N). Equation 17.1 is expressed as follows:

$$CR = \frac{CI}{RI}; \quad where\ CI = \frac{\lambda_{max}}{N-1}\ and \quad RI = 1.45\ (for\ 9\ criteria)\ldots\ldots \tag{17.1}$$

### 17.3.3 Selection and Classification of Flood Predictors

The physical process of flash flood mechanisms is controlled by the complex interactions among multiple variables that vary worldwide according to the geomorphological, hydrometerological, and other environmental settings. Therefore, in flood susceptibility mapping, the selection of suitable predictors that control flood generation is a difficult but crucial step. Considering a detailed literature study (Wang et al., 2011; Papaioannou et al., 2015; Costache and Bui, 2019; Dash and Sar, 2020) and environmental settings of the Bhagirathi River basin, nine criteria were selected. These are: elevation, slope angle, plan curvature, topographic position index (TPI), topographic wetness index (TWI), convergence index (CI), land use–land cover (LULC), vertical distance to channel networks, and geomorphology.

Each criterion was prepared in a GIS environment, mostly using SRTM DEM. The raster layers of elevation and slope angle (in degrees) were prepared in QGIS. The raster layers of plan curvature, TPI, TWI, and CI, were prepared in SAGA GIS using the Basic Terrain Analysis tool from the Terrain Analysis toolbox, and vertical distance to channel networks was prepared in SAGA GIS using the Morphometry tool from the Terrain Analysis toolbox. The LULC map was prepared from Landsat 8 data using the maximum likelihood algorithm-based supervized image classification technique in QGIS. The readily available shapefile of a geomorphology map was extracted for the study domain and reclassified into suitable classes.

After the data preparation, the value range of each parameter was categorized to increase the objectivity. Each raster layer was reclassified into groups, mostly in five classes: Elevation, slope angle, CI, and vertical distance to channel networks. Each criterion was classified into five groups based on the literature study and expert judgement for the study site. Plan curvature was classified into three classes based on the recommendation of previous studies (Costache and Bui, 2019). TPI and TWI were classified into five classes following the natural break method. Two qualitative parameters, LULC and geomorphology, were categorized according to local settings. Considering the influence on flooding, each class from each criterion has been assigned a class weight (i.e., new raster value) between 1 and 5 based on judgment. For a given predictor (or criterion), the highest class weight is assigned to a class that has maximum influence on flooding, while the lowest class weight is assigned that has a minimum influence for flooding. For example, the lowest slope angle class is more prone to flood; therefore, assigned a class weight of 5; while the highest slope angle class, by most, is negligibly prone to flood; hence, a class weight of 1 is assigned.

### 17.3.4 Flood Hazard Index

In this study, flood hazard index (FHI) was prepared to identify the potential flash flood hazard zone (Kazakis et al., 2015; Dash and Sar, 2020). After estimating the criteria weight through AHP, and class weight of each criterion, FHI was calculated. FHI is expressed as the sum of the product of criterion weight and class weight. FHI was prepared in QGIS from Raster Calculator using Equation 17.2:

$$FHI = \sum_{i=1 to 9}^{j=1 to 5} r_{ij} . w_i \tag{17.2}$$

where $r_{ij}$ is the class weight of $i$th parameter for $j$th class, $w_i$ is the score (weight) of the $i$th parameter obtained through AHP, $i$ represents the number of criteria used in the study (here total nine criteria used), and $j$ represents class weight that varies between 1 and 5 (considered for this study). Finally, the estimated FHI raster layer was classified into five classes by the Natural Break method: very low, low, moderate, high, and very high vulnerability zones.

### 17.3.5 Validation

To validate flood susceptibility mapping, several studies have applied receiver operating characteristics (ROC) as simple and robust techniques to check the accuracy of the model (Arora et al., 2019; Costache and Bui, 2019). The ROC curve is prepared by plotting 1-specificity on the $x$-axis and sensitivity on the $y$-axis. The area under the ROC curve reflects the accuracy of the model output. However, in this study, only flooded event points (true positive) were taken. For this purpose, 178 training sample points of past flooding were selected from the Google Earth image (after images of the massive flood 2013). A large number of points were taken, mostly near the boundary of the inundation extent. The calculated FHI map was reclassified into 100 classes at equal intervals and the raster value of the reclassified map was assigned in a descending order of FHI values. Then, the frequency (and percentage) of flooding was calculated for each reclass category using overlay operations. The cumulative percentage of frequency for each class is calculated to create an ROC curve by plotting cumulative frequency on the $y$-axis and reclass value (high FHI to low FHI) on the $x$-axis (Das, 2020). The corresponding area under thre curve (AUC) was calculated to estimate model accuracy.

## 17.4 Results and Discussion

### 17.4.1 Flood Controlling Factors

Each of the selected nine predictors were classified into suitable classes and assigned with class weight (Figure 17.2 and Table 17.1). Elevation and slope angle are considered important predictors for triggering flash flood, especially in mountainous topography. The areas lying at high elevation and high mountainous slope manifest runoff generation processes, while areas belonging to low elevation and flat to gentle slopes are prone to flooding. The elevation of the study domain was classified into five groups at an equal interval of 1000 m: <1500 m; 1500 to 2500 m; 2500 to 3500 m; 3500 to 4500 m; >4500 m (Figure 17.2a). Each

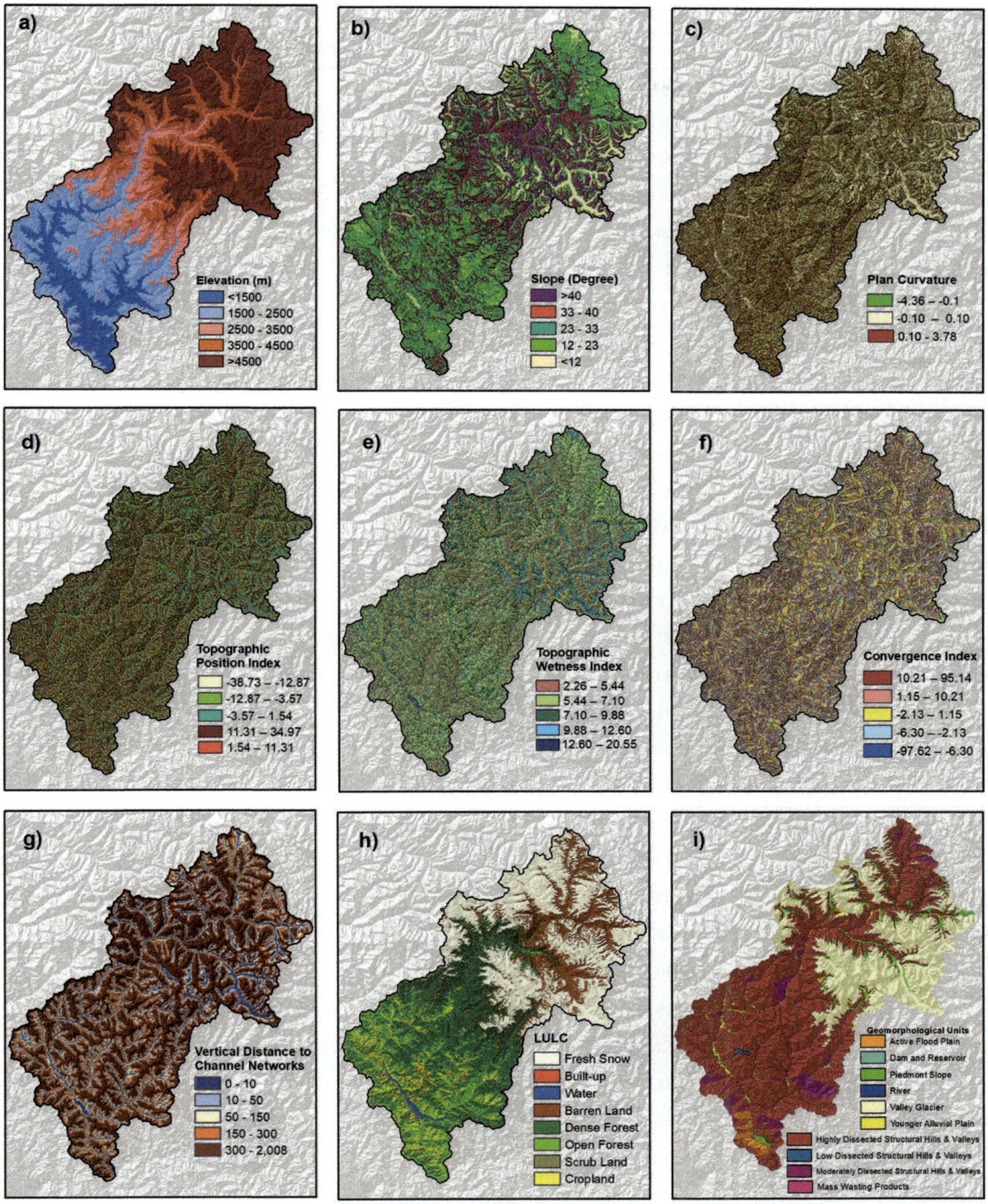


**Figure 17.2** Thematic layers of flood controlling factors used for AHP analysis: (a) elevation; (b) slope angle; (c) plan curvature; (d) topographic position index; (e) topographic wetness index; (f) convergence index; (g) vertical distance to channel networks; (h) land use land cover; and (i) geomorphology.

**Table 17.1** Class weight and criteria weight of flood predictors used for flood hazard index.

| Criteria | Class | Class Weight | Criteria Weight |
|---|---|---|---|
| Elevation | <1500 m | 5 | 0.157 |
| | 1500 to 2500 m | 4 | |
| | 2500 to 3500 m | 3 | |
| | 3500 to 4500 m | 2 | |
| | >4500 m | 1 | |
| Slope | <12° | 5 | 0.122 |
| | 12° to 23° | 4 | |
| | 23° to 33° | 3 | |
| | 33° to 40° | 2 | |
| | >40° | 1 | |
| Plan Curvature | −4.36 to −0.1 | 3 | 0.028 |
| | −0.1 to 0.1 | 5 | |
| | 0.1 to 3.78 | 1 | |
| Topographic Position Index (TPI) | −38.73 to −12.87 | 5 | 0.054 |
| | −12.87 to −3.57 | 4 | |
| | −3.57 to 1.54 | 3 | |
| | 1.54 to 11.31 | 2 | |
| | 11.31 to 34.97 | 1 | |
| Topographic Wetness Index (TWI) | 2.26 to 5.44 | 1 | 0.069 |
| | 5.44 to 7.10 | 2 | |
| | 7.10 to 9.88 | 3 | |
| | 9.88 to 12.60 | 4 | |
| | 12.60 to 20.55 | 5 | |
| Convergence Index (CI) | −97.62 to −6.30 | 5 | 0.034 |
| | −6.30 to −2.13 | 4 | |
| | −2.13 to 1.15 | 3 | |
| | 1.15 to 10.21 | 2 | |
| | 10.21 to 95.14 | 1 | |
| Land Use–Land Cover (LULC) | Snow cover | 1 | 0.034 |
| | Dense forest, Open forest, Scrubland | 2 | |
| | Cropland, Barren land | 3 | |
| | Builtup | 4 | |
| | Water | 5 | |

**Table 17.1** (Continued)

| Criteria | Class | Class Weight | Criteria Weight |
|---|---|---|---|
| Vertical Distance to Channel Networks | 0 to 10 m | 5 | 0.295 |
| | 10 to 50 m | 4 | |
| | 50 to 150 m | 3 | |
| | 150 to 300 m | 2 | |
| | 300 to 2000 m | 1 | |
| Geomorphology | River, lake, dam, other waterbodies | 5 | 0.207 |
| | Active flood plain, younger alluvial plains | 4 | |
| | Piedmont slope | 3 | |
| | Low, moderate, and highly dissected hill, mass wasting product | 2 | |
| | Glacier and glacial topography | 1 | |

elevation class was assigned a class weight between 1 (>4500 m) and 5 (<1500 m) (Table 17.1). The slope angle was categorized into five classes: <12°; 12° to 23°; 23° to 33°; 33° to 40°; >40°, and rated with a class weight between 1 (>40°) and 5 (<12°). (Figure 17.2b and Table 17.1). The plan curvature reflects the curvature of the surface slope, determined by the intersection of the horizontal plane with terrain (Wilson and Gallant, 2000). The negative value of plan curvature indicates a concave surface, whereas positive values represent a convex surface, and so values close to zero represent a flat surface. The plan curvature value ranged between −4.36 and 3.78 that was classified into three classes: −4.36 to −0.1; −0.1 to 0.1; and 0.1 to 3.78. The flat (−0.1 to 0.1), concave (−4.36 to −0.1), and convex (0.1 to 3.78) curvatures were assigned with class weights of 5, 3, and 1 respectively (Figure 17.2c and Table 17.1). TPI is considered as another important predictor for flash flood hazard mapping as it reflects the variation in elevation of a pixel from the average elevation of neighboring pixels (Zaharia et al., 2017). In this study, the TPI value range (−38.73 to 34.79) was reclassified into five classes: −38.73 to −12.87; −12.87 to −3.57; −3.57 to 1.54; 1.54 to 11.31; and 11.31 to 34.97 (Figure 17.2d). The highest TPI class was assigned with class weight 1 and the lowest TPI class with class weight 5 (Table 17.1). TWI reflects the influence of surface characteristics on runoff generation and water accumulation. It is calculated as the ratio between the cumulative upslope draining area and the slope gradient. The high TWI values are mostly associated with valley bottoms, gentle slopes, and terraced surfaces, while low values are assigned to high slope and mountain tops (Arabameri and Pourghasemi, 2019). The estimated TWI value range was reclassified into five classes: 2.26 to 5.44; 5.44 to 7.10; 7.10 to 9.88; 9.88 to 12.60; and 12.60 to 20.55 (Figure 17.2e). The weight was assigned 1 to 5 accordingly to the lowest TWI value class to the highest TWI value class (Table 17.1). CI provides an insight into convergence and divergence of flow in a terrain; therefore, an important parameter for flood susceptibility mapping. The negative value of CI is associated with convergence of flow, while a high value is associated with divergence of flow, mostly interfluvial area and ridge (Zaharia et al., 2017; Costache and Bui, 2019). After several iterations from expertise judgement based on the literature study and topographic characteristics of the study area, CI values were reclassified into five classes: −97.62 to −6.30; −6.30 to −2.13; −2.13 to 1.15; 1.15 to 10.21; and 10.21 to 95.14 (Figure 17.2f). The class weight of CI was assigned from highest class to lowest class, accordingly with 1 to 5 (Table 17.1). The vertical distance to channel networks expresses the difference in elevation of a pixel from the channel network. Several studies (Rahmati et al., 2016; Costache and Bui, 2019) have used a buffer distance from the main channel that could generalize the topographic variations at local scale. The high values of vertical distance to channel networks indicate the area located at high elevation from the main channel; therefore, less prone to flood; while low values indicate that an area lies close to the valley and highly prone to flood. This thematic layer was reclassified into five classes based on expert judgement: 0 to 10 m; 10 to 50 m; 50 to 150 m; 150 to 300 m; and 300 to 2000 m (Figure 17.2g); the weight of each reclassified class is given in Table 17.1.

Following supervized classification, eight LULC cases were identified: snow cover, dense forest, open forest, scrubland, cropland, barren land, builtup, and water (Figure 17.2h). The classified LULC was reclassified into five class based on their influence on flooding (Table 17.1). It is noteworthy that though snowmelt runoff and glacial lake outburst may magnify flooding in the study area, it hardly affects the upstream glaciated valley. Therefore, snow cover is assigned with class weight of 1. The major geomorphological units of the area are: river, lake and other waterbodies; old and new alluvial plains; flood plain; piedmont slope; low, moderate, and highly dissected hills; and glacier and glacial topography (Figure 17.2i). The class weight of each geomorphological unit is given in Table 17.1.

### 17.4.2 Multicriteria Analysis

The expert's suggested pairwise comparison matrix of nine criteria has been analyzsd through the AHP technique to calculate the relative weights or influence of the criteria on flash flood. Table 17.2 shows the pairwise comparison matrix used in AHP. The CR of the pairwise comparison is 6.6% that affirms the consistency of the pairwise comparison score assigned to each criterion in this study. The criteria weight obtained through AHP technique is presented in Table 17.1.

The highest weight is obtained for the criterion vertical distance to channel networks (29.5%), followed by geomorphology (20.7%), elevation (15.7%), slope (12.2%), TWI (6.9%), TPI (5.4%), CI (3.4%), LULC (3.4%), and plan curvature (2.8%). The weight value reflects the influence of each criterion for flash flood in the study area. A high value of criteria weight suggests high influence of the factor for flooding; therefore, highly sensitivity in flood generation. Vertical distance to channel networks in a hilly terrain like Uttarakhand Himalaya is a most important predictor for flash floods because the river valley accumulates a huge amount of water along with debris material after the cloudburst, heavy rainfall, glacial lake outbursts, etc. Therefore, areas lying close to the valley elevation are worse affected. The past events of flooding in the Bhagirathi basin reflected the influence of valley topography in flood susceptibility. Elevation was found to be the third-most important criteria after geomorphology. However, the AHP analysis suggests that four criteria relating to elevation, slope, vertical distance from river, and geomorphology share 78.1% influence, while the remaining five criteria share only 21.9% influence for flooding. Hence, flood susceptibility of the area largely depends on these factors that should be given priority in planning and decision-making.

### 17.4.3 Flood Susceptibility Mapping

Combining the criteria weight and corresponding class weight (Table 17.1), the flood susceptibility index, i.e., FHI was prepared. The calculated FHI value range (1 to 5) was further classified based on the Natural Break method into five classes: very low (1.00 to 1.77); low (1.77 to 2.29); moderate (2.29 to 2.82); high (2.82 to 3.56); and very high (3.56 to 5.00) susceptibility zones (Figure 17.3). The maximum part (80%) of the Bhagirathi basin is potentially low hazard prone in which very low, low, and moderate susceptibility zones cover 21.56%, 31.73%, and 26.36% of the basin area, respectively. The very high susceptibility zone covers only 5.23% and high susceptibility zone covers 15.13% area of the study basin. Most of the area close to the valley, especially along the main channel, is highly susceptible to flooding. Noticeably, major settlements in the Bhagirathi basin are located in this highly susceptible zone. Therefore, though the highly susceptible zone covers only 20% of the geographical area, the impact of flooding is significant.

**Table 17.2** Pairwise comparison matrix of criteria weights (averaged) used in AHP.

| | Elevation | Slope | Plan Curvaure | TPI | TWI | CI | LULC | VDCN | Geomorphology |
|---|---|---|---|---|---|---|---|---|---|
| Elevation | 1 | 2 | 4 | 5 | 3 | 3 | 7 | 0.33 | 0.5 |
| Slope | 0.5 | 1 | 3 | 2 | 1 | 4 | 7 | 0.5 | 1 |
| Plan curvature | 0.25 | 0.33 | 1 | 0.50 | 0.33 | 1 | 0.50 | 0.12 | 0.14 |
| TPI | 0.20 | 0.50 | 0.20 | 1 | 2 | 2 | 1 | 0.14 | 0.20 |
| TWI | 0.33 | 1 | 3 | 0.50 | 1 | 3 | 4 | 0.11 | 0.25 |
| CI | 0.33 | 0.25 | 1 | 0.50 | 0.33 | 1 | 2 | 0.12 | 0.17 |
| LULC | 0.14 | 0.14 | 2 | 1 | 0.25 | 0.5 | 1 | 0.20 | 0.25 |
| VDCN | 3 | 2 | 8 | 7 | 9 | 8 | 5 | 1 | 1 |
| Geomor-phology | 2 | 1 | 7 | 5 | 4 | 6 | 4 | 1 | 1 |

TPI: topographic position index; TWI: topographic wetness index; CI: convergence index; LULC: land use–land cover; VDCN: vertical distance to channel networks

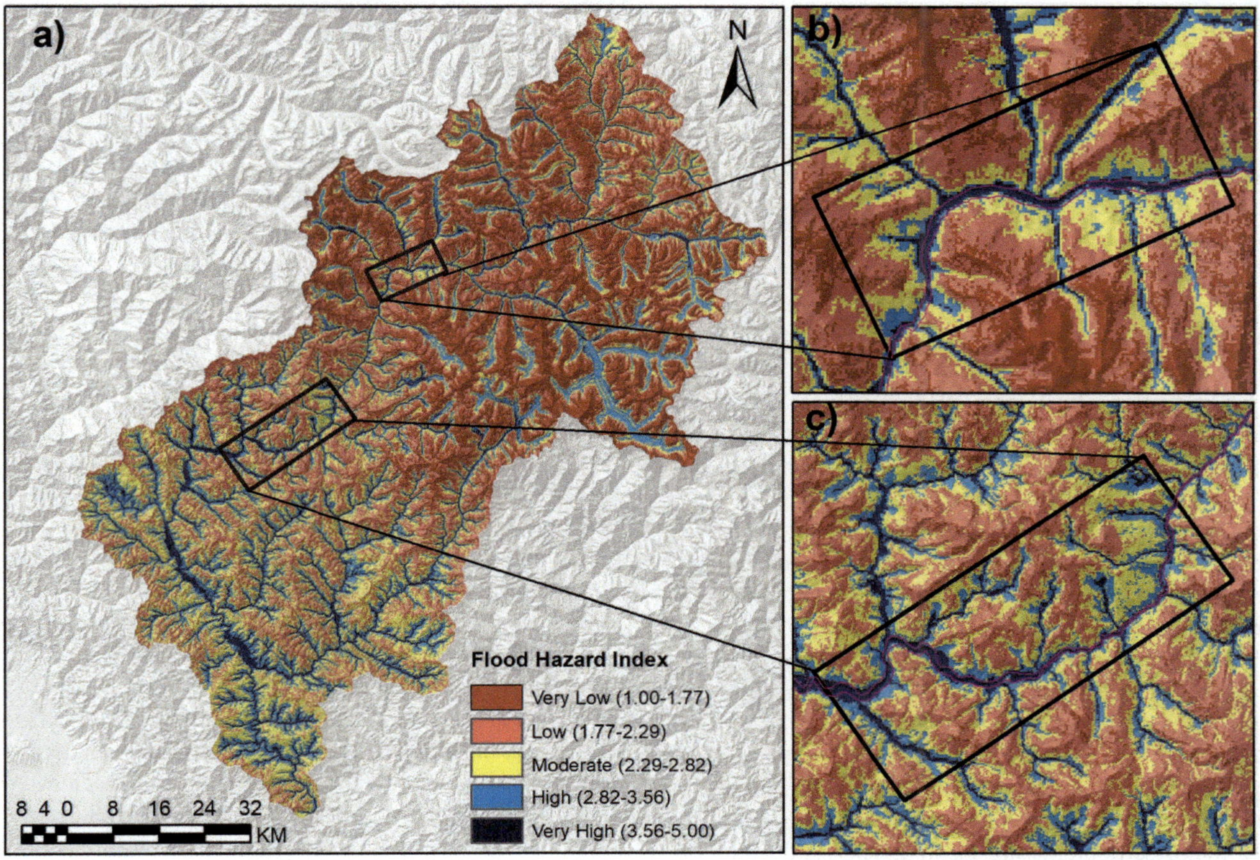


**Figure 17.3** Flood hazard map: (a) flood hazard zones of the study area; (b and c) close view of hazard map in two segments along the main channel that shows accurate matching of the very high susceptible zone with the area devastated during 16–18 June 2013 (drawn in red).

### 17.4.4 Validation

Estimation of model performance is another crucial part of multicriteria analysis. The precision of model performance is often validated by means of statistical analysis with field data such as historical inundation extent, and sample points of flood effected areas. ROC and AUC are well-established measures commonly used in flood susceptibility studies to test the prediction accuracy of model output (Costache and Bui, 2019; Das, 2020). The prepared ROC curve is presented in Figure 17.4 in which the AUC value is 0.78. The high AUC value suggests a good accuracy (77.73%) of the model performance (Tehrany et al., 2013).

Besides AUC, the classified hazard map is closely examined by overlapping the mask of flood affected area delineated from the Google Earth image that was captured after the massive flood event of 2013. Figures 17.3b and 17.3c show the selected few segments along the main channel of the Bhagirathi River where the very high hazard zone closely matches with the actual affected area delineated by Google Earth. Thus, it can be inferred that the present methodology is robust enough to identify the potential hazard area in hilly terrain.

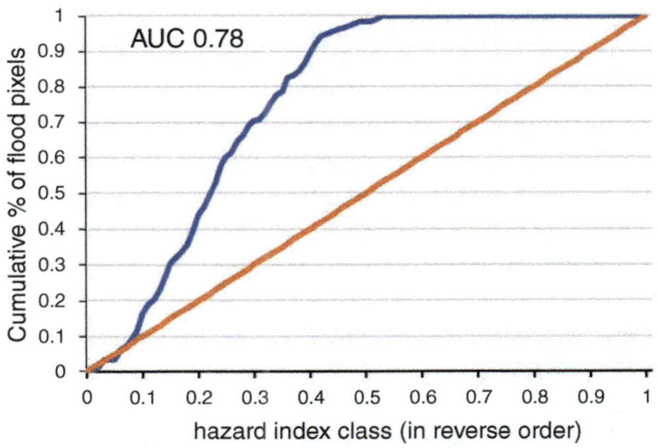


**Figure 17.4** Area Under the Curve (AUC) of the prepared hazard susceptibility map.

## 17.5 Conclusion

The Uttarakhand Himalaya, especially the Bhagirathi River basin, is often worse affected by high magnitude torrential floods that cause casualties and severe damage to infrastructure, land resources, economy, and livelihood of the area. This chapter tries to develop a flood susceptibility map of the Bhagirathi basin (Uttarakhand, India) through AHP-based multicriteria analysis using geospatial technology. The AHP analysis considered nine flood predictor criteria pertain to topography, morphometry, geomorphology, and surface characteristics. The spatial datasets of predictors, i.e., elevation, slope angle, plan curvature, TPI, TWI, CI, LULC, vertical distance to channel networks, and geomorphology were prepared, reclassified, and overlayed in GIS to obtain a flood hazard index (FHI). The criteria weight of each predictor was estimated through the AHP method. The accuracy of the present methodology was tested by receiver operating characteristics (ROC) and area under the curve (AUC)-based data validation technique.

The AHP analysis showed that vertical distance to channel networks is the most influential predictor (priority ~29.5%), followed by geomorphology (20.7%), elevation (15.7%), and slope (12.2%). The estimated flood susceptibility map shows that 20% of the area is highly susceptible to flooding, mostly the areas adjacent to the main channel where major settlements are located. The AUC value of 0.78 suggests high accuracy of the model performance which is highly acceptable. The findings of the study may help planners, administrators, and decision-makers to take preventive measure in the Bhagirathi River basin.

The study basin is characterized by a fragile mountainous ecosystem. The environmental conditions along with rampant development activities have increased the magnitude of flooding in this region. Hence, this study will help to prioritize the regions and conditions for taking structural and non-structural measures to minimize the devastating nature of flooding in the Bhagirathi basin. Multicriteria-based AHP analysis and GIS have been found to be robust for identifying potential flood-prone regions. This approach may be applied elsewhere in the world for flood hazard mapping, especially under data-limited conditions. This approach may help researchers in further study, as well as contributing to methodological development of multicriteria analysis, especially in criteria selection.

## References

Ahrens, J. and Rudolph, P.M. (2006). The importance of governance in risk reduction and disaster management. *Journal of Contingencies and Crisis Management* 14(4): 207–220.

Arabameri, A. and Pourghasemi, H.R. (2019). Spatial modeling of gully erosion using linear and quadratic discriminant analyses in GIS and R. In: *Spatial Modeling in GIS and R for Earth and Environmental Sciences* (ed. H.R. Pourghasemi and C. Gokceoglu), 299–321. Elsevier.

Arora, A., Pandey, M., Siddiqui, M.A et al. (2019). Spatial flood susceptibility prediction in Middle Ganga Plain: comparison of frequency ratio and Shannon's entropy models. *Geocarto International* 36(18): 2085–2116. doi:10.1080/10106049.2019.1687594.

Arora, M., Kumar, R., Singh, R.D. et al. (2016). Analysis of unusual meteorological conditions that led to recent floods in Bhagirathi Basin (Uttarakhand Himalayas). *Hydrological Sciences Journal* 61(7): 1238–1243.

Asthana, A.K.L. and Asthana, H. (2014). Geomorphic control of cloud bursts and flash floods in Himalaya with special reference to Kedarnath area of Uttarakhand, India. *International Journal of Advancement in Earth and Environmental Sciences* 2(1): 16–24.

Costache, R. (2019). Flash-Flood Potential assessment in the upper and middle sector of Prahova river catchment (Romania): a comparative approach between four hybrid models. *Science of the Total Environment* 659: 1115–1134.

Costache, R. and Bui, D.T. (2019). Spatial prediction of flood potential using new ensembles of bivariate statistics and artificial intelligence: a case study at the Putna river catchment of Romania. *Science of the Total Environment* 691: 1098–1118.

Das, S. (2020). Flood susceptibility mapping of the Western Ghat coastal belt using multi-source geospatial data and analytical hierarchy process (AHP). *Remote Sensing Applications: Society and Environment* 20: 100379.

Dash, P. and Punia, M. (2019). Governance and disaster: analysis of land use policy with reference to Uttarakhand flood 2013, India. *International Journal of Disaster Risk Reduction* 36. doi:10.1016/j.ijdrr.2019.101090.

Dash, P. and Sar, J. (2020). Identification and validation of potential flood hazard area using GIS-based multi-criteria analysis and satellite data-derived water index. *Journal of Flood Risk Management* 13(3): 1–14. doi:10.1111/jfr3.12620.

de Brito, M.M. and Evers, M. (2016). Multi-criteria decision-making for flood risk management: a survey of the current state of the art. *Natural Hazards and Earth System Sciences* 16(4): 1019–1033.

Dobhal, D.P., Gupta, A.K., Manish, M. et al. (2013). Kedarnath disaster: facts and plausible causes. *Current Science* 105(2): 171–174.

Kala, C.P. (2014). Deluge, disaster and development in Uttarakhand Himalayan region of India: challenges and lessons for disaster management. *International Journal of Disaster Risk Reduction* 8: 143–152. doi:10.1016/j.ijdrr.2014.03.002.

Kazakis, N., Kougias, I., and Patsialis, T. (2015). Assessment of flood hazard areas at a regional scale using an index-based approach and Analytical Hierarchy Process: application in Rhodope–Evros region, Greece. *Science of the Total Environment* 538: 555–563.

Khosravi, K., Pourghasemi, H.R., Chapi, K. et al. (2016). Flash flood susceptibility analysis and its mapping using different bivariate models in Iran: a comparison between Shannon's entropy, statistical index, and weighting factor models. *Environmental Monitoring and Assessment* 188(12): 1–21.

Kumar, A., Gupta, A.K., Bhambri, R. et al. (2018). Assessment and review of hydrometeorological aspects for cloudburst and flash flood events in the third pole region (Indian Himalaya). *Polar Science* 18: 5–20.

Li, X., Yan, D., Wang, K. et al. (2019). Flood risk assessment of global watersheds based on multiple machine learning models. *Water* 11(8): 1654.

Nandargi, S., Gaur, A., and Mulye, S.S. (2016). Hydrological analysis of extreme rainfall events and severe rainstorms over Uttarakhand, India. *Hydrological Sciences Journal* 61(12): 2145–2163.

Nkwunonwo, U.C., Whitworth, M., and Baily, B. (2020). A review of the current status of flood modelling for urban flood risk management in the developing countries. *Scientific African* 7: e00269.

Papaioannou, G., Vasiliades, L., and Loukas, A. (2015). Multi-criteria analysis framework for potential flood prone areas mapping. *Water Resources Management* 29(2): 399–418.

Rahmati, O., Pourghasemi, H.R., and Zeinivand, H. (2016). Flood susceptibility mapping using frequency ratio and weights-of-evidence models in the Golastan Province, Iran. *Geocarto International* 31(1): 42–70.

Rao, K.H.V.D., Rao, V.V., Dadhwal, V.K. et al. (2014). Kedarnath flash floods: a hydrological and hydraulic simulation study. *Current Science* 106(4): 598–603.

Saaty, T.L. (1980). *The Analytic Hierarchy Process: Planning, Priority Setting, Resource Allocation*. New York: McGraw-Hill.

Shen, G. and Hwang, S.N. (2019). Spatial–temporal snapshots of global natural disaster impacts revealed from EM-DAT for 1900–2015. *Geomatics, Natural Hazards and Risk*, 10(1): 912–934. doi:10.1080/19475705.2018.1552630.

Tehrany, M.S., Pradhan, B., and Jebur, M.N. (2013). Spatial prediction of flood susceptible areas using rule based decision tree (DT) and a novel ensemble bivariate and multivariate statistical models in GIS. *Journal of Hydrology* 504: 69–79.

Thakur, V.C. (2000). Cloudburst floods and flash floods caused by landslide and debris flow dam failures in the Indian Himalayas. *ICIMOD, Newsletter of the International Centre for Integrated Mountain Development* 38: 10–11.

Wang, Y., Li, Z., Tang, Z., and Zeng, G. (2011). A GIS-based spatial multi-criteria approach for flood risk assessment in the Dongting Lake Region, Hunan, Central China. *Water Resources Management* 25(13): 3465–3484.

Wilson, J.P. and Gallant, J.C. (2000), *Terrain Analysis: Principles and Applications*. New York: John Wiley & Sons.

Zaharia, L., Costache, R., Prăvălie, R. et al. (2017). Mapping flood and flooding potential indices: a methodological approach to identifying areas susceptible to flood and flooding risk. Case study: the Prahova catchment (Romania). *Frontiers of Earth Science* 11(2): 229–247.

# 18

# The Role of Himalayan Frontal Thrust in the Upliftment of Kimin Formation and the Migration of Sedimentary Basin in Arunachal Himalaya, Around Bandardewa, Papumpare District, Arunachal Pradesh

*Mondip Sarma[1,*], Sajeed Zaman Borah[2], Devojit Bezbaruah[1], Tapos Kumar Goswami[1], and Upendra Baral[3,4,5]*

[1] *Department of Applied Geology, Dibrugarh University, Dibrugarh 786004, Assam, India*
[2] *Techno Canada Inc., Barmer 344001, Rajasthan, India*
[3] *State Key Laboratory of Tibetan Plateau Earth System Science, Resources and Environment, Institute of Tibetan Plateau Research, Chinese Academy of Sciences, Beijing 100101, China*
[4] *Kathmandu Center for Research and Education, CAS-TU, Kirtipur 446001 Nepal*
[5] *Institute of Fundamental Research and Studies, Kathmandu 446001 Nepal*
* *Corresponding author*

## 18.1 Introduction

The collision of the Indian and Asian plates was initiated during the Early Eocene Epoch (Yin and Harrison, 2000; Dupont-Nivet et al., 2010), following severe shortening of crustal material (Goswami et al., 2021) and development of southward propagating (Goswami et al., 2018) thrust sheets. This collision resulted in development of the largest active collisional orogen of the world (Yin, 2006), the Himalayan mountain belt, as well as uprising of the Tibetan Plateau (Hodges, 2000). The subduction of the Indian plate underneath the Asian plate is still ongoing (Burgess et al., 2012), which is well specified by small to catastrophic earthquakes. Various geomorphic expressions have been documented by several studies along the frontal fault zone of the Himalayan range that distorted the Late Pleistocene and Holocene surfaces (Valdiya, 1984, 1992; Nakata, 1989; Yeats et al., 1992; Wesnousky et al., 1999; Lavé and Avouac, 2000; Malik and Nakata, 2003).

The rock sequence of the Himalayan region is older to younger successively from north to south. The youngest mountain, i.e., Siwalik rocks (Sub-Himalaya) of Neogene age are folded and uplifted (Kunte et al., 1983; Kumar, 1997, 2001) and separated from the alluvials of Brahmaputra by the Himalayan Frontal Thrust (HFT) (Sarma et al., 2020). The mountain front is tectonically active due to the southward progression of the Himalayan deformation (Lavé and Avouac, 2000). The fold/thrust belt of the sub-Himalaya of Arunachal Pradesh consists of Neogene lithostratigraphic units (Goswami et al., 2018). These units includes southward propagating thrust sheets, which are bounded by MBT in the north and HFT in the south. Toward the southward side of the MBT, the sedimentary sequence shows first-order ramp-flat geometry (Goswami et al., 2018). Three regionally significant faults (MBT, Tipi Thrust, and HFT) were developed from north to south (Kelty et al., 2004). These faults are striking E-W for about 200 km from Bhutan on the western side to the Dikrang River on the eastern side, and beyond that, the direction changes to NE to eastern syntaxsis (Yin et al., 2006).

Alluvial fans develop at the time of release of a solid load by the river channel at the base of drainage (Harvey et al., 2005; Blaire and McPherson, 2009; Adrian et al., 2005). An alluvial fan is a depositional landform developed when a channel flow is left behind in the mountain valley and the sediments are deposited at the valley mouth (Leeder, 1999). Miall (1985) suggested that the profile of rivers as well as fans changes with respect to base level change. Similarly, it also changes due to various conditions (Miall, 1985) like discharge of water, amount of sediment supply, and tectonic movement. Aggregation of sediments occurs at the positive accommodation space. This phenomenon does occur when the source area uplifts or base level rises resulting into the upward movement of a graded profile (Harvey, 1984, 1987). Alluvial fans present in an area are documented as a responsive recorder of the evolution of piedmont basins (Heyward, 1978; Nemec and Postma 1993; Blair and McPherson, 1994; Lloyd et al., 1998). Extensive facies of an alluvial fan are present in intermontane basins and are formed due to local upliftment and subsidence in regions of tectonically active orogens (Wagreich and Strauss, 2005). The tectonically active areas develop and preserve alluvial fans as tectonism creates and maintains relief (Chen et al., 2017).

*Advances in Remote Sensing Technology and the Three Poles*, First Edition. Edited by Manish Pandey, Prem C. Pandey, Yogesh Ray, Aman Arora, Shridhar D. Jawak, and Uma K. Shukla.

SRTM DEM Shuttle Radar Topography Mission (SRTM) and Digital Elevation Model (DEM) are widely used in geological investigations (Yang et al., 2011) as modes of remote sensing studies. The advancement of refining and applications of SRTM DEM has rapidly grown in the past few years (Rabus et al., 2003; Kiamehr and Sjöberg, 2005; Farr et al., 2007; Reuter et al., 2007). High-quality and high-resolution DEM data can be obtained from SRTM. The dataset has made the mapping of the Earth's surface an easier and more convenient task. In geological mapping, identification of surface fracture zones is very crucial, as they are the key elements of plate movement and earthquake distribution. Different remote sensing images and DEM are widely used as complementary information to conventional geological investigation. Additionally, these tools have great advantages in identification of regional linear features (Lee et al., 2002; Abduh et al., 2021; Elmahdy et al., 2021). Furthermore, the tectonic activities can be attributed by present geomorphological features, attitude, and slope of an area. Different authors used the SRTM DEM data to study major fault systems or tectonic boundaries (Fu et al., 2006).

Devi et al. (2010) have studied the channel forms and landforms of this area and their relationship with the neotectonics. Bezbaruah and Sarma (2013) discussed morphotectonic evolution of the area and proposed an evolutionary model. Recently, Singh and Bezbaruah (2021) discussed the active tectonic activity in the Siwalik basin along the Dikrang River using various geomorphic indices. Besides past studies, this study chiefly concentrates on tectono-sedimentological aspects of the Kimin Formation and Quaternary terrace deposits. Both of these units are interpreted as alluvial fan deposition. Different sedimentological studies have revealed that these sediments are part of alluvial fan deposits of the Dikrang River. From the observed lithological association of the Kimin Formation in this area, it can be concluded that these rocks were deposited in an alluvial fan environment. From the observed lithologies it can be inferred that a Paleoalluvial fan was present during deposition of the Kimin sediments, which is marked on the map. Study of lithologies of the study area indicates different parts of an alluvial fan, such as the distal, the mid, and the proximal fan. Geological mapping in the area indicates tectonic control over development and propagation of alluvial fans in the area. The Siwalik basin along the Dikrang River was interpreted to have been tectonically active and the basin had a tectonic influence by the HFT (Singh and Bezbaruah, 2021). Movement of a highly active Himalayan foothill fault is interlinked with the development of an alluvial fan in the present study area. A paleoalluvial fan and new alluvial fans have been identified from field observations and lithological associations.

## 18.2 Geology

### 18.2.1 Siwaliks of Arunachal Himalaya

The rocks of the Siwalik Group in Arunachal Pradesh are present as a series of belts from the eastern border of Bhutan to the Siluk River in the east (Kesari et al., 2010). On the eastern side, the Siwalik Group has faulted contact with rocks of the Lesser Himalaya (Sarma et al., 2020). The Siwalik Group of rocks is separated from the Lesser Himalayan sequence to the north and the Brahmaputra alluvium to the south by MBT and HFT, respectively. The Siwalik Group in Arunachal Himalaya has been subdivided into three formations (Karunakaran and Ranga Rao, 1979); Dafla, Subansiri, and Kimin formations, while Kumar (1997) included the Kimin Formation. More recently, the Siji Formation has also been proposed as a part of the Siwalik Group (Debnath et al., 2021).

Lithologically, the Dafla (Lower Siwalik) Formation dominantly consists of sandstones with alternating beds of shale and clay (Kumar, 1997). The Subansiri (middle Siwalik) Formation is well exposed along the Subansiri River (Karunakaran and Ranga Rao, 1979). Dominant lithology of this formation is sandstone. Salt and pepper texture is the characteristic feature of this sandstone. Irregular-shaped calcareous concretions, pebbles of quartzite, and fragments of carbonized and silicified wood are also frequent in this unit. Bands of friable sandstone, clay, and conglomerate are the lithological characteristics of the Kimin (Upper Siwalik) Formation. The sandstone of the Kimin Formation is characterized by a grey to orange color, is soft and massive, fine- to coarse-grained, and with some grit and pebbles (Kumar, 1997).

### 18.2.2 Geology of the Study Area

Rocks of the Siwalik Group and terrace deposit are mainly present in the study area. Toward the southernmost part of the region, sediments of the Kimin Formation are present on the western side of the Dikrang River (Bezbaruah and Sarma, 2013) (Figure 18.1), while a terrace deposit is present on the eastern side of the river. SRTM DEM (30 m) was used to identify linear features in the study area. Later these features were correlated with data obtained from field investigations and the geological map was prepared. DEM showing major structural features of the study area (Figure 18.2a) and a 3D model generated from DEM (Figure 18.2b) showing elevation difference of different lithounits were prepared.

At Bandardewa, the Kimin Formation comprises of beds of pebble with a sandy and gritty matrix and fine sand (Bezbaruah and Sarma, 2013) (Figure 18.3a), while on the western side of the Dikrang River, this formation is

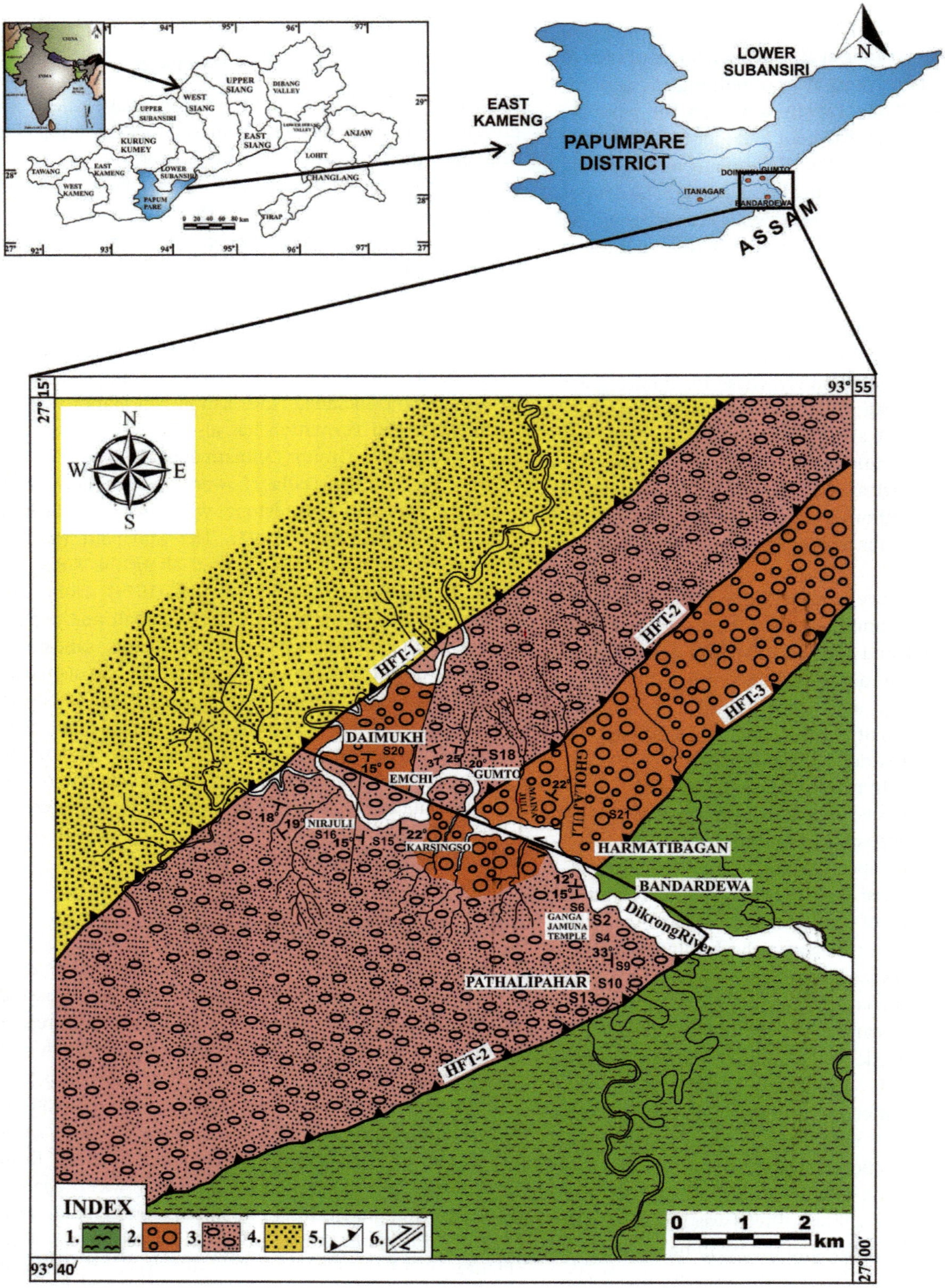


Figure 18.1 Geological map. 1. Alluvium. 2. Terrace deposit. 3. Kimin Formation. 4. Subansiri Formation. 5. Thrust. 6. Strike-slip fault (Adapted from Bezbaruah and Sarma, 2013).

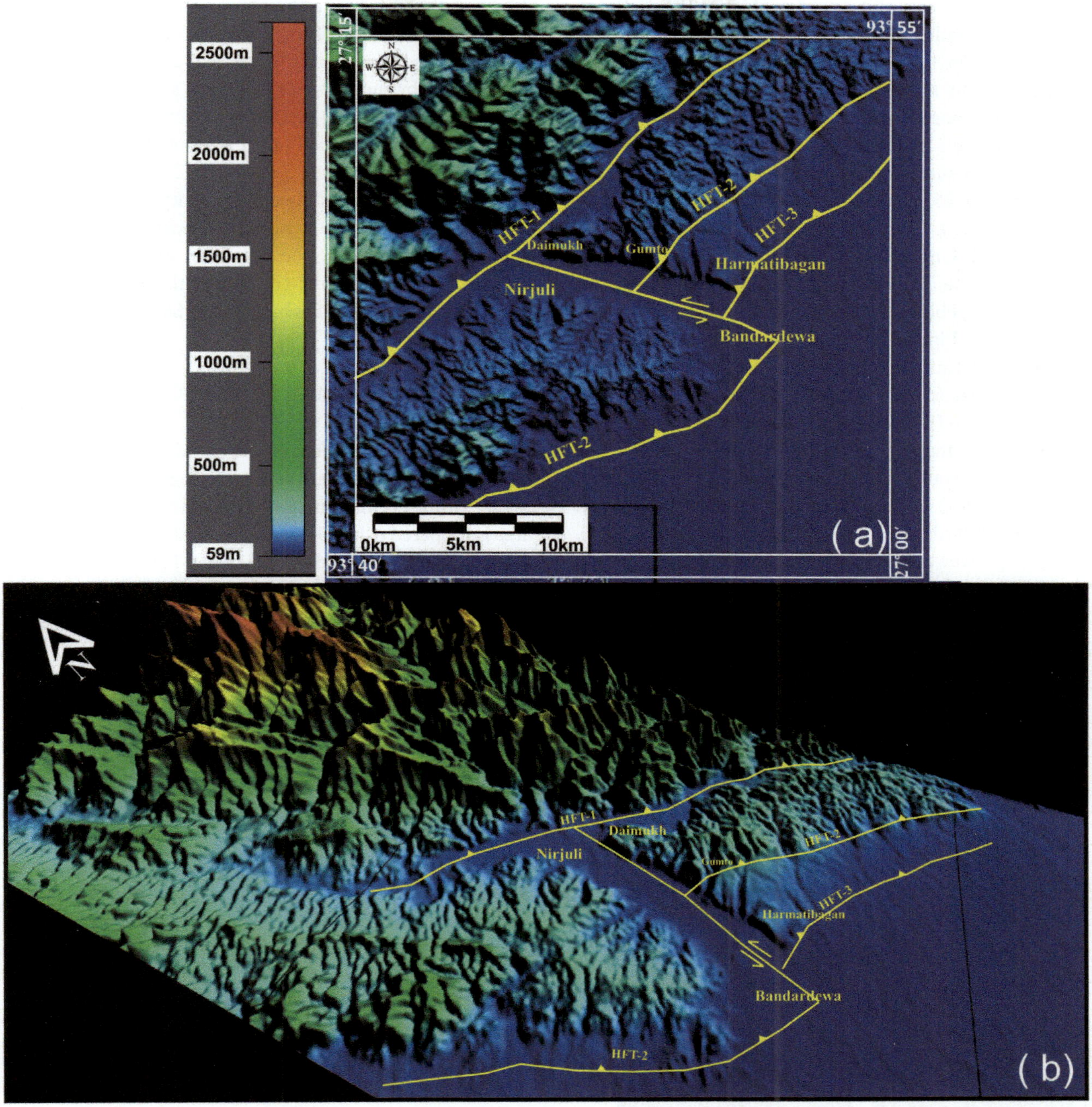


**Figure 18.2** (a) DEM image showing major structural features of the study area. (b) 3D image generated from DEM showing elevation and topographic difference of the study area.

characterized by the presence of gray-colored sandstone and an interbedded matrix supported by a pebble and cobble bed (Figure 18.3b). The Kimin Formation present on the eastern side of the Dikrang River dominantly comprises of fine- to medium-grained sandstone interbeded with gritty and pebble beds (Figure 18.3c), while there is dominancy of sandstone beds. The pebble beds contain clasts of gneiss, quartzite, and schist and the pebbles show preferred orientation. The sandstones are generally gray to yellowish orange, medium- to coarse-grained, and gritty and pebbly (Kumar, 1997).

Along the road section from Bandardewa to Naharlagan, both the Kimin Formation and terrace deposits are well exposed. The anticlinal structure near the Nirjuli area is identified by the variation of attitude of the Kimin Formation from a southerly to northwesterly direction (Bezbaruah and Sarma, 2013). More importantly, the horizontal disposition of the Kimin Formation is observed north of the Karsingso landslide zone (Figure 18.3d). Based on lineament and variation of bed attitude, a fault is marked here. On the eastern side of the Dikrang River, from Harmati to Mainajuli stream, terrace deposits are exposed, while north of the Mainajuli stream, the Kimin Formation is well exposed. The HFT uplifts the terrace deposit over recent alluvium, which is evidenced by the presence of a 20–25 m high scarp near the Harmati Tea garden (Figure 18.3e). The terraces are found to be tilted near Mainajuli and Gholajuli streams. The minor fault is mapped along the Gumtojuli stream. Terrace deposit with loosely-packed pebbles and cobbles are present around the Daimukh area (Figure 18.3f). Existence of sheared rocks and gouge material near Nunpu indicates the existence of HFT-1 that places the Subansiri Formation above the Kimin Formation (Bezbaruah and Sarma, 2013).

A cross fault is mapped along the Dikrong River, which is evidenced from southward advancement of the mountain front (Figure 18.1) on the west side of the Dikrang River, named the Bandardewa Fault (Bezbaruah and Sarma, 2013). The Dikrang River valley is wide, further indicating that the river was formed due to the effect of stike-slip movement along the Bandardewa Fault. The HFT-2 uplifts the Kimin Formation above recent alluvium near Bandardewa, which is evidenced by the presence of a steep scarp 20–30 m high in that area (Figure 18.4a). There is abrupt change in topography along HFT-2, as on the left bank of the Dikrang River near Gumto, there is a clear mark of HFT-2 that uplifts the Kimin Formation just above Quaternary terrace deposits. The inclination of the beds of the Kimin Formation is steeper on the eastern side than the western side of the Dikrang River, which might be differential movement along the Bandardewa Fault (Bezbaruah and Sarma, 2013).

**Figure 18.3** (a) Beds of pebble with sandy and gritty matrix and medium- to fine-grained sand of the Kimin Formation near Bandardewa; (b) Sand matrix-supported pebble and cobble bed with interbedded fine- to medium-grained friable sandstone of the Kimin Formation 500 m north of Banderdewa; (c) Alternate bands of boulder-cobble-pebble bed and friable sandstone with a 50–60 cm-thick shale layer of the Kimin Formation near Gumto; (d) Parallel disposition of mostly pebble-cobble bed of the Kimin Formation near Karsingso; (e) 20–25 m high scarp near Harmatibagan area; and (f) Loosely packed pebble and cobble of terrace deposit around the Daimukh area.

## 18.3 Materials and Method

The methodology of the study is based on field investigation and study of different sedimentological characteristics of the Kimin Formation and Terrace deposit. A geological map is prepared using Global Mapper 11.2 software by plotting field data on the scale of 1:25000. The Survey of India toposheet numbers 83 E/16 and 83 E/12 were used to prepare a base map to understand the topography, drainage, and location accessibility of the study area. SRTM DEM (30 m) is used to identify linear features in the study area. Later these features were correlated with data obtained from field investigation and the geological map was prepared. DEM showing major structural features of the study area (Figure 18.2a) and a 3D model generated from DEM (Figure 18.2b) showing elevation difference of different lithounits were prepared. A detailed geological investigation was carried out to record lithological variation, contacts of different lithologies, and identification of prominent structural features. Field investigation also involved collection of samples for grain size analysis. Samples were collected in a regular manner from the Kimin Formation and terrace deposit, mostly from road-cut sections. Lithologies were prepared from information collected in the field. Field sketches, and measurement of lithosection thickness were done while in the field, while grain size analysis was done by standard sieving methods at the Applied Geology Department, Dibrugrah University. In sieve analysis (Ingram, 1971), a compound sieve is made by placing various sieves one upon another with mesh size increasing from bottom to top. Screens of ½ $\Phi$ intervals of A.S.T.M. sieves were selected for the analysis. Data generated from the grain size analysis method are used to draw cumulative curves using arithmetic probability papers. Grain size analysis was done to understand the possible environment of deposition. Various studies by Passega (1957, 1964) and Friedman (1967) have revealed that a strong correlation exists among the depositional process and grain size frequency distribution. Moreover, the study emphasized that by proper selection, and combination of statistical parameters, it is possible to identify the depositional environment. For a better understanding of the characteristics and significance of the cumulative curves, they have been studied with the help of statistical parameters defined by Folk and Ward (1957) and Folk (1961). The present study also adopted various statistical parameters such as Graphic mean (Mz), Graphic kurtosis ($K_g$), Inclusive graphic standard deviation, and Inclusive graphic skewness (Ski).

## 18.4 Study of Alluvial Fan

Sediments of alluvial fans are generally coarse-grained; however, these sediments show a proximal to distal fining sequence reflecting loss of downstream power and transport capacity (Mountnet, 2005). An alluvial fan has three parts: Proximal, Mid, and Distal. The Proximal Facies is characterized by a thick framework of pebble bed in which the base is flat and convex upward in section transverse to the paleocurrent. Proximal fan deposits are generally disordered and without any definite structure (Grotzinger et al., 2006). Based on observed lithological association, the depositional environment of the Kimin Formation and terrace deposit is taken as an alluvial fan environment (Figure 18.5). Presence of unevenly distributed boulders is seen within the Kimin Formation (Figure 18.4b). As compared to a proximal fan, less gravel characterizes the mid fan facies and the bed bases are mostly scoured though some elongated and convex upward bases persist (Figure 18.4c). Distal facies are characterized by thin beds of gravels and lenses scattered in tabular and cross-bedded sandstones along with finer sediments. Planar and trough cross-beds are observed near Pathalipahar along the scarp of HFT-2 indicating the distal part of an alluvial fan (Figure 18.4d). In this study, vertical lithosection were prepared at different locations of the study area (Figure 18.5). The lithologies are compared with characteristics of different parts of an alluvial fan deposit.

**Figure 18.4** (a) 20–30 m high steep scarp near the Pathalipahar area; (b) Unevenly distributed boulder-cobble-pebble of Kimin Formation near Nirjuli; (c) Gray, friable, fine- to medium-grained sandstone with less amounts of pebble-cobble near Mainajuli stream representing a mid-fan facies; (d) Loose, friable, fine-grained sand with current bedding toward the mountain front area near Pathalipahar representing a distal fan facies; and (e) Boulders of Siwalik sandstone within the Kimin Formation.

### 18.4.1 Description of Lithosections

#### 18.4.1.1 Kimin Formation

The location is in front of the Ganga-Jamuna temple (Station 7). In this section, there is a pebble bed at the base indicating an active current zone of the river, which is followed by finer sequences. Moreover, the overlying sequences were developed because of the shifting of the river course. This deposit represents mid-fan deposits of an alluvial fan.

At a location 1 km upstream from the temple, a pebble bed at the base followed by a current bedded sand bed and then a swampy deposit is observed (Figure 18.3b). This sand bed is repeated at regular intervals, excluding the swampy deposit. From this lithology it is observed that the base of the section is deposited in an active channel of a river. This deposit is followed by a thin bed of carbonaceous shale indicating a back swamp deposit. Therefore, it can be inferred that there was avulsion of the river channel and this area become part of a backswamp in the flood plain. Active channel deposits follow this swampy deposit.

The study section is near Pathalipahar Area (Station 13). In this section, a current bedded sand is at the base followed by siltstone (Figure. 18.4d). This indicates that the energy condition is gradually decreasing from medium- to low-energy conditions. This type of sequence mainly represents the distal part of an alluvial fan.

Near the Gumto (Station 20), the pebbly deposit at the bottom indicates active channel deposit. A similar sedimentary environment prevailed during deposition of the upper part of the section.

At a location 300 m upstream from Harmatibagan (Station 21), near a road turning, a fine sand layer is overlain by alteration of sand and gritty layers (Figure 18.4b). These beds are overlain by a pebble bed which developed as a result of a high-energy current. This result concludes that during deposition of the lower fine sand layer, this area may be a part of a flood plain. During later stages, a high-energy current prevailed in this area.

The deposits near Nirjuli (Station 28) area comprise boulder and pebbles in a sandy and gritty matrix with a minor

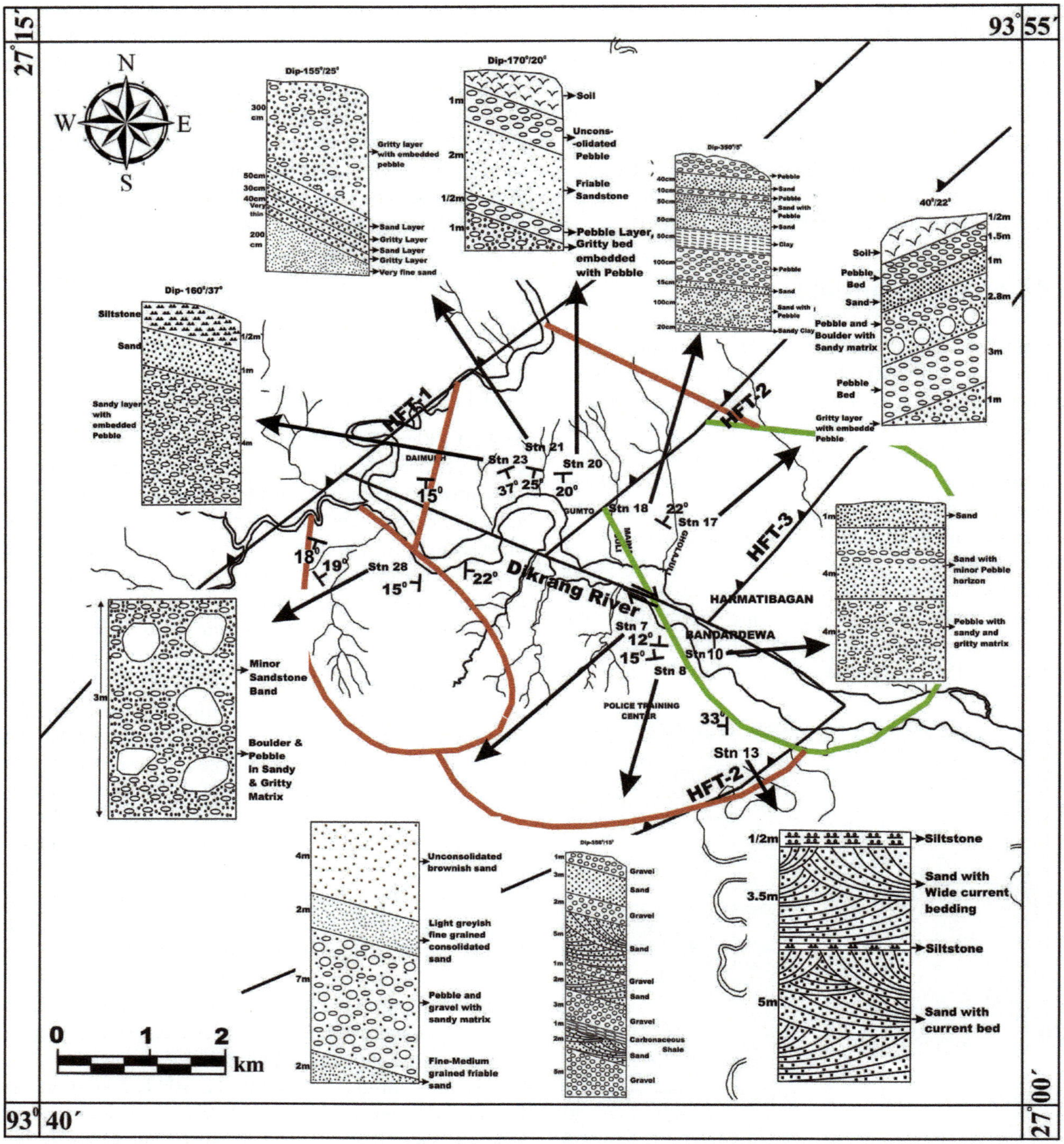


**Figure 18.5** Vertical sections of sedimentary structures at various locations of the alluvial fan, Bandardewa, Arunachal Pradesh.

sandstone band. It was developed by a very high-energy current. In this section, subrounded siltstone boulders of Siwalik rocks are found within the Kimin Formation (Figure 18.4e).

#### 18.4.1.2 Terrace Deposits

At the road turning near Bandardewa (Station 10), the beds are horizontal and comprised of unconsolidated pebbles and sand (Figure 18.3d). The horizontal disposition in this terrace deposit indicates that it is younger than the Kimin Formation.

In Harmatigaon area (Station 17), there is mainly pebbles and boulders, which indicates a high-energy current in the area. The dominance of coarser materials indicates that this is the proximal part of an alluvial fan.

From the lithological study 200 m downstream from Gumto (Station 18), it is observed that a layer of carbonaceous clay deposit is sandwiched between alternate layers of pebbly beds and sandy deposits (Figure 18.3c). Furthermore, during the deposition of the upper part of this section, there must be higher-energy conditions that prevailed in this area. Therefore, it can be inferred that during deposition of the carbonaceous clay, the area became a backswamp in the flood plain of the river channel from this area.

### 18.4.2 Grain Size Analysis

Analysis of size parameters of sediments is a very important process in understanding the basic property of clastic rocks and the clast and matrix relationship. The grain has a fundamental relationship with the physical forces involving transport mechanism and sediment deposition. Some of the legendary studies on a global scale by Passega (1957, 1964) and Friedman (1967) have established that there is a good relationship among size frequency distribution and the process of deposition. The appropriate selection as well as combination of results from various statistical parameters are helpful tools to identify the environment of deposition. The samples collected from various locations were further studied, mainly focusing on the various parameters for the grain size analysis.

### 18.4.3 Cumulative Curve

The log-probability curve is the illustration of cumulative grain-size distribution, and it is very helpful to discriminate the sediments according to their transportation mode (Visher, 1969). The samples collected near the Ganga-Yamuna temple, Bandardewa, have shown an S-shaped pattern with coarse truncation (CT) in between 0.0 and 1.0 $\varphi$ and fine truncation (FT) in between 4.0 and 4.5 $\varphi$. The samples collected near the Pathalipahar area also have shown an S-shaped pattern with CT in between 2.0 and 2.5 $\varphi$ and FT in between 8.0 and 8.5 $\varphi$. In addition, the samples collected from the Gumto and Daimukh area have shown an S-curve pattern with CT between –0.5 and 0.0 $\varphi$ and FT between 3.5 and 4.0 $\varphi$ (except samples 14 and 16). In most of the curves, slopes are smooth (Figure 18.6). In almost all the samples, the saltation population is predominant with little amount of rolling and suspension population. Thus, it can be inferred that the grains are moderately sorted.

### 18.4.4 Calculation of Size Parameters

#### 18.4.4.1 Graphic Mean

This mean (Mz) is a commonly used tools for the measurement of the central tendency. This mean can help in defining overall size and is calculated from the formula, $\varphi16 + \varphi50 + \varphi83/3$. In all the samples (except 7) the graphic mean ranges from 1.16 to 3.34 $\varphi$, which is the medium- to fine-sand size of Wentworth.

#### 18.4.4.2 Graphic Standard Deviations

This parameter ($\sigma1$) is the measurement of uniformity of sediment size distribution that is computed by the formula $\varphi84 - \varphi16/4 + \varphi95 - \varphi5/6.6$. From measurements of sorting value of many sediments, Folk (1966) has proposed a different sorting classification scale. The inclusive graphic standard deviation ranges from 0.6 to 2.0, which falls within the moderately well-sorted to poorly-sorted class.

#### 18.4.4.3 Graphic Skewness

The measurement of systematic distribution of all kinds of sediments, either coarse- or fine-grained, is possible by calculating graphic skewness (Ski) by the formula $\varphi 84 + \varphi 16 - 2\varphi50/(\varphi 84 - \varphi16) + \varphi95 + \varphi5 - 2\varphi50/(\varphi95 - 2\varphi5)$ (Folk and Ward, 1957). The zero value of Ski indicates symmetrical curves. Sediments having fine material as dominant constituents show positive skewness, while those having coarse material as dominant constituents show negative skewness. Samples 9, 15, and 20 have shown to be negatively skewed while others are positively skewed and graphic skewness ranges from 0.6 to 0.04, indicating that most of the samples have tails of finer materials.

#### 18.4.4.4 Graphic Kurtosis

Kurtosis ($K_G$) is a measurement that defines the parting from normality. This parameter measures sorting in the tails portion ratio. The value of normal kurtosis is 1.00. Leptokurtic condition prevails if the central portion is better sorted in comparison to the tail, whereas platykurtic condition prevails if tail portion is better sorted than central portion. Mesokurtic condition prevails if both are equally sorted.The graphic kurtosis in the present studied sample ranges from 0.7 to 2.2, which is within the range of the platykurtic to very leptokurtic class limit suggested by Folk (1966).

### 18.4.5 Inter-Relationship of Size Parameters

Relationship between different size parameters of sediments plays a key role to better understand various aspects of depositional environments (Passega, 1957; Friedman, 1961, 1967; Moiola and Weiser, 1968; Visher, 1969). Work of different authors on sediments for statistical size parameters (Folk and Ward, 1957; Mason and Folk, 1958; Sahu, 1964; Friedman, 1967; Moiola and Weiser 1968) have employed discrimination of the sedimentary environments by means of bivariate plots. Few plots have been attempted and have appeared to be important for various explanation

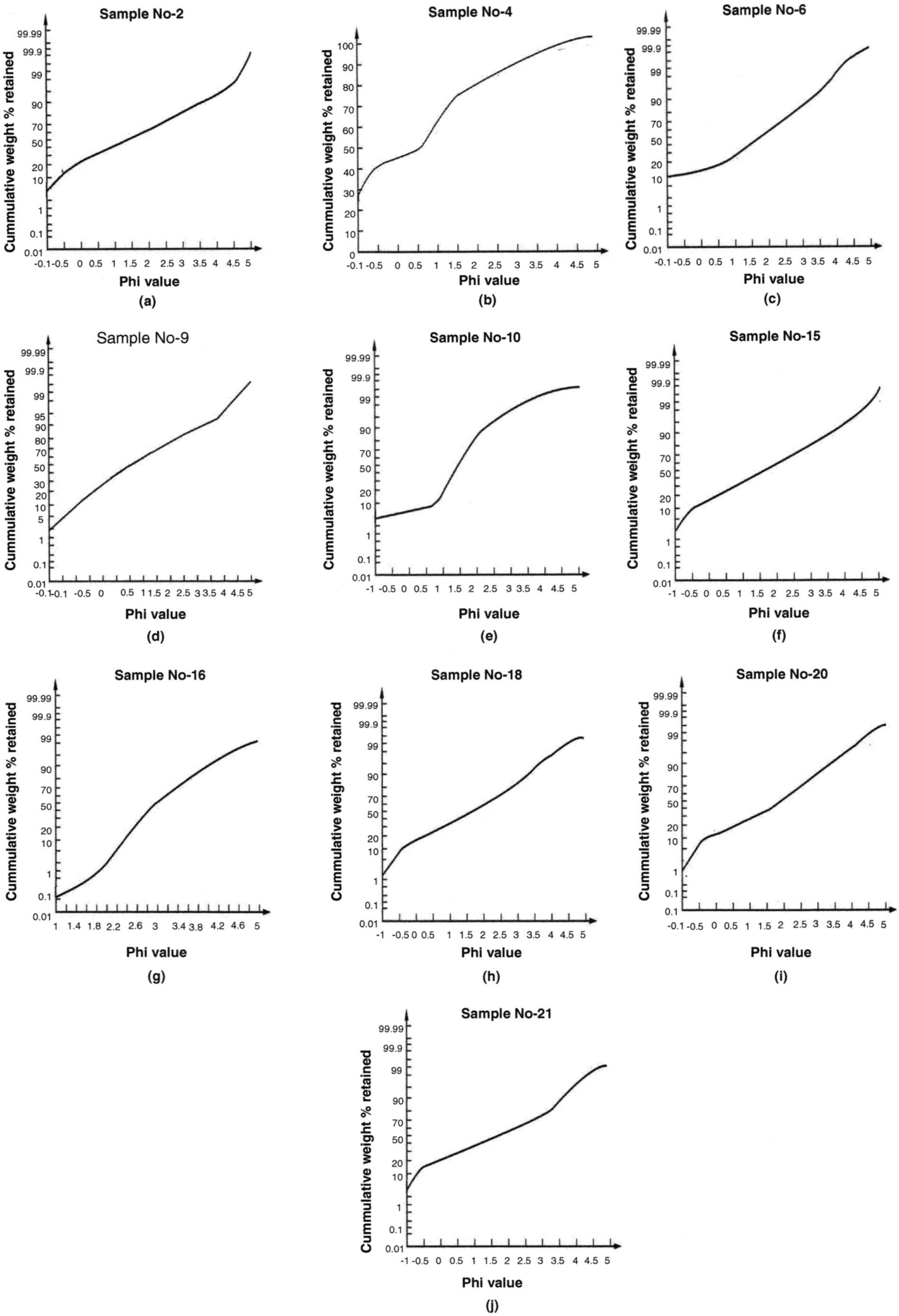


**Figure 18.6** Graphic representation of grain-size frequency data.

regarding depositional environment of sediments (Friedman, 1961, 1967). In the present study, four bivariate plots are drawn: 1) $\sigma_1$ vs. $SK_1$; 2) $\sigma_1$ vs. Mz; 3) $SK_1$ vs. Mz; and 4) $K_G$ vs. $SK_1$.The mean vs. standard deviation plot is one of the crucial tools to differentiate sands from beach and river (Friedman, 1967; Moiola and Weiser, 1968). Considering environmental fields of Friedman (1967), the river and beach sands (except one sample of the Kimin Formation) and all the other samples are scattered in the river field (Figure 18.7a). To distinguish between beach and river environments, Friedman (1967) and Moiola and Weiser (1968) contemplated plot of standard deviation vs. skewness (Figure 18.7b). The scatter of the samples shows a wide range. One sample point of the Kimin Formation is clustered in the beach field and the rest of the sample points fall in the river field. Most of the samples are positively skewed (Figure 18.7c). By means of discriminating fields of Friedman (1967), it is shown that the majority of the sample points are concentrated in the tidal current sands. The relationship of $Sk_1$ vs. $K_G$ can be analysed in the manner suggested by Thomas et al. (1972). The study designed four zones, A, B, C, and D, and related them to the conditions of hydraulic energy. High-energy conditions is represented by zone A, while other zones B, C, and D represents a decreasing trend. Almost all studied samples cluster around the normal distribution line, falling mainly in zone A and partly in zone B (Figure 18.7d). This may suggest the dominance of coarse- to medium-grained

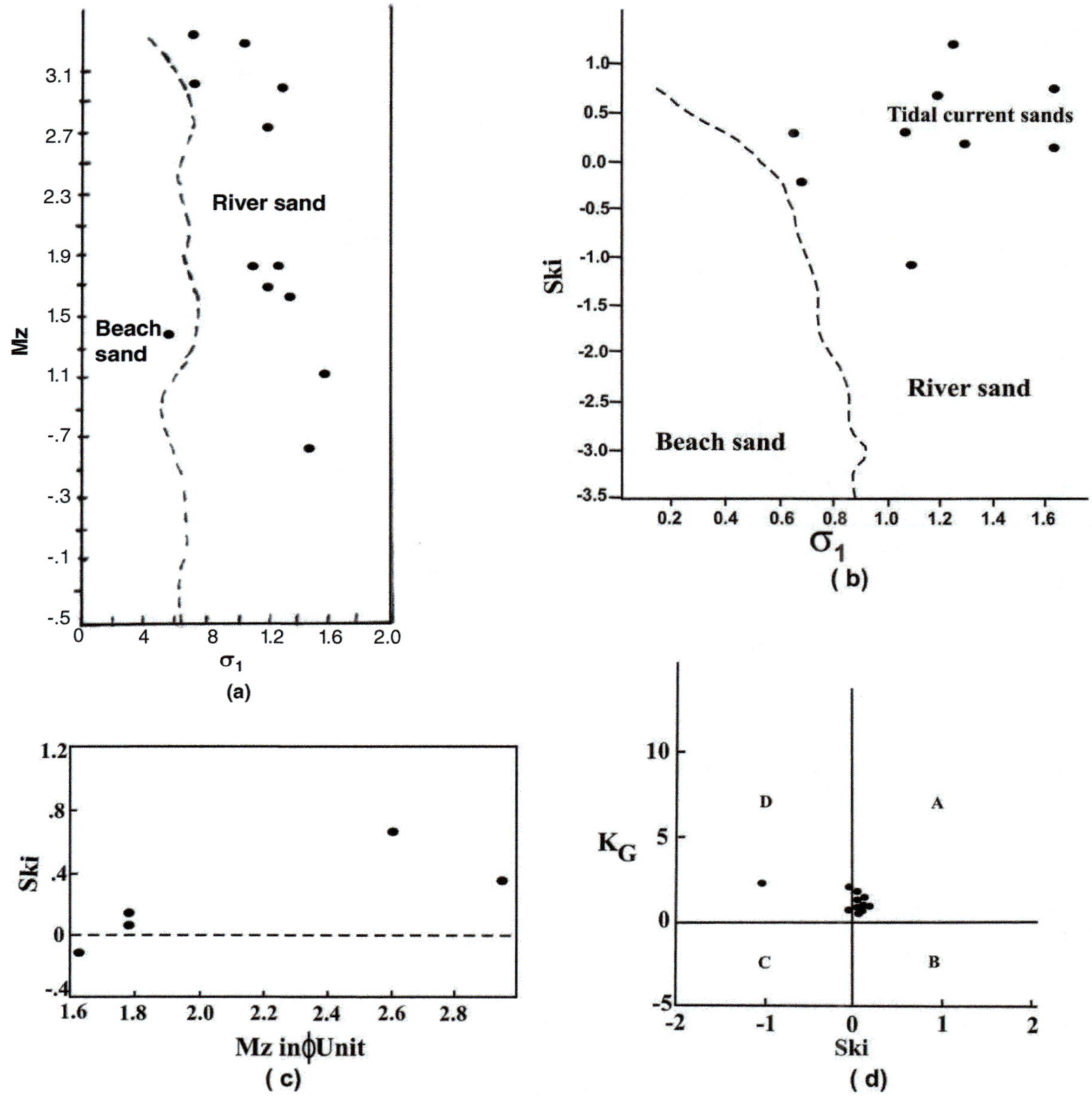


**Figure 18.7** (a) Plot of $\sigma1$ vs. Mz; (b) Plot of $\sigma_1$ vs. $SK_i$; (c) Plot of Ski vs. Mz.; and (d) Plot of KG vs. Ski (Table 18.1).

sediments and moderately higher hydraulic energy conditions during the deposition of the sediments.

### 18.4.6 CM Plot

The CM plot was first introduced by Passega (1957) to estimate different hydrodynamic forces that applied during the time of deposition (Srivastava and Mankar, 2009). Passega (1957) has also shown that there is a close association between the grain size distribution and their depositional process. Deposits in various environments can produce characteristic CM patterns. From the cumulative curves C, 1 percentile represents the coarsest fraction of the grain size distribution and M (median) represents 50 percentile values computed and their corresponding φ-values. Then the φ values are converted into microns. The obtained CM patterns are compared with basic patterns of Passega (1964). It reveals that most of the sample points lie above the line M = 100μ and cluster in the fields of I and II (Figure 18.8; Table 18.1). The spread of the sample points is limited within the O/P and R/S segments. Position of the points in the CM diagram indicates that some are rolled sediments along with some suspension sediments.

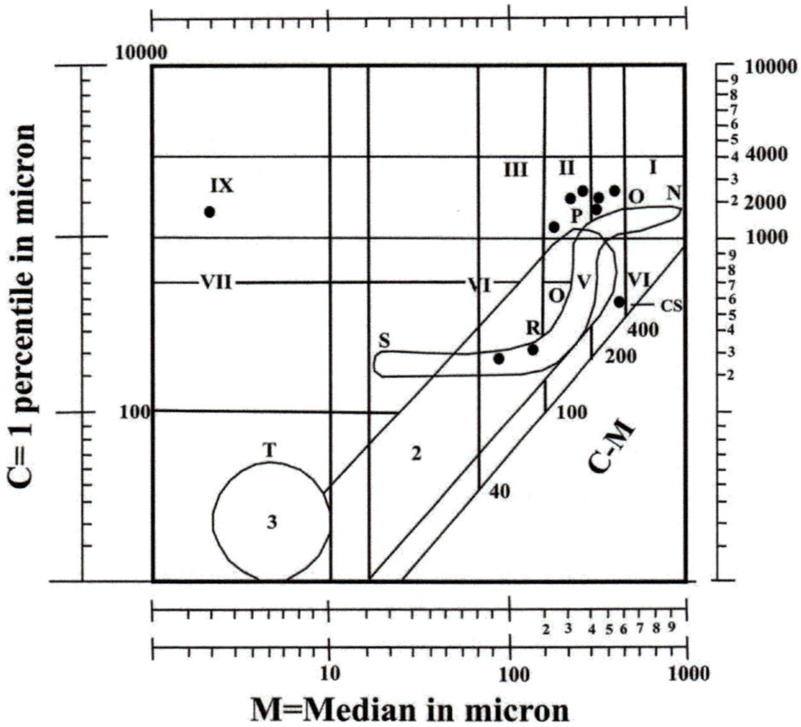


N/O: rolled sediments.
O/P: Rolled sediments with some suspension sediments.
P/Q:Graded suspension with some rolled sediments.
Q/R:Graded suspension(saltation) deposits.
R/S: uniform suspension.
CU: Maximum grain size transported as uniform suspension
CS: maximum grain size transported as graded suspension
T: Pelagic suspension.
1: Traction current deposits.
2: Turbidites.
3: Quiet water suspension deposits.
Segments I, II, III, & IX are denoted by C>1mm. Here mainly rolled sediments are incorporated; suspension is of minor importance. Segments IV, V, VI, & VII are denoted by C<1mm; mainly C<1mm are incorporated. The pattern of turbidity current deposits runs paralled limit C=M.

**Figure 18.8** CM plot of samples from and the details data is presented in Table 18.1.

**Table 18.1** Graphical statistical size parameters and their categories of Kimin Formation and terrace sediments.

| Serial No | Md | Mz | $\sigma_1$ | Ski | Kg | $\varphi$1 | C in μ | $\varphi$50 | M in μ |
|---|---|---|---|---|---|---|---|---|---|
| 2 | 1.3 | 1.16 | 1.70 | 0.018 | 1.208 | −1.4 | 2050 | 1.3 | 410 |
| 3 | 3.2 | 3.26 | 1.08 | 0.04 | 0.73 | −0.25 | 1200 | 3.2 | 110 |
| 9 | 3.4 | 3.34 | 0.73 | −0.057 | 0.98 | 1.8 | 290 | 3.4 | 96 |
| 10 | 1.45 | 1.48 | 0.6 | 0.13 | 1.71 | −2.5 | 5600 | 1.45 | 380 |
| 15 | 1.95 | 1.78 | 1.12 | −1.04 | 2.242 | −1.18 | 2200 | 1.95 | 260 |
| 16 | 2.9 | 3.07 | 0.7 | 0.35 | 0.79 | 1.75 | 300 | 2.9 | 135 |
| 18 | 1.9 | 1.78 | 1.34 | 0.15 | 1.29 | −1.3 | 2500 | 1.9 | 270 |
| 20 | 2.0 | 1.62 | 1.382 | −0.12 | 2.11 | −0.93 | 1900 | 2.0 | 250 |
| 21 | 0.9 | 1.21 | 1.69 | 0.67 | 0.98 | −1.32 | 2500 | 0.9 | 540 |

## 18.5 Discussion and Conclusions

From the observed lithological association, it can be concluded that the sediments of the Kimin Formation were deposited in an alluvial fan environment. From studied lithosections it can be inferred that a paleoalluvial fan of size approximately 120 $km^2$ was present in the area during deposition of the Kimin Formation. Furthermore, from the presence of siltstone boulders of Siwaliks within Kimin in some areas, it can be inferred that a part of Siwalik basin was uplifted during deposition of the Kimin Formation (Figure 18.4e). Therefore, the Kimin basin has developed in front of the HFT-1. Based on the study of terrace deposits in Harmatigaon, Gholajuli, and Bandardewa, the depositional environment of these sediments can be concluded to be of alluvial fan environment. In the lithologies, the dominance of boulders in the Harmotigaon area represents the proximal part of the fan, but near Banderdewa, a mid-fan facies is indicated.

The studied sediments are unconsolidated and thus it can be incidental that these sediments are younger than the Kimin Formation. Here the terrace sediments were deposited as a result of the uprise of the Kimin Formation along a sub-thrust (HFT-2). Because of this uprising phenomenon, the river emerges out of the mountain (south of the earlier alluvial fan, i.e., in front of HFT-2). This resulted in the formation of a new alluvial fan toward the south of the paleoalluvial fan (Kimin Formation). This phenomenon is clearly observed in the study area as older terrace deposits of the Harmatibagan area (Figure 18.1) and near Bandardewa.

From the cumulative curves, it is observed that almost in all the samples, the saltation population is predominant with a little amount of rolling and suspension population. The moderately-sorted grain in the studied localities further supports the results from the cumulative curves. It is worth mentioning that the predominance of saltation fraction is a characteristic of alluvial channel sands. Because of the lower energy condition in rivers, the sorting of river sand is poorer than the beach sand. Positive skewness distinguishes the river sand as it contains more fine particles, while the negative skewness supports the nature of beach sediments. Discrimination between beach and river sand can clearly be observed by the standard deviation (sorting) vs. skewness plot. From discrimination plots of fields of river and beach, it is observed that except for one sample point of the Kimin Formation, all the points fall in the river field. The plot of graphic standard deviation vs. graphic skewness, $\sigma_1$ vs. $SK_1$, shows that the majority of the sample points are concentrated in the tidal current sands. The $Sk_1$ vs. $K_G$ plot shows that almost all the samples cluster around zone A, indicating the presence of medium to coarse sands in the sediments. This is also indicated by the deposition energy condition of the sediments at moderate conditions. It is observed from the CM plot that the sediments are mostly rolled.

The important conclusions drawn from the present study are as follows:

i) The Kimin Formation and terrace deposits of the study area are interpreted to be deposited in an alluvial fan environment in the proximal, mid, and distal parts.
ii) The study highlighted that there was an uplift of the Kimin Formation by the Himalayan Frontal Fault. This uplift resulting into development of a younger alluvial fan toward the south of the paleoalluvial fan.
iii) Because of the uplift of the Kimin Formation, the river emerges out of the mountain (south of the earlier alluvial fan, i.e., in front of HFT-2), resulting in the formation of a new alluvial fan toward the south of the paleoalluvial fan (Kimin Formation).
iv) The grain size analysis of these sediments gives a fair idea about the process of deposition in which dominant sediments were deposited by saltation with minor

rolling and suspension. Characteristics of alluvial channel sands is also indicated by the predominance of a saltation fraction.

## Acknowledgments

The authors are highly grateful to the editors, especially editor Dr Manish Pandey of the University Center for Research & Development (UCRD), Chandigarh University, Mohali, India for giving the opportunity to contribute to the book "Advances in Remote Sensing Technology and the Three Poles" (multidisciplinary approach to the study of the Arctic, the Antarctic, and the Himalaya). Authors are grateful to the Department of Applied Geology, Dibrugarh University, Dibrugarh, India, for laboratory supports and field logistics. We, the authors, are thankful to two anonymous reviewers for their insightful comments and suggestions, which helps to finalize the manuscript in its present form.

## References

Abduh, A.G., Usman, F.C.A., Tampoy, W.M. et al. (2021). Remote sensing analysis of lineaments using Multidirectional Shaded Relief from Digital Elevation Model (DEM) in Olele Area, Gorontalo. *J. Phys. Conf. Ser.* 1783: 012095

Adrian, M., Harvey A.E., and Stokes, M. (2005). Alluvial fans: geomorphology, sedimentology, dynamics. Introduction: a review of alluvial-fan research. *Geological Society, London, Special Publications* 251: 1–7.

Bezbaruah, D. and Sarma, M. (2013). Morphotectonic evolution of the area in and around Bandardewa, Papumpare District, Arunachal Pradesh. *Journal of Earth Science* Special Volume: 223–232.

Blaire, T.C. and McPherson, J.G. (1994). Alluvial fan processes and forms. In: *Geomorphology of Desert Environments* (ed. A.D. Abrahams and A. Parsons), 354–402. London: Chapman & Hall.

Blaire, T.C. and McPherson, J.G. (2009). Processes and forms of alluvial fans. In: *Geomorphology of Desert Environments* (ed. A.J. Parsons and A.D. Abrahams), 354–402. Dordrecht: Springer. https://doi.org/10.1007/978-1-4020-5719-9_14.

Burgess, W.P., Yin, A., Dubey, C.S. et al. (2012). Holocene shortening across the Main Frontal Thrust zone in the eastern Himalaya. *Earth and Planetary Science Letters* 357–358: 152–167.

Chen, L., Ronald, J., Steel, F. et al. (2017). Alluvial fan facies of the Yongchong Basin: implications for tectonic and paleoclimatic changes during Late Cretaceous in SE China. *Journal of Asian Earth Sciences* 134: 37–54.

Debnath, A., Taral, S., and Mullick, S. (2021). The Neogene Siwalik Succession of the Arunachal Himalaya: a revised lithostratigrphic classification and its implications for the regional paleogeography. *Journal of the Geological Society of India* 97: 339–350.

Devi, M., Bhakuni, S.S., and Bora, P.K. (2010). Tectonic implications of drainage set-up in the Sub-Himalaya: a case study of Papumpare district, Arunachal Himalaya, India. *Geomorphology* 127(1–2): 14–31.

Dupont-Nivet, G., Lippert, P.C., Van Hinsbergen, D.J.J.et al.(2010). Paleolatitude and age of the Indo-Asia collision: paleomagnetic constraints. *Geophysical Journal International* 182: 1189–1198.

Elmahdy, S., Ali, T., and Mohamed, M. (2021). *Hydrological Modeling of Ar Rub Al Khali, Arabian Peninsula: A Modified Remote Sensing Approach Based on the Weight of Hydrological Evidence*. Geocarto International, Taylor & Francis. doi: 10.1080/10106049.2021.1936211

Farr, T.G., Rosen, P.A., Caro, E.et al.(2007). The shuttle radar topography mission.doi: doc: http://dx.doi.org/10.1029/2005RG000183

Folk R.L. (1966) A review of grain size parameters. *Sedimentology* 6(2): 73–93.

Folk, R.L. and Ward, W. (1957). Brazos river bar: a study in the significance of grains-size parameters. *Journal of Sedimentary Research* 27: 3–26.

Friedman, G.M. (1961). Distinction between dune, beach and river sands from their textural characteristics. *Journal of Sedimentary Research* 31(4): 514–529.

Friedman, G.M. (1967). Dynamic processes and statistical parameters compared for size frequency distribution of beach and river sands. *Journal of Sedimentary Petrology* 37(2): 327.

Friedman, G.M. (1979). Differences in size distributions of populations of particles among sands of various origins. *Sedimentology* 26: 3–32.

Fu, B.H., Zhang, S.L., Xie, X.P. et al. (2006). Late Quaternary tectonogeomorphic features along the Kangxiwar Fault, AltynTagh Fault system, northern Tibet. *Quaternary Sciences* 26: 228–235.

Goswami, T.K., Bezbaruah, D., Mahanta, B.N. et al. (2021). Structural framework, crustal shortening and Cenozoic exhumation in the eastern Himalaya: a review. *Physics and Chemistry of the Earth* 124(2): https://doi.org/10.1016/j.pce.2021.103048.

Goswami, T.K., Bezbaruah, D., Mukherjee, S. et al. (2018). Structures and morphotectonic evolution of the frontal fold-thrust belt, Kameng river section, Arunachal Himalaya, India. *Journal of Earth System Science* 127: 88.

Grotzinger, J., Bell, I., Herkenhoff, K. et al. (2006). Sedimentary textures formed by aqueous processes, Erebus crater Meridiani Planum. *Mars Geology* 34: 1085–1088.

Harvey, A.M. (1984). Aggradation and dissection sequences on Spanish alluvial fans: influence on morphological development. *Catena* 11: 289–304.

Harvey, A.M. (1987). Alluvial fan dissection: relationships between morphology and sedimentation. In: *Desert Sediments, Ancient and Modern* (ed. L. Frostick and I. Reid), 87–103. Geological Society of London, Special Publication 35. Oxford, UK: Blackwell.

Harvey, A.M., Mather, A.E., and Stokes, M. (2005). Alluvial fans: geomorphology, sedimentology, dynamics. Introduction: A review of alluvial-fan research. In: *Alluvial Fans: Geomorphology, Sedimentology, Dynamics* (ed. A.M. Harvey, A.E. Mather, and M.Stokes), 1–7. London: Geological Society Special Publications 251.

Heyward, A.P. (1978). Alluvial fan sequence and megasequence models, with examples for Westphalian D–Stephanian B coalfields, northern Spain. In: *Fluvial Sedimentology* (ed. A.D. Miall), 669–702. Memoir: Canadian Society of Petroleum Geologists, 5.

Hodges, K.V. (2000). Tectonics of the Himalaya and southern Tibet from two perspectives. *Geological Society America Bulletin* 112: 324–350.

Ingram, R.L. (1971). Sieve analysis. In: *Procedures in Sedimentary Petrology* (ed. R.E. Carver), 49–68. Wilson Interscience.

Karunakaran, C. and Ranga Rao, A. (1979). Status of exploration for hydrocarbons in the Himalayan region: contribution to stratigraphy and structure. *Him. Geol. Sem.* 1976, New Delhi Miscellaneous Publication. 41(Pt. V): 1–66.

Kelty, T.K., Dubey, C.S., and Yin, A. (2004). Structure and crustal shortening of the Subhimalayan fold and thrust belt, western Arunachal Pradesh, NE India. *Himalayan Journal of Sciences* 2: 175.

Kesari, G.K., Dasgupta, G., and Prakash, H.S.M. (2010). Geology and mineral resources of Arunachal Pradesh, Geological Survey of India. Miscellaneous Publication, No. 30 Part IV, Vol I(i). Arunachal Pradesh.

Kiamehr, R. and Sjöberg, L.E. (2005). Effect of the SRTM global DEM on the determination of a high-resolution geoid model: a case study in Iran. *Journal of Geodesy* 79: 540–551.

Kumar, G. (1997). *Geology of Arunachal Pradesh. Geological Society of India* 217. ISBN No: 81-85867-28-3

Kumar, S. (2001). Earthquake recurrence and rupture dynamics of Himalayan frontal thrust, India. *Science* 294: 2328–2331.

Kunte, S.V., Ganju, J.L., and Dutta, N.K. (1983). Geology and structure of the Tertiary belt between Barang and Pachin rivers, Arunachal Pradesh. *Geological Survey of India Miscellaneous Publication* 43: 124–129.

Lavé, J. and Avouac, J.P. (2000). Active folding of fluvial terraces across the Siwaliks Hills, Himalayas of central Nepal. *Journal of Geophysical Research*, 105: 5735–5770. https://doi.org/10.1029/1999JB900292.

Lee, T.H. and Moon, W.M. (2002). Lineament extraction from Landsat TM, JERS-1 SAR, and DEM for geological applications. In: *Proceedings of the International Geoscience and Remote Sensing Symposium (IGARSS)*, Toronto, ON, Canada, 24–28June 2002; 6: 3276–3278.

Leeder, M. (1999). *Sedimentology and Sedimentary Basins: From Turbulence to Tectonics*. Oxford, UK: Blackwell.

Lloyd, M.J., Nichols, G.J., and Friend, P.F. (1998). Oligo-Miocene alluvial fan evolution at the southern Pyrenean thrust front, Spain. *Journal of Sedimentary Research* 68: 869–878.

Malik, J.N. and Nakata, T. (2003). Active faults and related late Quaternary deformation along the northwestern Himalayan Frontal Zone, India. *Annals of Geophysics* 46: 917–936.

Mason, C.C. and Folk, R.L. (1958). Differentiation of beach, dune and aeolian flat environments by size analysis, Mustang Island, Texas. *Journal of Sedimentary Research* 28: 211–226.

Miall, A.D. (1985). Architecture-element analysis: a new method of facies analysis applied to fluvial deposits. *Earth-Science Reviews* 22: 261–308.

Moiola, R.J. and Weiser, D. (1968). Textural parameters: an evaluation. *Journal of Sedimentary Research* 38(1): 45–53. 539–549. Elsevier. ISBN 9780123693969. https://doi.org/10.1016/B0-12-369396-9/00176-3.

Nakata, T. (1989). Active faults of Himalaya of India and Nepal. *Geological Society of America* Special Paper no. 232: 243–264.

Nemec, W. and Postma, G. (1993). Quaternary alluvial fans in southwest Crete: sedimentation processes and geomorphic evolution. In: *Alluvial Sedimentation* (ed. M. Marzo, and C. Puigdefabrigas), 235–276. International Association of Sedimentologists, Special Publication 17. Wiley.

Passega, R. (1957). Texture as a characteristic of clastic deposition. *American Association of Petroleum Geologists* 41: 1952–1984.

Passega, R. (1964). Grain size representation by C-M pattern as a geological tool. *Journal of Sedimentary Research* 34: 830–847.

Passega, R. and Byramjee, R. (1969). Grain size image of clastic deposits. *Sedimentology* 13: 180–190.

Rabus, B., Eineder, M., Roth, A.et al. (2003). The shuttle radar topography mission: a new class of digital elevation models acquired by spaceborne radar. *ISPRS Journal of Photogrammetry and Remote Sensing* 57: 241–262.

Reuter, H. I., Nelson, A., and Jarvis, A. (2007). An evaluation of void filling interpolation methods for SRTM data. *International Journal of Geographical Information Science* 21: 983–1008.

Sahu, B.K. (1964). Depositional mechanisms from the size analysis of elastic sediments. *Journal of Sedimentary Petrology* 34: 73–83.

Sarma, M., Bezbaruah, D., Goswami, T.K.et al. (2020). Geologic and tectonic evolution of the Yinkiong Group and Abor volcanic rocks in the Eastern Himalaya: an overview of geologic data. *Geotectonics* 54: 395–409.

Singh, A. and Bezbaruah, D. (2021). Evaluation of active tectonics and geomorphic indices in Siwalik Basin around Dikrong River, Eastern Himalaya. *Journal of Scientific Research* 65(3): Institute of Science, Banaras Hindu University, Varanasi, India.

Srivastava, A.K. and Mankar, R.S. (2009). Grain size analysis and depositional pattern of upper Gondwana Sediments (Early Cretaceous) of Salbardi area, Districts Amravati, Maharashtra and Betul, Madhya Pradesh. *Journal Geological Society of India* 73: 393–406.

Thomas, R.L., Kemp, A.L.W., and Lewis, C.F.M. (1972). Distribution, composition and characteristics of the surficial sediments of Lake Ontario. *Journal of Sedimentary Research* 42(1): 66–84.

Valdiya, K.S. (1984). Evolution of the Himalaya. *Tectonophysics* 105(1–4): 229–248. ISSN 0040-1951.

Valdiya, K.S. (1992). The Main Boundary Thrust zone of the Himalaya, India. *Ann. Tecton.* 6: 54–84.

Visher, G.S. (1969). Grain size distributions and depositional processes. *Journal of Sedimentary Research* 39: 1074–1106.

Wagreich, M. and Strauss, P.E. (2005). Source area and tectonic control on alluvial-fan development in the Miocene Fohnsdorf intramontane basin, Austria. *Alluvial Fans: Geomorphology, Sedimentology, Dynamics.* London: Geological Society, Special Publications 251: 207–216. 0305–8719/05/$15 © The Geological Society of London.

Wesnousky, S.G., Kumar, S., Mohindra, R.et al. (1999). Uplift and convergence along the Himalayan Frontal Thrust of India. *Tectonics* 18: 967–976.

Yang, L., Meng, X., and Zhang, X. (2011). SRTM DEM and its application advances. *International Journal of Remote Sensing* 32(14): 3875–3896.

Yeats, R. S., Nakata, T., Farah, A.et al. (1992). The Himalayan Frontal Fault system. *Annales Tectonica* Special Issue: Supplement to Vol. VI: 85–98.

Yin, A. (2006). Cenozoic tectonic evolution of the Himalayan orogen as constrained by along-strike variation of structural geometry, exhumation history, and foreland sedimentation. *Earth Science Reviews* 76: 1–131.

Yin, A., Dubey, C.S., Kelty, T.K.et al. (2006). Structural evolution of the Arunachal Himalaya and implications for asymmetric development of the Himalayan orogen. *Indian Current Science Association* 90: 195–206.

Yin, A. and Harrison, T.M. (2000). Geologic evolution of the Himalayan-Tibetan orogen. *Annual Review of Earth and Planetary Sciences* 281: 211–280.

# 19

# Himalayan River Profile Sensitivity Assessment by Validating of DEMs and Comparison of Hydrological Tools

*Rahul Devrani[1,*], Rohit Kumar[2], Maneesh Kuruvath[3], Parv Kasana[4], Shailendra Pundir[5], Manish Pandey[6], and Sukumar Parida[7]*

[1] Delhi School of Climate Change & Sustainability, Institution of Eminence, University of Delhi, India 110007
[2] Discipline of Geology, School of Sciences, Indira Gandhi National Open University, New Delhi 110068, India
[3] Masters in Geology, Curtin University, Western Australia Perth.
[4] Department of Geology, University of Delhi, Delhi, India
[5] Wadia Institute of Himalayan Geology, 33 GMS. Road, Dehradun 248001, India
[6] University Center for Research & Development (UCRD), Chandigarh University 140413
[7] Physical Research Laboratory, Navrangpura, Ahmedabad 380009, India
* Corresponding author

## 19.1 Introduction

Rivers appear as physical substances, carrying water and sediments from the headwater to the sink. Since the dawn of civilization, humans have directly or indirectly depended on river resources for ecological growth and survival (Schumm, 2005). A river can be bedrock or alluvial and is fed by glaciers, snow, precipitation, groundwater, or lakes. The bedrock rivers are critical geomorphic markers as they play an essential role in shaping the landforms by responding rapidly and consistently to any change in tectonics and climate (Seeber and Gornitz, 1983; Ouchi, 1985; Holbrook and Schumm, 2000; Amos and Burbank, 2007; Jain and Sinha, 2009). The bedrock rivers set the base level, control the denudation rates (Whipple, 2004), and are always in the process of achieving "steady state" (Willett and Brandon, 2002). When the uplift and erosion are unbalanced, they adjust to changes by altering their width and slope, reflected in the landforms. Hence, fluvial landforms help our understanding of changes that may have occurred in the past.

Since the last century, geomorphologists have paid much attention to the geo-morphometrical parameters to study the effects of climatic–tectonic interaction on erosion rates in the rivers. The longitudinal river profile is always given special consideration among several geo-morphometrical parameters (Mudd et al., 2014, 2018). The interaction between climate and tectonics in a river can be expressed in different curvatures and a steady-state longitudinal profile. Theoretically, rivers should have a convex or straight profile along with knick points, but they generally have concave profiles. As they flow downstream, their gradient decreases, and there is a gradual increase in the area, which makes an upward concave shape. A river is sensitive to changes in the lithology, tectonics, and climate, which is reflected in their longitudinal profile. A river profile is one of the best geomorphic indices and is widely used in tectonically active areas as indicators of uplift (Whipple, 2004). The bedrock rivers show more diverse characteristics in hydrology, sediment flux, slope, and tectonics and will also show more geomorphic characteristics than other river types (Wohl, 2010).

The use of longitudinal profiles in the Himalayas has been carried out: i) to decipher the climatic and tectonic controls on the exhumation of deep crustal rocks (Baillie and Norbu, 2004; Clark et al., 2004; Jamieson et al., 2004; Thiede et al., 2004; Finnegan et al., 2008; Singh and Jain, 2009; Deeken et al., 2011; Devi et al., 2011, 2016; Dortch et al., 2011; Anoop et al., 2012; Godard et al., 2014; Munack et al., 2014; Scherler et al., 2014; Zhang et al., 2016; Jonell et al., 2017; Wang et al., 2018, 2019, 2020; Yang et al., 2018; Strong et al., 2019; Dey et al., 2021; Pundir et al., 2021; Tang et al., 2021); ii) to decipher the spatial and temporal pattern of erosion of the river catchments (Jain and Sinha, 2004; Sitaula et al., 2004; Vannay et al., 2004; Singh et al., 2007; Seong et al., 2008; Thakur et al., 2009; Tamrakar et al., 2011; Ghosh and Mishra, 2015; Nag and Phartiyal., 2015; Munack et al., 2016; Yang et al., 2016; Chaudhary et al., 2017; Parida et al., 2017; Schwanghart and Scherler, 2017; Sinclair et al., 2017; Wang et al., 2017; Chahal et al., 2019; Swarnkar et al., 2019; Kaushal et al., 2020; Prokop et al., 2020; Joshi et al., 2021); iii) to constrain the influence of tectonic, geomorphic, and anthropogenic activities on channel morphology and landform evolution, and to depict flash flood hazard impact (Brookfield, 1998; Ahmad and Bhat, 2012; Aiken and

*Advances in Remote Sensing Technology and the Three Poles*, First Edition. Edited by Manish Pandey, Prem C. Pandey, Yogesh Ray, Aman Arora, Shridhar D. Jawak, and Uma K. Shukla.

Brierley, 2013; Dar et al., 2014, 2019; Devrani et al., 2015; Borgohain et al., 2016); iv) to decipher the structural control on the landform evolution of the mountain belt and crustal deformation (Lavé and Avouac, 2000; Bishop et al., 2002; Wobus et al., 2003; Hodges et al., 2004; Robl et al., 2008; Singh and Tandon, 2010; Walsh et al., 2012; Thakur et al., 2014; Ahmad et al., 2015; King et al., 2016; Imsong et al., 2017; Mukul et al., 2017; Srivastava et al., 2017; Cannon et al., 2018; Nennewitz et al., 2018; Divyadarshini and Singh, 2019; Kothyari et al., 2019; Sharma et al., 2019; Wang et al., 2019; Joshi et al., 2020; Yu et al., 2020; Luirei et al., 2021; Wahyudi et al., 2021); v) to be used in the studies of paleo river discharges (Kumar et al., 2020); vi) to be used in the studies of landslide hazards, outburst floods, and mitigation and glacial and landslides feedbacks to landform evolution (Bhasin et al., 2002; Gupta and Sah, 2008; Korup et al., 2010); vii) to decipher the neo-tectonics and Himalayan seismic hazards (Philip and Sah, 1999; Malik et al., 2010; Singh and Awasthi, 2010; Duarah & Phukan, 2011; Mehta and Sanwal, 2011; Phartiyal and Kothyari, 2011; Kirby and Whipple, 2012; Bhakuni et al., 2013; Thakur, 2013; Cannon and Murphy, 2014; Morell et al., 2015; Imsong et al., 2016; Divyadarshini and Singh, 2017; Kaushal et al., 2017; Kothyari et al., 2017; Taloor et al., 2017; Talukdar et al., 2017; Dey et al., 2018; Joshi and Kotlia, 2018; Kothyari et al., 2019, 2020; Mishra, 2019; Stolle et al., 2019; Wani et al., 2019; Bhat et al., 2020; Chug et al., 2020; Shankar et al., 2020; Suresh and Kumar, 2020; Bhattarai et al., 2021; Joshi et al., 2021; Sinha and Sinha, 2021; Thakur et al., 2021); and viii) to study glacial landforms, river terraces and morphology, hill slope processes, and landscape evolution morphometry (Wager et al., 1937; Iwata et al., 1982; Yamanaka and Iwata, 1982; Burbank and Fort, 1985; Wells and Dorr, 1987; Froehlich et al., 1990; Fielding et al., 1994; Kirby and Whipple, 2001; Finlayson et al., 2002; Duncan et al., 2003; Finnegan et al., 2005; Korup, 2006; Craddock et al., 2007; Singh and Tandon, 2007, 2008; Suresh et al., 2007; Tandon and Sinha, 2007; Kothyari, 2014; Olen et al., 2015; Struck et al., 2015; Adams et al., 2016; Das et al., 2016; Sonam and Jain, 2017; Varay et al., 2017; Imsong et al., 2018, 2019; Chatterjee et al., 2019; Goswami et al., 2019; Sharma et al., 2019; DeCelles and Carrapa, 2021; Simoes et al., 2021; Wang et al., 2021).

In the present study, DEMs (Digital Elevation Model) from different sensors; e.g., ALOS-PALSAR, ASTER, COOPER, CARTOSAT, NASA, and SRTM, derive the sensitivity of the DEMs in the generation of longitudinal river profiles.

## 19.2 Study Area

The upper Alaknanda River catchment (UARC) comprises higher reaches of the mountainous catchment of the Ganga River basin, with its base level set as the Himalayan Frontal Thrust near Haridwar (Figure 19.1). The Alaknanda River originates from the Satopanth and the Bhagirath-Kharak glaciers (Nainwal et al., 2007, 2008) and traverses ~80 km through different lithostratigraphic units before joining the Dhauliganga River at Vishnuprayag. The Saraswati River is a major headwater tributary of the UARC and has a confluence with the Alaknanda River in the Keshavprayag. Apart from the Saraswati River, the Alaknanda River is also joined by several different sized and shaped tributaries. The UARC lies in the northwestern Himalayas, Garhwal Himalayas, and covers an area of 1540 km$^2$ in the Tethyan and Greater Himalayas region. The landforms and topography of the region are the result of the active tectonics, paleoglaciation, and active periglacial processes. Most of the region is covered by Quaternary glacial deposits, especially moraine and alluvial fans (Nainwal et al., 2007, 2008; Mehta et al., 2016; Rana et al., 2019).

## 19.3 Methodology (LSDTopoTools)

The extraction of river profiles is a very lengthy and tedious process. In a typical GIS software, the channel extraction requires multiple complicated steps. Each step generates some data that may or may not be helpful for the required genre of studies. Therefore, such software can be computationally costly and require advanced computer chipsets for hasty processing of DEM data. On the other hand, LSDTopoTools© (Mudd et al. 2014, 2018, https://github.com/LSDtopotools) is an Open-Source software package, and can be used for multiple topographical analyses. LSDTopoTools uses a combination of Python and C++ languages for calculating multiple topographic metrics. LSDTopoTools encourages reproducibility and transparency in the morphometric analysis, as it provides the users with the ability to regenerate the previously modeled data using similar parameters. It helps to better understand the previous problems and gives new users the ability to reason for their respective problems. So, for example, the data produced in this chapter can be reproduced by any curious worker in their computer systems using the similar parameters that the authors have used. The LSDTopoTools allows users to calculate various topographic metrics such as Hillslope metric calculation, Channel extraction, Chi analysis and Channel steepness analysis, Knick point analysis, Floodplain and Terrace extraction, and even Landscape evolution algorithms. We have considered the DEM resolution as an essential factor in channel extraction for this particular chapter. This work uses moderate resolution DEM (30 × 30 m), which can be readily used for channel network generation using the drainage area threshold method.

We have used the Chi analysis-based approach to extract the channels of the UARC. This approach uses the typical drainage area threshold method, which calculates

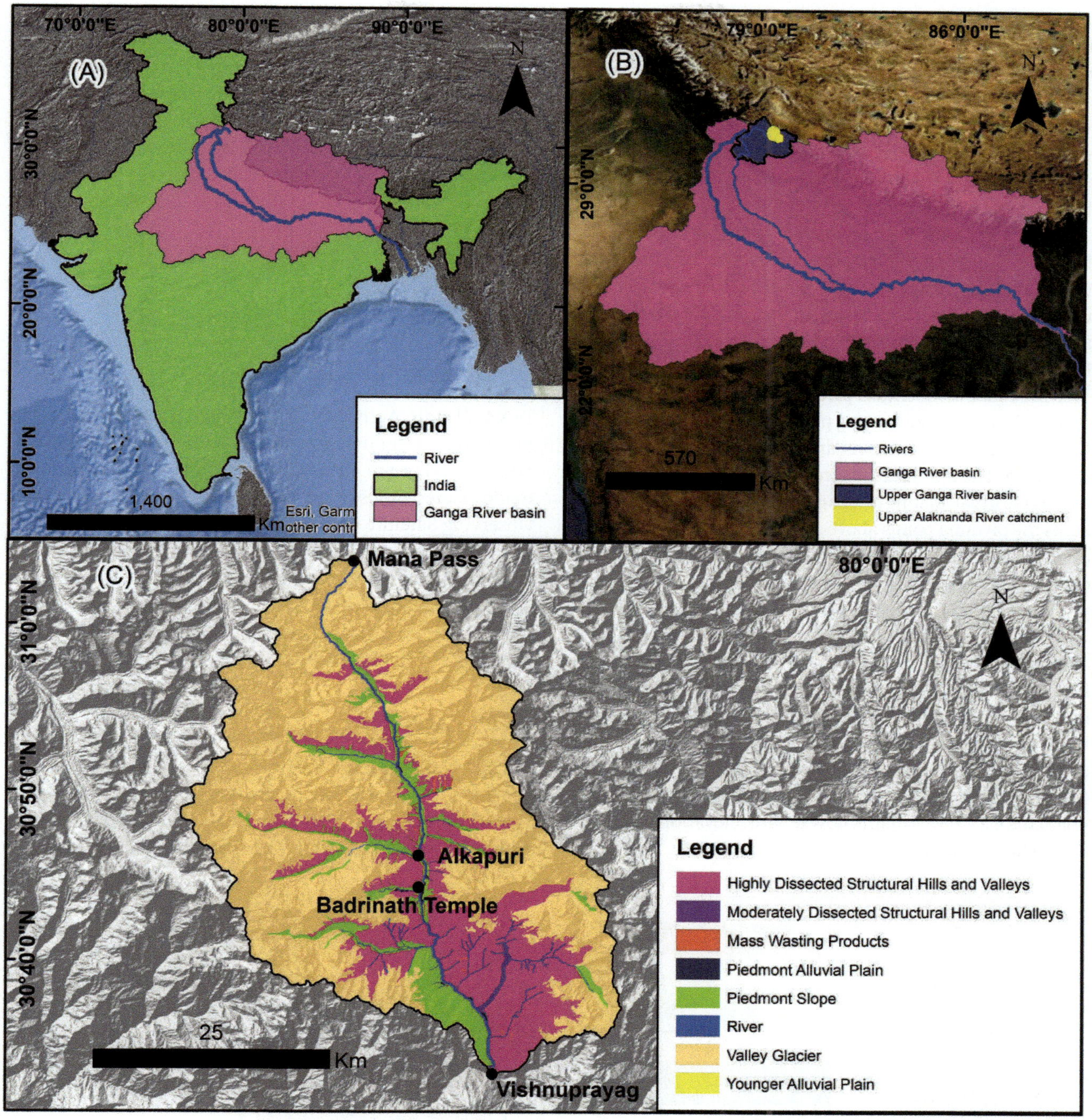


**Figure 19.1** (A) The Ganga River basin is draped over the hillshade map (courtesy – of ESRI online maps); (B) the upper Ganga River basin and upper Alaknanda River basin are represented by blue and yellow; and (C) the map represents regional-scale geomorphology in the upper Alaknanda River basin. *Source*: Based on Data from Bhukosh (https://bhukosh.gsi.gov.in/Bhukosh/Public).

drainages using contributing pixels in the steepest descent. It also uses the D8 algorithm, which calculates the steepest descent of the eight neighboring pixels to generate the fluvial network and drainage basin. The threshold contributing pixels are the primary factor in this approach because the channel network extraction depends directly upon it. For example, if we select a contributing threshold of 100 pixels for a 30 m DEM, a 0.09 $km^2$ area of drainage will be selected to construct a channel point. The culmination of these points will ultimately construct a drainage network using the D8 steepest descent algorithm. The threshold contributing pixels affect the drainage density since it depends on the number of pixels and DEM resolution. For example, a DEM with 30 m resolution with 1000 contributing pixels will have a lower drainage density than the same DEM with 100 contributing pixels. To maintain the homogeneous approach for channel extraction for different DEMs of 30 m resolution, we used a constant value of 500 pixels as threshold contributing pixels. This value provides us with almost identical drainage networks and successfully eliminates the areas affected by the hillslope processes.

Identifying basins from channel networks is also essential in calculating hydrological parameters. The LSDTopoTools uses the D8 algorithm for calculating the drainage area and, ultimately, the drainage basins. It uses a slightly different approach for calculating drainage basins from DEM. The user has to feed the compiled code with a driver/parameter file. This file contains all critical parameters required for analysis, such as threshold contributing pixels, the minimum and maximum drainage area of basins, and what kind of basin must be calculated. The minimum and maximum drainage areas play a vital role in calculating drainage basins because these values act as the pivoting points above and below which no drainage basin will be incorporated in the calculation. Similarly, for the case of "type of basin" to be calculated, it gives the user the power to control the basin calculation algorithm.

We validated the data by comparing our LSD basins with data generated from ArcGIS software. We found the similarity between the basins and used them for further comparisons and other analyses. After thoroughly assessing these parameters, we fed the software with this driver file and the selected DEM. The program ends by printing point datasets in .csv format, which can be used for further analysis. This CSV file contains flow length, elevation, drainage area, Chi value (if theta values are correctly provided), Mchi ($K_{sn}$ if $A0 = 1$), etc., for each data point of a tributary in a particular basin. Furthermore, we used a field correlated pour point/endpoint for the main channel. We considered this pour point as the ultimate endpoint of the channel and compared river profiles of Ganga for different DEMs. We have also used QGIS for sorting and splitting point datasets (.csv). The steps followed in extraction of different parameters used in this work are illustrated in Figure 19.2.

## 19.4 Details of DEM Datasets Used

### 19.4.1 ALOS-PALSAR

JAXA (Japan Aerospace Exploration Agency) has used more than 3 million scenes to develop ALOS-PALSAR DEM covering the entire globe through PRISM panchromatic stereo sensor placed on the Advanced Land Observing Satellite (ALOS) that worked from 2006 to 2011 (www.eorc.jaxa.jp). Its previous paid version was available at 5 m spatial resolution, but later they freely provided 30 m spatial resolution DEM. ALOS-PALSAR provides DEMs with a vertical accuracy of 5 m and a horizontal accuracy of 5 m (Table 19.1) (Figure 19.3A).

### 19.4.2 ASTER

NASA launched the Advanced Spaceborne Thermal Emission and Reflection Radiometer in 1999 with stereoscopic sensors, including a nadir observing telescope and a backward viewing telescope (https://lpdaac.usgs.gov/).

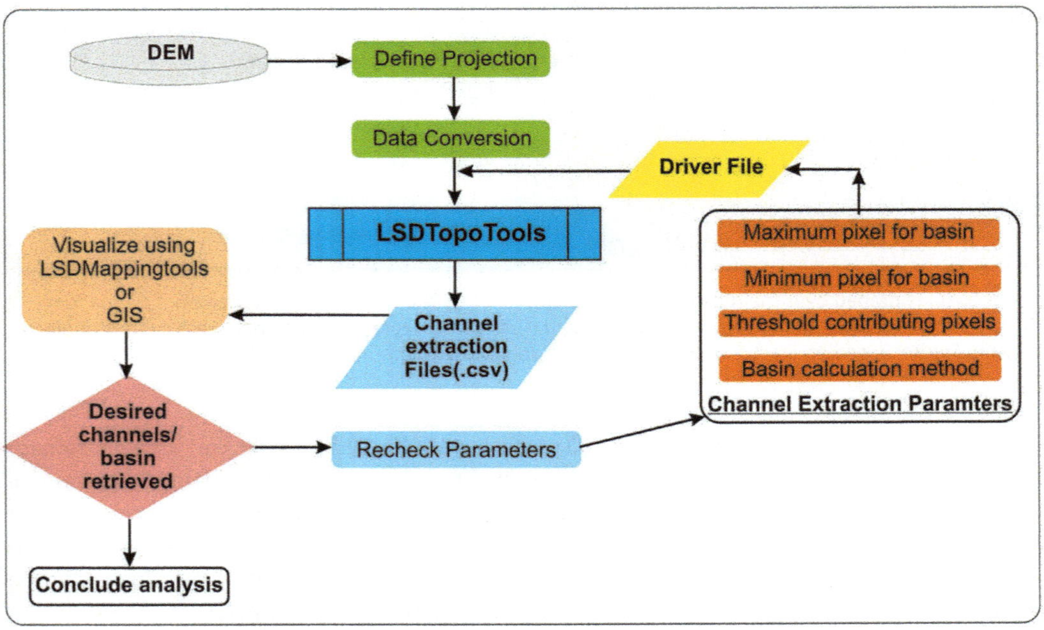


Figure 19.2 The flowchart represents the methodology used in the present study.

**Table 19.1** The accuracy and source of DEMs.

| Sensor | Satellite | Data Acquisition Time | Data Available Since | Data Resolution | Data Processed Using | Accuracy |
|---|---|---|---|---|---|---|
| PRISM panchromatic stereo mapping sensor | Advanced Land Observing Satellite "DAICHI" (ALOS) | 2006–2011 | 2015 | 30 m | Stereo Pan imagery | Vertical ~5 m, Horizontal ~5 m |
| ASTER with two telescopic for Nadir and backward viewing | NASA's Terra spacecraft in December 1999 | 2000–2013 | Publicly released by LP DAAC in August 2019 | 30 m | Stereo NIR imagery | Vertical ~20 m, horizontal ~50 m |
| CartoSAT two panchromatic in-track stereo cameras with 2.5 m spatial resolution | CartoSAT-1 | 2005 | Used in NRSC since 2011 | 30 m or coarser, generated using stereoscopic Images of resolution 2.5 m and GCPs (ground control points) | Augmented Stereo Strip Triangulation | Planimetric (absolute) ~15 m; Vertical (absolute) ~8 m; Vertical (Relative) >5 m |
| X-band Radar sensor (Copernicus DEM) | TanDEM-X and TerraSAR-X (TSX) radar satellite | 2007 | 2019 (90 m), 2021 (30 m) | GLO30-30 m, GLO90-90 m, generated using X-band Radar | | Horizontal (absolute) <6 m; Vertical (absolute) <4 m; Vertical (Relative) <2–4 m |
| C-band Radar sensor (NASA DEM) | Space Shuttle Endeavour | 11–21 February 2000 | LP DAAC in 2020 | 3 m, 90 m, generated using c-band Radar | Statistical-cost Network-flow Algorithm for Phase Unwrapping (SNAPHU) | Vertical 16 m; Horizontal 20 m |
| C-band Radar sensor (SRTM) | Space Shuttle Endeavour | 11–21 February 2000 | 2015 | 3 m, 90 m, generated using c-band Radar | Statistical-cost Network-flow Algorithm for Phase Unwrapping (SNAPHU) | Vertical accuracy 16 m; Horizontal accuracy 20 m |

ASTER used about 1,880,306 scenes to build the DEM through a stacking procedure, with aberrant data deleted and a water-body database included for enhanced accuracy. ASTER provide DEMs with the resolution of 30 m with 20 m vertical and 5 m horizontal accuracy (Table 19.1) (Figure 19.3B).

### 19.4.3 CartoDEM

To create the DEM, ISRO's Space Application Centre developed Augmented Stereo Strio Triangulation (ASST) software (Table 19.1). They used 20,000 scenes, ground control points, the Triangulated Irregular Network (TIN), and automatic mosaicing to create seamless DEMs. QV systems have quality checked each scene, and any irregularities such as water bodies, sinks, and leftover mosaics have been addressed. Its free version has a spatial resolution of 30 m and a vertical accuracy of 15 m (Table 19.1) (Figure 19.3C).

### 19.4.4 Copernicus DEM

The Copernicus (Cooper) DEM is a radar-based product acquired by the TanDEM-X mission of the German Aerospace Centre and Airbus Defence and Space. The mission was launched to generate global DEM and DSM using Synthetic Aperture Radar. The data were acquired in 2010 by TerraSAR-X and TanDEM-X satellites through bi-static strip map mode (www.help.emd.dk). It provides 30 m resolution data with a vertical accuracy of 4 m (Table 19.1) (Figure 19.3D).

### 19.4.5 NASA DEM

NASA DEMs were created from the SRTM data using hybrid processing techniques, and the data was released to the public in 2020. A hybrid approach produces 2D interferograms from radar pictures for improved accuracy.

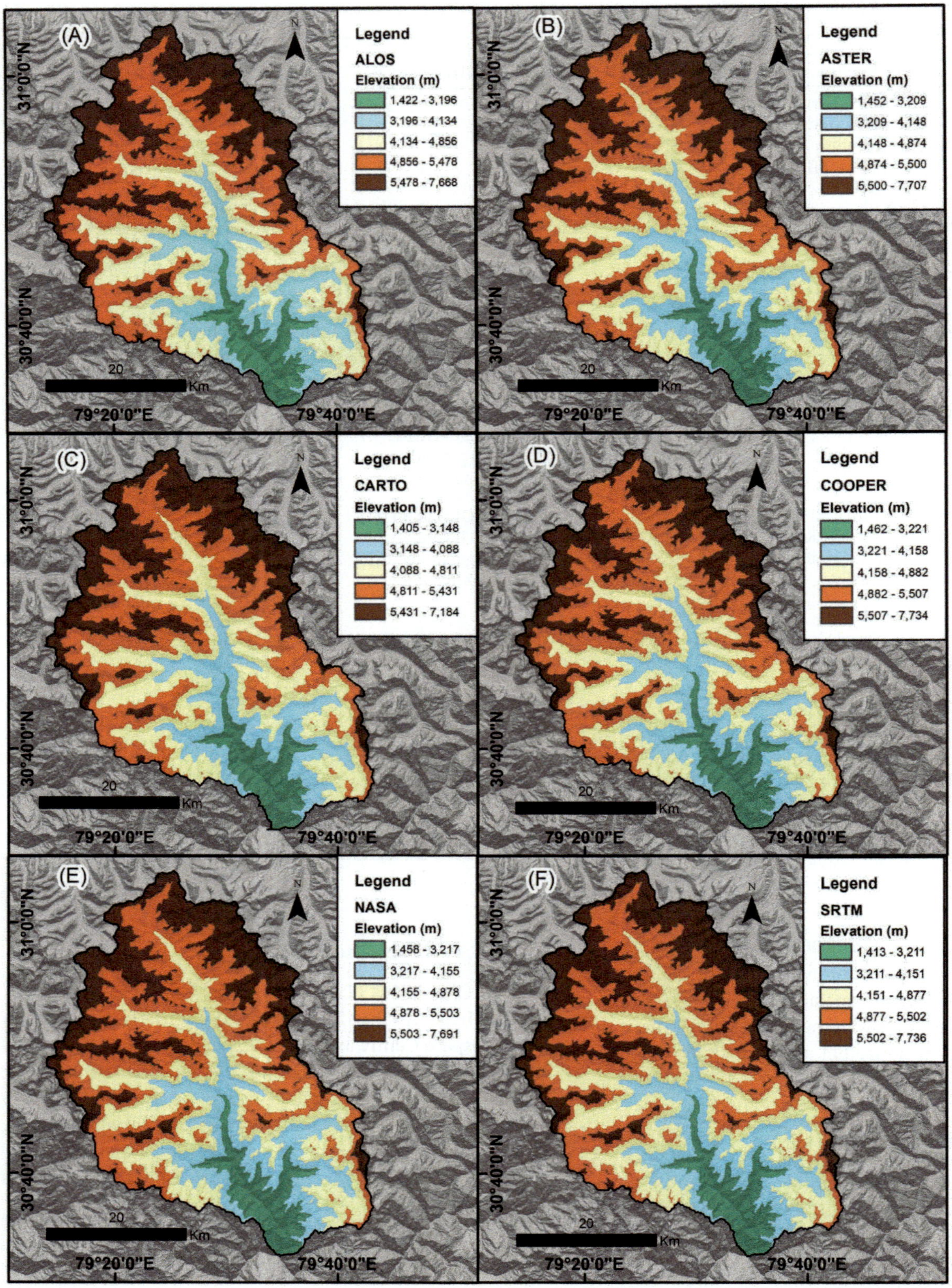


**Figure 19.3** Digital Elevation Models (DEM) of: (A) ALOS-PALSAR; (B) ASTER; (C) CARTOSAT; (D) COOPER; (E) NASA; and (F) SRTM.

Unwrapping techniques and supplementary data were not processed in SRTM, but they were included in NASA DEMs to remove voids and improve geographical coverage. Data from additional sources, such as ASTER, ALOS-PALSAR, PRISM, and Alaska DEMs, is used to fill the voids (https://lpdaac.usgs.gov/). Ground control points, laser profiles from glaciers, and ICESat were used to determine vertical accuracy. The vertical accuracy of NASA's DEM is 16 m, and the horizontal accuracy is 20 m (Table 19.1) (Figure 19.3E).

### 19.4.6 SRTM

The Shuttle Radar Topography Mission (SRTM) is led by NASA, the National Intelligence Agency, the German Aerospace Agency, and the Italian Space Agency (https://lpdaac.usgs.gov/). The mission, which uses C and X bands to cover the world from 60°N to 56°S, began in 2000. The C band has a wavelength of 5.6 cm, while the X band has a wavelength of 3.1 cm. As a result, particles with longer wavelengths and smaller sizes are transmitted. Its initial version, with a precision of 11 arc second, and its second version, with a precision of 31 arc second, are publicly available, with a vertical and horizontal accuracy of 161 m and 20 m, respectively (Table 19.1) (Figure 19.3F).

## 19.5 Result and Discussion

### 19.5.1 Assessment of DEMs Generated Watershed Boundary and Slope

The Digital Elevation Model (DEM) from different sensors is used to extract elevation gradient, watershed boundary, and slope gradient in degrees using semi-automatic approaches in GIS environment. The watershed boundaries are extracted though a fixed pour point to observe the variation in the DEMs and its outputs. The area of watershed boundary of the ALOS-PALSAR-DEM is 1537.88 $km^2$, 1537.04 $km^2$ through ASTER-DEM, 1546.52 $km^2$ through CartoDEM, 1537.9 $km^2$ through Cooper-DEM, 1539.59 $km^2$ through NASA-DEM, and 1539.39 $km^2$ through SRTM-DEM. The watershed boundaries are used to clip the DEMs from the downloaded DEMs. The elevation extracted from the DEMs show a slight different spatial distribution due to the attributes of sensors. Elevation in ALOS-PALSAR-DEM ranges from 7688 m to 1422 m with a mean elevation of 4774 m; in ASTER-DEM it ranges from 7707 m to 1452 m with a mean of 4795 m; in Carto-DEM it ranges from 7184 m to 1405 m with a mean of 4745 m; in Cooper-DEM elevation it ranges from 7734 m to 1462 m with a mean of 4798 m; in NASA-DEM it ranges from 7691 m to 1458 m with a mean of 4798 m; and in SRTM-DEM it ranges from 7736 m to 1413 m with a mean of 4801 m.

The slope gradient calculated from ALOS-PALSAR-DEM ranges from 0° to 79.33° having a mean value of 29.91°. Similarly, the slope gradient from the ASTER-DEM ranges from 0° to 78.53° with a mean of 30.33°; in Carto-DEM it ranges from 0° to 85.02° with a mean of 26.58°; in Cooper-DEM it ranges from 0° to 80.17° with a mean of 30.45°; in NASA-DEM it ranges from 0° to 79.22° with a mean of 30.07°; and in SRTM-DEM it ranges from 0° to 80.99° with a mean of 31.43°. Slope gradients (degrees) using all the digital elevation models is depicted in Figures 19.4A to 19.4F.

### 19.5.2 Sensivity of Longitudinal River Profiles Using Different DEMs

A longitudinal river profile shows the shape of the river's course from source to sink. It shows the changing width, depth, and gradient of the channel downstream. In longitudinal profile, the X coordinate represents the distance traveled from the origin (in km) and Y, which represents the elevation in meters. The longitudinal river profiles of the upper Alaknanda River (from Vishnuprayag) have been created from all DEMs (Figure 19.5).

The length of the river using ALOS-PALSAR DEM is 80.37 km, originating at an elevation of 5854 m. The river exhibits a steady state profile up to 32.75 km, and after that a knick point is identified and river shows a typical concave-up form up to 7 km. Then the river again shows comparatively gentle gradients and steady state profiles. The length of the river using ASTER DEM is 80.35 km and originates at the elevation of 5785 m. The river exhibits a steady state profile from source to sink, except for some areas where it shows as concave-up at 25.7 km and a knick point at 31.32 km with concave-up shape. Beyond this region, the river shows gentle gradients and steady state profiles. The river generated from CartoDEM is 83.64 km long and originates at the elevation of 5907 m. The river follows a constant state profile until it reaches a minor break (knick point) at 27.99 km. At 33.9 km, a major knick point is found and concave-up morphology is noted, after which the river resumes its steady state profile. The river profile shows extremely gentle gradients from 73.13 km to 81.42 km, followed by an abrupt break (perhaps knick point). The length of the river using COOPER DEM is 78.33 km and originated at an elevation of 5866 m. The river exhibits a steady state profile with some minor deviation up to 33 km then a knick point is identified and river shows a typical concave-up. After this region, the profile is in steady state. In the case of NASA DEM, the length of the river is 80.97 km originating at the elevation of 5923 m. Similarly, as with the COOPER DEM profile, this

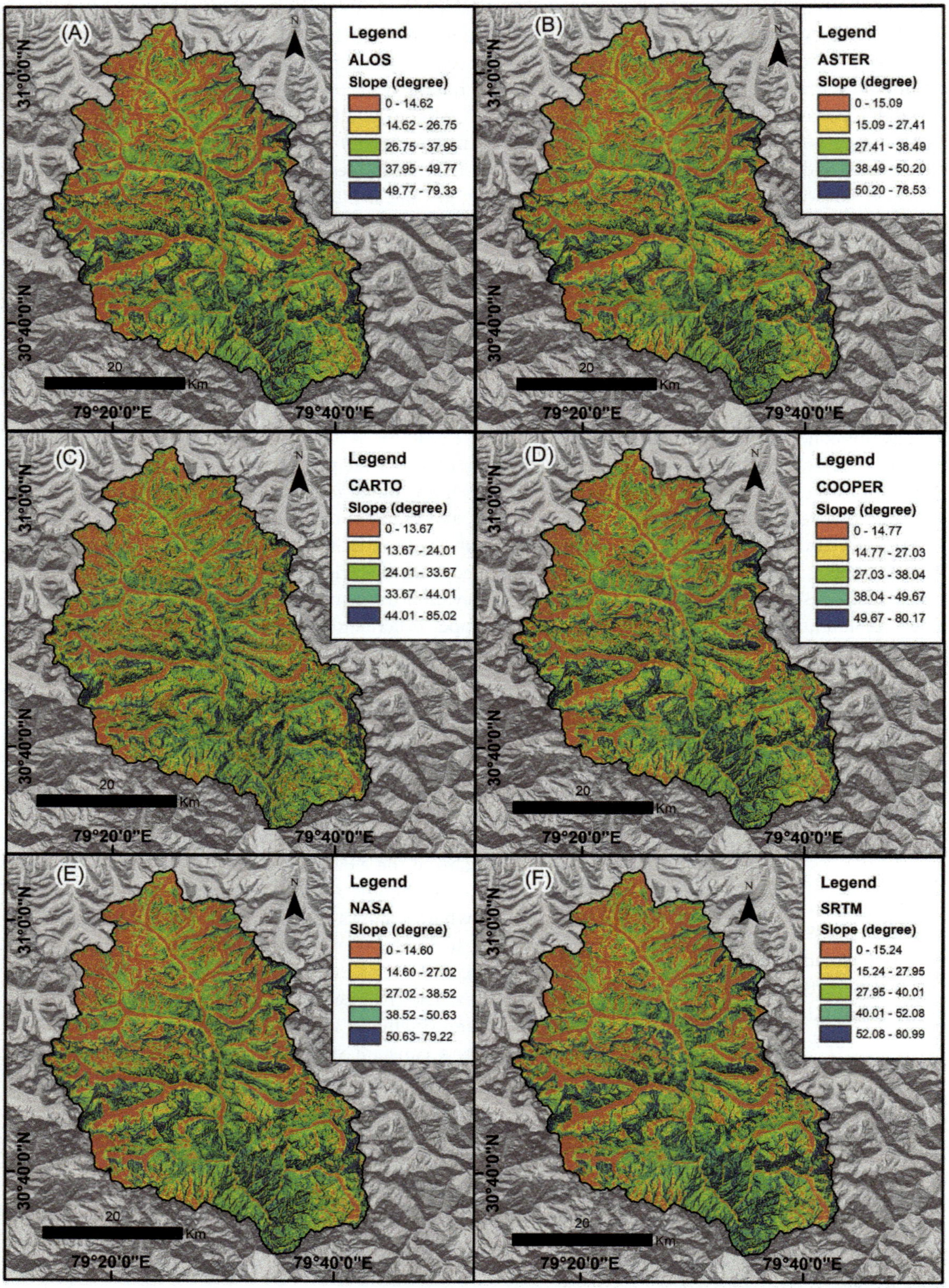


**Figure 19.4** Slope from different DEM: (A) ALOS-PALSAR; (B) ASTER; (C) CARTOSAT; (D) COOPER; (E) NASA; and (F) SRTM.

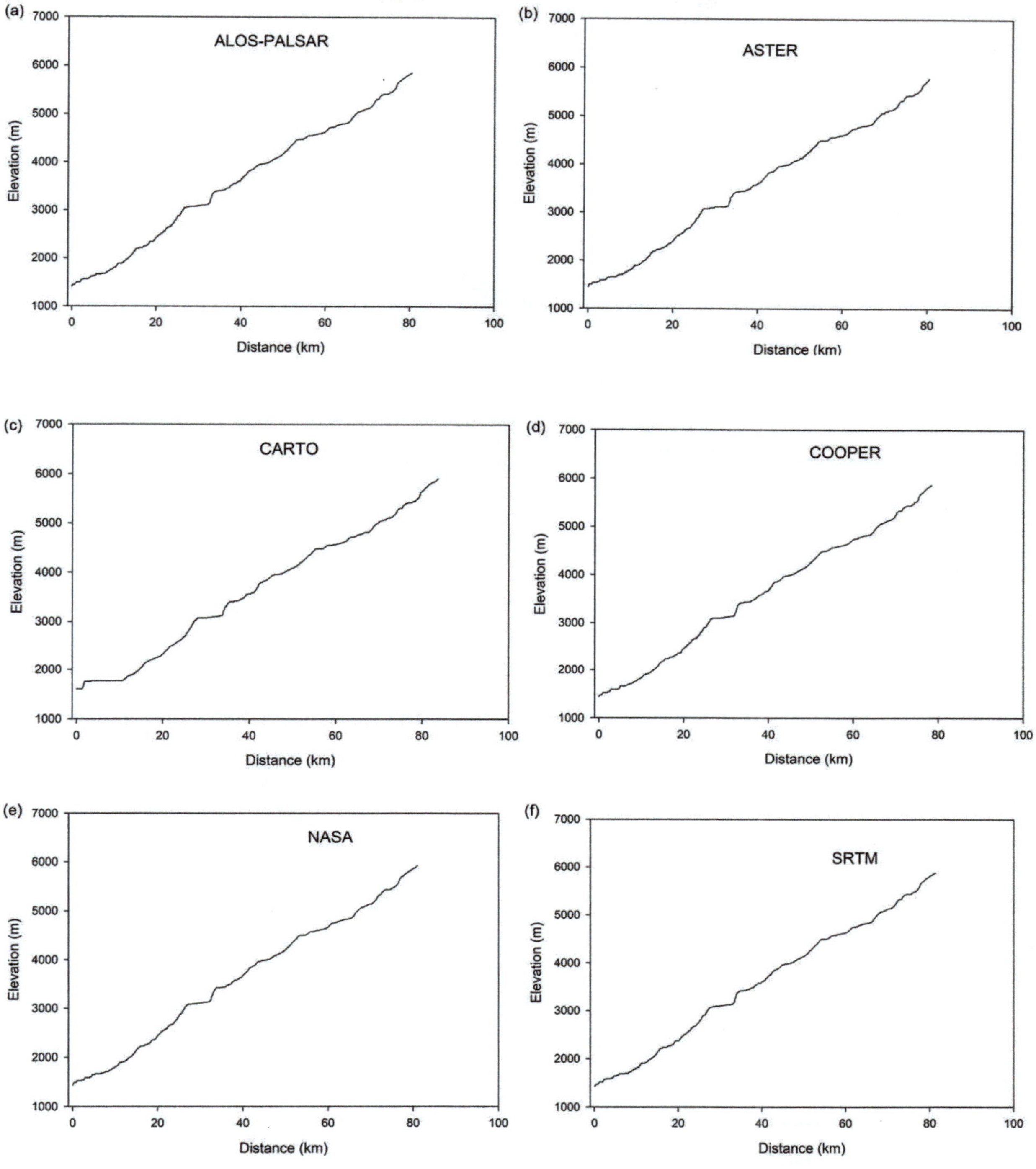


**Figure 19.5** Longitudinal river profile generated from different DEMs.

profile also shows a steady state profile with some minor deviation, but at 33.51 km a major knick point is identified and beyond this region the profile is in steady state. The length of the river using SRTM DEM is 81.35 km that originates at the elevation of 5878 m. The river flows in steady state throughout the session and only shows one major knick point at 34.93 km.

All the profiles generated from various DEMs exhibit that the river is in steady state and one major knick point is identified at between 30 and 35 km. This is considered as evidence of active deformation by tectonics, expected only in an area of rapid rock uplift and fluvial erosion due to tectonics high in that region. In Figure 19.6, we tried to compare the profiles collected from different sources of DEM.

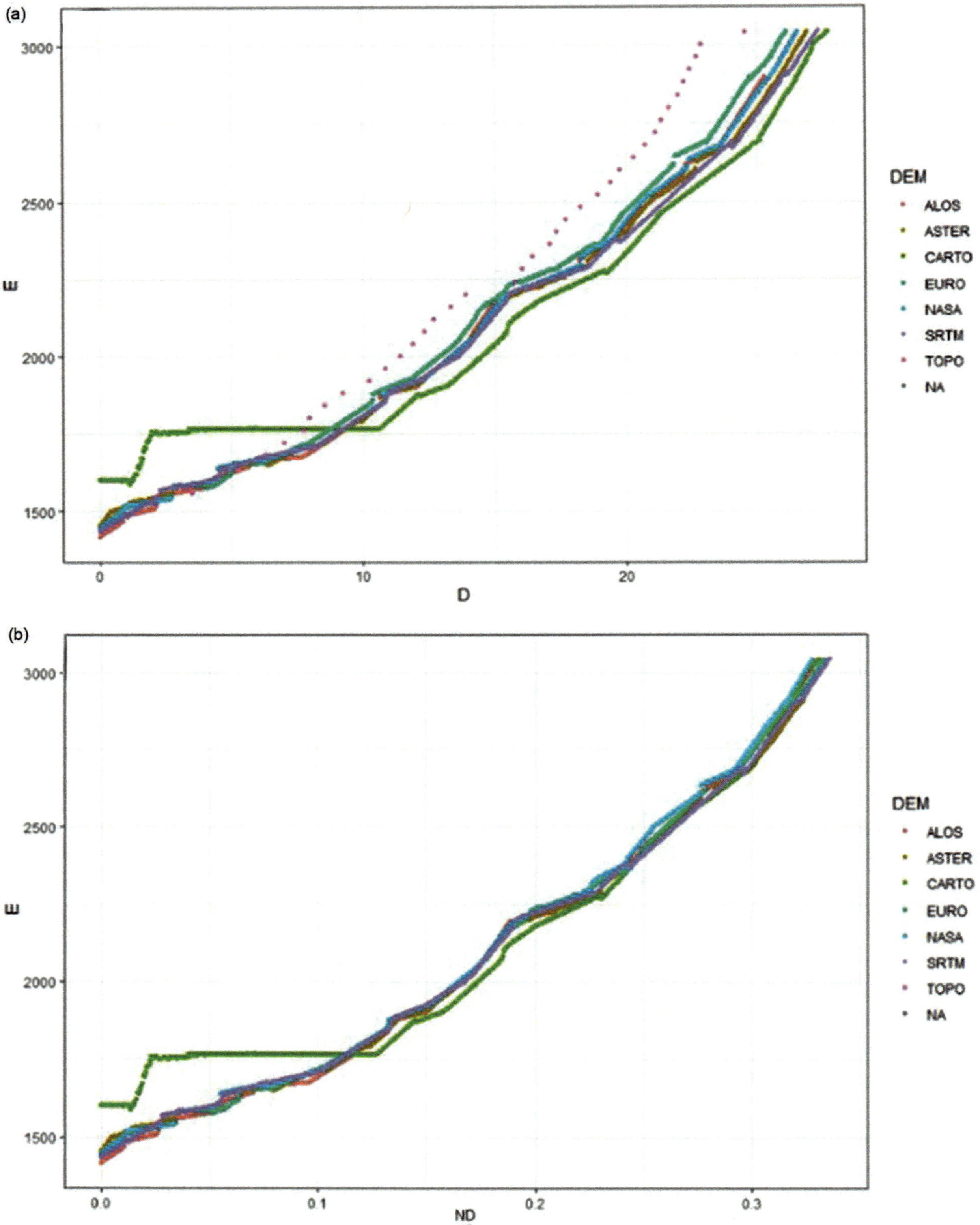


**Figure 19.6** Comparison of River Longitudinal profiles generated from different DEMs plotted at different polynomial orders of geometric progression in MS Excel.

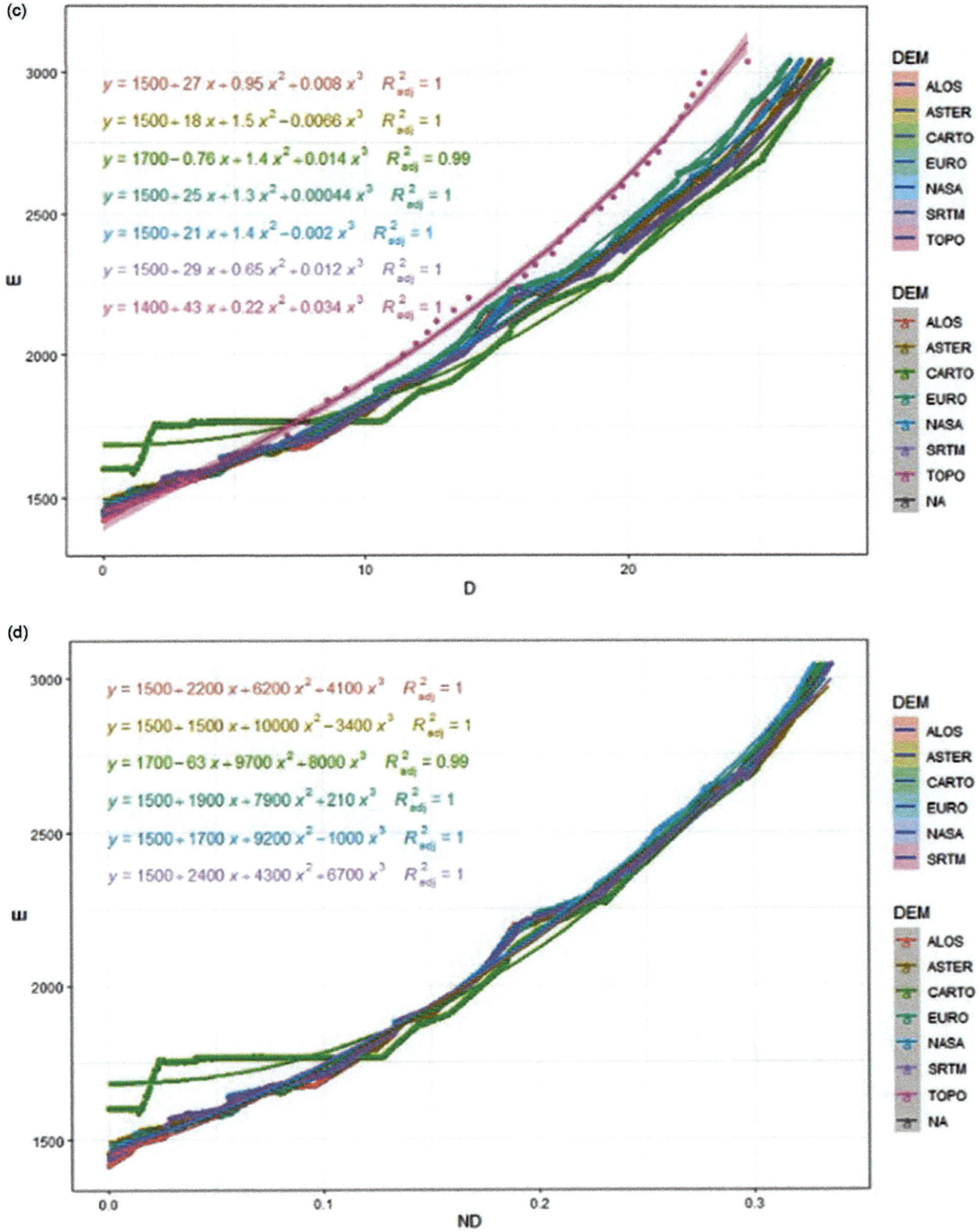


**Figure 19.6** (*Continued*)

(e)

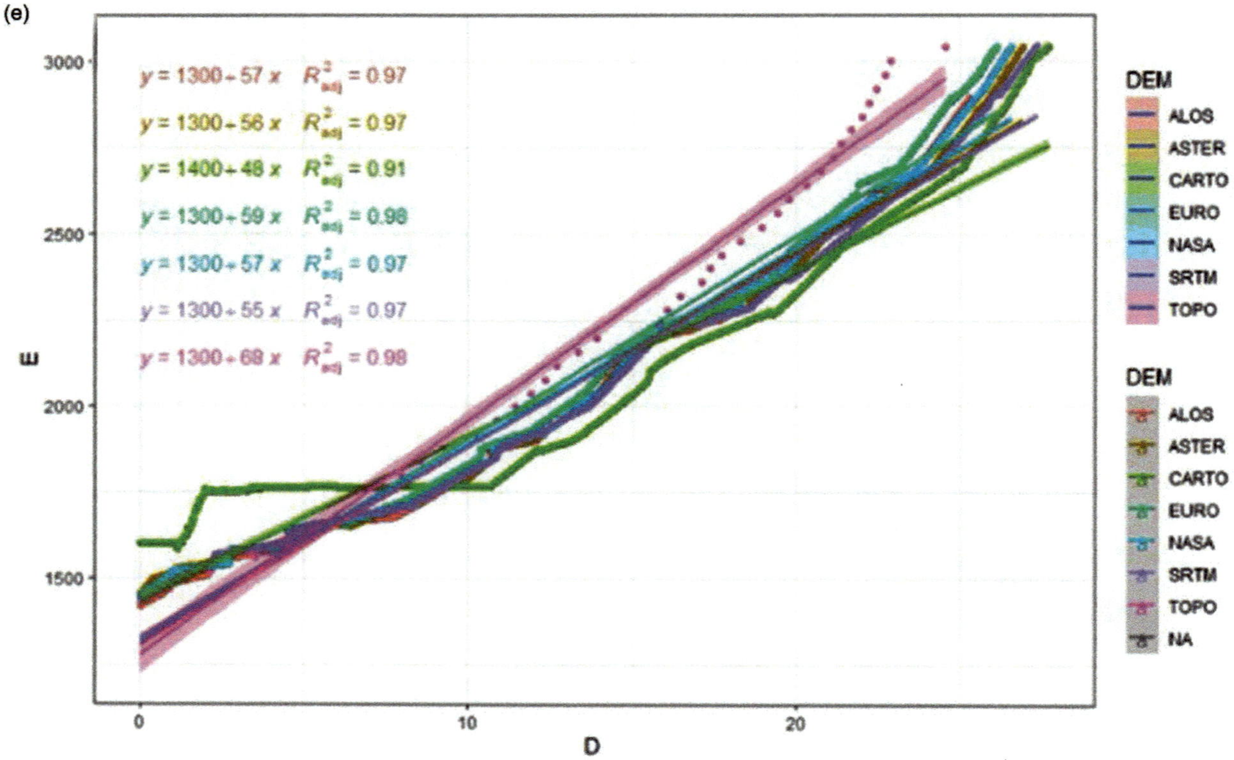


(f)

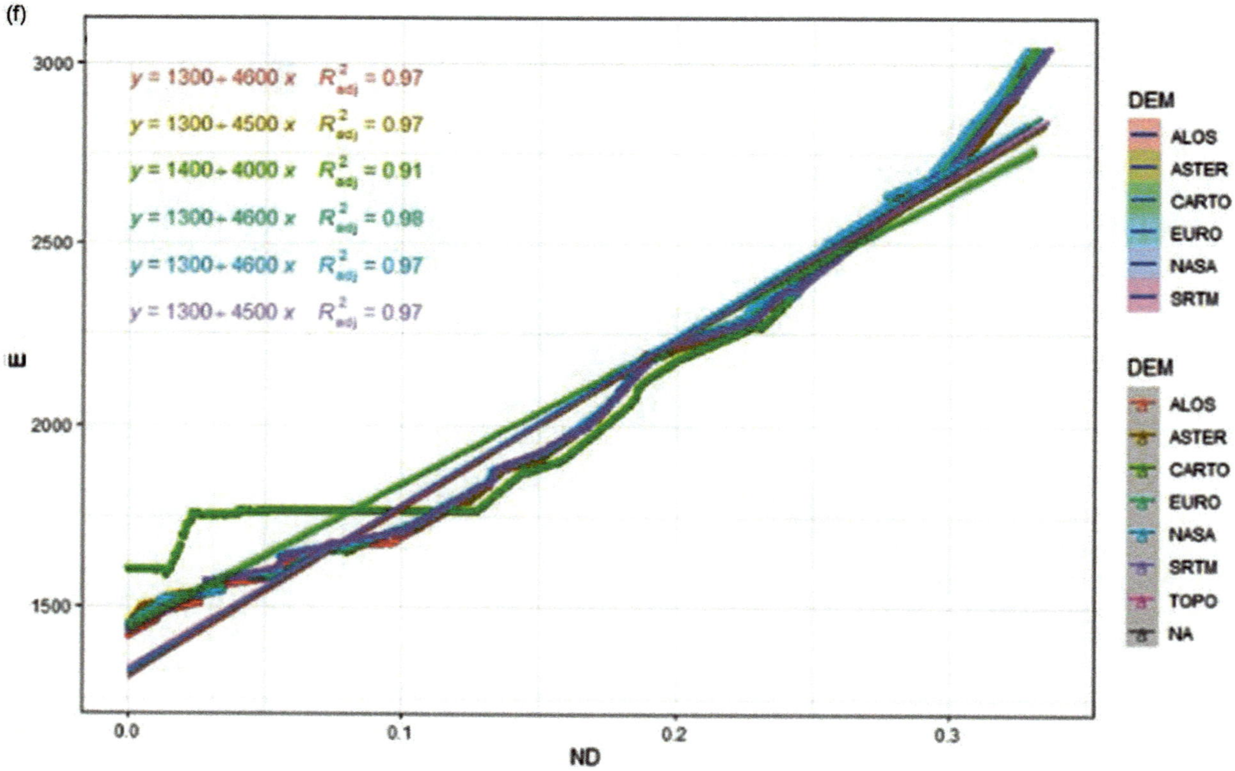


**Figure 19.6** (*Continued*)

## 19.6 Conclusion

Rivers have been identified as the primary agent in the erosional processes that shape landforms on a continent. Furthermore, rivers have always been important to human civilizations, so they are such an important feature. The assessment of river features such as longitudinal profile has become more exact and precise because of advances in spatially generated data and procedures. This study assesses the sensitivity of longitudinal river profiles obtained from ALOS-PALSAR, ASTER, COOPER, CARTOSAT, NASA, and SRTM Digital Elevation Models (DEMs). The extraction of river profiles in the traditional manner is time-consuming, as each step creates data that may or may not be useful for the genre of studies in question. To produce river profiles, LSDTopoTools, an open source software tool that uses a combination of Python and C++, is employed. When compared to toposheet at a scale of 1:50000, the results imply that SRTM provides a more reliable depiction of topographic characteristics. Furthermore, the river profile reveals a large variation, demonstrating that DEMs can be used to study the Himalayan river basin.

## Acknowledgments

Authors thank various sources for providing DEMs at the spatial resolution of 30 m. RD and PK thank the University of Delhi for providing infrastructure. RK acknowledges the Director, School of Sciences, IGNOU and MP thanks Chandigarh University for encouragement.

## References

Adams, B.A., Whipple, K.X., Hodges, K.V. et al. (2016). *In situ* development of high-elevation, low-relief landscapes via duplex deformation in the Eastern Himalayan hinterland, Bhutan. *Journal of Geophysical Research: Earth Surface* 121(2): 294–319.

Ahmad, S., Alam, A., Ahmad, B. et al. (2015). Geomorphic evidence of unrecognised Balapur fault segment in the southwest Kashmir basin of northwest Himalayas. *Geomorphology* 250: 159–172.

Ahmad, S. and Bhat, M.I. (2012). Tectonic geomorphology of the Rambiara basin, SW Kashmir Valley reveals emergent out-of-sequence active fault system. *Himalayan Geology* 33(2): 162–172.

Aiken, S J.B. and Brierley, G.J. (2013). Analysis of longitudinal profiles along the eastern margin of the Qinghai-Tibetan Plateau. *Journal of Mountain Science* 10(4); 643–657.

Anoop, A., Prasad, S., Basavaiah, N. et al. (2012). Tectonic versus climate influence on landscape evolution: a case study from the upper Spiti valley, NW Himalaya. *Geomorphology* 145: 32–44.

Baillie, I.C. and Norbu, C. (2004). Climate and other factors in the development of river and interfluve profiles in Bhutan, Eastern Himalayas. *Journal of Asian Earth Sciences* 22(5): 539–553.

Bhakuni, S.S., Luirei, K., and Kothyari, G.C. (2013). Neotectonic fault in the middle part of Lesser Himalaya, Arunachal Pradesh: a study based on structural and morphotectonic analyses. *Himalayan Geology* 34(1): 57–64.

Bhasin, R., Grimstad, E., Larsen, J.O. et al. (2002). Landslide hazards and mitigation measures at Gangtok, Sikkim Himalaya. *Engineering Geology* 64(4): 351–368.

Bhat, M.A., Dar, T., and Bali, B.S. (2020). Morphotectonic analysis of Aripal Basin in the North-Western Himalayas (India): an evaluation of tectonics derived from geomorphic indices. *Quaternary International* 568: 103–115.

Bishop, M.P., Shroder Jr, J.F., Bonk, R. et al. (2002). Geomorphic change in high mountains: a western Himalayan perspective. *Global and Planetary Change* 32(4): 311–329.

Borgohain, S., Das, J., Saraf, A.K. et al. (2016). Morphodynamic changes of Lohit River, NE India: GIS-based study. *Current Science* 110(9): 1810–1816.

Brookfield, M.E. (1998). The evolution of the great river systems of southern Asia during the Cenozoic India-Asia collision: rivers draining southwards. *Geomorphology* 22(3–4): 285–312.

Burbank, D.W. and Fort, M.B. (1985). Bedrock control on glacial limits: examples from the Ladakh and Zanskar ranges, northwestern Himalaya, India. *Journal of Glaciology* 31(108): 143–149.

Cannon, J.M. and Murphy, M.A. (2014). Active lower crustal deformation and Himalayan seismic hazard revealed by stream channels and regional geology. *Tectonophysics* 633: 34–42.

Chahal, P., Kumar, A., Sharma, C.P. et al. (2019). Late Pleistocene history of aggradation and incision, provenance and channel connectivity of the Zanskar River, NW Himalaya. *Global and Planetary Change* 178: 110–128.

Chakrabarti Goswami, C., Jana, P., and Weber, J.C. (2019). Evolution of landscape in a piedmont section of Eastern Himalayan foothills along India-Bhutan border: a tectono-geomorphic perspective. *Journal of Mountain Science* 16(12): 2828–2843.

Chatterjee, A., Ray, J.S., Shukla, A.D. et al. (2019). On the existence of a perennial river in the Harappan heartland. *Scientific Reports* 9(1): 1–7.

Chug, D., Pathak, A., Indu, J. et al. (2020). Observed evidence for steep rise in the extreme flow of western Himalayan rivers. *Geophysical Research Letters* 47(15): e2020GL087815.

Clark, M K., Schoenbohm, L M., Royden, L.H. et al. (2004). Surface uplift, tectonics, and erosion of eastern Tibet from large-scale drainage patterns. *Tectonics* 23(1): 1–20.

Craddock, W.H., Burbank, D.W., Bookhagen, B. et al. (2007). Bedrock channel geometry along an orographic rainfall gradient in the upper Marsyandi River valley in central Nepal. *Journal of Geophysical Research: Earth Surface* 112(F3): 1–17.

Dar, R.A., Romshoo, S.A., Chandra, R. et al. (2014). Tectono-geomorphic study of the Karewa Basin of Kashmir Valley. *Journal of Asian Earth Sciences* 92: 143–156.

Dar, R.A., Mir, S.A., and Romshoo, S.A. (2019). Influence of geomorphic and anthropogenic activities on channel morphology of River Jhelum in Kashmir Valley, NW Himalayas. *Quaternary International* 507: 333–341.

Das, N., Deka, J.P., Shim, J. et al. (2016). Effect of river proximity on the arsenic and fluoride distribution in the aquifers of the Brahmaputra floodplains, Assam, northeast India. *Groundwater for Sustainable Development* 2: 130–142.

DeCelles, P.G. and Carrapa, B. (2021). Coupled rapid erosion and foreland sedimentation control orogenic wedge kinematics in the Himalayan Thrust Belt of Central Nepal. *Journal of Geophysical Research: Solid Earth* 126(5): e2020JB021256.

Deeken, A., Thiede, R.C., Sobel, E.R. et al. (2011). Exhumational variability within the Himalaya of northwest India. *Earth and Planetary Science Letters* 305(1–2): 103–114.

Devi, R.M., Bhakuni, S.S., and Bora, P.K. (2011). Neotectonic study along mountain front of northeast Himalaya, Arunachal Pradesh, India. *Environmental Earth Sciences* 63(4): 751.

Devi, K., Khatua, K.K., and Das, B.S. (2016). Apparent shear in an asymmetric compound channel. In: *River Flow* 2016, 48–56.

Devrani, R., Singh, V., Mudd, S.M. et al. (2015). Prediction of flash flood hazard impact from Himalayan river profiles. *Geophysical Research Letters* 42(14): 5888–5894.

Divyadarshini, A. and Singh, V. (2017). Identifying active structures in the Chitwan Dun, Central Nepal, using longitudinal river profiles and SL index analysis. *Quaternary International* 462: 176–193.

Divyadarshini, A. and Singh, V. (2019). Investigating topographic metrics to decipher structural model and morphotectonic evolution of the Frontal Siwalik Ranges, Central Himalaya, Nepal. *Geomorphology* 337: 31–52.

Dortch, J.M., Dietsch, C., Owen, L.A. et al. (2011). Episodic fluvial incision of rivers and rock uplift in the Himalaya and Transhimalaya. *Journal of the Geological Society* 168(3): 783–804.

Duarah, B.P. and Phukan, S. (2011). Seismic hazard assessment in the Jia Bhareli river catchment in eastern Himalaya from SRTM-derived basin parameters, India. *Natural Hazards* 59(1): 367–381.

Duncan, C., Masek, J., and Fielding, E. (2003). How steep are the Himalaya? Characteristics and implications of along-strike topographic variations. *Geology* 31(1): 75–78.

Fielding, E., Isacks, B., Barazangi, M. et al. (1994). How flat is Tibet? *Geology* 22(2): 163–167.

Finlayson, D.P., Montgomery, D.R., and Hallet, B. (2002). Spatial coincidence of rapid inferred erosion with young metamorphic massifs in the Himalayas. *Geology* 30(3): 219–222.

Finnegan, N.J., Hallet, B., Montgomery, D.R. et al. (2008). Coupling of rock uplift and river incision in the Namche Barwa–Gyala Peri massif, Tibet. *Geological Society of America Bulletin* 120(1–2): 142–155.

Finnegan, N.J., Roe, G., Montgomery, D.R. et al. (2005). Controls on the channel width of rivers: implications for modeling fluvial incision of bedrock. *Geology* 33(3): 229–232.

Froehlich, W., Gil, E., Kasza, I. et al. (1990). Thresholds in the transformation of slopes and river channels in the Darjeeling Himalaya, India. *Mountain Research and Development* 10(4): 301–312.

Godard, V., Bourlès, D.L., Spinabella, F. et al. (2014). Dominance of tectonics over climate in Himalayan denudation. *Geology* 42(3): 243–246.

Gupta, V. and Sah, M.P. (2008). Impact of the trans-Himalayan landslide lake outburst flood (LLOF) in the Satluj catchment, Himachal Pradesh, India. *Natural Hazards* 45(3): 379–390.

Hodges, K.V., Wobus, C., Ruhl, K et al. (2004). Quaternary deformation, river steepening, and heavy precipitation at the front of the Higher Himalayan ranges. *Earth and Planetary Science Letters* 220(3–4): 379–389.

Imsong, W., Choudhury, S., and Phukan, S. (2016). Ascertaining the neotectonic activities in the southern part of Shillong Plateau through geomorphic parameters and remote sensing data. *Current Science* 110(10): 91–98.

Imsong, W., Choudhury, S., and Phukan, S. (2019). Mid–late Holocene fluvial aggradational landforms and morphometric investigations in the southern front of the Shillong plateau, NE India. *Journal of Earth System Science* 128(5): 1–26.

Imsong, W., Choudhury, S., Phukan, S. et al. (2018). Morphodynamics of the Kulsi River Basin in the northern front of Shillong Plateau: exhibiting episodic inundation and channel migration. *Journal of Earth System Science* 127(1): 1–15.

Iwata, S., Yamanaka, H., and Yoshida, M. (1982). Glacial landforms and river terraces in the Thakkhola regions, central Nepal. *Journal of Nepal Geological Society* 2: 81–94.

Jain, V. and Sinha, R. (2004). Fluvial dynamics of a branching river system in Himalayan foreland basin, Baghmati River, north Bihar plains, India. *Geomorphology* 60(1–2): 147–170.

Jamieson, R., Gordon, R., Joy, D. et al. (2004). Assessing microbial pollution of rural surface waters: a review of current watershed scale modeling approaches. *Agricultural Water Management* 70(1): 1–17.

Jonell, T.N., Carter, A., Böning, P. et al. (2017). Climatic and glacial impact on erosion patterns and sediment provenance in the Himalayan rain shadow, Zanskar River, NW India. *Bulletin* 129(7–8): 820–836.

Joshi, N., Kothyari, G.C., and Pant, C.C. (2021). Drainage conformation and transient response of river system in thrust segmentation of Northwest Himachal Himalaya, India. *Quaternary International* 575: 37–50.

Kaushal, R.K., Sarkar, A., Mishra, K. et al. (2020). Spatio-temporal variability in stream power distribution in the Upper Kosi River basin, Central Himalaya: controls and geomorphic implications. *Geomorphology* 350: 106888.

Kaushal, R.K., Singh, V., Mukul, M. et al. (2017). Identification of deformation variability and active structures using geomorphic markers in the Nahan salient, NW Himalaya, India. *Quaternary International* 462: 194–210.

King, G.E., Herman, F., and Guralnik, B. (2016).Northward migration of the eastern Himalayan syntaxis revealed by OSL thermochronometry. *Science* 353(6301): 800–804.

Kirby, E. and Whipple, K.X. (2001). Quantifying differential rock-uplift rates via stream profile analysis. *Geology* 29(5): 415–418.

Kirby, E. and Whipple, K.X. (2012). Expression of active tectonics in erosional landscapes. *Journal of Structural Geology* 44: 54–75.

Korup, O. (2006). Rock-slope failure and the river long profile. *Geology* 34(1): 45–48.

Korup, O., Montgomery, D.R., and Hewitt, K. (2010). Glacier and landslide feedbacks to topographic relief in the Himalayan syntaxes. *Proceedings of the National Academy of Sciences* 107(12): 5317–5322.

Kothyari, G.C. (2014). Morphometric analysis of tectonically active Pindar and Saryu River basins: central Kumaun Himalaya. *Zeitschrift für Geomorphologie* 59(4): 421–442.

Kothyari, G.C., Joshi, N., Taloor, A.K. et al. (2019). Landscape evolution and deduction of surface deformation in the Soan Dun, NW Himalaya, India. *Quaternary International* 507: 3021–323.

Kothyari, G.C., Kotlia, B.S., Talukdar, R. et al. (2020). Evidences of neotectonic activity along Goriganga River, higher central Kumaun Himalaya, India. *Geological Journal* 55(9): 6123–6146.

Kothyari, G.C., Shukla, A.D., and Juyal, N. (2017). Reconstruction of Late Quaternary climate and seismicity using fluvial landforms in Pindar River valley, Central Himalaya, Uttarakhand, India. *Quaternary International* 443: 248–264.

Kumar, A., Srivastava, P., and Devrani, R. (2020). Using clast geometries to establish paleoriver discharges: testing records for aggradation and incision from the upper Indus River, Ladakh Himalaya. *Geomorphology* 362: 107202.

Lavé, J. and Avouac, J.P. (2000). Active folding of fluvial terraces across the Siwaliks Hills, Himalayas of central Nepal. *Journal of Geophysical Research: Solid Earth* 105(B3): 5735–5770.

Mehta, J.S. and Sanwal, R. (2011). Evidence of active tectonics along oblique transverse normal fault in the Kosi River valley around Betalghat, Kumaun Lesser Himalaya, India. *Current Science* 101(4): 540–543.

Mehta, M., Dobhal, D. P., Shukla, T. et al. (2016). Instability processes triggered by heavy rain in the Garhwal Region, Uttarakhand, India. In: *Climate Change, Glacier Response, and Vegetation Dynamics in the Himalaya*, 219–234. Cham: Springer.

Mishra, K., Sinha, R., Jain, V. et al. (2019). Towards the assessment of sediment connectivity in a large Himalayan river basin. *Science of the Total Environment* 661: 251–265.

Morell, K.D., Sandiford, M., Rajendran, C.P. et al. (2015). Geomorphology reveals active décollement geometry in the central Himalayan seismic gap. *Lithosphere* 7(3): 247–256.

Mudd, S.M., Attal, M., Milodowski, D.T. et al. (2014). A statistical framework to quantify spatial variation in channel gradients using the integral method of channel profile analysis. *Journal of Geophysical Research: Earth Surface* 119(2): 138–152.

Mudd, S.M., Clubb, F.J., Gailleton, B. et al. (2018). How concave are river channels? *Earth Surface Dynamics* 6(2): 505–523.

Munack, H., Korup, O., Resentini, A. et al. (2014). Postglacial denudation of western Tibetan Plateau margin outpaced by long-term exhumation. *Bulletin* 126(11–12): 1580–1594.

Nainwal, H.C., Chaudhary, M., Rana, N. et al. (2007). Chronology of the Late Quaternary glaciation around Badrinath (upper Alaknanda Basin: preliminary observations. *Current Science* 90–96.

Nainwal, H.C., Negi, B.D.S., Chaudhary, M. et al. (2008). Temporal changes in rate of recession: evidences from Satopanth and Bhagirath Kharak glaciers, Uttarakhand, using Total Station Survey. *Current Science* 653–660.

Olen, S.M., Bookhagen, B., Hoffmann, B. et al. (2015). Understanding erosion rates in the Himalayan orogen: a case study from the Arun Valley. *Journal of Geophysical Research: Earth Surface* 120(10): 2080–2102.

Philip, G. and Sah, M.P. (1999). Geomorphic signatures of active tectonics in the Trans-Yamuna segment of the western Doon valley, northwest Himalaya, India. *International Journal of Applied Earth Observation & Geoinformation* 1(1): 54–63.

Pratt-Sitaula, B., Burbank, D.W., Heimsath, A. et al. (2004). Landscape disequilibrium on 1000–10,000 year scales Marsyandi River, Nepal, central Himalaya. *Geomorphology* 58(1–4): 223–241.

Prokop, P., Wiejaczka, Ł., Sarkar, S. et al. (2020). Morphological and sedimentological responses of small stream channels to extreme rainfall and land use in the Darjeeling Himalayas. *Catena* 188, 104444.

Rana, N., Sharma, S., Ali, S.N. et al. (2019). Investigating the sensitivity of glaciers to climate variability since the MIS-2 in the upper Ganga catchment (Saraswati valley), Central Himalaya. *Geomorphology* 346: 106854.

Robl, J., Hergarten, S., and Stüwe, K. (2008). Morphological analysis of the drainage system in the Eastern Alps. *Tectonophysics* 460(1–4): 263–277.

Schumm, S.A. (2005). *River Variability and Complexity, 1–219.* New York: Cambridge University Press.

Seeber, L. and Gornitz, V. (1983). River profiles along the Himalayan arc as indicators of active tectonics. *Tectonophysics* 92(4): 335–367.

Shankar, R., Singh, A.K., Satyam, G.P. et al. (2020). Active tectonics influences in the Satluj river basin in and around Rampur, Himachal Himalaya, India. *Arabian Journal of Geosciences* 13(14): 1–21.

Singh, T. and Jain, V. (2009). Tectonic constraints on watershed development on frontal ridges: Mohand Ridge, NW Himalaya, India. *Geomorphology* 106(3–4), 231–241.

Singh, V. and Tandon, S.K. (2007). Evidence and consequences of tilting of two alluvial fans in the Pinjaur dun, Northwestern Himalayan Foothills. *Quaternary International* 159(1): 21–31.

Sinha, S. and Sinha, R. (2021). Active tectonics, landscape evolution and sediment dynamics in Dehra Dun, Northwest Himalaya inferred from geomorphic indices and GIS tools. *Quaternary International* 585: 55–69.

Sonam, S. and Jain, V. (2017). Geomorphic effectiveness of long profile shape and role of inherent geological controls, Ganga River Basin, India. In: *EGU General Assembly Conference Abstracts*, 15464.

Swarnkar, S., Tripathi, S., and Sinha, R. (2019). Soil erosion and sediment yield modelling of Geomorphological Diverse Western (UGB) and Eastern (KB) River Basins of the Ganga River System, India. In: *Geophysical Research* Abstracts, vol. 21.

Taloor, A.K., Ray, P.K.C., Jasrotia, A.S. et al. (2017). Active tectonic deformation along reactivated faults in Binta Basin in Kumaun Himalaya of north India: inferences from tectono-geomorphic evaluation. *Zeitschriftfür Geomorphologie* 61(2): 159–180.

Thakur, P.K., Garg, R.D., Aggarwal, S.P. et al. (2013). Snow density retrieval using SAR data: algorithm validation and applications in part of North Western Himalaya. *The Cryosphere Discussions* 7(3): 1927–1960.

Thiede, R.C., Bookhagen, B., Arrowsmith, J.R. et al. (2004). Climatic control on rapid exhumation along the Southern Himalayan Front. *Earth and Planetary Science Letters* 222(3–4): 791–806.

Wager, L.R. (1937). The Arun River drainage pattern and the rise of the Himalaya. *The Geographical Journal* 89(3): 239–250.

Wang, W., Gao, Y., Anacona, P.I. et al. (2018). Integrated hazard assessment of Cirenmaco glacial lake in Zhangzangbo Valley, Central Himalayas. *Geomorphology* 306: 292–305.

Wells, N.A. and Dorr Jr., J.A. (1987). Shifting of the Kosi River, northern India. *Geology* 15(3): 204–207.

Whipple, K.X. (2004). Bedrock rivers and the geomorphology of active orogens. *Annu. Rev. Earth Planet. Sci.* 32: 151–185.

Willett, S.D. and Brandon, M.T. (2002). On steady states in mountain belts. *Geology* 30(2): 175–178.

Wohl, E. (2010). *A world of rivers*. In: *A World of Rivers*. Chicago, IL: University of Chicago Press.

Yamanaka, H. and Iwata, S.I. (1982). River terraces along the Middle Kali Gandaki and Marsyandi Khola, central Nepal. *Journal of Nepal Geological Society* 2: 95–111.

Zhang, H., Oskin, M.E., Liu-Zeng, J. et al. (2016). Pulsed exhumation of interior eastern Tibet: implications for relief generation mechanisms and the origin of high-elevation planation surfaces. *Earth and Planetary Science Letters* 449: 176–185.

# 20

# Glacier Ice Thickness Estimation in Indian Himalaya Using Geophysical Methods

## A Brief Review

*Aditya Mishra*[1,*], *Harish Chandra Nainwal*[1], *and R. Shankar*[2]

[1] *Department of Geology, H.N.B. Garhwal University, Srinagar Garhwal, Uttarakhand, India, 246174*
[2] *The Institute of Mathematical Sciences, Chennai, Tamil Nādu, 600113*
* *Corresponding author*

## 20.1 Introduction

The Himalayan cryosphere is a major reservoir of fresh water and is frequently referred to as the "third pole of the world" and "water tower of Asia" (Immerzeel et al., 2010). The glacier and snow-melt water contributes considerably to the runoff of the major river system (i.e., Indus, Ganga, and Bramhaputra) of High Mountain Asia and plays a crucial role in regulating the hydrology of the basins (Thayyen and Gergan, 2010; Lutz et al., 2018; Azam et al., 2021). According to Raina and Srivastava (2008), there are 9575 glaciers with an area of 37,466 $km^2$ in the Indian Himalayan Region (IHR). In contrast, other studies reported that the total glacier area lies between 22,800 and 25,041 $km^2$ in the Himalayan range, exclusive of Karakoram (Bolch et al., 2012; Kulkarni and Karyakarte, 2014) and 40,775 $km^2$ with the glaciers of Karakoram (Bolch et al., 2012). An updated basin-wise compilation of glacier data indicates that the Indus Basin has a maximum glacierized area of 26,530 $km^2$, followed by the Bramhaputra (11,381 $km^2$) and Ganga (8066 $km^2$) basins (Kulkarni et al., 2021). The discrepancy in glacier area has resulted from the differences in datasets used, mapped area and followed methodology, and time period by different researchers (Bolch et al., 2019). The recent studies reveal that most of the glaciers in the Himalayan region are retreating and losing mass at heterogeneous rates (Brun et al., 2017; Azam et al., 2018; Shukla et al., 2020; Kulkarni et al., 2021; Mehta et al., 2021; Kumar et al., 2021). Under the accelerating atmospheric warming, shrinking of the mountain glaciers is assumed to become enhanced in the future (Hock et al., 2019). A model analysis forecasts that if the global temperature rises to 1.5°C above pre-industrial levels, about 60% of the total ice mass of the present day in High Mountain Asia will be lost by the end of this century (Kraaijenbrink et al., 2017).

Ice thickness is one of the fundamental inputs in knowing the volume of the glacier. Glacier volume determines the amount of water available in frozen form, which is important for projecting river runoff and potential sea-level rise under the changing climate scenario (Farrinotti et al., 2019). However, volume estimates in the Himalayan regions are highly uncertain due to limited field-based ice-thickness measurements (Azam et al., 2018; Bolch et al., 2019; Kulkarni et al., 2021). Apart from the field-based ice-thickness measurements, volume can be determined by empirical relations (Chen and Ohmura, 1990; Bahr et al., 1997; Cogley, 2011) and numerical models (Farinotti et al., 2009, 2019; Huss and Farinotti, 2012; Linsbauer et al., 2012; Frey et al., 2014; Gantayat et al., 2014; Ramsankaran et al., 2018).

Similar to the consortium on compilation of a worldwide glacier inventory (Pfeffer et al., 2014), an initiative of a world glacier monitoring system (WGMS) was taken to compile the glacier thickness database (GlaThiDa) at global scale (Gärtner-Roer et al., 2014). The mean and maximum ice-thickness data of glaciers and ice caps are compiled in a systematic way with their metadata and source information (Gärtner-Roer et al., 2014). In the recent version (GlaThiDa v3), 3,854,279 thickness measurements distributed over roughly 3000 glaciers were compiled (Welty et al., 2020). Out of the total, only two glaciers (Chhota Shigri and Dokriani) of the entire HK range are added to the database. In earlier comprehensive reviews (Bolch et al., 2012; Kulkarni et al., 2021), the efforts of

*Advances in Remote Sensing Technology and the Three Poles,* First Edition. Edited by Manish Pandey, Prem C. Pandey, Yogesh Ray, Aman Arora, Shridhar D. Jawak, and Uma K. Shukla.

geophysical methods conducted between the 1960s and 1980s (Raina, 2009) to measure the glacier ice thickness in Himalayan region are overlooked.

In this chapter, we compile the geophysical surveys conducted in the Indian Himalayan region in the past six decades since the 1960s. Moreover, the modern geophysical method (cf. ground penetrating radar) done in the IHR in the last two decades has also been presented with our observations on one of the debris-covered Central Himalayan glacier. An alternative method to measure ice thickness of debris-covered glaciers and increase the ice thickness data base in Indian Himalayan region is suggested.

## 20.2 Geophysical Methods for Estimation of Glacier Ice Thickness

Geophysical techniques are an efficient way to investigate the inaccessible subsurface glacier topography including ice thickness, nature of ice-bed interface, and englacial features without causing damage (Reynolds, 2011; Siegert, 1999; Eisen et al., 2003). These geophysical methods use the physical properties of the substance to be investigated. Below are the most often applied geophysical methods in glaciology.

### 20.2.1 Gravity

Newton's Law of Gravitation is the foundation of gravity surveying. This method investigates the variations in the gravitational field arising from the differences in density of subsurface materials (Zeng and Wan, 2004; Nabighian et al., 2005). An uneven distribution of underground mass causes localized perturbations in the gravitational field, referred to as a gravity anomaly (Chapin, 1996). These variations are measured with the help of instruments and finally corrections to the gravity data are applied (Bull and Hardy, 1956; Hackney and Featurestone, 2003).

### 20.2.2 Magnetic

This magnetic geophysical method senses the magnetic field of the subsurface earth material and its variability over the surface. It is measured with different types of magnetometers. In glaciology, this method is not particularly useful for glacier ice thickness because of the non-magnetic nature of the ice bodies. While determining the subsurface geology, magnetism of the rocks play an important role. In estimation of glacier ice thickness, the results of the magnetic survey are inferred. The magnetic method can be conducted together with a gravity survey as the physical properties used in these two methods do not influence each other.

### 20.2.3 Resistivity

The electric resistivity method is based on the measurement of the potential produced by the direct (or low-frequency alternating) current introduced into the ground. This method involves production of an electric current into the ground between two current electrodes and simultaneously measuring the electric field between the two different electrodes, known as potential electrodes (Rothlisberger, 1967). The "current" and "potential" electrodes can be arranged in 102 different configurations (Szalai and Szarka, 2008); Schlumberger and Wenner are two well-known electrode configurations used in the electric resistivity method (Reynolds, 2011).

### 20.2.4 Seismic

The principle of seismic methods is based on elastic waves traveling through different materials at different velocities. The propagation of seismic waves through layered ground is determined by the reflection and refraction of the waves at the layer interface. When the wave reaches the interface, some energy is refracted into the deeper layer, while the reflected wave directs energy back into the upper layer (Schrott and Sass, 2008). A strong benefit of seismic methods over other geophysical methods is the penetration depth.

### 20.2.5 Ground Penetrating Radar

Ground Penetrating Radar is one of the efficient geophysical method for glaciological applications that utilizes the frequency range between 1 and 1000 MHz of the electromagnetic spectrum to gather subsurface glacier information (Plewes and Hu, 2001; Annan, 2002). The radiowaves propagate through the subsurface materials of different compositions having different dielectric properties and reflected from the discontinuities. These discontinuities are used to demarcate the subsurface features (Daniels, 2007). The penetration of the waves is the function of frequency; the lower the frequency, the deeper the penetration and vice-versa (Van Dam, 2012). The limited penetration of radiowaves in the moist medium is one of the demerits of GPR surveys.

## 20.3 Geophysical Methods in the Indian Himalaya Region

The Geological Survey of India (GSI) is the chief organization that systematically initiated the glaciological studies by selecting several glaciers in the Indian Himalayan region (Raina, 2009). The idea of measuring ice thickness through geophysical methods was conceived during the

International Geophysical Year, 1957. It took about 8 years to covert the idea into reality and during the year 1965, the Zemu Glacier in the Eastern Himalaya witnessed the first ever seismic surveys in the Himalayas (Bose et al., 1971). The estimated ice thickness near the snout region (lower ablation zone) was ~200 m.

After a decade, the Gara Glacier in the Western Himalaya was surveyed through seismic and resistivity methods by GSI (Table 20.1; GSI, 1998). The thickness of glacier ice was estimated to be 27–269 m based on the velocity of seismic waves and resistivity of the medium. Moreover, the thickness of the ice layer was confirmed by a higher resistivity than the layer above and below it. Subsequently, the thickness of Nehnar, Gorgarang, and Shaunegarang glaciers, in the Western Himalaya, were measured using the above described seismic and resistivity methods by GSI (Table 20.1; GSI, 1998). In Shaunegarang Glacier, the magnetic survey through the fluxgate magmatic method was also done in two longitudinal traverses. The observed ice thickness of the glacier varies from 16–80 m along the central flow-line.

In continuation, Tipra Glacier, Alaknanda Basin, Central Himalaya was surveyed using resistivity and magnetic methods during the field season of 1981 and resistivity and seismic in 1982 (GSI, 1992). The ice thickness could be interpreted only in two traverses in the ranges of 50–100 m and 40 to 80 m. Dunagiri Glacier, located in parts of the Dhauliganga Basin, was the second glacier after Tipra in the Central Himalaya, where geophysical investigations were carried out in August and September of 1986 (GSI, 1987). The results of two resistivity soundings show that ice thickness varies from 30 to 60 m and 80 to 120 m. The surveys conducted by GSI are tabulated in Table 20.1. An independent thickness estimation has been done for Chhota Shigri Glacier, Western Himalayas, through the gravity method (Dobhal et al., 1995). In this survey, 15 cross profiles were taken. The thickness of Chhota Shigri Glacier was estimated in the range of 15–130 m from the snout to accumulation zone.

The use of radio waves in demarcating the glacier thickness and other subglacial features started back to the 1950s in the Polar Regions (Robin, 1975; Annan, 2002). A recent compilation shows data acquisition, processing, and applications of radar waves in glaciological applications which has been done in the past five decades (Schroeder et al., 2020). The modern GPR has been commercially available since the 1970s and has undergone technological up-grade and system modifications (Annan and Davis, 1978). However, an effort to collect radar data to estimate glacier thickness in the Indian Himalaya started much later. The Dokriani glacier, Central Himalaya was the first glacier in Indian Himalaya, which saw the GPR survey in 1995 (Gergan et al., 1999). The results of GPR surveys having a central frequency of 12 MHz, show that the thickness varies from 15 to 25 m at the frontal part and 120 m in the zone of accumulation. Singh et al. (2010) have made an attempt to estimate the thickness of Patseo and Samudra Tapu glaciers using different frequency antennas of 50, 100, and 500 MHz. The depth of the Patseo Glacier at a single location was estimated to be 40 m, but thickness of the Samudra Tapu Glacier could not be measured using 50 and 100 MHz frequency antennae. The Chhota Shigri Glacier in Western Himalaya has seen two different GPR surveys with different antenna frequencies. In one of the studies, Singh et al. (2012), conducted a GPR survey using 16 MHz frequency antennas and showed that the thickness varies from 110 to 150 m for the surveyed locations of the glacier. Another study used low-frequency antenna (4.2 MHz) to map the bedrock of the same glacier (Azam et al., 2012). The total five transects at various elevation zones across the glacier were taken, the most successful attempt where the maximum number of transverse transects was covered. The reason may be a debris-free glacier and the use of

**Table 20.1** Details of geophysical investigations.

| S. No. | Glacier | State and Region | Survey Year | Geophysical Method (No. of Profiles) | Estimated Thickness (m) |
|---|---|---|---|---|---|
| 1 | Zemu | Sikkim, EH | 1965 | S (3) | 125–300 |
| 3 | Gara | HP, WH | 1975–76 | S (8), R (8) | 27–269 |
| 2 | Neh-Nar | J and K, WH | 1977 | S (6), R (7) | 45–160 |
| 4 | Gorgarang | HP, WH | 1978 | S (6), R (5) | 14–45 |
| 5 | Shaunegarang | HP, WH | 1983 | R (13), M (2) | 16–80 |
| 6 | Tipra | UK, CH | 1981–82 | S (5), R (2) | 40–200 |
| 7 | Dunagiri | UK, CH | 1986 | R (16), M (3) | 30–120 |
| 8. | Chhota Shigri | HP, WH | 1988 | G (15) | 15–130 |

(S: seismic; R: resistivity soundings; M: magnetic; G: gravity) in Indian Himalayan Regions (EH: Eastern Himalaya; WH: Western Himalaya; aCH: Central Himalaya). HP: Himachal Pradesh; J and K: Jammu and Kashmir; UK: Uttarakhand.

lower-frequency antenna in which signal dispersion and attenuation are relatively low. Swain et al. (2018) measured the depth of the Hamtah and Parang glaciers using 100 and 200 MHz antennas. The ice thickness of 40 to 140 m and 30 to 95 m has been calculated for the Parang and Hamtah glaciers, respectively. A GPR survey has been done on Saser La Glacier in the Karakoram Range (Singh et al., 2019) in point measuring mode. The thickness varies from 27 to 43 m at different locations near the frontal parts. Recently, the Menthosa Glacier, Western Himalaya observed the thickness through radar survey with a central frequency of 25 MHz (Prakash et al., 2021).

From the above studies, it is clear that the number of GPR surveys are very limited and sparse. In parts of the Central and Western Himalaya, only two and six glaciers respectively have been mapped by direct field-based radio techniques. We have summarized and tabulated (Table 20.2) all the glaciers in the Indian Himalaya where direct field-based GPR surveys have been successfully conducted. The locations of GPR surveyed glaciers in the Himalayan region are also shown in Figure 20.1.

## 20.4 GPR Surveys in the Debris Covered Glaciers

The majority of the glaciers in the Himalaya are covered by debris or supraglacial moraines in their ablation zones (Herried and Pelliccotti, 2018). The extent of debris over the glacier depends primarily on avalanches, bared area of valley walls in the accumulation zones, and their slopes (Scherler et al., 2011; Laha et al., 2017). Furthermore, debris thickness varies considerably from one location to another over a glacier, with typically more than a meter near the lower ablation zone and only a few centimeter in the upper reaches (Pratap et al., 2016; Sharma et al., 2016; Shah et al., 2019; Rounce et al., 2021). Of course, there are certain exceptions where large boulders sized at 5–10 m are present separately or in clusters. Glacier ice beneath the thick (>1 cm) debris retards melting through its insulating effects, while thin debris (<1 cm) accelerates surface-ice melting, even at higher altitudes (Pratap et al., 2016; Shah et al., 2019). This is due to the fact that the debris-covered glaciers respond differently than the clean-ice or debris-free

**Table 20.2** Details of ground penetrating radar (GPR) surveys in Indian Himalayas.

| S. No. | Glacier | Survey Year | Operating Frequency | Profile Length (km) | Profiles or Point Observations | Estimated Thickness (m) | Study |
|---|---|---|---|---|---|---|---|
| 1 | Dokriani | 1995 | 12.5 | 5.00 | 11 | 15–120 | (Gergan et al., 1999) |
| 2 | Patseo | 2004 | 50 | 0.08 | 2 | 40 | (Singh et al., 2010) |
| | | 2013 | 40 | – | 1 | 39 | (Singh et al., 2018) |
| | | 2017 | 35 | 0.93 | 4 | 36–103 | (Kumari et al., 2021) |
| 3 | Chhota Shigri | 2010 | 16 | 0.40 | 3 | 110–150 | (Singh et al., 2012) |
| | | 2012 | 4.2 | 4.20 | 5 | 124–270 | (Azam et al., 2012) |
| 4 | Satopanth | 2016 | 16 & 20 | 5.00 | 30 | 38–112 | (Mishra et al., 2018) |
| 5 | Hamtah | 2017 | 16 | 5.70 | 44 | 35–95 | (Swain et al., 2018) |
| 6 | Parang | 2017 | 100 | 1.30 | 20 | 40–140 | (Swain et al., 2018) |
| 7 | Saser La | 2018 | 40 | – | 17 | 28–62 | (Singh et al., 2019) |
| 8 | Menthosa | 2016–17 | 25 | 0.60 | 6 | 24–55 | (Prakash et al., (2021) |

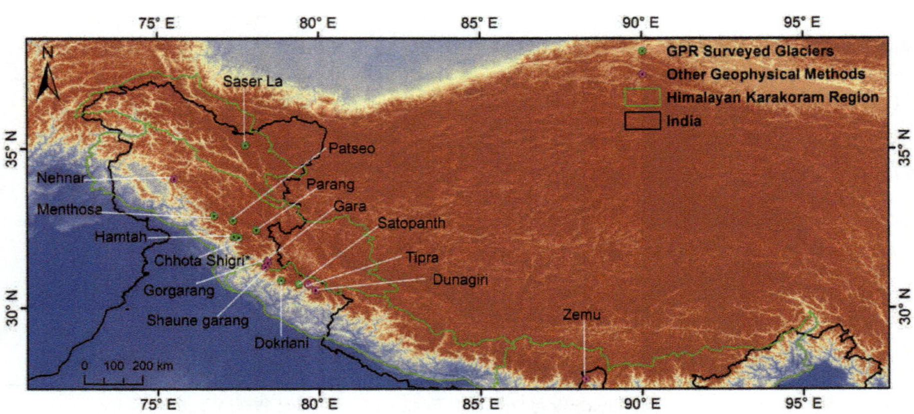


**Figure 20.1** Glaciers of Indian Himalaya where field-based glacier ice-thickness measurements have been carried out. *Source*: Google LLC.

glaciers (Scherler et al., 2011; Dobhal et al., 2013, 2021; Banerjee & Shankar, 2014).

The supraglacial debris does not only alter the surface processes and glacier mass balance (Brun et al., 2019), but also poses problems in acquiring geophysical (GPR) data due to the rough surface and presence of boulders (Figures 20.2b). To obtain deeper bedrock reflections, usually low-frequency antennas are preferred due to their high energy and wavelength and thus higher penetration in the medium (Robin, 1976) while the low frequency GPR antennas have long arms of about 2–10 m, and so cannot be properly coupled with the debris-mantled glacier surface. Moreover, uneven surfaces cause elevation differences between transmitting and receiving antennas; one to be higher or lower in elevation than the others. This causes the positional error in GPR data. In addition, the high-frequency (100–200 MHz) monostatic GPR systems in continuous mode with survey mode are probably not possible in debris-covered glaciers. Such system could be run on debris-free or sparsely-distributed smals fragments (10–20 cm) of rocks. Debris also transports englacially and poses problem in radar data interpretation due to overmasking of the bedrock.

The Himalayan glaciers, being temperate in nature, have enough water content subglacially to dampen the propagation of radio waves in the glacier ice as compared to the dry regions like Antarctica.

In view of the above, there are only two glaciers in Central Himalaya (Dokriani and Satopanth) with high debris cover fractions where ice thickness could be estimated. The debris distribution and behavior of Himalayan glaciers can be considered for low success rates of GPR surveys in the region (Gades et al., 2000). There may several reason but the following three may be considered as the most prominent, as in temperate glaciers, the water content is high compared to cold glaciers. Consequently, the radio wave dissipation is more. Second, in the regions of thick supra-glacial debris, the surface is extremely uneven and so the coupling between the antenna and the ground could be poor. Third, in debris-covered glaciers, there is a lot of englacial debris. This causes many reflections which weaken the signal due to scattering and could also mask bedrock reflections.

## 20.5 A Case Study on Debris-Covered Satopanth Glacier

Satopanth Glacier is a ~13 km-long well studied debris-covered compound valley glacier in the Central Himalaya, located at the head of Alaknanda River, Uttarakhand. The glacier lies in the elevation range between 3860 and 6200 m a.s.l. and occupies the debris covered area of ~11 $km^2$ (60%), out of the total ~18 $km^2$ glacier area. The recessional history shows that Satopanth Glacier retreated at a rate of ~6 m/year in the past 56 years, during 1936 to 2013 (Nainwal et al., 2016). Avalanches are the major contributor in the annual mass balance of the Satopanth Glacier (Laha et al., 2017). The debris of variable thickness is distributed over the glacier with a maximum thickness being 1.5 m near the glacier front (Shah et al., 2019).

Detailed GPR surveying has been conducted on the Satopanth Glacier during the ablation period of 2016 and 2017 (Mishra et al., 2018, 2021). The number of profiles (~30) were taken through the Geophysical Survey System Inc. (GSSI)-make GPR with a subsurface interface radar (SIR)–4000 control unit and multiple low frequency, bistatic, unshielded antennae. The transects (both along and across) were acquired in a thickly debris-covered region via 16 and 20 MHz frequency antenna. These GPR profiles were distributed in the elevations between 3900 and 4600 m a.s.l.

After processing of field acquired data with specified software (i.e., RADAN), we could distinguish the glacier bedrock from the overlying ice in radargrams at only three sites of the Satopanth Glacier. The detailed results are

**Figure 20.2** (a) Debris-covered glacier: Satopanth Glacier in the Central Himalaya. Photo taken in 2020; (b) GPR survey over the debris-covered parts of the Satopanth Glacier at an altitude of 4400 m a.s.l.

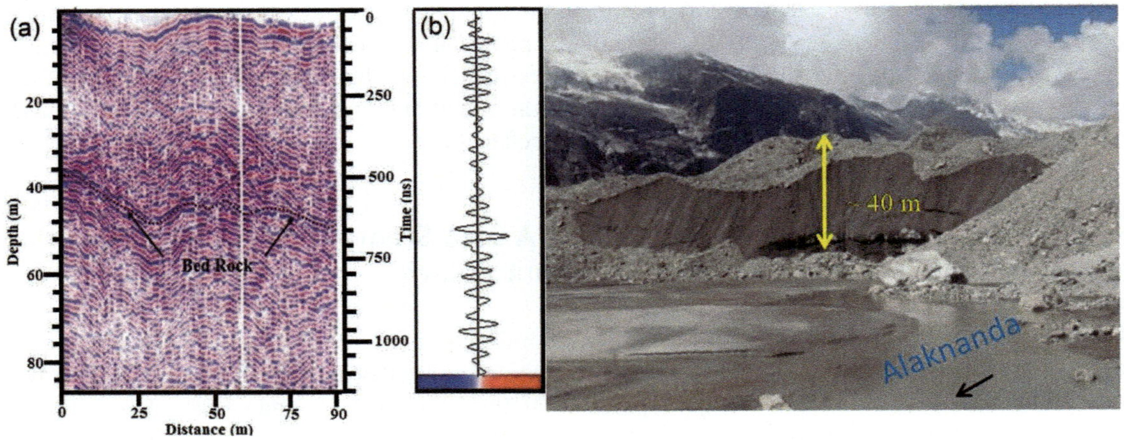


Figure 20.3 (a) Radargram (GPR profile) near the front of Satopanth Glacier Mishra et.al [2017] / Cambridge Univerisity Press / CC BY 4.0; (b) Corresponding to radargram ~40 m-thick ice cliff exposed at the glacier front. https://www.researchgate.net/profile/Aditya-Mishra-6/publication/324121343_Estimation_of_Ice_Thickness_of_the_Satopanth_Glacier_Central_Himalaya_Using_Ground_Penetrating_Radar/links/5abee08e0f7e9bfc0459adc6/Estimation-of-Ice-Thickness-of-the-Satopanth-Glacier-Central-Himalaya-Using-Ground-Penetrating-Radar.pdf.

presented in Mishra et al. (2018, 2021). The first location was near the glacier front where thickness was estimated to be 38–50 m (Figure 20.3a). The same was established through the exposed ice cliff at the glacier front in the proximity of the GPR profile (Figure 20.3b). The two other GPR profiles in which the ice–bedrock interface could be marked were located 9 km up glacier at an elevation of 4600 m a.s.l. During the data processing, we observed several reflections in the profile but demarcated the bedrock showing the highest and continuous reflections. Several other reflections were identified as englacial due to their hyperbolic shape and discontinuity.

Despite the thick debris over the temperate glacier ice, as discussed in section 20.3, we could mark the bedrock in 3 GPR profiles out of 35, possibly due to the following reasons. In the identified location (namely near the glacier snout) bedrock was located at a lower depth while in the two up-glacier profiles, debris thickness was relatively less compared to the other localities. Moreover, timing (diurnal and seasonal) of the data acquisition was found to be another important factor in temperate glaciers. Early morning (before sunrise) and late ablation period (October) recorded suitable GPR data in debris-covered glaciers, probably due to the least ice melting.

## 20.6 Conclusions and Future Prospects

All available field acquired thickness data in the Indian Himalaya region have not been paid attention in the previous global and regional glacier data compilations. This leads to the writing of the present chapter to compile all datasets. There are 15 glaciers in the IHR, where approximately 200 ice-thickness datasets were taken with the help of different geophysical methods. However, in view of the highest glaciated area the after poles, these are significantly lower than that of the other glaciated mountain ranges.

It is clear that many more direct measurements of ice volume of the glaciers are required to improve the reliability of the ice-volume estimates at basin and regional scales. Observations at these scales can only be done using remote-sensing methods, which only observe the glacier surface. The ice thickness and consequently the ice volumes have to inferred from this data using models. Direct field measurements are required to validate these models and choose one over the others.

Helicopter-borne ice radar systems can be deployed in the glaciers of the rugged Himalayan terrain to increase the number of field-based ice thickness data. Pritchard et al. (2020) proved the suitability of such radar systems in the debris-covered glaciers of Nepal Himalayas. They modified the dipole frequency antennas and developed a heli-borne radar system that could operate with low frequency antennas to gain deeper signal penetration over debris-covered and temperate glaciers. These airborne surveys take relatively less time to cover the entire glacier as compared to ground surveys.

## Acknowledgments

The GPR survey over Satopanth Glacier was funded by the Science & Engineering Research Board (SB/DGH-101/2015) and Department of Science and Technology (DST/CCP/NHC/158/2018). We are grateful to the editors, especially Dr. Manish Pandey and two anonymous reviewers, for their fruitful suggestions to improve the content. The Geological Survey of India (GSI) is acknowledged for making the reports available on its portal (www.gsi.gov.in).

## References


Annan, A.P. (2002). GPR: history, trends, and future developments. *Subsurface Sensing Technologies and Applications* 3(4): 253–270.

Annan, A.P. and Davis, J.L. (1978). Methodology for radar transillumination experiments: report of activities. *Geological Survey of Canada,* Paper 78-1B, 107–110.

Arendt, A., Echelmeyer, K., Harrison, W. et al. (2006). Updated estimates of glacier volume changes in the western Chugach Mountains, Alaska, and a comparison of regional extrapolation methods. *Journal of Geophysical Research: Earth Surface* 111(F3).

Azam, M.F., Wagnon, P., Berthier, E. et al. (2018). Review of the status and mass changes of Himalayan–Karakoram glaciers. *Journal of Glaciology* 64(243): 61–74.

Azam, M.F., Wagnon, P., Ramanathan, A. et al. (2012). From balance to imbalance: a shift in the dynamic behaviour of Chhota Shigri Glacier, western Himalaya, India. *Journal of Glaciology* 58(208): 315–324.

Azam, M.F., Kargel, J.S., Shea, J.M. et al. (2021). Glaciohydrology of the Himalaya-Karakoram. *Science* 373(6557).doi:10.1126/science.abf3668

Bahr, D.B., Meier, M.F., and Peckham, S.D. (1997). The physical basis of glacier volume-area scaling. *Journal of Geophysical Research: Solid Earth* 102(B9): 20355–20362.

Banerjee, A. and Shankar, R. (2013). On the response of Himalayan glaciers to climate change. *Journal of Glaciology* 59(215): 480–490.

Bolch, T., Shea, J.M., Liu, S. et al. (2019) *The Hindu Kush Himalaya Assessment*. Springer International Publishing. doi:10.1007/978-3-319-92288-1

Bolch, T., Kulkarni, A., Kääb, A. et al. (2012). The state and fate of Himalayan glaciers. *Science* 336(6079): 310–314.

Bose, R.N., Dutta, N.P., and Lahiri, S.M. (1971). Refraction seismic investigation at Zemu Glacier, Sikkim. *Journal of Glaciology* 10(58): 113–119.

Brun, F., Berthier, E., Wagnon, P. et al. (2017). A spatially resolved estimate of High Mountain Asia glacier mass balances from 2000 to 2016. *Nature Geoscience* 10(9): 668–673.

Brun, F., Wagnon, P., Berthier, E. et al. (2019). Heterogeneous influence of glacier morphology on the mass balance variability in High Mountain Asia. *Journal of Geophysical Research: Earth Surface* 124(6): 1331–1345.

Bull, C. and Hardy, J.R. (1956). The determination of the thickness of a glacier from measurements of the value of gravity. *Journal of Glaciology* 2(20): 755–763. http://doi.org/10.3189/S0022143000025028.

Chaohai, L. and Sharma, C.K. (1988). Report on first expedition to glaciers in the Pumqu (Arun) and Poiqu (Bhote-Sun Kosi) river basins, Xizang (Tibet), China.

Chapin, D.A. (1996). The theory of the Bouguer gravity anomaly: a tutorial. *The Leading Edge* 15(5): 361–363.

Chen, J. and Ohmura, A. (1990). Estimation of alpine glacier water resources and their change since the 1870s. *IAHS publ.* 193: 127–135.

Cogley, J.G. (2011). Present and future states of Himalaya and Karakoram glaciers. *Annals of Glaciology* 52(59): 69–73.

Daniels, D.J. (2007). An assessment of the fundamental performance of GPR against buried landmines. In: *Detection and Remediation Technologies for Mines and Minelike Targets XI*, vol. 6553, 103–117). SPIE.

DeBeer, C.M. and Sharp, M.J. (2007). Recent changes in glacier area and volume within the southern Canadian Cordillera. *Annals of Glaciology* 46: 215–221.

Dobhal, D.P., Kumar, S., and Mundepi, A.K. (1995). Morphology and glacier dynamics studies in monsoon–arid transition zone: an example from Chhota Shigri Glacier, Himachal-Himalaya, India. *Current Science* 936–944.

Dobhal, D.P., Mehta, M., and Srivastava, D. (2013). Influence of debris cover on terminus retreat and mass changes of Chorabari Glacier, Garhwal region Central Himalaya, India. *Journal of Glaciology* 59(217).

Dobhal, D.P., Pratap, B., Bhambri, R. et al. (2021). Mass balance and morphological changes of Dokriani Glacier (1992–2013), Garhwal Himalaya, India. *Quaternary Science Advance* 4. https://doi.org/10.1016/j.qsa.2021.100033.

Eisen, O., Nixdorf, U., Keck, L. et al. (2003). Alpine ice cores and ground penetrating radar: Combined investigations for glaciological and climatic interpretations of a cold Alpine ice body. *Tellus 55B*: 1007–1017.

Farinotti, D., Brinkerhoff, D.J., Clarke, G.K. et al. (2017). How accurate are estimates of glacier ice thickness? Results from ITMIX, the Ice Thickness Models Intercomparison eXperiment. *The Cryosphere* 11(2): 949–970.

Farinotti, D., Huss, M., Bauder, A. et al. (2009). A method to estimate the ice volume and ice-thickness distribution of alpine glaciers. *Journal of Glaciology* 55(191): 422–430.

Farinotti, D., Huss, M., Fürst, J.J. et al. (2019). A consensus estimate for the ice thickness distribution of all glaciers on Earth. *Nature Geoscience* 12(3): 168–173.

Frey, H., Machguth, H., Huss, M. et al. (2014). Estimating the volume of glaciers in the Himalayan–Karakoram region using different methods. *The Cryosphere* 8(6): 2313–2333.

Gades, A., Conway, H., Nereson, N. et al. (2000). Radio echo-sounding through supraglacial debris on Lirung and Khumbu Glaciers, Nepal Himalayas. IAHS Publication 264: 13–24.

Gantayat, P., Kulkarni, A.V., and Srinivasan, J. (2014). Estimation of ice thickness using surface velocities and slope: case study at Gangotri Glacier, India. *Journal of Glaciology* 60(220): 277–282.

Gärtner-Roer, I., Naegeli, K., Huss, M. et al. (2014). A database of worldwide glacier thickness observations. *Global and Planetary Change* 122: 330–344.

Geological Survey of India (GSI) (1987). *A report on geophysical investigation on Dunagiri Glacier District Chamoli*, Uttal Pradesh.

Geological Survey of India (GSI) (1992). *Report on glaciological studies on Tipra Bank Glacier District Chamoli*, Uttal Pradesh.

Gergan, J.T., Dobhal, D.P., and Kaushik, R. (1999). Ground penetrating radar ice thickness measurements of Dokriani Bamak (glacier), Garhwal Himalaya. *Current Science* 77(1): 169–173.

Hackney, R.I. and Featherstone, W.E. (2003). Geodetic versus geophysical perspectives of the "gravity anomaly." *Geophysical Journal International* 154(1): 35–43.

Haeberli, W. and Hölzle, M. (1995). Application of inventory data for estimating characteristics of and regional climate-change effects on mountain glaciers: a pilot study with the European Alps. *Annals of Glaciology* 21: 206–212.

Herreid, S. and Pellicciotti, F. (2020). The state of rock debris covering Earth's glaciers. *Nature Geoscience* 13(9): 621–627.

Hock, R., Rasul, G., Adler, C. et al. (2019). High mountain areas. In: *IPCC Special Report on the Ocean and Cryosphere in a Changing Climate* (ed. H.-O. Pörtner, D.C. Roberts, V. Masson-Delmotte, P. et al.), 131–202. Cambridge, UK and New York: Cambridge University Press. https://doi.org/10.1017/9781009157964.004.

Huss, M. and Farinotti, D. (2012). Distributed ice thickness and volume of all glaciers around the globe. *Journal of Geophysical Research: Earth Surface* 117(F4).

Immerzeel, W.W., Van Beek, L.P., and Bierkens, M.F. (2010). Climate change will affect the Asian water towers. *Science* 328(5984): 1382–1385.

Kraaijenbrink, P.D.A., Bierkens, M.F.P., Lutz, A.F. et al. (2017). Impact of a global temperature rise of 1.5 degrees Celsius on Asia's glaciers. *Nature* 549(7671): 257–260. doi:10.1038/nature2387

Kulkarni, A.V., Shirsat, T.S., Kulkarni, A. et al. (2021). State of Himalayan cryosphere and implications for water security. *Water Security* 14: 100101.

Kulkarni, A.V. and Karyakarte, Y. (2014). Observed changes in Himalayan glaciers. *Current Science* 106(2): 237–244.

Kumari, S., Pandit, A., Ramsankaran, R.A.A.J. et al. (2021). Modelling ice thickness distribution and volume of Patsio Glacier in Western Himalayas. *Journal of Earth System Science* 130(3): 1–14.

Kumar, V., Shukla, T., Mehta, M. et al. (2021). Glacier changes and associated climate drivers for the last three decades, Nanda Devi region, Central Himalaya, India. *Quaternary International* 575: 213–226.

Laha, S., Kumari, R., Singh, S. et al. (2017). Evaluating the contribution of avalanching to the mass balance of Himalayan glaciers. *Annals of Glaciology* 58(75, pt. 2): 110–118.

Linsbauer, A., Paul, F., and Haeberli, W. (2012). Modeling glacier thickness distribution and bed topography over entire mountain ranges with GlabTop: Application of a fast and robust approach. *Journal of Geophysical Research: Earth Surface* 117(F3).

Lutz, A.F., Immerzeel, W.W., Kraaijenbrink, P.D. et al. (2016). Climate change impacts on the upper Indus hydrology: Sources, shifts and extremes. *PloS one*, 11(11): e0165630.

Mehta, M., Kumar, V., Garg, S. et al. (2021). Little Ice Age glacier extent and temporal changes in annual mass balance (2016–2019) of Pensilungpa Glacier, Zanskar Himalaya. *Regional Environmental Change* 21(2): 1–18.

Mishra, A., Negi, B.D S., Banerjee, A. et al. (2018). Estimation of ice thickness of the Satopanth Glacier, Central Himalaya using ground penetrating radar. *Current Science* 114(4): 785.

Mishra, A., Nainwal, H.C., Dobhal, D.P. et al. (2021). Volume estimation of glaciers in Upper Alaknanda Basin, Garhwal Himalaya using numerical and scaling methods with limited field based data. *Himalayan Geology* 42(2): 336–344.

Nabighian, M.N., Ander, M.E., Grauch, V.J.S. et al. (2005). Historical development of the gravity method in exploration: historical development of gravity method. *Geophysics* 70(6): 63ND–89ND.

Nainwal, H. C., Banerjee, A., Shankar, R. et al. (2016). Shrinkage of Satopanth and Bhagirath Kharak glaciers, India, from 1936 to 2013. *Annals of Glaciology* 57(71): 131–139.

Pfeffer, W.T., Arendt, A.A., Bliss, A. et al. (2014). The Randolph Glacier Inventory: a globally complete inventory of glaciers. *Journal of Glaciology* 60(221): 537–552.

Plewes, L.A. and Hubbard, B. (2001). A review of the use of radio-echo sounding in glaciology. *Progress in Physical Geography* 25(2): 203–236.

Prakash, S., Sharma, M. C., Sreekesh, S. et al. (2021). Decadal terminus position changes and ice thickness measurement of Menthosa glacier in Lahaul region of north-western Himalaya. *Geocarto International* 1–20.

Pratap, B., Dobhal, D.P., Bhambri, R. et al. (2016). Four decades of glacier mass balance observations in the Indian Himalaya. *Regional Environmental Change* 16(3): 643–658.

Pritchard, H.D., King, E.C., Goodger, D.J. et al. (2020). Towards Bedmap Himalayas: Development of an airborne ice-sounding radar for glacier thickness surveys in high-mountain Asia. *Annals of Glaciology* 61(81): 35–45.

Raina, V.K. (2009). A state-of-art review of glacial studies, glacial retreat and climate change. Discussion Paper, MOEF, New Delhi. Available online at: http://www.moef.nic.in/sites/default/files/MoEF%20Discussion%20Paper%20_him.pdf

Raina, V.K. and Srivastava, D. (2008). *Glacier Atlas of India*. Bangalore: Geological Society of India.

Ramsankaran, R.A.A.J., Pandit, A., and Azam, M.F. (2018). Spatially distributed ice-thickness modelling for Chhota Shigri Glacier in western Himalayas, India. *International Journal of Remote Sensing* 39(10): 3320–3343.

Reynolds, J.M. (2011). *An Introduction to Applied and Environmental Geophysics*. John Wiley & Sons.

Robin, G.D.Q. (1975). Velocity of radio waves in ice by means of a bore-hole interferometric technique. *Journal of Glaciology* 15(73): 151–159.

Röthlisberger, H. (1967). Electrical resistivity measurements and soundings on glaciers: introductory remarks. *Journal of Glaciology* 6(47): 599–606.

Rounce, D.R., Hock, R., McNabb, R.W. et al. (2021). Distributed global debris thickness estimates reveal debris significantly impacts glacier mass balance. *Geophysical Research Letters* 48(8): e2020GL091311.

Scherler, D., Bookhagen, B., and Strecker, M.R. (2011). Hillslope-glacier coupling: the interplay of topography and glacial dynamics in High Asia. *Journal of Geophysical Research: Earth Surface* 116(F2).

Schroeder, D.M., Bingham, R.G., Blankenship, D.D. et al. (2020). Five decades of radioglaciology. *Annals of Glaciology* 61(81): 1–13.

Schrott, L. and Sass, O. (2008). Application of field geophysics in geomorphology: advances and limitations exemplified by case studies. *Geomorphology* 93(1–2): 55–73.

Shah, S.S., Banerjee, A., Nainwal, H.C. et al. (2019). Estimation of the total sub-debris ablation from point-scale ablation data on a debris-covered glacier. *Journal of Glaciology* 65(253): 759–769.

Sharma, P., Patel, L.K., Ravindra, R. et al. (2016). Role of debris cover to control specific ablation of adjoining Batal and Sutri Dhaka glaciers in Chandra Basin (Himachal Pradesh) during peak ablation season. *Journal of Earth System Science* 125(3): 459–473.

Shukla, A., Garg, S., Mehta, M. et al. (2020). Temporal inventory of glaciers in the Suru sub-basin, western Himalaya: impacts of regional climate variability. *Earth System Science Data* 12: 1245–1265.

Siegert, M.J. (1999). On the origin, nature and uses of Antarctic ice-sheet radio-echo layering. *Progress in Physical Geography* 23(2): 159–179.

Singh, K.K., Kulkarni, A.V., and Mishra, V.D. (2010). Estimation of glacier depth and moraine cover study using ground penetrating radar (GPR) in the Himalayan region. *Journal of the Indian Society of Remote Sensing* 38(1): 1–9.

Singh, S.K., Rathore, B.P., Bahuguna, I.M. et al. (2012). Estimation of glacier ice thickness using Ground Penetrating Radar in the Himalayan region. *Current Science* 68–73.

Singh, K.K., Singh, D.K., Negi, H.S. et al. (2018). Temporal change and flow velocity estimation of Patseo Glacier, Western Himalaya, India. *Current Science* 776–784.

Singh, K.K., Negi, H.S., and Singh, D.K. (2019). Assessment of glacier stored water in Karakoram Himalaya using satellite remote sensing and field investigation. *Journal of Mountain Science* 16(4): 836–849.

Swain, A.K., Mukhtar, M.A., Majeed, Z. et al. (2018). Depth profiling and recessional history of the Hamtah and Parang glaciers in Lahaul and Spiti, Himachal Pradesh, Indian Himalaya. *Geological Society, London, Special Publications* 462(1): 35–49.

Szalai, S. and Szarka, L. (2008). On the classification of surface geoelectric arrays. *Geophysical Prospecting* 56(2); 159–175.

Thayyen, R.J. and Gergan, J.T. (2010). Role of glaciers in watershed hydrology: a preliminary study of a "Himalayan catchment." *The Cryosphere* 4(1): 115.

Van Dam, R.L. (2012). Landform characterization using geophysics: recent advances, applications, and emerging tools. *Geomorphology*: 137(1): 57–73.

Welty, E., Zemp, M., Navarro, F. et al. (2020). Worldwide version-controlled database of glacier thickness observations. *Earth System Science Data* 12(4): 3039–3055.

Zeng, H. and Wan, T. (2004). Clarification of the geophysical definition of a gravity field. *Geophysics* 69(5): 1252–1254.

# 21

## Landscapes and Paleoclimate of the Ladakh Himalaya

*Anil Kumar*[1,*], *Rahul Devrani*[1,2], *and Pradeep Srivastava*[1,3]

[1] *Wadia Institute of Himalayan Geology, Dehradun 248001*
[2] *Delhi School of Climate Change & Sustainability,Institution of Eminence, University of Delhi, Delhi 110007*
[3] *Indian Institute of Technology, Roorkee 247667*
* *Corresponding author*

### 21.1 Introduction

The arid-glaciated mountainous landscape of the Ladakh Himalaya belongs to the western part of the Himalayan mountain range. In the exceptionally varied geomorphic landscapes of the Indian subcontinent, the Ladakh Himalaya exhibits a perfect assemblage of glaciated, fluvial, and lacustrine landforms and its exclusive human civilization history (Chadda et al., 2021). The well-defined first-order structures and topography of the Ladakh Himalaya can be traced to pre- to post-Himalayan orogeny processes. Since its upliftment in the Eocene, due to the collision of the Indian and the Eurasian plates, the Ladakh Himalaya has been interacting with tectonic, glacial, fluvial, and hillslope processes (Owen et al., 2006; Dortch et al., 2013; Sinclair et al., 2016; Kumar and Srivastava, 2017; Kumar et al., 2020b; Pundir et al., 2020; Nag et al., 2021). Such variety in the landforms in the Ladakh Himalaya is an added result of diverse structural and climatic characteristics and geomorphological processes working at different temporal and spatial scales.

In the Quaternary period, there was no persistency in the climatic conditions, and several cool, dry, and wet phases have been characterized by the period (Prell and Kutzbach, 1987; Clemens et al., 1996; Overpeck et al., 1996; Gupta et al., 2003; Dutt et al., 2015, and reference therein). The Ladakh Himalaya region has witnessed several phases of glacial maxima (Owen et al., 2006; Dortch et al., 2013; Ganju et al., 2018). These glaciations have produced large U-shaped valleys and multiple sets of moraines having various extent and preservations from north to south but had synchronous glaciation timing (Dortch et al., 2013). Hence, paleo-glacial and present-day glaciated valleys show parabolic-shaped cross profiles above the incised valley system of the Indus River and its tributaries. Despite being a glaciated-arid region, the Ladakh Himalaya is well sculpted by the Indus River and its tributaries (Blöthe et al., 2014; Munack et al., 2014; Kumar and Srivastava, 2017). On a geological scale, these rivers acted as a significant geomorphic agent to distribute sediments in the rivers of the Ladakh Himalaya, and in its present landscape, a large amount of valley-fill deposits are present on banks of these rivers (Sant et al., 2011a, 2011b; Blöthe et al., 2014; Munack et al., 2014; Kumar and Srivastava, 2017). The recurring periodic sedimentation and erosion in the Indus River and its tributaries during the Quaternary period is well connected to the different climatic cycles in the Ladakh Himalaya (Blöthe et al., 2014; Munack et al., 2014; Kumar and Srivastava, 2017). Extreme hydrological events like glacial lake outburst floods and flood deposits are some of the distinct geomorphic legacies in the region. In the present landscape of the Ladakh Himalaya, the slack water deposits (SWDs) are well-acknowledged and deposited in the waning phase of the flood. In the last few decades, the diverse landscape of the Ladakh Himalaya has become a focus of several researchers (Hobley et al., 2012; Srivastava et al., 2017; Ganju et al., 2018; Kumar et al., 2020a; Sharma et al., 2022). In the present chapter, we have tried to summarize the prominent landforms and paleoclimate present in the Ladakh Himalaya.

### 21.2 Geology of the Ladakh Himalaya

The Ladakh region is marked as the Indian and the Eurasian plate collision zone and termed as the Indus

*Advances in Remote Sensing Technology and the Three Poles,* First Edition. Edited by Manish Pandey, Prem C. Pandey, Yogesh Ray, Aman Arora, Shridhar D. Jawak, and Uma K. Shukla.

suture zone (ISZ) (Thakur and Misra, 1984), which is considered to have occurred at the beginning of the Eocene at ~60–50 Ma (Clift et al., 2002; Yin, 2006). The Lhasa and Karakoram blocks lie in the Eurasian plate; however, the Tethys sedimentary succession (TSS), High Himalayan crystalline (HCC) and Lesser Himalayan meta-sedimentary rocks are part of the Indian plate. The Karakoram fault (~800 km long) extends through Pamir in the north (Pakistan) to the Shyok Suture Zone (SSZ) and Gar Basin in southwest Tibet. The SSZ developed parallel to the ISZ by the Kohistan–Ladakh island arc collision with the southern fringe of the Eurasian plate during the Late Cretaceous (Clift et al., 2000). The calc-alkaline granitic intrusion and volcanism occurred by continuous subduction of the Indian oceanic crust below the Eurasian plate. This magmatic arc is represented in the Ladakh Batholith, an assemblage of granite, granodiorite, and tonalite suites of rocks (Thakur and Misra, 1984). The Ladakh plutonic complex exhibits a multiphase intrusive geological history during 100–90 Ma, 60–40 Ma, 20 Ma, and younger times (Honegger et al., 1982; Brookfield and Andrews-Speed, 1984; Searle et al., 1988; Chung et al., 2005; Henderson et al., 2010b, and references therein). The regions to the south of the Indus River include Paleozoic to Eocene siliciclastic and carbonates of the TSS and rocks of the HHC (Figure 21.1) (Brookfield and Andrews-Speed, 1984; Thakur and Misra, 1984; Chung et al., 2005). The rocks associated with the ISZ region are divided into two groups, the Tar Group and the Indus Group. The Tar Group characterizes marine successions that are considered the cessation of the neo-Tethys sea (Garzanti and Van Haver, 1988; Clift et al., 2002). The Tar Group lies on the southern verge of the ISZ and is composed of shallow marine to fluvial-deltaic deposits of early Eocene age (Searle et al., 1990; Henderson et al., 2010a, 2010b). Stratigraphically, the Indus Group lies conformably over the Tar Group and consists of both marine and terrestrial sedimentary rocks. These rocks are understood as the post-collisional deposits derived from the Ladakh

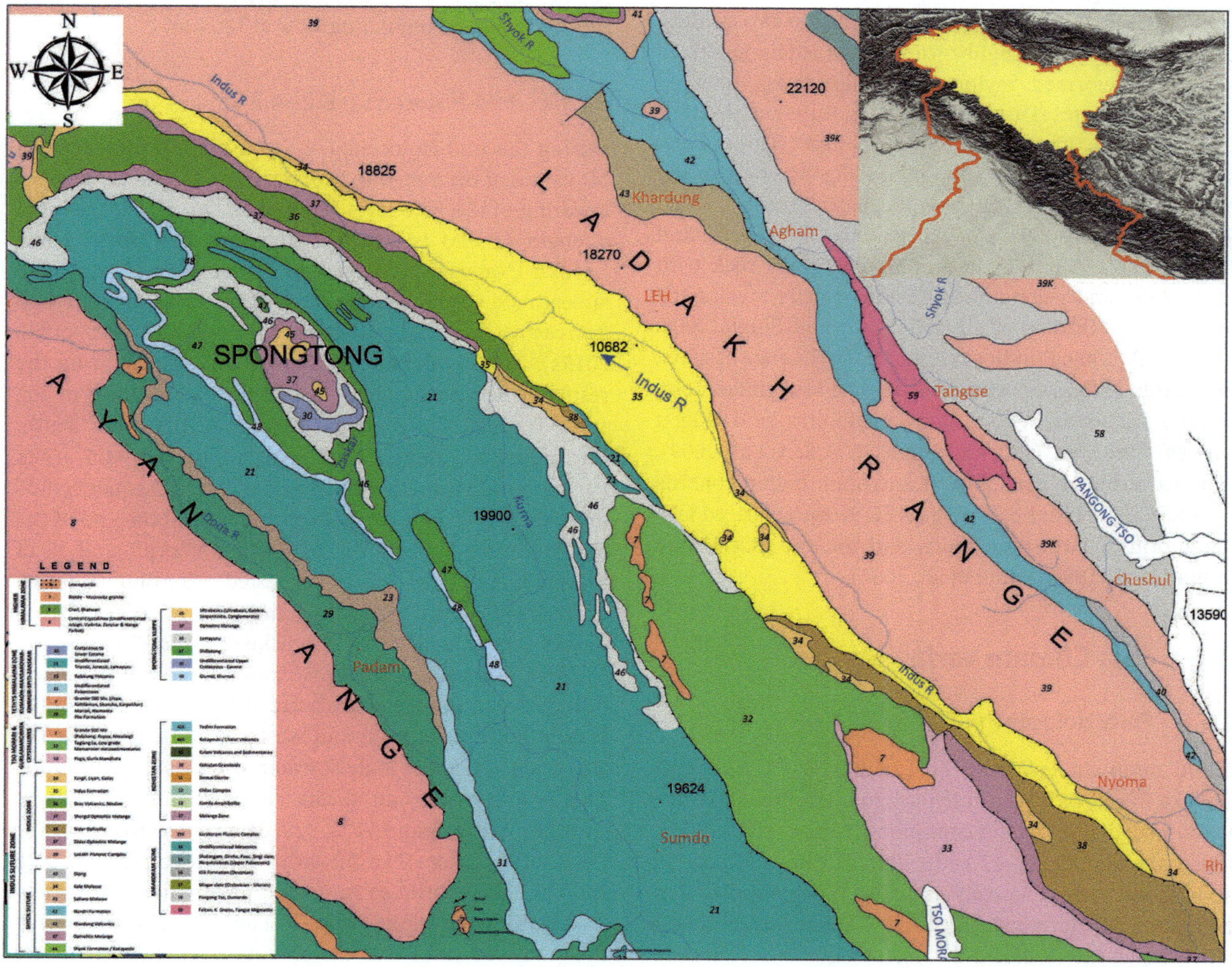


Figure 21.1 Geological map of the Ladakh region (Adapted from Thakur and Rawat, 1992).

Batholith (Thakur, 1983; Searle, 1986; Wu et al., 2007; Henderson et al., 2010a, 2010b, 2011). The Indus Group also contains rocks of fluvial-deltaic to alluvial facies and are considered as the paleo-Indus River deposits (Brookfield and Andrews-Speed, 1984; Garzanti and Van Haver, 1988; Henderson et al., 2010a).

Toward the west, sedimentary rocks of the Kargil Formation have an average bedrock thickness of ~1500 m, and rests unconformably over the Ladakh Batholith. The molassic rocks are composed of conglomerate, sandstone, and shale. Freshwater molluscs and plant fossils of Neogene Age were recovered from the sandstone beds of Kargil Molasse (Thakur and Misra, 1984).

### 21.2.1 Karakoram Region

The Karakoram region is subdivided into three major geological units: 1) Karakoram Fault Zone (KFZ); 2) Karakoram Metamorphic Complex (KMC); and 3) Karakoram Plutonic Complex (KPC). In the eastern Karakoram region, the Karakoram Fault (KF) passes through the Nubra Valley and bifurcates into two strands, i.e., the Tangtse and Muglib strands, and is characterized by ductile deformed mylonitic granite gneiss, sheared phyllites, and slates (Dunlap et al., 1998; Rutter et al., 2007; Pundir et al., 2020, and references therein). The oblique slip component of the KF along the Pangong controls the geomorphology of the Pangong Tso Basin (Dunlap et al., 1998; Searle et al., 1998; Chevalier et al., 2005; Searle and Phillips, 2007). A dominant amphibolite-grade rock with leucogranite and pegmatite intrusions occurs toward the Pangong Tso (Dunlap et al., 1998; Jain and Singh, 2008; Streule et al., 2009; Pundir et al., 2020). A sequence of schist and marble with slates characterizes the low-grade metamorphic sequence with granite mylonites and augen gneisses of the KMC (Jain and Singh, 2008; Sen et al., 2009).

The easternmost part of the Karakoram region consists of the KPC. The inner part of the KPC is characterized by a granodiorite, quartz monzonite, and tonalite assemblage (Dunlap et al., 1998; Rutter et al., 2007).

## 21.3 Past Climate Variability

### 21.3.1 Early Holocene (~11.7 to 8.2 ka)

A sand ramp studied from the Ladakh region at around ~12–7 ka BP shows fluvial gullying and inter-dunal lake formations, suggesting wetter conditions (Kumar et al., 2017). Prior to the period ~10–12 ka BP, the δD value of Pangong Tso is found to be relatively enriched. This suggests that penetration of the monsoon at the Tibetan Plateau commenced from 10–12 ka BP throughout the Late Holocene (Hou et al., 2017). Paleoflood records based on slack water deposits from the Ladakh Himalaya, reported by Srivastava et al. (2017), suggested megafloods during ~11–10 ka BP, inferring enhanced hydrological phenomena at the Western Himalaya. The pollen assemblages from Tso Kar, by Demske et al. (2009), indicate the most substantial monsoon influence during ~10.9–9.2 ka BP. However, a dryer phase has been revealed from the spread of desert steppe during ~12.2–11.8 ka BP (Demske et al., 2009). Enhanced precipitation of Tso Moriri Lake (NW Himalaya) during ~11.5–4.5 ka BP is also suggested from palynological study (Leipe et al., 2014). Carbonate isotope records from Tso Moriri Lake reveal that the highest monsoon-related precipitation occurred during this phase (Mishra et al., 2015b). It is followed by a weakening of the Indian Summer Monsoon (ISM) commencing from 8.5 ka BP, as inferred from the enrichment of $\delta^{18}O_{carb}$ and reduced Ti input, which is a proxy of surficial inflow (Mishra et al., 2015b). Rising of lake level from around 11.8 ka BP and maximum lake level around 8.5–7 ka BP at Tso Kar Lake is reported by Wunnemann et al. (2010). This study also suggests that the period ~12.5 ka BP to ~11.4–8.6 ka BP witnessed enhanced precipitation and regular outflow at Tso Kar.

### 21.3.2 Mid-Holocene (~8.2–4.2 ka)

During the mid-Holocene (~8.5–7.5 ka BP), the enriched δD values of the Pangong Tso represent an arid phase with lowering of the lake level (Hou et al., 2017). Deteriorated strength of ISM is reported from the Tso Moriri Lake, Ladakh Himalaya from ~8.5 to 5.5 ka BP, evident by the appearance of endogenetic carbonates and aragonite and decreasing siliciclastic titanium (Ti) influx (Mishra et al., 2015a). This weakening in ISM is understood to be the result of decreasing solar insolation and shifting of the Inter Tropical Convergence Zone (ITCZ) further south. Furthermore, enhanced hydrological activity at ~4.5–4.3 ka BP is reported from the Tso Moriri with a subsequent cold–dry phase at ~4.3 ka BP that continues until the end of the mid-Holocene (Dutt et al., 2018). Palynomorph analysis of the sediment of Tso Moriri suggested a continuous influx of freshwater until ~4.5 ka BP followed by arid and cold phases (Leipe et al., 2014). Demske et al. (2009), from the Tso Kar, suggested this period commenced with westerly influences and reduced moisture supply until ~4.8 ka BP. Due to both monsoonal and westerly influences, the Tso Kar experienced the highest water level at around 8 ka BP. This is followed by an abrupt climatic shift to aridity at ~4.8 ka. Wünnemann et al. (2010) reported that Tso Kar Lake shrank after ~7 ka BP with a reverse trend at ~5–6 ka BP. The lake continues to shrink, having its lowest stand ~4.2 ka BP suggesting perma-frost activity. The majority of the study indicates that the climate fluctuated rapidly in the mid-Holocene in the Western Himalaya.

### 21.3.3 Late-Holocene (~4.2 ka–Present)

After ~4 ka BP, Pangong Tso suffered a gradual reduction in the lake level, implying that weakening monsoon intensity subsequently limited the hydrologic budget of the region (Hou et al., 2017). Enriched δD values at ~3 and ~2 ka BP at Pangaog Tso suggest arid phases (Hou et al., 2017). At Tso Moriri, weakening ISM continued from ~4.3 ka to 3.5 ka BP; however, the period between ~2.8 and 0.55 ka BP witnessed enhanced discharge that suggests intense monsoonal rainfall. After this, a cold–arid phase coincided with the Little Ice Age (Dutt et al., 2018). Pollen records from the Gya peat section suggest that the region between ~2.6 and 1.9 cal kyr BP was relatively warm and suffered a cold and dry phase between ~1.8 and 1.1 cal kyr BP (Sharma et al., 2020). Weakening monsoon intensity is also inferred from low pollen invasion during ~2.8–1.3 ka BP at Tso Kar (Demske et al., 2009). Carbonate isotope records of Tso Moriri Lake witness persistent aridity throughout the entire Late Holocene (Mishra et al., 2015a). Palynological studies at Tso Moriri suggest the Late Holocene witnessed aridity and glaciation until ~1.8 ka BP (Leipe et al., 2014). Wünnemann et al. (2010) highlighted that Tso Kar experienced continued low lake levels in the entire Late Holocene.

## 21.4 Modern Climatic and Vegetation

Climatically, the Ladakh region lies in the cold semi-arid to cold arid climatic zone of the Himalaya. The average annual precipitation in the region is ~300 mm, with a maximum and minimum of ~735 and ~110 mm, respectively (Figure 21.2). Generally, the maximum precipitation occurs in a small zone near the southern boundary of the Ladakh region. The Kargil is the transitional zone where precipitation decreases toward the north. The western part of the Ladakh receives higher precipitation than its eastern counterpart (Figures 21.2a and 21.2b). The rainfall curve shown in Figure 21.2b, trending west to east, suggests that minimum precipitation in the western part is up 300 mm/a and 190 mm/a in the eastern part.

The region receives moistures (i.e., precipitation) by two major climatic systems, i.e., Indian Summer monsoon (ISM) and the Westerlies. The water budget in the Indus and its tributaries are dominated by ISM precipitation during June to September. From November to February, the Westerlies become active and precipitate as snow in the Kargil, Zanskar, Ladakh, and Karakoram regions. The average annual precipitation decreases to ~40 mm toward the Karakoram ranges, and in the Leh Valley it is ~150 mm (Lang and Barros, 2004). Strong winds with an average speed of ~6 km/hr dominate this region throughout the year (Kumar et al., 2017). During winter, the average diurnal temperature reaches –20°C, and in summer it is ~25°C.

The Zanskar catchment is situated in a similar climate domain with temperatures ranging from –14 to –3°C in winter and 10 to 25°C in summer (Taylor and Mitchell, 2000). There is scanty vegetation cover in the region and this is limited to the vicinity of streams (Baud et al., 1984; Osmaston 2001). The high-resolution Tropical Rainfall Measuring Mission (TRMM) indicates that the headwater of the Zanskar receives precipitation of ~200–400 mm/yr, whereas downstream it reduces to ~150 mm/yr (Bookhagen et al., 2009; Lee et al., 2014; Chahal et al., 2019; Tropical Rainfall Measing Mission 2011).

Very sparse vegetation and floral diversity are observed in Ladakh owing to its rugged terrain, low moisture, and high-frequency climatic oscillation. Due to these conditions, soil

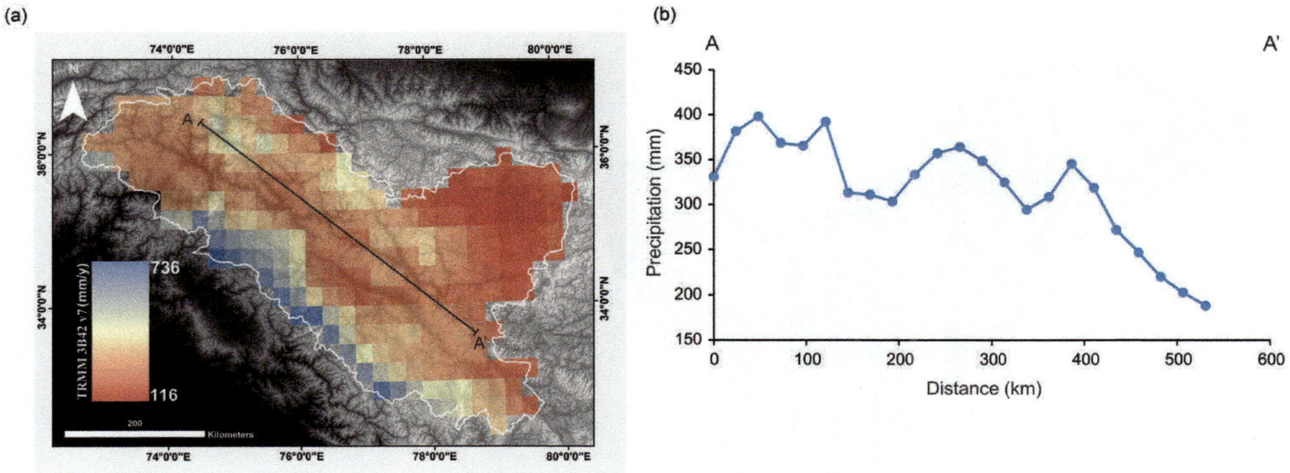


Figure 21.2 (a) Time averaged map of precipitation 3-hourly 0.25 degree shows precipitation gradient from south to north in the Ladakh region. (Dataset: TRMM 3B42 v7 over 1997-12-31 to 2019-12-31 downloaded from Earth-Data, NASA). A-A' is a precipitation profile given in (b). (b) The precipitation profile along A-A' shows a sharp decline in precipitation from NW to NE Ladakh.

fertility is very poor. Most likely places for vegetation cover are along perennial streams composed of sandy silt with high salinity and low nutrients. Highly tolerant plant species are abundant in *Artemisia macrocephala* (desert-steppes) and *Suaedaolufsenii* and other deep-rooted shrubs. In places where permanent freshwater is available, aquatic plant species such as *Potamogeton* occurs. Wide wetter plains, typical of southern Ladakh, are covered with grasses such as *Cyperaceae* and *Gramineae*, which function as grazing grounds. Such areas are relatively fertile due to the deposition of faecal matter from grazing animals. In such areas, species such as *Chenopodiaceae* are relatively abundant (Dvorský et al., 2011). Eastern Ladakh is dominated by *Gramineae, Cyperaceae, Brassicaceae,* and *Fabaceae*. At present, there are almost no naturally growing tree elements, except for those planted such as *Salyx*. Among shrubs, Junipers and Betula are found to be sparse (Kala and Mathur, 2002).

## 21.5 Landscapes in the Ladakh Region

The Indus River is a significant river of Ladakh that flows in a northwesterly direction through a valley and represents the first-order geomorphological feature of the Ladakh region. The Indus River is the backbone of the Ladakh and supports people from agricultural, livelihood, and socio-economic points of view (Dimri et al. 2020). It originates from the Mansarovar Lake at an elevation of ~5182 m a.s.l. in the vicinity of Mt Kailas and drains through the Karakoram zone, Ladakh Batholith, and tectonic units of ISZ. The Indus River is the main river with the Zanskar, Shyok, Nubra, and Suru Rivers as its major tributaries. In the present scenario, the Indus River is seventh-order, with the Shyok River as its major tributary being sixth-order (Figure 21.3).

Furthermore, the Zanskar, Suru, and Nubra Rivers are fifth-order and the drainage density of the Ladakh region is around 221 $mm^2$. The headwaters of all significant tributaries of the Indus River have large- and medium-sized glaciers that are often flanked by moraines of past glacial advances. The geomorphology of Ladakh Himalaya is marked by steep sedimentary cones made up of frost shattered rocks, glacial moraines, sand ramps, alluvial fills, alluvial fans, alluvial fan dammed paleo-lake deposits, debris flows, and filled- and strath-terraces (Figures 21.4, 21.5, and 21.6) (Jamieson et al., 2004; Pant et al., 2005; Phartiyal et al., 2005; Sant et al., 2011a, 2011b; Blöthe et al., 2014; Nag et al., 2016; Kumar and Srivastava, 2017; Kumar et al., 2017; Chahal et al., 2019). The average elevation of Ladakh is ~3000 m above the mean sea level (a.s.l.). All the prominent fluvial geomorphologic feature lie in the deeply incised valley of the Ladakh (Figure 21.7); however, U-shaped valleys and glacial moraines are situated at the first- to second-order channels. The tributaries contribute enormously to the sediment budget of the Indus, and phases of high sediment generation and transportation helped to form wide and thick valley fills that were subsequently transformed into terraces due to river incision. A ~350 km stretch of the Indus River in the Ladakh region exhibits different configurations of valley cross-sections, from wide U-shaped to deeply incised V-shaped gorges. Based on the valley configuration and channel gradient, the catchment of the Indus can be divided into four segments (Kumar and Srivastava, 2017).

Segment I (U-shaped valley) extends from Nyoma to Mahe, where the valley is vast (~4 km) with a braided pattern and a very gentle river slope, i.e., 0.75 m/km. The older tight meander of the river can be seen in the flood plain, and is flanked with sediment-filled terraces of a few meters in thickness.

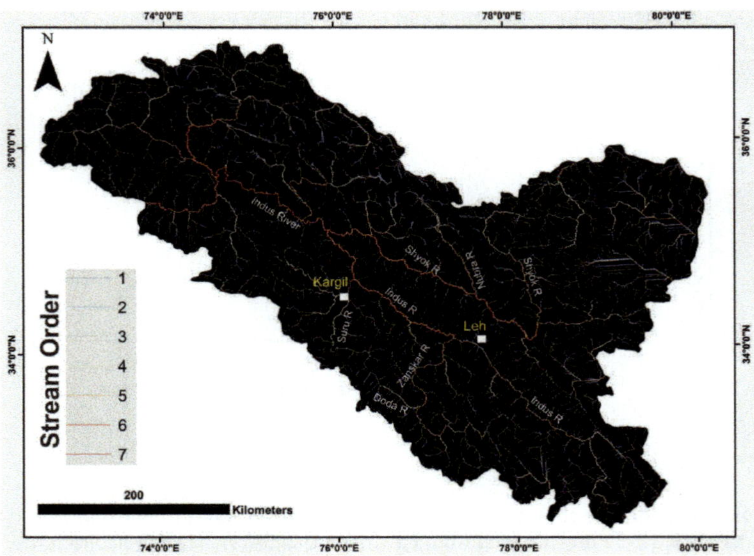


Figure 21.3 Drainage map of the Ladakh region showing significant rivers and their stream order.

**Figure 21.4** (a) and (b) are the panoramic views of a sand ramp along the Indus River at Nyoma and Saboo villages, respectively. *Source*: Geological Society of Korea; (c) and (d) are the panoramic views of paleolake deposits associated with the landslide dammed lakes in the Ladakh Himalaya. *Source*: Claudia Ioan / Flickr.

**Figure 21.5** (a) Valley fluvial terraces are the most common landforms in the Ladakh Himalaya. The valley filling in the Indus and Zanskar occurred in three pulses, at ~52, 28, and ~16 ka (Kumar et al. ( 2017) / ELSEVIER). (b) A panoramic view of a starth terrace near the Khalsi.

Segment II (V-shaped valley) extends from Mahe to Upshi (~107 km), where the valley narrows compared to other segments. Here the average valley width reduces to ~192 m, and the slope increases to ~7.2 m/km. The topographic slope of the river reduces from 4116 m to 3397 m a.s.l. This segment shows the development of thick-fill terraces that range in thickness from 10.5 m to 70 m (Kumar and Srivastava, 2017).

Segment III extends from Upshi to Leh (~60 km), with a large valley width (~1200 m) and low channel gradient

Figure 21.6 (a) Glacial moraine in the Suru valley and (b) lateral moraines associated Pensi La, DodaValley.

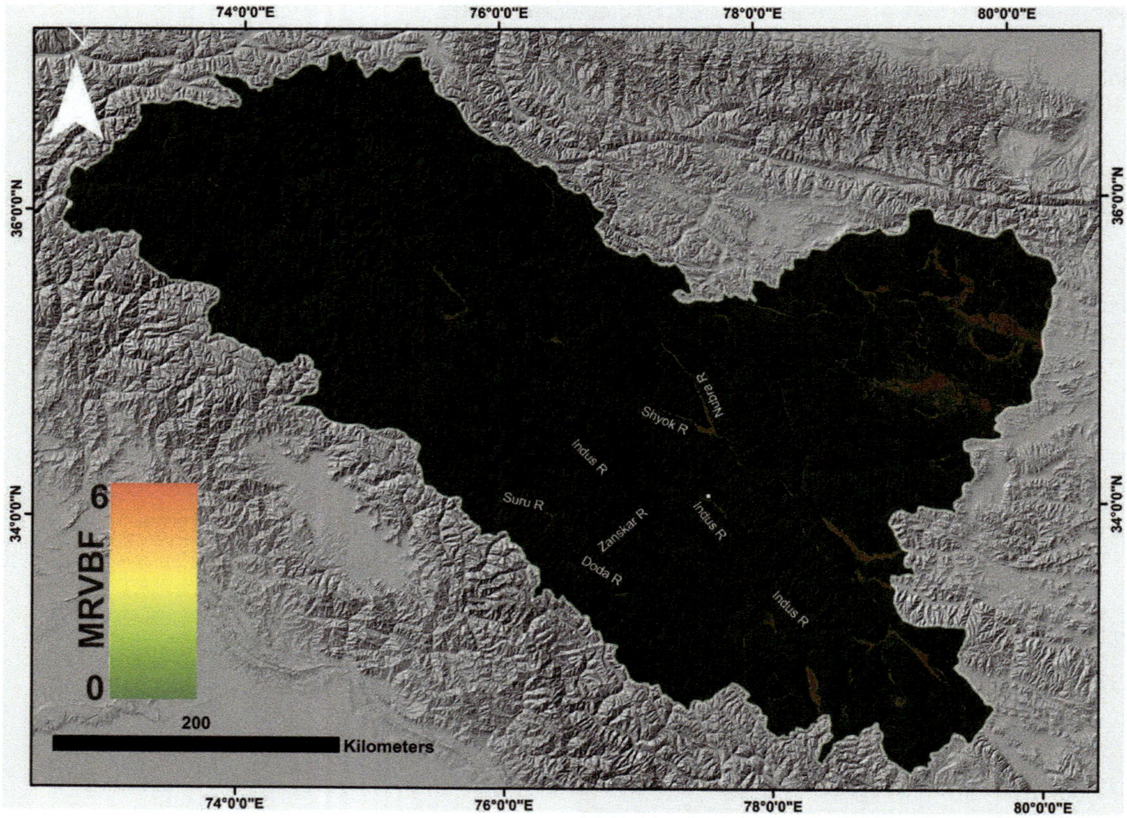


Figure 21.7 The valley bottom index separates the flat-wide and deep-narrow valleys in the Ladakh Himalaya.

(1.8 m/km). The meanders of Indus become braided in the middle of this segment. The valley is covered by alluvial fans both south and north of the trunk channel. Eleven fans emerge from de-glaciated amphitheatre valleys and are bounded by the Ladakh Batholith on three sides. The continuous sedimentation from north and south fans raises the local base level, leading to a decrease in local slope and results in expansion of aggradation into the tributaries coming from the north (Sant et al., 2011a, 2011b; Kumar and Srivastava, 2017).

Segment IV extends from Nimu to Dah-Hanu (~120 km). Strath terraces, deep gorges, and steep gradient characterize this segment. The Indus River entered the gorge by drifting its course from Segment III (Batholith-Molasse contact zone) to Segment IV (Molassic bedrock). Again, the valley has a V-shaped configuration and shows one to two levels of strath terraces at some places and one level of fill terrace. The valley width and channel gradient range from 100 to 500 m and 4 to 7.5 m/km, respectively. The strath terraces are situated at an average elevation of ~135 m above the river level (Kumar and Srivastava, 2017). Paleo-lake deposits are also an important geomorphologic feature of the Ladakh and are found sporadically in the upper Indus Valley (Figure 21.4). The paleo-lake deposits in Ladakh indicate past glacial melting and can be linked with past temperature rise (Dimri et al., 2020).

## 21.6 Glaciation and Associated Landforms

The Ladakh Himalaya and south-western Tibetan Plateau played a vital role during onset of Quaternary glaciations and influenced west winds and the Indian summer monsoon system (Scherler et al., 2010; Yan et al., 2020). The glacial landforms in the Ladakh Himalaya were studied to understand the glacial history and their forcing factors (Scherler et al., 2010; Sharma and Shulka, 2018; Yan et al., 2020; Kumar et al., 2021). For the northwestern Himalaya, a large volume of literature is available that establishes the teleconnections between the episodes of glaciation in the NW Himalaya and Northern Hemisphere climate variability (Mayewski et al., 2004; Rowan, 2017; Batbaatar et al., 2018; Shukla et al., 2018; Kumar et al., 2021). During marine isotopic stage 3 (MIS-3), glacial expansion occurred because of the strengthening of the Indian monsoon. During peak monsoon phases, moisture-laden wind penetrated the drier Himalaya and precipitated over the higher elevations of the mountains; as a result, positive glacier mass balance and glacieal advance occurred (Shukla et al., 2018; Kumar et al., 2021). Synchronous or asynchronous behavior of glacial advances with the monsoon has been debated and is still an essential question for research (Murari et al., 2014; Batbaatar et al., 2018; Yan et al., 2020).

In the Ladakh Himalaya, significant glacial advances and recessions are marked by various glacial landforms such as lateral, terminal, and recessional moraines (Figure 21.6), drumlines, cirques, outwash fans, glacial movement striations, and U-shaped valleys occurring mainly in the Indus, Zanskar, Nanga Parbat, and Karakoram regions during the MIS-4 to the Recent (Seong et al., 2007; Lee et al., 2014). The glaciation in the Nun-Kun massif and southern Zanskar occurred during 62–38 ka and 33–25 ka (MIS-4 to -3) (Lee et al., 2014; Sharma and Shukla, 2018; Kumar et al., 2021) that were associated with the strengthened mid-latitude Westerlies (Dortch et al., 2013). During the Last Glacial Maxima (LGM), the glaciers in the Ladakh Himalaya advanced (Scherler et al., 2010; Dong et al., 2018; Yan et al., 2020).

In the Suru and Doda valleys, glaciers advanced between 25 and 20 ka (Sharma and Shukla, 2018; Ali et al., 2020; Kumar et al., 2021). The southern Zanskar Glacial advances during the LGM are coeval to the Trans-Himalaya (Hedrick et al., 2011), Nubra, Karakorum (Nagar et al., 2013), Indus Valley (Dortch et al., 2013), and Lato massif (Orr et al., 2018). During the Younger Dryas period, glaciers advanced in the Suru and Doda valleys (Zanskar) between 17 and 11 ka; this advancement in glaciation was coeval with the events in Nanga Parbat, Nubra Valley, the Karakoram, and Lahul Himalaya (Phillips et al., 2000; Owen et al., 2001; Spencer and Owen, 2004; Seong et al., 2007; Lee et al., 2014; Ganju et al., 2018; Sharma and Shukla, 2018; Kumar et al., 2021). In the Holocene, two significant phases of glacial advances were recorded by the occurrence of extensive moraines, one at 10 to 7 ka and another at ~3 to 2 ka (Orr et al., 2018; Saha et al., 2018; Sharma and Shukla, 2018; Kumar et al., 2021).

## 21.7 Flood History and Disaster

It is vital to understand the floods in the Himalaya because of their enormous erosive energy to destroy or demolish the infrastructures in the valley (Devrani et al., 2015; Srivastava et al., 2017; Panda et al., 2020; Devrani et al., 2021) and carries substantial sediment load (mineral-rich sediment) deposited in the lowland areas which in turn helps the economy and agriculture (Goswami, 1985). The higher the relief, the higher will be sediment production, and vice-versa. In the modern time, floods in the drier Ladakh Himalaya are controlled by climatic systems such as: i) intense precipitation due to the Indian monsoon; ii) precipitation due to upper atmospheric circulation that interacts with the high peaks of the Himalayan orography; and iii) high winter precipitation (Westerlies).

The floods in Ladakh also are caused by lake outbursts such as glacial lake outburst floods (GLOFs) and landslide lake outburst floods (LLOFs) (Srivastava et al., 2013, 2017; Kumar et al., 2020a; Sharma et al., 2022). Previous studies also suggest that rainfall in Ladakh has been increasing in recent years, indicating that 40% of extreme rainfall events occur during July to September (monsoon months) (Bhan et al., 2015). The Indian Meteorological Department (IMD) monthly average precipitation for the years 2000 to 2018 from Leh Valley suggested that monsoon activity has increased recently in the valley; therefore, the recurrence of large flash floods has increased accordingly.

The studies on the paleoflood records and associated climatic forcing from the arid Ladakh Himalaya is sparse. Paleodischarge reconstruction of large floods in the Ladakh Himalaya through clast geometry and slack water deposits (SWDs) is the first-order dataset available (Srivastava et al., 2017; Chahal et al., 2020; Kumar et al., 2020a; Sharma et al., 2022). SWDs are identified as the flood deposits that contain material from the suspended load during the time of flooding at the location, where flood energy dissipates due to the sudden widening of the valley.

The Indus valley has recorded flood events between 13 and 5 ka. In the upper Zanskar regions (Padam Valley), a fill terrace dates back to 17 ± 2 ka. Above the fill terrace, 27 floods were identified as SWDs between 17 and 6 ka (Chahal et al., 2019; Sharma et al., 2022). In the Padam Valley, 11 large floods occurred between ~17 and 9 ka (Sharma et al., 2022). At the same time, in the lower

Zanskar (Chiling area), between 15 and 3 ka, two phases of large floods were recorded: older between 15 and 9 ka, with at least 14 large floods (Srivastava et al., 2017) and younger from 6 to 3 ka, with at least 10 large floods (Sharma et al., 2022). The frequency curve of flood chronologies of these two rivers suggested that there were three major phases of enhanced floods: i) phase-1, between 14 and 11 ka; ii) phase-2, between 10 and 8 ka, and iii) phase-3, from 7 to 4 ka. Therefore, it can be envisaged that 15 to 9 ka was a phase when Zanskar faced catchment-scale flooding. However, during 9 to 6 ka, the events were local; therefore, more data was needed to understand the forcing factors of these young floods (Sharma et al., 2022).

## 21.8 Conclusion

The present landscape of the Ladakh Himalaya is the result of long-term tectonic and subsequent glacial, hillslope, and fluvial processes. The arid-glaciated mountainous landscape of the the Ladakh Himalaya provides an exceptional natural laboratory to study landforms, landscape evolution, and paleoenvironments. The tectonic settings near the Ladakh Himalaya have shaped its outline, and in a few locations, we still observe tectonic activity in the recent past (Kumar et al., 2020b). The glacial history of the region has endowed the drainage network where glacial and hillslope deposits are reworked by the Indus River and its tributaries. Furthermore, the Ladakh Himalaya has been overwhelmed by both glacial and landslide lake outburst floods and has immensely contributed in the landscape evolution of the region.

## Acknowledgment

Director, WIHG is acknowledged for providing the necessary facility and infrastructure.

## References

Ali, S.N., Morthekai, P., Bajpai, S. et al. (2020). Redefining the timing of Tongul glacial stage in the Suru valley, NW Himalaya, India: new insights from luminescence dating. *Journal of Earth System Science* 129: 16. doi.org/10.1007/s12040-019-1280-9.

Bhan, S.C., Devrani, A.K., and Sinha, V. (2015). An analysis of monthly rainfall and the meteorological conditions associated with cloudburst over the dry region of Leh (Ladakh), India. *Mausam* 66(1): 107–122.

Batbaatar, J., Gillespie, A.R., Fink, D. et al. (2018). Asynchronous glaciations in arid continental climate. *Quaternary Science Reviews* 182: 1–19.

Baud, A., Gaetani, M., Garzanti, E. et al. (1984). Geological observations in southeastern Zanskar and adjacent Lahul area (northwestern Himalaya). *EclogaeGeologicaeHelvetiae* 77: 171–197.

Blöthe, J.H., Munack, H., Korup, O. et al. (2014). Late Quaternary valley infill and dissection in the Indus River, western Tibetan Plateau margin. *Quaternary Science Reviews* 94: 102–119.

Brookfield, M.E. and Andrews-Speed, C.P. (198). Sedimentology, petrography and tectonic significance of the shelf, flysch and molasse clastic deposits across the Indus Suture Zone, Ladakh, NW India. *Sedimentary Geology* 40: 249–286.

Chaddha, A.S., Mathews, R.P., Kumar, K. et al. (2021). Caves as interim-refugia: chemical signatures of human habitation under extreme environments of Ladakh, NW India. *Journal of Archaeological Science: Reports* 36: 102799.

Chahal, P., Kumar, A., Sharma, C.P. et al. (2019). Late Pleistocene history of aggradation and incision, provenance and channel connectivity of the Zanskar River, NW Himalaya. *Global and Planetary Change* 178: 110–128.

Chahal, P., Kumar, A., Sharma, P.C. et al. (2020). A preliminary assessment of the geological evidence of the mega floods in the upper Zanskar catchement, NW Himalaya. *Journal of the Palaeontological Society of India* 65(1): 64–72.

Chevalier, M.L., Ryerson, F.J., Tapponnier, P. et al. (2005). Slip-rate measurements on the Karakorum fault may imply secular variations in fault motion. *Science* 307(5708): 411–414.

Chung, S.L., Chu, M.F., Zhang, Y. et al. (2005). Tibetan tectonic evolution inferred from spatial and temporal variations in post-collisional magmatism. *Earth-Science Reviews* 68(3): 173–196.

Clemens, S.C., Murray, D.W., and Prell, W.L. (1996). Nonstationary phase of the Plio-Pleistocene Asian monsoon. *Science* 274(5289): 943–948.

Clift, P.D., Carter, A., Krol, M. et al. (2002). Constraints on India-Eurasia collision in the Arabian sea region taken from the Indus Group, Ladakh Himalaya, India. *Geological Society, London, Special Publications* 195(1): 97–116.

Clift, P.D., Degnan, P.J., Hannigan, R. et al. (2000). Sedimentary and geochemical evolution of the Dras forearc basin, Indus suture, Ladakh Himalaya, India. *Geological Society of America Bulletin* 112(3): 450–466.

Demske, D., Tarasov, P.E., Wünnemann, B. et al. (2009). Late glacial and Holocene vegetation, Indian monsoon and westerly circulation in the Trans-Himalaya recorded in the lacustrine pollen sequence from Tso Kar, Ladakh, NW India. *Palaeogeography, Palaeoclimatology, Palaeoecology* 279(3): 172–185.

Devrani, R., Singh, V., Mudd, S.M. et al. (2015). Prediction of flash flood hazard impact from Himalayan river profiles. *Geophysical Research Letters* 42(14): 5888–5894.

Devrani, R., Singh, V., Mehta, M. et al. (2021). Assessing sediment pulse during an extreme hydrological event in the Alaknanda Basin, Northwestern Himalaya, India. *Journal of the Geological Society of India* 97(1): 48–54.

Dimri, A.P., Bookhagen, B., Stoffel, M. et al. (eds) (2020). *Himalayan Weather and Climate and their Impact on the Environment*. Springer International Publishing.

Dong, G., Zhou, W., Yi, C. et al. (2018). The timing and cause of glacial activity during the last glacial in central Tibet based on 10Be surface exposure dating east of Mount Jaggang, the Xainza range. *Quaternary Science Reviews* 186: 284–297.

Dortch, J.M., Owen, L.A., Caffee, M.W. (2013). Timing and climatic drivers for glaciation across semi-arid western Himalayan Tibetan orogen. *Quaternary Science Reviews* 78: 188–208.

Dunlap, W.J., Weinberg, R.F., and Searle, M.P. (1998). Karakoram fault zone rocks cool in two phases. *Journal of the Geological Society* 155(6): 903–912.

Dutt, S., Gupta, A.K., Clemens, S.C. et al. (2015). Abrupt changes in Indian summer monsoon strength during 33,800 to 5500 years BP. *Geophysical Research Letters* 42(13): 5526–5532.

Dutt, S., Gupta, A.K., Wünnemann, B. et al. (2018). A long arid interlude in the Indian summer monsoon during ~4,350 to 3,450 cal yr BP contemporaneous to displacement of the Indus valley civilization. *Quaternary International* 482: 83–92.

Dvorský, M., Doležal, J., De Bello, F. et al. (2011). Vegetation types of East Ladakh: species and growth form composition along main environmental gradients. *Applied Vegetation Science* 14(1): 132–147.

Ganju, A., Nagar, Y.C., Sharma, L.N. et al. (2018). Luminescence chronology and climatic implication of the late quaternary glaciation in the Nubra valley, Karakoram Himalaya, India. *Palaeogeography, Palaeoclimatology, Palaeoecology* 502: 52–62.

Garzanti, E. and Van Haver, T. (1988). The Indus clastics: forearc basin sedimentation in the Ladakh Himalaya (India). *Sedimentary Geology* 59(3–4): 237–249.

Goswami, D.C. (1985). Brahmaputra River, Assam, India: physiography, basin denudation, and channel aggradation. *Water Resources Research* 21(7): 959–978.

Gupta, A.K., Anderson, D.M., and Overpeck, J.T. (2003). Abrupt changes in the Asian southwest monsoon during the Holocene and their links to the North Atlantic Ocean. *Nature* 421(6921): 354–357.

Henderson, A.L., Foster, G.L., and Najman, Y. (2010a). Testing the application of *in situ* Sm–Nd isotopic analysis on detrital apatites: a provenance tool for constraining the timing of India–Eurasia collision. *Earth and Planetary Science Letters* 297(1–2): 42–49.

Henderson, A.L., Najman, Y., Parrish, R. et al. (2010b). Geology of the Cenozoic Indus Basin sedimentary rocks: paleoenvironmental interpretation of sedimentation from the western Himalaya during the early phases of India–Eurasia collision. *Tectonics* 29: 1–35.

Henderson, A.L., Najman, Y., Parrish, R. et al. (2011). Constraints to the timing of India–Eurasia collision: a re-evaluation of evidence from the Indus Basin sedimentary rocks of the Indus–Tsangpo Suture Zone, Ladakh, India. *Earth-Science Reviews* 106(3): 265–292.

Hedrick, K.A., Seong, Y.B., Owen, L.A. et al. (2011). Towards defining the transition in style and timing of Quaternary glaciation between the monsoon-influenced Greater Himalaya and the semi-arid Transhimalaya of Northern India. *Quaternary International* 236(1–2), 21–33.

Hobley, D.E., Sinclair, H.D., and Mudd, S.M. (2012). Reconstruction of a major storm event from its geomorphic signature: the Ladakh floods, 6 August 2010. *Geology* 40(6): 483–486.

Honegger, K., Dietrich, V., Frank, W. et al. (1982). Magmatism and metamorphism in the Ladakh Himalayas (the Indus-Tsangpo suture zone). *Earth and Planetary Science Letters* 60(2): 253–292.

Hou, J., D'Andrea, W.J., Wang, M. et al. (2017). Influence of the Indian monsoon and the subtropical jet on climate change on the Tibetan Plateau since the late Pleistocene. *Quaternary Science Reviews* 163: 84–94.

Jain, A.K. and Singh, S. (2008). Tectonics of the southern Asian Plate margin along the Karakoram Shear Zone: constraints from field observations and U–Pb SHRIMP ages. *Tectonophysics* 451(1–4): 186–205.

Jamieson, S.S.R., Sinclair, H.D., Kirstein, L.A. et al. (2004). Tectonic forcing of longitudinal valleys in the Himalaya: morphological analysis of the Ladakh Batholith, North India. *Geomorphology* 58: 49–65.

Kala, C.P. and Mathur, V.B. (2002). Patterns of plant species distribution in the Trans-Himalayan region of Ladakh, India. *Journal of Vegetation Science* 13(6): 751–754.

Kumar, A. and Srivastava, P. (2017). The role of climate and tectonics in aggradation and incision of the Indus River in the Ladakh Himalaya during the late Quaternary. *Quaternary Research* 87(3): 363–385.

Kumar, A., Srivastava, P., and Meena, N.K. (2017). Late Pleistocene aeolian activity in the cold desert of Ladakh: a record from sand ramps. *Quaternary International* 443: 13–28.

Kumar, A., Srivastava, P., and Devrani, R. (2020a). Using clast geometries to establish paleoriver discharges: Testing records for aggradation and incision from the upper Indus River, Ladakh Himalaya. *Geomorphology* 362: 107202.

Kumar, A., Srivastava, P., Sen, K. et al. (2020b). Evidence for late Quaternary brittle deformation and back thrusting within the Indus Suture Zone, Ladakh Himalaya. *Tectonophysics* 792: 228597.

Kumar, V., Mehta, M., Shukla, A. et al. (2021). Late Quaternary glacial advances and equilibrium-line altitude changes in a semi-arid region, Suru Basin, western Himalaya. *Quaternary Science Reviews* 267: 107100.

Lang, T.J. and Barros, A.P. (2004). Winter storms in the central Himalayas. *Journal of the Meteorological Society of Japan* 82: 829–844.

Lee, S.Y., Seong, Y.B., Owen, L.A. et al. (2014). Late Quaternary glaciation in the Nun-Kun massif, northwestern India. *Boreas* 43(1): 67–89.

Leipe, C., Demske, D., Tarasov, P.E. et al. (2014). Potential of pollen and non-pollen palynomorph records from Tso Moriri (Trans-Himalaya, NW India) for reconstructing Holocene limnology and human–environmental interactions. *Quaternary International* 348: 113–129.

Mayewski, P.A., Rohling, E.E., Stager, J.C. et al. (2004). Holocene climate variability. *Quaternary Research* 62(3): 243–255.

Mishra, P.K., Prasad, S., Anoop, A. et al. (2015). Carbonate isotopes from high altitude Tso Moriri Lake (NW Himalayas) provide clues to late glacial and Holocene moisture source and atmospheric circulation changes. *Palaeogeography, Palaeoclimatology, Palaeoecology* 425: 76–83.

Mishra, P.K., Anoop, A., Schettler, G. et al. (2015a.) Reconstructed late quaternary hydrological changes from Lake Tso Moriri, NW Himalaya. *Quaternary International* 371: 76–86.

Mishra, P.K., Prasad, S., and Anoop, A. (2015b). Carbonate isotopes from high altitude Tso Moriri Lake (NW Himalayas) provide clues to late glacial and Holocene moisture source and atmospheric circulation changes. *Palaeogeography, Palaeoclimatology, Palaeoecology* 425: 76–83.

Munack, H., Korup, O., Resentini, A. et al. (2014). Postglacial denudation of western Tibetan Plateau margin outpaced by long-term exhumation. *Geological Society of America Bulletin* 126: 1580–1594.

Murari, M.K., Owen, L.A., Dortch, J.M. et al. (2014).Timing and climatic drivers for glaciation across monsoon-influenced regions of the Himalayan–Tibetan orogen. *Quaternary Science Reviews* 88: 159–182.

Nag, D., Phartiyal, B., and Joshi, M. (2021). Late Quaternary tectono-geomorphic forcing vis-a-vis topographic evolution of Indus catchment, Ladakh, India. *Catena* 199: 105103.

Nag, D., Phartiyal, B., and Singh, D.S. (2016). Sedimentary characteristics of palaeolake deposits along the Indus River valley, Ladakh, Trans-Himalaya: implications for the depositional environment. *Sedimentology* 63(6): 1765–1785.

Nagar, Y.C., Ganju, A., Satyawali, P.K. et al. (2013). Preliminary optical chronology suggests significant advance in Nubra valley glaciers during the Last Glacial Maximum. *Current Science* 96–101.

Orr, E.N., Owen, L.A., Saha, S. et al. (2018). Quaternary glaciation of the Lato Massif, Zanskar Range of the NW Himalaya. *Quaternary Science Reviews* 183: 140–156.

Osmaston, H. (2001). The geology, geomorphology and Quaternary history of Zanskar. *Himalayan Buddhist Villages* 1–36.

Overpeck, J., Anderson, D., Trumbore, S.et al. (1996). The southwest Indian Monsoon over the last 18,000 years. *Climate Dynamics* 12(3): 213–225.

Owen, L.A., Caffee, M.W., Bovard, K.R. et al. (2006). Terrestrial cosmogenic nuclide surface exposure dating of the oldest glacial successions in the Himalayan orogen. Ladakh Range, northern India. *Geological Society of America Bulletin* 118: 383–392.

Owen, L.A., Gualtieri, L.Y.N., Finkel, R.C. et al. (2001). Cosmogenic radionuclide dating of glacial landforms in the Lahul Himalaya, northern India: defining the timing of Late Quaternary glaciation. *Journal of Quaternary Science* 16(6): 555–563.

Panda, S., Kumar, A., Das, S. et al. (2020). Chronology and sediment provenance of extreme floods of Siang River (Tsangpo-Brahmaputra River valley), northeast Himalaya. *Earth Surface Processes and Landforms* 45(11): 2495–2511.

Pant, R.K., Phadtare, N.R., Chamyal, L.S. et al. (2005). Quaternary deposits in Ladakh and Karakoram Himalaya: a treasure trove of the palaeoclimate records. *Current Science* 88: 1789–1798.

Phartiyal, B., Sharma, A., Upadhyay, R. et al. (2005). Quaternary geology, tectonics and distribution of palaeo- and present fluvio/glacio lacustrine deposits in Ladakh, NW Indian Himalaya: a study based on field observations. *Geomorphology* 65: 241–256.

Phillips, W.M., Sloan, V.F., Shroder Jr, J.F. et al. (2000). Asynchronous glaciation at Nanga Parbat, northwestern Himalaya Mountains, Pakistan. *Geology* 28(5): 431–434.

Prell, W.L. and Kutzbach, J.E. (1987). Monsoon variability over the past 150,000 years. *Journal of Geophysical Research: Atmospheres* 92(D7): 8411–8425.

Pundir, S., Adlakha, V., Kumar, S. et al. (2020). Closure of India–Asia collision margin along the Shyok Suture Zone in the eastern Karakoram: new geochemical and zircon U–Pb geochronological observations. *Geological Magazine* 157(9): 1451–1472.

Rowan, A.V. (2017). The "Little Ice Age" in the Himalaya: a review of glacier advance driven by Northern Hemisphere temperature change. *The Holocene* 27(2): 292–308.

Rutter, E.H., Faulkner, D.R., Brodie, K.H. et al. (2007). Rock deformation processes in the Karakoram fault zone, Eastern Karakoram, Ladakh, NW India. *Journal of Structural Geology* 29(8): 1315–1326.

Saha, S., Owen, L.A., Orr, E.N. et al. (2018). Timing and nature of Holocene glacier advances at the northwestern end of the Himalayan-Tibetan orogen. *Quaternary Science Reviews* 187: 177–202.

Sant, D.A., Wadhawan, S.K., Ganjoo, R.K. et al. (2011a). Morphostratigraphy and palaeoclimate appraisal of the Leh valley, Ladakh Himalayas, India. *Journal of the Geological Society of India* 77: 499–510.

Sant, D.A., Wadhawan, S.K., Ganjoo, R.K. et al. (2011b.) Linkage of paraglacial processes from last glacial to recent inferred from Spituk sequence, Leh Valley, Ladakh Himalaya. *Journal of the Geological Society of India* 78(2): 147–156.

Scherler, D., Bookhagen, B., Strecker, M.R. et al. (2010). Timing and extent of late Quaternary glaciation in the western Himalaya constrained by 10Be moraine dating in Garhwal, India. *Quaternary Science Reviews* 29(7–8): 815–831.

Searle, M.P. (1986). Structural evolution and sequence of thrusting in the High Himalayan, Tibetan-Tethys and Indus suture zones of Zanskar and Ladakh, Western Himalaya. *Journal of Structural Geology* 8: 923–936.

Searle, M.P., Cooper, D.J.W., and Rex, A.J. (1988). Collision tectonics of the Ladakh-Zanskar Himalaya. In: *Tectonic Evolution of the Himalayas and Tibet* (ed. R.M. Shackleton, J.F. Dewey, and B.F. Windley) *Philosophical Transactions of the Royal Society of London* 326: 117–150.

Searle, M.P., Dewey, J.F., and Dunlap, W.J. (1998). Transpression tectonics along the Karakoram fault zone, northern Ladakh: constraints on Tibetan extrusion. In: *Continental Transpressional and Transtensional Tectonics* (ed. R.E. Holdsworth, R.A. Strachan, and J.E. Dewey). *Geological Society of London Special Publication* 135: 307–326.

Searle, M.P. and Phillips, R.J. (2007). Relationships between right-lateral shear along the Karakoram fault and metamorphism, magmatism, exhumation and uplift: evidence from the K2–Gasherbrum–Pangong ranges, north Pakistan and Ladakh. *Journal of the Geological Society* 164(2): 439–450.

Searle, M.P., Pickering, K., and Cooper, D. (1990). Restoration and evolution of the intermontane Indus molasse basin, Ladakh Himalaya, India. *Tectonophysics* 174: 301–314. https://doi.org/10.1016/0040-1951(90)90327-5.

Sen, K., Mukherjee, B.K., and Sachan, H.K. (2009). Field and microstructural analysis of the Pangong Granodiorite, Ladakh (NW India): implications for tectonics along the Karakoram Fault Zone. *Current Science* 96(8): 1124–1130.

Seong, Y.B., Owen, L.A., Bishop, M.P. et al. (2007). Quaternary glacial history of the Central Karakoram. *Quaternary Science Reviews* 26(2528): 3384–3405.

Sharma, S. and Shukla, A.D. (2018). Factors governing the pattern of glacier advances since the Last Glacial Maxima in the transitional climate zone of the Southern Zanskar Ranges, NW Himalaya. *Quaternary Science Reviews* 201: 223–240.

Sharma, C.P., Chahal, P., Kumar, A. et al. (2022). Late Pleistocene–Holocene flood history, flood-sediment provenance and human imprints from the upper Indus River catchment, Ladakh Himalaya. *GSA Bulletin* 134(1–2): 275–292. https://doi.org/10.1130/B35976.1

Sharma, C.P., Rawat, S.L., Srivastava, P. et al. (2020). High-resolution climatic (monsoonal) variability reconstructed from a continuous ~2700-year sediment record from Northwest Himalaya (Ladakh). *The Holocene* 30(3): 441–457.

Shukla, T., Mehta, M., Jaiswal, M.K. et al. (2018). Late Quaternary glaciation history of monsoon-dominated Dingad basin, central Himalaya, India. *Quaternary Science Reviews* 181: 43–64.

Sinclair, H.D., Mudd, S.M., Dingle, E. et al. (2016). Squeezing river catchments through tectonics: shortening and erosion across the Indus Valley, NW Himalaya. *Geological Society of America Bulletin* 129: 203–217.

Spencer, J.Q and Owen, L.A. (2004). Optically stimulated luminescence dating of Late Quaternary glaciogenic sediments in the upper Hunza valley: validating the timing of glaciation and assessing dating methods. *Quaternary Science Reviews* 23(1–2) 175–191.

Srivastava, P., Kumar, A., Mishra, A. et al. (2013). Early Holocene monsoonal fluctuations in the Garhwal higher Himalaya as inferred from multi-proxy data from the Malari paleolake. *Quaternary Research* 80(3): 447–458.

Srivastava, P., Kumar, A., Chaudhary, S. et al. (2017). Paleofloods records in Himalaya. *Geomorphology* 284: 17–30.

Streule, M.J., Phillips, R.J., Searle, M.P.et al. (2009). Evolution and chronology of the Pangong Metamorphic Complex adjacent to the Karakoram Fault, Ladakh: constraints from thermobarometry, metamorphic modelling and U–Pb geochronology. *Journal of the Geological Society* 166(5): 919–932.

Taylor, P.J. and Mitchell, W.A. (2000). The Quaternary glacial history of the Zanskar Range, north-west Indian Himalaya. *Quaternary International* 65–66: 81–99. https://doi.org/10.1016/S1040-6182(99)00038-5.

Thakur, V.C. (1983). Palaeotectonic evolution of Indus-Tsangpo Suture Zone in Ladakh and southern Tibet. In: *Geology of Indus Suture Zone of Ladakh* (ed. V.C. Thakur and K.K. Sharma), 195–204.

Thakur, V.C. and Misra, D.K. (1984). Tectonic framework of the Indus and Shyok suture zones in the eastern Ladakh, Northwest Himalaya. *Tectonophysics* 101: 207–220. Dehra Dun, India: Wadia Institute of Himalayan Geology.

Thakur, V.C. and Rawat, B.S. (1992). Geological map of northwestern Himalaya.

Tropical Rainfall Measuring Mission (TRMM) (2011). TRMM (TMPA) Rainfall Estimate L3 3 hour

0.25 degree × 0.25 degree V7, Greenbelt, MD, Goddard Earth Sciences Data and Information Services Center (GES DISC). https://doi.org/10.5067/TRMM/TMPA/3H/7.

Wu, F.Y., Clift, P.D., and Yang, J.H. (2007). Zircon Hf isotopic constraints on the sources of the Indus Molasse, Ladakh Himalaya, India. *Tectonics* 26: TC2014. doi:10.1029/2006TC002051.

Wünnemann, B., Demske, D., Tarasov, P. et al. (2010). Hydrological evolution during the last 15 kyr in the Tso Kar lake basin (Ladakh, India), derived from geomorphological, sedimentological and palynological records. *Quaternary Science Reviews* 29(9–10): 1138–1155.

Yan, Q., Owen, L.A., Zhang, Z. et al. (2020). Deciphering the evolution and forcing mechanisms of glaciation over the Himalayan-Tibetan orogen during the past 20,000 years. *Earth and Planetary Science Letters* 541: 116295. doi.org/10.1016/j.epsl.2020.116295

Yin, A. (2006). Cenozoic tectonic evolution of the Himalayan orogen as constrained by along-strike variation of structural geometry, exhumation history, and foreland sedimentation. *Earth-Science Reviews* 76(1–2): 1–131.

# 22

# A Review of Remote Sensing and GIS-Based Soil Loss Models With a Comparative Study From the Upper and Marginal Ganga River Basin

*Rohit Kumar*[1,*], *Rahul Devrani*[2,*], *and Benidhar Deshmukh*[1]

[1] *Discipline of Geology, School of Sciences, Indira Gandhi National Open University, New Delhi 110068, India.*
[2] *Delhi School of Climate Change & Sustainability, Institution of Eminence, University of Delhi, India*
* *Corresponding author*

## 22.1 Introduction

Soil erosion is a geomorphic process that occurs globally due to varying erosion rates and magnitude of precipitation, soil types and their characteristics, land use-landcover (LULC), and results in depletion of soil nutrients (Thomas et al., 2018). This geomorphic process is a significant aspect of land degradation in which top fertile organic matter-rich soil is displaced from its source region by various processes, like soil detachment, transportation, and deposition (Balasubramanian, 2017). Such land degradation processes are induced by the action of water, wind, and tillage (NRSC, 2014). Due to its implications over nature, agronomic efficiency, and food security, soil erosion is considered an imperative theme of the 21st century (Eswaran et al., 2001). Soil erosion is driven by natural events like droughts, floods, landslides, earthquakes, and anthropogenic activities. Anthropogenic activities like deforestation, farming, river damming, industrialization, mining, and development projects (human settlements, roads, and highways) increase soil erosion (Lal, 2003). On a worldwide scale, 1094 million hectares are affected by high erosion, and 224 million hectares suffer from severe erosion that cannot recover (Oldeman, 1994). The erosion of soil is categorized based on the method of transport and physical characteristics of erosional features like the splash erosion occurring due to raindrops, sheet erosion due to runoff flow, rill erosion due to rapid concentrated runoff flow, gully erosion due to expansion of rills and sheets, and channel erosion due to bank vegetation disturbance and increased stream power (Holy, 1980; Goldman, et al., 1986; Morgan, 1986; Balasubramanian, 2017). Sheet and rill erosion is categorized under the slight erosion category, shallow gullied areas as moderate erosion and ravines, or deep gullied regions as the severe category of erosion (SAC, 2016).

Healthy fertile soil is the foundation of agriculture and resource of food, fiber, and clean water (Borrelli et al., 2017); however, agricultural lands are decreasing across the globe despite food needs rising at an unprecedented rate (Bruinsma, 2003; FAO, 2005, 2015; Gelfand et al., 2013; Lambin et al., 2013). Soil erosion has become one of the most serious forms of soil degradation, resulting in a considerable drop in soil fertility and agricultural production as a result of inappropriate land use, and it needs to be monitored and managed. The mapping of erosion-affected soils, degraded land, and potential erosion sites, and information collected from field surveys and satellite data, will help to collect vital data for managing its effects (Shoshany et al., 2013). Several diverse models related to soil erosion have been used or tested at multi-scales, based on their appropriateness for various areas throughout the world (Thomas et al., 2018) (Table 22.1), yet there is uncertainty in the global erosion rates that vary from 20 to 200 petagrams per year (Doetterl et al., 2012). For instance, Fournier (1960) and Walling (1987) estimated the rate of soil loss as between 0.088 mm/yr and 0.30 mm/yr and Pimentel et al. (1995) suggests that every year ~75 billion metric tons of soil is eroded from the surface by fluvial (water), aeolian (wind), and ice activities. The FAO and ITPS (2015) suggested that 20–30 Gt of soil is worn down every year, and Montanarella (2015) claimed soil loss of 25–40 billion tons. Borrelli et al., (2017) estimated a soil erosion of 35 petagrams per year for 2001 and 35.9 petagrams per year in 2012 (an increase of 2.5%), and a recent study carried out by Borrelli et al., (2020) estimated a possible rate of $43 \pm 9.2$ petagrams per year that may increase to +30% to

*Advances in Remote Sensing Technology and the Three Poles*, First Edition. Edited by Manish Pandey, Prem C. Pandey, Yogesh Ray, Aman Arora, Shridhar D. Jawak, and Uma K. Shukla.

**Table 22.1** Description of some widely used soil erosion models (Adapted from Avwunudiogba and Hudson, 2014).

| Model | References | Comments/Parameters |
|---|---|---|
| **USLE**<br>Universal Soil Loss Equation | (Wischmeier and Smith, 1965) | Rainfall erosivity, erodibility of soil, topographic factor, crop management, support practices. In this model, all variables are represented by a single number predicted from rainfall, soil, or research data that are not geographic location based (Renard et al., 1997). |
| **RUSLE**<br>Revised Universal Soil Loss Equation | (Wischmeier and Smith, 1978) | Changes done in RUSLE: i) development of erosion index; ii) representation of erodibility of soil to compute the erodibility factor for various soil textures; iii) topographic factors for irregular slopes; and iv) improvement in erosion control support practices. |
| **ANSWERS**<br>The Areal Non-point Source Watershed Environment Response<br>Simulation | (Beasley et al., 1980) | This process-based model estimates soil erosion from a watershed. It includes a hydrology model, a sediment detachment model, and a water movement model. It may be connected to GIS and employs runoff, infiltration, subsurface drainage, and erosion characteristics. |
| **CREAMS**<br>Chemical Runoff and Erosion from Agricultural<br>Management System | (Knisel, 1980) | An empirically-based model with the ability to analyze nutrients in agricultural fields. It was created in the United States, and the inputs are precipitation, radiation, temperature, landuse, cultural practices, plant nutrients, and pesticides. |
| **SLEMSA**<br>Soil Loss Estimation Model for Southern Africa | (Stocking and Elwell, 1973) | Predicts long-term annual soil erosion through sheets on slopes and in agricultural fields in Southern Africa.<br>It can be expressed as:<br>$Z = K * C * X$<br>where Z represents the annual mean soil loss, K is annual soil loss from a plot of 30 m × 10 m with 4.5% slope, F is erodibility, C is cropped plot soil loss ratio and bare fallow plot, and X is the ratio of plot length and slope. |
| **EPIC**<br>Erosion Productivity<br>Impact Calculator | (Williams et al., 1983) | This model was created to work out the link between soil erosion and soil production across the United States. Its major components are temperature of soil, tillage, climate, hydro, nutrients, etc. |
| **GLEAMS**<br>Groundwater Loading<br>Effects of Agricultural<br>Management System | (Leonard et al., 1987) | A mathematical model for evaluating impact of agricultural management methods on field-size areas. The GLEAMS parameters are the hydrology component, erosion component, and pesticide component. |
| **MADALUS**<br>Mediterranean<br>Desertification and Land<br>Use | (Kirkby et al., 1988) | Process-based model for simulating erosion, plant development, and other processes that were created in Europe. This model had not been well tested in different climates or ecoregions, and it requires a number of variables. |
| **KYERMO**<br>Kentucky Erosion Model | (Hirschi and Barfield, 1988) | Process-based model created in United States, not as extensively utilized as others. |
| **WEPP**<br>Water Erosion Prediction<br>Project | (Nearin, 1989) | Process-based model that incorporates rill detachment (which comprises two parameters: baseline rill erodibility, and critical hydraulic shear of the soil), inter-rill detachment, and deposition parameters. It was created in 1985. |

Table 22.1 (Continued)

| Model | References | Comments/Parameters |
|---|---|---|
| **LISEM**<br>Limburg Soil Erosion Model | (De Roo et al., 1996) | Process and GIS-based hydrological mode that incorporates rainfall, infiltration, and vertical water movement in soil, surface storage in micro-depressions, and runoff. |
| **EUROSEM**<br>European Soil Erosion Model | (Morgan et al., 1998) | Estimates erosion and runoff though rills. It includes the following parameters: vegetation data, soil hydraulic properties, rainfall, soil mechanical properties, microtopographical data, and topographic data. It can be connected to GIS. |
| **GAMES**<br>Guelph Model for Evaluating the Effects of Agricultural Management Systems on Soil Erosion and Sedimentation | (Singh and Frevert, 2005) | With three components: erosion, sediment delivery, and issue categorization, this model was created as a screening tool to erosion and deposition in the field and watershed sizes. |
| **SWAT**<br>Soil and Water Assessment Tool | (Gassman et al., 2007) | Physical based-basin-scaled continuous-time model for analyzing water resource and nonpoint pollution concerns, as well as sediment and agricultural chemical output in a watershed. Landcover management, temperature of soil, and amount of nutrients are some major components. It may be connected to GIS. |
| **Erosion 2D/3D**<br>2D/3D Rainfall Erosion Model | (Lenz et al., 2018) | This is an eight-parameter physical model with moisture and texture of soil, density of bulk, surface roughness, erosion resistance, and skin factor. |

+66% in the future due to vigorous hydrological cycles. An average soil erosion rate of 30, 40, and 17 tons/ha/yr is estimated for Asia, South America, and Africa (Pimentel et al., 1995). In India, studies by Narayana and Babu (1983) and Kothyari (1996) estimated 5334 million tonnes of soil loss (average 16.4 tons/ha/yr) per year. Saha (2003) claimed 1650 million hectares of land are affected by erosion. In 2003–2005, 94.5 million hectares of land were estimated to have been impacted by soil erosion, which increased to 96.4 million hectares in 2011–2013 (SAC, 2016).

## 22.2 Geospatial Models

The erosion rate has primarily been estimated by field-based approaches employing different techniques, e.g., field photographs (Watson and Evans, 1991) and Microfilometer and Erosion pin techniques (Joshi, 2014). Estimating soil loss by field surveys is expensive, time-consuming, and with inaccessibility due to rugged terrain and other unfavorable scenarios. Modern technologies, i.e., aerocraft, remote sensing, and GIS, have advanced soil erosion assessment and permit testing and evaluation at a global or regional scale (Fistikoglu and Harmancioglu, 2002). Some recently-used models are SWAT (Soil and Water Assessment Tool), WEPP (Water Erosion Prediction), USLE (Universal Soil Loss Equation), RULSE (Revised Universal Soil Loss Equation, and MUSLE (Modified Universal Soil Loss Equation*)* (Pandey et al., 2007; Dabral et al., 2008; Benzer, 2010; Sheikh et al., 2011; Biswas, 2012; Kumar et al., 2014).

Cook (1936) sought to predict soil loss using characteristics such as soil texture, erosivity by rainfall and runoff, and vegetation cover (Table 22.2) (Renard et al., 1997). Zingg (1940) provided a quantitative estimating equation for erosion in the field, based on the length and slope percentage (Table 22.2) (Wischmeier and Smith, 1978; Renard et al., 1997). It was modified by Smith in 1941 by adding crop management and support practice factors. Browning et al. (1943) made the table for crop management and support practices for simple field use of the equation given by Smith. Smith and Whitt (1948) developed an equation for the estimation of erosion (A = C*LS*K*P) in late 1940, where "A," "C," "LS," "K," and "P" represent annual soil loss, crop management, slope length and steepness, erodibility of soil, and support practice, respectively. However, this equation needed to include the rainfall factor. Hence, Wischmeier and Smith developed the widely used USLE model in 1965, and its updated version, RUSLE, was introduced in 1978, which was recently modified and presented as MUSLE (Table 22.1).

**Table 22.2** History of soil loss estimation using geospatial techniques.

| S. No | Model | Researchers | Remarks |
|---|---|---|---|
| 1. | NO NOMENCLATURE | (Cook, 1936) | Based on soil texture, erosivity of rainfall and runoff, precipitation in the form of water, and vegetation cover |
| 2. | NO NOMENCLATURE | (Zingg, 1940) | First mathematical equation incorporating slope steepness and slope length (referred to as the slope practice method) |
| 3. | NO NOMENCLATURE | (Smith, 1941) | Zingg's equation was modified by adding crop management and support practice |
| 4. | NO NOMENCLATURE | (Browning et al.,, 1943) | Crop management and support practice factors tables were added to simplify field use of the equation given by Smith |
| 5. | NO NOMENCLATURE | (Smith and Whitt, 1948) | The equation (A = C*LS*K*P) was presented where, C is crop management, LS is topographic factor, K is erodibility of soil, and P is support practice. However, it lacks the rainfall erosivity factor |
| 6. | USLE | (Wischmeier and Smith, 1965) | Includes rainfall erosivity, erodibility of soil, slope length and steepness, crop management, and support practices |
| 7. | RUSLE-1 | (Wischmeier and Smith, 1978) | Includes erosion index, soil erodibility, topographic factors, support practices, and conservation factors |
| 8. | RUSLE-2 | (USDA, 2001) | Incorporates all soil-related information in soil factors and these factors are combined with topographic factors to define hillslope profile for erosion computation based on climatic conditions |
| 9. | MUSLE | (Williams and Berndt, 1972) | This model replaces the rainfall factor to the runoff characteristics and other factors are the same as RUSLE |

### 22.2.1 USLE (Universal Soil Loss Equation)

Wischmeier and Smith presented an empirical model of soil erosion to estimate average annual soil erosion in rills of small watersheds, based on 40 years of field observations gathered through the United State Department of Agriculture (USDA) (Novotny, 1994). The model equation can be written as

$$A = R * K * LS * C * P \tag{22.1}$$

where average annual soil loss in t/ha/yr is $A$, rainfall erosivity $R$ (MJ mm ha$^{-1}$ h$^{-1}$ yr$^{-1}$), soil erodibility $K$ (t ha h ha$^{-1}$ MJ$^{-1}$ mm$^{-1}$) slope length and steepness $LS$, crop management $C$, and support practice $P$.

USLE field use started in 1960, and in 1965 the primary reference manual *Agricultural Handbook* was released, which was later revised in 1978 (Renard et al., 1991). It estimates average annual soil erosion by rill and sheets (Renard et al., 1991; Jain et al., 2001). USLE results were used to guide farmers and conservationists in selecting different approaches that would control soil erosion, and it is also used in food security (Renard et al., 1991). However, the USLE cannot estimate deposition, channel erosion, sediment yield, and gully erosion (Jain et al., 2001). The USLE works best on a farmland scale or small area rather than a larger catchment area, which means that it will not give accurate estimation at a global scale and provides estimation over an extended time period, not for a single storm event (Fistikoglu and Harmancioglu, 2002). It compares estimated soil loss with the particular soil type (Erdogan et al., 2007). USLE lacks spatially distributed structure, thus not involving spatial resolution (Fistikoglu and Harmancioglu, 2002). As the USLE model is based on experimental erosion data of a minimal basin area or sub-watershed on a uniform hillslope, the results are restricted to comparable circumstances (Renard et al., 2010). The results failed to account for the influence of channels forming under natural conditions and the gully development process at steep boundaries (Renard et al., 2010). Hence, the USLE was revised to RUSLE for estimations.

### 22.2.2 RUSLE (Revised Universal Soil Loss Equation)

RUSLE can predict rate of soil erosion in all natural environmental conditions (Renard et al., 1997). It is used to monitor erosion through rills and sheets developed by surface runoff (Merritt et al., 2003). It can be present as:

$$A = \text{'}R\text{'} * \text{'}K\text{'} * \text{'}LS\text{'} * \text{'}C\text{'} * \text{'}P\text{'} \tag{22.2}$$

where $A$ is annual soil loss rate, $R$ is rainfall erosivity, $K$ is erodibility of soil, $LS$ is slope length and steepness, $C$ is crop management, and $P$ is support practice.

The first version of RUSLE was released in 1997, and it operated in a DOS-based computer environment which permitted only English input units, so conservation of output and input in metric units requires hand conservation (Renard et al., 2010). The most significant change was in the cover management factor, and now it can be used in any landuse class (Renard et al., 2010). There are some weaknesses in the model, such as: i) RUSLE-1 structure was based on science rather than user views: ii) user can manipulate the database; and iii) DOS-based interface in RUSLE-1 required windows-based graphics (Renard et al., 2010). The second version was released in 2001, but it was distributed and appropriately used in 2004 (Renard et al., 2010). Fundamental erosion science and process-based equations are used in RUSLE-2 (Renard et al., 2010). RUSLE-2 organizes information into objects and object-oriented organization applies to computer programming and how users put values of parameters (Renard et al., 2010). Soil-related information and management information is combined with the topographic description to develop to extract desire data to compute the erosion equation (Renard et al., 2010).

#### 22.2.2.1 Rainfall Erosivity Factor "R"

Rainfall Erosivity Factor "R" is a function of rainfall intensity, falling raindrops, and its intensity for 30 m, or R is the erosion index for the period of storm (Pandey et al., 2007; Devatha et al., 2015). This factor varies depending on the climate zone, e.g., tropical climate (around equatorial regions), tropical, wet, and dry zones, and semi-arid zones. Diverse equations are used to calculate factor R (Ghosal and Bhattacharya, 2020). In India, Prasannakumar et al. (2011) utilized the Arnoldus (1980) equation to compute the R factor for the Attapady Valley in Kerala (Ghosal and Bhattacharya, 2020). The equation is as follows:

$$R = \sum_{i=1}^{12} 13735 \times 10^{\left(1.5\log 10\left(\frac{Pi}{P}\right) - 0.08188\right)} \tag{22.3}$$

where R is erosivity of rainfall, *Pi* is rainfall for a month in mm, and *P* is rainfall for an annual period in mm. Similarly, for tropical wet and dry climatic zones, Kumar et al. (2014) and Kumar and Deshmukh (2018) applied the equation derived by Singh et al. (1981) in the case of the Himalayas and central India. This equation is commonly used throughout India:

$$R = 79 + 0.363 \times AAP \tag{22.4}$$

where *R* is the erosivity of rainfall and *APP* is the average of annual rainfall

#### 22.2.2.2 Soil Erodibility "K"

Soil Erodibility "K" is the soil loss rate determined by its characteristics, while all other factors remain constant (Wischmeier and Smith, 1978). Types of soil texture, organic matter, and erosion play a significant role in erosion, which may be assessed by field surveys (Devatha et al., 2015). Different equations can calculate it based on organic matter, variable size and structure, permeability, and color and type of soil (Ghosal and Bhattacharya, 2020). The K value can also be assigned by soil texture identification (Devatha et al., 2015).

#### 22.2.2.3 Slope Length and Steepness "LS"

Slope Length and Steepness "LS" measures the effect of water runoff on slope length and steepness of rills, sheet, or inter-rill regions. The choice of selecting the LS factor depends on the DEM spatial resolution, computing source availability, and scale of the study area (Benavidez et al., 2018). DEM with a spatial resolution greater than 100 m should be used with a technique that simply considers slope length and steepness. However, DEM with a spatial resolution of 15 m or higher should be used with an LS approach that considers flow accumulation (Schmitt, 2009; Benavidez et al., 2018). Factor L characterizes the overland stream beginning point to the point where the slope angle diminishes, and deposition takes place or water enters a channel that will be part of the drainage system (Wischmeier and Smith, 1978). Slope length is computed through the equation given by McCool et al. (1987), which is written as:

$$L = \left(\frac{\ell}{72.6}\right)^m \tag{22.5}$$

where L is slope length, $\ell$ is slope length in field, $m$ is exponent that depends on slope steepness (if the slope is <5% then the $m$ value 0.5 is used; if the slope = 4% then 0.5 for slope >3 m is 0.3).

Factor S is described as the inclination of the area from the beginning of the overland stream or hillslope to the deposition (Wischmeier and Smith, 1978). Although runoff will increase with an increase in slope gradient, it can also be influenced by crop types, surface roughness, and profile saturation (Wischmeier and Smith, 1978). If slope length is greater than 4 m, then the equation given by McCool et al. (1987) can be used:

$$S = 10.8 sin\theta + 0.03, s < 9\% \tag{22.6}$$

$$S = 16.8 sin\theta + 0.05, s \geq 9\% \tag{22.7}$$

where *S* is slope steepness and Θ is slope angle in degree.

Several equations have been developed by Wischmeier and Smith (1957), Moore and Burch (1986), and McCool et al. (1987) for estimating LS factors together based on topography and DEM resolution. Wischmeier and Smith (1957) developed this frequently-used equation by interchanging slope gradient (Ghosal and Bhattacharya, 2020). Recently, this equation was used in the Arc-Macro-based tool developed by Hickey (2001) with some modifications in the C++ programming language (Van Remortel et al., 2004). This equation is also frequently used in many studies (Kumar et al., 2014, 2021; Kumar and Deshmukh, 2018). The equation is as follows:

$$LS = \left(\frac{\ell}{72.6}\right)^{m} (65.41 sin^{22} + 4.56 sin^{2} + 0.065) \quad (22.8)$$

where, $\ell$ slope length, β is slope angle, and $m$ is variable of slope.

#### 22.2.2.4 Crop Management (C)

The Crop Management (C) factor estimates the type and density of the vegetation as it prevents the raindrops from impacting directly onto the ground and slowing runoff (Das et al., 2018). The C factor value depends on various factors such as category of vegetation, development, and density of that vegetation (Ghosal and Bhattacharya, 2020). Its value ranges from 0 to 1, indicating non-erodible to erodible conditions, respectively (Das et al., 2018). C values are calculated either by assigning previously defined values or by the equation given by Zhou et al. (2008) and Kouli et al. (2009), which is expressed as:

$$C = \exp\left[-\alpha(NDVI) / \beta(NDVI)\right] \quad (22.9)$$

where $NDVI$ is Normalized Difference Vegetation Index, and $\beta = 1$ and $\alpha = 2$ are constants (Vander-Knijff et al., 2000).

#### 22.2.2.5 Support Practices "P"

The Support Practices "P" factor is the soil loss ratio caused by some support method due to up-down slope farming (Pandey et al., 2007). When a slope or sloping soil is exposed, the system's protection against close-growing crops must be backed up by techniques that delay surface water flow and limit the carrying capacity (Wischmeier and Smith, 1978). The essential cropland practices are the terrace system, contour tillage, and strip cropping on the contour (Wischmeier and Smith, 1978; Ghosal and Bhattacharya, 2020). Its values range 0 to 1, where zero indicates good conservation and one indicate very less or poor conservation (Das et al., 2018).

### 22.2.3 MUSLE (Modified Universal Soil Loss Equation)

The RUSLE and USLE models do not take runoff into account; even transported sediments moving through streams and rainfall variation also affects sediment yield (Kinnell, 2005; Pandey et al., 2009). To address this problem, Williams and Berndt presented the MUSLE (Modified Universal Soil Loss Equation) in 1972, which substitutes runoff characteristics for the rainfall factor. Since the runoff factor is a function of rainfall energy and moisture condition, it ignores delivery ratios and enables the estimate of individual storm event sediment production (Zhang et al., 2009). The MUSLE is used to calculate the total quantity of sediment supplied to the stream network within a watershed by adding all of the sites where overland flow joins the stream (Zhang et al., 2009). It is expressed as

$$Y = 11.8 * (Q * qp)^{0.56} * K * LS * C * P \quad (22.\ 10)$$

where $Y$ is sediment yield (metric tons), $Q$ is runoff volume, $qp$ is peak flow rate ($m^3$/sec), $K$ is erodibility of soil, $LS$ is slope length and steepness, $C$ is crop management, and $P$ is support practice.

## 22.3 A Case Study in Upper and Marginal Ganga River Basins Using RUSLE Model

Numerous soil estimation models have been developed, with the RUSLE model being the most commonly used model. The RUSLE model estimates the average annual rate of erosion in different environments, e.g., forest, agriculture, mining area, construction sites (Stone and Hilborn, 2000; Dais, 2008). In many studies, the RUSLE model has been used in gullied areas (Wang et al., 2018; Yesuph and Dagnew, 2019; Senanayake et al., 2020). It can be used in hilly terrain (Garde and Kothyari, 1987; Kumar et al., 2014; Mahapatra et al., 2018; Swarnkar et al., 2018) as well as in plain areas (Biswas and Pani, 2015; Kumar and Deshmukh, 2018; Kumar et al., 2021). We have selected two sub-watersheds from the upper and marginal Ganga River Basin to observe the soil estimation through the RUSLE model in different topographic conditions.

### 22.3.1 Study Area (Upper and Marginal Ganga River Basins)

Two sites have been selected in different diverse locations in the upper and marginal Ganga River basins; the Rishiganga River Basin (RG) in the upper Ganga River Basin and the lower Chambal Basin in the marginal Ganga River Basin (Figures 22.1A and 22.1B). The RG is a tributary of the Dhauliganga River, which originates from the north and snout of the Rishi glacier (Kothyari et al., 2021; Kumar et al., 2021), and its basin area lies in the 79°39'31" to 80°3'0" longitude and 30°33'29.41" to 30°14'8.3" latitude and has a total area of 687.838 $km^2$ (Figure 22.1A). The RG

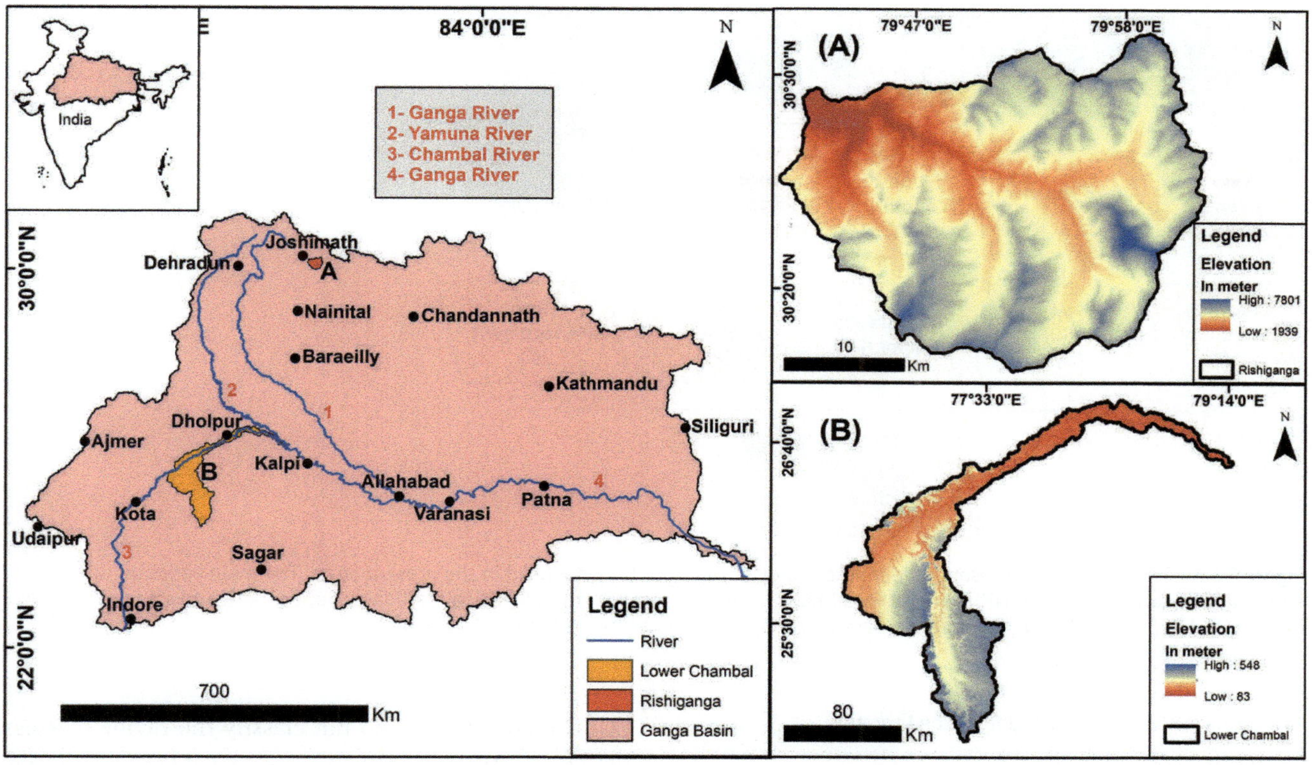


**Figure 22.1** The map shows the basin of Ganga River, the largest river basin in India, covering an area of ~988,727.8 km$^2$. In the present study, two locations, e.g., Rishiganga Basin and lower Chambal Basin are selected from the upper and marginal part of the Ganga River Basin. (A) and (B) show the topography of the selected basins developed from ALOS-PALSAR 12.5 m resolution DEM. Inset, the blue polygon, represents river channels and 1, 2, and 3 indicate Ganga, Yamuna, and Chambal Rivers and 4 represents the Ganga River after the Yamuna-Ganga confluence.

Basin is influenced by the Indian monsoon and highlands of Tibet (Kumar et al., 2021; Pandey et al., 2021), and its maximum and minimum elevation ranges from 7801 m.s.l. to 1939 m.s.l. with a mean elevation of 5033.49 m.s.l.

The frequent flash floods make this basin very prone to erosion as high flood discharge erodes soil rapidly The most recent flash flood occurred in February 2021 (Rana et al., 2021; Sain et al., 2021). The RG basin lies in high-grade metamorphic rocks, e.g., Kyanite gneiss, quartzite, staurolite, garnet, and biotite schists of the Vaikrata Group, bounded by the Trans Himadari Fault in the north-east and the Vaikrata Thrust in the southwestern (Valdiya, 1980; Pandey et al., 2021).

The second window is the Lower Chambal Basin (LC) lying in the marginal Ganga Basin. The Chambal River being a tributary of River Yamuna, originates from the Janapav Valley at the Malwa ridge (Kaushik and Ghosh, 2015). The LC basin is bounded between the 76°31'47.39" and 79°15'28.76" longitudes and the 26°53'41.44" and 24°43'22.39" latitudes, covering an area of 11,382.5 km$^2$ with the elevation ranging from 548 river m.s.l. to 83 river m.s.l. (with a mean elevation of 296 river m.s.l.) (Figure 22.1B). The LC basin is located in a humid to sub-humid climate with ~900 mm annual rainfall (Joshi, 2014). The rugged topography developed by gullies and ravines over the Bhander, Rewa, and Malwa groups is the distinct geomorphic unit that makes the LC Basin prone to soil erosion (Singh, 1996; Kumar et al., 2022b).

## 22.3.2 Dataset and Methodology

The RUSLE model (Equation 22.1) was utilized in this work to predict the rate of soil loss, and available high spatial resolution data were employed for a more detailed analysis (Figure 22.2). Soil texture maps for RG and LC river basins are procured from the National Bureau of Soil survey and Land Use planning (NBSS & LUP) as it provides 1:50,000-scale data by using mainly field surveys, remote-sensing analysis, and laboratory analysis (Shyampura and Sehgal, 1995; Tamgadge, 1996; Singh et al., 2004). Factor K values have been assigned from the literature (Table 22.3). Rainfall data of 2020 procured from Persiann-CCS (Cloud Classification System) developed by CHRS (Center for Hydrometeorology and Remote Sensing) at 4 km resolution have been used to create a yearly average rainfall map.

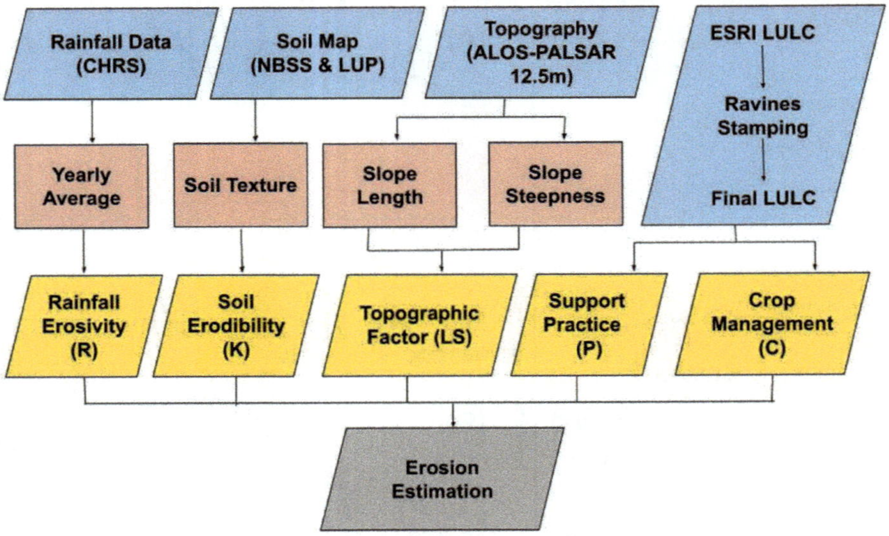


**Figure 22.2** The flowchart represents the methodology of the RUSLE model used in the present study. The main variables are soil erodibility calculated by using NBSS and LUP soil texture, rainfall erosivity from CHRS rainfall data, slope length and steepness factor from ALOS-PALSAR DEM, and crop management and support practices derived from LULC maps.

**Table 22.3** K values assigned from the literature.

| Soil Texture | K Value | Reference |
|---|---|---|
| Coarse loamy | 0.032 | (Jain et al., 2010) |
| Loamy | 0.02 | (Jain et al., 2010) |
| Loamy skeletal | 0.023 | (Jain et al., 2010) |
| Sandy skeletal | 0.042 | (Jain et al., 2010) |
| Silty | 0.75 | (Biswas and Pani, 2015) |
| Clayey | 0.1 | (Jones et al., 1996) |
| Fine loamy | 0.057 | (Jones et al., 1996) |

Raster layers of rainfall are converted into point data, and interpolation has been carried out to export data in 10 m resolution. Rainfall erosivity maps of the basins have been calculated by using Equation 22.4. Slope length and slope steepness (LS) of the RG and LCB have been calculated by utilizing 12.5 m resolution ALOS-PALSAR DEM procured from the ASF (Alaska Satellite Facility). ALOS DEM was further preprocced to fill the voids and then used for further analysis. The LS factor is calculated through a tool developed by Hickey (2001) in Arc Marco Langauge (Equation 22.8).

The LULC map for the year 2020 created by ESRI (Environmental Systems Research Institute) using Sentinel-2 imageries at 10 m resolution by Microsoft's Planetary Computer and scaled using Microsoft Azure Batch (Karra et al., 2021), has been used to calculate crop management and support practice factors. The LULC maps by ESRI do not classify the ravine classes; therefore, ravines of the LC basin have been identified and digitized in Google Earth and imported into the GIS platform. Later, this layer was combined with the LC Basin LULC map, and the modified LULC map was generated. P and C values are assigned from the literature (Table 22.4).

### 22.3.3 Rate of Soil Loss in Rishiganga Basin (RG)

The mean rate of soil erosion in the RG Basin is 9.82 ton/ha/yr for 2020, and these values are divided into six classes (Table 22.5), as proposed by Singh et al. (1992) for India with minor modifications (Figure 22.4). The RG Basin comprises coarse loamy, loamy, loamy skeletal, and sandy skeletal soil texture, covering an area of 81.14 $km^2$, 195.14 $km^2$, 5.05 $km^2$, and 406.51 $km^2$, respectively (Figure 22.3A). The mean soil loss rate in sandy skeletal texture soil is highest (11.12 ton/ha/yr), followed by 10.72 ton/ha/yr in coarse loamy, 10.30 ton/ha/yr in loamy skeletal, and 6.94 ton/ha/yr in loamy soil, indicating sandy texture as more prone to soil erosion. The RG Basin's mean annual rainfall erosivity in 2020 is 98.42 MJ mm/ha/hr/yr, which has been divided into five classes (Figure 22.3B). Class 1 ranging from 93.40 to 95.94 MJ mm/ha/hr/yr shows a mean rate of 8.9 ton/ha/yr; class 2 ranging from 95.95 to 97.79 MJ mm/ha/hr/yr shows a mean rate of 10.22 ton/ha/yr; class 3 ranging from 97.80 to 99.44 MJ mm/ha/hr/yr shows a mean rate of 9.52 ton/ha/yr; class 4 ranging from 99.45 to 101.05 MJ mm/ha/hr/yr shows a mean rate of 9.9 ton/ha/yr; and class 5

**Table 22.4** LULC classes and their assigned C and P values from the literature.

| LULC Class | C Value | P Value | Reference |
|---|---|---|---|
| Builtup | 0.01 | 1 | (Wanielista and Yousef, 1993) |
| Crop land | 0.08 | 0.5 | (Wanielista and Yousef, 1993) |
| Forest | 0.008 | 0.8 | (Pandey et al., 2007) |
| Scrub land | 0.1 | 0.8 | (Jain et al., 2001) |
| Water body | 0 | 0 | (Jain et al., 2001) |
| Barren land | 0.45 | 1 | (Wanielista and Yousef, 1993) |
| Gullies/Ravines | 1 | 1 | (Kumar et al., 2022a) |
| Sandy | 1 | 1 | (Wanielista and Yousef, 1993) |
| Glacier/Snow | 0 | 0 | (Wanielista and Yousef, 1993) |

**Table 22.5** Soil loss rate in different categories.

| Soil Loss Category | Rate of Soil Loss (t/ha/yr) | Area ($km^2$) |
|---|---|---|
| Slight erosion | 0 to 5 | 414.63 |
| Moderate erosion | 5–10 | 78.39 |
| High erosion | 10–20 | 82.25 |
| Very high erosion | 20–40 | 66.09 |
| Severe erosion | 40–80 | 38.47 |
| Very severe erosion | >80 | 8.02 |

ranging from 101.06 to 103.67 MJ mm/ha/hr/yr shows a mean rate of 10.98 ton/ha/yr.

The slope length and steepness factor (LS) are combined and divided into five classes to observe the change in soil loss rate corresponding to the LS value (Figures 22.3C and 22.3D). The LS values are classified from 0 to 8 under 1st class, 8 to 22 in 2nd class, 22 to 40 in 3rd class, 40 to 66 in 4th class, and 66 to 87 in class 5. Rate of soil loss highest category is observed in the 5th class having a mean of 49.01 tons/ha/yr. The RG Basin has a steep slope; hence, the velocity of the runoff is very high, thus eroding more soil. The rate of soil loss in classes 4, 3, 2, and 1 is 34.13 tons/ha/yr, 21.53 tons/ha/yr, 10.57 tons/ha/yr, and 2.12 tons/ha/yr, respectively. It shows that as the slope becomes gentler the soil loss rate also decreases. In LULC classes, the RG Basin is dominated by barren land with an area of 329.85 $km^2$, followed by snow cover/glaciers of 178.53 $km^2$, scrubland 154.59 $km^2$, forest 22.93 $km^2$, waterbodies 1.37 $km^2$, and builtup 0.20 $km^2$ (Figure 22.3E and 2.3F). The barren land region has the highest rate of soil loss (18.44 ton/ha/yr), as this class has no or very little vegetation and promotes more soil erosion than other classes, such as scrubland (4.44 ton/ha/yr) and forest (0.39 ton/ha/yr), and the RUSLE model cannot estimate soil loss in snow cover and water bodies, so these classes show zero values.

### 22.3.4 Rate of Soil Loss in Lower Chambal Basin (LC)

The mean rate of soil loss in the year 2020 for the LC Basin is 1.58 ton/ha/yr, and is divided into six classes as proposed by Singh et al. (1992) for India but with some modification based on the severity in plain areas (Table 22.6). We observe very less erosion in 81.93% of the total area as it is covered by forest, scrubland, and cropland. Six types of soil texture, e.g., loamy (6675.06 $km^2$), silty (1350.54 $km^2$), clayey (1332.65 $km^2$), fine loamy (1205.64 $km^2$), loamy skeletal (630.297 $km^2$), and coarse loamy (186.38 $km^2$) are identified in the Basin (Figure 22.5). The mean rate of soil loss is 0.18 ton/ha/yr in loamy, 8.18 ton/ha/yr in silty, 1.41 ton/ha/yr in clayey, 3.14 tons/ha/yr in fine loamy, 0.11 tons/ha/yr in loamy skeletal, and 0.30 ton/ha/yr in coarse loamy. Rainfall erosivity of the basin has been categorized into five classes: class 1 ranging from 100.48 to 105.55 MJ mm/ha/hr/yr shows 2.13 ton/ha/yr rate of soil loss, class 2 ranging 105.56 to 109.06 MJ mm/ha/hr/yr range contributes 1.1 ton/ha/yr; class 3 ranging from 109.07 to 112.29 MJ mm/ha/hr/yr contributes 1.37 ton/ha/yr; class 4 ranging from 112.30 to 115.71 MJ mm/ha/hr/yr contributes 1.95 ton/ha/yr; and class 5 ranging from 115.72 to 124.02 MJ mm/ha/hr/yr contributes 2.08 ton/ha/yr. The LC basin classifies into 7 LULC classes, with cropland accounting for 4212.07 $km^2$, scrub for 3490.72 $km^2$, open forest for 1936.76 $km^2$, ravines/gullies for 1371.56 $km^2$, water bodies for 168 $km^2$, builtup for 157.86 $km^2$, and sandy for 45.53 $km^2$. The mean rate of soil loss in cropland is 0.37 ton/ha/yr, 1.32 ton/ha/yr in scrubs areas, 0.04 ton/ha/yr from open forest, 8.43 ton/ha/yr from the ravines/gullies, and 4.93 ton/ha/yr from the sandy class.

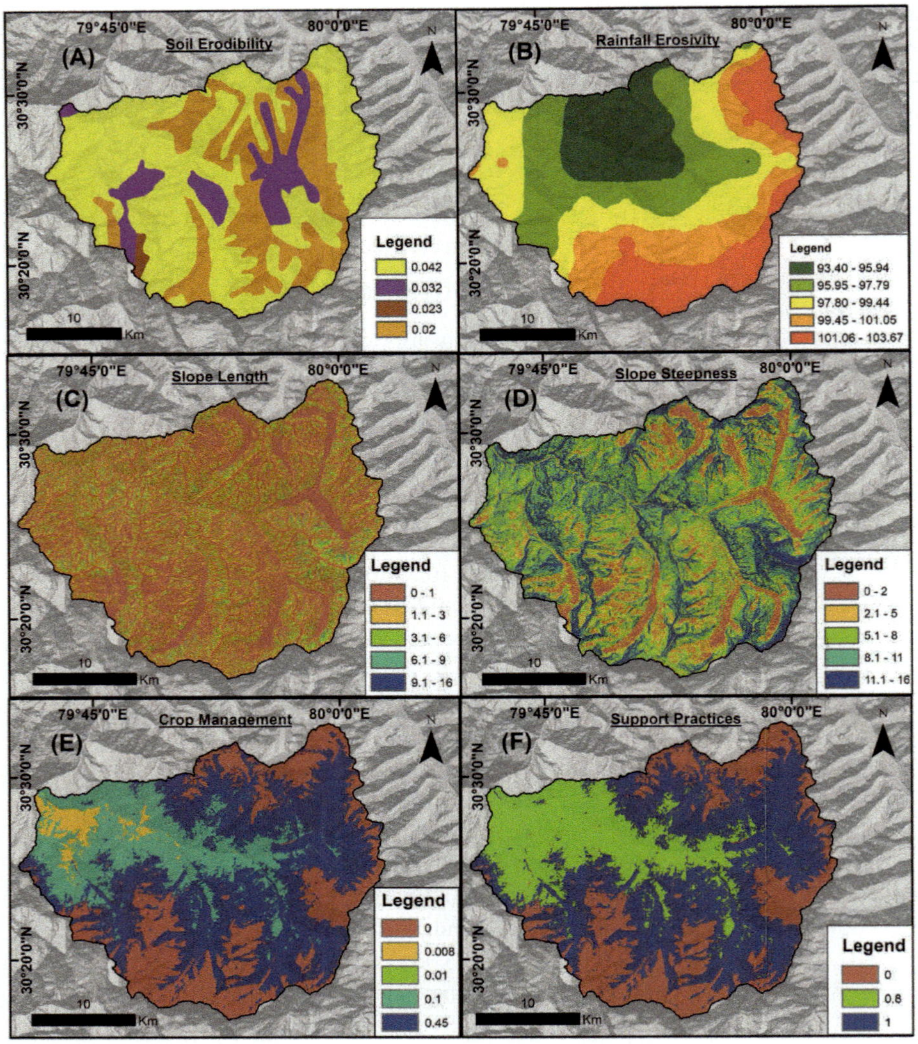


**Figure 22.3** The factors calculated for the Rishiganga Basin. (A) Soil Erodibility factor created by assigning appropriate values in the soil texture classes from the literature; (B) Rainfall Erosivity factor calculated from the CHRS average annual rainfall; (C) Slope length and (D) steepness factors calculated by tool developed in Arc Macro Language; (E) crop management; and (F) Support practice factor.

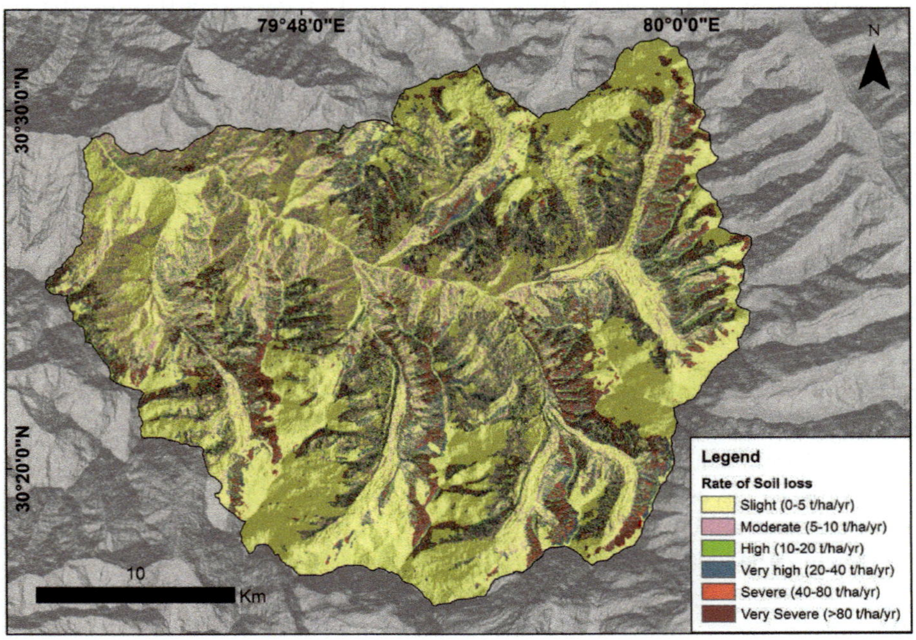


**Figure 22.4** The spatial pattern of soil loss rate in the RG Basin categorized into six classes based on severity. Slight erosion class covers 60.28% of the total area, moderate soil erosion covers 11.40% area, high erosion covers 11.96% area, very high erosion covers 9.61% area, severe erosion covers 5.59%, and very severe erosion class covers 1.17% of the total basin area.

**Table 22.6** Soil loss rate in different categories.

| Soil Loss Category | Rate of Soil Loss (t/ha/yr) | Area (km$^2$) |
|---|---|---|
| No erosion | 0 | 9325.89 |
| Slight erosion | 0–2 | 497.84 |
| Moderate erosion | 2–5 | 541.89 |
| High erosion | 5–10 | 671.302 |
| Severe erosion | 10–20 | 212.158 |
| Very severe erosion | >20 | 133.424 |

## 22.4 Discussion

The RUSLE is a globally accepted empirical soil erosion model that works in all environmental conditions and estimates on a large spatial scale (Renard et al., 1997). Its previous version (USLE), was applicable in small plots or sub-watersheds in a uniform hillslope, but did not work in concentrated flows, gullies, and sediment yields (Renard et al., 1997, 2010; Jain et al., 2001). This model best works at the farmland scale and can estimate over an extended time period, not only for a single storm. It does not involve any spatial resolution (Fistikoglu and Harmancioglu, 2002), and does not represent fundamental hydrologic and erosional

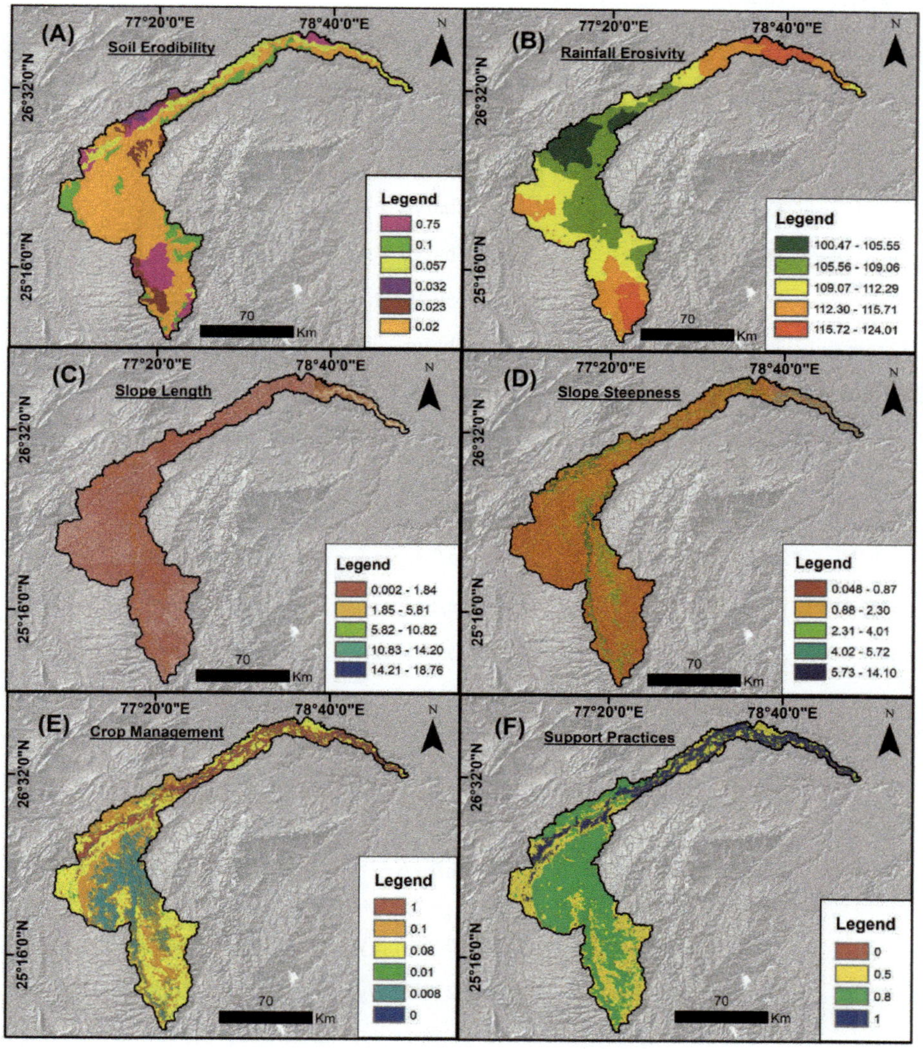


**Figure 22.5** The factors generated and employing the RUSLE model in Lower Chambal Basin. (A) Soil erodibility factor classified on the basis of K values assigned to soil texture. (B) Rainfall Erosivity factor calculated from the CHRS average annual rainfall. (C) Slope length and (D) steepness factors calculated by tool developed in Arc Macro Language, and (E) crop management, and (F) support practice.

processes explicitly (Renard et al., 1991). The RUSLE model with enhanced capabilities predicts long-term soil loss in all conditions, i.e., scrubland, forest, cropland, fallow land, and mining sites, and it can also be used in gullies-affected areas (Renard et al., 1991; Kong, 2020; Senanayake et al., 2020).

For 40 years (1977 to 2017), the RUSLE model has been widely used to estimate soil loss and in 1556 studies related to USLE/RUSLE with an average citation rate of 18.8 calculated by the Science Direct tool (Alewell et al., 2019). For the 250 m spatial resolution data, Borrelli et al. (2017) calculated a 35 Pg/year worldwide rate of soil erosion in 2001 and a 35.9 Pg/year global rate of soil erosion in 2012 that increased to 43 ± 9.2 Pg/year in 2015 (Borrelli et al., 2020). Pirnentel et al. (1995) estimated 30 to 40 ton/ha/yr rate of soil loss in Asia, South America, and Africa and 17 tons/ha/yr for the USA and Europe. Many RUSLE-based soil loss estimations in the Himalayas, Indo-Gangetic plains, and Peninsular plateaus have been carried out and explained in Table 22.7.

**Table 22.7** Recent studies in the different regions of India using the RUSLE model.

| S. No | Topography | Region | Study Area | Rate of Soil Loss | Author |
|---|---|---|---|---|---|
| 1 | Western Himalaya | Himalayan Mountains | Kangra region, Western Himalaya | 25.63 ton/ha/yr | (Kumar et al., 2014) |
| 2 | Mid Himalaya | | Tehri Garhwal | 28.38 ton/ha/yr | (Gupta and Kumar, 2017) |
| 3 | Mid Himalaya | | Tehri Garhwal | 65.84 ton/ha/yr | (Kalambukattu and Kumar, 2017) |
| 4 | Northeastern Himalaya | | Tista River | 3.53 ton/ha/yr | (Chatterjee, 2020) |
| 5 | Mid Himalaya | | Kosi River Basin | 40 ton/ha/yr | (Uddin et al., 2016) |
| 6 | Central Himalaya | | Nepal | 25 ton/ha/yr | (Koirala et al., 2019) |
| 7 | Northeastern Himalaya | | Tirap district | 1.38 to 59.05 ton/ha/yr | (Das et al., 2018) |
| 8 | Northeastern Himalaya | | Ri-Bhoi district | 36 ton/ha/yr | (Jena et al., 2018) |
| 9 | Upper Gangetic plains | Gangetic plains | Garra River Basin | 23 ± 4.7 ton/ha/yr | (Swarnkar et al., 2018) |
| 10 | Upper Gangetic plains | | Ghaghara River Basin | 21.39 ton/ha/yr | (Kumar and Singh, 2021) |
| 11 | Marginal Gangetic plains | | Parbati River Basin | 1.2 ton/ha/yr | (Kumar et al., 2021) |
| 12 | Central Indo Gangetic | | Madhya Pradesh | 6.64 ton/ha/yr | (Suryawanshi et al., 2021) |
| 13 | Central Indo Gangetic | | Jhagrabaria watershed | 0.76 ton/ha/yr | (Rawat et al., 2016) |
| 14 | Southern Peninsular | Peninsular plateaus | Pamba River Basin, Kerala | 17.73 ton/ha/yr | (Prasannakumar et al., 2012) |
| 15 | Southern Peninsular | | Andipatti watershed, Tamil Nadu | 6.73 ton/ha/yr | (Balasubramani et al., 2015) |
| 16 | Southern Peninsular | | Siruvani watershed, Kerela | 14.917 ton/ha/yr | (Prasannakumar et al., 2011) |
| 17 | Eastern Peninsular | | Sanjal watershed, Jharkhand | 61.4 ton/ha/yr | (Dutta et al., 2015) |
| 18 | South-East Peninsular | | Daltonganj watershed | 26 to 30 ton/ha/yr | (Tirkey et al., 2013) |
| 19 | Eastern Peninsular | | Barakar River Basin, Jharkhand | 0 to 282 ton/ha/yr | (Biswas and Pani, 2015) |
| 20 | Southern Peninsular | | Southern Western Ghats, Kerala | 14.36 ton/ha/yr | (Thomas et al., 2018) |
| 21 | Central Peninsular | | Bhandara District, Maharashtra | 20 to 40 ton/ha/yr | (Kashiwar et al,, 2021) |

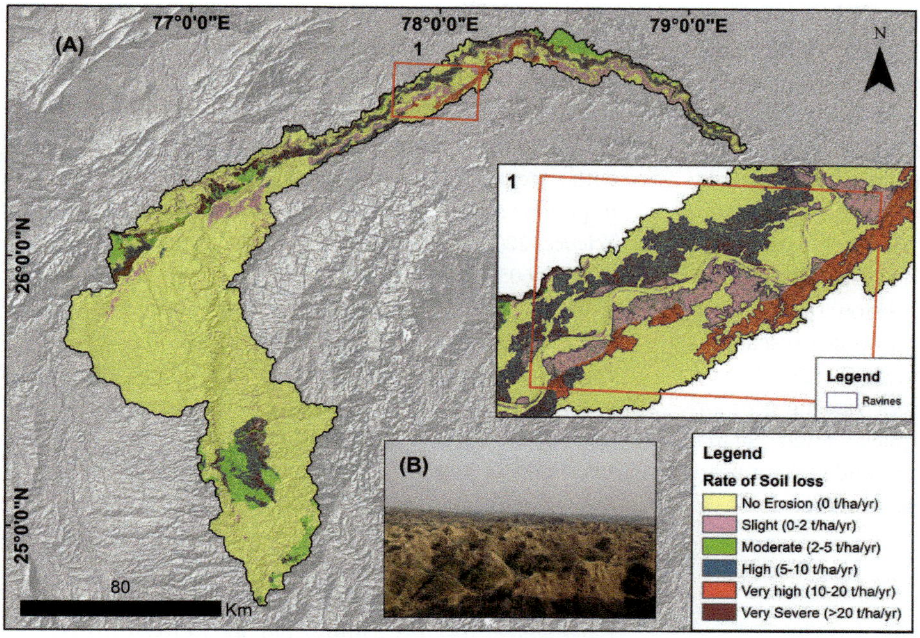


**Figure 22.6** The spatial pattern of soil loss in the LC Basin classified into six classes based on severity. No erosion occurred in 81.93% of the area, slight erosion covered 4.37%, moderate erosion covered 4.76%, high erosion covered 5.90%, very high erosion covered 1.86%, and very severe covered 1.17% of the Basin. (A) The zoomed area to show better classification of the rate of soil erosion in gully/ravine affected areas. (B) The field photograph of the gully/ravine showing steep slopes.

We used the RUSLE model for the estimation of soil loss of two sub-basins in the Ganga River Basin, i.e., Rishiganga (RG) in the upper Ganga River Basin and lower Chambal basin (LC) in the marginal Ganga River Basin. These river basins have diverse topographic and climatic conditions, e.g., RG lies in the Central Himalayas affected by Indian monsoons and the cold and dry highlands of Tibet (Pandey et al., 2021). In contrast, the LC Basin lies in plain areas with semi-arid climatic conditions (Joshi, 2014), and is mainly distinguished by the ravines/badlands. The average rate of soil erosion in the RG Basin is 9.82 ton/ha/yr, whereas the rate in the LC Basin is 1.58 ton/ha/yr, which is between the soil loss tolerable limit of 4.5 ton/ha/yr to 11 ton/ha/yr in India (Mannering, 1981). Rate of soil loss is high in the RG Basin due to higher slopes ranging from 0° to 79.5° (30.93° mean slope) and barren land (which covers 47.96% of the basin) with no to very little vegetation, promoting direct rainfall–soil interaction, weakening the basic structure of the soil, and resulting in higher surface runoff, both to accelerate the rate of soil loss in the basin. The slope is very low in the LC Basin, and ranges from 0° to 56.23° (with a mean slope of 4.08°). However, the area is dominated by rugged topography (covering ~12.04% of the basin), e.g., gully, ravines and badlands. These features are deeply dissected erosional landforms developed over soft rock terrain and generally associated with the uplifted semi-arid settings (Goudie, 2004). The high drainage density with an incised network of narrow channels (Bryan and Yair, 1982; Campbell, 1989) and steep slope (Ranga et al., 2016) of the ravines/badlands makes it very prone to soil erosion (Kumar et al., 2021). The rate of soil loss in this rugged topography is 8.43 ton/ha/yr, the highest in the basin. These two studies in different zones are an addition to the previous RUSLE-based studies in nearby areas and support the role of the RUSLE model to estimate the rate of soil loss with better accuracy. These two studies in separate topographic and climatic zones are in addition to earlier RUSLE-based research in adjacent locations and support its function in accurately estimating the rate of soil loss.

## 22.5 Conclusion

Several geospatial-based soil erosion models developed since the 1930s are discussed in this present study, emphasising the widely used models, such as USLE and its successor RUSLE. The RUSLE empirical soil erosion model is explored for estimating long-term soil loss rates in various environmental settings and topography. The Upper (Rishiganga) and Marginal (Lower Chambal) Ganga River Basin is selected for the case study to validate the RUSLE model in two different, i.e., Himalayan and Indo-Gangetic plains, topographic conditions. The factor R is created by CHRS average rainfall data, factor K is produced by providing appropriate values to the NBSS-LUP soil texture classes, factor LS as from the ALOS-PALSAR DEM by an Arc

Macro Langauge tool, and C and P factors are produced by assigning values in LULC classes developed by ESRI using sentinel-2 satellite imageries. In 2020, the average rate of soil loss in the Rishiganga Basin was 9.82 ton/ha/yr, whereas the lower Chambal basin is 1.58 ton/ha/yr. The highest soil erosion is observed in the rugged topography (8.43 ton/ha/yr) and barren land (18.44 ton/ha/yr) as there is very little vegetation that protects the direct interaction of soil and water with steep slopes. The RUSLE model has very accurately quantified the soil erosion rate in both regions, showing very similar figures to other researches carried out in adjacent areas. The results point out the versatility of the RUSLE model in terms of varying geomorphic conditions and morphodynamics. High spatial resolution at the Indian scale is recommended for a more in-depth study of erosion-affected areas in India.

## Acknowledgments

Authors acknowledge ESRI for LULC maps, Alaska Satellite Facility for ALOS-PALSAR DEM and IGNOU library for providing NBSS-LUP soil texture maps. RK and BD acknowledge the Director, School of Sciences, IGNOU for infrastructure and encouragement. RD thanks the Delhi School of Climate Change & Sustainability, Institution of Eminence, University of Delhi, India, for providing infrastructure.

## References

Alewell, C., Borrelli, P., Meusburger, K. et al. (2019). Using the USLE: chances, challenges and limitations of soil erosion modelling. *International Soil and Water Conservation Research* 7(3): 203–225. https://doi.org/10.1016/j.iswcr.2019.05.004.

Arnoldus, H.M.J. (1980). An approximation of the rainfall factor in the Universal Soil Loss Equation. 127–132.

Avwunudiogba, A. and Hudson, P.F. (2014). A review of soil erosion models with special reference to the needs of humid tropical mountainous environments. *European Journal of Sustainable Development* 3(4): 299–299. https://doi.org/10.14207/ejsd.2014.v3n4p299.

Balasubramani, K., Veena, M., Kumaraswamy, K.et al. (2015). Estimation of soil erosion in a semi-arid watershed of Tamil Nadu (India) using revised universal soil loss equation (RUSLE) model through GIS. *Modeling Earth Systems and Environment* 1(3): 1–17. https://doi.org/10.1007/s40808-015-0015-4.

Balasubramanian, A. (2017). Soil erosion: causes and effects. Centre for Advanced Studies in Earth Science, 1–7. Mysore, India: University of Mysore.

Beasley, D.B., Huggins, L.F., and Monke, A. (1980). ANSWERS: a model for watershed planning. *Transactions of the ASAE* 23(4): 938–0944. https://doi.org/10.13031/2013.34692.

Benavidez, R., Jackson, B., Maxwell, D.et al. (2018). A review of the (Revised) Universal Soil Loss Equation ((R) USLE): with a view to increasing its global applicability and improving soil loss estimates. *Hydrology and Earth System Sciences* 22(11): 6059–6086. https://doi.org/10.5194/hess-22-6059-2018.

Benzer, N. (2010). Using the geographical information system and remote sensing techniques for soil erosion assessment. *Polish Journal of Environmental Studies* 19(5): 881–886.

Biswas, S. (2012). Estimation of soil erosion using remote sensing and GIS and prioritization of catchments. *International Journal of Emerging Technology and Advanced Engineering* 2(7): 124–128.

Biswas, S.S. and Pani, P. (2015). Estimation of soil erosion using RUSLE and GIS techniques: a case study of Barakar River basin, Jharkhand, India. *Modeling Earth Systems and Environment* 1(4): 1–13. https://doi.org/10.1007/s40808-015-0040-3.

Borrelli, P., Robinson, D.A., Fleischer, L.R. et al. (2017). An assessment of the global impact of 21st century land use change on soil erosion. *Nature Communications* 8(1): 1–13. https://doi.org/10.1038/s41467-017-02142-7.

Borrelli, P., Robinson, D.A., Panagos, P.et al. (2020). Land use and climate change impacts on global soil erosion by water (2015–2070). *Proceedings of the National Academy of Sciences* 117(36): 21994–22001. https://doi.org/10.1073/pnas.2001403117.

Browning, G.M., Parish, C.L., and Glass, J. (1943). A method for determining the use of limitations of rotation and conservation practices in the control of soil erosion in Iowa. *Journal of the American Society of Agronomy* 23: 246–249.

Bruinsma, J. (ed.) (2003). *World Agriculture: Towards 2015/2030: An FAO Perspective*. London: Earthscan.

Bryan, R. and Yair, A. (eds.) (1982). *Badland Geomorphology and Piping*. Norwich, UK: Geo Books .

Campbell, L.A. (1989). Badlands and badland gullies. In: *Arid Zone Geomorpholog (ed.* D.S.G. Thomas), 159–193. New York: Halstead Press.

Chatterjee, N. (2020). Soil erosion assessment in a humid, Eastern Himalayan watershed undergoing rapid land use changes, using RUSLE, GIS and high-resolution satellite imagery. *Modeling Earth Systems and Environment* 6(1): 533–543. https://doi.org/10.1007/s40808-019-00700-0.

Cook, H.L. (1936). The nature and controlling variables of the water erosion process. *Soil Science Society of America Journal* 1(C): 487–494. https://doi.org/10.2136/sssaj1937.03615995000100000085x.

Dabral, P.P., Baithuri, N., and Pandey, A. (2008). Soil erosion assessment in a hilly catchment of north eastern India using USLE, GIS and remote sensing. *Water Resources Management* 22: 1783–1798. https://doi.org/10.1007/s11269-008-9253-9.

Dais (2008). Directorate Agricultural Information Services. Directorate Agricultural Land and Resources Management, 1–7. Directorate Agricultural Information Services, Pretoria.

Das, B., Paul, A., Bordoloi, R. et al. (2018). Soil erosion risk assessment of hilly terrain through integrated approach of RUSLE and geospatial technology: a case study of Tirap District, Arunachal Pradesh. *Modeling Earth Systems and Environment* 4(1): 373–381. https://doi.org/10.1007/s40808-018-0435-z.

De Roo, A.P.J., Wesseling, C.G., and Ritsema, C.J. (1996). LISEM: a single-event physically based hydrological and soil erosion model for drainage basins. I: Theory, input and output. *Hydrological Processes* 10(8): 1107–1117. https://doi.org/10.1002/(SICI)1099-1085(199608).

Devatha, C.P., Deshpande, V., and Renukaprasad, M.S. (2015). Estimation of soil loss using USLE model for Kulhan Watershed, Chattisgarh: a case study. *Aquatic Procedia* 4: 1429–1436. https://doi.org/10.1016/j.aqpro.2015.02.185.

Doetterl, S., Van Oost, K., and Six, J. (2012). Towards constraining the magnitude of global agricultural sediment and soil organic carbon fluxes. *Earth Surface Processes and Landforms* 37(6): 642–655. https://doi.org/10.1002/esp.3198.

Dutta, D., Das, S., Kundu, A. et al. (2015). Soil erosion risk assessment in Sanjal watershed, Jharkhand (India) using geo-informatics, RUSLE model and TRMM data. *Modeling Earth Systems and Environment* 1(4): 1–9. https://doi.org/10.1007/s40808-015-0034-1.

Erdogan, E.H., Erpul, G., and Bayramin, İ. (2007). Use of USLE/GIS methodology for predicting soil loss in a semiarid agricultural watershed. *Environmental Monitoring and Assessment* 131(1–3): 153–161. https://doi.org/10.1007/s10661-006-9464-6.

Eswaran, H., Lal, R., and Reich, P.F. (2001). Land degradation: an overview. In: *Responses to Land Degradation* (ed. E.M. Bridges, I.D. Hannam, and L.R. Oldeman), 20–35. New Delhi, India: Oxford Press.

FAO (Food and Agriculture Organization) (2005). The state of food insecurity in the world 2005: eradicating world hunger e-key to achieving the millennium development goals.

FAO (Food and Agriculture Organization) (2015). Healthy soils are the basis for healthy food production. (Food and Agriculture Organization of the United Nations), 1–4. http://www.fao.org/documents/card/en/c/645883cd-ba28-4b16-a7b8-34babbb3c505.

FAO and ITPS (2015).Status of the world's soil resources (SWSR): Main report. Food and agriculture. Organization of the United Nations and Intergovernmental Technical Panel on Soils, Rome, Italy. Available online: http://www.fao.org/3/a-i5199e.pdf.

Fistikoglu, O. and Harmancioglu, N.B. (2002). Integration of GIS with USLE in assessment of soil erosion. *Water Resources Management* 16(6): 447–467. https://doi.org/10.1023/A:1022282125760.

Fournier, F. (1960). Climate Erosion; la relation entre l'érosion du sol par l'eau et les précipitations atmosphériques, 302–551. Presses universitaires de France.

Garde, R.J. and Kothyari, U.C. (1987). Flood estimation using regional flood frequency analysis. *Hydraul. Eng. Rep.* No. HYD-8702, Civ. Eng. Dep., Roorkee: University of Roorkee.

Gassman, P.W., Reyes, M.R., Green, C.H. et al. (2007). The soil and water assessment tool: historical development, applications, and future research directions. *Transactions of the ASABE* 50(4): 1211–1250. https://doi.org/10.13031/2013.23637.

Gelfand, I., Sahajpal, R., Zhang, X. et al. (2013). Sustainable bioenergy production from marginal lands in the US Midwest. *Nature* 493(7433): 514–517. https://doi.org/10.1038/nature11811.

Ghosal, K. and Bhattacharya, S.D. (2020). A review of RUSLE model. *Journal of the Indian Society of Remote Sensing* 48(4): 689–707. https://doi.org/10.1007/s12524-019-01097-0.

Goldman, S.J., Jackson, K., and Bursztynsky, T.A. (1986). *Erosion and Sediment Control Handbook*.New York: Routledge Taylor & Francis Group.

Goudie, A. (2004). *Encyclopedia of Geomorphology*, vol. 2. London and New York: Psychology Press.

Gupta, S. and Kumar, S. (2017). Simulating climate change impact on soil erosion using RUSLE model: a case study in a watershed of mid-Himalayan landscape. *Journal of Earth System Science* 126(3): 43. https://doi.org/10.1007/s12040-017-0823-1.

Hickey, R. (2001). Slope angle and slope length solutions for GIS. *Cartography* 29(1): 1–8. https://doi.org/10.1080/00690805.2000.9714334.

Hirschi, M.C. and Barfield, B.J. (1988). KYERMO: a physically based research erosion model part I. model development. *Transactions of the ASAE* 31(3): 804–0813. https://doi.org/10.13031/2013.30785.

Holy, M. (1980). *Erosion and Environment*. vol. 9, 25–215. Oxford, UK: Pergamon Press.

Jena, R.K., Padua, S., Ray, P. et al. (2018). Assessment of soil erosion in sub-tropical ecosystem of Meghalaya, India using remote sensing, GIS and RUSLE. *Indian Journal of Soil Conservation* 46(3): 273–282.

Jain, S.K., Kumar S., and Varghese J. (2001). Estimation of soil erosion for a Himalayan watershed using GIS technique. *Water Resources Management* 15(1): 41–54. https://doi.org/10.1023/A:1012246029263.

Jain, M.K., Mishra, S.K., and Shah R.B. (2010). Estimation of sediment yield and areas vulnerable to soil erosion and deposition in a Himalayan watershed using GIS. *Current Science* 98(2): 213–221.

Jones, D.S., Kowalski, D.G., and Shaw, R.B. (1996). Calculating revised universal soil loss equation (RUSLE) estimates on department of defense lands: a review of RUSLE factors and US army land condition-trend analysis (LCTA) data gaps. Center for Ecological Management of Military Lands, Colorado State University.

Joshi, V.U. (2014). The Chambal badlands. In: *Landscapes and Landforms of India*, 143–149. Dordrecht: Springer. https://doi.org/10.1007/978-94-017-8029-2_13.

Kalambukattu, J. and Kumar, S. (2017). Modelling soil erosion risk in a mountainous watershed of mid-Himalaya by integrating RUSLE model with GIS. *Eurasian Journal of Soil Science* 6(2): 92–105. https://doi.org/10.18393/ejss.286442.

Karra, K., Kontgis, C., Statman-Weil, Z. et al. (2021). Global land use/land cover with Sentinel 2 and deep learning. In: *2021 IEEE International Geoscience and Remote Sensing Symposium IGARSS*, 4704–4707. https://doi.org/10.1109/IGARSS47720.2021.9553499.

Kashiwar, S.R., Kundu, M.C., and Dongarwar, U.R. (2021). Soil erosion estimation of Bhandara region of Maharashtra, India, by integrated use of RUSLE, remote sensing, and GIS. In: *Natural Hazards*, 1–23. https://doi.org/10.1007/s11069-021-04974-5.

Kaushik, P. and Ghosh, P. (2015). Geomorphic evolution of Chambal river origin in Madhya Pradesh using remote sensing and GIS. *International Journal of Advanced Remote Sensing & GIS* 4(1): 1130–1141.

Kinnell, P.I.A. (2005). Why the universal soil loss equation and the revised version of it do not predict event erosion well. *Hydrological Processes: An International Journal* 19(3): 851–854. https://doi.org/10.1002/hyp.5816.

Kirkby, M.J., Abrahart, R., McMahon, M.D. et al. (1988). MEDALUS soil erosion models for global change. *Geomorphology* 24: 35–39. https://doi.org/10.1016/S0169-555X(97)00099-8.

Knisel, W.G. (1980). CREAMS: a field scale model for chemicals, runoff, and erosion from agricultural management systems. United States: Agricultural Research Service, US Department of Agriculture.

Koirala, P., Thakuri, S., Joshi, S. et al. (2019). Estimation of soil erosion in Nepal using a RUSLE modeling and geospatial tool. *Geosciences* 9(4): 1–19. https://doi.org/10.3390/geosciences9040147.

Kong, H. (2020). Evaluation of soil erosion in hilly gully area of the Loess Plateau based on remote sensing survey and RUSLE model. *Geoscience and Remote Sensing* 3(1): 9–16. https://doi.org/10.23977/geors.2020.030102.

Kothyari, G.C., Joshi, N., Taloor, A.K. et al. (2021). Reconstruction of active surface deformation in the Rishi Ganga basin, Central Himalaya using PSInSAR: a feedback towards understanding the 7th February 2021 Flash Flood. In: *Advances in Space Research*. https://doi.org/10.1016/j.asr.2021.07.002.

Kothyari, U.C. (1996). Erosion and sedimentation problems in India. In: *IAHS Publications-Series of Proceedings and Report: Intern Association Hydrological Sciences* 236: 531–540.

Kouli, M., Soupios, P., and Vallianatos, F. (2009). Soil erosion prediction using the Revised Universal Soil Loss Equation (RUSLE) in a GIS framework, Chania, Northwestern Crete, Greece. *Environment Geology* 57(3): 483–497. https://doi.org/10.1007/s00254-008-1318-9.

Kumar, A., Devi, M., and Deshmukh, B. (2014). Integrated remote sensing and geographic information system based RUSLE modelling for estimation of soil loss in western Himalaya, India. *Water Resources Management* 28(10): 3307–3317. https://doi.org/10.1007/s11269-014-0680-5.

Kumar, N. and Singh, S.K. (2021). Soil erosion assessment using earth observation data in a trans-boundary river basin. *Natural Hazards* 107(1): 1–34. https://doi.org/10.1007/s11069-021-04571-6.

Kumar, R. and Deshmukh, B. (2018). Soil loss assessment using geospatial technology: challenges and opportunities with reference to the Chambal basin, central India. Paper presented at the Environmental and Ecological Sustainability, IGNOU, November 2018.

Kumar, R., Deshmukh, B., and Kumar, A. (2022a). Using Google Earth Engine and GIS for basin scale soil erosion risk assessment: A case study of Chambal river basin, central India. *Journal of Earth System Science* 131(4): 1–16. https://doi.org/10.1007/s12040-022-01977-z.

Kumar, R., Deshmukh, B., and Sathunuri, K. (2021). RUSLE model based assessment of soil erosion in Parbati River Basin, Central India using Google Earth Engine and GIS. In: *EGU General Assembly Conference Abstracts*, EGU21-1524. https://doi.org/10.5194/egusphere-egu21-1524.

Kumar, R., Naqvi, H. R., Devrani, R. et al. (2022b). Sediment Yield Assessment, Prioritization and Control Practices in Chambal River Basin Employing SYI Model. *Journal of the Geological Society of India* 98(11): 1585–1594. https://doi.org/10.1007/s12594-022-2215-7

Lal, R. (2003). Soil erosion and the global carbon budget. *Environment International* 29(4): 437–450. https://doi.org/10.1016/S0160-4120(02)00192-7.

Lambin, E.F., Gibbs, H.K., Ferreira, L. et al. (2013). Estimating the world's potentially available cropland using a bottom-up approach. *Global Environmental Change* 23(5): 892–901. https://doi.org/10.1016/j.gloenvcha.2013.05.005.

Lenz, J., Yousuf, A., Schindewolf, M. et al. (2018). Parameterization for EROSION-3D model under simulated rainfall conditions in lower Shivaliks of India. *Geosciences* 8(11): 396.

Leonard, R.A., Knisel, W.G., and Still, D.A. (1987). GLEAMS: groundwater loading effects of agricultural management systems. *Transactions of the ASAE* 30(5): 1403–1418.

Mahapatra, S.K., Obi Reddy, G.P., Nagdev, R. et al. (2018). Assessment of soil erosion in the fragile Himalayan ecosystem of Uttarakhand, India using USLE and GIS for sustainable productivity. *Current Science* 115(1): 108–121. https://doi.org/10.18520/cs/v115/i1/108-12.

Mannering, J.V. (1981). Use of soil loss tolerances as a strategy for soil conservation. In: *Soil Conservation Problems and Prospects. Proceedings of Conservation 80, the International Conference on Soil Conservation*, Silsoe, Bedford, UK, 21–25 July 1980.

McCool, D.K., Brown, L.C., Foster, G.R. et al. (1987). Revised slope steepness factor for the Universal Soil Loss Equation. *Transactions of the ASAE* 30(5): 1387–1396.

Merritt, W.S., Letcher, R.A., and Jakeman, A.J. (2003). A review of erosion and sediment transport models. *Environmental Modelling & Software* 18(8–9): 761–799. https://doi.org/10.1016/S1364-8152(03)00078-1.

Montanarella, L. (2015). Agricultural policy: govern our soils. *Nature News* 528(7580): 1–3. https://doi.org/10.1038/528032a.

Moore, I.D. and Burch, G.J. (1986). Physical basis of the length slope factor in the Universal Soil Loss Equation. *Soil Science Soc. Am.* 50: 1294–1298.

Morgan, R.P.C. (1986). *Soil Erosion and Conservation*, 1–298. UK: Blackwell Publishing Ltd. Longman Scientific & Technical,

Morgan, R.P.C., Quinton, J.N., Smith, R.E. et al. (1998). The European Soil Erosion Model (EUROSEM): a dynamic approach for predicting sediment transport from fields and small catchments. *Earth Surface Processes and Landforms: The Journal of the British Geomorphological Group* 23(6): 527–544.

Narayana, D.V. and Babu, R. (1983). Estimation of soil erosion in India. *Journal of Irrigation and Drainage Engineering* 109(4): 419–434. https://doi.org/10.1061/(ASCE)0733-9437(1983)109:4(419).

Nearing, M.A., Foster, G.R., Lane, L.J. et al. (1989). A process-based soil erosion model for USDA-Water Erosion Prediction Project technology. *Transactions of the ASAE* 32(5): 1587–1593.

Novotny, V. (1994). *Water Quality: Prevention, Identification and Management of Diffuse Pollution*. New York: Van Nostrand-Reinhold Publishers.

NRSC Report (2014). *Erosion Map of India. Soil & Land Resources Assessment Division*, 1–22. Balanags, Hyderabad: National Remote Sensing Centre.

Oldeman, L.R. (1994). An international methodology for an assessment of soil degradation, land georeferenced soils and terrain database. RAPA Publication, 1994(3).

Pandey, A., Chowdary, V.M., and Mal, B.C. (2007). Identification of critical erosion prone areas in the small agricultural watershed using USLE, GIS and remote sensing. *Water Resources Management* 21(4): 729–746. https://doi.org/10.1007/s11269-006-9061-z.

Pandey, A., Chowdary, V.M., and Mal, B.C. (2009). Sediment yield modelling of an agricultural watershed using MUSLE, remote sensing and GIS. *Paddy and Water Environment* 7(2): 105–113. https://doi.org/10.1007/s10333-009-0149-y.

Pandey, P., Chauhan, P., Bhatt, C.M. et al. (2021). Cause and process mechanism of rockslide triggered flood event in Rishiganga and Dhauliganga River Valleys, Chamoli, Uttarakhand, India Using satellite remote sensing and *in situ* observations. *Journal of the Indian Society of Remote Sensing* 49(5): 1011–1024. https://doi.org/10.1007/s12524-021-01360-3.

Pirnentel, D., Harvey, C., Resosudanno, F. et al. (1995). Environmental and economic costs of soil erosion and conservation benefits. *Science* 267(5201): 1117–1123. https://doi.org/10.1126/science.267.5201.1117.

Prasannakumar, V., Shiny, R., Geetha, N. et al. (2011). Spatial prediction of soil erosion risk by remote sensing, GIS and RUSLE approach: a case study of Siruvani river watershed in Attapady valley, Kerala, India. *Environmental Earth Sciences* 64(4): 965–972. https://doi.org/10.1007/s12665-011-0913-3.

Prasannakumar, V., Vijith, H., Abinod, S. et al. (2012). Estimation of soil erosion risk within a small mountainous sub-watershed in Kerala, India, using Revised Universal Soil Loss Equation (RUSLE) and geo-information technology. *Geoscience Frontiers* 3(2): 209–215. https://doi.org/10.1016/j.gsf.2011.11.003.

Rana, N., Sundriyal, Y., Sharma, S. et al. (2021). Hydrological characteristics of 7th February 2021 Rishi Ganga Flood: implication towards understanding flood hazards in higher Himalaya. *Journal of the Geological Society of India* 97(8): 827–835. https://doi.org/10.1007/s12594-021-1781-4.

Ranga, V., Poesen, J., Van Rompaey, A. et al. (2016). Detection and analysis of badlands dynamics in the Chambal River Valley (India), during the last 40 (1971–2010) years. *Environmental Earth Sciences* 75(3): 1–12. https://doi.org/10.1007/s12665-015-5017-z.

Rawat, K.S., Mishra, A.K., and Bhattacharyya, R. (2016). Soil erosion risk assessment and spatial mapping using LANDSAT-7 ETM+, RUSLE, and GIS, a case study. *Arabian Journal of Geosciences* 9(4): 288. https://doi.org/10.1007/s12517-015-2157-0.

Renard, K.G., Foster, G.R., Weesies, G.A. et al. (1997). *Predicting Soil Erosion by Water: A Guide to Conservation Planning with the Revised Universal Soil Loss Equation (RUSLE)*. 703. Washington, DC: United States Department of Agriculture.

Renard, K.G., Foster, G.R., Weesies, G.A. (1991). RUSLE: revised universal soil loss equation. *Journal of Soil and Water Conservation* 46(1): 30–33.

Renard, K.G., Yoder, D.C., Lightle, D.T. et al. (2010). Universal soil loss equation and revised universal soil loss equation. *Handbook of Erosion Modelling* 8: 135–167.

SAC Report (2016). Desertification and land degradation atlas of India (based on IRS AWiFS data of 2011–13 and 2003–05), 1–219. Ahmedabad, India: Space Applications Centre, ISRO.

Saha, S.K. (2003). Water and wind induced soil erosion assessment and monitoring using remote sensing and GIS. In: Satellite Remote Sensing and GIS Applications in Agricultural Meteorology, 315–330.

Sain, K., Kumar, A., Mehta, M.et al. (2021). A perspective on Rishiganga-Dhauliganga flash flood in the Nanda Devi Biosphere Reserve, Garhwal Himalaya, India. *Journal of the Geological Society of India* 97(4): 335–338. https://doi.org/10.1007/s12594-021-1691-5.

Schmitt, L.K. (2009). Developing and applying a soil erosion model in a data-poor context to an island in the rural Philippines. *Environment, Development and Sustainability* 11(1): 19–42. https://doi.org/10.1007/s10668-007-9096-1.

Senanayake, S., Pradhan, B., Huete, A. et al. (2020). Assessing soil erosion hazards using land-use change and landslide frequency ratio method: a case study of Sabaragamuwa Province, Sri Lanka. *Remote Sensing* 12(9): 1–18. https://doi.org/10.3390/rs12091483.

Sheikh, A.H., Palria, S., and Alam, A. (2011). Integration of GIS and universal soil loss equation (USLE) for soil loss estimation in a Himalayan watershed. *Recent Research in Science Technology* 3(3): 51–57.

Shoshany, M., Goldshleger, N., and Chudnovsky, A. (2013). Monitoring of agricultural soil degradation by remote-sensing methods: a review. *International Journal of Remote Sensing* 34(17): 6152–6181. https://doi.org/10.1080/01431161.2013.793872.

Shyampura, R.L. and Sehgal, J. (1995). Soils of Rajasthan for optimizing land use. *Soils of India Series*. NBSS Publication, 51(b):

Singh, V.P. and Frevert, D.K. (2005). GAMES: the Guelph model for evaluating the effects of agricultural management systems on erosion and sedimentation. In: *Watershed Models*, 507–524. Boca Raton, FL: CRC Press.

Singh, G., Babu, R., Narain, P. et al. (1992). Soil erosion rates in India. *Journal of Soil and Water Conservation* 47(1): 97–99.

Singh, G., Chandra, S., and Babu, R. (1981). Soil loss and prediction research in India, Central Soil and Water Conservation Research Training Institute. Bulletin No T-12 D: 9.

Singh, I.B. (1996). Geological evolution of Ganga Plain an overview. *Journal of the Palaeontological Society of India* 41: 99–137.

Singh, S.P., Ram, J., Walia, C.S. et al. (2004). Soils of Uttar Pradesh for optimizing land use. *NBSS Publication* 68(b): 1–91.

Singh, V.P. and Frevert, D.K. (2005). GAMES: The Guelph model for evaluating the effects of agricultural management systems on erosion and sedimentation. In: *Watershed Models* (ed. V.P. Singh and D.K. Frevert), 507–524. Boca Raton, FL: CRC Press.

Smith, D.D. (1941). Interpretation of soil conservation data for field use. *Agricultural Engineering* 22: 173–175.

Smith, D.D. and Whitt, D.M. (1948). Estimating soil losses from field areas of claypan soil. *Soil Science Society of America Journal* 12(C): 485–490. https://doi.org/10.2136/sssaj1948.036159950012000C0108x.

Stone, R.P. and Hilborn, D. (2000). Universal Soil Loss Equations Fact Sheet (USLE). Ministry of Agriculture, Food and Rural Affairs Version 2 (RUSLE2).

Suryawanshi, A., Nema, A.K., Jaiswal, R.K. et al. (2021). Identification of soil erosion prone areas of Madhya Pradesh using USLE/RUSLE. *Journal of Agricultural Engineering* 58(2): 177–191. https://doi.org/10.52151/jae2021581.1744.

Swarnkar, S., Malini, A., Tripathi, S. et al. (2018). Assessment of uncertainties in soil erosion and sediment yield estimates at ungauged basins: an application to the Garra River basin, India. *Hydrology and Earth System Sciences* 22(4): 2471–2485. https://doi.org/10.5194/hess-22-2471-2018.

Tamgadge, D.B. (1996). Soils of Madhya Pradesh for optimising land use. National Bureau of Soil Survey & Land Use Planning Publication 59(b).

Thomas, J., Joseph, S., and Thrivikramji, K.P. (2018). Assessment of soil erosion in a tropical mountain river basin of the southern Western Ghats, India using RUSLE and GIS. *Geoscience Frontiers* 9(3): 893–906. https://doi.org/10.1016/j.gsf.2017.05.011.

Tirkey, A.S., Pandey, A.C., and Nathawat, M.S. (2013). Use of satellite data, GIS and RUSLE for estimation of average annual soil loss in Daltonganj watershed of Jharkhand, India. *Journal of Remote Sensing Technology* 1(1): 20–30.

Uddin, K., Murthy, M.S.R., Wahid, S.M. et al. (2016). Estimation of soil erosion dynamics in the Koshi basin using GIS and remote sensing to assess priority areas for conservation. *PloS One* 11(3): 1–19. https://doi.org/10.1371/journal.pone.0150494.

Valdiya, K.S. (1980). *Geology of Kumaun Lesser Himalaya*, 1–291. Dehradun, India: Wadia Institute of Himalayan Geology.

Van Remortel, R., Maichle, R., and Hickey, R. (2004). Computing the RUSLE LS factor based on array-based slope length processing of digital elevation data using a C++ executable. *Computers and Geosciences* 30(9–10): 1–43.

Vander-Knijff, J.M., Jones, R.J.A., and Montanarella, L. (2000). Soil erosion risk assessment in Europe. EUR 19044

EN. Office for official publications of the European Communities, Luxembourg, 1–34.

Walling, D.E. (1987). Methods of computing sedimentation in lakes and reservoirs. UNESCO technical report, 1985. Available from IHP Secretariat, Division of Water Sciences, 1–224. Paris: Unesco.

Wang, L., Qian, J., Qi, W.Y. (2018). Changes in soil erosion and sediment transport based on the RUSLE model in Zhifanggou watershed, China. *Proceedings of the International Association of Hydrological Sciences* 377: 9–18. https://doi.org/10.5194/piahs-377-9-2018.

Wanielista, M.P. and Yousef, Y.A. (1993). Stormwater Management 21(2). Weinheim, Germany: John Wiley & Sons. https://doi.org/10.1002/aheh.19930210213.

Watson, A. and Evans, R. (1991). A comparison of estimates of soil erosion made in the field and from photographs. *Soil and Tillage Res.* 5(19): 17–27. https://doi.org/10.1016/0167-1987(91)90106-8.

Williams, J.R. and Berndt, H.D. (1972). Sediment yield computed with universal equation. *Journal of the Hydraulics Division* 98(12): 2087–2098. https://doi.org/10.1061/JYCEAJ.0003498.

Williams, J.R., Renard, K.G., and Dyke, P.T. (1983). EPIC: a new method for assessing erosion's effect on soil productivity. *Journal of Soil and Water Conservation* 38(5): 381–383.

Wischmeier, W.H. and Smith, D.D. (1957). Factors affecting sheet and rill erosion. *Eos, Transactions American Geophysical Union* 38(6): 889–896.

Wischmeier, W.H. and Smith, D.D. (1965). Predicting rainfall-erosion losses from cropland east of the Rocky Mountains: a guide for selecting practices for soil and water conservation. USDA Agricultural Handbook No. 282. http://www.ars.usda.gov/SP2UserFiles/Place/64080530/RUSLE/AH_282.pdf.

USDA-Agricultural Research Service (ARS) (2001). A guide to conservation planning. In: Predicting Rainfall Erosion Losses. Science and Education Administration, US Dept. of Agriculture Publisher, Agriculture Handbook (US Dept. of Agriculture); no. 537, USDA Publications.

Yesuph, A.Y. and Dagnew, A.B. (2019). Soil erosion mapping and severity analysis based on RUSLE model and local perception in the Beshillo Catchment of the Blue Nile Basin, Ethiopia. *Environmental Systems Research* 8(1): 1–21. https://doi.org/10.1186/s40068-019-0145-1.

Zhang, Y., Degroote, J., Wolter, C. et al. (2009). Integration of Modified Universal Soil Loss Equation (MUSLE) into a GIS framework to assess soil erosion risk. *Land Degradation & Development* 20(1): 84–91. https://doi.org/10.1002/ldr.893.

Zhou, P., Luukkanen, O., Tokola, T. et al. (2008). Effect of vegetation cover on soil erosion in a mountainous watershed. *Catena* 75(3): 319–325. https://doi.org/10.1016/j.catena.2008.07.010.

Zingg, A.W. (1940). Degree and length of land slope as it affects soil loss in run-off. *Agricultural Engineering* 21: 59–64.

# 23

# Wetlands as Potential Zones to Understand Spatiotemporal Plant-Human-Climate Interactions

## A Review on Palynological Perspective from Western and Eastern Himalaya

*Sandhya Misra*, Anupam Sharma, Ravi Shankar Maurya, and Krishna G. Misra*

*Birbal Sahni Institute of Palaeosciences, 53 University Road, Lucknow 226007, India*
** Corresponding author*

## 23.1 Introduction

The land or part of the land which is partially or completely covered by water during any/some part of the year is known as wetlands. They act as the passage between the land and the water ecosystem. Based on their origin, geomorphology, hydrology, water depth, floral/faunal composition, physical and chemical nature, and water chemistry/substratum characteristics, these areas support a variety of living forms and are widely defined/classified by many workers across the globe. Varied definitions from diversified parts of the globe can be seen in Sharma and Singh (2021 and references therein). In the Indian context, Gopal and Sah (1995) classified the wetlands on the basis of vegetation types that represent the prevailing hydrological state in these regions. India, with 15.3 million hectares of wetland area, has over 757,000 wetlands in the form of ponds, lakes, rivers, swamps, ditches, estuaries, mudflats, etc. covering 4.7% of the nation's total geographical area (Global Wetland Outlook, 2018). Varying ecoclimatic zones in India facilitate a diverse variety of wetland systems ranging from cold desert wetlands at high altitude to warm and wet wetlands along the coastal area (Prasad et al., 2002).

Third pole/Indian Himalaya is considered as the "Water Tower of Asia," and plays a significant role in influencing hydroclimatic conditions, defining floristic wealth, flourishing/development of civilizations, and agriculture-based economy of the nation. This ~18% of India's geographical area with varied orography, topography, precipitation changes, altitudinal variation, glaciers, etc. houses a wide variety of wetlands. These wetlands also define the water regimes in the major and minor streams and water bodies flowing through this part of the Himalaya. Various plant taxa found in the wetlands of the Indian subcontinent are categorized into marshes, swamps, and peatland taxa, depending on the type of environmental conditions they prefer for their establishment, growth, and proliferation (Gopal et al., 1995). Emergent vegetation like species of *Typha*, *Scirpus*, *Carex*, *Sagittaria latifolia*, *Nuphar variegate*, *Nymphaea odorata*, *Iris versicolor*, and floaters, *Lemna minor*, *Potamogeton spp.*, and *Elodea canadensis*, represent the marshy vegetation. Non-native invasive plants such as *Lythrum salicaria* and *Phragmites australis* impose risk to marshes (Bandyopadhyay and Mukherjee, 2005).

## 23.2 Importance of Wetlands

Wetlands, popularly called as the "Kidney of the Earth" (Mitsch and Gosselink, 1993), serve as an important part of the social and ecological system. They help in regulating several processes in the Earth's biosphere and are among the important sources of carbon. They serve a range of purposes like food, fuel, fiber, aesthetics, economical, ecological, and recreational services to mankind (Keddy, 2010). They also aid in regulating biogeochemical cycle, purification of surface and groundwater, recharging of the water table, and recycling wastewater. They are evaluated to play a critical role in protecting watersheds, monitoring climate and disasters, stabilizing shorelines, and desynchronizing flood flows.

Natural climatic changes along with the varying anthropogenic activities in the form of urbanization, population pressure, industrialization, and poor sewage disposal have massively imposed critical conditions for the survival and sustainability of these ecologically fragile and economically

*Advances in Remote Sensing Technology and the Three Poles,* First Edition. Edited by Manish Pandey, Prem C. Pandey, Yogesh Ray, Aman Arora, Shridhar D. Jawak, and Uma K. Shukla.

important ecosystems (Gosselink and Turner, 1978). The sensitivity of these ecosystems can be realized from the fact that the biota present at these sites responds even to the minor fluctuations in the prevailing environment, which is often reflected in the form of unprecedented changes in species composition and richness and also productivity (Houlahan et al., 2006; Keddy, 2010). Many of the tangible services provided by wetlands are subjected to the anthropogenic pressures either to meet out a livelihood or as a result of enhanced population pressure (Reis et al., 2017, 2019). According to a study by Xu et al. (2019), except for the 2303 number of wetlands he discussed in his paper, in other areas >50% of wetlands are impacted by contamination from various industrial and domestic wastes, biological use, alteration to the existing natural system, fishing, paddy field cultivation, aquaculture, and so on.

According to recent reports under the potential impact of urban projects and uncontrolled urban expansion, the country is losing its wetlands at an accelerated rate of nearly 25 hectares with every one square kilometer increase in built-up areas (Khandekar, 2020). Even the GIS-based reports on the status of wetlands in India documented the changes in the wetlands areas. Based on the observation and assessment of data from wetlands Global Wetland Outlook (2018), its report evidenced that since 1700 the world had lost 35% of its wetlands at a pace three times faster than the loss of other natural vegetations. On the other hand, in the same report, it mentions that the period from 1970 to 2014 has witnessed a two-fold rise in human-made wetlands indicating anthropogenic influence on natural ecosystem balance. Palynological investigation and some recent GIS-based studies carried out on various aspects of wetlands from the diversified locations in the Himalayan region raised the alarm toward the deteriorating environmental conditions, change in catchment areas, vegetation composition variation, increased human pressures, etc., creating a grave danger to their existence and a call to academicians, politicians, and scientists to come up with rescue/conservational plans for sustainable future developments.

## 23.3 Climate of Himalaya

On the basis of monthly and annually mean temperature changes and precipitation variation recorded in the diversified locations of Himalayan region, it is found that the area enjoys a cold-humid winter tundra and polar type climate. Areas with winter temperatures nearly or equal to 10°C and summer temperature of less than 18°C, witness humid summers and cold humid winters (Siwalik Hills, Lesser Himalaya. and some parts in northeast India). Higher altitude sites experience a tundra-type climate, while the regions that are covered by snow for the maximum part of the year experience a polar-type climate. In such regions, the warmest month recorded a temperature less than 10°C and snowfall as the most characteristic form of precipitation. The mean annual temperature and precipitation recorded in the western Himalayas is 13.92°C and 717.6 mm, respectively. In the eastern part, recorded mean annual temperature is 23.27°C, while reported mean annual precipitation is 2471.61 mm. The western region receives rainfall from both Indian summer monsoon (ISM) and Westerlies and in the eastern part ISM contributes the major part (Kar and Quamar, 2019).

## 23.4 Vegetation Types in the Himalayan Region

According to Reddy et al., 2015, vegetation of a region reflects the structural and functional attributes of regional/local climatic conditions, which in turn are found playing a very dramatic role in assessing the carbon stock, biodiversity, and sustainability scenarios at the global scale. Champion and Seth (1968) categorized the vegetation in this region into nine classes representing the taxa from sub-tropical broadleaved to tropical pine forest to dry evergreen to montane wet temperate forest to dry temperate to sub-alpine, moist-alpine to dry alpine forests (Table 23.1).

## 23.5 Wetlands as Sites for Floristic Analysis

To interpret fossil pollen records and to gain an insight into past vegetation that provides valuable information for past climatic changes, modern pollen data under varying vegetational and eco-climatic conditions is required. Realizing and relying on the importance of modern pollen rain studies and assumptions that the pollen morpho-types present in modern samples represent the number of plant species present in the surrounding vegetation (Xu et al., 2007), a good number of attempts have been made at varying locations in the Indian Himalayas using archives such as surface sediment/soil, moss cushions, top sediments from lake beds and wetlands, tree bark, spider's nests, leaf/rock surfaces, and locally produced honey (Kar and Quamar, 2018, 2019) to understand the extant vegetation in the region. Judicious use of such data helps to rectify the uncertainties in the interpretation of fossil pollen assemblage for the current geological period (Wright, 1967). An overview of the studies conducted on modern pollen-rain aspects (mentioned in Quamar and Kar, 2019), it is realized that under the varying preservation, pollen production, and mode and means of pollen dispersal, there is no direct relationship observed

**Table 23.1** Types of Vegetation in the Himalayan region.

| Type of Forest | Remarks |
|---|---|
| Sub-tropical broad-leaved hill forests | These forests with taxa like *Alnus, Betula, Juglans, Carpinus, Corylus, Quercus*, and *Ulmus* occur in lesser and greater Himalaya in the western Himalayan region. |
| Sub-tropical pine forests | Such forests with *Pinus roxburghii* and *Pinus kesiya* are found in central, western, and north-eastern hills. |
| Sub-tropical dry evergreen forests | These forests are composed of conifers like species of *Pinus* and *Abies*, and broad-leaved taxa like *Betula, Alnus* and *Corylus* etc., and are often encountered in Lesser Himalaya in the NW Himalayan region. |
| Montane wet temperate forests | These forests with species of *Pinus, Picea wallichiana*, and *Abies* are found from 1800–3000 m altitude in the eastern Himalaya. While in the Western Himalayas, except for places of rainfall below 1000 mm, they occur at altitudes between 1500 m and 3000 m. |
| Himalayan wet/moist temperate forests | These forests are encountered between an elevation of 1500 m and 3000 m in the eastern Himalaya. In these forests, the understorey vegetation is found growing luxuriantly. |
| Himalayan dry temperate forests | Distributed in the lesser Himalayan in the eastern and western Himalayan regions and consists of both conifers, i.e., *Abies, Juniperus, Podocarpus, Picea*, and *Pinus* and broad-leaved forests like *Alnus, Betula, Carpinus, Corylus, Juglans, Salix*, and *Ulmus*. |
| Sub-alpine forests | Found throughout the Himalaya up to the treeline above 3000 m altitude. The characteristic vegetation is composed of *Betula, Cedrus, Juniperus, Pinus, Quercus*, and *Rhododendron*. |
| Moist alpine forests | Such forests are usually encountered on the windward side above treelines in the greater Himalayan range. They are distributed in regions with annual rainfall of over 2500 mm. In context to the western Himalaya, they are found above 3500 m, while in the eastern Himalaya above 4000 m is represented by diverse herbaceous ground flora in alpine meadows. |
| Dry alpine forests | They are usually present on the leeward side where precipitation/rainfall is much less than that experienced in the rain shadow areas. In the northwestern part of the Trans-Himalayas, such forests dominantly occur giving the region an alpine meadow look with few ephemeral herbs. |

between the taxa represented in the contemporary pollen spectra and the present-day vegetation in the study areas. This non-linear correlation between the modern pollen rain and the actual number of taxa present in the vegetation is also reported by several workers (Fagerland, 1952; Davis, 1963; Xu et al., 2006, 2007; Bajpai and Kar, 2018; Quamar and Kar, 2019). It is noted that the pollen dispersed by wind, i.e., *Anemophilous*, is well/highly represented in comparison with the taxa which are dispersed by insects, i.e., *Entomophilou*. There are certain plant taxa whose pollens are not recorded in the modern pollen spectra despite their good presence in the surrounding vegetation. Such taxa marking their presence in the extant vegetation but absent in the pollen assemblages, popularly called palynologically silent taxa, this is attributed to the individual species interaction with its environment and surrounding climate (Hicks, 2001; Spieksma et al., 2003; Bajpai and Kar, 2018; Quamar and Kar, 2019).

Synthesis of the published records reviewed in the present chapter shows that the taxa like Pinus (*P. roxburghii, P. wallichiana, Alnus)* is often over/highly represented and is often encountered as extra local elements along with *Betula, Carpinus, Corylus, Elaeocarpus, Ilex*, and *Ulmus*. *Abies, Aesculus, Alnus, Betula, Carpinus, Cedrus, Corylus, Elaeocarpus, Ilex, Juglans, Juniperus, Mallotus, Podocarpus, Picea, Quercus,Tsuga, and Ulmu*, which are not well represented or sometimes underrepresented due to several factors owing to their inefficient production, dispersal, and/or preservational conditions. While these factors do not influence *Juniperus, Podocarpus*, and *Tsuga*, they are significantly marked/recorded in their respective regions of growth. Plant taxa belonging to families such as *Asteraceae, Amaranthaceae, Chenopodiaceae, Malvaceae* and *Poaceae* (grasses) reveals their true composition in the background flora.

## 23.6 Wetlands as Sites for Past Vegetation-Climate-Human Interaction

Wetlands with their capacity of exchanging heat, water, and energy with their surroundings are observed to play a salient role in regulating biogeochemical cycles, through carbon sequestration. Therefore, they are found to contribute in modulating the local/regional atmospheric constituents and environment, which in turn affects the respective ecosystem and its biotic constituents (Fan et al., 2010; Russi et al., 2013; Sharma and Singh, 2021).

In comparison to other geological/archaeological sites, being located in low land areas or natural depressions, wetlands have relatively high preservation potential for spore/pollen and other floral/faunal remains. Hence, they serve as a repository for the fossil record of terrestrial and aquatic biodiversity found in the vicinity of the wetlands-associated habitats. These factors make the sites ideal for palynological investigations to test the actual temporal changes in plant biodiversity in spatial chrono sequences over thousands of years under varied climate, vegetation, and human influence. To understand the underlying causal mechanisms, whether its natural or increased anthropogenic pressures, we need long-term data through time and space. Analysis of temporal variation of wetland biodiversity provides data on controlling factors of the variability of the biodiversity including climatic changes, which had a pivotal role either directly or indirectly. In the absence/sparsely available instrumental data for climatic records, long climate data is generated through analyses of various proxy records (Bradley, 1999). Among the available catalogue of proxies, spore/pollen from dated sediments is found promising for the understanding of the long-term changes in biodiversity such as climate change. Pollen driven vegetational/floristic analysis also documents the amplitude, frequency, and magnitude of precipitational/monsoonal variations in the wetland areas, depicting the change in the lake levels, and its expansion or change in the catchment area (Faegri and Iversen, 1964; Gaussen et al., 1965; Gasse et al., 1991; Gunnell, 1997; Bonnefille et al., 1999; Karet al., 2002; Chen et al., 2006; Prasad et al., 2014; Ghosh et al., 2015; Rawat et al., 2015; Quamar et al., 2017; Kar and Quamar, 2018).

An overview of palynological analysis carried out from the wetlands areas in the Himalayan region (Figure 23.1, Table 23.2) shows that the wetlands serve as potential sites that preserve a long-term response of vegetation to natural and human influences and thus makes them ideal sites for palaeoclimate and palaeovegetation dynamics study. Analysis of pollen/spores from the wetland sites reviewed in this chapter recorded the signals and response of vegetation not only to the global climatic events (Last Glacial Maximum: LGM; Younger Dryas: YD; Holocene Climatic Optimum: HCO; Medieval Warm Period: MWP; and Little Ice Age: LIA); but also deciphered the role of the regional and local environment in controlling the vegetation dynamics and state of the environment in these wetlands systems. Anthropogenic impacts in the form of grazing of animals, agriculture, and litter decompositions are also recorded in these studies. Studies show that agricultural activities in these areas are deciphered with the record of a good frequency of pollen taxa of *Amaranthaceae, Caryophyllaceae, Brassicaceae, Artemisia, and Cerealia*, i.e., plants which are cultured by humans for their use. Pastoral activities are inferred from the raised/high frequency of pollens of *Tubuliflorae* and *Liguliflorae* (*Asteraceae*). Wet/humid/moist conditions in an area are observed if there is an increase in the frequency of *Cyperaceae*, *Polygonum*, *Hygrophila*, and *Apiaceae* pollens. Availability of water in study areas is deciphered based on pollen/spores representing the aquatic taxa, such as *Lemna* (duckweed), *Potamogeton* (pond weed), *Typha* (reed), and *Utricularia*

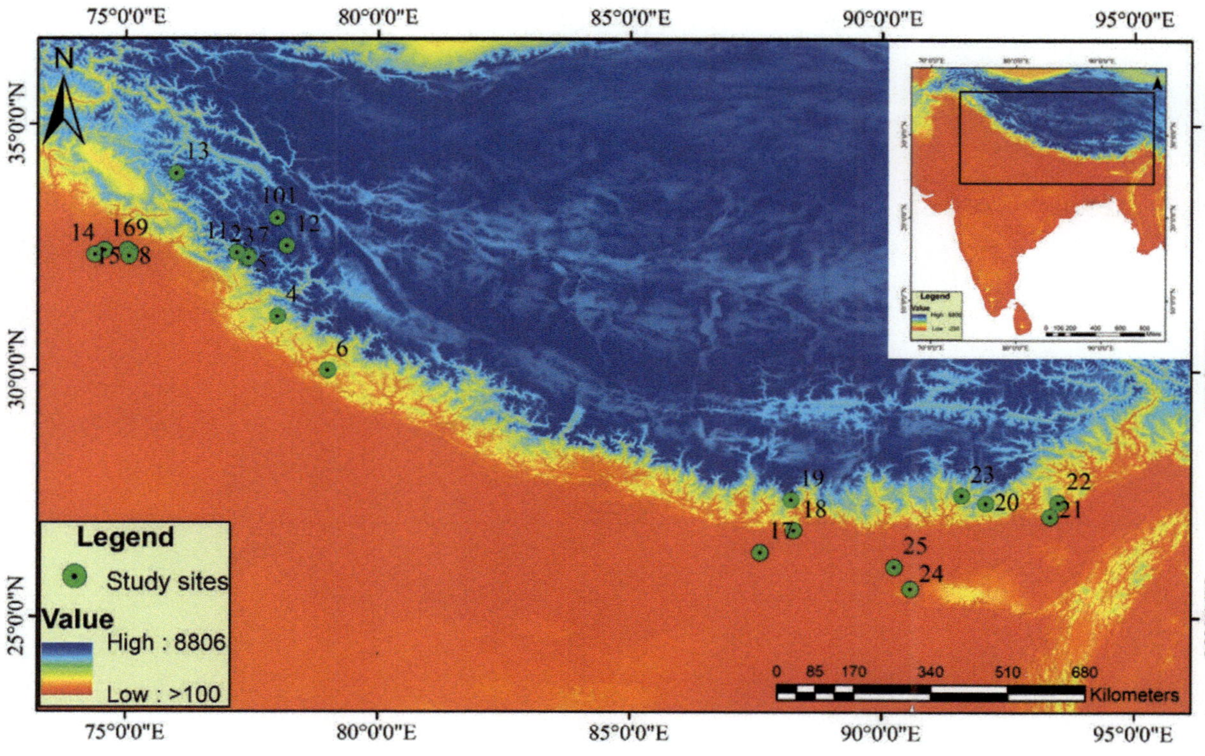


**Figure 23.1** Map locating the wetland sites referred to for this study.

**Table 23.2** Salient features of vegetation and climate recorded from wetlands sites in the Himalayan region.

| | Location (as in Chapter) | | Remark | | | |
|---|---|---|---|---|---|---|
| **Area** | **Lat.** | **Long.** | **Period** | **Inferred Vegetation** | **Climate** | **References** |
| Tso-Kar Lake, Ladakh | 33°10′N | 78°00′E | 77–10 ka BP<br>30–28,<br>21–18.375,<br>16 and 10 ka | *Chenopodiaceae-Artemisia* steppe phases of increase in pollen of *Betula, Juniperus, Hippophae*, and *Lonicera* | Cold and arid Warm and moist fluctuations | (Bhattacharyya, 1983) |
| Kunzum La, Batal, HP | 32°33′N | 77°42′E | Around or before 2–1.8 ka BP<br>1.8 and 1.37 ka BP<br>Until 800 year BP<br>800 and 496 year BP<br>From 496 year BP to Present | Dry alpine steppe elements<br>Expansion of *Juniperus* within the steppe<br>Expansion of dry steppe elements<br>Expansion of *Juniperus*<br>Expansion of dry steppe elements | Cold and dry<br>Warm and moist<br>Cold and dry<br>Warm and moist<br>Drier | (Bhattacharyya, 1988) |
| Takche lake, lahual-spiti, HP | 32°27′N | 77°38′ E | 2 ka<br>Until 1 ka<br>1 ka–400 year BP<br>From 400 year BP to Present | Dominance of grasses and sedges over shrubby elements<br>Continuation of the dominance of grasses and sedges over shrubs<br>Decline of grasses and sedges<br>Increase of steppe taxa | Cool and dry<br>Cool and dry<br>Warm, wet and moist<br>Cool and dry | (Mazari et al., 1995) |
| Chharaka Tal, Uttarakhand | 31°10′N | 78°E | 2800–1900 year BP<br>1900–1200 year BP<br>1200 year BP to Present | Mixed temperature conifer/broad-leaved forests<br>Decline of *Pinus, Betula, Alnus, Quercus*, and sedges<br>Expansion of *Abies, Artemisia, Pinus, Cedrus, Grasses, Ranunculaceae*, and sedges | Cool and moist<br>Deterioration in the climate (less cool and moist)<br>Further deterioration of climate (cold and dry) | (Chauhan et al., 1997) |
| Sitikher Bog, HP | 32°30′N | 77°40′E | 2300–1500 year BP<br>1500–900 year BP<br>900 year BP to Present | In the lower elevations, advancement in the glaciers is experienced<br>Retreat of glaciers and shift of treeline toward the higher elevations, glacial retreat accompanied with the treeline migration was encountered<br>Glaciers/treeline descended | Cold and dry<br>Warm and moist<br>Cold climate which continues until Present | (Chauhan et al., 2000) |

**Table 23.2** (Continued)

| | Location (as in Chapter) | | Remark | | | |
|---|---|---|---|---|---|---|
| **Area** | **Lat.** | **Long.** | **Period** | **Inferred Vegetation** | **Climate** | **References** |
| Dewar Tal, Uttarakhand | 30°N | 79°E | 2500–2300 year BP<br>2300–2000 year BP<br>2000–1400 year BP<br>1400–400 year BP<br>400 year BP to present | Chirpine-oak forest<br>Forests expanded<br>Decrease in frequency of pollen of chirpine and oak, while an enhancement in the frequency of members of *Poaceae*<br>Development of oak forests<br>Decrease in oak in expansion of chirpine | Cool and dry<br>Warm, wet and moist<br>Cool and dry<br>Favorable climatic (warm and moist)<br>Cool and dry | (Chauhan and Sharma, 2000) |
| Naychhudwari bog, HP | 32°30'N | 77°43'E | 1300–750 year BP<br>750–450 year BP<br>450 year BP to Present | Alpine vegetation<br>Glacier receded and treeline migrated to higher elevation, while glaciers receded<br>Treeline migration to lower altitude while glaciers experienced an advancement | Warm and moist<br>Intermittent deterioration and amelioration of climate<br>Cold and dry | (Chauhan, 2006) |
| Mansar lake, J&K | 32°41'28"N | 75°8'52" E | 9000–8000 year BP<br>8000–7000 year BP<br>7000–300 year BP<br>5500–4250 year BP<br>3000–750 year BP<br>750 year BP to Present | Mixed chirpine (*Pinus* ef. Roxburghii), oak (*Quercus*) forests<br>Forest of oak, chirpine forests<br>Broad-leaved taxa declined<br>Presence of sandy deposits<br>Oak and broad-leaved taxa expanded<br>Mixed chirpine, oak forests | Cool and dry<br>Warm, wet and humid<br>Cool and dry<br>Brief spell of pluvial activity<br>Warm and more humid<br>Deterioration of climate | (Trivedi and Chauhan, 2008) |
| Surinsar lake, J&K | 32°45'N | 75°2'E | 9500–7700 year BP<br>7700–6125 year BP<br>6125–4330 year BP<br>4330–4000 year BP<br>4000–2100 year BP<br>2100–800 year BP<br>800 year BP to Present | Mixed oak, broad-leaved/chirpine forests<br>Mixed chirpine/oak-broad-leaved forests<br>Expansion of oak and its broad-leaved associates<br>Presence of sandy layer<br>Decline in oak and other broad-leaved taxa and a concurrent increase in chirpine<br>Presence of sandy deposits<br>Slight advance in oak | Warm and humid<br>Cool and dry<br>Moderately warm and humid<br>A brief spell of pluvial environment<br>Cool and dry<br>Pluvial episode<br>Amelioration of climate | (Trivedi and Chauhan, 2009) |

*(Continued)*

**Table 23.2** (Continued)

| | Location (as in Chapter) | | Remark | | | |
|---|---|---|---|---|---|---|
| **Area** | **Lat.** | **Long.** | **Period** | **Inferred Vegetation** | **Climate** | **References** |
| Tso-kar lake, Ladakh | 33°10'N | 78°00'E | 15.2 to 14 kyr BP<br>After 14 kyr BP<br>12.9 and 12.5 kyr BP<br>12.2–11.8 kyr BP<br>Transition between Late Glacier and Early Holocene period<br>10.9 and 9.2 kyr BP<br>9.2 to 4.8 kyr BP<br>8 kyr BP<br>After 4.8 kyr BP<br>2.8–1.3 kyr BP | Alpine desert vegetation<br>alpine meadows and aquatic taxa increased<br>Rich meadows expanded along with *Artemisa*<br>Cheno/Ams were abundant in vegetation<br>Wet alpine meadows dominated by *Artemisia*<br>Growth of the alpine meadow, steppe and desert-steppe vegetation<br>Highest water levels to Tso-Kar<br>Expansion of *Chenopodiaceae*-dominated desert steppe<br>Scarce vegetation cover | Dry and cold<br>Weak sign of climate amelioration<br>Improved moisture conditions<br>Extremely weak monsoon<br>Strengthening of summer monsoon<br>Maximum monsoon<br>Weaker summer monsoon and enhancement of the winter westerlies flow<br>Combined effect of both monsoonal and westerly influence<br>An abrupt shift toward aridity<br>Weak monsoon | (Demske et al., 2009) |
| Chandra Tal, Lahul | 32°40' N | 77°20'E | Prior to 12,800 year BP<br>Till 11,640 year BP<br>YD terminates indicating initiation of the Holocene | Alpine meadow vegetation<br>Notable decrease in local (meadow) and regional (desert steppe) vegetation<br>Gradual reappearance of local and regional flora | Wet and warm<br>Cool and dry climate marking the onset of the YD<br>Wet and warm conditions | (Rawat et al., 2012) |
| Tso-Moriri Lake, Ladakh | 32°54'N | 78°19'E | 11–9.6 cal ka BP<br>5.2 cal ka BP<br>4.5–4.3 cal ka BP<br>4–3.6 cal ka BP<br>3.2 cal ka BP<br>3.5–3 cal ka BP | | Gradual decline in MAP<br>Changes in climatic conditions affected the ancient Harappan Civilization<br>Increased dryness<br>Increased aridity associated with North Atlantic cooling<br>Increased aridity associated with North Atlantic cooling<br>Collapse of Harappan Civilization | (Leipe et al., 2013) |
| Lamayuru palaeolake, Ladakh | 34°N | 76°E | Prior to 3 ka year BP<br>35 ka year BP<br>22 ka year BP<br>Subsequent phase | *Chenopodiceae-Ephrdra-Artemisia*<br>Migration of *Betula* into steppe<br>*Ephedra-Artemisia-Chenopodiaceae*<br>Expansion of steppe taxa | Semi-arid<br>Comparatively less arid than before<br>Comparatively favorable climatic<br>Coolar and drier | (Ranhotra et al., 2007) |

Table 23.2 (Continued)

| | Location (as in Chapter) | | Remark | | | |
|---|---|---|---|---|---|---|
| Area | Lat. | Long. | Period | Inferred Vegetation | Climate | References |
| Gharana Wetland, Jammu and Kashmir | 32°36′ 51.52′N | 74°38′ 58.15′E | 10,696–8536 cal year BP<br>8536–5296 cal year BP<br>5296–2776 cal year BP<br>2776–1376 cal year BP<br>1376 cal year BP to Present | Insignificant pollen recovery<br>Mixed conifer/ broad-leaved forests<br>Mixed broad-leaved/ conifer forests<br>Dense mixed broad-leaved/conifer forests<br>Mixed conifer/ broad-leaved forests | Pluvial environment<br>Cool and dry<br>Warm and humid<br>Further increase in monsoon precipitation<br>Deterioration of climate | (Quamar, 2019a) |
| Nanga wetland, Samba, Jammu and Kashmir | 32°33′ 15.44′N | 75°06′ 55.70′E | 3984–2784 cal year BP<br>2784–1584 cal year BP<br>1584–320 cal year BP<br>320 cal year BP to Present | Poor recovery of pollen taxa. Sporadic presence of *Pinus* and *Poacae* is recorded<br>Mixed broad-leaved/ conifer forests<br>Mixed conifer/ broad-leaved forests<br>Mixed broad-leaved/ conifer forests | Pluvial environment<br>Warm and humid with increased monsoon<br>Cool and dry<br>Warm and humid | (Quamar, 2019b) |
| Bajalta Lake, Jammu and Kashmir | 32°45.621′ N | 74°57.026′ E | 3205–2485 cal year BP<br>2485–1585 cal year BP<br>1585–865 cal year BP<br>865 cal year BP to Present<br>(CE 1085 onwards) | Mixed broad-leaved/ conifer forests<br>Mixed conifer/ broad-leaved forests<br>Further increase in conifers<br>Increased in broad-leaved taxa and simultaneous<br>reduction of conifers | Warm and humid with increased monsoon precipitation<br>Cool and dry, with reduced monsoon precipitation<br>Deterioration of climate, reduced monsoon precipitation<br>Warm and humid, with increased monsoon precipitation (partly corresponding to the Medivial Warm Period (MWP) between AD 740 and 1150) | (Quamar, 2019a, 2019b) |

(bladderwort). Algae, zoospores, and zygospores, such as *Spirogyra*, *Zygnema*, *Pseudoschizaea*, and *Botryococcus* in the sub-surface sediment samples also was proved to show similar results. The record of fern spores of both monolete and trilete reflects the damp and shady conditions at the study sites. Records of coprophilous fungal spores of *Delitschia*, *Glomus*, *Pleospora*, *Sordaria*, *Sporormiella*, *Tripterospora*, and *Urocystis* species in samples indicate the presence of domisticated animals, thus human habitation in the vicinity of the study areas.

## 23.7 Conclusions

In view of the studies done from the Himalayan region, we summarize that pollen/spores recorded in both the parts of Himalayan regions document the impact and response of vegetation toward both the monsoonal systems, i.e., Indian Summer Monsoon (ISM) and Westerlies/Western Disturbances (WD). The vegetation in the western Himalayas responded well to the precipitation-driven changes, while in the eastern part the role of temperature is found to be very significant in determining vegetation dynamics. This study also highlights that in the western Himalaya, the impact of ISM and WD is recorded in proxy records, while from the eastern part, mainly the role of ISM is recorded. Hence, records characterized by the evidence of prolonged wet/humid phases are reported from the eastern sector, whereas the arid/dry episodes are comparatively well recorded in the western part. The varied response of proxy toward the varying temperature and precipitation driven changes in these regions is also reflected in the form of response to the Global Climatic events like Last Glacial Maximum (LGM), Younger Dryas (YD), Holocene Climatic Optimum (HCO), Roman Warm Period (RWP), Dark Ages Cold Period (DACP), Medieval Warm Period (MWP), and Little Ice Age (LIA).

Analysis of pollen/spore studies at both extant and extinct vegetation scenarios show that for the past few centuries the wetlands in these regions are facing potential human impact leading to deteriorating conditions and floristic changes. Surprisingly, economic benefits of wetland and associated services have not been recognized until now by the policy/decision-makers, at any level, whether its local, regional, or global. Serious concerns are raised and expressed by scientific and societal communities and have become a matter of great concern and immediacy from academic to political fronts encompassing scientists, ecologists, policy-makers/planners, environmentalists, governing bodies, ruling parties, and economists to conserve and preserve the natural resources of the world. Lack of scientific data in this regard imposes challenges for the decision-makers to come up with a holistic approach to rescue, restoration/management plans. For this, applied research is urgently required to check the restoration, monitoring, and management at regular intervals through the use of GIS, remote sensing, and other geospatial data. In this regard, the use of Land-Use/Land-Cover Change (LULC), Landsat data (Multispectral Scanner (MSS), Thematic Mapper (TM), and Enhanced Thematic Mapper (ETM+) proved to be very useful to access the forest cover change and anthropogenic impacts at annual or seasonal scale. This requires environmentalists and social scientists to work together with strict legal action/legislation and science-based participatory regulations to save these rapidly vanishing ecosystems.

## Acknowledgments

We are grateful to Dr Vandana Prasad, Director Birbal Sahni Institute of Palaeosciences, Lucknow for the constant support, necessary facilities, and permission (BSIP/RDCC/Publication no 93/2021-22) to carry out and publish this work. We acknowledge all the distinguished researchers whose work have been cited in this chapter. We also acknowledge the anonymous reviewers for their valuable comments and suggestions which greatly improved the earlier version of the manuscript.

## References

Bajpai, R. and Kar, R. (2018). Modern pollen deposition in glacial settings in the Himalaya (India): abundance of *Pinus* pollen and its significance. *Palynology* 42(4): 475–482. https://doi.org/10.1080/01916122.2017.1407835.

Bandyopadhyay, S. and Mukherjee, S.K. (2005). Diversity of aquatic and wetland vascular plants of Koch Bihar district, West Bengal. In: *Plant taxonomy: Advances and Relevance* (ed. A.K. Pandey, Jun Wen, and J.V.V. Dogra), 223–244. New Delhi: CBS Publishers & Distributors.

Bhattacharyya, A. (1983). Studies in the vegetational history of the alpine region of North-West Himalaya (unpubl. PhD thesis). Lucknow University, Lucknow, 278.

Bhattacharyya, A. (1988). Vegetation and climate during post-glacial period in the vicinity of Rohtang Pass, Great Himalayan Range. *Pollen et Spores* 30(3 and 4): 417–427.

Bonnefille, R., Anupama, K., and Barboni, D. (1999). Modern pollen spectra from tropical South India and Sri Lanka: altitudinal distribution. *Journal of Biogeography* 26: 1255–1280. https://doi.org/10.1046/j.1365-2699.1999.00359.x.

Bradley, R.S. (1999). *Paleoclimatology: Reconstructing Climates of the Quaternary*. Elsevier.

Champion, H.G. and Seth, S.K. (1968). *A Revised Survey of Forest Types of India*. Delhi: Manager of Publications.

Chauhan, M.S. (2006). Late Holocene vegetation and climate change in the alpine belt of Himachal Pradesh. *Current Science* 91: 1562–1567.

Chauhan, M.S., Mazari, R.K., and Rajagopalan, G. (2000). Vegetation and climate in upper Spiti region, Himachal Pradesh during Late Holocene. *Current Science* 79(3): 373–377.

Chauhan, M.S., Sharma, C., and Rajagopalan, G. (1997). Vegetation and climate during the Late Holocene in Garhwal Himalaya. *The Palaeobotanist* 46(1,2): 211–216.

Chen, F.H., Huang, X.Z., and Zhang, J.W. (2006). Humid little ice age in arid central Asia documented by Bosten Lake, Xinjiang, China. *Science in China Series D: Earth Sciences* 49(12): 1280–1290. https://doi.org/10.1007/s11430-006-2027-4.

Davis, M.B. (1963). On the theory of pollen analysis. *American Journal of Science* 261(10): 897–912. doi: https://doi.org/10.2475/ajs.261.10.897

Demske, D., Tarasov, P.E., Wünnemann, B. et al. (2009). Late glacial and Holocene vegetation, Indian monsoon and westerly circulation in the Trans-Himalaya recorded in the lacustrine pollen sequence from Tso Kar, Ladakh, NW

India. *Palaeogeography, Palaeoclimatology, Palaeoecology* 279(3–4): 172–185. https://doi.org/10.1016/j.palaeo.2009.05.008.

Faegri, K. and Iversen, J. (1964). *Text Book of Pollen Analysis*. Waltham, MA: Chronica Botanica Co., 239p.

Fagerland, F. (1952). The real significance of pollen diagrams. *Botaniska Notiser* 105: 185–224.

Fan, J., Comstock, J.M., and Ovchinnikov, M. (2010). The cloud condensation nuclei and ice nuclei effects on tropical anvil characteristics and water vapor of the tropical tropopause layer. *Environmental Research Letters* 5: 044005. https://doi.org/10.1088/1748-9326/5/4/044005.

Gardner, R.C. and Finlayson, C. (2018). *Global Wetland Outlook: State of the World's Wetlands and Their Services to People*. 2020–2025. Ramsar Convention Secretariat, Stetson University College of Law Research Paper No. 2020-5, Available at SSRN: https://ssrn.com/abstract=3261606. Switzerland: Secretariat of the Ramsar Convention.

Gasse, F., Arnold, M., and Frontes, J.C. (1991). A 13,000-year climate record from western Tibet. *Nature* 353: 742–745. https://doi.org/10.1038/353742a0.

Gaussen, H., Legris, P., Viart, M. et al. (1965). *Godavari and Mahanadi maps and booklets*. International map of vegetation and of environmental conditions. Sheet Bombay with notice. New Delhi: ICAR.

Ghosh, R., Bera, S., and Sarkar, A. (2015). A ~50 ka record of monsoonal variability in the Darjeeling foothill region, Eastern Himalayas. *Quaternary Science Reviews* 114: 100–115. https://doi.org/10.1016/j.quascirev.2015.02.002.

Gopal, B. and Sah, M. (1995). Inventory and classification of wetlands in India. *In: Classification and Inventory of the World's Wetlands* (ed. C.M. Finlayson A. van der Valk), 39–48. Dordrecht, Netherlands: Springer. doi: 10.1007/978-94-011-0427-2_5.

Gosselink, J.G. and Turner, R.E. (1978). The role of hydrology in fresh water wetland ecosystems. In: *Freshwater Wetlands: Ecological Processes and Management Potential* (ed. R.E. Good, D.F. Whigham, and R.L. Simpson), 63–78. New York: Academic Press.

Gunnell, Y. (1997). Relief and climate in South Asia: the influence of the Western Ghats on the current climate pattern of Peninsular India. *International Journal of Climatology* 17: 1169–1182. https://doi.org/10.1002/(SICI)1097-0088(199709)17:11<1169::AID-JOC189>3.0.CO;2-W.

Hicks, S. (2001). The use of annual arboreal pollen deposition values for delimiting tree-lines in the landscape and exploring models of pollen dispersal. *Review of Palaeobotany and Palynology* 117(1–3): 1–29. https://doi.org/10.1016/S0034-6667(01)00074-4.

Houlahan, J.E., Keddy, P.A., Makkay, K. et al. (2006). The effects of adjacent land use on wetland species richness and community composition. *Wetlands* 26(1): 79–96.

Kar, R. and Quamar, M.F. (2018). Pollen-based Quaternary palaeoclimatic studies in India: an overview of recent advances. *Palynology*. Epub ahead of print 14 March. https://doi.org/10.1080/01916122.2017.1410502.

Kar, R. and Quamar, M.F. (2019). Pollen-based quaternary paleoclimatic studies in India: an overview of the recent advances. *Palynology* 43(1): 76–93. https://doi.org/10.1080/01916122.2017.1410502.

Kar, R., Ranhotra, P.S., and Bhattacharyya, A. (2002). Vegetation vis-à-vis climate and glacial fluctuations of the Gangotri glacier since last 2000 years. *Current Science* 82: 347–351.

Keddy, P.A. (2010). *Wetland Ecology: Principles and Conservation*. Cambridge, UK: Cambridge University Press. http://doi.org/10.1017/CBO9780511778179.

Khandekar, N. (2020). https://thewire.in/environment/world-wetlands-day-ramsar-convention-catchment-water-pollution-urbanisation.

Leipe, C., Demske, D., and Tarasov, P.E. (2013). Fossil pollen record of composite sediment core TMD from Tso Moriri, analysis of modern surface pollen samples, mean annual precipitation reconstruction, and digitisation and recalibration of different discussed palaeo climate proxy records. https://doi.org/10.1594/PANGAEA.808958.

Mazari, R.K., Bagati, T.N., Chauhan, M.S. et al. (1995). Palaeoclimatic record of last 2000 years in Trans Himalayan Lahaul-Spiti Region. In: *Nagoya IGBP-PAGES/PEP-II Symposium*, 262–269.

Mitsch, W.J. and Gosselink, J.G. (1993). *Wetland*, 2e. New York: Van Nostrand Reinhold Press.

Prasad, S., Anoop, A., Riedel, N. et al. (2014). Prolonged monsoon droughts and links to Indo-Pacific warm pool: a Holocene record from Lonar Lake, central India. *Earth and Planetary Science Letters* 391: 171–182. https://doi.org/10.1016/j.epsl.2014.01.043.

Prasad, S.N., Ramachandra, T.V., Ahalya, N. et al. (2002). Conservation of wetlands of India: a review. *Tropical Ecology* 43(1): 173–186.

Quamar, F.M. and Kar, R. (2019). Modern pollen dispersal studies in India: a detailed synthesis and review. *Palynology* 44(2): 217–236. https://doi.org/10.1080/01916122.2018.1557274.

Quamar, M.F. (2019). Vegetation dynamics in response to climate change from the wetlands of Western Himalaya, India: Holocene Indian summer monsoon variability. *The Holocene* 29(2): 345–362. https://doi.org/10.1177/0959683618810401.

Quamar, M.F. (2020). Surface pollen distribution from Akhnoor of Jammu District (Jammu and Kashmir), India: implications for the interpretation of fossil pollen records. *Palynology* 44(2): 270–279. https://doi.org/10.1080/01916122.2019.1568317.

Quamar, M.F., Nawaz Ali, S., and Nautiyal, C.M. (2017). Vegetation and climate reconstruction based on a ~4 ka pollen record from north Chhattisgarh, central India. *Palynology* 41(4): 504–515. https://doi.org/10.1080/01916122.2017.1279236.

Ranhotra, P.S., Bhattacharyya, A., and Kotlia, B.S. (2007). Vegetation and climatic changes around Lamayuru, Trans-Himalaya during the last 35 kyr BP. *The Palaeobotanist* 56(1–3): 117–126. http://hdl.handle.net/123456789/1119.

Rawat, S., Gupta, A.K., and Sangode, S.J. (2015). Late Pleistocene–Holocene vegetation and Indian summer monsoon record from the Lahaul, Northwest Himalaya, India. *Quaternary Science Reviews* 114: 167–181. https://doi.org/10.1016/j.quascirev.2015.01.032.

Rawat, S., Phadtare, N.R., and Sangode, S.J. (2012). The Younger Dryas cold event in NW Himalaya based on pollen record from the Chandra Tal area in Himachal Pradesh, India. *Current Science* 102: 1193–1198. http://www.jstor.org/stable/24107763.

Reddy, K.R. and DeLaune, R.D. (2008).*Biogeochemistry of Wetlands: Science and Applications*, 1e. CRC Press. https://doi.org/10.1201/9780203491454

Reis, V., Hermoso, V., Hamilton, S.K., Bunn, S.E. et al. (2019). Characterizing seasonal dynamics of Amazonian wetlands for conservation and decision-making. *Aquatic Conservation: Marine and Freshwater Ecosystems* 29(7): 1073–1082. https://doi.org/10.1002/aqc.3051.

Reis, V., Hermoso, V., Hamilton, S.K. et al. (2017). Global assessment of inland wetland conservation status. *Bioscience* 67: 523–533. https://doi.org/10.1093/biosci/bix045.

Russi, D., ten Brink, P., Farmer, A. et al. (2013). The economics of ecosystems and biodiversity for water and wetlands. IEEP, London and Brussels; Ramsar Secretariat, Gland.

Sharma, S. and Singh, P. (eds) (2021). *Wetlands Conservation*. John Wiley & Sons. https://doi.org/10.1002/9781119692621.ch1.

Spieksma, F.T.M., Corden, J.M., Detandt, M. et al. (2003). Quantitative trends in annual totals of five common airborne pollen type (*Betula, Quercus, Pinaceae, Urtica*, and *Artemisia*), of five pollen monitoring stations in western Europe. *Aerobiologia* 19(3–4): 171–184.

Trivedi, A. and Chauhan, M.S. (2008). Pollen proxy records of Holocene vegetation and climate change from Mansar Lake, Jammu region, India. *Current Science* 95(9): 1347–1354. https://www.jstor.org/stable/24103247.

Trivedi, A. and Chauhan, M.S. (2009). Holocene vegetation and climate fluctuations in northwest Himalaya, based on pollen evidence from Surinsar Lake, Jammu region, India. *Journal of the Geological Society of India* 74(3): 402–412. https://doi.org/10.1007/s12594-009-0142-5.

Wright, H.E., Jr. (1967). The use of surface samples in Quaternary pollen analysis. *Review of Palaeobotany and Palynology* 2(1–4): 321–330. https://doi.org/10.1016/0034-6667(67)90162-5.

Xu, Q.H., Li, Y.C., and Li, Y. (2006). A discussion about modern pollen and study of Quatrenary environment (in Chinese). *Progress in Natural Science* 16: 647–656.

Xu, Q.H., Li, Y.C., Yang, X.L. et al. (2007). Quantitative relationship between pollen and vegetation in northern China. *Science China Earth Sciences* 50(4): 582–599. https://doi.org/10.1007/s11430-007-2044-y.

Xu, T., Weng, B., Yan, D. et al. (2019). Wetlands of international importance: status, threats, and future protection. *International Journal of Environmental Research and Public Health* 16(10): 1818. https://doi.org/10.3390/ijerph16101818.

# 24

# Investigation of Land Use/Land Cover Changes in Alaknanda River Basin, Himalaya During 1976–2020

*Varun Narayan Mishra**

*Amity Institute of Geoinformatics & Remote Sensing, Amity University, Sector 125, Noida 201313, Uttar Pradesh, India*
** Corresponding author*

## 24.1 Introduction

In general, the land-use term mentions the utilization of land resources by many anthropogenic activities, while land cover is the habitat or natural surface of the Earth (Di Gregorio and Jansen, 2005; Mishra and Rai, 2016). Accurate and up-to-date land use/land cover (LULC) information is vital for the proper utilization, planning, and inventory of natural resources in a sustainable manner (Sheeja et al., 2011; Mishra and Rai, 2016). At the same time, changes in LULC are assumed to be a major threat for the biodiversity, ecosystem services, and ecological balance in future scenarios. Timely information of the LULC change (LULCC) is of increasing significance in several fields of studies like environmental impact and vulnerability assessment (Nguyen and Liou, 2019; Talukdar and Pal, 2020), monitoring of hazards and natural disasters (Liou et al., 2012; Zhang et al., 2019), and regional and urban planning (Mishra et al., 2019; Arora et al., 2021) at different scales. The effects of LULCC on both the function of the Earth's system and the majority of ecosystems are widely accepted phenomena worldwide (Lambin et al., 2001; Wu et al., 2017; Muleta et al., 2021). The spatio-temporal dynamics of LULCC is strongly influenced by anthropogenic interferences. In recent years, LULCC has emerged as a significant underlying driver due to rapid urbanization and increased anthropogenic activities that lead to environmental changes at both regional and global levels (Mishra et al., 2018; Gogoi et al., 2019; Li et al., 2020).

The mountain ecosystems, especially the Himalayas, are most unstable and strongly affected by the global drivers of changes such as LULCC and climate change that have harmful impacts on the ecological and social developments in these regions (Shrestha et al., 2012; Batar et al., 2017). It is also challenging to achieve sustainable development of the ecosystems of Himalayan mountains. So, it is essential to analyse the response between LULCC and quantify the ongoing changes on regional levels for recommending sustainable land-use management systems.

With the development of satellite-based remote-sensing, data at various spatial and temporal scales are now available (Batar et al., 2017; Wu et al., 2017; Mishra et al., 2019). It provides cost-effective acquisition of multi-temporal LULC information in a quick manner over large areas in comparison to ground surveys (Chen and Wang, 2010; Wittke et al., 2019). In the last few years, several studies have been carried out to map and monitor the dynamics of LULCC using earth observations from satellites (Guidigan et al., 2019; Gupta and Sharma, 2020; Arora et al., 2021). A number of studies have been reported to describe the sources like anthropogenic, economic activities, population growth, and agricultural expansion causing LULCC at different observational scales (Mishra et al., 2019, 2018; Cao et al., 2020).

The Alaknanda River Basin of the Himalayan region experienced rapid population growth and urbanization processes in the last few decades. This area consists of many dams and hydropower plants that have created pressure on the changes in land-use patterns. In this context, it is much needed to assess the changes in LULC of the Alaknanda Basin over this period of time. The objective of the present study is to investigate the spatial-temporal changes in LULC between the years 1976 and 2020 using remote-sensing datasets in the Alaknanda River Basin, one

*Advances in Remote Sensing Technology and the Three Poles*, First Edition. Edited by Manish Pandey, Prem C. Pandey, Yogesh Ray, Aman Arora, Shridhar D. Jawak, and Uma K. Shukla.

of the most heterogenous parts in the Himalayan region, India. This work can be useful to quantify ongoing changes that is vital for the sustainable planning and management of the study area.

## 24.2 Materials and Methods

### 24.2.1 Study Area

The Alaknanda River Basin located in the eastern part of the Garhwal Himalayan region of India is chosen as the study area, having center latitude 30º31′56.92″N and 79º17′47.08″E and total geographical area of 8956.25 $km^2$. The study area comprises the Bhagirath Kharak and Satopanth glaciers, which are the sources of the Alaknanda River, a major tributary of the Ganga. The location map of the study area is shown in Figure 24.1. It is mainly characterized by hilly terrain, deep gorges, and river valleys. The Alaknanda River and its tributaries are among the major rivers of India, which originates and flows from this region. Climatologically, the basin experiences variations due to the altitudinal differences and diverse physiography. There is variation in temperature from season to season and from valley to high altitude regions.

### 24.2.2 Data Used

In this study, cloud-free images acquired by Landsat 2-Multispectral Scanner System (MSS), and Landsat 8 Operational Land Imager (OLI) sensors in years 1976 (19 November) and 2020 (8 and 17 November), respectively

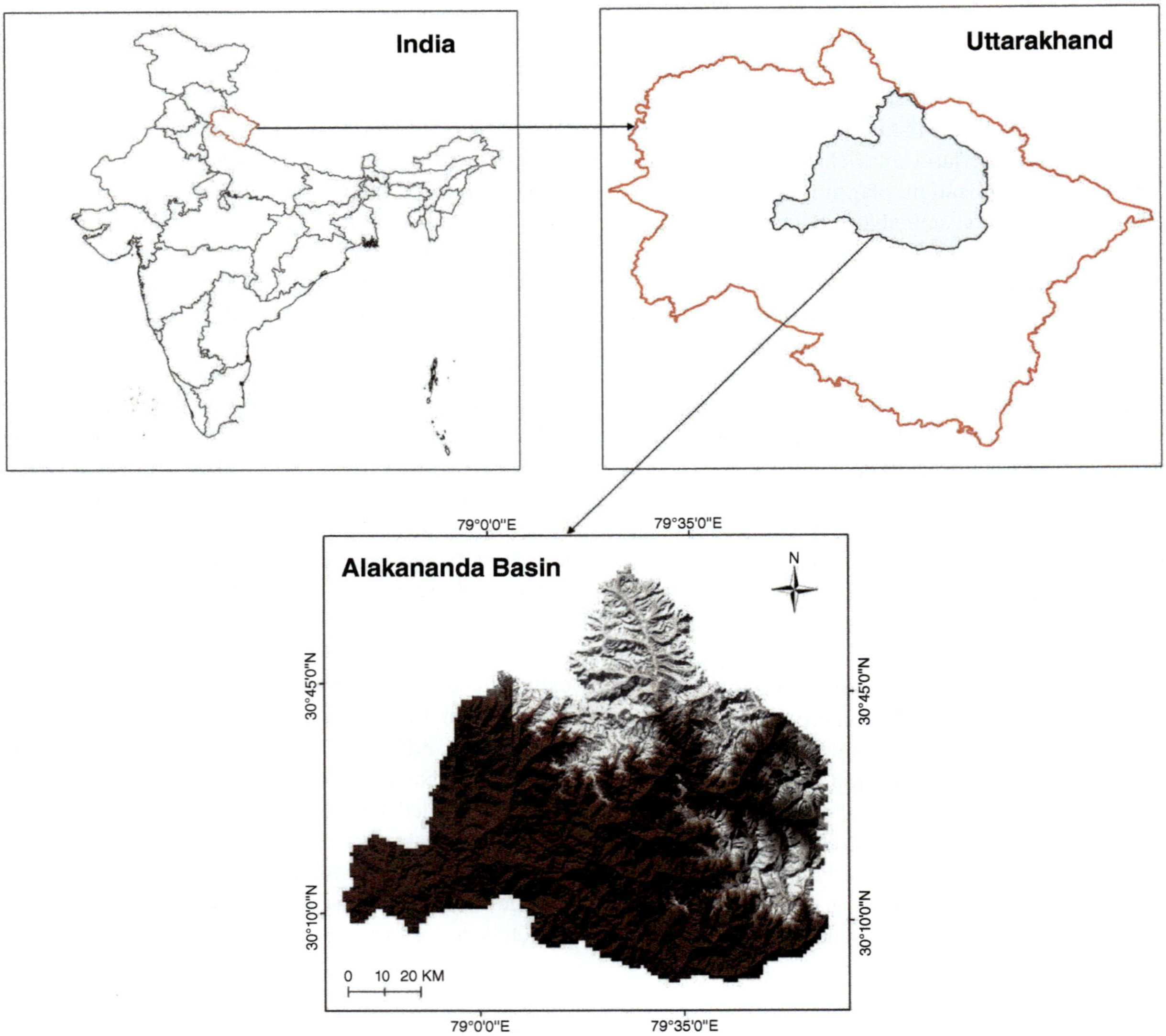


Figure 24.1 Location map of the study area.

were used to prepare the LULC maps of the area under investigation. The satellite images were downloaded from USGS Earth Explorer website (http://earthexplorer.usgs.gov).

### 24.2.3 Methods

The present investigation was based on the LULC maps derived by classifying the Landsat images of the years 1976 and 2020, using the supervized classification method.

#### 24.2.3.1 LULC Classification Scheme

The study area falls under the mountainous landscape. So, it is required to perform the pre-processing of the satellite images to reduce the differences caused by the atmospheric or sensor variations of the two dates. The pre-processing of satellite images includes geometric rectification, radiometric calibration, atmospheric rectification, mosaicking, and extraction of study area prior to classification (Gashaw et al., 2017; Lin et al., 2018). Afterwards, the images were resampled using the nearest neighbor method to the pixel size of 30 m with the Universal Transverse Mercator (UTM) coordinate system (Zone 44, North). An appropriate band combination was used to generate false color composite (FCC) for both the remote-sensing images. A supervized maximum-likelihood classification (MLC) was applied for extracting LULC information. The error matrix approach was used to evaluate the accuracies of classified maps obtained by the MLC method. The accuracy assessment of maps was performed by computing the overall classification accuracy (OA), producer's accuracy (PA), user's accuracy (UA), and kappa coefficient (Kc) (Congalton and Green, 1999).

#### 24.2.3.2 LULC Change Investigation

The investigation of LULCC was carried out using the post-classification change detection method. The area statistics of classified maps was compared in order to determine the quantitative information of the changes between the time period of 1976 and 2020. The annual rate of change (rt) for each LULC class is calculated, as well using Equation 24.1 (Puyravaud, 2003; Mishra et al., 2019) given as

$$rt = \frac{1}{(t_2 - t_1)} \times \ln\left(\frac{A_2}{A_1}\right) \times 100 \qquad (24.1)$$

where $A_1$ and $A_2$ are the areas (in km$^2$) of individual LULC at years t1 (initial time) and t2 (later time), respectively. The positive and negative rt values show increase and decrease in the area of a specific LULC category.

## 24.3 Results and Discussion

The classified LULC results corresponding to the Landsat satellite images based on MLC method are shown in Figures 24.2a and 24.2b. The class-wise area distribution for the years 1976 and 2020 is given in Table 24.1.

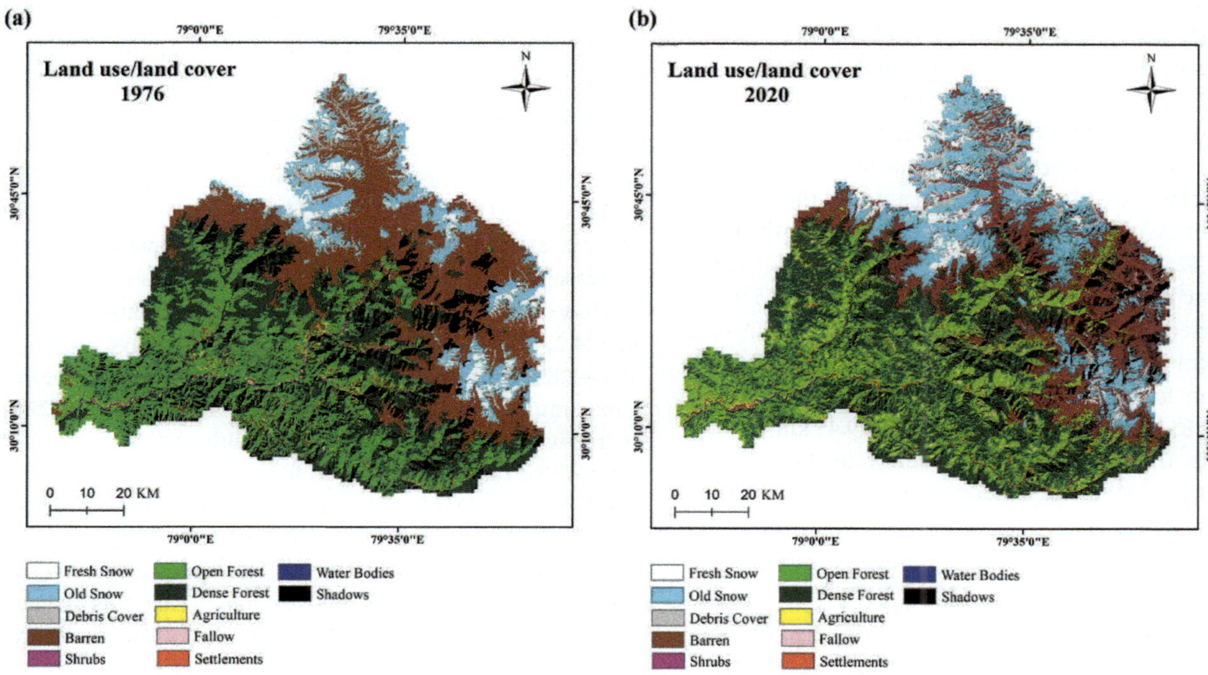


Figure 24.2 LULC status of the study area for year (a) 1976; and (b) 2020 based on MLC.

Table 24.1 Area and amount of change in LULC classes during 1976 to 2020.

| Year | 1976 | | 2020 | | Change 1976–2020 | | |
|---|---|---|---|---|---|---|---|
| LULC Category | Area (km$^2$) | Area (%) | Area (km$^2$) | Area (%) | Area (km$^2$) | Area (%) | Rate of change (%) |
| Fresh Snow | 299.94 | 3.35 | 359.00 | 4.01 | 59.06 | 0.66 | 0.41 |
| Old Snow | 623.62 | 6.96 | 1040.68 | 11.62 | 417.06 | 4.66 | 1.16 |
| Debris Cover | 324.64 | 3.62 | 268.35 | 3.00 | −56.29 | −0.63 | -0.43 |
| Barren | 2655.09 | 29.65 | 1782.35 | 19.90 | −872.74 | −9.74 | -0.91 |
| Shrubs | 68.38 | 0.76 | 98.57 | 1.10 | 30.20 | 0.34 | 0.83 |
| Dense Forest | 2154.33 | 24.05 | 2653.37 | 29.63 | 499.04 | 5.57 | 0.47 |
| Open Forest | 1812.98 | 20.24 | 1920.99 | 21.45 | 108.01 | 1.21 | 0.13 |
| Agriculture | 115.46 | 1.29 | 49.51 | 0.55 | −65.94 | −0.74 | -1.92 |
| Fallow | 69.60 | 0.78 | 110.85 | 1.24 | 40.95 | 0.46 | 1.05 |
| Settlements | 16.69 | 0.19 | 119.43 | 1.33 | 102.73 | 1.15 | 4.47 |
| Water Bodies | 28.58 | 0.32 | 28.92 | 0.32 | 0.34 | 0.00 | 0.03 |
| Shadows | 786.65 | 8.78 | 524.22 | 5.85 | −262.43 | −2.96 | -0.92 |
| **Total** | **8956.25** | **100** | **8956.25** | **100** | **0.00** | **00.00** | |

## 24.3.1 LULC Status

The accuracy assessment of LULC classification results showed an OA of 80.53% (Kc = 0.774) for 1976 and 85.82% (Kc = 0.849) for 2020, respectively. Figure 24.2 represents the spatial pattern of the LULC distributional of the Alaknanda Basin for the years 1976 and 2020, respectively. From Table 24.1, the class-wise area statistics reveal that in 1976, 3.35% (299.94 km$^2$) area was under fresh snow, 6.96% (623.62 km$^2$) under old snow, 3.62% (324.64 km$^2$) under debris cover, 29.65% (2665.09 km$^2$) under barren, 0.76% (68.38 km$^2$) under shrubs, 24.05% (2154.33 km$^2$) under dense forest, 20.24% (1812.98 km$^2$) under open forest, 1.29% (115.46 km$^2$) under agriculture, 0.78% (69.60 km$^2$) under fallow, 0.19% (16.69 km$^2$) under settlements, 0.32% (28.58 km$^2$) under water bodies, and 8.78% (786.65 km$^2$) under shadow areas. In 2020, the area under these LULC categories was found to be 4.01% (359.00 km$^2$) for the fresh snow, 11.62% (1040.68 km$^2$) under old snow, 3.00% (268.35 km$^2$) under debris cover, 19.90% (1782.35 km$^2$) under barren, 1.10% (98.57 km$^2$) under shrubs, 29.63% (2653.37 km$^2$) under dense forest, 21.45% (1920.99 km$^2$) under open forest, 0.55% (49.51 km$^2$) under agriculture, 1.24% (110.85 km$^2$) under fallow, 1.33% (119.43 km$^2$) under settlements, 0.32% (28.92 km$^2$) under water bodies, and 5.85% (524.22 km$^2$) under shadow regions.

## 24.3.2 LULC Change

Table 24.1 shows both the positive and negative changes occurring in LULC condition of the Alaknanda Basin during the period of 1976–2020. The fresh snow increased by 59.06 km$^2$ during the period under investigation, which accounts for 0.66% of the total study area. The old snow increased by 417.06 km$^2$, which accounts for 4.66% of the total study area. The debris cover decreased by 56.29 km$^2$, which accounts for 0.63%. There was a decrease of 872.74 km$^2$ area in the barren category. The shrubs class increased by 30.20 (0.34%) during the period under investigation. The dense forest increased from 2154.33 km$^2$ in 1976 to 2653.37 km$^2$ in 2020, which accounts for 5.57% of the total area. The open forest increased from 1812.98 km$^2$ in 1976 to 1920.99 km$^2$ in 2020, which accounts for 1.21% of the total area. There was a decrease of 65.94 km$^2$ area in the agriculture class. The fallow land increased by 40.95 km$^2$ during the period under investigation, which accounts for 0.46% of the total study area. The settlements increased from 16.69 km$^2$ in 1976 to 119.43 km$^2$ in 2020, which accounts for 1.15% of the total area of the basin. The water bodies have very slight increase from 28.58 km$^2$ in 1976 to 28.92 km$^2$ in 2020. The shadow covered area has decreased from 786.65 km$^2$ in 1976 to 524.22 km$^2$ in 2020, which accounts for 2.96%. During 1976–2020, the rate of loss of agriculture was found to be at a maximum with −1.94% followed by shadows with −0.92%, barren with −0.91%, and debris cover with −0.43%. The highest positive rate of change was found for settlements with 4.47% followed by old snow (1.16%), fallow (1.05%), shrubs (0.83%), fresh snow (0.41%), dense forest (0.47%), open forest (0.13%), and water bodies (0.03%). The settlements showed the highest positive rate of change while the agriculture had the highest negative rate of change in the Alaknanda Basin during 1976–2020.

Moreover, the key sources of uncertainties in this study are the sensor's characteristics and classification method. The erroneous description of LULC classes is another cause of uncertainty in the classification method. The results of image classification may also be affected by interpretation skills and shadow regions of hilly terrain. As a result, the changing pattern of LULC may vary at different scales and depend on the resolution of the images used.

## 24.4 Conclusions

This study reveals the effectiveness of multi-date satellite images in investigating LULCC for the Alaknanda Basin of the Himalayan region, India. It experienced a decrease in barren land and agriculture and an increase in shrubs, dense forest, open forest, fallow land, and settlements between 1976 and 2020. The study indicates that expansion in settlements are the major drivers of change in agriculture land and other categories. It also suggested that land cover changes may lead to agriculture loss with implications for food security and people's livelihoods. The settlements have continued to develop along the Alaknanda River. These areas are subject to the construction of infrastructure and hydropower plants. The present study establishes the worth of Landsat images for understanding the spatio-temporal dynamics of LULCC at lower cost with reasonable accuracy. Locating the areas with ongoing changes will help to contribute toward the sustainable development and land-use planning of the study site.

## References


Arora, A., Pandey, M., Mishra, V.N. et al. (2021). Comparative evaluation of geospatial scenario-based land change simulation models using landscape metrics. *Ecol. Indic.* 128: 107810. https://doi.org/10.1016/j.ecolind.2021.107810.

Batar, A.K., Watanabe, T., and Kumar, A. (2017). Assessment of land-use/land-cover change and forest fragmentation in the Garhwal Himalayan region of India. *Environments* 4: 34. https://doi.org/10.3390/environments4020034.

Cao, Y., Zhang, X., Fue, Y. et al. (2020). Urban spatial growth modeling using logistic regression and cellular automata: a case study of Hangzhou. *Ecol. Indic.* 113: 106200. https://doi.org/10.1016/j.ecolind.2020.106200.

Chen, Z.and Wang, J. (2010). Land use and land cover change detection using satellite remote sensing techniques in the mountainous Three Gorges area, China. *Int. J. Remote Sens.* 31: 1519–1542. https://doi.org/10.1080/01431160903475381.

Congalton, R.G. and Green, K. (1999). *Assessing the Accuracy of Remotely Sensed Data: Principles and Practices.* Boca Raton, FL: CRC/Lewis Press.

Di Gregorio, A. and Jansen, L.J.M. (2005). *Land Cover Classification System: LCCS: Classification Concepts and User Manual.* Rome: Food and Agriculture Organization of the United Nations.

Gashaw, T., Tulu, T., Argaw, M. et al. (2017). Evaluation and prediction of land use/land cover changes in the Andassa watershed, Blue Nile Basin, Ethiopia. *Environ. Syst. Res.* 6: 17. https://doi.org/10.1186/s40068-017-0094-5.

Gogoi, P.P., Vinoj, V., Swain, et al. (2019). Land use and land cover change effect on surface temperature over Eastern India. *Sci. Rep.* 9: 8859. https://doi.org/10.1038/s41598-019-45213-z.

Guidigan, M.L.G., Sanou, C.L., Ragatoa, D.S. et al. (2019). Assessing land use/land cover dynamic and its impact in Benin Republic using land change model and CCI-LC products. *Earth Syst. Environ.* 3(1): 127–137. https://doi.org/10.1007/s41748-018-0083-5.

Gupta, R.and Sharma, L.K. (2020). Efficacy of Spatial Land Change Modeler as a forecasting indicator for anthropogenic change dynamics over five decades: a case study of Shoolpaneshwar Wildlife Sanctuary, Gujarat, India. *Ecol. Indic.* 112: 106171. https://doi.org/10.1016/j.ecolind.2020.106171.

Lambin, E.F., Turner, B.L., Geist, H.J. et al. (2001). The causes of land-use and land-cover change: moving beyond myths. *Glob. Environ. Change* 11: 261–269. https://doi.org/10.1016/S0959-3780(01)00007-3.

Li, Z.T., Li, M., and Xia, B.C. (2020). Spatio-temporal dynamics of ecological security pattern of the Pearl River Delta urban agglomeration based on LUCC simulation. *Ecol. Indic.* 114: 106319. https://doi.org/10.1016/j.ecolind.2020.106319.

Lin, X., Xu, M., Cao, C. et al. (2018). Land-use/land-cover changes and their influence on the ecosystem in Chengdu City, China during the period of 1992–2018. *Sustainability* 10: 3580. https://doi.org/10.3390/su10103580.

Liou, Y.A., Sha, H.C., Chen, T.M. et al. (2012). Assessment of disaster losses in rice paddy field and yield after tsunami induced by the 2011 great East Japan earthquake. *J. Mar. Sci. Technol.* 20: 618–623. http://dx.doi.org/10.6119%2fJMST-012-0328-2.

Mishra, V.N., Prasad, R., Kumar, P. et al. (2019). Assessment of spatio-temporal changes in land use/land cover over a decade (2000–2014) using earth observation datasets: a case study of Varanasi district, India. *Iran J. Sci. Technol. Trans. Civ. Eng.* 43: 383–401. https://doi.org/10.1007/s40996-018-0172-6.

Mishra, V.N. and Rai, P.K. (2016). A remote sensing aided multi-layer perceptron-Markov chain analysis for land use and land cover change prediction in Patna district (Bihar), India. *Arab. J. Geosci.* 9(4): 1–18. https://doi.org/10.1007/s12517-015-2138-3.

Mishra, V.N., Rai, P.K., Prasad, R. et al. (2018). Prediction of spatio-temporal land use/land cover dynamics in rapidly developing Varanasi district of Uttar Pradesh, India using Geospatial approach: a comparison of hybrid models. *Appl. Geomat.* 10(3): 257–276. https://doi.org/10.1007/s12518-018-0223-5.

Muleta, T.T., Kidane, M., and Bezie, A. (2021). The effect of land use/land cover change on ecosystem services values of Jibat forest landscape, Ethiopia. *GeoJournal* 86: 2209–2225. https://doi.org/10.1007/s10708-020-10186-4.

Nguyen, K.A. and Liou, Y.A. (2019). Mapping global eco-environment vulnerability due to human and nature disturbances. *MethodsX* 6: 862–875. https://doi.org/10.1016/j.mex.2019.03.023.

Puyravaud, J.P. (2003). Standardizing the calculation of the annual rate of deforestation. *For. Ecol. Manag.* 177: 593–596. https://doi.org/10.1016/S0378-1127(02)00335-3.

Sheeja, R.V., Joseph, S., Jaya, D.S. et al. (2011). Land use and land cover changes over a century (1914–2007) in the Neyyar River Basin, Kerala: a remote sensing and GIS approach. *Int. J. Digit. Earth* 4(3): 258–270. https://doi.org/10.1080/17538947.2010.493959.

Shrestha, U.B., Gautam, S., and Bawa, K.S. (2012). Widespread climate change in the Himalayas and associated changes in local ecosystems. *PLoS ONE* 7: 1–10. https://doi.org/10.1371/journal.pone.0036741.

Talukdar, S. and Pal, S. (2020). Wetland habitat vulnerability of lower Punarbhaba River Basin of the uplifted Barind region of Indo-Bangladesh. *Geocarto Int.* 35(8): 857–886. https://doi.org/10.1080/10106049.2018.1533594.

Wittke, S., Yu, X., Karjalainen, M. et al. (2019). Comparison of two-dimensional multitemporal Sentinel-2 data with three-dimensional remote sensing data sources for forest inventory parameter estimation over a boreal forest. *Int. J. Appl. Earth Obs. Geoinf.* 76: 167–178. https://doi.org/10.1016/j.jag.2018.11.009.

Wu, M., Schurgers, G., Ahlström, A. et al. (2017). Impacts of land use on climate and ecosystem productivity over the Amazon and the South American continent. *Environ. Res. Lett.* 12: 054016. http://dx.doi.org/10.1088/1748-9326/aa6fd6.

Zhang, Y., Ge, T., Tian, W. et al. (2019). Debris flow susceptibility mapping using machine-learning techniques in Shigatse Area, China. *Remote Sens.* 11(23): 2801. https://doi.org/10.3390/rs11232801.

# Section IV

## The Arctic: The Northernmost Ocean Having the North Pole Environment and Remote Sensing

# 25

# Hydrological Changes in the Arctic, the Antarctic, and the Himalaya

## A Synoptic View from the Cryosphere Change Perspective

*Shyam Ranjan*[1,*], *Manish Pandey*[2,3,*], *and Rahul Raj*[4]

[1] *School of Environmental Sciences, Jawaharlal Nehru University, New Delhi 110067, India*
[2] *University Center for Research and Development, Chandigarh University, Mohali 140413, Punjab, India*
[3] *Department of Civil Engineering, Chandigarh University, Mohali 140413, Punjab, India*
[4] *Centre for Korean Studies, School of Language, Literature & Culture Studies Jawaharlal Nehru University, New Delhi 110067, India*
* *Corresponding authors*

## 25.1 Introduction

A glacier is a robust tool to understand the climate change on the Earth's system. The formation of the Earth's climate is a long process. It can be defined as a long-term average of a particular location's weather conditions (a combination of various meteorological conditions, i.e., temperature, precipitation, humidity, wind speed, and wind direction). Different model results have shown the rise of near-surface temperature in recent decades (IPCC, 2013). Since glaciers are highly sensitive to temperature variations, an increase in temperature negatively impacts the cryospheric system, especially on glaciers. Furthermore, at the spatio-temporal scale, it affects the glacio-hydrology of glacierised stream catchments (Barnett et al., 2005). The term "glacio-hydrology" is a combination of hydrological as well as glaciological aspects of science that incorporates the water movement, its distribution, and its linkage to physical factors during in-phase and phase change of the hydrological cycle, as well as understanding the distinct behavior of glaciers and its association with different meteorological conditions (Subramanya, 2013).

Understanding the glacio-hydrology component of the cryospheric system is very important because studies suggest that anthropogenic factors played a significant role in climate change that elevated the global mean temperature (GMT) by 0.9°C since 1850 (Masson-Delmotte et al., 2018). This temperature increase has severely impacted the Earth's cryosphere by reducing the stored polar ice volumes (IPCC, 2013). Recent trends, especially from the late 1970s to the mid-2000s, show that the North Polar region's summer sea-ice area has shrunk by >10% per decade (Stroeve et al., 2012). Assuming the continuity of this trend remains the same, the Arctic region could face an ice-free summer by the end of the 21st century (Wunderling et al., 2020). Furthermore, field observation records show that the North Polar Region summer sea ice is declining faster than expected through the projected scenario from GCMs (General Circulation Models) (Stroeve et al., 2012).

Scientific observations indicate that both the Greenland Ice Sheet and the West Antarctic Ice Sheet have been losing cryospheric mass at an accelerating pace in the past few decades (Zwally et al., 2011; Khan et al., 2014; Shepherd et al., 2018) and so expected to increase the sea level with progressing global warming (Clark et al., 2016; Steffen et al., 2018). Apart from the polar regions, mountain glaciers also show a similar retreating trend worldwide with an average weight equivalent ice loss of around 250 ± 30 Gt/year in the last 100 years (Leclercq et al., 2011; Gardner, 2013).

According to the IPCC (2013), it is estimated that nearly 600 glaciers have disappeared, and many more will follow the same trend in the future. Moreover, it has been found that several mountain glaciers are in disequilibrium and may likely loss cryospheric mass in the near future (Christian et al., 2018).

Understanding glacier loss in high mountain ranges of the Himalaya, also known as the Third Pole, has become more critical due to its ability to store an enormous amount of fresh water in the form of solid (glacier, permafrost) and meltwater (lakes, wetlands, and springs). Meltwater storage in the high mountainous regions of the Himalaya is the second largest after polar areas; therefore, it is also considered "Asia's Water Tower" (Singh et al., 2016). In the high-altitude Himalayan

*Advances in Remote Sensing Technology and the Three Poles,* First Edition. Edited by Manish Pandey, Prem C. Pandey, Yogesh Ray, Aman Arora, Shridhar D. Jawak, and Uma K. Shukla.

rages of the Hindu Kush Himalayan region, there are approximately 54,000 glaciers that spread over an area of about 60,000 $km^2$ with nearly 6000 $km^3$ of ice volume reserves (Williams, 2013). Various meltwater channels of these glaciers mingle and form a glacial river that provides fresh water to billions of people in downstream regions (ICIMOD, 2009; Immerzeel et al., 2010; Lutz et al., 2016). Himalayan glaciers' unique location (lower latitude with a high amount of heat exchange compared to the Arctic and the Antarctic) makes them vulnerable to changes in temperature or precipitation patterns (Immerzeel et al., 2010) and therefore it becomes essential to monitor them continuously.

Various *in-situ* and satellite based observations confirm that most Himalayan glaciers are losing their ice mass. Still, the rate of retreat is variable over different spatial and temporal spaces as well as topography and local climatic conditions (Barry, 1990; Gerrard et al., 1993; Shrestha et al., 1999; Barnett et al., 2005; Oerlemans, 2005; Bolch et al., 2012; Kääb et al., 2012; Schmidt et al., 2012; Field et al., 2014; Vashisht et al., 2017). The primary concern about glacial retreat in the high mountain regions is limited to the physical processes and ecological, social, and economic aspects, especially near the mountainous hillslope zone and in the downstream regions. Therefore, in this study, we will explore and understand the changes in the cryospheric system of the Arctic, the Antarctic, and the Third Pole (Himalaya), together with their association with the hydrological ecological, societal, and economic conditions.

## 25.2 Cryosphere and Its Influence on Socio-Ecological-Economical (GLASOECO) System

Glaciers, a part of the cryospheric system, are well connected with the Earth's ecological and societal system through its meltwater discharge and glacial river system that flows downstream and sustains the societal, ecological, and economical (SOECO) components. Therefore, any result in the cryospheric system (glacier, snow cover, glacial ice, permafrost) results in disequilibrium between SOECO components. Strong association of the cryospheric system to SOECO requires an in-depth understanding of the complex linkage between SOECO components and their feedback at different climatic conditions (Warner, 2010; Derksen et al., 2012; Huggel et al., 2015). It is crucial to understand that even a minute change at the level of the cryospheric system in the sense of its size or characteristics can affect the micro- as well as the macroclimate of the nearby ecological system as well as livelihood related to them (Warner, 2010; Fountain et al., 2012; Berman and Schmidt, 2019). These changes are non-linear and highly complex and often cascade through the domino effect (one effect followed by another, and so on) on the spatio-temporal scale (Galaz et al., 2011; Cramer et al., 2015). For instance, glacier melting combined with a heavy influx of precipitation can create situations that may lead to glacial lake outburst floods (GLOFs). That can trigger other events, including land degradation, increase in the meltwater discharge, high rate of sedimentation, degradation of biodiversity, and ultimately socio-economic vulnerabilities (McDowell et al., 2013; Carey et al., 2017). Food security risk, livestock, fisheries, tourism, energy production, ecosystem vulnerability, species migration, etc., are major affected sectors from cryospheric change and are discussed below. Changes in cryosphere and related effects in various domains of our physical and social environmental variables is portrayed in the sketch diagram presented in Figure 25.1.

### 25.2.1 Cryospheric Change and Its Influence on Agriculture and Livestock

A high mountain range like the Himalaya is a significant source of water availability to farmers for their agriculture and daily needs. Significant populations living near upstream regions are dependent on glacial meltwater (especially during the spring and summer runoff) for their livelihood, vegetation growth, groundwater recharge, and food security (Bury et al., 2013; Clouse, 2016; Parveen et al., 2015; Paudel and Andersen, 2013). Therefore, any adverse impact on the cryospheric system will directly influence upstream agriculture and agro-pastoralism (Dame and Nüsser, 2011; Parveen et al., 2015; Clouse, 2016).

The unique topography of the western Himalayan region (northwestern India, northern Pakistan, and Afghanistan) compel farmers to entirely depend on glacial meltwater for their agriculture production and food security (Parveen et al., 2015; Clouse, 2016; Mukherji et al., 2019). The situation is even more alarming in northern India, where scientific observation has found the gradual thinning of low-lying glaciers in the last three decades, reducing meltwater supply for irrigation and leading to drought in many agricultural villages (Grossman, 2015). Cold desert-like Ladakh, India, entirely relies on glacier meltwater-based water sources for its agricultural production. Similarly, Melamchi Valley, Nepal is also wholly dependent on glacier meltwater for its 100% water needs, while in Gilgit-Baltistan, glacier and snowmelt water contributes to nearly 66% for irrigation (Khadka et al., 2008; Dame and Nüsser, 2011; McDowell et al., 2013; Vincent et al., 2019). Not only in the upstream catchments, but these glacier-fed rivers also become vital for a downstream agricultural area. For instance, near the Himalayan foothills, the Bhakhra Nangal dam provides enough irrigation water to three major

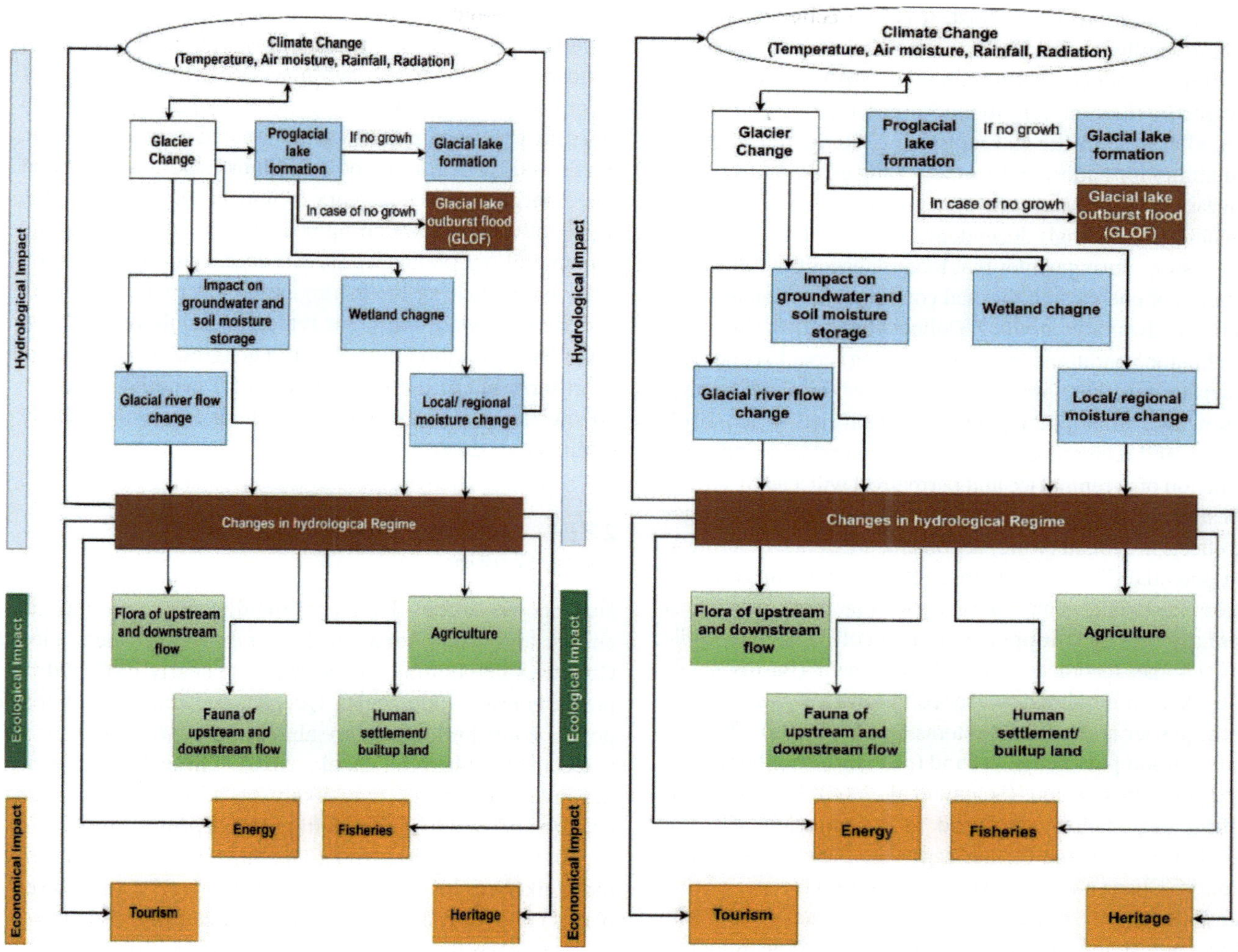


**Figure 25.1** Flow diagram showing climate change and its association with hydrological, ecological, and economic balance.

Indian states to help India become and remain self sufficient in food grain availability (Malik, 2008). Similarly, meltwater in the Indus Basin mainly originates from six major glaciers and irrigates nearly 80% of the total cultivated area of Pakistan (ICIMOD, 2010). As an output, it produces more than 80% of rice and wheat sugar (Qureshi, 2011).

Livestock and animal husbandry is an integrated part for farmers in mountain areas that sustains each other by providing food and manure. Additionally, animals are a prime source of food during extreme climatic conditions (no agriculture during snowfall and winter) and for skins, wool, and transportation. Therefore, any adverse impacts on water availability directly affect livestock production and food security (Rasul and Molden, 2019). For example, in the Nagqu, Tibet, He and Richards (2015) observed that the decreasing density of plants in meadows is partly due to change in the cryosphere; this has negatively affected the animal herders with increasing grazing intensity. Thus, any change in the cryospheric system not only affects agricultural production and food security but animal husbandry also.

### 25.2.2 Cryospheric Change and Its Influence on Ecosystem and Environment

The sensitivity of cryospheric system to temperature variations makes it a matter of serious concern for the anthropogenically induced climate change that potentially affects the vegetation, carbon balance, water availability, and micro-climate (Fountain et al., 2012; Williams et al., 2015). Alteration in glacier mass (e.g., ice loss) directly influences the ecosystem through habitat loss and changes in the thermal condition. In contrast, it is indirectly affected by altering the light absorption and thus nutrient availability to primary producers (Fountain et al., 2012). Additionally, changes in thermal conditions induce melting of permafrost, alteration in the soil moisture content, and nutrient availability for habituated species (Yang et al., 2010).

In fact, in the cryospheric system, plant growth and food availability are strongly dependent on the duration and thickness of snow during winter that too influences the soil water, an influx of energy, and thermal conditions for biomass production (Edwards et al., 2007; DeBeer et al., 2016). Changing the climatic conditions (climate warming) could reduce the snowpack and, therefore, alter the length of the plant growing season and, thus, plant phenology (Gottfried et al., 2012; Warren and Lemmen, 2014). In the worst-case scenario, extinction of perennial ice and permafrost will lead to surface exposure and decrease in vegetation cover, which will trigger the effects of aeolian (wind) action and accelerate desertification potential (Wang and French, 1994, 1995). Furthermore, worldwide treeline data analysis shows that more than 50% of sites have observed the uphill movement of the treeline, while only 1% has recorded downhill movement (Barros et al., 2016). A warming climate will not only have a negative impact on the glacier and treeline systems, but it will also affect the stored carbon pool in the wetland (peatland) and the sequestered permafrost regions (Schuur et al., 2008). Therefore, permafrost degradation could lead to the emission of major greenhouse gases and further aggravate the climate change scenario (Cheng and Wu, 2007). Ma et al. (2011) reported that many pollutants, especially persistent organic pollutants (POPs), are stored in Arctic glacier ice in sizeable amounts. They could release them back into the atmosphere and increase the concentration due to climate warming.

### 25.2.3 Cryospheric Change and Its Influence on the Economy

Climate warming not only changes the cryospheric ice volume, agriculture, and ecological equilibrium, but also affects the fisheries (Lehodey et al., 2006; Brander, 2007). For instance, it has been observed that rapid glacier melting and glacier outburst floods (GLOF) in the Kenai River, Alaska, have severely affected fish production and annual income worth USD 70 million (Milner et al., 2017). Similarly, fish production from the Himalayan rivers is also under threat due to changing climate (Allison et al., 2009). Furthermore, climate warming also influences river and lake water warming, which leads to acceleration in the rise of invasive species. The study has found that in Arctic and Continental Great Lakes ecoregions, commercial fishing has increased but at the cost of the food security risk for the indigenous people, dependent on subsistence fishing (Warren and Lemmen, 2014) for centuries.

Tourism is another crucial sector for the source of income and livelihoods for the mountainous regions that supports multi-million dollar recreation services (Burakowski and Magnusson, 2012). A new trend of polar tourism (PT) wherein exploration and leisure related activities in the Arctic and the Antarctic polar regions has been witnessed has emerged in recent years (Shijin et al., 2020). In PT, main activities are watching polar bears, whale watching, snowmobiling, photography, reindeer sledding, watching polar lights, fishing/ice fishing, and boat tours. Changing climate with warming polar regions possibly open a new route for exploration purposes and generate revenue in the short term but at the same time create negative feedback on the native species and bring them into vulnerability (Shijin et al., 2020).

### 25.2.4 Cryospheric Change as a Risk to Energy Security

Hydropower is considered eco-friendly, clean, and natural energy, and it supplies about 16% of the world's electricity. Countries like Bhutan and Norway use nearly 100% hydropower energy while Nepal, Quebec, and Canada's hydropower contributions go to almost 93%, 90%, and 59%, respectively (Shrestha et al., 2016). These hydropower-dependent countries may likely be affected by climate change and rapid glacier melting (IPCC, 2001; Boehlert et al., 2016; Turner et al., 2017). The impact of climate warming is likely to be more on "run of the river" hydropower plants (Kopytkovskiy et al., 2015) that can face little to no water storage issues (Boehlert et al., 2016; Turner et al., 2017). Glacier melting initially can provide an ample amount of water volume for hydropower, but that does not lead to an increase in the energy production due to storage limitation of the reservoir (Warren and Lemmen, 2014; Tarroja et al., 2016). But the situation turns even worse in the dry season when runoff volume is reduced for hydropower production and the initial surplus amount in annual flow from melted glaciers is not able to compensate for decreased water volume during the dry season (Rees et al., 2004; Tarroja et al., 2016). Warming climate increases the frequency of mountain disasters that also affect hydropower production from a flash flood, debris, and dam-structure damage. For instance, heavy rainfall and lake outbursts near the Chorabari Glacier, India, in June 2013 resulted in heavy casualties and damage to the hydropower project. Similarly, Dig Tsho GLOF in Nepal destroyed a multi-million dollar hydropower project in 1985 (Schwanghart et al., 2016). Therefore, the energy sector, specially hydroelectric power sector is likely to suffer heavy losses in the long run due to climate change, ultimately increasing the stress for energy demand (Berman and Schmidt, 2019).

## 25.3 Hydrological Changes in the Arctic and the Antarctic Regions

### 25.3.1 Hydrological Changes in the Arctic

Long-term observational data from 1971 to 2017 show that the annual average Arctic surface air temperature has increased by 2.7°C, surprisingly 2.4 times faster than average for the Northern Hemisphere (AMAP, 2019). A recent study by Rantanen et al., (2022) presents observational dataset based evidence of four times temperature increase in the Arctic region as compared to previous reports of 2-3 times increase only. The projected and observed annual average warming of the Arctic is scarier. It shows the continuity with two times more than the global mean level and the increase in frequency and duration of Arctic winter warming (Serreze and Barry, 2011; Graham et al., 2017). Annual precipitation also shows an increasing trend with an estimate of 1.5%–2.0%/decade, with the strongest in October to May. The warming impact can also be seen through a change in the precipitation phase, such as the Baltic Sea Basin and Scandinavia have observed lower precipitation in snow fall, but more precipitation has been experienced in liquid rain form.

In contrast, the trend in spring snow in the Arctic lands was found to decrease in volume (AMAP, 2019). In direct form, the evidence of Arctic warming is visible through the reduction in the extent of sea ice, snow cover areas, permafrost regions, and river and lake ice content (Shijin et al., 2020). Satellite observation of the last decade shows that Arctic winter sea-ice maximums in the years 2015 to 2018 were at record low levels (AMAP, 2019). It has been estimated that from 1850 to 2000, the percentage contribution of Arctic melting to sea-level rise is 48% (10 cm) and significantly 30% of the total sea-level rise has happened only over the last 25 years (the years 1992 to 2017) (AMAP, 2019). The permafrost layer is also under warming stress, and an observed average temperature of the permafrost active layer reached a record high, especially in the cold permafrost regions of the northern Arctic, resulting in the vulnerability of river and lake ice subsidence and thawing of the permafrost areas (Hori et al., 2018; AMAP, 2019). For instance, because of thawing, the situation of ground stability of the Russian Arctic is under threat. Still, these regions are economically crucial. Nearly 45% of the hydrocarbon extraction fields are located here and are under threat due to climate warming, leading to severe damage to the economy and environment (Hjort et al., 2018).

### 25.3.2 Hydrological Changes in the Antarctic

Antarctica, the world's largest "cold source" (temperature in most areas is below 0ºC), is sensitive to global climate change. Changing climate in the polar region is primarily due to global warming, cryosphere melting, and changes in the equilibrium condition of the Antarctic environment. However, changes in the Antarctic cryosphere are not uniform. For instance, southeastern Antarctic hardly shows any changes in the surface temperature since 1950, while the southwestern Antarctic has increased by about 0.7°C/10a (Hodgson et al., 2010; Orsi et al., 2012; Turner et al., 2014). In the last 25 years, it has been estimated that the Antarctic ice sheet lost a total of 2,720 ± 139 billion tons of ice, which is enough for the 7.6 ± 3.9 mm global sea-level rise (SLR) (Shepherd et al., 2018). In the projected emission scenario, the total cryospheric volume of Antarctica is enough to raise the sea level by 1 m by 2100 and >15 m by 2500 (DeConto and Pollard, 2016). Furthermore, rapid warming significantly affects the ice shelves that disintegrate with time from the main body, either partly or entirely (Cook and Vaughan, 2010; Shepherd et al., 2018). Regional differences and changes cannot be ignored despite a slight increase in the Antarctic Sea ice. Interestingly, changes in the sea-ice extent are insignificant in the Indian Ocean and the Western Pacific, but the Bellingshausen–Amundsen seas region has undergone significant changes and decrease of its sea ice from −5.1 ± 1.6%/10a to −2.9 ± 1.4%/10a, due to the air temperature cooling effect (De Santis et al., 2017). The recent trend (2014 to 2017) of sea ice extent in the Southern Hemisphere shows an unprecedented annual decline together with a significant reduction in the sea-ice covered area (Parkinson, 2019). Antarctic permafrost, another essential part of the polar cryospheric system, accounts for a small fraction of its total size (~0.36%). Still, it is considered the primary controlling factor for the South Pole terrestrial ecosystem (Vieira et al., 2010). Permafrost is sensitive to temperature change, and climate warming can accelerate permafrost degradation (Bockheim et al., 2013; Guglielmin and Vieira, 2014).

## 25.4 Hydrological Changes in the Third Pole (Himalaya)

Himalayan glaciers, especially in the Karakoram region (Hindu-Kush, HK), are high-altitude and high-volume glaciers. Out of six river basins of the Himalayan range, HK glaciers contribute to about 100%, 93%, 31%, and 22% of the total ice volumes in the Ganges, Indus, Brahmaputra, and Tarim river basins, respectively (Nie et al., 2021). The formation of a glacier in the Himalayan region is primarily due to major precipitation sources: western disturbances that decline from west to east, Indian summer monsoon that decline from east to the west (Nie et al., 2021) and local moisture recycling which is yet to be fully understood in the Himalayan region (Ranjan et al., 2021).

Rapid climate warming rings an alarm for Himalayan glaciers and glacier-hydrology as glacier mass loss and a decrease in the precipitation at high altitudes is reported by studies (Nie et al., 2017), although Karakoram glaciers show relatively stable conditions (the Karakoram anomaly) (Bolch et al., 2012; Farinotti et al., 2020). The stability of the Karakoram glacier, even in the rising global temperature scenario, is a topic of debate among glaciologists. Some propose that it could be due to interchange between regional warming that accentuates glacier mass loss rate and increases localized mass gain in winter and summer cooling as a result of enhanced snowfall conditions (Treydte et al., 2006; Bolch et al., 2012; Mukhopadhyay and Khan, 2014; Azam, 2018; de Kok et al., 2018; Farinotti et al., 2020). Physical characteristics of a glacier, i.e., debris-free or debris-covered, also play a crucial role in the melting process. For instance, debris-covered (thin layered debris cover) glaciers directly affect the glacier melt rate by increasing absorption of solar radiation and reduction in the surface albedo and thus increasing the glacial response to climate change (Scherler et al., 2011). The supraglacial lakes formed over the glacier surface equally contribute to changing the glacier melt rate because these lakes or ponds transmit the thermal energy to glacier ice (Sakai et al., 2009; King et al., 2018) and promoting cliff collapse or calving (Tsutaki et al., 2019; Liu et al., 2020; Watson et al., 2020). Studies have observed that Himalayan glaciers are losing their mass rapidly due to supraglacial lake/pond formation (Brun et al., 2019; King et al., 2019; Maurer et al., 2019).

Changes in glacier volume directly affect the meltwater runoff and, consequently, the seasonal river flow from upstream to downstream (Nie et al., 2021). In a high runoff scenario, all these channels decide the fate of incidence and lake outburst-flood intensity in the mountainous regions. Meltwater from glaciers contributes to total runoff as it is one of the components of the total-runoff equation, including snowfall, rainfall, basal flow, groundwater, and glacier meltwater (Nie et al., 2021). The percentage contribution of glacier runoff is significant in the changing climatic scenario because presently, in the HK region, it contributes nearly 42% and 33% in Tarim and upper Indus basins, respectively. It is noteworthy that the HK region primarily receives precipitation from the western disturbances (Lutz et al., 2016), while Brahmaputra and Upper Ganga basins, whose glacier-runoff contribution is ~16% and ~12%, respectively, receive precipitation primarily from the monsoon system (Lutz et al., 2014). It has been reported that, between 2000 and 2018, the Indus, Ganga, and Brahmaputra river basins experienced the most significant meltwater runoff of 4.55, 3.26, and 5.23 Gt per year due to the loss of glacier ice (Shean, 2020). Therefore, a warming climate is likely to increase the melting of Himalayan glaciers and lead to massive mass loss in the long run but increased glacier runoff in the short run; and intensifies the conditions conducive for runoff flooding related catastrophic losses (Nie et al., 2021).

### 25.4.1 Runoff Flooding

The Himalaya is prone to different types of flooding. For instance, the 2010 flooding in Pakistan was due to excessive rainfall and had a massive economic setback, apart from around 2000 fatalities (Atta-ur-Rahman and Khan, 2013). Similarly, the Uttarakhand, India flooding in June 2013, caused by catastrophic heavy rainfall, resulted in the death of nearly 6000 people (Allen et al., 2016; Mishra et al., 2019). Most of the floods in these regions are generally caused by either excessive rainfall or temperature-induced glacier melting (Atta-ur-Rahman and Khan, 2013; Mishra et al., 2019; Nie et al., 2021). Recent studies report that Himalayan regions, especially Ganga and Brahmaputra basins, are more prone to rainstorm-related flooding. In contrast, the Indus and Tarim basins (especially at high altitude) are more associated with temperature-induced flooding (Zhang, 2016). Combined effects of these two flood generating factors worsen and intensify the flood hazard situation in the downstream areas, especially near the foothills and floodplain zones. For example, combined effects of ongoing accelerated glacier-melting-induced water discharge and heavy catastrophic rainfall had been witnessed in the case of the June 2013 floods in Uttarakhand, India (Martha et al., 2015; Houze et al., 2017; Rounce et al., 2017). The intensity of flood disasters is very high during the monsoon peak season (June to September), while melt-water-induced flood intensity is high during May to September (Goswami et al., 2006). The magnitude and frequency of floods in the HK region (Tarim, Ganges, Brahmaputra, and Indus basins) have increased from the 1950s to 2009 (Elalem and Pal, 2015; Zhang, 2016). Changes in the weather pattern and alteration in the long-term weather systems (i.e., monsoonal precipitation) are the major causes of intensely extreme flooding, weather storms, and flash flooding (Goswami et al., 2006; Atta-ur-Rahman and Khan, 2013). Furthermore, studies using climate prediction models project that the magnitude and frequency of flooding will most likely increase by the end of this century, and with the 2°C global temperature increment, and this scenario is likely to exacerbate (Dankers et al., 2014; Mohammed et al., 2017; Lutz et al., 2019).

### 25.4.2 Future Hydrological Change in the Third Pole

The future prediction for the fate of glaciers of the HK region is crucial to understand the economy and livelihood of the population that directly or indirectly rely on them. RCPs4 model projection predicts the increasing temperature and precipitation in the HK region, leading to glacier mass loss (Lutz et al., 2014; Kraaijenbrink et al., 2017). Even if we can restrict the global mean temperature to 1.5°C, projected warming of 2.1 ± 0°C through the

RCP2.6 scenario for Himalaya will result in an overall 36 ± 7% mass loss of glaciers in the region by the end of the 21st century (Kraaijenbrink et al., 2017). This projected scenario indicates the change in debris covered glaciers and the increase in the number of supraglacial ponds and lakes (Irvine-Fynn, 2017; Jiang, 2018; Scherler et al., 2018; King et al., 2019). Similarly, the RCP8.5 model predicts more severe floods. It has been estimated that the peak flow of a 50-year-return flood is likely to increase by about 51%, 108%, and 80% for the Upper Indus, Upper Ganges, and Upper Brahmaputra basins, respectively (Wijngaard et al., 2017). The model prediction also indicates that the Brahmaputra and the Ganges basins will be highly vulnerable to flooding due to the predicted heightened peak discharges. (Dankers et al., 2014; Lutz et al., 2019). The Upper Indus and the Tarim basins will receive high flow not only during summer but also during winter with possible shifting of peak flow (Luo et al., 2018; Khan et al., 2020), resulting in more intense floods. Therefore, regional cooperation with more discharge monitoring station networks in the HK region would be ideal for precise prediction (Vashisht et al., 2017). That will help policymakers to make proper risk assessments and find a sustainable solution (Mishra et al., 2019).

## 25.5 Conclusion

This study addressed the following points of Earth's cryosphere system and the impact of climate warming on the cryospheric system:

- Rapid warming of the climate is significantly changing the glacier volume of Third Pole glaciers, which is very important for the sustainability of billions of people's livelihood.
- Rapidly losing glaciers directly affect the Himalayan cryospheric system and freshwater availability locked in solid form (glacier, permafrost) and its release through meltwater (lakes, wetlands, and springs).
- Warming climate also increases the chances of mountain disasters and damage and destruction of hydropower production plants from flash floods, debris flows, etc. An example is June 2013 due to heavy rainfall and lake outbursts near the Chorabari Glacier. It has been predicted that more GLOFs in the Himalayan region are likely to increase with the warming climate.
- Unlike the Himalaya, the warming climate in the Arctic region is opening new routes for transportation, exploration, and revenue purposes and creating negative feedback on the native species and bringing them into vulnerability.
- Rapid warming also significantly impacts the Antarctic ice shelves with breakage and disintegration with time. It not only alarms us about the sea-level rise threats but also attracts our attention to the changing ocean salinity and nutrient availability to the marine ecosystem.
- Well researched and preplanned mitigation framework and measures are necessary for maintaining the equilibrium of the cryosphere system by reducing greenhouse gas emissions. Furthermore, addressing the impacts of cryosphere shrinkage requires global cooperation and immediate steps to monitor and facilitate adaption and mitigation measures.

## Acknowledgments

The authors thank CSIR-SRA (13-9059-A-2019 Pool) and Science and Engineering Board (SERB) for funding this project (Project no: SR/FTP/ES-134/2014).

## References

Allen, S.K., Rastner, P., Arora, M. et al. (2016). Lake outburst and debris flow disaster at Kedarnath, June 2013: hydrometeorological triggering and topographic predisposition. *Landslides* 13: 1479–1491. doi.org/10.1007/s10346-015-0584-3

Allison, E.H., Perry, A.L., Badjeck, M.C. et al. (2009). Vulnerability of national economies to the impacts of climate change on fisheries. *Fish* 10: 173–196. doi.org/10.1111/j.1467-2979.2008.00310.x

AMAP: Arctic Monitoring and Assessment Programme (2019). Arctic climate change update 2019: an update to key findings of Snow, Water, Ice, and Permafrost in the Arctic (SWIPA) 2017. Assessment Report, 12.

Atta-ur-Rahman and Khan, A.N. (2013). Analysis of 2010-flood causes, nature, and magnitude in the Khyber Pakhtunkhwa, Pakistan. *Nat. Hazard.* 66: 887–904. doi.org/10.1007/s11069-012-0528-3

Azam, M.F., Wagnon, P., Berthier, E. et al. (2018). Review of the status and mass changes of Himalayan-Karakoram glaciers. *Journal of Glaciology* 64: 61–74. doi.org/10.1017/jog.2017.86

Barnett, T P., Adam, J. C., and Lettenmaier, D.P. (2005). Potential impacts of a warming climate on water availability in snow-dominated regions. *Nature* 438: 303–309. doi.org/10.1038/nature04141

Barros, C., Guéguen, M., Douzet, R., Douzet, R. et al. (2017). Extreme climate events counteract the effects of climate and land-use changes in Alpine tree lines. *Journal of Applied Ecology* 54: 39–50. doi.org/1J0.1111/1365-2664.12742

Barry, R.G. (1990). Evidence of recent changes in global snow and ice cover. *GeoJournal* 20: 121–127. doi.org/10.1007/BF00196739

Berman, M. and Schmidt, J.I. (2019). Economic effects of climate change in Alaska. *Weather Clim. Soc.* 11: 245–258. doi.org/10.1175/WCAS-D-18-0056.1

Bockheim, J., Vieira, G., Ramos, M. et al. (2013). Climate warming and permafrost dynamics in the Antarctic Peninsula region. *Glob. Planet. Chang.* 100: 215–223. doi.org/10.1016/j.gloplacha.2012.10.018

Boehlert, B., Strzepek, K., Gebretsadik, Y. et al. (2016). Climate change impacts and greenhouse gas mitigation effects on US hydropower generation. *Appl. Energy* 183: 1511–1519. doi.org/10.1016/j.apenergy.2016.09.054

Bolch, T., Kulkarni, A., Kääb, A. et al. (2012). The state and fate of Himalayan glaciers. *Science 80 336: 310–314.* doi.org/10.1126/science.1215828

Brander, K.M. (2007). Global fish production and climate change. *Proc. Natl. Acad. Sci. USA.* 104: 19709–19714. doi.org/10.1073/pnas.0702059104

Brun, F., Wagnon, P., Berthier, E. et al. (2019). Heterogeneous influence of glacier morphology on the mass balance variability in High Mountain Asia. *J. Geophys. Res. Earth Surf.* 124: 1331–1345. doi.org/10.1029/2018JF004838

Burakowski, E. and Magnusson, M. (2012). *Climate impacts on the winter tourism economy in the United States.* NRDC's Policy Pap. Sch. Repos. University of Hampshire, UK, 1–33.

Bury, J., Mark, B.G., Carey, M. et al. (2013). New geographies of water and climate change in Peru: coupled natural and social transformations in the Santa River watershed. *Ann. Assoc. Amer. Geogr.* 103: 363–374. doi.org/10.1080/00045608.2013.754665

Carey, M., Molden, O.C., Rasmussen, M.B. et al. (2017). Impacts of glacier recession and declining meltwater on mountain societies. *Ann. Am. Assoc. Geogr.* 107: 350–359. doi.org/10.1080/24694452.2016.1243039

Cheng, G. and Wu, T. (2007). Responses of permafrost to climate change and their environmental significance, Qinghai-Tibet Plateau. *J. Geophys. Res. Earth Surf.* 112: F02S03. doi.org/10.1029/2006JF000631

Christian, J.E., Koutnik, M., and Roe, G. (2018). Committed retreat: controls on glacier disequilibrium in a warming climate. *J. Glaciol.* 64: 675–688. doi.org/10.1017/jog.2018.57

Clark, P.U., Shakun, J.D., Marcott, S.A. et al. (2016). Consequences of twenty-first-century policy for multi-millennial climate and sea-level change. *Nat. Clim. Chang.* 6: 360–369. doi.org/10.1038/nclimate2923

Clouse, C. (2016). Frozen landscapes: climate-adaptive design interventions in Ladakh and Zanskar. *Landsc. Res.* 41: 821–837. doi.org/10.1080/01426397.2016.1172559

Cook, A.J. and Vaughan, D.G. (2010). Overview of areal changes of the ice shelves on the Antarctic Peninsula over the past 50 years. *Cryosphere* 4: 77–98. doi.org/10.5194/tc-4-77-2010

Cramer, W., Yohe, G.W., Auffhammer, M. et al. (2015). Detection and attribution of observed impacts. In: *Climate Change 2014 Impacts, Adaptation and Vulnerability: Part A: Global and Sectoral Aspects* (ed. C.B. Field, V.R. Barros, D.J. Dokken, et al.), 979–1038. Cambridge, UK: Cambridge University Press. doi.org/10.1017/CBO9781107415379.023

Dame, J. and Nüsser, M. (2011). Food security in high mountain regions: agricultural production and the impact of food subsidies in Ladakh, Northern India. *Food Secur.* 3: 179–194. doi.org/10.1007/S12571-011-0127-2

Dankers, R., Arnell, N.W., Clark, D.B. et al. (2014). First look at changes in flood hazard in the Inter-Sectoral Impact Model Intercomparison Project ensemble. *Proc. Natl. Acad. Sci. USA* 111: 3257–3251. doi.org/10.1073/pnas.1302078110

de Kok, R.J., Tuinenburg, O.A., Bonekamp, P.N.J. et al. (2018). Irrigation as a potential driver for anomalous glacier behavior in High Mountain Asia. *Geophys. Res. Lett.* 45: 2047–2054. doi.org/10.1002/2017GL076158

De Santis, A., Maier, E., Gomez, R. et al. (2017). Antarctica, 1979–2016 sea ice extent: total versus regional trends, anomalies, and correlation with climatological variables. *Int. J. Remote Sens.* 38: 7566–7584. doi.org/10.1080/01431161.2017.1363440

DeBeer, C.M., Wheater, H.S., Carey, S.K. et al. (2016). Recent climatic, cryospheric, and hydrological changes over the interior of western Canada: a review and synthesis. *Hydrol. Earth Syst. Sci.* 20: 1573–1598. doi.org.10.5194/hess-20-1573-2016

DeConto, R.M. and Pollard, D. (2016). Contribution of Antarctica to past and future sea-level rise. *Nature.* 531: 591–597. doi.org/10.1038/nature17145

Derksen, C., Smith, S., Sharp, M. et al. (2012). Variability and change in the Canadian cryosphere. *Climage Change* 115: 59–88. doi.org/10.1007/s10584-012-0470-0

Edwards, A.C., Scalenghe, R., and Freppaz, M. (2007). Changes in the seasonal snow cover of alpine regions and its effect on soil processes: a review. *Quat. Int.* 162–163: 172–181. doi.org/10.1016/j.quaint.2006.10.027

Elalem, S. and Pal, I. (2015). Mapping the vulnerability hotspots over Hindu-Kush Himalaya region to flooding disasters. *Weather Clim. Extrem.* 8: 46–58. doi.org/10.1016/j.wace/2014.12.001

Farinotti, D., Immerzeel, W.W., de Kok, R.J. et al. (2020). Manifestations and mechanisms of the Karakoram Glacier Anomaly. *Nat. Geosci.* doi.org/10.1038/s41561-019-0513-5

Field, R.D., Kim, D., LeGrande, A.N. et al. (2014). Evaluating climate model performance in the tropics with retrievals of water isotopic composition from Aura TES. *Geophys. Res. Lett.* 41: 6030–6036. doi.org/10.1002/2014gl060572

Fountain, A.G., Campbell, J.L., Schuur, E.A.G. et al. (2012). The disappearing cryosphere: impacts and ecosystem responses to rapid cryosphere loss. *Bioscience* 62: 405–415. doi.org/10.1525/bio.2012.62.4.11

Galaz, V., Moberg, F., Olsson, E.K. et al. (2011). Institutional and political leadership dimensions of cascading ecological crises. *Public Admin.* 89: 361–380. doi.org/10.1111/j.1467-9299.2010.01883.x

Gardner, A.S., Moholdt, G., Cogley, J.G. et al. (2013). A reconciled estimate of glacier contributions to sea-level rise: 2003 to 2009. *Science 80*: 340: 852–857. doi.org/10.1126/science.1234532

Gerrard, J., Stone, P.B., and Agenda, M. (1993). The state of the world's mountains: a global report. *Geogr. J.* 159: 101. doi.org/10.2307/3451524

Goswami, B.N., Venugopal, V., Sengupta, D. et al. (2006). Increasing trend of extreme rain events over India in a warming environment. *Science* 80: 314: 1442–1445. doi.org/10.1126/science.1132027

Gottfried, M., Pauli, H., Futschik, A. et al. (2012). Continent-wide response of mountain vegetation to climate change. *Nat. Clim. Chang.* 2: 111–115. doi.org/10.1038/nclimate1329

Graham, R.M., Cohen, L., Petty, A.A. et al. (2017). Increasing frequency and duration of Arctic winter warming events. *Geophys. Res. Lett.* 44: 6974–6983. doi.org/10.1002/2017GL073395

Grossman, D. (2015). As Himalayan glacier melts, two towns face the fall out. Yale Environment 360 (WWW document). Yale School of Forestry and Environment Studies. e360.yale.edu/features/as_himalayan_glaciers_melt_two_towns_face_the_fallout (accessed 25 February 2022).

Guglielmin, M. and Vieira, G. (2014). Permafrost and periglacial research in Antarctica: new results and perspectives. *Geomorphology* 225: 1–3. doi.org/10.1016/j.geomorph.2014.04.005

He, S. and Richards, K. (2015). Impact of meadow degradation on soil water status and pasture management: a case study in Tibet. *L. Degrad. Dev.* 26: 468–479. doi.org/10.1002/ldr.2358

Hjort, J., Karjalainen, O., Aalto, J. et al. (2018). Degrading permafrost puts Arctic infrastructure at risk by mid-century. *Nat. Commun.* 9: 5147. doi.org/10.1038/s41467-018-07557-4

Hodgson, D.A., Convey, P., Verleyen, E. et al. (2010). The limnology and biology of the Dufek Massif, Transantarctic Mountains 82° south. *Polar Sci.* 4: 197–214. doi.org/10.1016/j.polar.2010.04.003

Hori, Y., Cheng, V.Y.S., Gough, W.A. et al. (2018). Implications of projected climate change on winter road systems in Ontario's Far North, Canada. *Clim. Change.* 148: 109–133. doi.org/10.1007/s10584-018-2178-2

Houze, R.A., Mcmurdie, L.A., Rasmussen, K.L. et al. (2017). Multiscale aspects of the storm producing the June 2013 flooding in Uttarakhand, India. *Mon. Weather Rev.* 145: 4447–4466. doi.org/10.1175/MWR-D-17-0004.1

Huggel, C., Carey, M., Clague, J.J. et al. (2015). *The High-Mountain Cryosphere: Environmental Changes and Human Risks*. Cambridge, UK; Cambridge University Press. doi.org/10.1017/CBO9781107588653

ICIMOD (2009). Biodiversity and Climate Change in the Himalaya: Sustainable Mountain Development No. 55. Kathmandu.

ICIMOD (2010). Glacial Melt and Downstream Impacts on Indus Dependent Water Resources and Energy. Kathmandu.

Immerzeel, W.W., van Beek, L.P.H., and Bierkens, M.F.P. (2010). Climate change will affect the Asian water towers. *Science* 80: 328: 1382–1385. doi.org/10.1126/science.1183188

IPCC (2001). Intergovernmental Panel on Climate Change. Impacts, Adaptation and Vulnerability. Working Group II Contribution to the Intergovernmental Panel on Climate Change (IPCC). IPCC Fourth Assessment Report: Climate Change 2007 (AR4).

IPCC (2013). Working Group 1: I. Stocker, T.F., Qin, D., Plattner, G.-K. et al. (2013). *Climate Change 2013: The Physical Science Basis*. Contribution of Working Group I to the Fifth Assessment Report of the Intergovernmental Panel on Climate Change. IPCC.

Irvine-Fynn, T.D.L., Porter, P.R., Rowan, A.V. et al. (2017). Supraglacial ponds regulate runoff from Himalayan debris-covered glaciers. *Geophys. Res. Lett.* 44: 11894–11904. doi.org/10.1002/2017GL075398

Jiang, S., Nie, Y., Liu, Q. (2018). Glacier change, supraglacial debris expansion and glacial lake evolution in the Gyirong River Basin, central Himalayas, between 1988 and 2015. *Remote Sens* 10:986. doi.org/10.3390/rs10070986

Kääb, A., Berthier, E., Nuth, C. et al. (2012). Contrasting patterns of early twenty-first-century glacier mass change in the Himalayas. *Nature* 488: 495–498. doi.org/10.1038/nature11324

Khadka, R. and Khanal, E. (2008). Environmental management plan (EMP) for Melamchi water supply project, Nepal. *Environmental Monitoring and Assessment* 146: 225–234. doi.org/10.1007/s10661-007-0074-8

Khan, A.J., Koch, M., and Tahir, A.A. (2020). Impacts of climate change on the water availability, seasonality and extremes in the Upper Indus Basin (UIB). *Sustainability* 12: 1283. doi.org/10.3390/su12041283

Khan, S.A., Kjær, K H., Bevis, M. et al. (2014). Sustained mass loss of the northeast Greenland ice sheet triggered by regional warming. *Nat. Clim. Chang.* 4: 292–299. doi.org/10.1038/nclimate2161

King, O., Bhattacharya, A., Bhambri, R. et al. (2019). Glacial lakes exacerbate Himalayan glacier mass loss. *Sci. Rep.* 9: 18145. doi.org/10.1038/s41598-019-53733-x

King, O., Dehecq, A., Quincey, D. et al. (2018). Contrasting geometric and dynamic evolution of lake and land-terminating glaciers in the central Himalaya. *Glob. Planet. Chang.* 167: 46–60. doi.org/10.1016/j.gloplacha.2018.05.006.

Kopytkovskiy, M., Geza, M., and McCray, J.E. (2015). Climate-change impacts on water resources and hydropower potential in the Upper Colorado River Basin. *J. Hydrol. Reg. Stud.* 3: 473–493. doi.org/10.1016/j.ejrh.2015.02.014

Kraaijenbrink, P.D.A., Bierkens, M.F.P., Lutz, A.F. et al. (2017). Impact of a global temperature rise of 1.5 degrees Celsius on Asia's glaciers. *Nature* 549: 257–260. doi.org/10.1038/nature23878

Leclercq, P.W., Oerlemans, J., and Cogley, J.G. (2011). Estimating the glacier contribution to sea-level rise for the period 1800–2005. *Surv. Geophys.* 32: 519–535. doi.org/10.1007/s10712-011-9121-7

Lehodey, P., Alheit, J., Barange, M. et al. (2006). Climate variability, fish, and fisheries. *J. Clim.* 19: 5009–5030. doi.org/10.1175/JCLI3898.1

Liu, Q., Mayer, C., Wang, X. et al. (2020). Interannual flow dynamics driven by frontal retreat of a lake-terminating glacier in the Chinese Central Himalaya. *Earth Planet. Sci. Lett.* 546: 116450. doi.org/10.1016/j.epsl.2020.116450

Luo, Y., Wang, X., Piao, S. et al. (2018). Contrasting streamflow regimes induced by melting glaciers across the Tien Shan–Pamir–North Karakoram. *Sci. Rep.* 8: 16470. doi.org/10.1038/s41598-018-34829-2

Lutz, A.F., Immerzeel, W.W., Kraaijenbrink, P.D. et al. (2016). Climate change impacts on the upper Indus hydrology: sources, shifts and extremes. *PLoS One* 11: e0165630. doi.org/10.1371/journal.pone.0165630

Lutz, A.F., Immerzeel, W.W., Shrestha, A.B. et al. (2014). Consistent increase in High Asia's runoff due to increasing glacier melt and precipitation. *Nat. Clim. Chang.* 4: 587–592. doi.org/10.1038/nclimate2237

Lutz, A F., ter Maat, H.W., Wijngaard, R.R. et al. (2019). South Asian river basins in a 1.5°C warmer world. *Reg. Environ. Chang.* 19: 833–847. doi.org/10.1007/s10113-018-1433-4

Ma, J., Hung, H., Tian, C., and Kallenborn, R. (2011). Revolatilization of persistent organic pollutants in the Arctic induced by climate change. *Nat. Clim. Chang.* 1: 255–260. doi.org/10.1038/nclimate1167

Malik, R. (2008). Growth impacts of development and management of water resources. In: Proceedings of the 7th Annual Partners Meeting, IWMI TATA Water Policy Research Program, ICRISAT, Patancheru, Hyderbad, India, 2–4 April 2008. Patancheru, Andhra Pradesh, India, 859–870. Hyderabad: International Water Management Institute. Available at: https://publications.iwmi.org/pdf/H042935.pdf

Martha, T.R., Roy, P., Govindharaj, K.B. et al. (2015). Landslides triggered by the June 2013 extreme rainfall event in parts of Uttarakhand state, India. *Landslides* 12: 135–146. doi.org/10.1007/s10346-014-0540-7

Maurer, J.M., Schaefer, J.M., Rupper, S. et al. (2019). Acceleration of ice loss across the Himalayas over the past 40 years. *Sci. Adv.* 5: eaav7266. doi.org/10.1126/sciadv.aav7266

McDowell, G., Ford, J.D., Lehner, B. et al. (2013). Climate-related hydrological change and human vulnerability in remote mountain regions: a case study from Khumbu, Nepal. *Reg. Environ. Chang.* 13: 299–310. doi.org/10.1007/S10113-012-0333-2

Milner, A.M., Khamis, K., Battin, T.J. et al. (2017). Glacier shrinkage driving global changes in downstream systems. *Proc. Natl. Acad. Sci. USA* 114: 9770–9778. doi.org/10.1073/pnas.1619807114

Mishra, A., Appadurai, A.N., Choudhury, D. et al. (2019). Adaptation to climate change in the Hindu Kush Himalaya: stronger action urgently needed. In: *The Hindu Kush Himalaya Assessment*, 1st ed. (ed. P. Wester, A. Mishra and A.B.S. Aditi), 457–490. Cham: Springer. doi.porg/10.1007/978-3-319-92288-1_13

Mohammed, K., Islam, A.S., Tarekul Islam, G. et al. (2017). Extreme flows and water availability of the Brahmaputra River under 1.5 and 2°C global warming scenarios. *Clim. Change* 145: 159–175. doi.org/10.1007/s10584-017-2073-2

Mukherji, A., Sinisalo, A., Nüsser, M. et al. (2019). Contributions of the cryosphere to mountain communities in the Hindu Kush Himalaya: a review. *Reg. Environ. Change* 19: 1311–1326. doi.org/10.1007/s10113-019-01484-w

Mukhopadhyay, B. and Khan, A. (2014). Rising river flows and glacial mass balance in central Karakoram. *J. Hydrol.* 513: 192–203. doi.org/10.1016/j.jhydrol.2014.03.042

Nie, Y., Pritchard, H.D., Liu, Q. et al. (2021). Glacial change and hydrological implications in the Himalaya and Karakoram. *Nat. Rev. Earth Environ.* 2: 91–106. doi.org/10.1038/s43017-020-00124-w

Nie, Y., Sheng, Y., Liu, Q. et al. (2017). A regional-scale assessment of Himalayan glacial lake changes using satellite observations from 1990 to 2015. *Remote Sens. Environ.* 189: 1–13. doi.org/10.1016/j.rse.2016.11.008

Oerlemans, J. (2005). Extracting a climate signal from 169 glacier records. *Science* 80: 308: 675–677. doi.org/10.1126/science.1107046

Orsi, A.J., Cornuelle, B.D., and Severinghaus, J.P. (2012). Little Ice Age cold interval in West Antarctica: evidence from borehole temperature at the West Antarctic Ice Sheet (WAIS) Divide. *Geophys. Res. Lett.* 39: L09710. doi.org/10.1029/2012GL051260

Parkinson, C.L. (2019). A 40-y record reveals gradual Antarctic sea ice increases followed by decreases at rates far exceeding the rates seen in the Arctic. *Proc. Natl. Acad. Sci. USA* 116: 14414–14423. doi.org/10.1073/pnas.1906556116

Parveen, S., Winiger, M., Schmidt, S. et al. (2015). Irrigation in Upper Hunza: evolution of socio-hydrological interactions in the Karakoram, northern Pakistan. *Erdkunde*. 69: 69–85. doi.org/10.3112/erdkunde.2015.01.05

Paudel, K.P. and Andersen, P. (2013). Response of rangeland vegetation to snow cover dynamics in Nepal Tramns Himalaya. *Clim. Change* 117: 149–162. doi.org/10.1007/s10584-012-0562-x

Qureshi, A.S. (2011). Water management in the Indus Basin in Pakistan: challenges and opportunities. *Mt. Res. Dev.* 31: 252–260. doi.org/10.1659/MRD-JOURNAL-D-11-00019.1

Ranjan, S., Ramanathan, Al., Keesari, T., Singh, V.B., Kumar, N., Pandey, M. and Leuenberger, M.C. (2021). Triple Water Vapour–Isotopologues Record from Chhota Shigri, Western Himalaya, India: A Unified Interpretation based on δ17O, δ18O, δD and Comparison to Meteorological Parameters. *Front. Earth Sci.* 8: 599–632. Available at: https://doi.org/10.3389/feart.2020.599632.

Rantanen, M., Karpechko, A.Yu., Lipponen, A., Nordling, K., Hyvärinen, O., Ruosteenoja, K., Vihma, T. and Laaksonen, A. (2022) 'The Arctic has warmed nearly four times faster than the globe since 1979', *Communications Earth & Environment*, 3(1), p. 168. Available at: https://doi.org/10.1038/s43247-022-00498-3.

Rasul, G. and Molden, D. (2019). The global social and economic consequences of mountain cryospheric change. *Front. Environ. Sci.* 7: 91. doi.org/10.3389/FENVS.2019.00091

Rees, H.G., Holmes, M.G.R., Young, A.R. et al. (2004). Recession-based hydrological models for estimating low flows in ungauged catchments in the Himalayas. *Hydrol. Earth Syst. Sci.* 8: 891–902. doi.org/10.5194/hess-8-891-2004

Rounce, D.R., Byers, A.C., Byers, E.A. et al. (2017). Brief communication: observations of a glacier outburst flood from Lhotse Glacier, Everest area, Nepal. *Cryosphere* 11: 443–449. doi.org/10.5194/tc-11-443-2017

Sakai, A., Nishimura, K., Kadota, T. et al. (2009). Onset of calving at supraglacial lakes on debris-covered glaciers of the Nepal Himalaya. *J. Glaciol.* 55: 909–917. doi.org/10.3189/002214309790152555

Scherler, D., Bookhagen, B., and Strecker, M.R. (2011). Spatially variable response of Himalayan glaciers to climate change affected by debris cover. *Nat. Geosci.* 4: 156–159. doi.org/10.1038/ngeo1068

Scherler, D., Wulf, H., and Gorelick, N. (2018). Global assessment of supraglacial debris-cover extents. *Geophys. Res. Lett.* 45: 11798–11805. doi.org/10.1029/2018GL080158

Schmidt, S. and Nüsser, M. (2012). Changes of high altitude glaciers from 1969 to 2010 in the Trans-Himalayan Kang Yatze Massif, Ladakh, northwest India. *Arct. Antarct. Alp. Res.* 44: 107–121. doi.org/10.1657/1938-4246-44.1.107

Schuur, E.A.G., Bockheim, J., Canadell, J.G. et al. (2008). Vulnerability of permafrost carbon to climate change: implications for the global carbon cycle. *Bioscience* 58: 701–714. doi.org/10.1641/B580807

Schwanghart, W., Worni, R., Huggel, C. et al. (2016). Uncertainty in the Himalayan energy-water nexus: estimating regional exposure to glacial lake outburst floods. *Environ. Res. Lett.* 11: 074005. doi.org/10.1088/1748-9326/11/7/074005

Serreze, M C. and Barry, R.G. (2011). Processes and impacts of Arctic amplification: a research synthesis. *Glob. Planet. Chang.* 77: 85–96. doi.org/10.1016/j.gloplacha.2011.03.004

Shean, D.E., Bhushhan, S., Montesano, P. et al. (2020). A systematic, regional assessment of High Mountain Asia glacier mass balance. *Front. Earth Sci.* 7: 1–19. doi.org/10.3389/feart.2019.00363

Shepherd, A., Ivins, E., Rignot, E. et al. (2018). Mass balance of the Antarctic Ice Sheet from 1992 to 2017. *Nature* 558: 219–222. doi.org/10.1038/s41586-018-0179-y

Shijin, W., Yaqiong, M., Xueyan, Z. et al. (2020). Polar tourism and environment change: opportunity, impact and adaptation. *Polar Sci.* 25: 100544. doi.org/10.1016/j.polar.2020.100544

Shrestha, A.B., Wake, C.P., Mayewski, P.A. et al. (1999). Maximum temperature trends in the Himalaya and its vicinity: an analysis based on temperature records from Nepal for the period 1971–94. *J. Clim.* 12: 2775–2786. doi.org/10.1175/1520-442(1999)012<2775:MTTITH>2.0.CO;2

Shrestha, F., Uddin, K., Bikash Maharjan, S. et al. (2016). Application of remote sensing and GIS in environmental monitoring in the Hindu Kush Himalayan region. *AIMS Environ. Sci.* 3: 646–662. doi.org/10.3934/environsci.2016.4.646

Singh, S., Kumar, R., Bhardwaj, A. et al. (2016). Changing climate and glacio-hydrology in Indian Himalayan Region: a review. *Wiley Interdiscip. Rev. Clim. Change* 7: 393–410. doi.org/10.1002/WCC.393

Steffen, W., Rockström, J., Richardson, K. et al. (2018). Trajectories of the earth system in the anthropocene. *Proc. Natl. Acad. Sci. USA* 115: 8252–8259. doi.org/10.1073/pnas.1810141115

Stroeve, J.C., Serreze, M.C., Holland, M.M. et al. (2012). The Arctic's rapidly shrinking sea ice cover: a research synthesis. *Clim. Change* 1005–1027. doi.org/10.1007/s10584-011-0101-1

Subramanya, K. (2013). *Engineering Hydrology*. New Delhi: McGraw Hill Education (India) Pvt. Ltd.

Tarroja, B., AghaKouchak, A., and Samuelsen, S. (2016). Quantifying climate change impacts on hydropower generation and implications on electric grid greenhouse gas emissions and operation. *Energy* 111: 295–305. doi.org/10.1016/j.energy.2016.05.131

Treydte, K.S., Schleser, G.H., Helle, G. et al. (2006). The twentieth century was the wettest period in northern Pakistan over the past millennium. *Nature* 440: 1179–1182. doi.org/10.1038/nature04743

Tsutaki, S., Fujita, K., Nuimura, T. et al. (2019). Contrasting thinning patterns between lake- and land-terminating glaciers in the Bhutanese Himalaya. *Cryosphere* 13: 2733–2750. doi.org/10.5194/tc-13-2733-2019

Turner, J., Barrand, N.E., Bracegirdle, T.J. et al. (2014). Antarctic change and the environment. *Polar Record (Gt. Brit.)* 50: 237–259. doi.org/10.1017/S0032247413000296

Turner, S.W.D., Ng, J.W., and Galelli, S. (2017). Examining global electricity supply vulnerability to climate change using a high-fidelity hydropower dam model. *Sci. Total Environ.* 590–591: 663–675. doi.org/10.1016/j.scitotenv.2017.03.022.

Vashisht, P., Pandey, M., Ramanathan, AL., Tayal, S. and Jackson, M. (2017). Comparative Assessment of Volume Change in Kolahoi and Chhota Shigri Glaciers, Western Himalayas, Using Empirical Techniques. *J. Clim. Change* 3(1): 37–8. Available at: https://doi.org/10.3233/JCC-170004.

Vieira, G., Bockheim, J., Guglielmin, M. et al. (2010). Thermal state of permafrost and active-layer monitoring in the Antarctic: advances during the international polar year 2007–2009. *Permafr. Periglac. Process.* 21: 182–197. doi.org/10.1002/ppp.685

Vincent, A., Violette, S., and Aðalgeirsdóttir, G. (2019). Groundwater in catchments headed by temperate glaciers: a review. *Earth-Science Rev.* doi.org/10.1016/j.earscirev.2018.10.017

Wang, B. and French, H.M. (1994). Climate controls and high-altitude permafrost, Qinghai-Xizang (Tibet) Plateau, China. *Permafr. Periglac. Process.* 5: 87–100. doi.org/10.1002/ppp.3430050203

Wang, B. and French, H.M. (1995). Permafrost on the Tibet Plateau, China. *Quat. Sci. Rev.* 14: 255–274. doi.org/10.1016/0277-3791(95)00006-B

Warner, K. (2010). Global environmental change and migration: governance challenges. *Glob. Environ. Change* 20: 402–413. doi.org/10.1016/j.gloenvcha.2009.12.001.

Warren, F.J. and Lemmen, D.S. (2014). Synthesis. In: *Canada in a Changing Climate: Sector Perspectives on Impacts and Adaptation* (ed. F.J. Warren and D.S. Lemmen), 1–18 Ottawa, ON: Government of Canada.

Watson, C.S., Kargel, J.S., Shugar, D.H. et al. (2020). Mass loss from calving in Himalayan proglacial lakes. *Front. Earth Sci.* 7. doi.org/10.3389/feart.2019.00342

Wijngaard, R.R., Lutz, A.F., Nepal, S. et al. (2017). Future changes in hydro-climatic extremes in the Upper Indus, Ganges, and Brahmaputra River basins. *PLoS One* 12: e0190224. doi.org/10.1371/journal.pone.0190224

Williams, M.W. (2013). The status of glaciers in the Hindu Kush–Himalayan Region. *Mt. Res. Dev.* 33: 114. doi.org/10.1659/mrd.mm113

Williams, M.W. and Seastedt, T. et al. (2015). An overview of research from a high elevation landscape: the Niwot Ridge, Colorado Long-Term Ecological Research programme. *Plant Ecology and Diverity.* 8(5–6): 597–605. doi.org/10.1080/17550874.2015.1123320

Wunderling, N., Willeit, M., Donges, J.F. et al. (2020). Global warming due to loss of large ice masses and Arctic summer sea ice. *Nat. Commun.* 11: 5177. doi.org/10.1038/s41467-020-18934-3

Yang, Z., Ou, Y.H., Xu, X. et al. (2010). Effects of permafrost degradation on ecosystems. *Acta Ecol. Sin.* 30: 33–39. doi.org/10.1016/j.chnaes.2009.12.006

Zhang, Q., Gu, X., Singh, V.P. et al. (2016). Magnitude, frequency and timing of floods in the Tarim River Basin, China: changes, causes and implications. *Glob. Planet. Chang.* 139: 44–55. doi.org/10.1016/j.gloplacha.2015.10.005

Zwally, H.J., Li, J., Brenner, A.C. et al. (2011). Greenland ice sheet mass balance: distribution of increased mass loss with climate warming: 2003–07 versus 1992–2002. *J. Glaciol.* 57: 88–102. doi.org/10.3189/002214311795306682

# 26

# High-Resolution Remote Sensing for Mapping Glacier Facies in the Arctic

*Shridhar Digambar Jawak[1], Sagar Filipe Wankhede[2,*], Alvarinho J. Luis[3], and Keshava Balakrishna[2]*

[1] *Svalbard Integrated Arctic Earth Observing System (SIOS), SIOS Knowledge Centre, PO Box 156, N-9171, Longyearbyen, Svalbard, Norway*
[2] *Department of Civil Engineering, Manipal Institute of Technology, Manipal Academy of Higher Education, Manipal 576104, India*
[3] *Earth System Sciences Organization, National Centre for Polar and Ocean Research (NCPOR), Ministry of Earth Sciences, Government of India, Headland Sada, Vasco-da-Gama, Goa 403804, India*
[*] *Corresponding author*

## 26.1 Introduction

Glaciers have shaped the Earth's landscapes throughout its history. Eroding, abrading, and depositing material throughout the traversing regime of a glacial system, they continue to morph present-day continents. In combination with ice sheets, frozen lakes, and sea ice, they form the planet's cryosphere. The interchange of mass and energy between the cryosphere and other components of the earth systems make glaciers sensitive indicators of climate change (Benn and Evans, 2013). Alpine glaciers provide perennial sources of drinking water to millions of people across the world. Supporting life through meltwater discharge and threatening famine through potential depletion of glacial bodies, they are fundamental resources of both scientific and economic significance. The deposition of snowflakes followed by their accumulation, compaction, and ageing into firn, morphing into ice, mixing with sediments and debris, while flowing along the slope of the land, is a summary of a glacier's birth and life. While in motion, this mass of snow and ice releases heat on account of friction and seasonal variations. This produces meltwater, which often causes the snowpack to become wet or saturated. These motions, movements, and processes are a result of the accumulation and melt/ablation cycles of a glacial system. These also characterize the two broad zones on a glacier. The zone where accumulation dominates melt is known as the accumulation zone, and the zone where melt dominates is known as the ablation zone. Both are separated by an equilibrium line where accumulation and melt are net zero. Alternating cycles of accumulation and melt, along with the natural transformation of snow to ice, lead to the formation of stratigraphic facies in the two zones (Benson, 1962). When these facies are observed from the surface, the mixing with debris, dust, surface impurities, and crevasses lead to the determination of surface classes (Pope and Rees, 2014a), also known as surface facies (Yousuf et al., 2019). These surface facies are supraglacial representatives of the stratigraphic glaciological zones of a glacier. Therefore, understanding and monitoring the evolution of surface facies can provide an estimate of the glacier's health (Jawak et al., 2019). One of the primary reasons for mapping facies is their capacity to assist in mass balance estimations, by calibrating distributed mass balance models (Braun et al., 2007).

Methods of monitoring facies include *in-situ* diagenetic analysis (Müller, 1962), field spectra analysis (Gore et al., 2017), remote sensing (RS) analysis of airborne imagery (Pope and Rees, 2014a), satellite-based optical (Hall et al., 1987) and synthetic aperture radar (SAR) data (Adam, 1997). Techniques of remote observations of facies have been reviewed by König et al. (2001) and Pellikka and Rees (2010). Given that a large variety of sensors operate with the purpose of monitoring the Earth's surface, the specific calibrations of a single sensor may not always favor detection of snow and ice (SI) types (Pope and Rees, 2014b). Hence, rigorous analysis of SI features with respect to their properties affecting satellite observations have provided a plethora of methods that could be used for mapping facies. Some of the methods include image classification, spectral band ratios (Jawak et al., 2019), backscatter thresholding (Brown, 2012), incorporation of ancillary materials (Shukla and Ali, 2016), and integration of ground penetrating radar (GPR) into SAR measurements (Barzycka et al., 2020). The response of facies to incoming radiation, when measured in the optical range, yields the spectral signature for each

*Advances in Remote Sensing Technology and the Three Poles,* First Edition. Edited by Manish Pandey, Prem C. Pandey, Yogesh Ray, Aman Arora, Shridhar D. Jawak, and Uma K. Shukla.

facies. Dozier (1984) measured the reflectance pattern of snow using Landsat thematic mapper. Since then, progress in the assessment of facies has improved the understanding of spectral properties and sensor calibrations. Reflectance characteristics of facies are important inputs when mapping is performed using optical data. The output of research attempts aiming to map facies is usually in the form of thematic maps (Braun et al., 2007; Tran et al., 2008; Keshri et al., 2009; Ghosh et al., 2014; Zhang et al., 2019). A combination of appropriate mapping methods with geospatial data with higher spatial and spectral resolution and therefore better spectral variability should yield accurate thematic maps. However, classification of satellite data for land-cover mapping often requires operator assistance for the division of classes and presentation of results. The skill of an operator comes into play here when thematic classes must be identified. Often, a single operator trains a classification algorithm using either prior knowledge or reference material through the visual and spectral assessment of an image. It is sporadically acknowledged that human performance will impact RS image analysis results (Gardin et al., 2011). This can be applied to the mapping of surface facies as well. An operator aiming to map surface facies using an optical satellite image will have to train the classifier to the best of their knowledge. This depends upon the operator's knowledge of the glacial body, derived spectral response patterns, and comparative assessment against previous attempts. To date, no study on surface facies has aimed at identifying the impacts of different operators on the classification of a single image.

This chapter aims to test this variability. We assigned four anonymous operators the task of selecting training data for classification of surface facies. Each operator was provided with the same base pre-processed image. The operators identified potential surface facies using their own understanding of the image and after assessment of relevant literature. These training samples were then used to train two pixel-based classifiers to obtain the differences in mapped facies.

### 26.1.1 Glacier Facies Mapping Using Multispectral Data

Multispectral (MSS) data exploits the spectral separability of Earth features for mapping procedures. The different wavelength ranges encapsulated by MSS sensors provide the spectral signatures for a target feature. The spectral response pattern of facies in different spectral bands are investigated by Dozier (1989), Hall et al. (1987), Keshri et al. (2009), and Gore et al. (2017). Snow grain size, moisture content, and sediment mixture affects the spectral signature recorded by an MSS sensor (Dozier et al., 1981). One way of exploiting these sensitivities is by band ratioing (Hall et al., 1988). An example of custom band ratioing is discussed by Pope and Rees (2014a). They obtained airborne multispectral images of the Midtre Lovénbreen glacier in Ny-Ålesund and assessed *in-situ* data to generate linear combinations (LCs) for mapping glacier facies. Another example is reported by Jawak et al. (2019), who generated customized spectral index ratios for mapping facies using the multispectral bands of WorldView-2 (WV-2) MSS data. Interestingly, Pope and Rees (2014b) studied the spatial and radiometric characteristics of resampled imagery on the detection of facies. While their results do not suggest the influence of spatial resolution on facies, it is important to note that the study did not test supervised classification. The test was conducted using unsupervised methods. Higher spatial resolution does enable an operator to map more distinct surface features and subsequently more efficient training samples can be generated. Utilization of classification algorithms has been shown to provide good results for detecting glacier facies (Pandey et al., 2013; Luis and Singh, 2020). Furthermore, comparative studies between classification techniques suggest that while methods such as the object-based image analysis deliver high accuracies, they can come at a cost of time and efficiency (Rastner et al., 2014; Nijhawan et al., 2016). This is also observed by Jawak et al. (2019). Facies have been mapped using a variety of optical data driven methods (Paul et al., 2016; Shukla and Ali, 2016, Yousuf et al., 2019; Luis and Singh, 2020). Supervised classification techniques stand out from other methods because of their ability to use multiple classifying algorithms through the input of a single set of training samples. However, prior to performing supervised classification, it is necessary to identify factors governing its process and the effects that different training samples may have on the final classification.

### 26.1.2 Image Classification

Digital images are a $n$-Dimensional assortment of pixels. In the case of MSS images, $n$ refers to the number of spectral bands. The value of a pixel, known as the digital number (DN) or digital brightness is a measure of the radiation sensed by an acquiring sensor from a target object. This radiative energy is measured in terms of bits, wherein 8 bits retains brightness values between 0 and 255. These DNs in each respective band of a multispectral image, house pertinent information related to the entire captured scene. After conversion of these raw DNs to at-sensor radiance and then to surface reflectance for removing atmospheric effects and image distortions, the image is ready for classification procedures (Cracknell and Hayes, 1991). The process of grouping homogenous pixels into categories or classes is known as image classification. The criterion for homogeneity usually corresponds to spectral similarity with respect to the application, the objects within the scene, the technique applied for classification, and the end goal of the classification method. These techniques rely on categorizing each individual pixel into specified classes; hence, they are known as pixel-based methods of image analysis. When several pixels with homogeneous properties are clubbed to form objects within an image, and then these objects are categorized into classes, it is known as object-based method of image analysis. Therefore, all classification procedures rely either on the basic pixel or its transformed forms. Jawak et al. (2015) divide classification procedures into three broad categories according to the procedural basis: i) classification based on the kind of learning; ii) classification based on presumptions of data distribution; and iii) classification based on the number of outputs for a given spatial unit. Among these, learning-based classification is of two types, supervised and unsupervised. Unsupervised classification localizes the natural differences in spectral signature of pixels in the $n$-D space and groups similar pixels into clusters. An operator then assigns a thematic value to each class or adjusts them according to some reference of the target scene. Popular mechanisms of this method include the iterative self-organizing data analysis (ISODATA) and the K-means algorithms (Lu and Weng, 2007). Supervised classification on the other hand requires an operator to provide a classifying algorithm (called classifier henceforth) a set of samples for each class depending upon prior knowledge and reference data. These samples are often identified using visual and spectral analysis of a given image. As these samples are used to "train" the classifier, they are called training samples or sets. The training samples denote specific spectral attributes for each target which the classifier numerically matches for the whole scene in each band, thereby accomplishing classification. The procedure for supervised classification is as follows: i) selection of training samples; ii) application of a classifier; iii) post-classification adjustments (if necessary); and iv) accuracy assessment of the classification according to reference data. The final accuracy of this technique relies heavily on the training sample set provided by the operator and the choice of classifier for a given image. This warrants a closer look at the requirements for training data selection by an operator.

### 26.1.3 Training Samples and Operator Skill

A training sample is a set of pixels that are delineated using points (single pixels), linear arrays (linear set of pixels), or polygonal enclosures (non-linear group of pixels) of an identifiable feature in a RS image. A set of training samples of different features provides the training data for classification. The aim of an operator while assigning training samples is to gather an ensemble of statistics that can be attributed to the spectral response pattern for each target in the image (Lillesand et al., 2014). The goal is to provide adequate data which is representative of all the targets. In terms of surface facies, it is the representation of all facies within a set of training samples. Low to moderate spatial resolution data can often impede the allocation of training sets because of low feature differentiation and spectral mixing (Bosdogianni et al., 1994; Lu and Weng, 2007). However, in the case of VHR data, where spatial resolution is typically finer than 2 m, spectral mixing is less of an impedance to training samples. The onus of accurate classification then lies upon the skill of the operator. An operator's ability to identify accurate feature details for classification would rely upon visual acuity and perception, reference data/prior knowledge, sample set size, and choice of classifier. Gardin et al. (2011) reviewed the performance of operators based on psychological factors to outline a research concept for determining human performance variability. This is in line with the observations made by Congalton (1991) and Foody (2008), who suggest that human errors influence a classification analysis. Documentation of operator variability (Leckie, 2003; Powell et al., 2004; Zhou et al., 2010) suggest that subjectivity plays a ruling factor in the interpretation of data. Within the context of glacier facies, this presents a unique opportunity to understand the subjectivity introduced into sample collection for classification based on both spectral and visual interpretation of the image. Jawak et al. (2019) elaborate upon the spatial and spectral components used while determining facies on a Himalayan glacier. They further correlate the identified facies with respect to previous attempts in near-by areas and published spectral reflectance curves. To the best of our knowledge, the differences in mapped facies using the same classification algorithms, but by employing training data from different operators, have not been tested.

### 26.1.4 The Test of Operator Influence

In this chapter we set up an experiment incorporating the VHR WV-3 satellite data for classification of glacier facies. Pre-processing of the data follows standard radiometric calibrations and boundary delineations. This base image is then supplied to four different operators who are tasked to select training samples of the facies they identify on any glacier through their understanding of the glacial terrain. The resultant four training sets are used as input into two classification algorithms. Thus, the different facies identified, the thematic results obtained, and the performance of the classifiers are evaluated.

## 26.2 The Geographical Area and Geospatial Data

Ny-Ålesund is a premier research town, located on the Brogger peninsula, on the island of Spitsbergen, in the Svalbard Archipelago of Norway (Figure 26.1). The research facilities maintained here are part of an international consortium of countries, institutes, and organizations. The region consists of sub-polar type glaciers, categorized into cirque, ice caps, piedmont, tidewater and continental glaciers (Table 26.1). The area is of immense importance as the region is experiencing warming at double the rate of the global mean (Isaksen et al., 2016). The glaciers visible in Figure 26.1 are Vestre Brøggerbreen, Austre Lovénbreen, Austre Brøggerbreen, Midtre Lovénbreen, Edithbreen, Botnfjellbreen, Pedersbreen, and Uvérsbreen. This region is selected for examination as existing literature consisting of *in-situ* and MSS data on facies of this area (Pope and Rees, 2014a, 2014b) will direct the operators when selecting training samples.

The data used in this study consists of LV2A WV-3 imagery (WorldView-3 © 2016). The image was acquired on 10 August 2016 by Digital Globe, USA. Its MSS spatial resolution is 1.24 m, and the short-wave infrared bands (SWIR) are of 3.7 m resolution. The spectral bands of the data consist of coastal (0.40–0.45 μm), blue (0.45–0.51 μm), green (0.51–0.58 μm), yellow (0.585–0.625 μm), red (0.63–0.69 μm), red edge (0.705–0.745 μm), near infrared 1 (NIR1) (0.770–0.895 μm), near infrared 2 (NIR2) (0.86–1.04 μm), SWIR1 (1.99–1.22 μm), SWIR2 (1.55–1.59 μm), SWIR3 (1.64–1.68 μm), SWIR4 (1.71–1.75 μm), SWIR5 (2.14–2.18 μm), SWIR6 (2.18–2.22 μm), SWIR7 (2.23–2.28 μm), and SWIR8 (2.29–2.36 μm).

## 26.3 Methodology

The data processing methodology is segmented into radiometric calibration, digitization, and then supply of data to operators for glacier selection and training sample collection. The samples are then used as input for two classification algorithms, the Minimum Distance classifier (MD) and the Mahalanobis Distance classifier (MHD). Figure 26.2 displays the methodology adopted. The protocol was executed in ENVI 5.3.

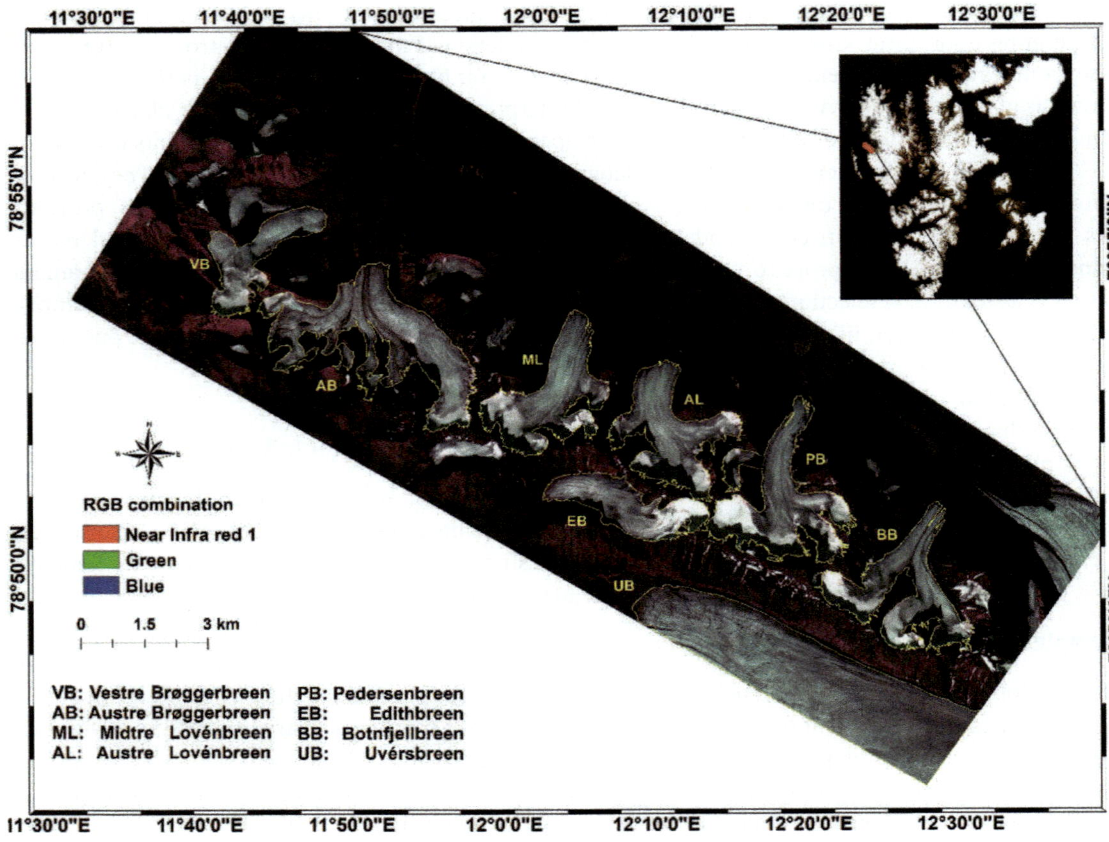


Figure 26.1 The geographical region selected for analysis (WorldView-3 © 2016).

**Table 26.1** Glaciers selected by each operator and the facies identified.

| Operator | Glacier Selected | Facies |
|---|---|---|
| Operator 1 (O1) | Austre Brøggerbreen | Fresh Snow, Wet Snow, Glacier Ice, Melting Ice, Dirty Ice, Debris, Shadow Ice |
| Operator 2 (O2) | Midtre Lovénbreen | Fresh Snow, Shadowed Snow, Off Glacier, Low Debris, Medium Debris, High Debris, Dry Snow, Semi-dry Snow |
| Operator 3 (O3) | Austre Brøggerbreen | Dry Snow, Percolation Snow, Wet Snow, Melting Ice, Dirty Ice, Debris, Off Glacier, Water Stream, Crevasses, Shadow |
| Operator 4 (O4) | Midtre Lovénbreen | Dry Snow, Wet Snow, Saturated Snow, Melting Snow, Melting Glacier Ice, Glacier Ice, Dirty Ice, Shadowed Snow, Streams and Crevasses |

### 26.3.1 Radiometric Calibration and Digitization

Separability of spectral signatures is necessary for classification (Mausel et al., 1990). Therefore, removal of atmospheric effects via radiometric calibration is an important precursor to any reflectance-based mapping method. The protocol for radiometric calibration follows Jawak et al. (2018) and Rastner et al. (2014). It is a two-step process consisting of: a) conversion of DN values to at-sensor spectral radiance; and b) obtaining surface reflectance by implementing atmospheric correction modules. Selection of atmospheric correction method is usually image and application specific. However, this protocol uses the Fast Line-of-Sight Atmospheric Analysis of Spectral Hypercubes (FLAASH) method after

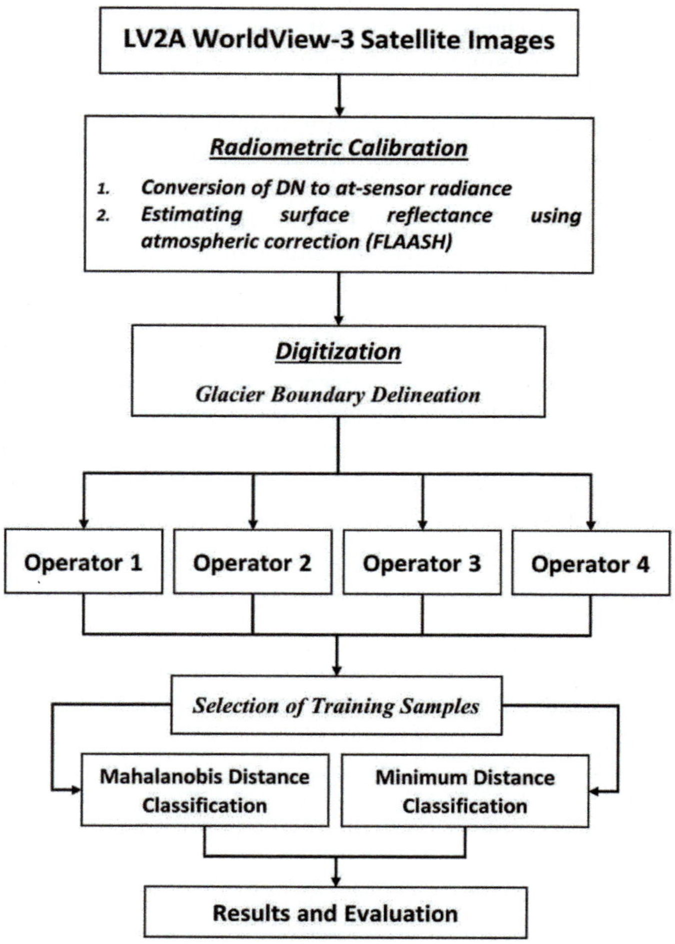


**Figure 26.2** Workflow of the research concept.

reviewing its successful application in alpine areas (Nolin et al., 1993; Guo et al., 2021) and recommendations of comparative studies between various atmospheric correction methods (Dewi and Trisakti, 2016). FLAASH is developed on the moderate resolution atmospheric transmission 4 (MODTRAN4) radiative transfer code (Kaufman et al., 1997) and utilizes information from the image metadata to estimate the atmospheric conditions at the time of scene capture. A radiative transfer equation then uses the modeled atmosphere to convert radiance to reflectance (https://www.harrisgeospatial.com/docs/backgroundflaash.html). Parameters for the FLAASH module were assigned according to Abreu and Anderson (1996).

Glacial boundaries and ice divides were manually digitized in ArcGIS. Therefore, each of the operators received an atmospherically corrected VHR WV-3 image with delineated glacial boundaries. The operators could extract the glacier of choice and then subject the subset of the glacier image to classification procedures.

### 26.3.2 Operator Selections

The operators selected the glaciers and identified facies to the best of their knowledge and skill. The training samples were selected accordingly.

### 26.3.3 Classification and Reference Point Selection

The MHD and MD classifiers are selected for classification as their performance for mapping facies have been tested elsewhere (Jawak et al., 2019; Luis and Singh, 2020). These are available under the Terrain Categorization (TERCAT) workflow in ENVI. Post-classification processing was avoided to prevent further inclusion of operator bias. Each operator then identified reference sample points for accuracy assessment for the facies and the glacier under consideration. The directive was to follow an equalized random sampling process based on their interpretation of spectral plots and visual analysis (Keshri et al., 2009). Accuracy assessment was then conducted using the overall accuracy and kappa statistic (κ) (Foody, 2010).

## 26.4 Results and Discussion

Table 26.2 displays the accuracies achieved by each operator for their classification schemes using each classifier. O2 and O4 achieved similar results for the MHD classification and MD. O1 achieved average overall accuracy and κ value. The results of O3, however, were below average.

It appears that O2 and O4 have achieved similar results based on accuracies for Midtre Lovénbreen. However, the only common "facies" that they have is the dry snow facies. Shadowed snow facies are common for all; hence, it will not be considered as identified facies, but rather a common surface class for these glaciers. O2's schema for identifying facies appears to be based on the level of debris on the ice. O1, O3, and O4 appear to follow a similar pattern of accumulation and melt facies visible on glaciers. These facies resemble some of the classes identified by Pope and Rees (2014a). However, one prominent distinction is the presence of fresh snow. The image was acquired at the end of the melt season. This implies that all the facies of the glacier would be ideally visible. But the availability of fresh snow is rather questionable. Both O1 and O2 seem to have identified fresh snow during the training sample allocation process. O3 shows percolation snow, which is more aptly identified using SAR data (Pellikka and Rees, 2010). O4's classification scheme appears to follow the prescribed accumulation and melt facies usually associated at the end of the ablation season.

Upon first glance, it seems that O2's scheme is the most accurate (Figure 26.3); however, close examination of the kinds of facies portray a nuanced picture. Based upon the accuracies alone, the trend of operator reliability looks like O2>O4>O1>O3. However, after examining the facies identified and the accuracies together, the trend is modified as O4>O3>O1>O2.

**Table 26.2** Overall Accuracy and Kappa Statistic obtained.

| | Overall Accuracy | | Kappa Statistic | |
|---|---|---|---|---|
| **Operators** | **MHD** | **MD** | **MHD** | **MD** |
| Operator 1 | 76 | 75 | 0.71 | 0.7 |
| Operator 2 | 85 | 74 | 0.82 | 0.68 |
| Operator 3 | 67 | 55 | 0.64 | 0.5 |
| Operator 4 | 84 | 72 | 0.8 | 0.66 |

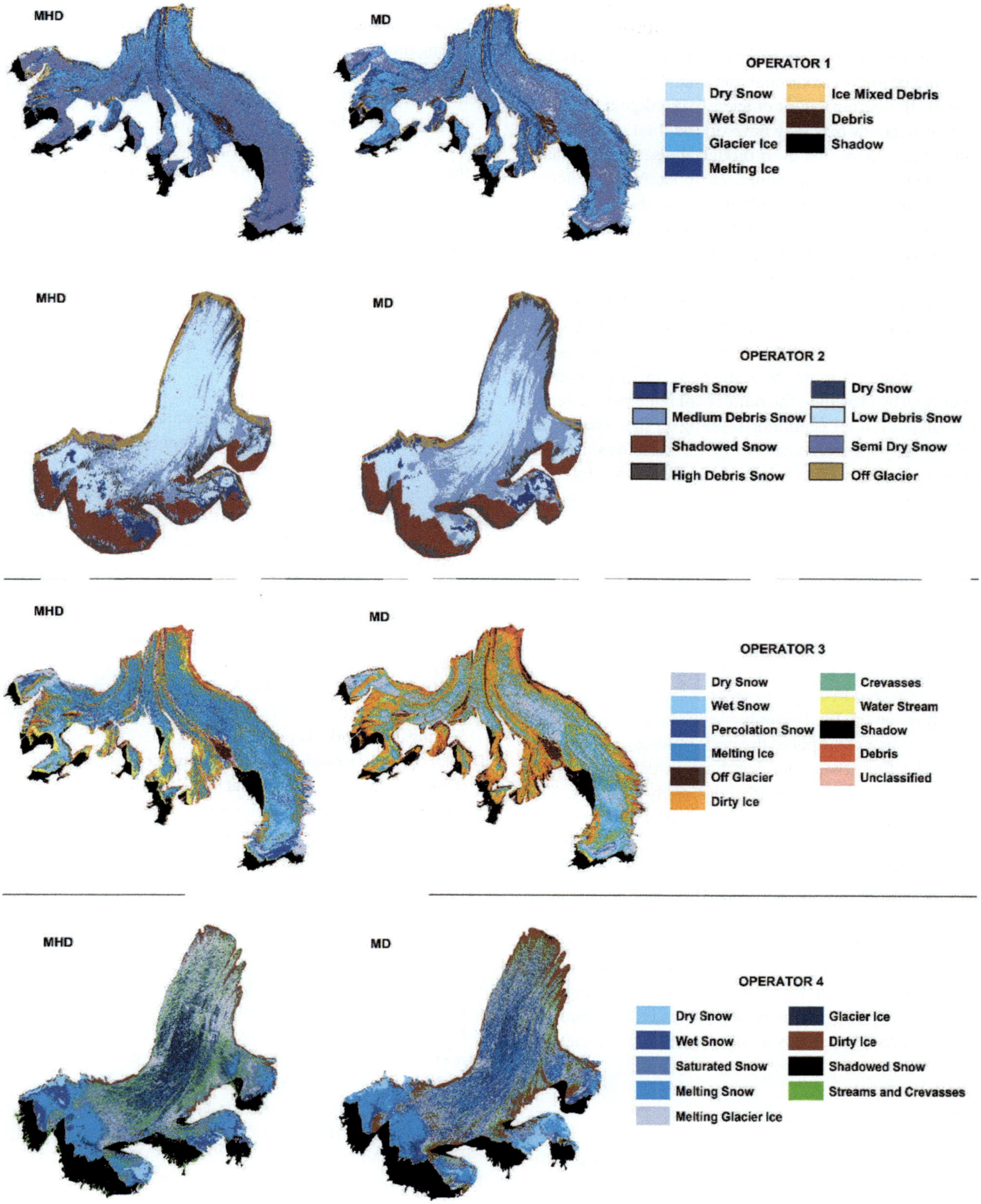


**Figure 26.3** Classified outputs from all the respective operators. MHD: Mahalanobis Distance, MD: Minimum Distance.

## 26.5 Inferences and Recommendations

Detailed examination of the success and failures of classification would be improved through the availability of field data. However, the current work efficiently highlights the influence that operator bias may have on the final classification results. It is evident that even though the same base image was provided to the operators, each identified different classes. A primary reason would be the psychological variability described by Gardin et al. (2011). Another factor would be the allocation of different training set size (Ramezan et al., 2021). The classifiers have shown a generically similar performance, where the MHD has performed better than MD, to previous works (Jawak et al., 2019). However, it is the influence of the operators that require more detailed analysis. Future research must check the performance variability of operators by providing the same set of reference materials and perhaps even set a common training set size. That would highlight only the differences in facies solely based on the understanding and knowledge of the operators. Moreover, glacier facies mapping would be improved if field data is to be included in the validation process.

## 26.6 Conclusion

The study of glacier facies is an important aspect of glaciology. Using RS data, it is easier and more efficient now than in much of recent history. The impact that a trained operator can have on the final mapping procedure is tremendous. The current work compared the performance of four operators on the final thematic classification of glacier facies. Different facies were identified by each operator. O2's schema was focused on the level of debris on the ice, O1 and O2 show fresh snow at the end of the ablation season, and O3 identified percolation snow. O4 outlined summer facies to the maximum; however, the resultant accuracy achieved was less than that of O2. Thus, it appears that a significant amount of effort must be exerted in the training of operators to correctly identify feature classes, as the entire classification protocol depends on the training data. Moreover, reference points must be allocated in accordance with field data if the goal is accurate mapping of facies.

## References


Abreu, L.W. and Anderson, G.P. (1996). The MODTRAN 2/3 report and LOWTRAN 7 model. *Contract* 19628: 0132.

Adam, S. (1997). Glacier snow line mapping using ERS-1 SAR imagery. *Remote Sensing of Environment* 61(1): 46–54. doi: 10.1016/s0034-4257(96)00239-8.

Barzycka, B., Grabiec, M., Błaszczyk, M. et al. (2020). Changes of glacier facies on Hornsund glaciers (Svalbard) during the decade 2007–2017. *Remote Sensing of Environment* 251: 112060. doi: 10.1016/j.rse.2020.112060.

Benn, D. and Evans, D. (2013). *Glaciers and Glaciation*. London: Routledge.

Benson, C. (1962). Stratigraphic studies in the snow and firn of the Greenland ice sheet. No. RR70: Cold Regions Research and Engineering Lab: Hanover NH, USA. Available online: http://acwc.sdp.sirsi.net/client/en_US/search/asset/100139 2;jsessionid=351D596A6CE87F45BAEB04E7B9ECE897. enterprise-15000 (accessed 17 January 2018).

Bosdogianni, P., Petrou, M., and Kittler, J. (1994). Mixed pixel classification in remote sensing. *Image and Signal Processing for Remote Sensing* 2315: 494–505. doi: 10.1117/12.196750.

Braun, M., Schuler, T.V., Hock, R. et al. (2007) Comparison of remote sensing derived glacier facies maps with distributed mass balance modelling at Engabreen, northern Norway. *IAHS Publications: Series of Proceedings and Reports* 318: 126–134.

Brown, I. (2012). Synthetic aperture radar measurements of a retreating firn line on a temperate icecap. *IEEE Journal of Selected Topics in Applied Earth Observations and Remote Sensing* 5(1): 153–160. doi: 10.1109/jstars.2011.2167601.

Congalton, R.G. (1991). A review of assessing the accuracy of classifications of remotely sensed data. *Remote Sensing of Environment* 37(1): 35–46. doi: 10.1016/0034-4257(91)90048-B.

Cracknell, A.P. and Hayes, L. (1991). *Introduction to Remote Sensing*. New York: Taylor & Francis.

Dewi, E.K. and Trisakti, B. (2016). Comparing atmospheric correction methods for Landsat OLI data. *International Journal of Remote Sensing and Earth Sciences* 13(2): 105–120. Available online: https://core.ac.uk/download/pdf/298933391.pdf.

Dozier, J. (1984). Snow reflectance from LANDSAT-4 thematic mapper. *IEEE Transactions on Geoscience and Remote Sensing* GE-22(3): 323–328. doi: 10.1109/tgrs.1984.350628.

Dozier, J. (1989). Spectral signature of alpine snow cover from the Landsat Thematic Mapper. *Remote Sensing of Environment* 28: 9–22. doi: 10.1016/0034-4257(89)90101-6.

Dozier, J., Schneider, S., and McGinnis, D. (1981). Effect of grain size and snowpack water equivalence on visible and near-infrared satellite observations of snow. *Water Resources Research* 17(4): 1213–1221. doi: 10.1029/wr017i004p01213.

Foody, G.M. (2008). Harshness in image classification accuracy assessment. *International Journal of Remote Sensing* 29(11): 3137–3158. doi: 10.1080/01431160701442120.

Foody, G.M. (2010). Assessing the accuracy of land cover change with imperfect ground reference data. *Remote Sensing of Environment* 114(10): 2271–2285. doi: 10.1016/j.rse.2010.05.003.

Gardin, S., Van Laere, S.M.J., Van Coillie, F.M.B. et al. (2011). Remote sensing meets psychology: a concept for operator performance assessment. *Remote Sensing Letters* 2(3): 251–257. doi: 10.1080/01431161.2010.516280.

Ghosh, S., Pandey, A., and Nathawat, M. (2014). Mapping of debris-covered glaciers in parts of the Greater Himalaya Range, Ladakh, western Himalaya, using remote sensing and GIS. *Journal of Applied Remote Sensing* 8(1): 083579. doi: 10.1117/1.jrs.8.083579.

Gore, A., Mani, S., Shekhar, C. et al. (2017). Glacier surface characteristics derivation and monitoring using Hyperspectral datasets: a case study of Gepang Gath glacier, Western Himalaya. *Geocarto International* 34(1): 23–42. doi: 10.1080/10106049.2017.1357766.

Guo, Z., Geng, L., Shen, B. et al. (2021). Spatiotemporal variability in the glacier snowline altitude across high mountain Asia and potential driving factors. *Remote Sensing* 13(3): 425. doi: 10.3390/rs13030425.

Hall, D., Chang, A., and Siddalingaiah, H. (1988). Reflectances of glaciers as calculated using Landsat-5 Thematic Mapper data. *Remote Sensing of Environment* 25(3): 311–321. doi: 10.1016/0034-4257(88)90107-1.

Hall, D., Ormsby, J., Bindschadler, R. et al. (1987). Characterization of snow and ice reflectance zones on glaciers using Landsat Thematic Mapper Data. *Annals of Glaciology* 9: 104–108. doi: 10.1017/s0260305500000471.

Harrisgeospatial.com. Available online: https://www.harrisgeospatial.com/docs/backgroundflaash.html (accessed 21 June 2018).

Isaksen, K., Nordli, Ø., Førland, E.J. et al. (2016). Recent warming on Spitsbergen: influence of atmospheric circulation and sea ice cover. *Journal of Geophysical Research* 121(20): doi: 10.1002/2016JD025606.

Jawak, S., Devliyal, P., and Luis, A. (2015). A comprehensive review on pixel oriented and object-oriented methods for information extraction from remotely sensed satellite images with a special emphasis on cryospheric applications. *Advances in Remote Sensing* 4(3): 177–195. doi: 10.4236/ars.2015.43015.

Jawak, S., Wankhede, S., and Luis, A. (2019). Explorative study on mapping surface facies of selected glaciers from Chandra Basin, Himalaya using WorldView-2 data. *Remote Sensing* 11(10): 1207. doi: 10.3390/rs11101207.

Jawak, S.D., Wankhede, S.F., and Luis, A.J. (2018). Exploration of glacier surface facies mapping techniques using very high-resolution WorldView-2 satellite data. *Proceedings* 2(7): 339.

Kaufman, Y.J., Wald, A.E., Remer, L.A. et al. (1997). The MODIS 2.1-/spl mu/m channel-correlation with visible reflectance for use in remote sensing of aerosol. *IEEE Transactions on Geoscience and Remote Sensing* 35(5): 1286–1298. doi: 10.1109/36.628795.

Keshri, A., Shukla, A., and Gupta, R. (2009). ASTER ratio indices for supraglacial terrain mapping. *International Journal of Remote Sensing* 30(2): 519–524. doi: 10.1080/01431160802385459.

König, M., Winther, J., and Isaksson, E. (2001). Measuring snow and glacier ice properties from satellite. *Reviews of Geophysics* 39(1): 1–27. doi: 10.1029/1999rg000076.

Leckie, D. (2003). Stand delineation and composition estimation using semi-automated individual tree crown analysis. *Remote Sensing of Environment* 85(3): 355–369. doi: 10.1016/S0034-4257(03)00013-0.

Lillesand, T., Kiefer, R.W., and Chipman, J. (2014). *Remote Sensing and Image Interpretation*. Hoboken, NJ: John Wiley & Sons.

Lu, D. and Weng, Q. (2007). A survey of image classification methods and techniques for improving classification performance. *International Journal of Remote Sensing* 28(5): 823–870. doi: 10.1080/01431160600746456.

Luis, A. and Singh, S. (2020). High-resolution multispectral mapping facies on glacier surface in the Arctic using WorldView-3 data. *Czech Polar Reports* 10(1): 23–36. doi: 10.5817/cpr2020-1-3.

Mausel, P.W., Kramber, W.J., and Lee, J.K. (1990). Optimum band selection for supervised classification of multispectral data. *Photogrammetric Engineering and Remote Sensing* 56: 55–60.

Müller, F. (1962). Zonation in the accumulation area of the glaciers of Axel Heiberg Island, NWT, Canada. *Journal of Glaciology* 4(33): 302–311. doi: 10.1017/s0022143000027623.

Nijhawan, R., Garg, P., and Thakur, P. (2016). A comparison of classification techniques for glacier change detection using multispectral images. *Perspectives in Science* 8: 377–380. doi: 10.1016/j.pisc.2016.04.080.

Nolin, A., Dozier, J., and Mertes, L. (1993). Mapping alpine snow using a spectral mixture modeling technique. *Annals of Glaciology* 17: 121–124. doi: 10.3189/S0260305500012702.

Pandey, P., Kulkarni, A., and Venkataraman, G. (2013). Remote sensing study of snowline altitude at the end of melting season, Chandra-Bhaga Basin, Himachal Pradesh, 1980–2007. *Geocarto International* 28(4): 311–322. doi: 10.1080/10106049.2012.705336.

Paul, F., Winsvold, S., Kääb, A. et al. (2016). Glacier remote sensing using Sentinel-2. Part II: Mapping glacier extents and surface facies, and comparison to Landsat 8. *Remote Sensing* 8(7): 575. doi: 10.3390/rs8070575.

Pellikka, P. and Rees, G. (2010). *Remote Sensing of Glaciers*. Boca Raton, FL: CRC Press.

Pope, A. and Rees, G. (2014a). Using *in situ* spectra to explore Landsat classification of glacier surfaces. *International*

*Journal of Applied Earth Observation and Geoinformation* 27: 42–52. doi: 10.1016/j.jag.2013.08.007.

Pope, A. and Rees, W. (2014b). Impact of spatial, spectral, and radiometric properties of multispectral imagers on glacier surface classification. *Remote Sensing of Environment* 141: 1–13. doi: 10.1016/j.rse.2013.08.028.

Powell, R., Matzke, N., de Souza, C. et al. (2004). Sources of error in accuracy assessment of thematic land-cover maps in the Brazilian Amazon. *Remote Sensing of Environment* 90(2): 221–234. doi: 10.1016/J.RSE.2003.12.007.

Ramezan, C.A., Warner, T.A., Maxwell, A.E. et al. (2021). Effects of training set size on supervised machine-learning land-cover classification of large-area high-resolution remotely sensed data. *Remote Sensing* 13(3): 368. doi: 10.3390/rs13030368.

Rastner, P., Bolch, T., Notarnicola, C. et al. (2014). A comparison of pixel- and object-based glacier classification with optical satellite images. *IEEE Journal of Selected Topics in Applied Earth Observations and Remote Sensing* 7(3): 853–862. doi: 10.1109/jstars.2013.2274668.

Shukla, A. and Ali, I. (2016). A hierarchical knowledge-based classification for glacier terrain mapping: a case study from Kolahoi Glacier, Kashmir Himalaya. *Annals of Glaciology* 57(71): 1–10. doi: 10.3189/2016aog71a046.

Tran, N., Rémy, F., Feng, H., and Féménias, P. (2008). Snow facies over ice sheets derived from Envisat active and passive observations. *IEEE Transactions on Geoscience and Remote Sensing* 46(11): 3694–3708. doi: 10.1109/tgrs.2008.2000818.

Yousuf, B., Shukla, A., Arora, M.K.., et al. (2019), Glacier facies characterization using optical satellite data: impacts of radiometric resolution, seasonality, and surface morphology. *Progress in Physical Geography: Earth and Environment* 43(4): 473–495. doi: 10.1177/0309133319840770.

Zhang, J., Jia, L., Menenti, M., and Hu, G. (2019). Glacier facies mapping using a machine-learning algorithm: the Parlung Zangbo Basin case study. *Remote Sensing* 11(4): 452. doi: 10.3390/rs11040452.

Zhou, W., Schwarz, K., and Cadenasso, M. (2010) Mapping urban landscape heterogeneity: agreement between visual interpretation and digital classification approaches. *Landscape Ecology* 25(1): 53–67. doi: 10.1007/S10980-009-9427-8.

# 27

# Supraglacial Lake Filling Models

## Examples From Greenland

*Prateek Gantayat**

*Lancaster Environment Centre, Lancaster University, Lancaster, UK*
** Corresponding author*

## 27.1 Introduction

Due to the acceleration in the rate of loss of ice mass, the Greenland Ice Sheet (GrIS) will become a potential contributor to global mean sea level rise for the rest of this century (Mouginot et al., 2019; Smith et al., 2020). Between 2010 and 2018, Mouginot et al. (2019) observed a six-fold increase in the rate of mass loss as compared to that in the 1980s.

Mass loss of the ice sheet is due to an imbalance between mass gain due to snowfall and mass loss due to surface melting and dynamic ice discharge. The surface melting of an ice sheet may accelerate if there is a decrease in surface albedo due to the formation of a large number of supraglacial lakes (SGLs) (Perovich et al., 2002). These lakes also act as temporary reserves of meltwater, thereby reducing the rate of the runoff from the ice sheet. As the melt season progresses, these lakes may either drain (via crevasses) thereby transferring the lake melt to the base of the ice sheet (Selmes et al., 2011), or become filled and subsequently overflow at slower rates (Das et al., 2008).

When a lake drains rapidly and consequently the lake melt content flows into the ice sheet bed, the ice sheet's velocity shows short-term spikes due to an increase in lubrication at its base (Christofferson et al., 2018). Due to this phenomenon, the ice sheet's hypsometry changes, thereby increasing the mass loss (Leeson et al., 2012). In a warming climate, the abundance of meltwater in inland areas helps the propagation of these short-term velocity spikes (Sole et al., 2011; Bartholomew 2011), which consequently might result in a net acceleration. Knowledge of SGL lake evolution over shorter time periods is crucial to pinpoint not only the conditions required for lake drainage but also the timing of the occurrence of such events.

In this chapter, three methods that simulate the evolution of SGLs in the GrIS are discussed. These models are called SLING (Leeson et al., 2012), SRLF (Banwell et al., 2012a), and SRLFCI (Koziol et al., 2017). Both these methods were applied over different regions located in the southwest GrIS.

## 27.2 Methods

### 27.2.1 Supraglacial Lake FilLING (SLING)

Leeson et al. (2012) developed the SLING model. This is a fully transient 2D hydrology model that is based on the flow of meltwater through a porous medium (Chow, 1988) and Manning's open channel flow (Manning, 1891). The model consists of three main components, namely:

1) *Flow direction component*: For a given DEM cell called "source," this component determines the appropriate "destination" cell. In every time step, the model routes meltwater produced in the source cell to its destination cell. In the model, the destination cell is the lowest lying cell in the 3-by-3 neighborhood of the source cell.
2) *Flux transfer component*: In every timestep, this component calculates the displacement of meltwater runoff between a source and its destination cell. This is calculated depending on whether the meltwater flows through a porous medium or an open channel.
   a) *Flow through porous medium*: The volumetric flow rate or flux "Q" of meltwater through a snowpack can be modeled according to Darcy's law (Chow, 1988):

$$Q = \frac{kAS}{\mu} \tag{27.1}$$

*Advances in Remote Sensing Technology and the Three Poles*, First Edition. Edited by Manish Pandey, Prem C. Pandey, Yogesh Ray, Aman Arora, Shridhar D. Jawak, and Uma K. Shukla.

where $k$ is defined as the permeability of snow which is estimated using the model proposed by Shimizu (1969), $\mu$ is viscosity of water assumed to be $1.763 \times 10^{-3}$ Pa s, $A$ is cross-sectional area of flow (Figure 27.1a), and $S$ is free surface slope.

b) *Open channel flow*: When meltwater flows over bare ice, the volumetric flow rate or flux $Q$ is modeled according to Manning (1891):

$$Q = \frac{A}{n} R^{2/3} S^{1/2} \tag{27.2}$$

where $R$ is the hydraulic radius and is approximated by the depth of the flow for wide, shallow channels, $S$ is the free surface slope of the channel, $A$ is the cross-sectional area of flow (Figure 27.1b), and $n$ is Manning's roughness coefficient assumed to be 0.01 for ice (Lotter et al., 1932).

Let $i$ denote a source cell, $j$ denote its corresponding destination cell, $z_i$ and $z_j$ represent the surface elevations of the source and destination cells, and $d_i$ and $d_j$ represent the depth of standing water at the source and destination cell respectively (Figure 27.1b). Based on whether the source and destination cells share a border or a diagonal path, the path length ($P$) between the two cells can be $L$ or $(2)^{1/2}L$ respectively. The free surface slope is defined as:

$$S = \frac{(z_i + d_i) - (z_j + d_j)}{P} \tag{27.3}$$

The rate of change of water depth ($d'$) due to flux leaving the cell is formulated as:

$$Q = d'^{FLUX} L^2 \tag{27.4}$$

where $L^2$ is assumed to be the area of the cell. For a source cell $i$, the rate of change of water depth ($d_i'^{OUT}$) is equal to the sum of the rate of change of water depth due to the outgoing flux ($d_i'^{FLUX}$) and meltwater runoff ($d_i'^{RUN}$) respectively:

$$d'^{OUT}_i = d'^{FLUX}_i + \frac{d'^{RUN}_i}{86400} \tag{27.5}$$

Using the fact that $d'^{FLUX}_i = -d'^{FLUX}_j$, the rate of change of water depth at cell $j$ is:

$$d'^{OUT}_j = -d'^{FLUX}_i + \frac{d'^{RUN}_j}{86400} \tag{27.6}$$

Thus, we have a system of two differential equations, Equation 27.5 and Equation 27.6. This system is solved using the 4th-order Runge Kutta method.

Finally, in case a cell receives water from multiple cells, the new depth of water at a given cell $n$ at time $t$ is given by:

$$d_n^t = d_n^0 + \Sigma d_n^{in} + \int_0^t d'^{RUN}_n - \int_0^t d'^{OUT}_n \tag{27.7}$$

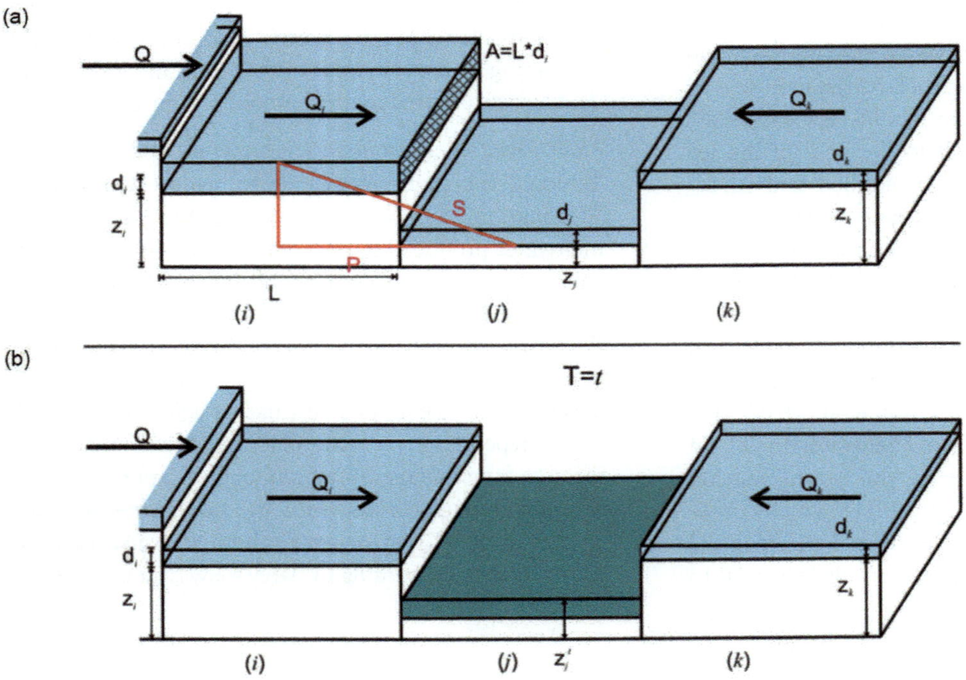


Figure 27.1 (a) $j$ is the destination cells for cells marked as $i$ and $k$ respectively. The slope of the ice surface is represented as S. The flow path is represented as P. Qi and Qj represent the outgoing volumetric flux from cells $i$ and $k$. (b) A schematic to show the working of the lake accumulation component. *Source:* Leeson et al., 2012 / European Geosciences Union / CC BY 3.0.

3) *Lake accumulation component*: At the end of every timestep, after the flow from each source cell to its respective destination cell has been calculated, an iterative algorithm is used to bring the water into the lakes. This iterative algorithm is the lake accumulation component. Like the flow direction component, for every cell the corresponding 3-by-3 neighborhood is scanned to determine the existence of a "sink" cell. A sink is defined as the DEM cell whose elevation is lowest in its corresponding 3-by-3 neighborhood. In case a sink cell is located, if the height of the cell plus water is less than the height of the lowest-lying nearest neighbor, all the water in the sink cell is incorporated into the lake surface of the DEM. In case the sum of cell elevation, height of standing water, and the meltwater runoff is less than the same for its lowest lying neighbor, the difference between the heights of the two cells plus 1 mm is incorporated into the lake surface for that sink cell. The algorithm loops around all the DEM cells until no sinks are encountered. The algorithm can be visualized from Figure 27.1b.

### 27.2.2 Surface Routing and Lake Filling Model (SRLF)

Banwell et al. (2012a) developed the SRLF model. This model assumes that all the lakes form in the topographic hollows on the ice sheet surface. It consists of two components that are run in every timestep. The components are:

1) *Lake and Catchment Identification Algorithm (LCIA)*: This algorithm is based on the work done by Arnold et al. (2010). Before starting the discussion about the algorithm, it is important to define sink cells. A sink is defined as a DEM cell with no lower neighbors. Every sink cell is a potential nucleus for a lake. At first, the LCIA calculates the flow direction matrix for the DEM. For every DEM cell, the flow direction matrix gives the direction of flow based on the concept of steepest descent. For every sink cell, the LCIA loops over the flow direction matrix to identify the cells that ultimately flow into the sink cell. The area spanned by those DEM cells is the catchment for the sink cell. The LCIA then searches the catchment boundary for the lowest possible cell over which the water would pour as the sink (and any surrounding cells lower than their neighbors) floods with water. This lowest possible cell is the outlet cell for the catchment. Once the outlet cell has been located, the lake hypsometry can be calculated based on the quantity of meltwater present at the sink cell. An example of modeled supraglacial catchments is shown in Figure 27.2.

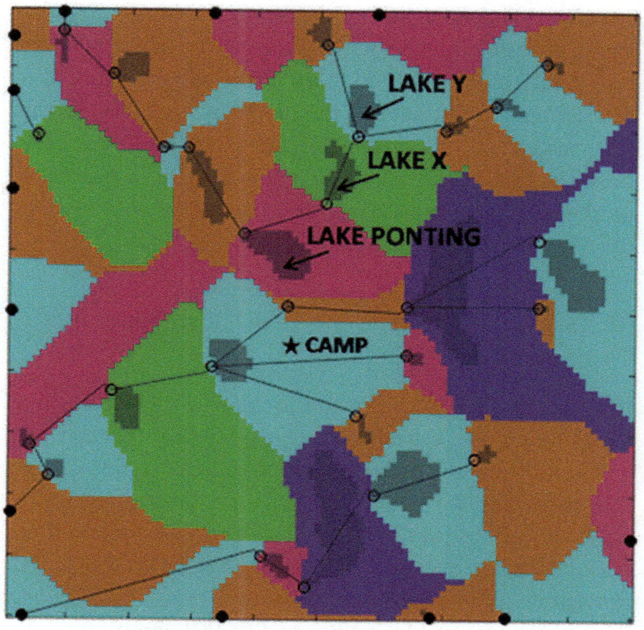


**Figure 27.2** The colored regions are the topographical catchments as delineated by LCIA. The hollow black circles represent the location of the outlet of every catchment. The thin black lines represent the direction of flow of overflowing water from an upstream lying catchment to another catchment located in the downstream. Maximum possible lake extents are shown in gray. The solid black circles represent the outlet cells that lie at the domain margin. *Source:* Banwell et al., 2012a / John Wiley & Sons.

2) *Flow Delay Algorithm (FDA)*: The flow direction matrix calculated by LCIA delineates the flow potential path from every DEM cell to the corresponding sink (or lake) cell. However, the correct location of the cell at which the runoff would end up in a given timestep is determined by the FDA component of the model. The FDA uses the flow direction of the matrix with assumptions about the physical processes to estimate the flow delay time between each DEM cell and its corresponding sink (or lake) cell. Between two DEM cells, the meltwater runoff may have a Darcian (Colbeck, 1978) or Manning's flow (Manning, 1891), depending upon the presence or absence of a snowpack. The relevant equations have been discussed in Arnold et al. (1998). In simpler terms, for every DEM cell, the FDA splits out the travel time taken by water to cross that cell depending on whether the water flows through a snowpack or flows through an open channel over bare ice. By integrating the travel times downslope, we calculate the total time taken by water to flow from a DEM cell to the corresponding sink (or lake cell).

Thus, at the end of every timestep, distinct hydrographs are produced at every sink (or lake) cell. The total accumulated volume of meltwater at each sink or lake cell can be calculated and from this calculation we can determine at which timestep the lake will be filled to the brim. Once the lake is full, any further addition of meltwater is routed to downstream catchments as calculated by the LCIA. In this way, water can potentially flow in a series of "cascades" from the initial source cell, through a series of full lakes, until it either reaches a lake which has not yet overflown or the edge of the DEM domain.

### 27.2.3 Surface Routing and Lake Filling With Channel Incision (SRLFCI)

Koziol et al. (2017) developed the SRLFCI model. Just like SRLF, this model assumes that every surface depression is a potential site for the formation of an SGL. In this model, the meltwater is routed and SGLs are formed using the same concept as that for SRLF. However, the major difference between SRLF and SRLFCI is that SRLFCI also has provisions to simulate rapid lake and non-lake drainage through crevasses. In addition, SRLFCI simulates lake overflow in a more realistic manner, unlike SRLF. In the following few paragraphs, we will explain how the above new processes were simulated into the model:

1) *Rapid lake and non-lake drainage through crevasses*: This mechanism simulates rapid drainage of meltwater from lake as well as non-lake areas in the following ways:
   a) *Non-lake drainage*: Surface stresses derived from winter velocities (Joughin et al., 2010) and satellite imagery were used to visually determine the location of crevassed areas. Based on the delineated regions, a threshold stress value of 132.5 kPa was determined. All the DEM cells that had a stress value more than the threshold, were classified as crevassed cells. In every timestep, while routing surface meltwater from a source cell to its corresponding sink or destination cell, if a crevassed cell is encountered, the entire meltwater runoff from the source cell is transferred to the crevassed cell, rather than the sink or destination cell. The propagation of crevasses (due to the deposition of meltwater) in these cells was determined by solving the equation of Linear Elastic Fracture Mechanics (LEFM) (van der Veen, 2008; Clason et al., 2012, 2015).
   b) *Lake drainage*: This occurred when the lake volume was found to be enough to fill an underlying crevasse penetrating the local ice thickness to the ice sheet's base. The model uses the concept of fracture area (Fa). Fa is assumed to be constant across the study area and is found to vary between 4000 and 8000 m$^2$ (Arnold et al., 2014). In every timestep, for every lake, the deepest point within the lake is located and the ice thickness at that point is noted. If the volume of the lake is more than the product of Fa and the ice thickness, the lake is marked as draining and the lake water is emptied in either a single or a number of successive timesteps.
2) *Slow lake drainage through channel incision*: Selmes et al. (2013) showed that around a third of lakes located in southwest Greenland drain slowly over the surface of the ice sheet. While overflowing, the outgoing meltwater incises a channel into the ice at the edge of the lake. This phenomenon is called channel incision and is simulated by using the spillway model developed by Raymond and Nolan (2000). A detailed description has been given in Koziol et al. (2017). Briefly, for a given lake, the spillway model is a system of two coupled differential equations that gives the rate of evolution of lake depth as well as the channel height. Whenever the lake fills up and overflows through the lake's outlet, the overflowing water erodes the underlying ice due to frictional heating. Consequently, the surface elevation of the lake's outlet decreases, thereby aiding lake overflow.

## 27.3 Study Area

SLING was applied over a domain that measures 6753 km$^2$ in area and is located in the western GrIS. The study area is shown in Figure 27.3. The ice sheet elevation in this region ranges from 1100 to 1752 m a.s.l.

SRLF is applied over the Paakitsoq region located in west central Greenland (Figure 27.4). This region spans over an area of 2300 km$^2$. In this study, the focus is on the Ponting Lake area, as shown in Figure 27.4.

SRLFCI was applied over the Paakitsoq region, as shown in Figure 27.5. The region spanned over an area of 2368 km$^2$.

## 27.4 Data Used

For producing results using the SLING model, InSAR data derived DEM of the corresponding study area was used (Palmer et al., 2011). The DEM was produced at 100 m spatial resolution. Data regarding snow pack properties and meltwater runoff was derived from the MAR Regional Climate Model (Fettweis et al., 2011). For comparing results with observed data, MODIS observed annual lake extents for the period 2003 to 2007 were used.

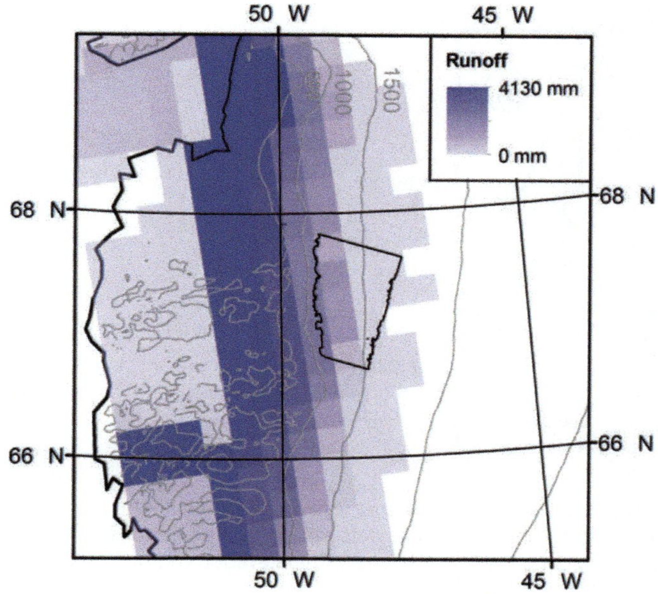


**Figure 27.3** The study area is the domain contained inside the thin black boundary. The elevation contours are represented in gray. These contours are drawn at 500 m altitudinal intervals. The daily runoff which is derived from the MAR climate model is shown in blue. *Source:* Leeson et al., 2012 / European Geosciences Union / CC BY 3.0.

For producing results using the SRLF model, ASTER GDEM of the study area was used. For all model simulations, the GDEM was resampled to 100 m spatial resolution and then smoothed using an s-by-6 median filter. Data regarding daily runoff and snow properties were derived using a surface mass balance model (Banwell et al., 2012b) and meteorological records derived from weather stations installed at the study area, as shown by the black stars in Figure 27.4. For validating modeled results, observed lake volume data for Lake Ponting were used.

To produce the results using the SRLFCI model, GIMP DEM at 90 m spatial resolution was used (Howat et al., 2015). Before all model simulations, the DEM was smoothed using a 2-by-2 median filter to remove small-scale noise. Then, a 11-by-11 Gaussian filter was used to remove the terracing effects of the DEM. The local ice thickness data was taken from the BedMachine dataset (Morlighem et al., 2015). To locate the crevassed areas, WorldView imagery acquired during the 2009 and 2010 melt seasons was used. Daily surface melt runoff data was extracted from the RACMO2.3 regional climate model simulations (Noël et al., 2015)

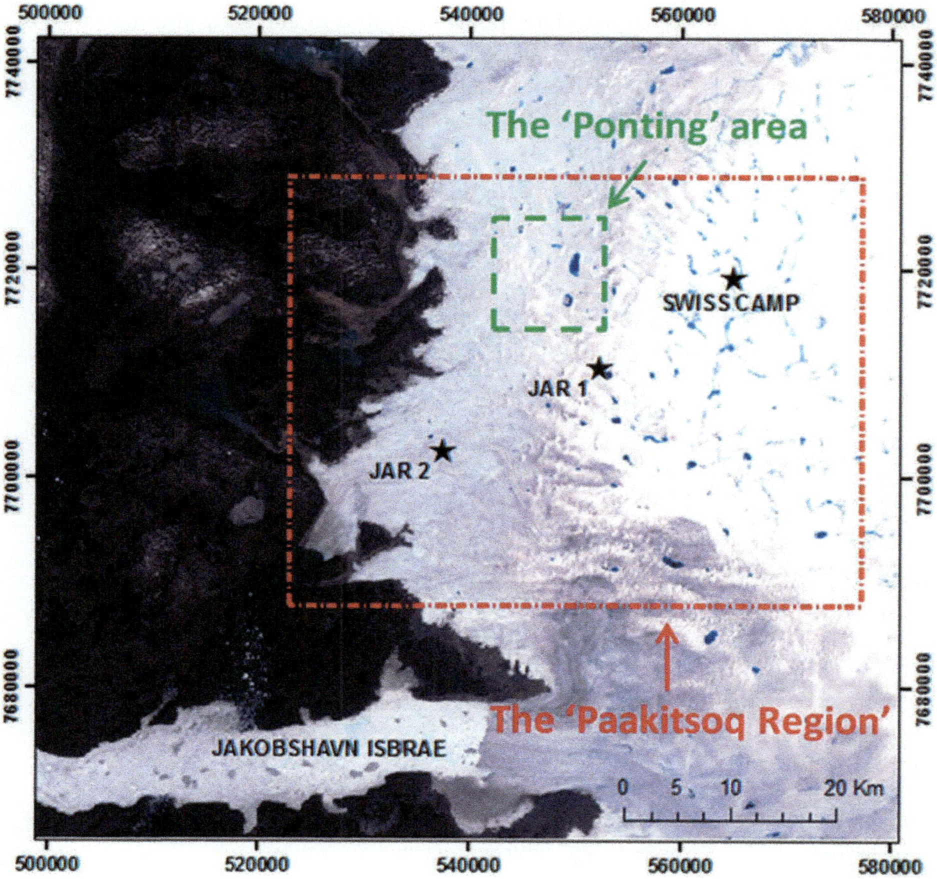


**Figure 27.4** Map of the study site. The Ponting area is delineated by the green box, located within the larger Paakitsoq region (red box). The base Landsat 7 ETM + image is dated 7 July 2001. *Source:* Banwell et al. (2012a), John Wiley & Sons.

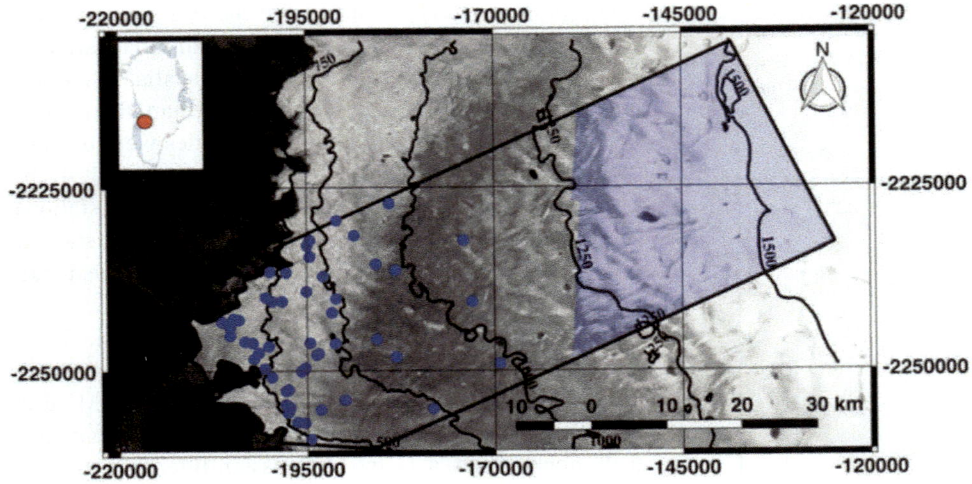


Figure 27.5 Map showing the study area used in Koziol et al. (2017). The location of the study area is shown as a red dot on the map of Greenland (top left corner). The study area is bounded by the rectangular shape. The blue dots represent the locations of the moulins as derived from WorldView imagery. The background image is a Grey scale Landsat 8 imagery. The elevation contours were derived from the GIMP DEM. *Source:* Koziol et al., 2017 / Cambride University Press / CC BY 4.0.

## 27.5 Results

### 27.5.1 Results For SLING Model

The model performance is assessed by analyzing its skill in predicting the lake location on a lake-by-lake basis. Figure 27.6 shows a comparison between observed and simulated lake extents. There are also a large number of instances where the observed and modeled extents overlap. In such instances, the overlapped area has been marked as violet in Figure 27.6. In general, modeled and observed lake extents are in good agreement with each other. The model was able to correctly simulate 66% of observed lakes that had an aerial extent greater than 0.125 $km^2$. In addition, various lake cross-sections are also shown in Figure 27.6.

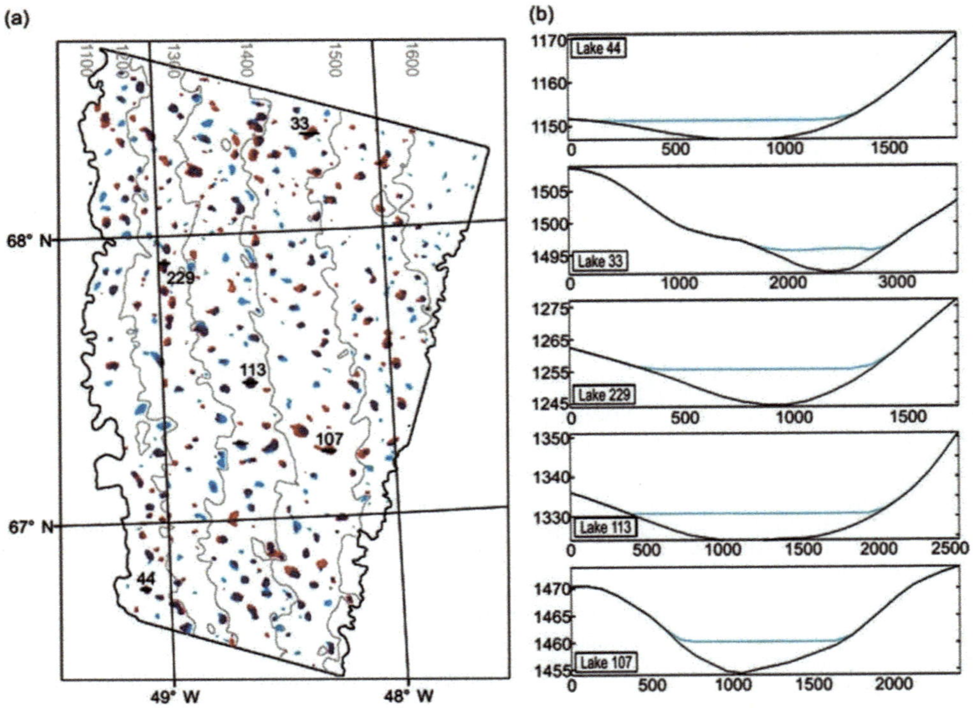


Figure 27.6 (a) Composite plot showing observed lakes (red), modeled lakes (blue), and the regions where modeled and observed lakes overlap (violet). (b) 1D cross-sections of some modeled SGLs. *Source:* Leeson et al., 2012 / European Geosciences Union / CC BY 3.0.

### 27.5.2 Results For SRLF Model

Figure 27.7 shows a comparison between modeled and observed Lake Ponting volumes. Two major cases have been considered. In the first case, the rate of change of volume of Lake Ponting was assumed to be a function of only the meltwater generated in the corresponding catchment (the green line plot in Figure 27.7). Consequently, the lake volume was modeled without taking account of the overflowing lake water from lake X and lake Y (Figure 27.2). In the second case, the rate of change of volume of Lake Ponting was assumed to be a function of the meltwater generated in the corresponding catchment as well as the meltwater contribution from overflowing lakes X and Y (the blue and purple line plot in Figure 27.7). The results demonstrate that the model is able to correctly model the volume of Lake Ponting only when the contribution of meltwater due to lake overflow from X and Y is taken into account. The dip in the observed curve after 19 June 2011 is because the lake starts to drain from that point. For further details, the reader is requested to refer to Banwell et al. (2012a).

### 27.5.3 Results For SRLFCI Model

This model was able to identify which lakes drain rapidly or slowly and which lakes survive the melting season and will now undergo freezing in the coming winter. This information is shown in Figure 27.8.

The above information is also shown in Table 27.1 and this metric is compared with that observed by Selmes et al. (2013). From the table we can see that the modeled metrics broadly match with those observed. However, we must keep in mind that the observed numerical figures are with respect to the entire southwest Greenland, while the modeled numerics are with respect to the study area only.

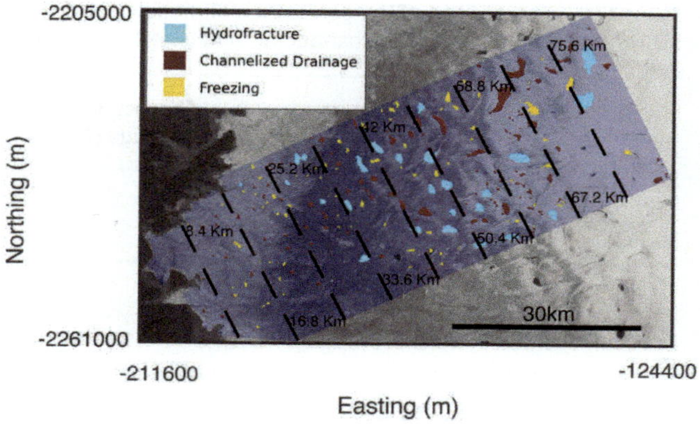


Figure 27.8 The modeled lakes have been classified into three categories, namely: i) ones that drained via hydrofracture (blue polygons); ii) ones that drained via channel incision (brown polygons); and iii) ones that survived the melting season and will now undergo freezing in the coming winter (yellow polygons). *Source:* Koziol et al., 2017 / Cambridge University Press / CC BY 4.0.

## 27.6 Discussion

The SLING model is the simplest of all the models. However, in a given timestep, it routes surface meltwater only within the 3-by-3 neighborhood of the source cell. Consequently, the meltwater runoff is unable to travel between more than two DEM cells in a single timestep. This becomes an issue when either a coarse timestep is used for model simulations or the study area produces a large amount of meltwater runoff. In addition, the model

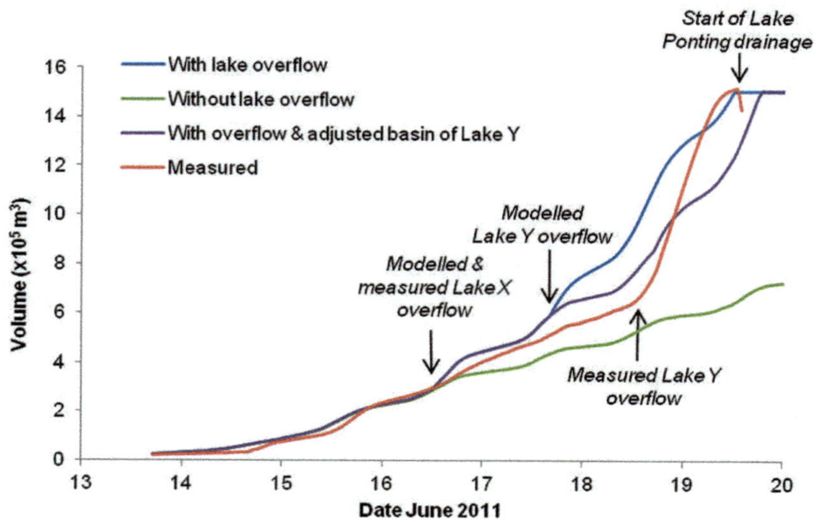


Figure 27.7 Comparison of modeled and observed volumes of Lake Ponting. *Source:* Banwell et al., 2012a / John Wiley & Sons.

**Table 27.1** Comparison of the percentage of modeled and observed lakes that drain via: i) hydrofracture (rapidly); ii) via channel incision (slowly); or iii) neither of the two.

| | Hydrofracture Drainage (%) | Channel Incision Drainage (%) | None (%) | Unknown (%) |
|---|---|---|---|---|
| Observed | 11.7 | 38.9 | 43.8 | 5.6 |
| Modeled | 10.7 | 37.8 | 51.5 | 0.0 |

does not simulate rapid drainage of SGLs or meltwater runoff through crevasses. Also, the phenomenon of slow lake drainage via channel incision is not simulated.

The SRLF model is more complex as compared to SLING. This model takes care of one of the shortcomings of SLING, i.e., unlike SLING, in a single timestep, from a source cell, the SRLF model is able to route the meltwater runoff to a destination which may not be in the 3-by-3 neighborhood of the source cell. However, like SLING, this model also does not simulate rapid drainage of SGLs or meltwater runoff through crevasses. In addition, the phenomenon of slow lake drainage via channel incision is not simulated.

The SRLFCI model is the most complex of the three models discussed here. It overcomes the shortcomings of the SLING and SRLF models. This model not only routes meltwater from a source cell over multiple cells in a given timestep but also simulates the phenomenon of rapid, SGL, and meltwater drainage via hydrofractures and slow drainage of SGL via channel incision. From the above results it is apparent that the third model, i.e., SRLFCI, is better able to model SGL evolution.

## 27.7 Conclusions

In this chapter, three supraglacial lake filling and meltwater routing models have been discussed. One model used climate model derived daily runoff, snow depth, and snow density data. The second used station derived meteorological records and an SMB model for routing meltwater and simulating the evolution of supraglacial lakes. The third model not only simulated meltwater routing and formation of SGLs, but also modeled their drainage and overflow in a realistic manner.

## Acknowledgments

The author is thankful to the *Journal of Glaciology* and *Journal of Geophysical Research* for their open data use policy. The author is grateful to the co-authors of this book for reviewing and giving valuable suggestions for improving this chapter.

## References

Arnold, N., Richards, K., Willis, I. et al. (1998). et al.Initial results from a distributed, physically based model of glacier hydrology. *Hydrol. Process.* 12: 191–219. doi: 10.1002/(SICI)1099-1085(199802)12:2<91::AIDHYP571>3.0.CO;2-C.

Arnold, N.S. (2010). A new approach for dealing with depressions in digital elevation models when calculating flow accumulation values. *Prog. Phys. Geogr.* 34(6): 781–809. doi: 10.1177/0309133310384542.

Arnold, N.S., Banwell, A.F., and Willis, I.C. (2014). High-resolution modelling of the seasonal evolution of surface water storage on the Greenland ice sheet. *Cryosphere* 7(8): 1149–1160. doi: 10.5194/tc-8-1149-2014

Banwell, A.F., Arnold, N., Willis, I. et al. (2012a). Modelling supraglacial water routing and lake filling on the Greenland Ice Sheet. *J. Geophys. Res.* 117: doi: 10.1029/2012JF002393.

Banwell, A.F., Willis, I.C., Arnold, N.S. et al. (2012b). Calibration and validation of a high resolution surface mass balance model for Paakitsoq, west Greenland. *J. Glaciol.* 58(212): 1047–1062. doi: 10.3189/2012JoG12J034.

Bartholomew, I. (2011). Seasonal variations in Greenland Ice Sheet motion: inland extent and behaviour at higher elevations. *Earth Planet. Sci. Lett.*, 307(3–4): 271–278. ISSN 0012821X (doi: 10.1016/j.epsl.2011.04.014)

Chow, V.T. (1988). *Applied Hydrology. McGraw-Hill International Editions, Civil Engineering Series*. Singapore: McGraw-Hill Book Company .

Christoffersen, P., Bougamont, M., Hubbard, A. et al. (2018). Cascading lake drainage on the Greenland Ice Sheet triggered by tensile shock and fracture. *Nature Comm.* 9: 1064. doi: 10.1038/s41467-018-03420-8.

Clason, C., Mair, D.W.F., Burgess, D.O. et al. (2012). Modelling the delivery of supraglacial meltwater to the ice/bed interface: application to southwest Devon Ice Cap, Nunavut, Canada. *J. Glaciol.* 58(208): doi: 10.3189/2012JoG11J129.

Clason, C., Mair, D.W.F., Nienow, P.W. et al. (2015). Modelling the transfer of supraglacial meltwater to the bed of Leverett Glacier, Southwest Greenland. *Cryosph.* 9: 123–138. doi: 10.5194/tc-9-123-2015

Colbeck, S.C. (1978). The physical aspects of water flow through snow. In: *Advances in Hydroscience* (ed. V.T. Chow), vol. 11, 165–206: San Diego, CA: Academic.

Das, S.B., Joughin, I., Behn, M.D. et al. (2008).Fracture propagation to the base of the Greenland Ice Sheet during supraglacial lake drainage. *Science* 320(5877): 778–781. doi: 10.1126/science.1153360.

Fettweis, X., Tedesco, M., van den Broeke, M. et al. (2011). Melting trends over the Greenland ice sheet (1958–2009) from spaceborne microwave data and regional climate models. *Cryosph.* 5: 359–375. doi: 10.5194/tc-5-359-2011.

Howat, I., Negrete, A., and Smith, B. (2015). Measures Greenland Ice mapping project (GIMP) digital elevation model, version 1. 90 m resolution. NASA National Snow and Ice Data Center Distributed Active Archive Center, Boulder, CO. doi: 10.5067/NV34YUIXLP9W.

Joughin, I., Smith, B.E., Howat, I.M. et al. (2010). MEaSUREs Greenland Ice Velocity Map from InSAR Data. NASA DAAC at the National Snow and Ice Data Center, Boulder, CO. doi: 10.5067/MEASURES/ CRYOSPHERE/ nsidc-0478.001.

Koziol, C., Arnold, N., Pope, A. et al. (2017). Quantifying supraglacial meltwater pathways in the Paakitsoq region, West Greenland. *J. Glaciol.* 63(239): 464–476.

Leeson, A.A., Shepherd, A., Palmer, S. et al. (2012). Simulating the growth of supraglacial lakes at the western margin of the Greenland ice sheet. *Cryosph.* 6: 1077–1086. doi: 10.5194/tc-6-1077-2012.

Lotter, G.K. (1932). *Considerations on hydraulic design of channels with different roughness of walls*. Transactions, All-Union Scientific Research Institute of Hydraulic Engineering, Leningrad, 1932

Manning, R. (1891). On the flow of water in open channels and pipes. Transactions. *Institution of Civil Engineers of Ireland* 20: 161–207.

Morlighem, M., Rignot, E., Mouginot, J. et al. (2015). IceBridge bedMachine Greenland, Version 1, thickness. *NASA DAAC at the National Snow and Ice Data Center*, Boulder, CO. doi: 10.5067/AD7B0HQNSJ29, 2015.

Mouginot, J., Rignot, E., Bjork, A.A. et al. (2019). Forty-six years of Greenland Ice Sheet mass balance from 1972 to 2018. *PNAS* 116(19): 9239–9244. doi: 10.1073/pnas. 1904242116.

Noël, B., van de Berg, W.J., van Meijgaard, P. et al. (2015). Summer snowfall on the Greenland ice sheet: a study with the updated regional climate model racmo2.3. *Cryos. Discuss.* 9(1): 1177–1208. doi: 10.5194/tcd- 9-1177-2015.

Palmer, S., Shepherd, A., Nienow, P. et al. (2011). Seasonal speedup of the Greenland Ice Sheet linked to routing of surface water. *Earth Planet. Sci. Lett.* 302: 423–428. doi: 10.1016/j.epsl.2010.12.037.

Perovich, D.K., Tucker, W.B., and Ligett, K.A. (2002). Aerial observations of the evolution of ice surface conditions during summer. *J. Geophys. Res.* 107(C10): 8048. doi: 10.1029/2000JC000449.

Raymond, C. and Nolan, M. (2000). Drainage of a glacial lake through an ice spillway. *Intl. Assoc. Hydrol. Sci. Publ.* 264: 199–210.

Selmes, N., Murray, T., and James, T.D. (2011). Fast draining lakes of Greenland ice sheet. *Geophys. Res. Lett.* 38(15): 5 p. doi:10.1029/2011GL047872.

Selmes, N., Murray, T., and James, T.D. (2013). Characterizing supraglacial lake drainage and freezing on the Greenland Ice Sheet. *Cryos. Discuss.* 7(1): 475–505. ISSN 1994-0440 (doi:10.5194/tcd-7-475-2013)

Smith, B., Fricker, H.A., Gardner, A.S. et al. (2020). Pervasive ice sheet mass loss reflects competing ocean and atmosphere processes. *Science* 368(6496): 1239–1242. doi: 10.1126/science.aaz5845.

Sole, A.J., Mair, D.W.F., Nienow, P.W. et al. (2011). Site characteristics and ice speed at four sites on Kangiata Nunata Sermia glacier, southwest Greenland. PANGAEA, https://doi.org/10.1594/PANGAEA.837202, Supplement to: Sole, A.J. et al. (2011): Seasonal speedup of a Greenland marine-terminating outlet glacier forced by surface melt–induced changes in subglacial hydrology. *J. Geophys. Res.*116(F3): F03014. https://doi.org/10.1029/2010JF001948

van der Veen, C.J. (2008). Fracture propagation as means of rapidly transferring surface meltwater to the base of glaciers. *Geophys. Res. Lett.* 34(1): L01501. doi: 10.1029/2006GL028385.

# 28

## Arctic Sea Level Change in Remote Sensing and New Generation Climate Models

*S. Chatterjee[1,*], R.P. Raj[2], A. Bonaduce[2], and R. Davy[2]*

[1] National Centre for Polar and Ocean Research, Ministry of Earth Sciences, Goa 403 804, India
[2] Nansen Environmental and Remote Sensing Centre, and Bjerknes Center for Climate Research, Bergen, Norway
* Corresponding author

## 28.1 Introduction

Sea level change is a natural indicator of both anthropogenic forcing and natural variability induced climate change (Church et al., 2013; Oppenheimer et al., 2019). Variations in dynamic and thermodynamic processes in several earth system components, such as ocean, atmosphere, cryosphere, and hydrosphere, are integrated into sea level change. Quantification of contributions from the individual factors to sea level change thus needs to be done cautiously. Nonetheless, the global mean sea level (GMSL) rapidly increased at a rate of 3.05 ± 0.24 mm/yr during 1993–2016 (Horwath et al., 2022). The steric and mass component respectively contributed 38% and 57% to the GMSL trend. Furthermore, the GMSL has increased more rapidly in the most recent decade (2003–2016; 3.64 ± 0.26 mm/yr) with an increased contribution from ocean mass change due to Greenland ice sheet melting (Horwath et al., 2022). However, regional sea level changes are also of potential significance owing to its variety of governing forcing mechanisms, that may differ from its GMSL counterpart (Stammer et al., 2013; Raj et al., 2020). For example, changes in salinity are known to significantly impact regional sea level change (Munk, 2003). These regional sea level changes are particularly important for infrastructure, population, and coastal ecosystems. In this context, sea level change in the Arctic Ocean (AO) and surrounding seas are of particular interest for several reasons but not limited to: i) a rapid increase in oceanic heat content in recent decades (Skagseth and Mork 2012; Timmermans et al., 2018); ii) an increased glacier and ice-sheet melting (Shepherd et al., 2012; Gardner et al., 2013); and iii) a changing dynamic of atmospheric and oceanic circulation (Zhang et al., 2008; Screen et al., 2018; Armitage et al., 2020; Raj et al., 2020). In recent decades, the Arctic region experienced the most profound impacts of climate change compared to the entire globe. The warming rate in the Arctic is three times faster than the rest of the globe (AMAP Arctic Climate Change Update, 2021). This warming signature is also present in the adjacent seas, mainly in the Nordic Seas, which are the main gateway to the AO (Timmermans et al., 2018; Mork et al., 2019). Climate models predict a continuation of this warming trend in the AO, even in a low emission scenario (IPCC AR6, 2021). Accelerated glacier and ice-sheet melting (Shepherd et al., 2012; Gardner et al., 2013; Hofer et al., 2020) can significantly influence the sea level of the AO, which is becoming more accessible for exploration, shipping, and fishing. The majority of the climate models predict a "near-ice-free Arctic," also known as the "Blue-Ocean," in the near future.

## 28.2 Remote Sensing of Arctic Ocean Sea Level Changes

Determination of sea-surface height in the AO is challenging compared to other global oceans. Seasonal sea-ice cover makes *in-situ* measurement coverage poor in the AO, compared to other regional domains. Tide gage observations since the 1950s from the Norwegian and Russian coasts help partly in this aspect but they are spatially limited and do not provide an overall picture of sea level changes in the whole AO. Nonetheless, consistent with the rise in global mean sea level, most of these station records show significant positive

*Advances in Remote Sensing Technology and the Three Poles*, First Edition. Edited by Manish Pandey, Prem C. Pandey, Yogesh Ray, Aman Arora, Shridhar D. Jawak, and Uma K. Shukla.

trends (Proshutinsky et al., 2001, 2004). During the early 1990s, estimation of sea level through satellite altimetry began and since then it has improved the knowledge on spatial characteristics of AO sea level significantly. Observation-based studies on the pan-Arctic sea level have been made possible with satellite altimetry, and especially with the Cryosat-2 mission (2010) that provided a latitudinal coverage until 88°N. Notably, the latitudinal extents of all other satellite missions leave a larger polar gap (Table 28.1).

Sea level estimates from over the sea-ice covered regions is determined through detection of leads and water between the sea-ice floes by the altimeter. Challenges associated with the determination of sea level in the AO are explained in detail by Rose et al. (2019). Technological solutions are continuously needed and pursued to overcome the challenges and to enhance the spatial resolution of the altimetric signal, which can further enable the solution of the mesoscale dynamics, the role of which on the Arctic sea levels has not yet been studied. The improvement of satellite sensors (e.g., Delay-Doppler altimetry) and algorithms to retrieve sea level signals from remote sensing allow for obtaining sea level observations both in the open ocean and near the coast at the same accuracy (Passaro et al., 2014). Enhanced altimetry, based on the re-tracking of the altimetric waveforms to avoid echoes from bright targets (Gomez-Enri et al., 2010), has proved to accurately detect sea level and sea state in coastal regions (Benveniste et al., 2020, Passaro et al., 2021a). These dedicated developments, supported by several research projects during the last few years (e.g., ESA Climate Change Initiatives, Legeais et al., 2018; Horwath et al., 2022), allowed classification of radar altimetry signals at high latitudes, expanding our capabilities of observing sea level in sea-ice covered areas (Passaro et al., 2021b). In the latest completed ESA initiative in 2020, the ESA. Sea Level Budget Closure project is a cross-Essential Climate Variable (ECV) project focused on the closure of the sea level budget for the global ocean including the AO. As part of this initiative, sea level estimates in the AO are obtained using a physical retracker (ALES+). From the leads in the sea-ice covered regions it allows retrieval of specular waveforms. The reprocessing of the data is done using dedicated Arctic retracking, updated geophysical corrections, and improved leads/open ocean discrimination (Rose et al., 2019). However, there still exist major challenges (Raj et al., 2020), which include the retrieval of waveform data during summer in the presence of surface melt layers over sea ice, melt ponds, and the impact of waves (sea-state bias) on sea level estimates and associated retracking challenges in the marginal ice zones. An ongoing ESA funded project (Cryo-Tempo) is expected to address these still existing challenges.

Radar interferometry will be used in the near future for wide-swath measurements of sea-surface elevation. This will take the existing capability of nadir altimeters beyond and provide two-dimensional (2D) sea surface mapping at a very high-resolution. The first wide-swath altimetry

**Table 28.1** List of satellite altimeter missions launched to date.

| Satellite | Launch Year | Latitude Coverage (Deg.) |
|---|---|---|
| GEOSAT (NASA) | 1985 | 72 |
| ERS 1 (ESA) | 1991 | 81.5 |
| TOPEX/POSEIDON (NASA/CNES) | 1992 | 66 |
| ERS 2 (ESA) | 1995 | 81.5 |
| GFO (US Navy/NOAA) | 1998 | 72 |
| JASON 1 (CMES/NASA) | 2001 | 66 |
| ENVISAT (ESA) | 2002 | 81.5 |
| JASON 2 (CNES/NASA/NOAA/Eumetsat) | 2008 | 66 |
| CRYOSAT 2 (ESA) | 2010 | 88 |
| HY-2 (CNSA) | 2010 | 80.7 |
| SARAL (ISRO/NASA) | 2013 | 88 |
| SENTINEL 3 (ESA) | 2016 | 81.5 |
| JASON 3(CNES/NASA/NOAA/Eumetsat) | 2016 | 66 |
| SENTINEL 6 (ESA/NASA/NOAA/Eumetsat) | 2020 | 66 |

mission NASA/CNES Surface Water and Ocean Topography mission (SWOT; Durand et al., 2010; Fu et al., 2012) will be launched in 2022. Recent studies, based on synthetic observations, have proven the high potential of wide-swath altimetry observations at the regional scales (Bonaduce et al., 2018; D'Addezio et al., 2019, Souopgui et al., 2020), in the global ocean and at the high latitudes (Benkiran et al., 2021; Tchonang et al., 2021).

Along with the satellite altimeter missions, the Gravity Recovery and Climate Experiment (GRACE) has also played a major role in estimating the ocean mass change contribution to sea level. Since its launch in 2002, the GRACE mission has provided improved understanding of the Earth's gravity field (Tapley et al., 2004; Raj, 2017; Prandy et al. 2012). The minute change in the distance between the pair of satellites is measured and translated into changes in the Earth's gravitational pull. This gravitational anomaly information is further used to estimate ocean mass change. Generation of high-resolution data covering the coastal regions from the GRACE mission using the new Global tailored-kernel solutions (Groh and Horwath, 2016) is under progress.

These remotely-sensed satellite data products provide valuable sources for validation of general circulation and ocean models, which are often helpful to understand the forcing mechanisms behind these observed variabilities in sea level changes. Modeling of sea level changes requires mainly taking into consideration physical processes contributing to changes in the existing ocean volume due to temperature, salinity (steric changes), and changes in ocean mass by glaciers and ice sheets. A major source of uncertainty in the model performances, particularly in the AO, comes from the poor representation of the Arctic hydrography (Shu et al., 2019). The improper representation of the Arctic and Atlantic water layers in the models can lead to biased oceanic heat mixing in the water column and in turn, affect sea level estimates. Furthermore, the strong halosteric component of AO sea level changes requires proper representations of the freshwater budget in the models, which again is highly sensitive to model resolution (Wang et al., 2018) and complex interactions between different processes (Lique et al., 2016). However, by continuously increasing observational networks and improving model performances, significant progress has been achieved in understanding sea level variability in the AO.

## 28.3 Results and Discussion

### 28.3.1 Observed Trend and Variability

The AO sea level change for 66N-82N shows a positive trend of 2.2 ± 1.1 mm/yr during 1993–2015 (Andersen and Piccioni, 2016). The spatial distribution of the trend in annual mean sea level anomaly reveals that the Nordic Seas and the Beaufort Sea exhibit large positive trends, while negative trends are observed in the Russian shelf region (Figure 28.1). However, it should be noted that due to strong interannual and seasonal variability, these trends may be significantly different when considered for a different time period. It has been found that seasonal changes in freshwater fluxes (Solomon et al., 2021) can induce seasonality in sea-surface height variability of up to ~5 cm, with a peak in autumn-early winter and minimum in spring (Armitage et al., 2016). The strong positive trends over the Nordic Seas and the Beaufort Sea are not uniform in time due to strong interannual variability. Temporal evolutions of the sea level anomalies averaged over these regions depict a contrasting behavior in both cases for the periods 2003–2009 and 2010–2016 (Figure 28.2, Raj et al. 2020). The Beaufort gyre in the Beaufort Sea, with a mean anticyclonic circulation, accumulates the freshwater within it, whereas the Nordic Seas are largely influenced by the advection of warm and saline Atlantic Water. This explains the key control of halosteric (thermosteric) component on the steric sea level variability in Beaufort Sea (Nordic Seas) (Raj et al., 2020). Interannual variability of sea level in both the regions thus can be potentially influenced by the steric sea level changes. During 2003–2009 and 2010–2016, the increasing trend of sea level anomaly in the Beaufort gyre (Nordic Seas) region is mostly due to increasing halosteric (thermosteric) sea level change. The variable trends of halosteric sea level changes in the Beaufort sea during 2003–2009 and 2010–2016 can be explained by the interannual variability of dominant large-scale atmospheric circulation patterns, which controls the strength of Beaufort gyre circulation and thus freshwater accumulation within it (Giles et al., 2012; Raj et al., 2020). The thermosteric sea level changes in the Nordic Seas are mostly affected by volume and heat transported by Atlantic Water.

### 28.3.2 Arctic Ocean Sea Level and Large-Scale Atmospheric and Ocean Circulation

The robustness of large-scale atmospheric circulation variability-induced basin-wide sea level changes in the AO can be further realized with coupled sea ice–ocean models that help in the reconstruction of sea-surface height records beyond the satellite era. The long-term interannual variability of the annual mean AO sea-surface height is strongly influenced by the dominant large-scale atmospheric circulation patterns and associated changes in freshwater distribution within the AO basin (Xiao et al., 2020). The dominant atmospheric circulation patterns in the Arctic are the Arctic Oscillation and Arctic Dipole (Wu et al., 2006). While the Arctic Oscillation features as a sea level pressure (SLP) anomaly centered in the central Arctic, the Arctic Dipole is an opposite phase

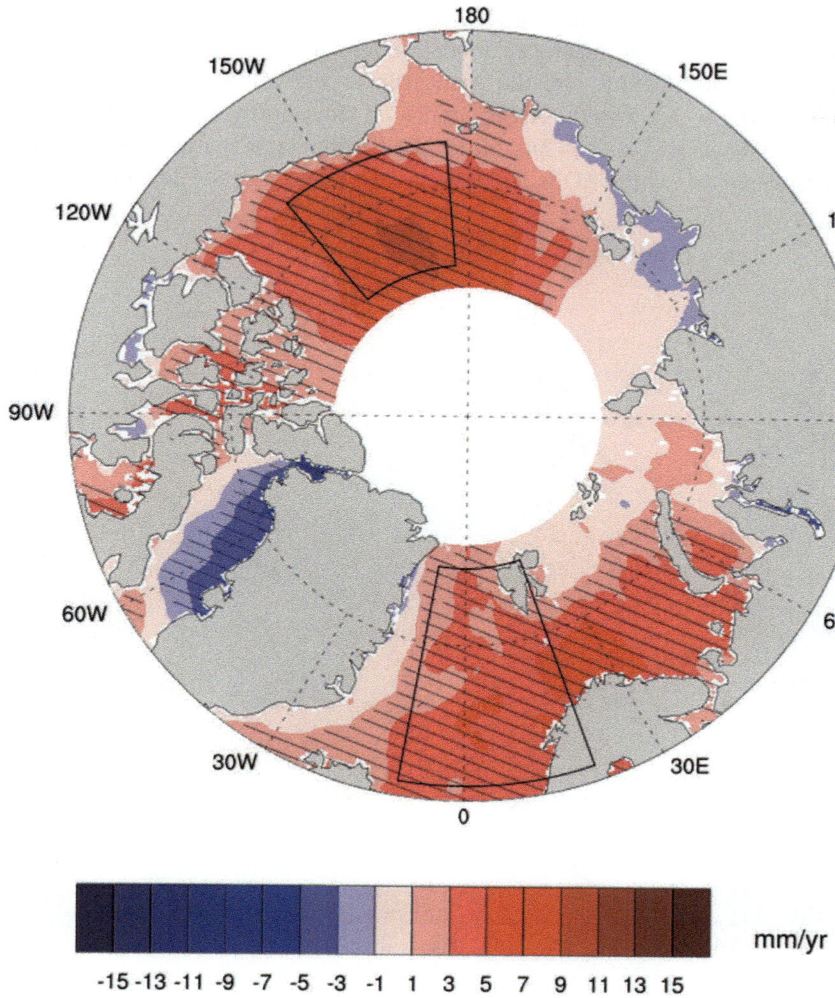


**Figure 28.1** Trend in annual mean sea level anomaly (mm/yr) during 1996–2016. The marked regions indicate Beaufort Sea and Nordic Seas (data: DTU).

distribution of SLP anomalies between the east and west Arctic. Arctic Oscillation-induced SLP anomalies and changes in wind-driven transport alters circulation and spatial distribution of freshwater in the central AO basin. As a result, the deep basin and shelf region of the AO exhibit an antiphase distribution of sea-surface height (Xiao et al., 2020). The contrasting shelf-deep basin sea-surface height pattern emerges as the first mode of variability in the annual mean sea surface height (see figure 4 in Xiao et al., 2020). Furthermore, the Arctic Dipole-associated atmospheric circulation pattern alters the cross-Arctic transport of freshwater and results in dipolar sea level spatial distribution between the Amerasian and Eurasian basins. The rapid increase in sea level in the Beaufort Sea during 2003–2009 (Figure 28.2) can be due to a dominant positive Arctic Dipole pattern (Figure 28.3a), which strengthens the Beaufort gyre

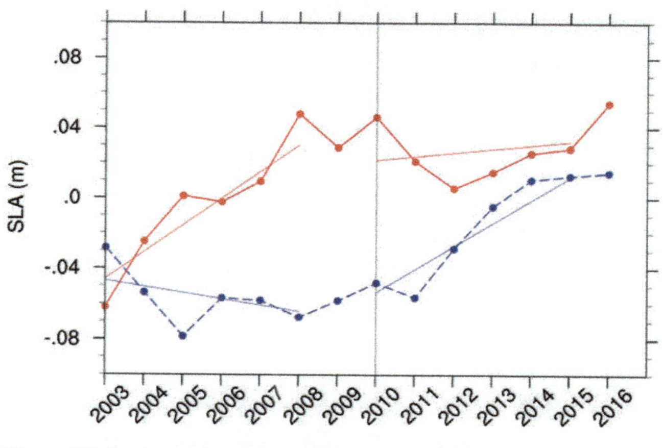


**Figure 28.2** Annual mean sea level anomaly (m) for Nordic Seas (66–80N, –10:20E; blue) and Beaufort Sea (72–80N, 140: –175W; red).

circulation and thus accumulated large amounts of freshwater in this region (Raj et al., 2020). The slowdown of this increasing trend during 2010–2016 (Figure 28.2) was due to the dominance of a positive Arctic Oscillation-like atmospheric circulation pattern (Figure 28.3b), which weakened the anticyclonic Beaufort gyre and thus sea level changes induced by freshwater accumulation (Raj et al., 2020). This can be further realized through co-variability of Beaufort Sea SLP with the Arctic Dipole index during 2003–2009 and with the Arctic Oscillation index during 2010–2016 (Figure 28.3c). Thus, the spatial distribution of sea-surface height in the AO is largely governed by the large-scale atmospheric

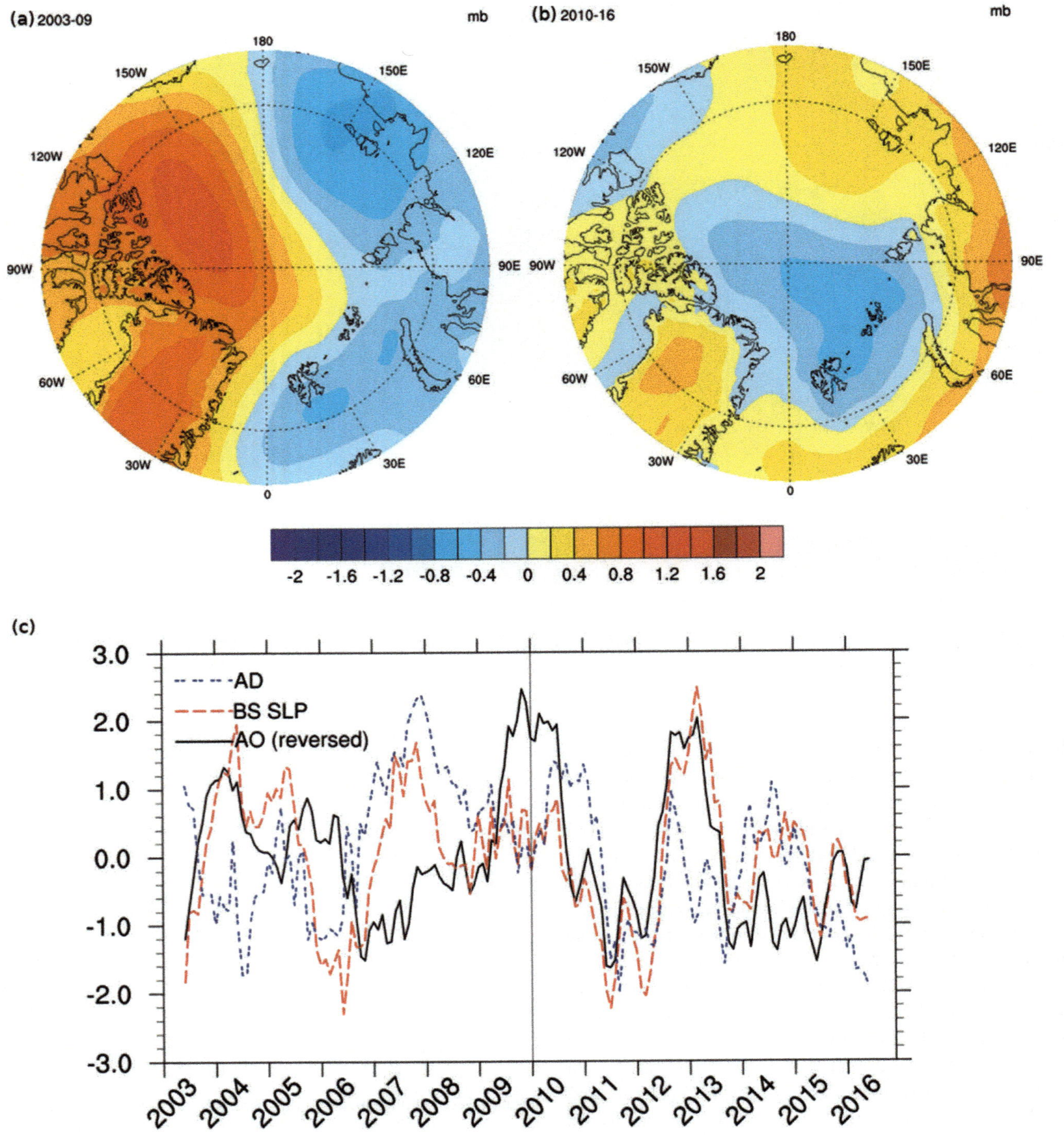


**Figure 28.3** Monthly mean sea level pressure anomaly (mb) averaged for (a) 2003–09 and (b) 2010–2016 (Raj et al., 2020). (c) Monthly indices of Arctic Oscillation (black) and Arctic Dipole (blue) pattern (1st and 2nd Principal Component of SLP anomaly north of 70N) smoothed by 11-month running mean and SLP averaged in Beaufort Sea (72–80N, 140–175W, in red).

circulation patterns. For details readers are advised to refer Raj et al., (2020).

The amount of heat that ends up reaching the interior AO largely depends on the two-branched flow of Atlantic water (AW) in the Nordic Seas. The Norwegian Atlantic Slope Current (NwASC) carries the warm and saline AW and bifurcates into the Fram Strait (FS) branch and Barents Sea branch near the Barent Sea Opening (BSO), north of the Norwegian coast. The FS branch retains most of its heat and takes it to the interior AO, circulating anticlockwise along the continental Arctic slope, whereas a large amount of the heat from the Barents Sea branch is lost to the atmosphere over the shallow Barents Sea basin (Skagseth et al., 2020; Skagseth and Mork, 2012). Thus, the amount of heat that goes into the AO is largely determined by the relative strength of these two branches. Wind-induced Ekman transport and resulting sea-height anomalies in the northern Barents Sea play a key role in determining the relative strength of these branches (Lien et al., 2013). A low SLP anomaly and associated Ekman transport induces a negative sea-surface height anomaly in the northern Barents Sea. This, in turn, creates a slope of sea-surface height gradient around the northern Barents Sea and strengthens (weakens) the AW inflow with the Barents Sea (Fram Strait) branch (Figure 28.4). The strength of this Barents Sea AW inflow largely determines the Barents Sea climate, which is one of the regions in the AO with the largest warming trend.

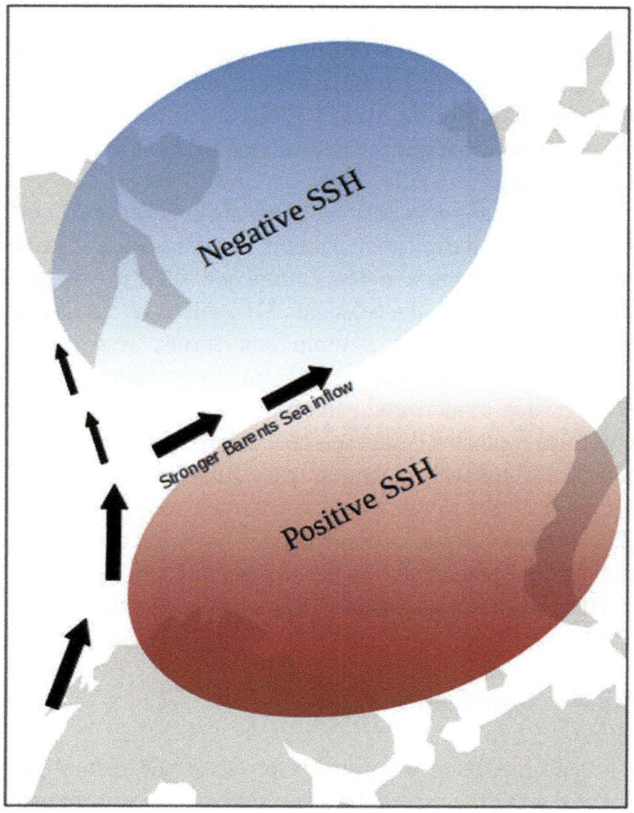


**Figure 28.4** Schematic representation of sea level influence on Barents Sea inflow as suggested by Lien et al. (2013).

Over the central AO, the surface-ocean circulation plays a key role in the distribution of sea ice and freshwater not only in the interior basin but also the export of those out of the AO basin in the Nordic Seas. This in turn can determine the strength of deep-water formation and potentially the thermohaline circulation. To better understand the impact of changes in AO circulation, Proshutinsky and Johnson (1997) defined an index, i.e., Arctic Ocean Oscillation (AOO), based on annual mean sea surface height over the AO basin. Subsequently, AOO has been found to be a robust measure for the variability in the Arctic environment (Polyakov et al., 1999; Proshutinsky et al., 1999; Overland, 2009; Krishfield et al., 2014; Rabe et al., 2014). Based on the sign of the AOO index, two distinct regimes of ocean circulation were proposed. During years with a positive AOO index, characterized as the anti-cyclonic circulation regime, the Arctic experiences a high sea level pressure (SLP) anomaly and associated wind-driven anti-cyclonic flow of sea ice accumulates freshwater in the Beaufort Gyre and Canadian basin. On the other hand, during the cyclonic circulation regime, the low SLP anomaly in the central Arctic weakens the anticyclonic Beaufort Gyre and thus higher amounts of sea ice and freshwater are released into the Nordic Seas, resulting in a strong salinity anomaly in the region (Dickson et al., 1988).

### 28.3.3 Arctic Ocean Sea Level in CMIP6

Coupled Model Intercomparison Project (CMIP) is an initiative of the World Climate Research Programme (WCRP), that has coordinated climate model experiments from multiple modeling teams worldwide. Since its launch in 1995, the CMIP initiative has been developed in six phases and continues to provide improved understanding of the past, present, and future climate change in a multi-model framework. Here we use 11 different CMIP6 climate models (Table 28.2) for assessing the sea level variability in the AO during 1993–2014. Analyzing the individual models, Figure 28.5 shows differences among models in representing the main circulation features of the AO and its adjacent seas. ACCESS-CM2, developed by CSIRO, captures the Beaufort Sea Gyre (BSG) and the cyclonic Nordic Seas (NS) circulation, but not the full spatial feature of the Subpolar Gyre (SPG) in the North Atlantic. Notably the ACCESS-ESM1-5 does resolve the SPG as well. CMCC-CM2-HR4, with the highest resolution among the selected models (Table 28.2), reveals finer details of the mean SSH pattern in the NS and the SPG. In particular, all three major ocean basins (the Norwegian Basin, the Lofoten Basin, the Greenland Basin) in the Nordic Seas are well represented. However, the high

**Table 28.2** List of CMIP6 models used here.

| Model | Institution | Nominal Resolution (The Resolution of the Grid on Which Data Are Reported) |
|---|---|---|
| ACCESS-CM2 | CSIRO, Australia | 250 km |
| ACCESS-ESM | CSIRO, Australia | 250 km |
| CMCC-CM2-HR4 | Euro-Mediterranean Center on Climate Change, Italy | 25 km |
| CMCC-CM2-SR5 | Euro-Mediterranean Center on Climate Change, Italy | 100 km |
| EC-Earth3-Veg-LR | EC-Earth-Consortium | 100 km |
| MPI-ESM-1-2-HAM | Max Planck Institute for Meteorology, Germany | 250 km |
| MPI-ESM-1-2-HR | Max Planck Institute for Meteorology, Germany | 50 km |
| NorCPM1 | | 100 km |
| NorESM2-LM | Norwegian Climate Consortium (NCC), Norway | 100 km |
| NorESM2-MM | Norwegian Climate Consortium (NCC), Norway | 100 km |
| IPSL-CM6A-LR | IPSL, France | 100 km |

mean sea level in the BSG is not found. Notably the lower-resolution model developed by the same institute captures the high mean sea level of the BSG. It also represents the mean sea level of the Nordic Seas and the SPG but with comparatively less details. In an opposite behavior, the high-resolution model (MPI-ESM-1-2-HR) of the Max Planck Institute captures all the mean SSH features of the AO and its adjacent seas in comparison to the lower resolution model (MPI-ESM-1-2-HAM), mainly the BSG. The EC-Earth Consortium model also captures the main features, but not the details of the Nordic Seas sea level, where the model does not capture the low mean sea level in the Lofoten and Norwegian Basin. The Norwegian Climate Prediction model (NorCPM), even though it represents the mean low sea level in the SPG, falls short of representing the mean sea level in the Nordic Sea and BSG. The Norwegian Earth System Model (NorESM), both the low- and medium-resolution, represents the Nordic Seas and BSG; however, the sea level of the Nordic Seas is underestimated. In comparison, the IPSL model better represents all three main features of the AO. These similarities or discrepencies among the models may arise from the representation of hydrography, dynamics, freshwater input etc., which may vary across the models and thus needs to be investigated in further details.

The spatial distribution of annual-mean sea-surface height (SSH) in the ensemble mean is similar to the observed pattern and captures major circulation features (Figure 28.6). The spatial pattern of annual mean SSH clearly marks the well-known dipole circulation pattern with cyclonic circulation over the Nordic Seas and Russian sector of the AO (negative SSH values) and anticyclonic circulation over the Canadian basin with maximum SSH over the Beaufort Gyre region due to Ekman convergence (positive SSH values). The cyclonic circulation over the Nordic Seas features two major Gyre circulations, i.e., Greenland Sea gyre and Lofoten Basin gyre (Hansen and Nansen, 1909). The strength of these gyre circulations and associated changes in Atlantic Water inflow modifies the sea-ice concentration in the Greenland Sea (Chatterjee et al., 2021) and Atlantic water temperature at the Fram Strait (Chatterjee et al., 2018). The Atlantic water, after entering the central AO basin through the Fram Strait, continues to move cyclonically around the AO basin and reaches the Eurasian basin, where it can cause a significant impact on the sea-ice formation (Polyakov et al., 2017). Over the Nordic Seas, the higher negative values in ensemble mean (Figure 28.6a), compared to observation (Figure 28.6b), indicates somewhat weaker gyre circulations (convective mixing) in the ensemble mean compared to observation. However, a more detailed analysis of the AO dynamics and its impact on sea-surface height is required in CMIP6 models to confirm these features.

Seasonal changes in SSH in the AO are largely determined by SLP and wind circulation patterns, river run-offs, and steric effects due to changes in temperature and salinity. Averaged over the whole AO, the seasonal SSH exhibits its maximum and minimum during autumn and spring, respectively. However, in the coastal areas, SSH can also exhibit a peak in summer due to increased run-offs. In both observation and ensemble mean, the largest seasonal difference is found in the Siberian sector with maximum river run-offs (Figure 28.7). However, in the Nordic Seas, the ensemble mean shows less difference compared to observation. This could result from differences in the models in the representation of Nordic Sea circulation and/or its response to atmospheric forcing.

## 28.4 Conclusions

The AO has been warming faster than the rest of the globe during the recent decades and based on future projections, it is expected that the warming will continue, even in a low emission scenario. The cumulative effect of these changes can cause significant sea level changes in

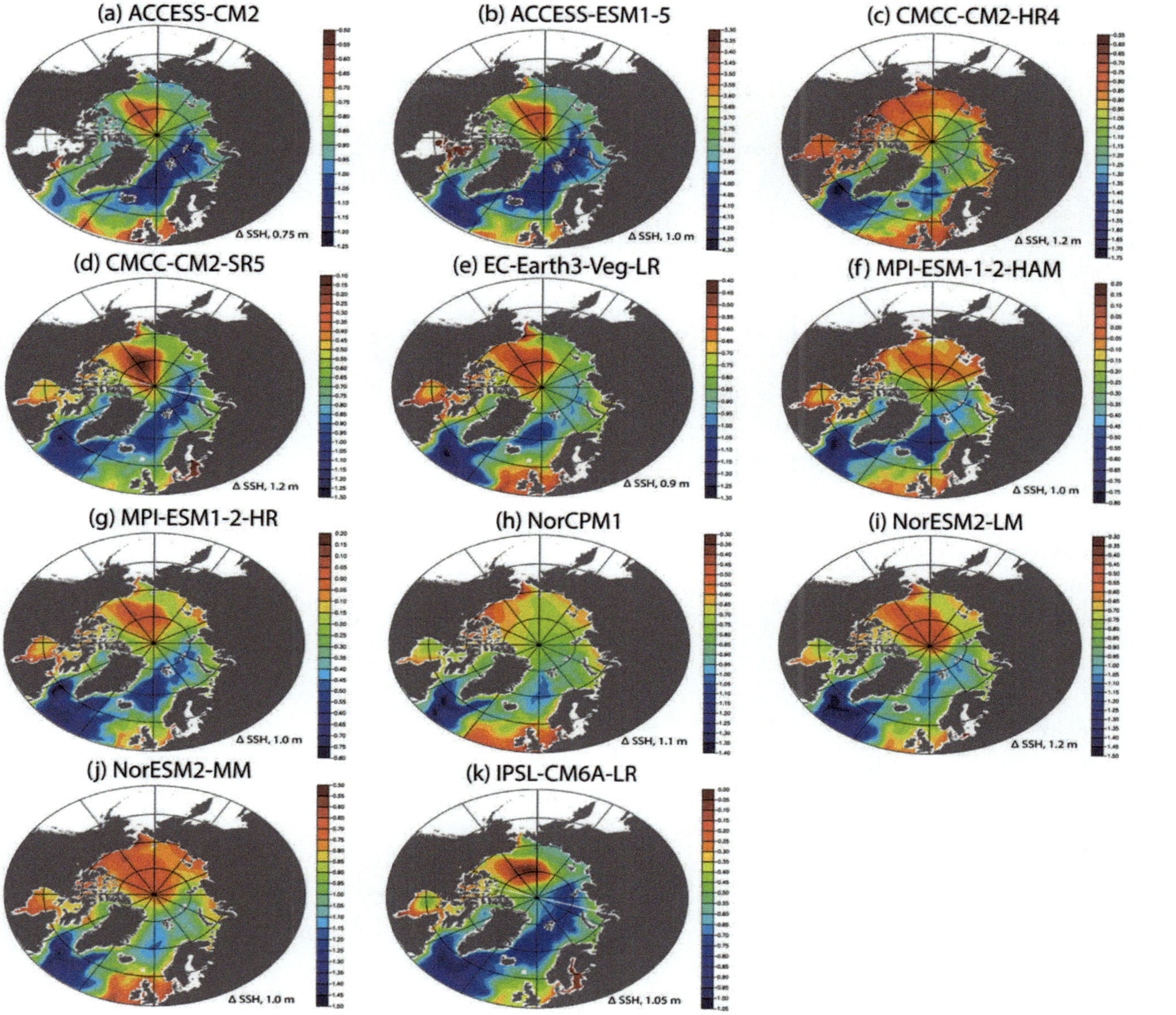


**Figure 28.5** Mean (1993–2014) sea-surface height (m) of 11 CMIP6 models.

AO, the monitoring of which needs to be prioritized. Besides the many challenges in retrieving accurate sea level data, satellite altimetry provides the best opportunity to study the pan-Arctic sea level. Ongoing research projects targeting accurate retrieval of sea level estimates in the AO, together with the innovative forthcoming next-generation altimeter missions, shall provide further information on Arctic sea level variability. It is now well established that the atmospheric variability significantly influences the sea level in the AO. Here, altimeter data is used to validate the performance of 11 CMIP6 model-derived sea level variabilities in the Arctic for the time period 1993–2014. Individual CMIP6 models are found to show differences in representing the mean circulation of the AO. However, the spatial distribution annual-mean sea-surface height in the AO estimated from all ensembles mean of 11 models successfully captures major circulation features as derived from satellite data. Averaged over the whole AO, the seasonal SSH is maximum during autumn and minimum in spring. In both observation and ensemble mean, the largest seasonal difference is found in the Siberian sector. Further understanding of the ocean and atmospheric dynamics affecting the sea level in the AO and improved representation of those in the climate models, accompanied with more accurate sea level estimates from forthcoming altimeter missions, are expected to improve our knowledge on AO sea level variability and its future course.

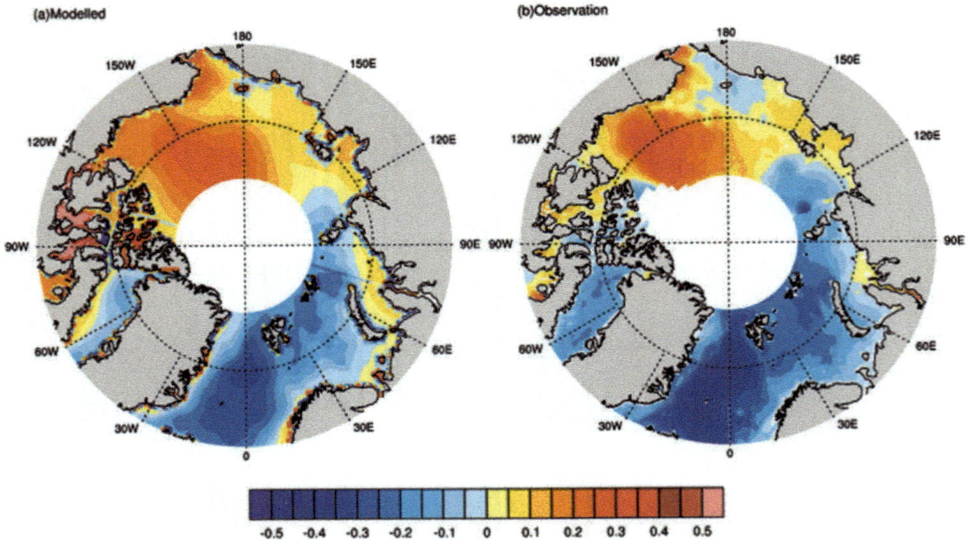


**Figure 28.6** Annual mean sea-surface height (m), averaged for 1993–2014 (a) model ensemble mean (b) Observation. Area averaged values are removed in each case for comparison.

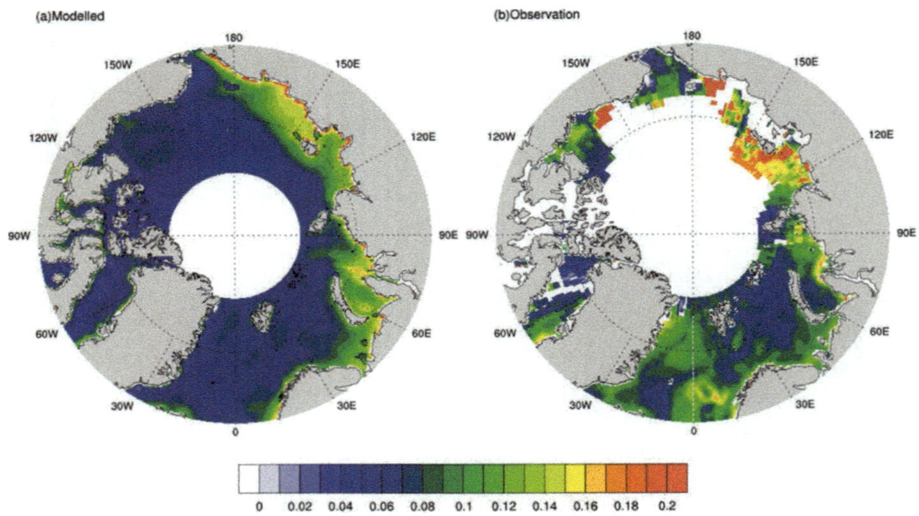


**Figure 28.7** Difference in sea surface height climatology (1993–2014) between September to October and March to April. (a) Model ensemble mean; and (b) Observation.

## Acknowledgments

Author SC would like to acknowledge NCPOR, fully funded by Ministry of Earth Science, Govt. of India. Authors AB and RPR are supported by Sea Level Predictions and Reconstructions (SeaPR) project funded by the Bjerknes Center for Climate Research (BCCR) initiative for strategic projects. Authors also acknowledge the European Space Agency's DRAGON 5 project.

## References

AMAP (2021). Arctic Climate Change Update 2021: key trends and impacts: summary for policy-makers. Arctic Monitoring and Assessment Programme (AMAP), Tromsø, Norway. 16.

Andersen, O.B. and Piccioni, G. (2016). Recent Arctic sea-level variations from satellites. *Front. Mar. Sci.* 3: 76.

Armitage, T.W.K., Bacon, S., Ridout, A.L. et al. (2016). Arctic sea surface height variability and change from satellite radar altimetry and GRACE, 2003–2014. *J. Geophys. Res. Oceans* 121: 4303–4322. https://doi.org/10.1002/2015JC011579.

Armitage, T.W.K., Manucharyan, G.E., Petty, A.A. et al. (2020). Enhanced eddy activity in the Beaufort Gyre in response to sea ice loss. *Nat. Commun.* 11: 761. https://doi.org/10.1038/s41467-020-14449-z.

Benkiran, M., Ruggiero, G., Greiner, E. et al. (2021). Assessing the impact of the assimilation of SWOT observations in a global high-resolution analysis and forecasting system. Part 1: Methods. *Front. Mar. Sci.* 8: 691955. https://doi.org/10.3389/fmars.2021.691955.

Benveniste, J., Birol, F., Calafat, F. et al. (2020). Coastal sea-level anomalies and associated trends from Jason satellite altimetry over 2002–2018. *Sci. Data* 7: 357. https://doi.org/10.1038/s41597-020-00694-w.

Bonaduce, A., Benkiran, M., Remy, E. et al. (2018). Contribution of future wide-swath altimetry missions to ocean analysis and forecasting. *Ocean Sci.* 14: 1405–1421. https://doi.org/10.5194/os-14-1405-2018.

Chatterjee, S., Raj, R.P., Bertino, L. et al. (2018). Role of Greenland Sea gyre circulation on Atlantic Water temperature variability in the Fram Strait. *Geophysical Research Letters* 45: 8399–8406. https://doi.org/10.1029/2018GL079174.

Chatterjee, S., Raj, R.P., Bertino, L. (2021). Combined influence of oceanic and atmospheric circulations on Greenland sea ice concentration. *The Cryosphere* 15: 1307–1319. https://doi.org/10.5194/tc-15-1307-2021.

Church, J., Clark, P., Cazenave, A. et al. (2013). Sea-level change. In: *Climate Change 2013: The Physical Science Basis. Contribution of Working Group I to The Fifth Assessment Report of the Intergovernmental Panel on Climate Change* (ed. T.F. Stocker, D. Qin, and G.-K.Plattner), 1137–1216. Cambridge, UK and New York: Cambridge University Press.

D'Addezio, J.M., Smith, S., Jacobs, G.A. et al. (2019). Quantifying wavelengths constrained by simulated SWOT observations in a submesoscale resolving ocean analysis/forecasting system. *Ocean Modelling* 135: 40–55. https://doi.org/10.1016/j.ocemod.2019.02.001.

Dickson, R.R., Meincke, J., Malmberg, S.-A. et al. (1988). The "Great Salinity Anomaly" in the northern North Atlantic 1968–82. *Prog. Oceanogr.* 20: 103–151.

Durand, M., Fu, -L.-L., Lettenmaier, D.P. et al. (2010). The surface water and ocean topography mission: observing terrestrial surface water and oceanic submesoscale eddies. *Proc. IEEE* 98: 766–779.

Fu, L.-L., Alsdorf, D., Morrow, R. et al. (eds) (2012). *SWOT: The Surface Water and Ocean Topography Mission; Wide-Swath Altimetric Measurement of Water Elevation on Earth*. Pasadene, CA: Jet Propulsion Laboratory, National Aeronautics and Space Administration. 12-05: 228.

Gardner, A S. (2013). A reconciled estimate of glacier contributions to sea-level rise: 2003 to 2009. *Science* 340(6134): 852–857. https://doi.org/10.1126/science.1234532.

Giles, K., Laxon, S., Ridout, A. et al. (2012). Western Arctic Ocean freshwater storage increased by wind-driven spin-up of the Beaufort Gyre. *Nat. Geosci.* 5: 194–197. https://doi.org/10.1038/ngeo1379.

Gomez-Enri, J., Vignudelli, S., Quartly, G.D. et al. (2010). Modeling Envisat RA-2 waveforms in the coastal zone: case study of calm water contamination. *IEEE Geosci. Remote Sens. Lett.* 7: 474–478. https://doi.org/10.1109/LGRS.2009.2039193.

Groh, A. and Horwath, M. (2016). The method of tailored sensitivity kernels for GRACE mass change estimates. *Geophysical Research Abstracts* 18: EGU2016-12065, EGU General Assembly.

Helland-Hansen and Nansen (1909). The Norwegian Sea: its physical oceanography based upon the Norwegian reseaches 1900–1904. *Report on Norwegian Fishery and Marine Investigations* 2: 39.

Hofer, S., Lang, C., Amory, C. et al. (2020). Greater Greenland Ice Sheet contribution to global sea-level rise in CMIP6. *Nat. Commun.* 11: 6289. https://doi.org/10.1038/s41467-020-20011-8.

Horwath, M., Gutknecht, B.D., Cazenave, A. et al. (2022). Global sea-level budget and ocean-mass budget, with a focus on advanced data products and uncertainty characterisation. *Earth Syst. Sci. Data* 14: 411–447, https://doi.org/10.5194/essd-14-411-2022.

IPCC AR6 (2021). Climate change 2021. In: *The Physical Science Basis. Contribution of Working Group I to the Sixth Assessment Report of the Intergovernmental Panel on Climate Change* (ed. V. Masson-Delmotte, P. Zhai, S.L. Pirani, et al.). Cambridge, UK and New York: Cambridge University Press.

Krishfield, R.A., Proshutinsky, A., Tateyama, K. et al. (2013).Deterioration of perennial sea ice in the Beaufort Gyre from 2003 to 2012 and its impact on the oceanic freshwater cycle, *J. Geophys. Res. Oceans* 119: 1271–1305. doi:10.1002/2013JC00899.

Legeais, J.-F., Ablain, M., Zawadzki, L.eta;l. et al. (2018). An improved and homogeneous altimeter sea-level record from the ESA Climate Change Initiative. *Earth Syst. Sci. Data* 10: 281–301. https://doi.org/10.5194/essd-10-281-2018.

Lien, V., Vikebø, F., and Skagseth, Ø. (2013). One mechanism contributing to co-variability of the Atlantic inflow branches to the Arctic. *Nat Commun* 4: 1488. https://doi.org/10.1038/ncomms2505, 2013

Lique, C., Holland, M.M., Dibike, Y.B. et al. (2016). Modeling the Arctic freshwater system and its integration in the global system: lessons learned and future challenges. *J. Geophys. Res. Biogeosci.* 121: 540–566. https://doi.org/10.1002/2015JG003120.

Mork, K.A., Skagseth, Ø., and Søiland, H. (2019). Recent warming and freshening of the Norwegian Sea observed by Argo data. *J. Clim.* 32(12): 3695–3705. (Accessed 6 September 2021.) https://doi.org/10.1175/JCLI-D-18-0591.1.

Munk, W. (2003). Ocean freshening, sea-level rising. *Science* 300: 2041–2043. https://doi.org/10.1126/science.1085534.

Oppenheimer, M., Glavovic, B., Hinkel, J. et al. (2019). Sea-level rise and implications for low-lying islands, coasts and communities. In: *IPCC Special Report on the Ocean and Cryosphere in a Changing Climate* (ed. H.-O. Pörtner, V. Roberts, P. Masson-Delmotte, et al.), 321–445. Cambridge, UK and New York: Cambridge University Press. https://doi.org/10.1017/9781009157964.006.

Overland, J.E. (2009). Meteorology of the Beaufort Sea. *J. Geophys. Res.* 114: C00A07. doi:10.1029/2008JC004861.

Passaro, M., Cipollini, P., Vignudelli, S. et al. (2014). ALES: a multi-mission adaptive subwave form retracker for coastal and open ocean altimetry. *Remote Sens. Environ.* 145: 173–189. https://doi.org/10.1016/j.rse.2014.02.008.

Passaro, M. (2021a). Recent advances in coastal altimetry and implications for sea level monitoring closer to the coast. *Ocean Decade Laboratories, Laboratory 2: "A Predicted Ocean" satellite activity, designing observing systems for ocean boundaries.*

Passaro, M. (2021b). Observing sea level and climate change at the coast and at the polar latitudes with reprocessed altimetry: a review. *1st Workshop of Inter-Commission Committee on Geodesy for Climate Research (ICCC) of the International Association of Geodesy (IAG).*

Polyakov, I., Proshutinsky. A., Johnson, M. (1999). The seasonal cycles in two regimes of Arctic climate. *J. Geophys. Res.* 104(C11): 25761–25788.

Polyakov, I.V., Pnyushkov, A.V., Alkire, M.B. et al. (2017). Greater role for Atlantic inflows on sea-ice loss in the Eurasian Basin of the Arctic Ocean. *Science* 356(6335): 285–291. https://doi.org/10.1126/science.aai8204.

Prandi, P., Ablain, M., Cazenave, A. et al. (2012). A new estimation of mean sea-level in the Arctic Ocean from satellite altimetry. *Mar. Geod.* 35: 61–81. https://doi.org/10.1080/01490419.2012.718222.

Proshutinsky, A.Y. and Johnson, M.A. (1997). Two circulation regimes of the wind driven Arctic Ocean. *J. Geophys. Res.* 102: 12493–12514.

Proshutinsky, A., Ashik, I.M., Dvorkin, E.N. et al. (2004). Secular sea-level change in the Russian sector of the Arctic Ocean. *J. Geophys. Res.* 109: C03042. https://doi.org/10.1029/2003JC002007.

Proshutinsky, A., Pavlov, V., and Bourke, R.H. (2001). Sea-level rise in the Arctic Ocean. *Geophys. Res. Lett.* 28(11): 2237–2240.

Rabe, B. (2014). Arctic Ocean basin liquid freshwater storage trend 1992–2012. *Geophys. Res. Lett.* 41: 2014.

Raj, R.P. (2017). Surface velocity estimates of the North Indian Ocean from satellite gravity and altimeter missions. *International Journal of Remote Sensing* 38(1): 2017.

Raj, R.P., Andersen, O.B., Johannessen, J.A. et al. (2020). Arctic sea-level budget assessment during the GRACE/Argo time period. *Remote Sens.* 12: 2837. https://doi.org/10.3390/rs12172837.

Rose, S.K., Andersen, O.B., Passaro, M. et al. (2019). Arctic Ocean sea-level record from the complete radar altimetry era: 1991–2018. *Remote Sens.* 11(14): 1672. https://doi.org/10.3390/rs11141672.

Screen, J.A., Deser, C., Smith, D.M. et al. (2018). Consistency and discrepancy in the atmospheric response to Arctic sea-ice loss across climate models. *Nat. Geosci.* 11: 155–163. https://doi.org/10.1038/s41561-018-0059-y.

Shepherd, A. et al. (2012). A reconciled estimate of ice-sheet mass balance. *Science* 338(6111): 1183–1189. https://doi.org/10.1126/science.1228102.

Shu, Q., Wang, Q., Su, J. et al. (2019). Assessment of the Atlantic water layer in the Arctic Ocean in CMIP5 climate models. *Clim. Dyn.* 53: 5279–5291. https://doi.org/10.1007/s00382-019-04870-6.

Skagseth, Ø. and Mork, K.A. (2012). Heat content in the Norwegian Sea, 1995–2010. *ICES J. Mar. Sci.* 69(5): 826–832. https://doi.org/10.1093/icesjms/fss026.

Skagseth, Ø., Eldevik, T., Årthun, M. et al. (2020). Reduced efficiency of the Barents Sea cooling machine. *Nat. Clim. Chang.* 10: 661–666. https://doi.org/10.1038/s41558-020-0772-6,2020

Solomon, A., Heuzé, C., Rabe, B. et al. (2021). Freshwater in the Arctic Ocean 2010–2019. *Ocean Sci.* 17: 1081–1102. https://doi.org/10.5194/os-17-1081-202.

Souopgui, I., D'Addezio, J.M., Bowley, C. et al. (2020). Multi-scale assimilation of simulated SWOT observations. *Ocean Model.* 154: 101683. doi: 10.1016/j.ocemod.2020.101683.

Stammer, D., Cazenave, A., Ponte, R.M. et al. (2013). Causes for contemporary regional sea level changes. *Annual Review of Marine Science* 5(1): 21–46.

Tapley, B.D., Bettadpur, S., Ries, J.C. et al. (2004). GRACE measurements of mass variability in the Earth system. *Science* 305(5683): 503–505. doi: 10.1126/science.1099192. PMID: 15273390, 2004

Tchonang, B.C., Benkiran, M., Le Traon, P.-Y. et al. (2021). Assessing the impact of the assimilation of SWOT observations in a global high-resolution analysis and forecasting system. Part 2: Results. *Front. Mar. Sci.* 8: 687414. https://doi.org/10.3389/fmars.2021.687414.

Timmermans, M.-L. et al. (2018). Warming of the interior Arctic Ocean linked to sea ice losses at the basin margins. *Sci. Adv.* 4(8): eaat6773. https://doi.org/10.1126/sciadv.aat6773.

Wang, Q., Wekerle, C., Danilov, S. et al. (2018). A 4.5 km resolution Arctic Ocean simulation with the global multi-resolution model FESOM 1.4. *Geosci. Model Dev.* 11: 1229–1255.

Wu, B., Wang, J., and Walsh, J.E. (2006). Dipole anomaly in the Winter Arctic atmosphere and its association with sea ice motion. *J. Clim.* 19: 210–225.

Xiao, K., Chen, M., Wang, Q. et al. (2020). Low-frequency sea-level variability and impact of recent sea ice decline on the sea-level trend in the Arctic Ocean from a high-resolution simulation. *Ocean Dyn.* 70: 787–802. https://doi.org/10.1007/s10236-020-01373-5.

Zhang, X., Sorteberg, A., Zhang, J. et al. (2008). Recent radical shifts of atmospheric circulations and rapid changes in Arctic climate system. *Geophys. Res. Lett.* 35: L22701. https://doi.org/10.1029/2008GL035607.

# 29

# Spatio-Temporal Variations of Aerosols Over the Polar Regions Based on Satellite Remote Sensing

*Rohit Srivastava**

*National Centre for Polar and Ocean Research, Ministry of Earth Sciences, Government of India, Goa, India 403804*
* *Corresponding author*

## 29.1 Introduction

Atmospheric aerosols are tiny solid particles suspended in the air medium having a size range of 0.001 to 100 μm. Aerosols are mainly produced by both natural as well as anthropogenic activities. Black carbon (soot), organic carbon (OC), sulfates, nitrates, mineral dust, and sea salt are the major aerosol species. Aerosols are mostly formed in nature by wind-blown dust, sea salt from breaking sea waves, forest fires, volcanic activities (directly emitted and formed due to gas to particle conversion), biogenic aerosol production, and natural gas to particle conversion products. e.g., sulfate aerosols formed by dimethyl sulfide emitted from the ocean surface. Vehicular, industrial exhausts, soil erosion in agriculture, open mining, and crop residue burning are the main anthropogenic activities that give rise to this particulate matter (aerosols) in the atmosphere.

Different removal processes reduce aerosols from the atmosphere. For larger aerosols, dry deposition (sedimentation) is a more effective removal procedure in which the aerosols settle down owing to gravity. The most efficient sink process for particles smaller than 1 μm is wet deposition, which is classified into three categories: rainout, washout, and sweep out. In the rainout process, aerosols act as cloud condensation nuclei and subsequently fall to the surface as raindrops. If an aerosol is already present in a cloud drop and it expands in size to fall as rain, the aerosols are washed out. The raindrop can collide with the aerosols present below the base of a rain cloud, causing the aerosols to be incorporated into the raindrop, which is subsequently swept from the atmosphere (Seinfeld and Pandis, 1998). Aerosols in the size range of 0.01 to 10 m have about a week of residence duration in the lower troposphere, which increases with altitude. In the stratosphere, the residence time of aerosols ranges from a few months to years.

Aerosols interact with the cryosphere and clouds and significantly affect the Arctic's radiation balance. The aerosol–cryosphere interaction may lead through complicated mechanisms, like the melting of sea ice. This can have an additional impact on the radiation budget by reducing surface reflectivity. The positive feedback might lead to fast ice loss. Anthropogenic aerosols, such as Black Carbon (BC), absorb radiation leading to warming. BC was reported as the second-leading contributor to global warming after $CO_2$ (Ramanathan and Carmichael, 2008). Aerosol-radiation interactions control three primary processes that can contribute to Arctic warming. The first is the absorption of incoming solar energy by aerosols (e.g., BC), which can cause tropospheric warming and surface cooling. As a result, air stratification changes, affecting the local circulation pattern (Shindell and Faluvegi, 2009). Second, lower-latitude absorbing aerosols may amplify latitudinal temperature gradients, leading to greater poleward heat transmission (Sand et al., 2013). Warming may give rise to the reduction in ice/snow albedo caused by the deposition of absorbing aerosols in the third process (Flanner et al., 2009).

Arctic amplification is a phenomenon in which the Arctic area has warmed more than double, as rapidly as the global mean (Cohen et al., 2014). Rapid Arctic warming has resulted in catastrophic loss of Arctic sea ice and spring snow cover, far faster than climate models predicted (Cohen et al., 2014). Antarctica in the South Pole is surrounded by the Southern Ocean and the air is normally pristine in the absence of any large local aerosol sources. Antarctica is

*Advances in Remote Sensing Technology and the Three Poles*, First Edition. Edited by Manish Pandey, Prem C. Pandey, Yogesh Ray, Aman Arora, Shridhar D. Jawak, and Uma K. Shukla.

mainly influenced by the long-range transport of aerosols from the other continents and sub-continents. Hansen and Nazarenko (2004) estimated a 2.5% reduction in albedo across the Arctic as a result of BC deposition over the snow. During the period 1980 to 2010, recent decreases in scattering aerosol (e.g., anthropogenic sulfate) loading in the Arctic resulted in a net surface warming of +0.27 ± 0.04 K over the Arctic region (Breider et al., 2017). The global annual mean air temperature may rise from 0.10 to 0.15°C when the snow is considered to be influenced by BC in the general circulation modeling simulations. The yearly Arctic warming, on the other hand, is stated to have risen from 0.50 and 1.61°C (Flanner et al., 2009). The variabilities of the aerosol properties over the Arctic and Antarctic atmosphere are still not understood well. In the present study, the seasonal and spatial variabilities in the aerosol optical property over the polar (Arctic and Antarctic) region have been investigated. The study presented in the chapter will be useful in improving the understanding of polar aerosol variability using satellite remote-sensing data.

## 29.2 Data and Methodology

Monthly mean Aerosol Optical Depth (AOD) at a mid-visible wavelength (0.55 μm) retrieved from MODerate resolution Imaging Spectroradiometer (MODIS) are investigated from 2002 to 2018. MODIS is installed on Terra and Aqua Earth Observing System (EOS) satellites, which orbit at 705 km altitude. The reflected solar radiation and terrestrial emission in the wavelength regime of 0.41–14.4 μm were measured using MODIS. The wavelength range is divided into 36 spectral channels comprising the visible, near, middle, and thermal infrared bands with 12-bit radiometric resolution. MODIS aerosol retrieval technique uses two well-known algorithms, depending from which surface the aerosol properties are retrieved. The Dark Target (DT) algorithm was used in retrieval of aerosol properties over land and ocean, the while Deep Blue (DB) technique was utilized for retrieval over the land. Over the darker surface (e.g., ocean), the DT technique was used in the retrieval of aerosol properties. In order to reduce the uncertainties in aerosol retrieval by the DT algorithm over a brighter surface, such as the desert regions, the DB technique was used. The latest combined DT and DB dataset in Collection-6.1 MODIS aerosol properties are developed based on the second-generation Dark Target technique (Levy et al., 2013) and enhanced DB algorithm (Hsu et al., 2013).

Heavy smoke can be detected by the Collection 6.1 DB algorithm from MODIS. The C6.1 method has decreased artifacts in diverse terrain, better land and ocean surface modeling in the high terrain; and modified spatial and temporal aerosol modeling over land and oceanic regions (Wei et al., 2019). Based on the DT and DB algorithms, MODIS has given two levels of aerosol characteristics products. The first is the 3 and 10 km horizontally resolved Level-2 atmosphere daily (swath) product and the second consists of Level-3 daily, weekly (or 8-day), and monthly aerosol properties at a 1° × 1° horizontal grid resolution. A look-up table approach was utilized in the MODIS retrieval algorithm. Surface reflectance measured by MODIS in various spectral bands is given to clear-sky pixels based on region, season, and kind of land cover. After the matchup of radiance values according to geo-location and season, the look-up table approach based on the aerosol model is used for aerosol retrievals (Remer et al., 2008).

MODIS aerosol products have been examined and contrasted with *in situ* and/or other satellite/ground-based remote-sensing data, and they have been improved regularly (Chu, 2002; Remer et al., 2008; Li et al., 2009; Wei et al., 2019). The variations in MODIS L 3 collection 6.1 mid-visible (0.55 μm) AOD from Terra are investigated over the polar regions in the northern and southern hemispheres, i.e., over the Arctic and Antarctic regions at a horizontal resolution of 1° × 1° from 2001 to 2018. Comparison of the collection 6.0 aerosol optical depth over land with Aerosol Robotic Network (AERONET)-derived AOD suggested that globally, AOD retrieved by MODIS are reported to have a good correlation with the AERONET-derived AOD measurements (Wei et al., 2019); while the aerosol properties, retrieved using combined DT and DB techniques, performed well at most of the regions. However, the DB product was claimed to be better at specific locations (Wei et al., 2019). Level 3 MODIS AODs at mid-visible wavelengths are examined for the values lying between 0.0 and 1.0, which is on the premise that AOD values larger than 1.0 are probably influenced owing to cloud contamination (Chung et al., 2005). The retrieval uncertainty in MODIS-derived AODs was claimed as ±(0.05 + 0.15 AOD) and ±(0.03 + 0.05 AOD) over the land and oceanic regions respectively (Remer et al., 2008; Li et al., 2009).

The fire count products obtained from MODIS provide the locations of fires that are burning. The fire counts products are accessible from the NASA Earth Observations website. The fire product was utilized for the years 2002–2018 at a grid resolution of 1°, which is produced from the MODIS product (MOD14A1) having a resolution of 1 km. The level 3 tile-based global fire counts are utilized from the MODIS Collection 6. Each pixel designated "fire" receives a count of the number of fires in the range of 0 to 30 counts. During each of the satellite overpasses, infrared anomalies compared to nearby pixels are used to detect them. The technique compares the signals in recognized fire pixels to those in the surrounding, non-fire pixels using brightness temperatures taken from the mid-IR (4 μm) as well as thermal IR (11 μm) wavelengths of the MODIS radiation measurements (Giglio et al., 2016). The meteorological parameters such as near-surface relative humidity (RH) and winds at

1000 and 850 hPa pressure levels are obtained from the ERA (ECMWF (European Centre for Medium-Range Weather Forecasts) Reanalysis) interim datasets from 2002 to 2018. The ERA meteorological data was generated using Integrated Forecast System (IFS) with 60-altitude levels at $0.75° \times 0.75°$ grid resolution (Dee et al., 2011).

## 29.3 Results and Discussion

### 29.3.1 Seasonal Variations of Relative Humidity (RH) Over Northern and Southern Polar Regions

#### 29.3.1.1 Arctic

The seasonal and spatial variations of RH over the Arctic region for boreal winter (December-January-February (DJF)), boreal spring (March-April-May (MAM)), boreal summer (June-July-August (JJA)) and boreal autumn (September-October-November (SON) are shown in Figure 29.1. The majority of the areas in the Arctic exhibit higher (>85%) RH values. The central Greenland regions show even higher (>95%) values of RH. During the rising air-mass, RH achieves saturation as it approaches the interior of Greenland; hence, the ice sheet's interior has greater RH values (Srivastava and Ravichandran, 2021). During boreal spring and summer seasons, the low latitude Arctic regions exhibit lower RH values. The higher RH values can lead to the growth of hydrophilic aerosols and affect the aerosol optical and radiative properties.

#### 29.3.1.2 Antarctic

The spatial variations of RH over the Antarctic region for austral summer (December-January-February (DJF)), austral autumn (March-April-May (MAM)), austral winter (June-July-August (JJA)), and austral spring (September-October-November (SON)) are shown in Figure 29.2. The Antarctic region exhibits higher (>90%) RH values, while surrounding oceanic regions show lesser RH values (<85%). In austral autumn and winter, the RH values are greater than 95% over Antarctica. Radiative

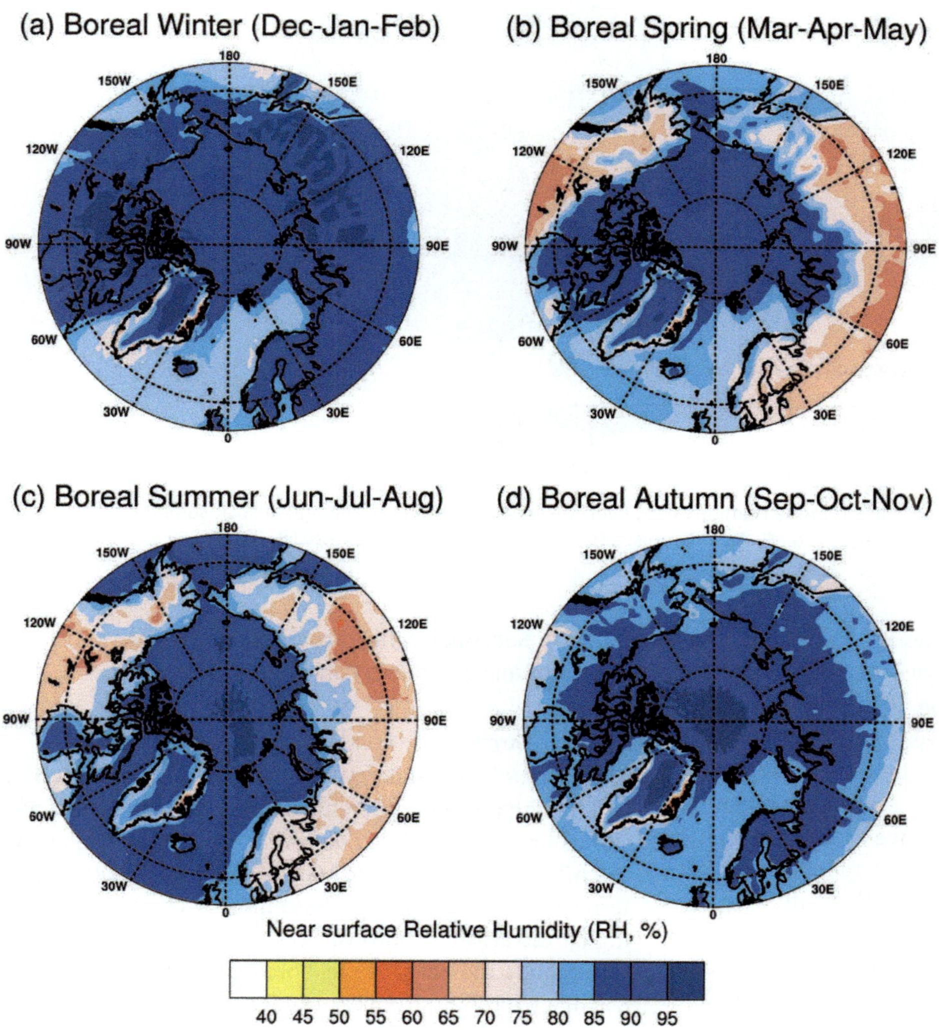


**Figure 29.1** Spatial variations of near surface relative humidity (RH) over the Arctic region during: (a) boreal winter (DJF); (b) boreal spring (MAM); (c) boreal summer (JJA); and (d) boreal Autumn (SON) seasons of years 2002–2018.

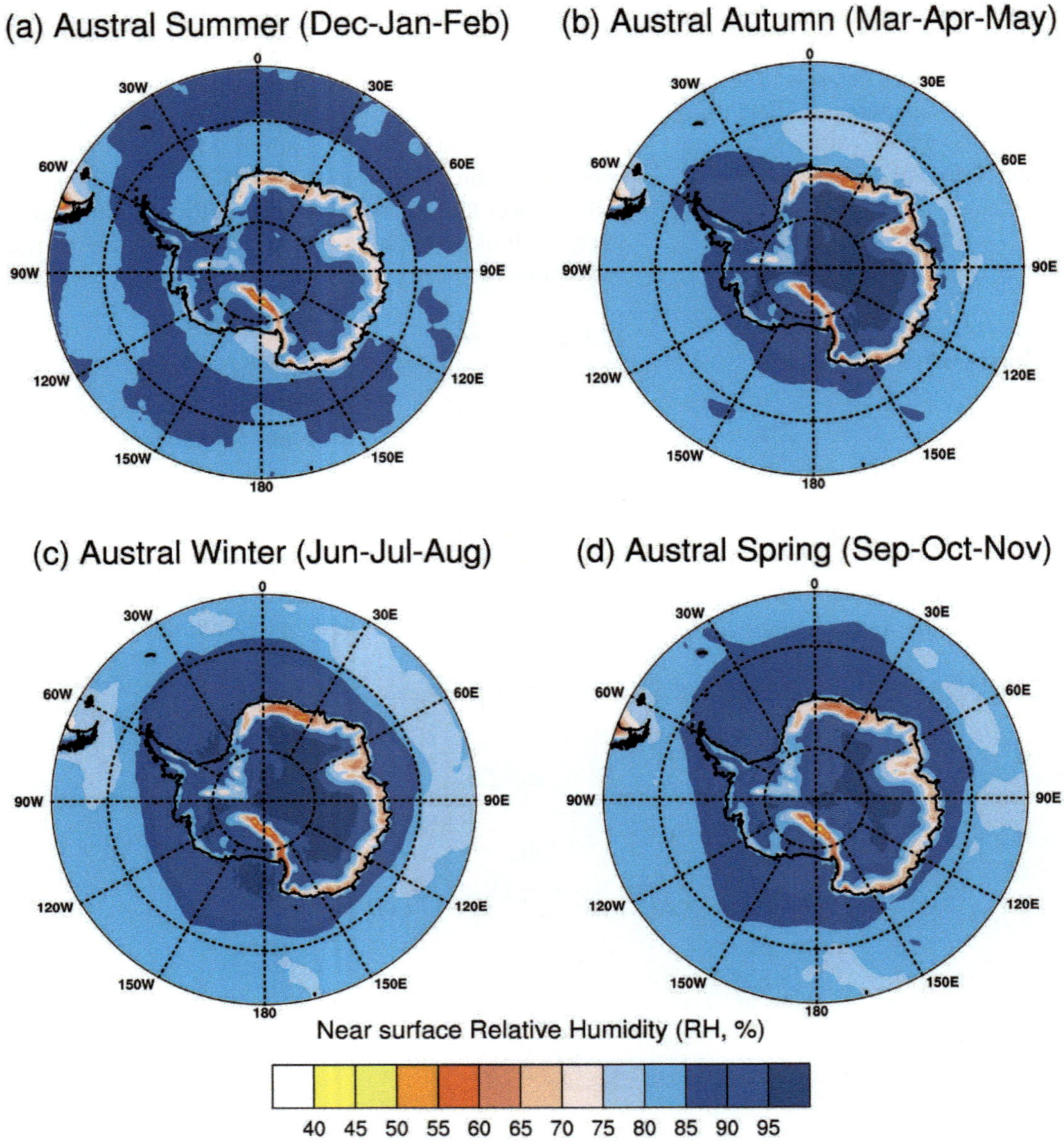


**Figure 29.2** Spatial variations of near surface relative humidity (RH) over the Antarctic region during: (a) austral summer (DJF); (b) austral autumn (MAM); (c) austral winter (JJA); and (d) austral spring (SON) seasons of years 2002–2018.

cooling gives rise to higher RH values and more condensation of moisture near the surface over the Antarctic plateau. This results in the inversion of specific humidity (SH) and creates SH inversion layer. As a result, the air masses which are dry with extremely low SH are generated. These dry air masses descend from the plateau by katabatic winds (fall winds) and give rise to adiabatic warming. Due to adiabatic warming, a reduction in RH gives rise to a downward sensible heat flow, allowing for effective surface evaporation on coastal slopes (Naakka et al., 2021). Higher RH levels can cause the growth of hydrophilic aerosols, which can influence the optical and radiative properties of the aerosol.

### 29.3.2 Seasonal Variations of Winds over Northern and Southern Polar Regions

#### 29.3.2.1 Arctic

The climatology winds at two pressure levels, 1000 hPa and 850 hPa during different seasons over the Arctic, are shown in Figure 29.3. The wind speeds are higher at 850 hPa when compared to 1000 hPa. A small cyclonic pattern can be observed near the central Arctic regions during all the seasons; however, their positions are different throughout the seasons. In boreal winter and autumn, 850 hPa winds are stronger over Siberia. The wind speed is higher over the Greenland regions at both pressure levels.

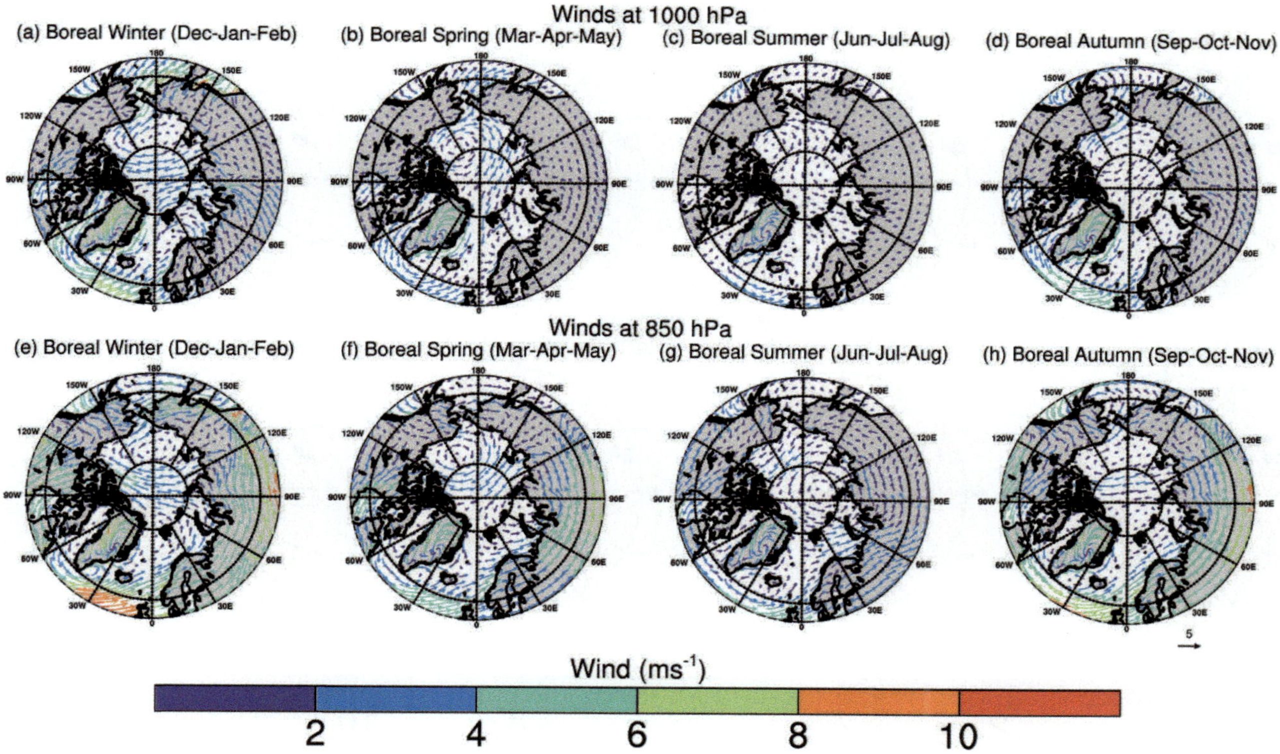


**Figure 29.3** Spatial variations of wind vectors over the Arctic region during: (a,e) boreal winter (DJF); (b,f) boreal spring (MAM); (c,g) boreal summer (JJA); and (d,h) boreal Autumn (SON) seasons of years 2002–2018 at 1000 hPa at 850 hPa pressure levels.

In winter, the winds are mostly easterly, coming from the east direction. During boreal summer and autumn, the wind speed is lower (<4 ms$^{-1}$) over the majority of the Arctic region at 1000 hPa (Figures 29.3c and 29.3d). The winds play a very crucial role in transporting aerosols from the source regions.

#### 29.3.2.2 Antarctic

The climatology winds at two pressure levels, 1000 hPa and 850 hPa, during different austral seasons over the Antarctic, are shown in Figure 29.4. The circumpolar current over the Antarctic can be seen during all the seasons at different pressure levels. The circumpolar winds are stronger (with wind speed greater than 8 ms$^{-1}$) at a height of 850 hPa, when compared to those at 1000 hPa where wind speed is relatively lower (<8 ms$^{-1}$). When radiative cooling occurs over the elevated Antarctic ice sheet, very cold, dense air is produced. Antarctica is usually encircled by a low-pressure belt with many low centers which is termed as the "circumpolar trough." However, high pressure dominates in the interior of the continent which gives rise to katabatic winds. Cold, thick air moving out from the interior's polar plateau down the sheer vertical dips along the Antarctic coast also forms the strong katabatic winds that blow across Antarctica. Strong katabatic winds arise along the sharp border of Antarctica when cold air flows over the landmass.

### 29.3.3 Seasonal Variations of Global Fire Activities

The fire counts were computed by adding daily active fire count data at 0.1° pixel, from 2002 to 2018, annually and for each season. These data are gridded at 1° resolution and quantify the fire counts per 1000 km$^2$. Figure 29.5 shows the seasonal mean globally distributed fire activities that occurred from 2002 to 2018. The coarser global fire maps are comparable to the wide global fire distribution shown in the 0.1° fire count data (Oom and Pereira, 2013; Andela et al., 2017). The fires mostly due to natural activities (forest fires) during the dry/summer season can be observed in less economically developed and less populated areas such as the African, Southeast Asian, and Siberian regions. Due to the heavy use of fossil fuels throughout the years, including wetter/cooler non-fire seasons, the fire levels in developed nations such as the United States and Australia are more consistent throughout the year, especially in densely populated regions

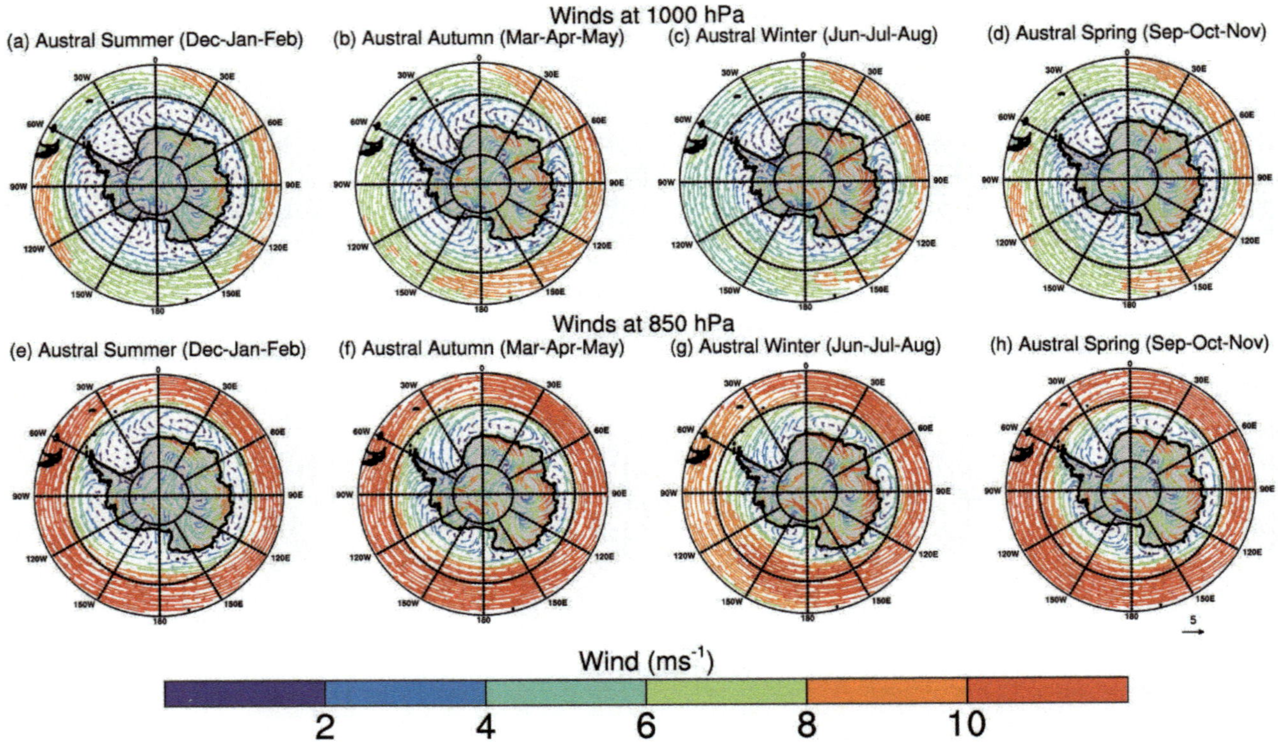


**Figure 29.4** Variations of wind vectors over the Antarctic region during: (a,e) austral summer; (b,f) austral autumn; (c,g) austral winter; and (d,h) austral spring seasons of years 2002–2018 at pressure levels of 1000 hPa and 850 hPa.

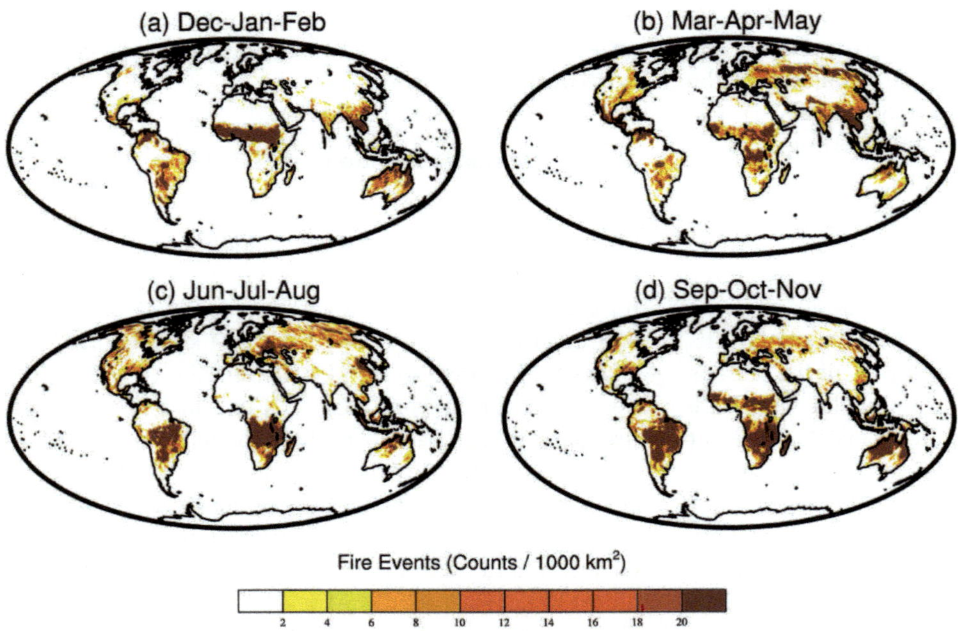


**Figure 29.5** Spatial distribution of the global fire activities: (a) Dec-Jan-Feb; (b) Mar-Apr-May; (c) Jun-Jul-Aug; and (d) Sep-Oct-Nov months of years 2002–2018.

(Figure 29.5). The Amazon Basin is responsible for 14% of all global fires (Earl and Simmonds, 2018). Siberian forests exhibit high fire counts during the boreal summer season, while south African and Madagascar regions show high fire counts during the boreal summer and autumn seasons. The fire activities produce black carbon aerosols during the burning of woods and can be transported to the polar regions.

### 29.3.4 Aerosol Variations Over the Northern and Southern Polar Region

The annual mean variations of aerosol optical depth (AOD) at mid-visible (0.55 μm) wavelength from 2002 to 2018 over the Arctic and Antarctic regions are shown in Figure 29.6. AOD is the measure of the columnar burden of aerosols, which indicates the presence of aerosols in the air column. AOD is found to be higher over the Arctic when compared to the Antarctic region. The higher values of AOD over the Arctic can be due to anthropogenic activities in surrounding regions, while Antarctica is surrounded by the Southern Ocean, which has much less anthropogenic activities. The aerosols present near the Antarctic coast can be due to the transport of aerosols from long distances. AOD data over the snow-covered Arctic and Antarctica regions cannot be retrieved from the satellite data due to the very high surface reflectance of ice. The Canadian Arctic Archipelago, Alaska regions show higher values of AOD, suggesting the presence of high aerosols due to anthropogenic activities. The part of the Siberian region also shows higher AOD values, which could be due to high fire activities in this region. The seasonal distribution of AOD over both polar regions is discussed in the following sections.

### 29.3.5 Seasonal Aerosol Variations Over the Northern and Southern Polar Regions

#### 29.3.5.1 Arctic

The seasonal and spatial variations of mid-visible AOD over the Arctic region for boreal winter, spring, summer, and autumn are depicted in Figure 29.7. AOD values cannot be retrieved over most of the locations of the Arctic region during boreal winter due to the absence of reflected short-wave radiation in the season. AOD over the central Arctic and other ice-covered regions can also not be retrieved by MODIS due to high surface reflectance values. During the boreal spring and summer, AOD values are observed to be higher (Figures 29.7b and 29.7c). The higher values can be owing to anthropogenic activities and fire events over the region.

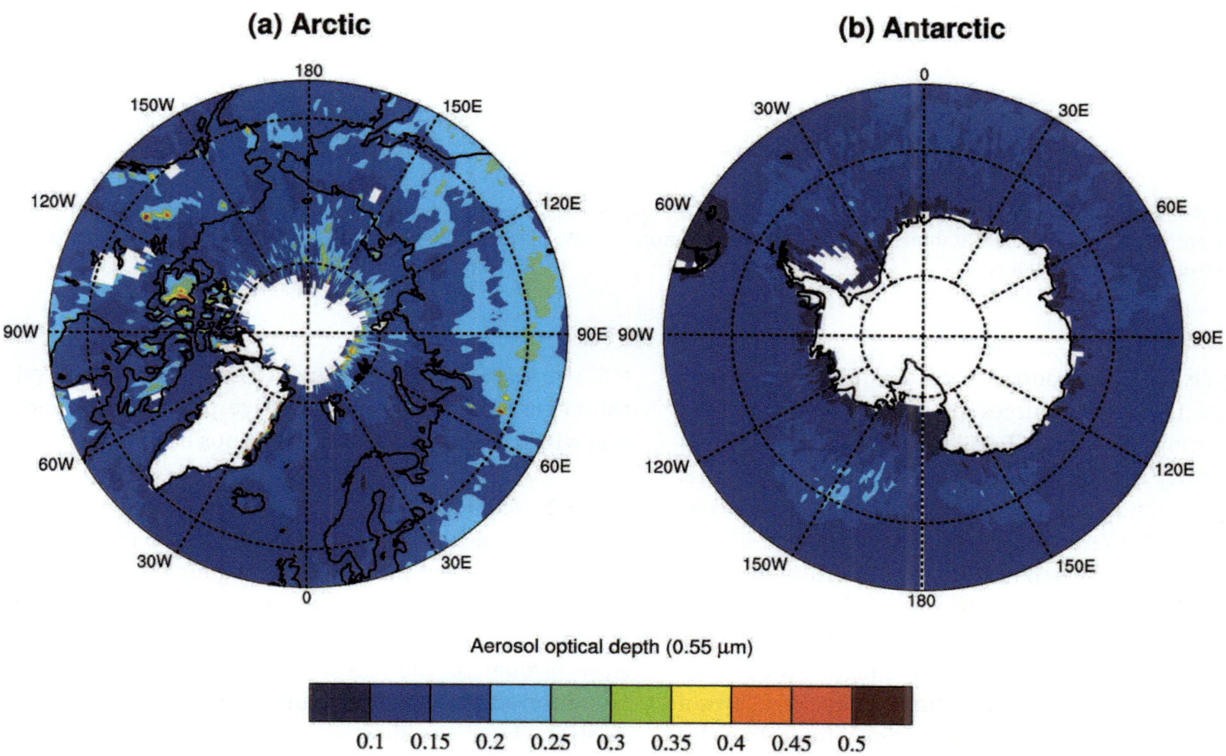


**Figure 29.6** Spatial variations of mid-visible aerosol optical depth (AOD): (a) over the Arctic and (b) Antarctic regions during years 2002–2018. AOD data could not be retrieved over the white regions.

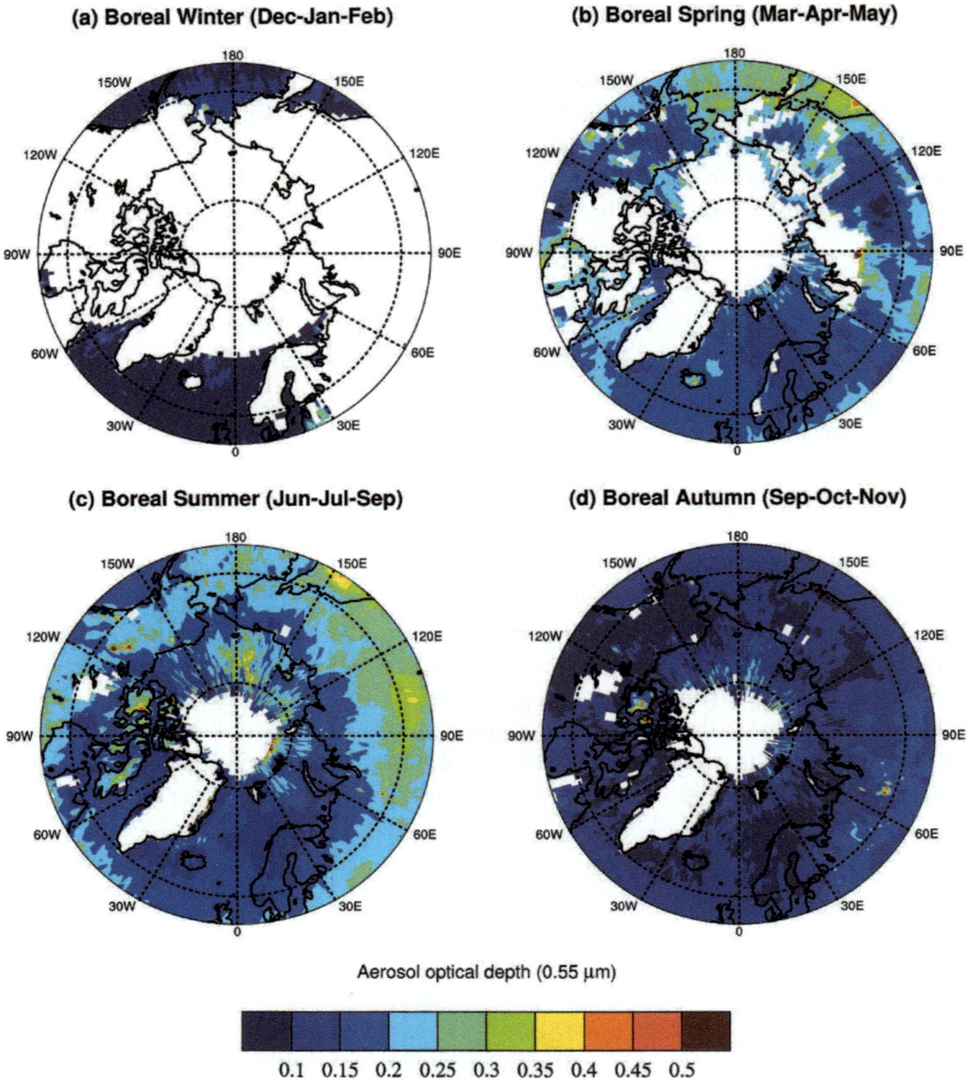


**Figure 29.7** Spatial variations of mid-visible (at 0.55 μm) aerosol optical depth over the Arctic during: (a) boreal winter (DJF); (b) boreal spring (MAM); (c) boreal summer (JJA); and (d) boreal Autumn (SON) seasons of years 2002–2018. AOD data could not be retrieved over the white regions.

Long-distance transport of natural and manmade aerosols are the main sources of aerosols reaching the Arctic during the winter (Breider et al., 2017). Late winter and early spring have an accumulation of aerosols known as Arctic Haze (Law and Stohl, 2007). The concentrations of different aerosol species are lower during the boreal summer when natural Arctic aerosol predominates (Breider et al., 2017). The mid-visible AOD values cannot be retrieved in boreal winter due to the unavailability of shortwave radiations. This suggests the importance of ground-based measurements of aerosol properties over the polar regions during winter. In order to address such issues, India is developing the *in-situ* network of aerosol measurements, i.e., POLar AERosol NETwork (POLAERNET). In POLAERNET, the physical, chemical, optical and radiative properties of aerosols will be measured on a contentious basis.

#### 29.3.5.2 Antarctic

The spatial variations of mid-visible (at 0.55 μm) aerosol optical depth over the Antarctic region for the austral summer, autumn, winter, and spring seasons are shown in Figure 29.8. The Antarctic landmass is covered with snow throughout the year; therefore, AOD values cannot be retrieved due to high surface reflectance. During Austral winter, the shortwave radiation is not available, which also causes the non-retrieval of AOD from satellite remote sensing.

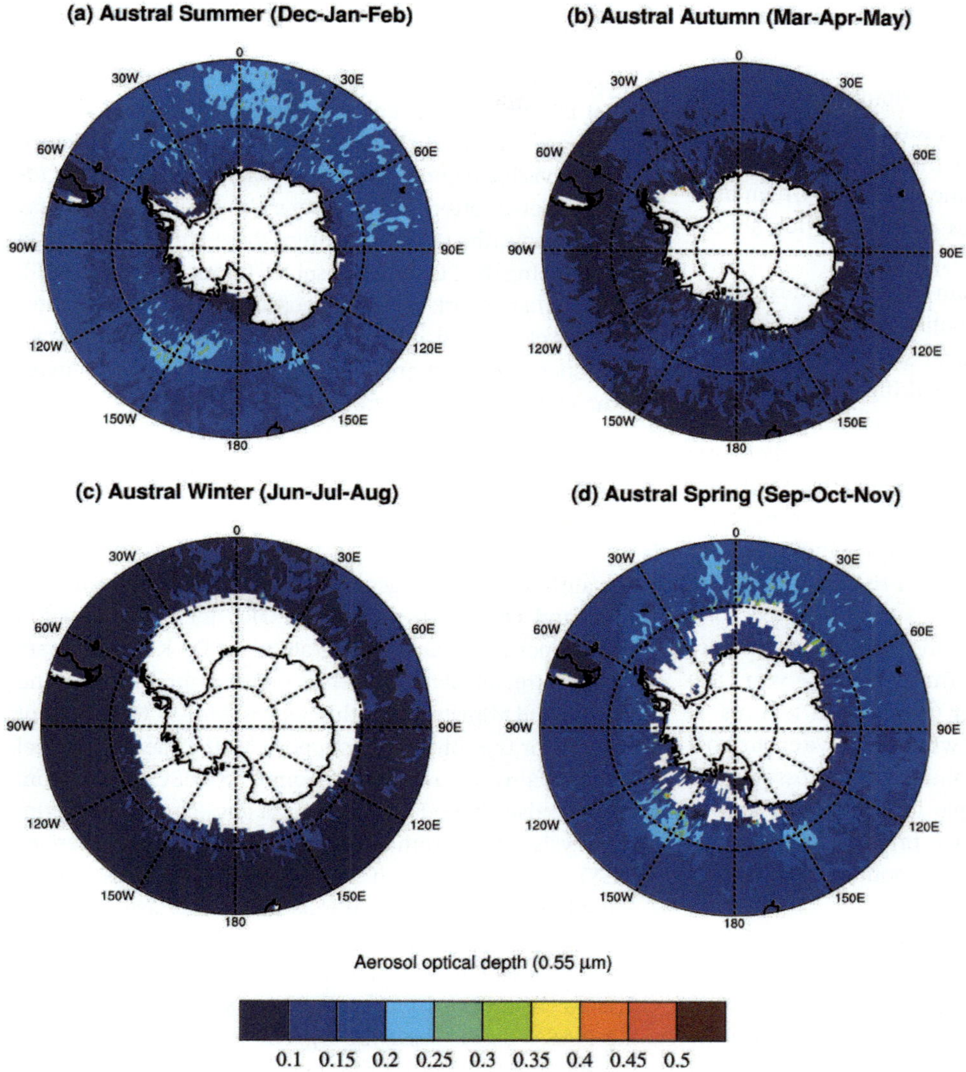


**Figure 29.8** Spatial variations of mid-visible (at 0.55 μm) aerosol optical depth over the Antarctic region during: (a) austral summer (DJF); (b) austral autumn (MAM); (c) austral winter (JJA); and (d) austral spring (SON) seasons of years 2002–2018. AOD data from MODIS could not be retrieved over the white colored regions.

The coastal Antarctic and circumpolar southern oceanic regions show higher AOD values during the austral summer and spring seasons, owing to the transport of manmade and forest fire aerosols originating over Madagascar and southern African regions (Srivastava et al., 2021). The AOD values over the Antarctic regions are lower in all the seasons when compared to the Arctic region, which is due to the lower influence of anthropogenic activities near the Antarctic region. The unavailability of satellite remote-sensing data during the winter season and the ice-covered region of Antarctica suggests the importance of *in-situ* measurements of aerosol properties under India's POLAERNET program.

## 29.4 Conclusions

Atmospheric aerosols absorb and scatter incoming solar and outgoing terrestrial radiations and have a crucial impact on polar as well as global climate. This present chapter highlights the spatial and seasonal variations of mid-visible aerosol optical depth obtained from the MODIS sensor installed on the Terra satellite. In addition, variations of near-surface meteorological parameters such as relative humidity and winds at 1000 and 850 hPa pressure levels are also discussed. These parameters are very important in the growth and transport of aerosols. The global distribution of fire counts derived from MODIS measurements is also discussed. The study revealed the following:

- The majority of the Arctic area exhibits higher (>85%) RH values. Central Greenland region shows even higher (>95%) RH values. Because of the rising airmass, RH reaches saturation as it enters the interior of Greenland, resulting in higher RH values within the ice sheet's interior.
- RH over Antarctica is more than 95% in the austral autumn and winter. Radiative cooling over the Antarctic plateau causes the increase in near-surface RH and leads to condensation of moisture, resulting in a practically permanent specific humidity inversion layer.
- At both pressure levels, the wind speed is higher over the Greenland areas. During the winter, the winds are primarily easterly. The wind speed is low throughout the boreal summer and autumn, with wind speeds at 1000 hPa being less than 4 $ms^{-1}$ over the majority of the Arctic area.
- The circumpolar circulation over the Antarctic may be noticed at various pressure levels throughout the year. The circumpolar winds are stronger (wind speed > 8 $ms^{-1}$). Very cold, dense air is produced when radiative cooling occurs over the high Antarctic ice sheet. A low-pressure zone around Antarctica with multiple low centers is known as the "circumpolar trough." High pressure, on the other hand, dominates the interior of the continent, resulting in katabatic winds.
- Fires in African, Southeast Asian, and Siberian regions tend to follow more natural fire seasons, occurring throughout the austral summer/boreal winter, whereas fire counts in the United States and Australia are more consistent throughout the year, especially in densely populated areas, during the cooler/wetter non-fire season.
- When compared to Antarctic regions, AOD is observed to be greater in the Arctic. Anthropogenic activity in the surrounding regions may account for the elevated AOD values over the Arctic. The Southern Ocean surrounds Antarctica, which has very little anthropogenic activity.
- Higher AOD levels are seen in the Canadian Arctic Archipelago and Alaska regions, indicating the presence of high aerosols owing to anthropogenic activities. AOD levels are also greater in the Siberian area, which might be attributable to the region's strong fire activity.
- During the austral summer and spring seasons, the AOD values in the coastal Antarctic and circumpolar southern oceanic regions are higher. This might be attributed to anthropogenic and forest fire aerosols being transported from South Africa and Madagascar regions.
- Because shortwave radiations are unavailable in boreal winter in the Arctic and austral winter in the Antarctic, the AOD values cannot be retrieved. In addition, AOD can also not be retrieved through satellite-derived remote sensing over snow-covered regions due to high surface reflectance in the visible regime. These emphasize the necessity of *in-situ* observations of aerosol properties over the polar areas throughout the year. To address these concerns, India is developing the POLar AERosol NETwork (POLAERNET), an *in-situ* network of aerosol measurements. The measurements of physical, optical, chemical, and radiative characteristics of aerosols are planned on continuous basis under the POLAERNET program.

## Acknowledgments

The author acknowledges the Director, National Centre for Polar and Ocean Research (NCPOR), for the motivation and support. The author is grateful to Dr K.P. Krishnan, Group Director, Arctic Science division for the help and support and Ministry of Earth Sciences (MoES), Government of India for the continuous support. MODIS aerosol optical depth data were downloaded from GES-DISC, NASA from https://ladsweb.modaps.eosdis.nasa.gov. The ERA-Interim reanalysis dataset from Copernicus Climate Change Service (C3S) is utilized, which is available from https://www.ecmwf.int/en/forecasts/datasets/archive-datasets/reanalysis-datasets/era-interim. We acknowledge the use of data from NASA's Fire Information for Resource Management System (FIRMS) (https://earthdata.nasa.gov/firms), part of the NASA Earth Observing System Data and Information System (EOSDIS). The author thanks the Editors of the book and the reviewers for their valuable comments and suggestions. This is NCPOR contribution no. B3–/2022–23.

## References

Andela, N., Morton, D.C., Giglio, L. et al. (2017). A human-driven decline in global burned area. *Science* 356(6345): 1356–1362. doi: 10.1126/science.aal4108.

Breider, T.J., Mickley, L.J., Jacob, D.J. et al. (2017). Multidecadal trends in aerosol radiative forcing over the Arctic: contribution of changes in anthropogenic aerosol to Arctic warming since 1980. *Journal of Geophysical Research: Atmospheres* 122(6): 3573–3594. doi: 10.1002/2016JD025321.

Chu, D.A. (2002). Validation of MODIS aerosol optical depth retrieval over land. *Geophysical Research Letters* 29(12): 8007. doi: 10.1029/2001GL013205.

Chung, C.E. (2005). Global anthropogenic aerosol direct forcing derived from satellite and ground-based observations. *Journal of Geophysical Research* 110(24): 1–17.

Cohen, J., Ramanathan, V., Kim, D. et al. (2014). Recent Arctic amplification and extreme mid-latitude weather. *Nature Geoscience* 7(9): 627–637. doi: 10.1038/ngeo2234.

Dee, D.P., Uppala, S.M., Simmons, A.J. et al. (2011). The ERA-Interim reanalysis: configuration and performance of the data assimilation system. *Quarterly Journal of the Royal Meteorological Society* 137(656): 553–597. doi: 10.1002/qj.828.

Earl, N. and Simmonds, I. (2018). Spatial and temporal variability and trends in 2001–2016 global fire activity. *Journal of Geophysical Research: Atmospheres* 123(5): 2524–2536. doi: 10.1002/2017JD027749.

Flanner, M.G., Zender, C.S., Hess, P.G. et al. (2009). Springtime warming and reduced snow cover from carbonaceous particles. *Atmospheric Chemistry and Physics* 9(7): 2481–2497.

Giglio, L., Schroeder, W., and Justice, C.O. (2016). The collection 6 MODIS active fire detection algorithm and fire products. *Remote Sensing of Environment* 178: 31–41. doi: 10.1016/j.rse.2016.02.054.

Hansen, J. and Nazarenko, L. (2004). Soot climate forcing via snow and ice albedos. *Proceedings of the National Academy of Sciences of the United States of America* 101: 423–428. doi: 10.1073/pnas.2237157100.

Hsu, N.C., Jeong, M.-J., Bettenhausen, C. et al. (2013). Enhanced Deep Blue aerosol retrieval algorithm: the second generation. *Journal of Geophysical Research: Atmospheres* 118(16): 9296–9315. doi: 10.1002/jgrd.50712.

Law, K.S. and Stohl, A. (2007). Arctic air pollution: origins and impacts. *Science* 315(5818): 1537–1540. doi: 10.1126/science.1137695.

Levy, R.C., Mattoo, S., Munchak, L.A. et al. (2013). The Collection 6 MODIS aerosol products over land and ocean. *Atmospheric Measurement Techniques* 6(11): 2989–3034. doi: 10.5194/amt-6-2989-2013.

Li, Z., Zhao, X., Kahn, R. et al. (2009). Uncertainties in satellite remote sensing of aerosols and impact on monitoring its long-term trend: a review and perspective. *Annales Geophysicae* 27: 2755–2770.

Naakka, T., Nygård, T., and Vihma, T. (2021). Air moisture climatology and related physical processes in the Antarctic on the basis of ERA5 reanalysis.*Journal of Climate* 34(11): 4463–4480. doi: 10.1175/JCLI-D-20-0798.1.

Oom, D. and Pereira, J.M.C. (2013). Exploratory spatial data analysis of global MODIS active fire data. *International Journal of Applied Earth Observation and Geoinformation* 21: 326–340. doi: 10.1016/j.jag.2012.07.018.

Ramanathan, V. and Carmichael, G. (2008). Global and regional climate changes due to black carbon. *Nature Geoscience* 1: 221–227.

Remer, L.A., Kliedman, R.G., Levy, R.C. et al. (2008). Global aerosol climatology from the MODIS satellite sensors. *Journal of Geophysical Research* 113: doi: 10.1029/2007JD009661.

Sand, M., Berntsen, T.K., Seland, Ø. et al. (2013). Arctic surface temperature change to emissions of black carbon within Arctic or mid-latitudes. *Journal of Geophysical Research* 118: 7788–7798. doi: 10.1002/jgrd.50613.

Seinfeld, J. and Pandis, S. (1998). *Atmospheric Chemistry and Physics: From Air Pollution to Climate Change*. Hoboken, NJ: John Wiley.

Shindell, D. and Faluvegi, G. (2009). Climate response to regional radiative forcing during the twentieth century. *Nature Geoscience* 2(4): 294–300. doi: 10.1038/ngeo473.

Srivastava, R., Asutosh, A., Sabu, P., Anilkumar, N.(2021). Investigation of Black Carbon characteristics over southern ocean: *Contribution of fossil fuel and biomass burning* (2021). Environmental Pollution 276: 116645. doi: 10.1016/j.envpol.2021.116645

Srivastava, R. and Ravichandran, M. (2021). Spatial and seasonal variations of black carbon over the Arctic in a regional climate model. *Polar Science* 30: 100670. doi: 10.1016/j.polar.2021.100670.

Wei, J., Feng, Y., Gup., J. et al. (2019). Performance of MODIS Collection 6.1 Level 3 aerosol products in spatial-temporal variations over land. *Atmospheric Environment* 206: 30–44. doi: 10.1016/j.atmosenv.2019.03.001.

# Section V

## The Research Institutions on the "Three Poles," Data Pools, Data Sharing Policies, Career in Polar Science Research and Challenges

# 30

# Multi-Disciplinary Research in the Indian Antarctic Programme and Its International Relevance

*Anand K. Singh*[1,*], *Yogesh Ray*[1], *Shailendra Saini*[1], *Rahul Mohan*[1], *and M. Javed Beg*[1]

[1] *National Centre for Polar and Ocean Research, Ministry of Earth Sciences, Goa, India*
[*] *Corresponding author*

## 30.1 Introduction

The Indian Antarctic endeavor started in the austral summer of 1981/82 with the expedition vessel, *MV Polar Circle*, and it has continued ever since. During the first and second expeditions, the scientific experiments and fieldwork were carried out from the vessel and ice-shelf in the campaign mode. It also included inland penetration to the Schirmacher Oasis and Wohlthat mountain. With India joining the Antarctic Treaty in 1983, the first permanent station, Dakshin Gangotri station, was commissioned on the ice-shelf in Central Dronning Maud Land in the season 1983/84 to facilitate year-round multi-disciplinary scientific work.

As the Antarctic Programme developed further, the second station, Maitri station, was constructed on the rocky terrain of Schirmacher Oasis in 1988/89 as a replacement for Dakshin Gangotri, which was completely buried under snow and eventually decommissioned in 1989/90. The site of Dakshin Gangotri has been declared a historic site and monument (HSM-44) under the provisions of the Antarctic Treaty (ATCM XXVI CEP VI, 2015). Meteorology and space weather, geological and magnetic surveys, ice-shelf dynamics, regional biology, oceanography, environment, lacustrine studies, etc., have been the major focus areas of research from Maitri (Ravindra, 2006).

For exploration of West Antarctica, the dedicated Weddell Sea Expedition was launched in 1989/90 for oceanographic survey, geological mapping of Nunataks, and study of the Filchner ice shelf (Ravindra and Kaul, 1995). The maiden Indian South Pole expedition was launched from Maitri by land in 2010/11 (Ravindra and Mohan, 2011). In addition to the glacial studies en-route and around the pole, atmospheric data acquisition and snow-core sampling were carried out during the South Pole expedition.

To further expand scientific horizon, India commissioned the third research station, Bharati station in the Larsemann Hills, in 2012. About 3000 km further east of Maitri, Bharati offered new opportunities to the Indian scientists. The location of Bharati, along the Indian longitude, has proved to be highly effective for data reception and tele-commanding to Indian remote-sensing satellites. Unique geology and mineral deposits in the Larsemann Hills, connection of India and Antarctica in the mesozoic era (Pant and Dasgupta, 2017), and ultra-modern infrastructure and facilities, collectively added a new chapter to the Antarctic Programme.

The Indian Antarctic Programme is managed by the National Centre for Polar and Ocean Research (NCPOR), an autonomous organization under the Ministry of Earth Sciences, Government of India.

This chapter is intended to introduce the reader to the Indian Antarctic Programme and the broad area of multi-disciplinary research undertaken in the expeditions. In the following sections, a brief introduction to the international bodies for regulation of Antarctic policies, and scientific and logistics collaboration, are presented, which is followed by an introduction to the multi-disciplinary research undertaken at Indian stations and their international relevance.

## 30.2 India in the International Bodies for Antarctica

Several international bodies have been constituted for peaceful scientific presence and cooperation in Antarctica for the rest of the world. The Antarctic Treaty, in force since 1961, is the main regulatory body and has been acceded to by 54 nations to date. India joined the Antarctic Treaty on 19 August 1983 and holds a consultative

*Advances in Remote Sensing Technology and the Three Poles*, First Edition. Edited by Manish Pandey, Prem C. Pandey, Yogesh Ray, Aman Arora, Shridhar D. Jawak, and Uma K. Shukla.

membership, which authorizes India to participate in decision-making of the Treaty.

As the human presence started to increase in Antarctica, the environmental monitoring and related guidelines became essential. The implementation of environment protocols for Antarctica is overseen and regulated by the Committee for Environmental Protection (CEP), which started functioning in 1998. India has been a member of CEP since the beginning. The Council of Managers of National Antarctic Programs (COMNAP), an international association of 31 countries including India, provides a platform to the governments for mutual cooperation in Antarctica (Wratt, 2013).

Scientific Committee on Antarctic Research (SCAR), created in 1958, helps initiate, develop, and coordinate international scientific research in Antarctica and Southern Ocean. SCAR includes 45 members from around the world. India joined the full membership of SCAR in 1984. With the objective of conserving Antarctic marine life, the Commission for the Conservation of Antarctic Marine Living Resources (CCAMLR) was established in 1982, and India is a member of the commission.

Resupply to Antarctic stations even today mostly depend on ships; however, for the movement of persons and light cargo, non-commercial flights are operated from different parts of the world. Nations actively involved in Dronning Maud Land (DML) of Antarctica formed a consortium in 2002, Dronning Maud Land Air Network (DROMLAN) for the air connectivity between Cape Town, South Africa to Antarctica and different stations. At present, the consortium has representation from 12 countries including India.

Antarctic bodies, such as CEP, COMNAP, SCAR, CCAMLR, and DROMLAN, independently advise the Antarctic Treaty on the relevant issues in Consultative Meetings.

## 30.3 Multi-Disciplinary Antarctic Research in the Last Decade

In the austral summer 2021/22, the 41st Indian scientific expedition to Antarctica (41-ISEA) was launched using the resupply vessel *MV Vasiliy Golovnin* and DROMLAN flights. A wide spectrum of scientific research has been undertaken at Indian Antarctic stations in the past four decades. Broadly, the research projects fall in the categories of: i) Atmospheric Sciences and Space Weather; ii) Biological Sciences; iii) Environmental Monitoring; iv) Earth Sciences and Glaciology; v) Medicine and Human Physiology; and vi) Student Participation and Outreach.

Research organizations and universities working in the relevant fields, carry out instrument- and field-based work at Indian Antarctic stations. In the summer season, the area of work expands to a few hundred kilometers around Maitri and Bharati stations, while during the winter months, work mostly continues in the vicinity of the stations.

Figure 30.1 depicts the number of scientific projects and participating organizations in the Antarctic Programme in the past 12 expeditions during 2010–22. Commissioning of the new station, Bharati, in 2012, opened a new avenue for Indian scientists, not only in the number of projects but also participation of the total number of institutions doubled. However, after 36-ISEA in 2016/17, various smaller projects were merged to address some of the pressing questions of the SCAR horizon scan. Year 2020/21 was an exception due to the Covid pandemic, when the scientific component was restricted to meteorological and space weather observations.

For the last 12 expeditions, the distribution of a total of 285 scientific projects led by principal investigators of 193 national organizations are shown in Figure 30.2. Earth Sciences and Glaciology are the major areas of interest in the Indian Antarctic Programme. Geological mapping,

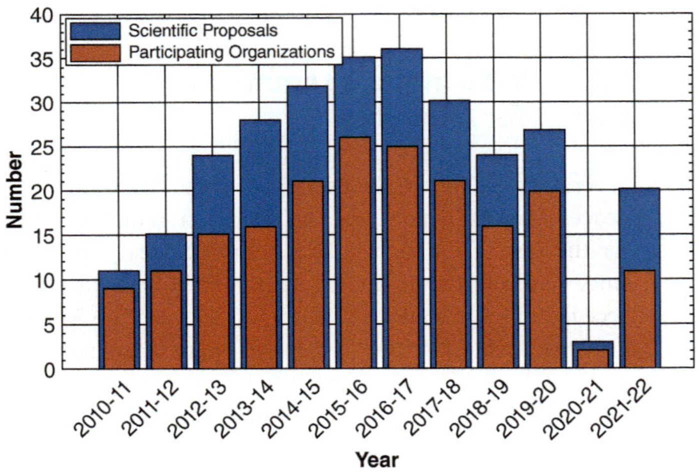


**Figure 30.1** Scientific projects and participating national organizations in Antarctic expeditions in the last 12 years.

glacier monitoring (Shrivastava et al., 2017), paleoclimatic studies using lake and ice cores (Thamban, 2017; Warrier et al., 2017) and sediment samples (Pant and Dasgupta, 2017) have been at the focus.

Atmospheric and Space Weather studies comprise of about one-third of the total projects in Antarctic expeditions. A suite of experiments for *in-situ* monitoring and remote sensing of the atmosphere, starting from the surface of the Earth to the exosphere and beyond, are operational at Maitri and Bharati stations. Uninterrupted monitoring and data acquisition of the meteorological parameters and periodic forecasting have continued (Dhote et al., 2021). Space Weather studies using magnetometers, riometer, radiosonde, all-sky camera, VHF radar, etc., have been key to the understanding of the solar-terrestrial interaction and vertical coupling in the atmospheric regions (Singh et al., 2019; Sinha et al., 2017).

Biological sciences and environmental monitoring contribute equally to the number of scientific projects in the Antarctic Programme. Wildlife monitoring (Pande et al., 2017) and algal and bacterial studies (Jani et al., 2021), etc., are regularly undertaken. Environmental monitoring with regard to local anthropogenic activities and long-range transport is the year-round activity at both stations (Botsa et al., 2021; Tiwari, 2017). Moreover, natural radiation levels are regularly monitored using multiple instruments (Prajith et al., 2019).

As a part of capacity building, merit-based projects of masters students are occasionally supported for the Antarctica expeditions. New approaches to study the human physiology in response to the extreme Antarctic conditions have also been introduced in the recent expeditions (Nirwan et al., 2021).

The Indian National Committee of SCAR produced a special issue on the recent Antarctic research in India (Nayak, 2017), which has several publications and overviews in the fields of paleoclimate, glacier monitoring, geology, biodiversity, environmental and wildlife monitoring, sea-ice, aerosol, meteorology, and space weather. In addition, a special volume of the *Polar Science* highlighted the new research in the Antarctic as well as Arctic expeditions (Kumar et al., 2021). Interested readers are referred to the special volumes (Nayak, 2017; Kumar et al., 2021) and references therein.

Apart from the Antarctic expeditions, dedicated Indian Scientific Expeditions to the Southern Ocean (ISESO) started in 2004 to pursue multi-disciplinary research in field of hydrodynamics, biogeochemistry, biodiversity, air–sea interaction, atmospheric processes, palaeoclimatology, etc. In addition to numerous scientific publications from the last 11 Southern Ocean Expeditions, special volumes of *Current Science* (Ramesh et al., 2010) and *Deep Sea Research* (Anilkumar and Achuthankutty, 2015; Anilkumar et al., 2020) highlight achievements and new research.

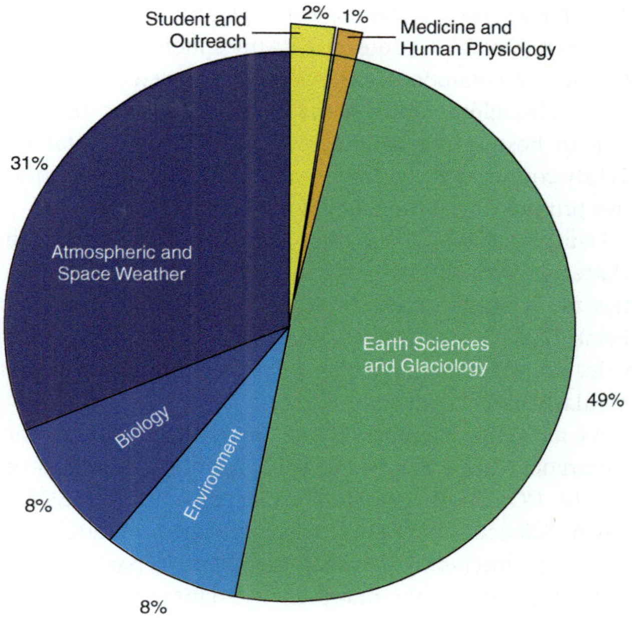


**Figure 30.2** Discipline-wise distribution of the scientific projects in the Indian Antarctic Programme from 2010–11 to 2021–22. Earth Sciences and Glaciology, and Atmospheric and Space Weather, attract the maximum participation. The projects include summer as well as winter components.

## 30.4 International Relevance

Antarctica is considered as a natural laboratory of the world and provides great opportunities for research in a variety of disciplines of science. However, the practical difficulty for access to the farthest continent and the hostile weather conditions are the greatest challenges, and call for international cooperation, more than anywhere else. Through an international initiative involving scientists and policy-makers, the Scientific Committee on Antarctic Research (SCAR) prioritized Antarctic research for the next two decades and beyond (Kennicutt et al., 2014, 2015). Out of six broad themes, a total number of 80 priority scientific questions have been identified for coordinated international efforts.

The scientific projects within the Indian Antarctic Programme have been aligned to the SACR priorities. Ongoing research at the Indian stations and Southern Ocean expeditions cover all the broad themes of the Science Horizon Scan. Recently, various multi-national projects, such as Mass balance, dynamics, and climate of the central Dronning Maud Land coast, East Antarctica (MADICE); Schirmacher Oasis Nippon (Japan), India Coring (SONIC); Sea Ice and Westerly winds during the Holocene in coastal

Antarctica (SIWHA); Geological Exploration of the Amery Ice Shelf (GeoEAIS); Integrated Atmospheric Observation Facility for Antarctica (IAOFA), etc., have been completed/ initiated to address some of the priority questions. In addition to hosting the several researchers of the Antarctic Treaty countries at the Indian stations, bilateral collaborative projects have been taken with the programme.

Data collected during the Antarctic expeditions are shareable with the national and international communities for scientific research. A dedicated portal, National Polar Data Center (https://npdc.ncaor.gov.in/npdc) provides information and access to the (meta)data, which are regularly updated and archived.

As an active member of the international bodies for Antarctica, India has hosted and organized meetings of ATCM, COMNAP, DROMLAN, etc. Also, the 10th SCAR Open Science Conference was organized in 2022. To encourage interaction between the scientists participating in the expeditions, the first-ever International Conference on Antarctic Research was organized at Bharati in 2020 with participation of over 50 scientists from India and neighboring stations: Davis (Australia), Progress (Russia), and Zhongshan (China).

## 30.5 Concluding Remarks

The India Antarctic endeavor, started in the austral summer of 1981/82, has expanded to commissioning of three research stations, and completion of hundreds of multi-disciplinary research projects in the past four decades. The Indian Antarctic Programme has been continuously expanding with international involvement and cooperation.

## References


Anilkumar, N. and Achuthankutty, C.T. (2015). Ecosystem survey in the Indian Ocean sector of the Southern Ocean-results from Indian expeditions. *Deep-Sea Research Part II* 118: 137–141.

Anilkumar, N., Ravichandran, M., and Jena, B. (2020). Indian scientific expeditions to the Southern Ocean: comprehensive surveys to understand atmospheric, physical, and biogeochemical processes. *Deep Sea Research Part II: Topical Studies in Oceanography* 178: 104860.

Botsa, S.M., Tara, D.L., Magesh, N.S. et al. (2021). Characterization of black carbon aerosols over Indian Antarctic station, Maitri and identification of potential source areas. *Environmental Science: Atmospheres* 1(6): 416–422.

Dhote, P.R., Thakur, P.K., Shevnina, E. et al (2021). Meteorological parameters and water balance components of Priyadarshini Lake at the Schirmacher Oasis, East Antarctica. *Polar Science* 30: 100763.

Jani, K., Kajale, S., Shetye, M. et al. (2021). *Marisediminicola senii* sp. nov. isolated from Queen Maud Land, Antarctica. *International Journal of Systematic and Evolutionary Microbiology* 71(2): 004641.

Kennicutt, M.C., Chown, S.L., Cassano, J.S. et al. (2014). Polar research: six priorities for Antarctic science. *Nature* 512(7512): 23–25.

Kennicutt, M.C., Chown, S.L., Cassano, J.J. et al. (2015). A roadmap for Antarctic and Southern Ocean science for the next two decades and beyond. *Antarctic Science* 27(1): 3–18.

Kumar, A., Imura, S., Kim, S.-J. et al. (2021). Polar studies-window to the changing earth. *Polar Science* 30: 100767.

Madrid ATCM XXVI CEP VI. (2015). Revised list of historic sites and monuments.

Nayak, S. (2017). Recent Antarctic research in India: the national committee report to SCAR. *Proceedings of the Indian National Science Academy* 83(2): 245–248.

Nirwan, M., Halder, K., Saha, M. et al. (2021). Improvement in resilience and stress-related blood markers following ten months yoga practice in Antarctica. *Journal of Complementary and Integrative Medicine* 18(1): 201–207.

Pande, A. Sivakumar, K., Sathyakumar, R. et al. (2017). Monitoring wildlife and their habitats in the Southern Ocean and around Indian research stations in Antarctica. *Proceedings of the Indian National Science Academy* 83: 483-496483–496

Pant, N.C. and Dasgupta, S. (2017). An introduction to the crustal evolution of India and Antarctica: the supercontinent connection. 457(1): 1–6. doi: 10.1144/SP457.14.

Prajith, R., Rout, R.P., Kumbhar, D. et al. (2019). Measurements of radon (222rn) and thoron (220rn) exhalations and their decay product concentrations at Indian Stations in Antarctica. *Environmental Earth Sciences* 78: 1–12.

Ramesh, R., Sudhakar, M., and Ravindra, R. (2010). Indian contribution in Southern Ocean. *Current Science* 99(10): 1378–1424.

Ravindra, R. (2006). The ninth Indian scientific expedition to Antarctica: events and achievements. Scientific Report, Ninth Indian Expedition to Antarctica, Technical Publication, No. 6: 1–20.

Ravindra, R. and Kaul, M.K. (1995). Scientific report, *Indian expedition to Weddell Sea*. Technical Publication No. 7: 1–69.

Ravindra, R. and Mohan, R. (2011). Three decades of polar science in India. *Journal of the Geological Society of India* 78(1–3): 5–6. https://doi.org/10.1007/s12594-011-0060-1.

Shrivastava, P.K., Roy, S.K., and Bhai, H.Y. (2017). Antarctic glacier monitoring. *Proceedings of the Indian National Science Academy* 83(2): 255–267.

Singh, A.K., Dhar, A., Gudade, A.L. et al. (2019). Study of solar-terrestrial interaction based on magnetic field observations from Bharati Station, Larsemann Hills, Antarctica. Scientific report. *Twenty Seventh Indian Antarctic Expedition*, Technical Publication No. 25: 121–131.

Sinha, A.K., Dhar, A., Singh, A.K. et al. (2017). India's contribution to geomagnetism and allied studies in Antarctica: a review. *Proceedings of Indian Academy of Sciences* 83(2): 299–326. doi: 10.16943/ptinsa/2017/48955.

Thamban, M. (2017). Antarctic palaeoclimatic reconstruction using ice cores: Indian initiatives during 2008–2016. *Proceedings of the Indian National Science Academy* 83(2): 249–254.

Tiwari, A.K. (2017). Environmental monitoring around Indian Antarctic stations. *Proceedings of Indian National Science Academy* 83: 399–413.

Warrier, A.K., Mahesh, B.S., and Mohan, R. (2017). Lake sediment studies in ice-free regions of East Antarctica: an Indian perspective. *Proceedings of Indian National Science Academy* 83: 289–297.

Wratt, G.S. (2013). *A Story of Antarctic Co-operation: 25 Years of the Council of Managers of National Antarctic Programs.* Council of Managers of National Antarctic Programs (COMNAP).

# 31

## Indian and International Research Coordination in the Arctic

*Archana Singh*[1,*], *Divya T. David*[1], *and K.P. Krishnan*[1]

[1] *National Centre for Polar and Ocean Research, Ministry of Earth Sciences, Vasco-Da-Gama 403804, Goa, India*
* *Corresponding author*

### 31.1 The Changing Arctic and Inherited Interest

With the whole world experiencing warming trends, the Arctic has warmed up almost four times faster than the rest of the globe since 1979 (Rantanen et al., 2022). The region is also experiencing intensified precipitation and winds.

Increasing air and sea surface temperatures are leading to a decrease in the sea ice extent (Overland et al., 2019; Docquier et al., 2021). Many of the Arctic regions have become sea ice free in summers, which has brought Arctic sea ice to its minimum levels in recent years (Parkinson and DiGirolamo, 2021). Similarly, the winter sea ice is also on decline (Park et al., 2015). Loss of sea ice is paving the way for more open waters that absorb more heat, as the darker color of water reflects less and thus has less albedo. The warmer waters further melt more ice, pushing the Arctic into positive ice-albedo feedback (Kashiwase et al., 2017). Sea ice-free waters are also prone to strong polar winds influencing the mixing and stratification of the water column (Davis et al., 2016). Overall, these changes are affecting the Arctic marine ecosystem and life in a significant way (Solan et al., 2020). The sea ice decline is distressing for the species that thrive on the sea ice like polar bears, seals, and sea lions (Gulland et al., 2022). They are forced to switch to terrestrial areas and thus drastically impact the food web.

On the terrestrial front, retreating of glaciers and thawing of permafrost have enhanced the susceptibility of coastal erosion and landslides. The landmasses are becoming greener and permafrost areas are releasing greenhouse gasses as thawing kicks-off decomposition of earlier frozen organic matter. Settlements on the land are seeing damage due to increasing active layer of permafrost (Hjort et al., 2018). The soil-associated and soil-dependent biota get affected in a major way as their habitat is evolving. Apart from impacting lives and the environment over land, the resultant run-off forces hydrological changes, especially in the coastal regions and fjords of the Arctic (Andresen et al., 2020). The intensified precipitation also contributes to the runoff-related hydrological changes (Bintanja, 2018).

The changes are of immediate concern to the indigenous communities in the Arctic (Huntington et al., 2022). The Arctic comprises the Arctic Ocean and surrounding landmasses of Alaska (USA), Canada, Greenland (Denmark), Iceland, Norway, Sweden, Finland. and Russia. Indigenous communities spread across the landmasses are the Inuits (Greenland, Canada, and Alaska), Sámi (northern parts of Norway, Sweden, Finland, and of the Murmansk Oblast, Russia), and Russian indigenous people. The multinational nature of the Arctic with its climatic significance demands close cooperation and coordination between the countries. In addition, the Arctic Ocean is well connected with the other parts of the globe via the Atlantic Ocean, the Pacific Ocean, and the global conveyor belt. These oceanic connections along with the atmospheric pathways link the changes happening in the Arctic and the global climate. Due to sea ice decline, new sea routes are opening up, enhancing tourism and resource exploration activities in the Arctic. Furthermore, policy-makers and researchers have recognized the importance of taking the indigenous communities into the loop (Taagholt, 1988; Kankaanpää and Young, 2012). With all this going on in parallel, the Arctic demands international cooperation and understanding among

*Advances in Remote Sensing Technology and the Three Poles,* First Edition. Edited by Manish Pandey, Prem C. Pandey, Yogesh Ray, Aman Arora, Shridhar D. Jawak, and Uma K. Shukla.

researchers, policy-makers, and other interested bodies. Such global coordination carries a great significance when the world is moving forward in adopting resilient and adaptation strategies toward climate change.

## 31.2 International Research Coordination

On the international front, the Arctic Council (https://arctic-council.org) is the intergovernmental forum that operates to take cognition and document the activities and status of the Arctic which are of concern to social and scientific forms of life. This includes people, biodiversity, pollution, climate, ocean, and emergencies. The Arctic Council was established in 1996 in Canada by signing the Ottawa Declaration, with Canada as the first chair. The council's work accommodates seeking cooperation, coordination, and interaction among the Arctic countries, indigenous people, and other inhabitants on first-hand issues such as environmental protection and sustainable development. The council gives equal importance to research data and knowledge sharing, assessments and documentation, and finally to advise policy-makers of the Arctic countries and the Intergovernmental Panel on Climate Change (IPCC). The Arctic Council has representatives from the Arctic countries and indigenous people of the Arctic as permanent participants, and 38 observers which include non-Arctic countries, intergovernmental and interparliamentary organizations, and non-governmental organizations. India became one of the non-Arctic observers in 2013.

There are six working groups under which the council's work is being carried out:

i) *Arctic Contaminants Action Program (ACAP)*: deals with the pollutants and environmental risks
ii) *Arctic Monitoring and Assessment Programme (AMAP)*: handles pollution studies and its climate change effect
iii) *Conservation of Arctic Flora and Fauna (CAFF)*: works toward biodiversity of Arctic species and habitats
iv) *Emergency Prevention, Preparedness and Response (EPPR)*: takes care of the environmental and other emergencies and accidents
v) *Protection of the Arctic Marine Environment (PAME)*: for the sustainable use of the Arctic environment including resource exploration, shipping, protected areas. and pollution
vi) *Sustainable Development Working Group (SDWG)*: a multi-subject working group to includes both social and environmental science to improve the environmental, economic, and social conditions of the indigenous and Arctic communities

Apart from these working groups, there are some expert groups and task forces that carry out additional work for the council. One such active expert group works on global black carbon and methane emissions. Another was functional from 2011 to 2013, which dealt with ecosystem-based management. Besides these expert groups, the Arctic Council has 11 task forces which were formed to handle specific issues and tasks.

One of the observers in the Arctic Council, the International Arctic Science Committee (IASC: https://iasc.info) is a non-governmental organization that deals with the coordination of the Arctic science. The IASC was founded in 1990 by the eight Arctic countries to encourage and facilitate cooperation in all aspects of Arctic Science. The IASC is governed by a council that meets every year in their annual meeting named "Arctic Science Summit Week" (ASSW). The IASC operates its activities with five working groups: atmosphere, cryosphere, marine, terrestrial, and social and human. India is a council member of the IASC. With its member countries and institutions, the IASC brings together the scientific updates in the Arctic research, logistical needs, and data sharing to one table. The committee also conducts workshops and conferences to communicate and share ideas and needs, and to cater to other necessities of the community. The IASC operates in collaboration with other coordination councils and committees working in the Arctic and the Antarctic. The affiliates and partners include the Arctic Council, International Science Council, Scientific Committee on Antarctic Research (SCAR), and many more.

Talking of international coordination in the Arctic, Svalbard encompasses all possible aspects due to round the year international presence. Svalbard is a hub of Arctic Science activities due to its location and research facilities coordinated by Norway, under which this region of the Arctic falls. Geographically, Svalbard is located north of the Atlantic Ocean. The warm saline Atlantic water mass enters the Arctic Ocean via the west Spitsbergen current to the west of Svalbard, where the process is known as "Atlantification." On the eastern side, a part of the cold current from the Arctic recirculates around Svalbard and the other part interacts with the Barents Sea. In the purview of enhanced Atlantification, the Svalbard region has become a hotspot for scientific studies, and research bases are being set on its land for long-term as well as short-term climate studies.

With the rise of this important and international presence, it became pertinent to have an organization to coordinate the research activities in the Svalbard region. Svalbard Science Forum (SSF: https://www.forskningsradet.no/en/svalbard-science-forum) and Svalbard Integrated Arctic Earth Observing System (SIOS: https://sios-svalbard.org) are two such bodies that fulfil the role of coordination in the region. India is a member of both. The SSF and SIOS are looking into these aspects and thus coordinate the activities, encouraging collaboration and sharing among different countries and research groups to utilize the infrastructure and facilities to their fullest.

Regulating and streamlining research activities is necessary for a sustainable Arctic. The SSF is one such regulating body that works under the coordination of the Research Council of Norway and is observed by the Governor of Svalbard. The forum helps researchers to get necessary approvals from the governor and register their activities on an online open portal, Research in Svalbard (RiS), especially the ones concerning the conserved areas.

The main role of the SIOS is to bring together all research infrastructure and data on the same platform to detect overlaps and gaps, and to optimize the observing system in Svalbard as a whole. This observing system also encourages and facilitates sharing of the infrastructure and expertise. The SIOS became operational in 2018 to handle the increased scientific presence and data generation in Svalbard. The organization receives host funding from the Research Council of Norway, and contributions from its member institutions for annual activities, which include funding projects under annual access calls. Recently, the SIOS has emphasized the role of remote sensing in Arctic research which can help in keeping the research in the Arctic more sustainable, as it will help to minimize the carbon footprint of the research per se. The SIOS has various working groups which separate as well as integrate the different components of the work. The working groups have representation from the member institutions and operate under the direction of the General Assembly or the Board of Directors:

i) *Science Optimisation Advisory Group (SOAG)*: the principal working group that advises the director of the SIOS on the science, new ideas and services, and supports the review process of project proposals for access call
ii) *Research Infrastructure Coordination Committee (RICC)*: handles operational relationships among institutions and the members
iii) *Remote Sensing Working Group (RSWG)*: deals with remote-sensing data and related activities of the SIOS
iv) *SIOS data management system working group (SDMS WG)*: coordinates data curation and sharing
v) *Information Advisory Group (IAG)*: handles SIOS communication and outreach

The SIOS is publishing annual reports on science-based development of the observing systems in and around Svalbard since 2018 (*The State of Environmental Science in Svalbard Reports*) and offers recommendations on research needs and future planning.

Arctic research is also being coordinated from the regional levels. For example, Ny-Ålesund in Svalbard, a northernmost permanent settlement, is inhabited by research communities from different countries and contributes significantly to Arctic research. The activities in the area are being coordinated by the Ny-Ålesund Science Managers Committee (NySMAC: https://nyalesundresearch.no/nysmac). The function of NySMAC is to encourage and facilitate coordination and sharing of information about the activities among all who are working in the region. NySMAC has set up four flagship programs based on the key areas of research activity and ongoing long-term measurements with active discussion and cooperation among the community. The four flagship programmes are:

i) *Atmosphere*: related to atmospheric studies in Ny-Ålesund and Svalbard seeking collaboration and joint research actions
ii) *Glaciology*: dealing with glaciological research and monitoring in and around Ny-Ålesund
iii) *Kongsfjorden system*: to discuss research in Kongsfjorden, a fjord near Ny-Ålesund in Svalbard, in relation to the increasing Atlantification
iv) *Terrestrial ecosystems*: coordinates research in terrestrial realms of life and ecosystem processes

Apart from these flagship programs, there is also a cross-cutting flagship group to deal with interdisciplinary subjects. For instance, the nutrient cycle was identified by the flagships as a link between the atmosphere, cryosphere, and biosphere.

Thus, Arctic research is well-coordinated under different international and geographically designated bodies. Some forums deal with the different regional parts of the Arctic and some are exclusive to address data or logistics issues. The list is long and also includes the Asian Forum of Polar Sciences (AFoPS), European Polar Board (EPB), EU-PolarNet, NordForsk, Pacific Arctic Group, Sustaining Arctic Observing Networks (SAON), Forum of Arctic Research Operators (FARO), etc. The Arctic Circle Assembly (https://www.arcticcircle.org) is a non-profit and non-partisan organization operated by the Icelandic government. The assembly is held annually inviting governments, research and academic institutions, think tanks, environmental organizations, indigenous communities, and concerned people. With diverse participation, the assembly is a democratic platform to discuss ideas, and showcase and celebrate anything that concerns the Arctic and the planet.

## 31.3 Arctic Research Coordination at the National Level

India's research activities in the Arctic are facilitated through the National Centre for Polar and Ocean Research (NCPOR: https://ncpor.res.in) under the Ministry of Earth Science, Government of India. The NCPOR is a central government autonomous institute located in Goa that coordinates and handles funding and logistics of national expeditions to the polar regions: the Arctic and Antarctica. The institute also facilitates research in the Himalaya, the Third Pole.

India's major Arctic research share comes from its annual activities in the Svalbard region. India's Arctic program launches expeditions to Ny-Ålesund, Svalbard every year. India has a research base named "Himadri" in Ny-Ålesund and is a part of NySMAC. The annual expedition process starts with the NCPOR seeking project proposals from institutes and universities across the country to carry out research in the Arctic region. The proposals are invited specifically in the following areas:

i) *Atmospheric Science*: with special reference to the study of aerosols, trace gasses, and precipitation over the Arctic
ii) *Marine Science*: dynamics and functioning of Arctic fjords (Kongsfjorden and Krossfjorden, the nearby fjord)
iii) *Environmental Chemistry*: natural contaminants in food webs and long-range pollutants
iv) *Cryospheric studies*: snow and ice chemistry, and glaciology

The proposals undergo two-level screening by subject experts. The first step includes a detailed review of the project proposals by the experts after which the proposals are selected for a second-level screening. In the second level, project investigators need to present their proposals before an expert committee and an open audience. The expert committee in consultation with the logistic team and the director of NCPOR finally selects projects for implementation based on the merits and practicability of the proposals. The selected projects are then sent for approval from the Ministry of Earth Sciences (MoES). An overview of the major research fields of the projects implemented under the Indian Arctic research program from 2011 to 2019 is presented in Figure 31.1.

Preparation for the expedition starts with briefing the participants about the study area, route of journey, do's and don'ts in line with the in-force regulations, and preparation of necessary documents for deputation. The expeditions are launched in batches from May to October with around 8 to 9 participants in each batch. Apart from the summer excursions, one research team is sent in spring during March-April. The NCPOR also supports Indian researchers to become part of different Arctic cruises and expeditions hosted by other countries and institutions. Apart from the regular expeditions to Ny-Ålesund, the NCPOR is dedicatedly working to expand India's research and logistic hand in the other parts of the Arctic.

In addition to field access support, the NCPOR is also funding projects in the Arctic and Antarctic from 2017 under the polar research initiative for a long-term and continuous commitment from Indian researchers. This initiative funds projects for three years to carry out research in polar sciences and land-ocean-atmosphere processes. NCPOR's outreach activities exhaustively cover India's research in the Arctic to educate, discuss, and spread awareness among interested students, educators, and scientists, as well as general public.

At the ministerial level, the NCPOR comes under the MoES (https://www.moes.gov.in), which handles national research on the Earth System and Climate Sciences. The ministry encompasses two subordinate offices, three attached offices, and five autonomous bodies, including NCPOR. Concerning Arctic research, the MoES reports to the parliamentary committee to advise on the activities and plans with a scientific background. The MoES has signed collaboration and MoUs with Norway, Sweden, and Canada to strengthen India's Arctic research. Recently, India released its Arctic policy in March 2022, emphasizing a sustainable, responsible, and transparent strategic framework

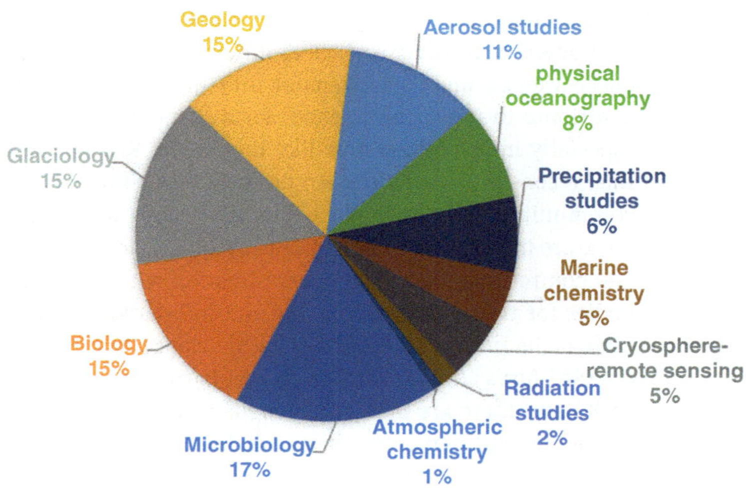


**Figure 31.1** Major research fields of the projects implemented under the Indian Arctic research program from 2011 to 2019. (*Source:* Adapted from NATIONAL CENTRE FOR POLAR AND OCEAN RESEARCH); "Biology" includes biological studies other than microbiology.

(https://www.moes.gov.in/sites/default/files/2022-03/compressed-SINGLE-PAGE-ENGLISH.pdf). The policy laid down its objectives under a multi-angle stand including science and research, climate and environmental protection, economic and human development, transportation and connectivity, governance and international cooperation, and national capacity building. The highlights of the policy are as follows:

i) Strengthen research base and activities at Himadri, Ny-Ålesund
ii) Channelize research activities in line with international Arctic research priorities including socio-economic, political, and traditional knowledge, and contribute toward the UN sustainable goals
iii) Acquire a dedicated ice-class polar research vessel for India
iv) Strengthen the funding support and collaborations at national and international levels
v) Increase participation in the work of the Arctic Council, IASC, NySMAC, SIOS, Arctic Circle assembly, SAON, etc.
vi) Work toward research and conservation of biodiversity, improve earth system modeling, environmental management, emergency preparedness, and actions in the Arctic
vii) Strengthen remote-sensing observations and fill gaps in Arctic research
viii) Work toward natural resource exploration, e-commerce, renewables, and bioenergy, with mutual collaboration and investment in a sustainable manner
ix) Investment in Arctic infrastructure, healthcare services, sustainable tourism, maritime services, and exploration
x) Work for the welfare of indigenous communities, and cultural and educational exchange
xi) Engage in shipbuilding, sea excursions, and transport corridors in the Arctic
xii) Capacity building from the research, exploration, and socio-economic angles.
xiii) Pursue cooperation and peace processes

## 31.4 Coordination Among Students, Young Researchers, and Educators

One of the important social responsibilities of research is to teach and educate. To support doctoral research in the polar areas, including the Arctic, the MoES and the NCPOR have MoUs at international, national, and state levels. At the international level, the MoES has an ongoing PhD program for Arctic and Antarctic research in collaboration with Norway. NCPOR supports in-house PhDs working on the Arctic. In Addition, the NCPOR has collaborated with various universities for producing PhDs in polar research. There is the CUSAT-NCPOR Centre for Polar Sciences and NCPOR is also recognized as a research center for PhDs at Goa, Mangalore, and Bharathidasan Universities. Recently, an MoU with Savitribai Phule Pune University was established for collaborative research programs for MSc, MPhil, and PhD degrees in Earth Sciences. Apart from PhDs, NCPOR hosts master's and bachelors' internships in the field of Earth Sciences that also include Arctic research, which acts as a gateway for the students to gain hands-on experience and exposure.

The Arctic research has impressive participation from the international bodies connecting the gap between students, researchers, and teachers. The International Polar Year (IPY) and the Arctic Council laid down the foundation of the Association of Polar Early Career Scientists (APECS: https://www.apecs.is) and the University of Arctic (UArctic: https://www.uarctic.org), respectively. The APECS works to support young researchers working in the Arctic and Antarctic. The organization considers undergraduate and postgraduate students, postdoctoral researchers and faculty members, early career professionals and educators, or anyone interested in polar research. The Indian counterpart of APECS is the Indian Polar Research Network (IPRN), which disseminated APECS's work and opportunities. These associations facilitate networking, early career development, and outreach to the concerned people. UArctic, on the other hand, is a network of universities, colleges, and research institutes working toward Arctic research and education. The consortium aims to build and strengthen the collective resources and infrastructure through research, education, and outreach. Besides, other international organizations work to bring polar research and academia as well as the general public on a common frame to emphasize polar research and education in the present scenario of climate change. For example, there is Polar Educators International (PEI: https://polareducator.org) that connects researchers, educators, and the global community fulfilling the broad outreach purpose. A summary of coordination organizations about India's Arctic research is presented in Figure 31.2.

Though we consider science as different from social science, science must complement and more importantly contribute to the social sciences for a sustainable future, especially in a sensitive area like the Arctic. Recognizing this, recent projects in the Arctic have taken indigenous communities and knowledge into the loop. Moreover, there are international and regional social science councils and committees that work in coordination with scientific bodies for the overall benefit. On the international level, there is the International Arctic Social Science Association (IASSA: https://iassa.org), which is a partner institute of IASC, and there are many indigenous councils, both governmental and non-governmental, which take active cognizance of Arctic research activities. However, India is yet to implement the social science and commerce-related

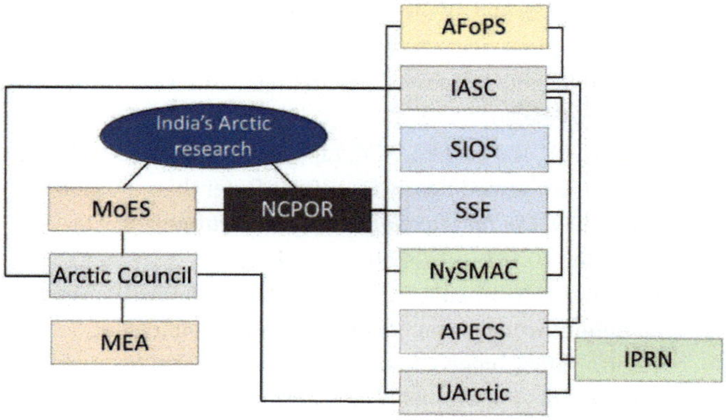


**Figure 31.2** Coordination organizations and their interrelations pertaining to India's Arctic research: red = ministerial level; light gray = international; dark gray = institutional level; yellow = Asian; light blue = Svalbard region; green = regional level; MoES = Ministry of Earth Sciences (India); NCPOR = National Centre for Polar and Ocean Research: MEA = Ministry of External Affairs (India); AFoPS = Asian Forum of Polar Sciences; IASC = International Arctic Science Committee: SIOS = Svalbard Integrated Arctic Earth Observing System: SSF = Svalbard Science Forum: NySMAC = Ny-Ålesund Science Managers Committee: APECS = Association of Polar Early Career Scientists; IPRN = Indian Polar Research Network; UArctic = University of the Arctic.

components in the Arctic research, which are now a part of India's Arctic Policy.

At the ground point, India has made good use of the above-mentioned coordination committees and working groups for the progress of its Arctic research activities. Lately, during the Covid-19 pandemic crisis, the connections were well used to continue some of the long-term measurements in Ny-Ålesund. The data collected by Indian scientists are getting their due in international environmental assessment reports and being used by the international community. Over the last five years, India has expanded its research in the emerging areas of concern such as microplastics, pollution, and metagenomic studies. To expand its research, India has signed an MoU with the Canadian High Arctic Research Station as a step toward its pan-Arctic vision. In addition, Indian researchers have been part of expeditions outside Ny-Ålesund, such as the Svalbard coastal cruise and IBRV ARON Chukchi Sea cruise (an AFoPS initiative), and international collaborative projects such as EU-Horizon and Arctic PASSION.

With all the expertise gained so far and available support at the international level, India is now ready to move forward with project-based major expeditions and cruises to address the big questions in the Arctic. At the national level, Arctic research needs participation from major research and academic institutes in India. At the administrative level, India needs to strengthen the Arctic logistic framework and hopes to own the polar research vessel soon.

## Acknowledgments

The authors thank the Director, National Centre for Polar and Ocean Research (NCPOR), to facilitate the Indian Arctic program, which has helped in formulating this chapter. The authors are grateful to the Ministry of Earth Sciences, Government of India, to fund and implement the program through the NCPOR. Special thanks to all the national and international collaborators whose cooperation has helped in the successful Indian research and logistical activities in the Arctic. This is NCPOR contribution number B-4/2022-23.

## Conflict of Interest Statement

The authors have no conflicts of interest to declare. All co-authors have seen and agree with the contents of the manuscript and there is no financial interest to report.

## References

Andresen, C.G., Lawrence, D.M., Wilson, C.J. et al. (2020). Soil moisture and hydrology projections of the permafrost region: a model intercomparison. *Cryosphere* 14(2): 445–459. doi: 10.5194/tc-14-445-2020.

Bintanja, R. (2018). The impact of Arctic warming on increased rainfall. *Scientific Reports* 8(1): 16001. doi: 10.1038/s41598-018-34450-3.

Davis, P.E.D., Lique, C., and Johnson, H.L. (2016). Competing effects of elevated vertical mixing and increased freshwater input on the stratification and sea ice cover in a changing Arctic Ocean. *Journal of Physical Oceanography* 46(5): 1531–1553. doi: 10.1175/JPO-D-15-0174.1.

Docquier, D., Koenigk, T., Fuentes-Franco, R. et al. (2021). Impact of ocean heat transport on the Arctic sea-ice decline: a model study with EC-Earth3. *Climate Dynamics* 56(5–6): 1407–1432. doi: 10.1007/s00382-020-05540-8.

Gulland, F.M.D., Baker, J.D., Howe, M. et al. (2022). A review of climate change effects on marine mammals in United States waters: past predictions, observed impacts, current research and conservation imperatives. *Climate Change Ecology* 3: 100054. doi: 10.1016/j.ecochg.2022.100054.

Hjort, J., Karjalainen, O., Aalio, J. et al. (2018). Degrading permafrost puts Arctic infrastructure at risk by mid-century. *Nature Communications* 9(1): 5147. doi: 10.1038/s41467-018-07557-4.

Huntington, H.P.. Zaporsky, A., Kaltenbormn, B.P. et al. (2022). Societal implications of a changing Arctic Ocean. *Ambio* 51(2): 298–306. doi: 10.1007/s13280-021-01601-2.

Kankaanpää, P. and Young, O.R. (2012). The effectiveness of the Arctic Council. *Polar Research* 31(1): 17176. doi: 10.3402/polar.v31i0.17176.

Kashiwase, H., Ohsima, K.I., Nihahi, S. et al. (2017). Evidence for ice-ocean albedo feedback in the Arctic Ocean shifting to a seasonal ice zone. *Scientific Reports* 7(1): 8170. doi: 10.1038/s41598-017-08467-z.

Overland, J., Dunleab, E., Box, J.E. et al. (2019). The urgency of Arctic change. *Polar Science* 21: 6–13. doi: 10.1016/j.polar.2018.11.008.

Park, D.-S.R., Lee, S., and Feldstein, S.B. (2015). Attribution of the recent winter sea ice decline over the Atlantic sector of the Arctic Ocean. *Journal of Climate* 28(10): 4027–4033. doi: 10.1175/JCLI-D-15-0042.1.

Parkinson, C.L. and DiGirolamo, N.E. (2021). Sea ice extents continue to set new records: Arctic, Antarctic, and global results. *Remote Sensing of Environment* 267: 112753. doi: 10.1016/j.rse.2021.112753.

Rantanen, M., Karpechko, A.Y., Lipponen, A. et al. (2022). The Arctic has warmed nearly four times faster than the globe since 1979. *Communications Earth & Environment* 3(1): 1–10. doi: 10.1038/s43247-022-00498-3.

Solan, M., Archambault, P., Renaud, P.E. et al. (2020). The changing Arctic Ocean: consequences for biological communities, biogeochemical processes and ecosystem functioning. *Philosophical Transactions of the Royal Society A: Mathematical, Physical and Engineering Sciences* 378(2181): 20200266. doi: 10.1098/rsta.2020.0266.

Taagholt, J. (1988). International Arctic Science Committee. *Polar Record* 24(150): 248–248. doi: 10.1017/S0032247400009256.

# Index



*Advances in Remote Sensing Technology and the Three Poles*, First Edition. Edited by Manish Pandey, Prem C. Pandey, Yogesh Ray, Aman Arora, Shridhar D. Jawak, and Uma K. Shukla.

## *h*

## *t*

## u



## v



## w